$44.50

X-Ray and Atomic Inner-Shell Physics–1982

(International Conference, University of Oregon)

International Conference on X-Ray and Atomic Inner-Shell Physics
August 23-27, 1982
University of Oregon
Eugene, Oregon
USA

AIP Conference Proceedings
Series Editor: Hugh C. Wolfe
Number 94

X-Ray and Atomic
Inner-Shell Physics–1982
(International Conference, University of Oregon)

Edited by
Bernd Crasemann
University of Oregon

American Institute of Physics
New York   1982

L.C. Catalog Card No. 82–74075
ISBN 0–88318–193–2
DOE CONF- 820855

Committees

Conference Organizing Committee

B. Crasemann, co-chairman, University of Oregon
R. D. Deslattes, co-chairman, National Bureau of
Standards
R. C. Elton, Naval Research Laboratory
C. P. Flynn, University of Illinois
M. O. Krause, Oak Ridge National Laboratory
E. Merzbacher, University of North Carolina
R. H. Pratt, University of Pittsburgh
P. Richard, Kansas State University
F. Wuilleumier, LURE, Université Paris-Sud

International Advisory Board

T. Åberg, Helsinki University of Technology, Finland
W. N. Asaad, American University in Cairo, Egypt
W. Bambynek, Joint Research Centre, Commission
of the European Communities, Belgium
D. Berényi, Institute of Nuclear Research of the
Hungarian Academy of Sciences, Hungary
J. P. Briand, Université Pierre et Marie Curie, France
C. E. Brion, University of British Columbia, Canada
J. Drahokoupil, Fysikalni ustav, Czechoslovakia
D. Fabian, University of Strathclyde, U.K.
J. G. Ferreira, Universidade de Lisboa, Portugal
V. Florescu, University of Bucharest, Rumania
M. Gavrila, FOM Institute for Atomic and Molecular
Physics, The Netherlands
Z. Iwinski, University of Warsaw, Poland
P. P. Kane, Indian Institute of Technology, Bombay,
India

H. Kleinpoppen, University of Stirling, U.K.
A. Kodre, Univerza Edvarda Kardelja v Lubljani,
Yugoslavia
K. Kohra, KEK National Laboratory for High Energy
Physics, Japan
C. Kunz, Deutsches Elektronen-Synchrotron, BRD
F. P. Larkins, Monash University, Australia
R. Manne, University of Bergen, Norway
W. Mehlhorn, Albert-Ludwigs-Universität, BRD
A. Meisel, Karl-Marx-Universität, DDR
V. I. Nefedov, Institute of General and Inorganic
Chemistry, Academy of Sciences of the USSR
C. Nordling, Uppsala Universitet, Sweden
H. Paul, Johannes-Kepler Universität, Linz, Austria
A. Ron, Racah Institute for Physics, Israel
H. K. Tseng, National Central University, China

Honorary Board

J. A. Bearden, Johns Hopkins University, USA
M. Blokhin, Rostov State University, USSR
Y. Cauchois, Université Pierre et Marie Curie, France
A. Faessler, Universität München, BRD
E. L. Jossem, Ohio State University, USA
L. G. Parratt, Cornell University, USA

Local Committee

B. Crasemann, chairman, University of Oregon
Sandra Hill, secretary, University of Oregon
Sharon Mullins, treasurer, University of Oregon
Mau Hsiung Chen, University of Oregon
John T. Moseley, University of Oregon
T. Darrah Thomas, Oregon State University
Larry H. Toburen, Battelle Northwest Laboratories

Foreword

The 1982 International Conference on X-Ray and Atomic Inner-Shell Physics, "X-82," was held on the University of Oregon campus in Eugene on August 23-27. The Conference succeeded meetings held in Stirling, Scotland, in 1980 and in Sendai, Japan, in 1978. The present series has resulted from the coalescence of two earlier series of conferences: on x rays (from Ithaca and Leipzig in 1965 through Washington in 1976), and on inner-shell ionization (Atlanta, 1972, and Freiburg, 1976).

The five-day Conference was attended by 222 participants from 26 countries. The truly international character of the meeting, its size, the high quality of the carefully prepared presentations, and the sunny Oregon summer weather all contributed to making X-82 a very fruitful scientific gathering.

The program incorporated an even representation of x-ray physics and atomic inner-shell physics. Each day, plenary sessions were followed by two parallel sessions pertaining to the two major areas covered by the Conference. There were invited Reviews and Progress Reports, and contributed papers presented in poster sessions.

These Proceedings contain the texts of invited papers presented at the Conference. The material has been reproduced directly from camera-ready copy provided by the authors. The papers have been grouped by subjects, in somewhat different order than in the Conference program. We have attempted to make the Proceedings available as soon as possible after the Conference, and are grateful to AIP Conference Proceedings Series Editor Hugh C. Wolfe for his help in attaining this aim. G. B. Armen kindly assisted with the compilation.

We gratefully acknowledge Conference sponsorship and support from the International Union of Pure and Applied Physics, the National Science Foundation, the U. S. Army Research Office, the U. S. Department of Energy Division of Chemical Sciences, the National Bureau of Standards, and the Department of Physics and Chemical Physics Institute of the University of Oregon.

Bernd Crasemann

TABLE OF CONTENTS

INNER SHELLS AS A LINK BETWEEN ATOMIC AND NUCLEAR PHYSICS

Eugen Merzbacher
University of North Carolina at Chapel Hill
Chapel Hill, N.C. 27514

ABSTRACT

Nuclear decay and reaction processes generally take place in
neutral or partially ionized atoms. The effects of static nuclear
properties (size, shape, moments) on atomic spectra are well known,
as are electronic transitions accompanying nuclear transitions, e.g.
K capture and internal conversion. Excitation or ionization of
initially filled inner shells, really or virtually, may modify
nuclear Q values, will require correction to measured beta-decay
endpoint energies, and can permit the use of inner-shell transitions
in the determination of nuclear widths Improvements in resolution
continue to enhance the importance of these effects. There is also
beginning to appear experimental evidence of the dynamical effects
of atomic electrons on the course of nuclear reactions.

The dynamics of a nuclear reaction, which influences and may in
turn be influenced by atomic electrons in inner shells, offers
instructive examples of the interplay between strong and
electromagnetic interactions and raises interesting questions about
coherence properties of particle beams. A variety of significantly
different collision regimes, depending on the atomic numbers of the
collision partners and the collision velocity, will be discussed and
illustrated.

INTRODUCTION

The border between atomic collision and nuclear reaction
physics is at present a developing region, with a number of
scattered successes to its credit and a future potential that has
yet to be fully assessed. This lecture is intended to provide a
survey of some of the major accomplishments and of the basic
concepts underlying the experiments that have been carried out so
far.

The reasons for the recent upsurge of interest in collision
processes at the interface of atomic and nuclear physics are easily
appreciated: Particle accelerators in the MeV range, especially
single-ended and tandem Van de Graaff accelerators, have found
increasing use for experiments involving atomic excitation and
ionization; nuclear physicists are turning more and more toward such
experiments, repaying a debt to atomic physics that was incurred
fifty years ago when, following the discovery of the neutron, many
experimental atomic physicists became nuclear physicists almost
overnight; the energy resolution of beams and detectors in
accelerator-related atomic and nuclear physics is now approaching

the eV level, making atomic energy differences discernible in nuclear reactions; and – last but not least – theoretical atomic physicists are prepared to deal with the complications that must be faced when the relatively weak and long-range atomic and the stronger short-range nuclear interactions are simultaneously considered in a many-body problem.

Inner shells of atoms constitute a natural bridge between nuclear interactions, which correspond to energies in the MeV and cross sections in the millibarn to barn range, and the bulk of atomic electrons with interaction energies in the eV and cross sections in the 10^{-16} cm^2 range. Although the cross sections for atomic inner-shell processes are generally far smaller than this – but still very large in comparison with nuclear cross sections – the energy transfers and transition rates involving inner-shell electrons are often more nearly comparable to characteristic nuclear quantities, accounting for the role that inner shell electrons play in the processes which are the subject of this lecture. In the interest of simplicity, the symbol K will, in this paper, be used as a generic subscript denoting any inner shell.

NUCLEAR REACTION WIDTHS

The issues are conveniently introduced by a brief review of an imaginative idea put forward by Gugelot[1] for measuring the width, Γ_n, of a compound nuclear resonance, or its reciprocal, the time τ(sec) $= 0.658 \times 10^{-15}/\Gamma_n$, by the use of an "atomic clock." Assuming that in the ("in" part of the) collision of a nuclear projectile Z_1 with a target nucleus Z_2, which leads to the formation of a compound nucleus Z_1+Z_2, an inner-shell electron vacancy is formed with known probability P_K^{in}, the nuclear reaction width Γ_n can be compared with the width or decay rate, Γ_K, of the excited atomic state, if the two rates are of similar magnitude. This method relies on the possibility of experimentally identifying the decay of the atomic vacancy state by the decay energy characteristic of the transient "united" atom (Z_1+Z_2) rather than the target atom (Z_2), and it assumes that the two decay branches, atomic and nuclear, for a compound nuclear state with a vacancy in its atomic K shell are independent of each other. Several recent experiments which have used this technique to determine, by coincidence measurements of x rays (or Auger electrons) from the united atom and purely nuclear scattering, are reviewed in this volume by W. Meyerhof.[2] The collision is schematically depicted in Fig.1. If $N_{P'}$ is the "singles" rate for the nuclear reaction and $N_{P',K(Z_1+Z_2)}$ the coincidence rate, then the relation between nuclear and atomic widths is given by

$$\frac{N_{P',K(Z_1+Z_2)}}{N_{P'}} = P_K^{in} \frac{\Gamma_{K(Z_1+Z_2)}}{\Gamma_{K(Z_1+Z_2)} + \Gamma_n} \tag{1}$$

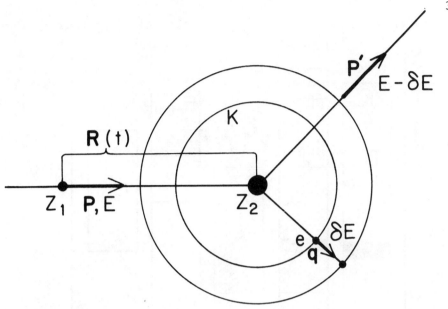

Fig.1. Schematic sketch indicating a nuclear collision with near-zero impact parameter between projectile Z_1, with momentum **P** and energy E, and target Z_2. Straight-line trajectories are assumed, and momentum **q** and energy δE are transferred to an atomic K-shell electron. The nuclear ejectile has final momentum **P'** and energy $E-\delta E$. The internuclear vector is **R**(t).

ATOMIC AND NUCLEAR ENERGY AND TIME SCALES

In judging the usefulness of equation (1) for a practical determination of nuclear widths, one must bear in mind that the atomic width Γ_K, which is of order 10^8 sec^{-1} for hydrogen, is proportional to Z^4, whereas Γ_n depends on the mass number A of the nucleus and its excitation energy, E_n, roughly as $\exp[-(A/E_n)^{1/2}]$, and P_K^{in} has values between 10^{-5} and 10^{-1}, depending on the strength of the vacancy-producing atomic interactions. In practice, it has been possible with this method to verify that compound nuclear states in medium-A nuclei at excitation energies of 10-15 MeV have $\Gamma_n \sim 10$ eV. In Fig.2, the approximate range of atomic and nuclear widths, to which this comparison method may be applicable, has been indicated.[3] Since the overlap is limited, Γ_K can serve as an atomic yardstick for Γ_n only under rather favorable circumstances.

Fig.2 also shows the approximate ranges of other relevant atomic and nuclear energy parameters (or their reciprocal time equivalents). These include typical nuclear excitation energies, E_n, atomic collision times T_{coll}, and atomic excitation energies ω_K,

4

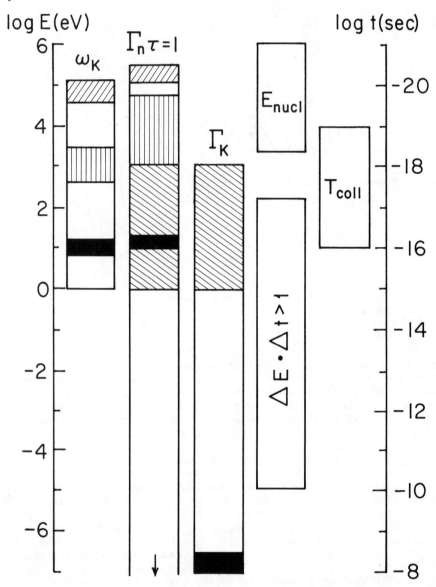

Fig.2. Relation between characteristic energies and times in nuclear reactions and ion-atom collisions. The symbols are explained in the text. The four shaded regions on the Γ_n column refer, starting from the bottom, respectively to the Gugelot-type experiments (Refs. 1, 2), Staub's experiment (Ref. 13), Blair's experiment (Ref. 7), and the proposed heavy-ion experiments (Ref. 5). The diagram is adapted from Ref. 3.

which are in the eV range for hydrogen and proportional to Z^2. Also indicated are some crudely estimated limits for the energy spread, ΔE, and time resolution, Δt, of the wave packets which describe the projectile motion in the usual experimental situation, but it should be remembered that usually for such wave packets $\Delta E \times \Delta t$ is much greater than unity. It is interesting to speculate on the possibility of observing physical effects which require a more detailed understanding of the structure of these wave packets.[4]

EFFECTS OF NUCLEAR REACTIONS ON ATOMIC COLLISIONS

If a nuclear reaction is initiated in a particular collision, the impact parameter on the atomic scale is obviously very small, and it is usually (except for very high-Z collision partners) permissible to assume that b = 0. The occurrence of the nuclear reaction signals that an atomic head-on collision has taken place!

For purposes of orientation, it is useful to think of a narrow compound nuclear resonance state as quasistationary with a lifetime $\tau \sim \Gamma$ and to describe the atomic collision by a semiclassical treatment. The prescribed classical orbital motion of the heavy collision partners is responsible for the time-dependent perturbations which cause transitions of the atomic electrons, and it is reasonable to inquire how the atomic transition amplitudes are affected by the time delay introduced into the heavy-particle motion by the nuclear reaction. The atomic target electrons are subject to the time-varying Coulomb fields of the projectile and the recoiling target nucleus. In first-order perturbation theory, the amplitude $a_n(+\infty)$ for an atomic transition to an excited (discrete or contiuum) state n from an initial (t=-∞) state K is, except for some irrelevant constants, given by

$$a_n(+\infty) \sim \int V(\mathbf{q}) \, F_{nK}(\mathbf{q}) \, d^3q \int_{-\infty}^{+\infty} \exp\{i[\omega_{nK}t - \mathbf{q} \cdot \mathbf{R}(t)]\} \, dt \qquad (2)$$

where $V(\mathbf{q})$ represents the Fourier transform of the perturbation and $F_{nK}(\mathbf{q})$ is an inelastic atomic form factor. In the high-velocity limit, the collision is impulsive, and $\mathbf{q}$ is the momentum transfer from the heavy collision partners to the electron. Most important for the present discussion is the classical orbital Fourier integral,

$$I(\omega_{nK}, \mathbf{q}) = \int_{-\infty}^{+\infty} \exp\{i[\omega_{nK}t - \mathbf{q} \cdot \mathbf{R}(t)]\} \, dt \qquad (3)$$

The orbital integral depends on the atomic energy transfer ω_{nK} and the moving position vector $\mathbf{R}(t)$ which describes the time variation of the nuclear configuration. If there is a time delay τ, as in the case of a compound nuclear reaction, this will affect the t-dependence of $\mathbf{R}(t)$. In the particular case of charged-particle capture or decay, only a "half-collision" occurs, and $\mathbf{R}(t)$ is set equal to zero after or before the nuclear process.

These considerations make it evident that atomic transition amplitudes can be significantly influenced by the time delay which a nuclear reaction between so-called "sticky" nuclei introduces into the Fourier integral of equation (3). The observation of such effects on the amplitudes implies the use of the atomic transition energies ω_K, rather than the widths Γ_K, as yardstick for the nuclear width parameter to be determined. Fig.2 shows that this approach to the problem extends considerably the potential range of values of Γ_n that is accessible to determination by atomic collision techniques.

For example, it has been suggested that oscillations in the energy spectrum of the emitted delta electrons might convey information about compound nuclear delay times in heavy-ion reactions.[5] Similar oscillations, with a period comparable to Γ_n, in the energy spectrum of positrons from heavy-ion collisions are currently under investigation.[6] In these instances, the dependence of the transition amplitude (2) on the energy of the final state of ionization of the atom is studied.

Blair et al[7] have shown that one may also study the dependence of the transition amplitude for atomic inner-shell vacancy production in the course of a nuclear resonance reaction on the energy of the incident particle. A consistent analysis in terms of stationary states of total energy E shows that the amplitude for K-shell excitation to a final atomic state n can be expressed as[8]

$$T_{nK} \sim \frac{1}{E-\delta E-E_r+i(\Gamma_n/2)} \, a^{in}_{nK} + a^{out}_{nK} \, \frac{1}{E-E_r+i(\Gamma_n/2)} \qquad (4)$$

The structure of this expression is suggestive: The first term describes atomic excitation (or ionization) on the "in" leg of the reaction, followed by the nuclear resonance scattering, with the incident energy reduced by the energy transfer δE to the atom, and the second term corresponds to atomic excitation on the "out"leg, following the nuclear reaction. The details of the pioneering proton-x ray coincidence experiment by Blair et al[7] verifying the interference between the two terms of equation (4) are reviewed in Meyerhof's lecture in these Proceedings.[2]

THE EFFECTS OF ATOMIC EXCITATION ON NUCLEAR REACTIONS

Since nuclear reaction experiments usually involve ions and atoms, rather than bare nuclei, one expects in nuclear reaction experiments to observe effects caused by real or virtual excitations of the atomic electrons, provided that the experimental resolution is good enough. Many years ago, following the prediction by Lewis[9], evidence of energy transfer to atoms was observed in the thick-target yields of certain narrow nuclear resonances produced by incident charged particles, such as the familiar $^{27}Al(p,\gamma)^{28}Si$ resonance at 992 keV.[10] This so-called "Lewis effect" consists of

slight oscillations of the thick-target yield, above the threshold
and the sharp initial rise, as a function of the energy of the
incident projectile. The oscillations occur because the reaction
yield is lowered when the incident projectile energy exceeds the
resonance energy by an amount that is less than (a multiple of) the
minimum energy lost by the charged particle in inelastic collisions
to the atoms of the target medium. Thus, the Lewis effect is a
manifestation of the quantum nature of the atomic energy transfers
during the passage of the projectile through the target, as it slows
down until it reaches the resonance energy.

The Lewis effect is distinctly different, however, from the
process considered in this lecture, because in the case of the Lewis
effect the atomic energy losses are mostly due to so-called
"distant" collisions at large impact parameters to outer electrons,
and these collisions do not occur in the same atom as the nuclear
reaction. When Christy[11] inquired whether atomic excitation effects
had to be taken into account in the experimental determination of
nuclear reaction Q values [which are equivalent to the change in the
total masses of the reactants, $Q=(M_{before}-M_{after})c^2$], he concluded
that such effects would be difficult to notice unless the resolution
were to reach the 100 eV level. Contemporary accelerator beam and
particle detection technology has brought atomic and nuclear
collision experiments into this range.

For example, precision measurements of endpoints in beta ray
spectra are of great current interest (e.g. for determining limits
on the neutrino mass). An accurate method for obtaining the
endpoint energies of positron spectra consists of measuring the Q
values of the inverse nuclear transformations, such as (p,n) and
(^{3}He,^{3}H) reactions, which tend to be endoergic (Q<0) and therefore
have a threshold energy. In such nuclear reactions, several
mechanisms can be invoked to describe atomic excitation, although
the underlying forces are of course the electrostatic particle
interactions in all instances. The energy transfer can be
attributed to Coulomb excitation of the atomic electrons by the
charged projectile as well as by the recoiling target nucleus, but
the excitation can also be considered as arising from the rather
abrupt change in nuclear charge and the ensuing shake-off and shake-
up processes. In some common examples among light nuclei, these
atomic effects require cumulative corrections of nuclear Q values by
amounts of the order of 100-300 eV.[12]

Since a compound nuclear state is in reality a quasistationary
state of an entire compound nuclear atom, one expects that in a
nuclear collision it might be possible to observe replicas of very
sharp nuclear resonances, displaced in energy by the atomic
excitation energies. A very difficult early heroic experiment by
Staub et al[13] was the first attempt to exhibit such atomic replicas,
or satellites, or echoes[14], of a very narrow nuclear resonance. The
particular reaction chosen was resonant elastic scattering of alpha
particles on ^{4}He in the neighborhood of 184 keV. For this resonance
in the ^{4}He(α,α)^{4}He reaction, the width ($\Gamma_n \sim 15$ eV) is comparable to
and generally smaller than the atomic transition energies $\omega_K \sim 100$ eV

8

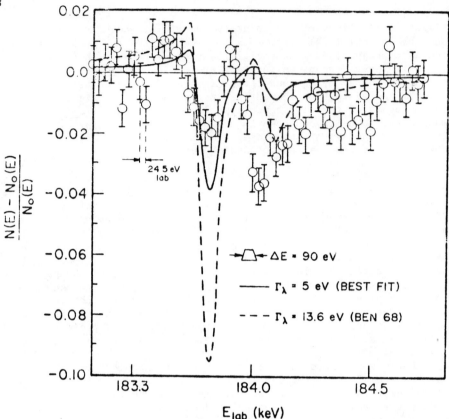

Fig.3. Relative yield of elastic scattering of α particles from ^{4}He near the resonance at 184 keV. The yield $N(E)$ is plotted relative to the yield $N_0(E)$ from Rutherford scattering and shows anomalies attributed to atomic excitation. The data and the dotted theoretical curve are from Ref.13. The solid curve is a recent analysis (Ref. 15).

(in the compound ion $^8Be^+$). The atomic excitations to the first few levels in Be show up as anomalies in the nuclear excitation function. These anomalies are displayed in Fig.3.[13,15]

Quite recently, Duinker et al[16] have found indications of a weak K-shell excitation satellite accompanying the same narrow $^{27}Al(p,\gamma)^{28}Si$ resonance at 992 keV that was mentioned above in the discussion of the Lewis effect. The resonance is narrow (about 100 eV) and the K-shell binding energy is substantially greater (1.6 keV), providing favorable conditions for the observation of such a satellite, but the experimental resolution is not good enough for a clean separation of the two lines. From the results, shown in Fig.4, Duinker estimated $P_K^{in} \sim 10^{-2}$.

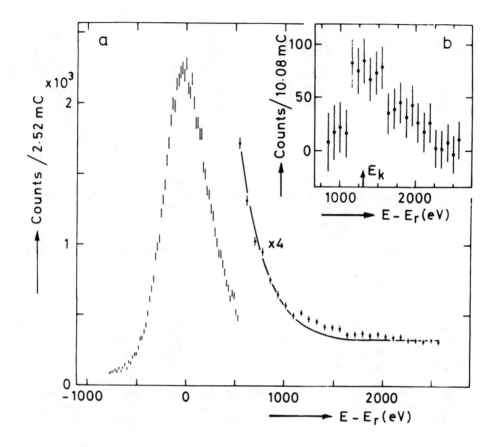

Fig.4. Data from Ref. 16 showing a small secondary peak, or satellite, approximately 1.6 keV above the 992 keV resonance in the ^{27}Al$(p,\gamma)^{28}$Si reaction. Insert (b) is a close-up of the excess counts due to the satellite.

Many years ago, Bilpuch[17] observed a conspicuous tendency of many of the Breit-Wigner resonances in very high-resolution experiments with gas targets at low temperature to exhibit high-energy tails, as a function of proton energy, suggesting the presence of atomic satellites. Fig.5 shows the data of Park[18] for the ^{40}Ar$(p,n)^{40}$K reaction. Coleman[19] made a preliminary R-matrix calculation for the many-body multi-level compound reaction system which represents the nucleus with its electronic surroundings. In this calculation, the atomic excitations are assumed to take place in the external region of nuclear configuration space. Since, unlike Blair's coincidence experiment, the inelastic energy transfer to the atom is not observed in any of these experiments, atomic Coulomb excitation and ionization in the <u>intermediate</u> state is the mechanism for the appearance of satellites in nuclear resonance excitation functions. Remarkably, virtual atomic excitation can occur even if the projectile and ejectile are neutral, as in an (n,n) reaction, provided only that at least one charged-particle

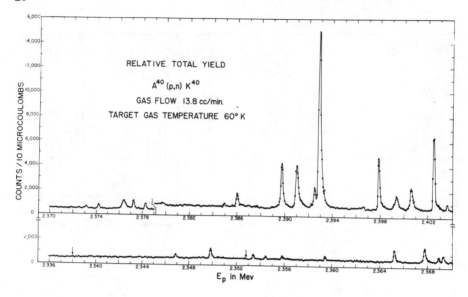

Fig.5. High-resolution resonance data in the ^{40}Ar(p,n)^{40}K reaction. The data, from Ref. 18, show many resonances to be asymmetric, with a suggestion of secondary (satellite) peaks on the high-energy side.

channel is open. This channel mediates the excitations through the strong nuclear interaction. It must be open so that an interaction with the atomic electrons can take place in the region external to the nucleus.

Yet another nuclear reaction regime is represented by the example of a relatively very narrow resonance in ^{12}C(p,p)^{12}C scattering with polarized protons at 14.2 MeV.[20] Measurements of the analyzing power, as well as the differential cross section, gave a resonance width of Γ_n=1,600 eV, after the necessary corrections for various instrumental causes of line broadening were made. This width is almost twice the width (900 eV) of the same state, as was determined from decay branching ratios.[21] Efforts have been made to understand this difference as a consequence of atomic excitations, adding an energy spread δE to the resonance width. The dominant mechanism for this broadening attributed to atomic excitation is not yet clearly understood, but the available interpretations have suggested the importance of kinematic excitation, either by recoil of the target nucleus in this energetic reaction,[20] or through the atomic center-of-mass motion of the nucleus.[14]

CONCLUSIONS

Coincidence and high-resolution experiments are beginning to reveal a number of manifestations of the physics at the interface between the atom and the nucleus. No comprehensive conceptual framework has yet been developed to understand these effects from a unified point of view. A number of difficult experiments in this area have, however, been recently completed. Useful new information for an appreciation of the dynamics of both atomic and nuclear collision processes may be forthcoming.

ACKNOWLEDGEMENTS

This paper was prepared with partial support from the Department of Energy under contract No.DE-AS05-76ER02408. It is dedicated to the memory of my friend and colleague John S. Blair, whose experimental and theoretical work advanced so greatly our understanding of nuclear effects in atomic physics.

REFERENCES

1. P. C. Gugelot, in Proceedings of the International Conference on Direct Interactions and Nuclear Reaction Mechanisms, Padua 1962, ed. E Clementel and C. Willi, Vol. II, p. 382, Gordon and Breach, London 1963.
2. W. Meyerhof, in Proceedings of the International Conference on X-Ray and Atomic Inner-Shell Physics, Eugene 1982, ed. B.Crasemann, p. 13 (AIP Conf. Proc. 94)
3. E. Merzbacher, in Proceedings of the Divisional Conference of the European Physical Society on Nuclear and Atomic Physics with Heavy Ions, Bucharest 1981, ed. M. Ivaşcu, D. Bucurescu, and I. A. Dorobanţu, p. 293, Central Institute of Physics 1982.
4. K. W. McVoy, X. T. Tang, and H. A. Weidenmüller, Z. Phys. A299, 195 (1981); P. von Brentano and M. Kleber, Phys. Lett. 92B, 5 (1980).
5. G. Soff, J. Reinhardt, B. Müller, and W. Greiner Phys. Rev. Lett. 43, 1981 (1979); J. Reinhardt, B. Müller, W. Greiner, and G. Soff, Z. Phys. A292, 211 (1979); U. Müller, J. Reinhardt, G. Soff, B. Müller, and W. Greiner, Z. Phys. A297, 357 (1980).
6. B. Müller, in Proceedings of the International Conference on X-Ray and Atomic Inner-Shell Physics, Eugene 1982, ed. B. Crasemann, p. 206 (AIP Conf. Proc. 94)
7. J. S. Blair, P. Dyer, K. A. Snover, and T. A. Trainor, Phys. Rev. Lett. 41, 1712 (1978).
8. J. S. Blair and R. Anholt, Phys. Rev. A25, 907 (1982); J. M. Feagin and L. Kocbach, J. Phys. B14, 4349 (1981); K. W. McVoy and H. A. Weidenmüller, Phys. Rev. A25, 1462 (1982).
9. H. W. Lewis, Phys. Rev. 125, 937 (1962).
10. W. L. Walters, D. G. Costello, J. G. Skofronick, D. W. Palmer, W. E. Kane, and R. G. Herb, Phys. Rev. 125, 2012 (1962).
11. R. F. Christy, Nucl. Phys. 22, 301 (1961).
12. J. M. Feagin, E. Merzbacher, and W. J. Thompson, Phys. Lett. 81B, 107 (1979).
13. J. Benn, E. B. Dally, H. H. Müller, R. E. Pixley, H. H. Staub, and H. Winkler, Nucl. Phys. A106, 296 (1968).
14. J. S. Briggs and A. M. Lane, Phys. Lett. 106B, 436 (1981).
15. J. M. Feagin, Ph. D. dissertation, University of North Carolina at Chapel Hill, 1979.

12

16. W. Duinker and C. R. Boersma, Phys. Lett. 100B, 13 (1981);
    W. Duinker, dissertation, Utrecht, 1981.
17. E. G. Bilpuch and H. W. Newson, private communications.
18. P. B. Parks, Ph. D. dissertation, Duke University, 1963.
19. L. A. Coleman, Ph. D. dissertation, University of North Carolina
    at Chapel Hill, 1969.
20. W. J. Thompson, J. F. Wilkerson, T. B. Clegg, J. M. Feagin,
    E. J. Ludwig, and E. Merzbacher, Phys. Rev. Lett. 45, 703
    (1980).
21. R. E. Marrs, E. G. Adelberger, and K. A. Snover, Phys. Rev. C16,
    61 (1977).

COMPOUND-NUCLEUS EFFECTS ON INNER-SHELL IONIZATION[*]

W. E. Meyerhof

Department of Physics, Stanford University, Stanford, CA 94305, USA

ABSTRACT

If, in ion-atom collisions, only those events are experimentally selected in which the projectile nucleus strikes the target nucleus to form a compound-nuclear (CN) system, K-shell ionization may be affected <u>during</u> CN (i.e. united-atom, UA) formation and <u>after</u> CN formation. In the first case, if the K-shell is ionized on the way into the collision, a UA x-ray may be emitted during the CN life-time. In the second case, if the CN formation is resonant, the target K-ionization probability can fluctuate as the resonance is traversed. This paper discusses mainly the second case and shows that the theory of inner-shell ionization accompanied by CN formation, first developed by Blair, agrees with experiment over a wide range of experimentally chosen parameters. Once this has been assured, these concepts may be extended to studies of other nuclear delay time effects, e.g. in deep-inelastic or superheavy nuclear reactions, as discussed by Müller and by Greenberg at this conference.

INTRODUCTION

In a preceding lecture, Merzbacher has drawn attention to a field of overlap between atomic and nuclear physics, which might provide new information about nuclear and atomic collisions: study of dynamic effects between atomic electrons and nuclear reactions.

Gugelot[1] first suggested that it should be possible to detect compound nucleus (CN) x-rays, i.e. atomic x-rays resulting from the formation of a relatively long-lived compound-nuclear system in a nuclear reaction. The interest in the detection of CN x-rays is twofold. From a nuclear-physics point of view, the possibility exists to determine the life-time of the compound nucleus from the magnitude of the x-ray production cross section. From an atomic-physics viewpoint, the possibility exists of determining the zero-impact-parameter ionization probability on the way into a collision,[2] a quantity of great theoretical interest, since normally only the way-in and -out probability is determined.[3] Also, if the energy of x-rays from (short-lived) super-heavy CN systems could be determined, a new spectroscopic tool would be available to study atomic systems in which extreme relativistic effects are expected.[4,5]

The emission of CN x-rays from super-heavy systems, not yet detected, is related to positron emission, which <u>has</u> been detected

---

* Supported in part by the National Science Foundation under Grant No. PHY 80-15348. Work done in collaboration with R. Anholt, J. F. Chemin, and Ch. Stoller.

(see the lectures by Greenberg, and by Müller). The search for
nuclear time-delay effects in positron emission[6,7] is presently under
way.

Besides the possible formation of compound-nucleus x-rays in a
nuclear reaction, there is another, more subtle, influence of nuclear
time-delay effects on the production of target x-rays in an elastic
or inelastic nuclear scattering reaction.[8] In principle, the target
x-ray production probability can be used to estimate the nuclear time
delay if the x-ray production process is understood, or, if the
nuclear time delay is known, properties of the atomic vacancy-
production amplitude can be investigated.[9] Theoretically, inner-shell
target ionization in nuclear reactions induced by light projectiles
has similarities to bremsstrahlung emission[10-12] by these projectiles:
the former process involves (delta-) electron ejection, the latter
photon production. A nuclear time-delay effect on projectile
bremsstrahlung has been detected.[13]

The following sections discuss, in turn, the formation of CN
x-rays produced by light projectiles and target x-ray production in
nuclear resonance reactions with light projectiles. We should note
that, from an atomic point-of-view, the requirement, that a nuclear
reaction take place simultaneously with electronic effects, restricts
these atomic collisions to an effectively zero atomic impact
parameter.

EXPERIMENTS WITH COMPOUND-NUCLEUS X-RAYS

Theoretical background. Consider a proton or alpha particle bombarding a target nucleus and forming a compound nucleus with a
cross section σ_R. Call $P_{\frac{1}{2}}$ the probability of forming a K vacancy on
the way into the collision (i.e. one-half of a complete atomic
collision). Then $\sigma_R P_{\frac{1}{2}}$ is the cross section for having a K vacancy
in the united atom with atomic number $Z_{UA} = Z_p + Z_t$, where Z_p and Z_t
are the projectile and target atomic numbers, respectively. Each
such united atom (with a K vacancy) can decay by electronic filling
of the K vacancy with a probability per unit time λ_K and by disin-
tegration of the compound nucleus with a decay probability $\lambda_c = 1/\tau_c$.
Hence the probability of K x-ray emission from the original united
atom is equal to[1,2]

$$\omega_K \frac{\lambda_K}{\lambda_c + \lambda_K} \ , \tag{1}$$

where ω_K is the K fluorescence yield. The cross section for K x-ray
production in the entire nuclear reaction is

$$\sigma_{cx} = \sigma_R P_{\frac{1}{2}} \omega_K \frac{\lambda_K \tau_c}{1 + \lambda_K \tau_c} \ , \tag{2}$$

a formula first given by Gugelot[1] and reexamined by von Brentano and
Kleber,[14] by McVoy et al,[15] and by Anholt.[16]

Equation (1) assumes a simple exponential law for compound nucleus (CN) decay, which may not be correct in all situations.[17] Also, it is assumed that K x-ray emission and CN decay are independent of each other, which is not true in all cases, as discussed below. The exact calculation of $P_{\frac{1}{2}}$ presents an interesting theoretical problem.[3,18] It appears that, approximately,

$$P_{\frac{1}{2}} \simeq \frac{1}{2} P_K(0), \tag{3}$$

where $P_K(b)$ is the K-vacancy production probability in a complete collision (way-in and way-out) at an atomic impact parameter b. Experimental results. The experimental difficulties in applying Gugelot's proposal to the determination of the CN lifetime τ_c can be appreciated by the following example. Consider the 12-MeV p + $^{106}_{48}$Cd reaction.[2,19,20] Here, the CN formation cross section is $\sigma_R \simeq 0.7b$ and $P_K(0) \simeq 7 \times 10^{-4}$, so that $\sigma_K P_{\frac{1}{2}} \omega_K \simeq 2 \times 10^{-4}b$ represents the maximum cross section for the production of the united-atom $^{107}_{49}$In K x-rays. The latter cross section must be compared to an expected cross section of $\sim 60b$ for the production of $_{48}$Cd K x-rays by the ordinary Coulomb ionization process. Furthermore, even the detection of In x-rays does not unambiguously signal the formation of the ^{107}In compound nucleus, because the ^{106}Cd (p,n) reaction can form ^{106}In in excited states as a reaction product. Internal conversion in the gamma decay of ^{106}In* will produce K x-rays which are indistiquishable from the K x-rays from ^{107}In CN formation. This problem can be avoided by detecting only those In K x-rays which are in coincidence with inelastically scattered protons from the ^{106}Cd (p,p') reaction. But, still, the energy interval of the inelastic protons must be judiciously chosen so that the decay of the ^{107}In compound nucleus is signaled unambiguously.[2,20]

Figure 1 shows the x-ray spectrum in true coincidence with inelastic protons from the 12-MeV p + ^{106}Cd reaction. Although the coincidence requirement reduces the Cd K x-rays enormously, it does not eliminate them, because they can be produced also by internal-conversion decay of ^{106}Cd excited states which are populated in the ^{106}Cd (p,p') reaction. Analysis of these and other[21,22] experiments gives the results shown in Table I. Agreement with the statistical theory of CN decay appears to be good. For completeness, we mention an experiment of Hardy et al.[23] in which the compound atom $^{69}_{33}$As was formed, with a K vacancy, by electron capture from $^{69}_{34}$Se. In this decay, $^{69}_{33}$As can be produced in proton-unstable excited states. The K vacancy is then filled while the nucleus is $^{69}_{33}$As or while it is $^{68}_{32}$Ge. An expression similiar to (1) leads to τ_c. A search for CN x-rays from elastic nuclear reactions has not been successful so far.[22]

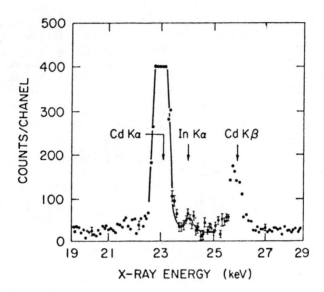

Fig. 1. X-ray spectrum from 12-MeV proton bombardment of ^{106}Cd in coincidence with inelastic protons (energy window 5.5 to 10 MeV). Accidental coincidences have been subtracted. The In Kα lines are due to CN formation. (From Ref. 20.)

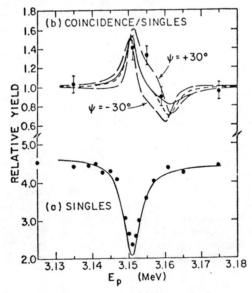

Fig. 2 (a). Yield of elastically scattered protons at 90 degree lab. angle from the reaction ^{58}Ni(p,p$_0$) as a function of bombarding energy. (b). Ratio of proton yield in coincidence with nickel K x-rays, divided by the proton yield. The ratio is normalized to unity off resonance. The curves correspond to different phases of the amplitude $a_{\frac{1}{2}}$ in Eq. (7). (From Ref. 9.)

TABLE I

Reaction	Compound nucl. & excit. energy	τ_c(measured) (sec)	τ_c(calculated) (sec)	Ref.
10-MeV p + ^{106}Cd	^{107}In*(14 MeV)	$(6.5 \pm 4) \; 10^{-17}$	11×10^{-17}	2
12-MeV p + ^{106}Cd	^{107}In*(16 MeV)	$(5.0 \pm 2.5) \; 10^{-17}$	5×10^{-17}	2
10-MeV p + ^{112}Sn	^{113}Sb*(13 MeV)	$(4.0 \pm 3.8) \; 10^{-17}$	a	21
12-MeV p + ^{112}Sn	^{113}Sb*(15 MeV)	$(3.4 \pm 2.0) \; 10^{-17}$	a	21

a A comparison of various measurements of τ_c with theoretical estimates can be found in Ref. 22.

EXPERIMENTS WITH TARGET X-RAYS

Theoretical background. We now turn to time-delay effects on target x-rays, i.e. on the vacancy-production probability in the target atom. In sufficiently asymmetric collisions, target K vacancies are produced predominately by ionization, with ejection of "delta electrons" whose energy is denoted by ε. At zero impact parameter, the K-vacancy production probability is given by the semiclassical approximation as[24]

$$P_K(0) = 2 \int_0^\infty d\varepsilon \sum_{\lambda\mu} |a_{\lambda\mu}(\varepsilon)|^2 \tag{4a}$$

where λ,μ are the orbital and magnetic quantum numbers of the ejected electron. A factor of 2 has been added to take into account the two K electrons. The amplitude $a_{\lambda\mu}$ can be written in the form[24]

$$a_{\lambda\mu} = \int_{-\infty}^\infty dt \, M_{\lambda\mu}(\varepsilon,t) e^{i\omega t}, \tag{4b}$$

where $M_{\lambda\mu}$ is a Coulomb matrix element connecting the initial and final electron states and $\hbar\omega \equiv |E_K| + \varepsilon$ is the energy which has to be imparted to the K electron with a binding energy E_K in order that it be ejected with the final kinetic energy ε. For projectiles of velocity considerably less than the initial Bohr velocity of the K electron, the $\lambda = 0$ term is dominant in Eq. (4a).[24] Hence, for simplicity, we drop all higher terms in Eq. (4a) and we drop the subscript $\lambda\mu$ in Eq. (4b). Breaking the integration in Eq. (4b) into the time intervals $-\infty$ to 0, 0 to T, and T to ∞, one can show[8] that for an elastic collision the ionization amplitude can be written

$$a = a_{\frac{1}{2}} - e^{i\omega(0)T} a_{\frac{1}{2}}^*, \tag{5}$$

where $a_{\frac{1}{2}}$ is the ionization amplitude on the way into the collision and $\omega(0)$ is the value of ω in the UA state.*

Conceptually, Eq. (5) is clear: looking at $a_{\frac{1}{2}}$ as the ionization amplitude for a time-reversed nuclear decay, one must, after a time lapse T and a phase change $\omega(0)T$, add the ionization amplitude for the outgoing nuclear decay.[8] The probability $P_K(0)$, Eq.(4a), must still be averaged over the distribution function $p(T)$ of the sticking time:

$$\overline{P}_K(0) = \int_0^\infty dT \; p(T) \; P_K(0). \tag{6}$$

This semiclassical treatment of a nuclear collision is not suited for resonance reactions, nor does it include such effects as Coulomb-nuclear interference. A proper quantum mechanical treatment must be made.[9,25-27] This yields instead of Eq. (6), for $\lambda = 0$ and elastic scattering,[9]**

$$\overline{P}_K(0) = 2 \int_0^\infty d\varepsilon \; |a_{\frac{1}{2}} \; f[\theta, E-\hbar\omega(0)] - a_{\frac{1}{2}}^* \; f(\theta,E)|^2 / |f(\theta,E)|^2. \tag{7}$$

Here, $f(\theta,E)$ is the elastic scattering amplitude for the projectile at a center-of-mass scattering angle θ and kinetic energy E. For the later discussion, we note that f can be decomposed into a Coulomb scattering amplitude f_c and a nuclear (resonant) amplitude f_n:

$$f = f_c + f_n. \tag{8}$$

One can interpret Eq. (7) as follows. If the K electron is ionized on the way into the collision, the projectile has only an energy $E-\hbar\omega$ left to initiate the nuclear reaction. If the K vacancy is produced only on the way out, the full energy E was available to the nuclear reaction. Hence, if f has a resonant behavior, the two terms in the integrand of Eq. (7) will change in relative magnitude and phase as the resonance is traversed. The magnitude of $\overline{P}_K(0)$ will reflect this variation. Of course, in a resonance reaction, the time delay is known from the width of the resonance.[13,17] So, the interest of Eq. (7) lies in the fact that one has here the rare opportunity to determine the phase of $a_{\frac{1}{2}}$ and compare it to theories of ionization.[9]

In comparing the semiclassical expressions (4) and (5) with the quantum mechanical expression (7), one sees that the phase factor $e^{-i\omega(0)T}$ [suitably averaged by Eq. (6)] in the former theory is equivalent to the ratio $f[\theta, E-\hbar\omega(0)]/f(E)$ in the latter theory. Expanding to first order in $\omega(0)$, the averaged semiclassical phase factor becomes

---

* The probability $P_{\frac{1}{2}}$ in Eq. (2) is related to $a_{\frac{1}{2}}$ by $P_{\frac{1}{2}} = 2 \int_0^\infty d\varepsilon |a_{\frac{1}{2}}|^2$.

** The term proportional to $a_{\frac{1}{2}}^*$ differs in sign from Ref. 9, because the factor i is included in our definition of the amplitude $\underline{a}$.

$$1 - i\omega(0)\tau \,, \tag{9}$$

assuming exponential decay with a mean life τ. The ratio of the scattering amplitudes is, to first order in $\omega(0)$,

$$1 - \hbar\omega(0)\partial\ell nf/\partial E\big|_E = 1 - i\omega(0)\tau' \,, \tag{10}$$

where τ' is the reaction time.[28] As discussed by Blair,[29] τ' can be imaginary and energy dependent; hence, the similiarly between expressions (9) and (10) is partly superficial.

In the case of a simple resonant behavior of $f(\theta,E)$, if $\hbar\omega(0)$ is much smaller than the width of the resonance, $f(\theta,E)$ cancels out in Eq. (7). In that case, no variation in $P_K(0)$ is expected as the resonance is traversed.[25-27]

Experimental results. Experimental tests of Eq. (7) consist in determining the variation of the magnitude of the ionization probability as a nuclear resonance is traversed. Away from the resonance, the ionization probability has a (nearly energy independent) value, as given by the usual semiclassical or other theories. The magnitude of the excursion at resonance dependes mainly on two parameters, the ratio E_K/Γ of the K-electron binding energy to the resonance width and the ratio $|f_n|/|f_c|$ of the nuclear to the Coulomb scattering amplitudes at resonance.[30] The excursion is expected to be largest for $E_K/\Gamma \simeq 1$ and $|f_n|/|f_c| \gg 1$, i.e. at backward scattering angles. The detailed shape of the excursion depends also on the projectile orbital angular momentum ℓ_p brought into the resonance. So far, all tests of Eq. (7) have used proton projectiles. A large range of the parameter E_K/Γ has been investigated, as we now describe.

Case $E_K/\Gamma \simeq 1$. The first determination of $\overline{P}_K(0)$ across a resonance was made by Blair et al.[9] for the elastic scattering of protons on ^{58}Ni near 3.15 MeV, at a lab. scattering angle of 90°. For this resonance, $E_K/\Gamma \simeq 1.5$ and $\ell_p = 0$. Figure 2(a) shows the resonance shape as a function of proton energy. Figure 2(b) gives the experimental points and the prediction of Eq. (7) (solid line). The sensitivity of $\overline{P}_K(0)$ to the phase of $a_{\frac{1}{2}}$ is shown by setting the phase of $a_{\frac{1}{2}}$ arbitrarily to +30° and -30° (long dashed lines) and to zero (short dashed line). Figures 3(a) and (b) give the relative cross section (90° lab.) and absolute ionization probability in the neighborhood of the 5.06-MeV resonance in ^{88}Sr(p,p_0) for which $E_K/\Gamma \simeq 0.8$ and $\ell_p = 2$.[31] The solid curve in Fig. 3(b) is the prediction of Eq. (7).

Case $E_K/\Gamma < 1$. Figures 4(a) and (b) are for a wider resonance in ^{88}Sr(p,p_0) at 6.06 MeV, again at a 90° lab. scattering angle.[30] Here, $E_K/\Gamma \simeq 0.2$ and $\ell_p = 0$. Experimentally, it is very difficult to determine ionization probabilities to better than ±10%. Hence, only consistency with Eq. (7) can be claimed in this case.

Case $E_K/\Gamma \ll 1$. Duinker et al.[32] measured the ionization probability across the 0.46-MeV resonance in ^{12}C(p,p_0) for which $E_K/\Gamma \simeq 0.01$ and

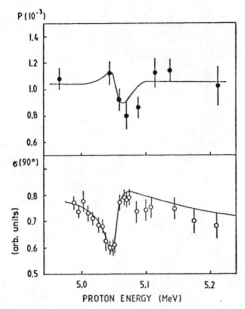

Fig. 3.(a) Cross section of elastically scattered protons at 90°
lab. angle from the reaction ^{88}Sr(p,p$_0$), as a function of bombarding
energy. (b) Ionization probability of ^{88}Sr in coincidence with
elastically scattered protons (90° lab.). The curve has been computed
using Eq. (7). (From Ref. 31.)

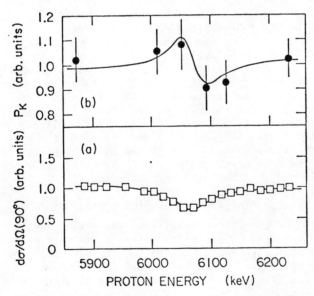

Fig. 4. Same as Fig. 3, but near the resonance at 6.06 MeV. (From
Ref. 30.)

$\ell_p = 0$. They found a ∿70% excursion in the ionization probability near resonance (at a 125° lab. scattering angle), whereas Eq. (7) predicts only a few percent excursion.[25-27] Duinker et al could fit their results by introducing an ad hoc phase difference between f_c and f_n in Eq. (8). This experiment has now been repeated[33] at a mean lab. scattering angle of 142°, using a CH_4 gas target instead of a solid target, used by Duinker. Figures 5(a) and (b) give the new results, as well as the predictions of Eq. (7) (solid line)[25] and of the ad hoc modification of Duinker et al (dashed line). It appears that the experimental results are consistent with Eq. (7). Most likely, the findings of Duinker et al[32] can be attributed to scattering effects;[33] also, the ad hoc modification of Eq. (7) has not found a theoretical basis.[25-27]

Case $E_K/\Gamma \gg 1$. Anholt et al[34] have proposed a method of applying Eq. (7) to the other extreme case $E_K/\Gamma \gg 1$, by measuring the ionization probability across an isobaric analog resonance (IAR) at lab. scattering angles near 180°. In theories of compound-nucleus reactions involving IAR's, it is customary to write the nuclear reaction amplitude as[35]

$$f(\theta,E) = \overline{f} + f^{fl} , \qquad (11)$$

where $\overline{f}$ consists of f_c [Eq. (8)] and the part of f_n which varies slowly with energy, whereas f^{fl} is the part of f_n which varies rapidly with energy due to the influence of CN levels of width Γ_{CN} in the IAR. The slowly varying part of f_n has a typical resonance shape with a width Γ_{IAR}. Indeed, the resonances in $^{58}Ni(p,p_0)$ and $^{88}Sr(p,p_0)$ shown in Figs. 2-4 are all IARs. Normally, it is assumed that the beam energy spread is much larger than the spacing between CN levels and one chooses $\overline{f}$ such that

$$\langle f^{fl} \rangle = 0 \qquad (12)$$

over the beam energy spread. The energy averaged elastic scattering cross section is then

$$\sigma(\theta,E) \equiv \langle |f|^2 \rangle = |\overline{f}|^2 + \langle |f^{fl}|^2 \rangle \equiv \sigma_D + \sigma_{CE} , \qquad (13)$$

where σ_D is called the direct scattering cross section and σ_{CE} the compound elastic cross section, a quantity of considerable interest in nuclear physics.[35] On substituting Eq. (11) into Eq. (7), one finds that the fluctuating part of f destroys the coherence between the incoming and outgoing amplitudes.[34] The final result can be written

$$\overline{P}_K(0) = P_D + P_{CE} , \qquad (14)$$

where, under the same assumptions that were made for Eq. (7),

22

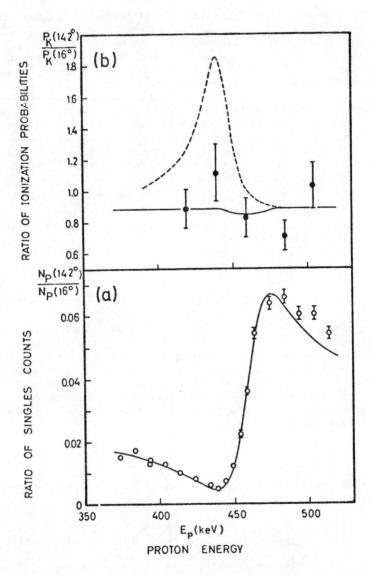

Fig. 5.(a) Ratio of 142° and 16° (lab.) elastic cross sections from the reaction $^{12}C(p,p_0)$, as a function of bombarding energy. (b) Ratio of ^{12}C ionization probabilities in coincidence with elastically scattered protons at 142° and 16° lab. angles. Solid curve computed from Eq. (7). Dashed curve computed from ad hoc modification of Eq. (7) by Duinker et al (Ref. 32.). (From Ref. 33.)

$$P_D = 2 \int_0^\infty d\varepsilon \ |a_{\frac{1}{2}} \overline{f}[\theta, E-\hbar\omega(0)] - a_{\frac{1}{2}}^* \overline{f}(\theta, E)|^2 / \sigma(\theta, E), \qquad (15)$$

$$P_{CE} = 2 \int_0^\infty d\varepsilon \ |a_{\frac{1}{2}}|^2 [\sigma_{CE}(\theta, E-\hbar\omega) + \sigma_{CE}(\theta, E)]/\sigma(\theta, E). \qquad (16)$$

The important result of this analysis is that, if E_K/Γ (= E_K/Γ_{CN}, here) $\gg 1$, the coherence between incoming and outgoing amplitudes in Eq. (7) is destroyed. A preliminary experimental test of Eq. (14) has been made,[36] by measuring the ionization probability across the 5.06-MeV resonance in $^{88}Sr(p,p_0)$ at the lab. scattering angle of 160°. Figure 6(a) shows the scattering cross section σ, as well as the compound-elastic cross section σ_{CE} determined by the use of polarized protons.[37] Figure 6(b) gives the measured ionization probability and values of $P_K(0)$ computed from Eqs. (14)-(16) assuming various magnitudes for the peak cross section of σ_{CE}. One sees that the experimental magnitude of 40mb/sr for the latter cross section gives a good fit for the energy dependence of the ionization probability. Theoretical calculations also show that at scattering angles of 90° or smaller, σ_{CE} has only a negligible influence on $P_K(0)$, so that the previous analyses shown in Figs. 2-4 are essentially unchanged.

In summary, the measurements show that over a wide range of the parameter E_K/Γ Eq. (7) is in agreement with experiment. Therefore, the concepts embodied in this equation can be extended to studies of other nuclear delay time effects by inner-shell vacancy production.

OUTLOOK

The successful detection of delay time effects on x-rays produced in nuclear reactions will inspire further work in this field. The incentive is high, because, in principle, the measurement of CN x-rays can determine nuclear lifetimes in the region of $\sim 10^{-15}$ to $\sim 10^{-18}$ sec and target x-ray methods are sensitive to nuclear delays from $\sim 10^{-18}$ to $\sim 10^{-21}$ sec.[2] Hence, the method has a potentially wide range of applicability. In particular, no other method is presently available to measure nuclear delay times in deep-inelastic heavy-ion reactions.[38] From an atomic physics point of view, one has here the opportunity to study details of the ionization process, not accessible by other means. As experimental techniques develop further, one may expect to see much activity in this fascinating area of overlap between atomic and nuclear physics.

REFERENCES

1. P.C. Gugelot, in Direct Interaction and Nuclear Reaction Mechanisms, Padua, E. Clementel and C. Villi, Eds. (Gordon and Breach, New York, 1962), p. 382.
2. J.F. Chemin, Doctor-of-Science Thesis, University of Bordeaux I (1978).
3. L. Kocbach, in Notes from the Nordic Spring Symposium on Atomic Inner Shell Phenomena, Geilo, Norway, J.M. Hansteen and R. Gundersen, Eds. (Department of Physics, University of Bergen, Norway, 1978), vol. II, p. 65.

24

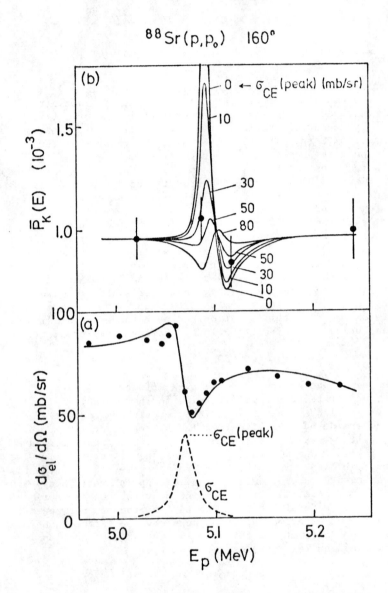

Fig. 6. (a) Cross section of elastically scattered protons at 160° lab. angle from the reaction ^{88}Sr(p,p$_0$), as a function of bombarding energy. (From Ref. 36.) Dashed curve gives compound-elastic cross section obtained by Kretchmer and Graw (Ref. 37.) (b) Ionization probability of ^{88}Sr in coincidence with elastically scattered protons (160° lab.). Curves are computed from Eq. (14) for various assumed peak values of σ_{CE}. (From Ref. 36.)

4. J. Rafelski, private communication (1978).
5. J. Reinhardt and W. Greiner, Repts. Prog. Phys. 40, 219 (1977).
6. J. Rafelski, B. Müller, and W. Greiner, Z. Physik A 285, 49 (1978).
7. U. Müller, Diploma Thesis, Institute of Theoretical Physics, University of Frankfurt (1981).
8. G. Ciochetti and A. Molinari, Nuovo Cimento 40B, 69 (1965).
9. J.S. Blair, P. Dyer, K.A. Snover, and T.A. Trainor, Phys. Rev. Lett. 41, 1712 (1978).
10. R.M. Eisberg, D.R. Yennie, and D.H. Wilkinson, Nucl. Phys. 18, 338 (1960).
11. H. Feshbach and D.R. Yennie, Nucl. Phys. 37, 150 (1962).
12. M.K. Liou, C.K. Liu, P.M.S. Lesser, and C.C. Trail, Phys. Rev. C 21, 518 (1980).
13. C. Maroni, I. Massa, and G. Vannini, Nucl. Phys. A 273, 429 (1976); C.C. Trail, P.M.S. Lesser, A.H. Bond, Jr., M.K. Liou, and C.K. Liu, Phys. Rev. C 21, 2131 (1980); P.M.S. Lesser, C.C. Trail, C.C. Perng, and M.K. Liou, Phys. Rev. Lett. 48, 308 (1982).
14. P. von Brentano and M. Kleber, Phys. Lett. 92B, 5 (1980).
15. K.W. McVoy, X.T. Tang, and H.A. Weidenmüller, Z. Physik A 299, 195 (1981).
16. R. Anholt, Z. Physik A (in press).
17. S. Yoshida, Ann. Rev. Nucl. Sci. 24, 1 (1974).
18. G. Ciochetti, A. Molinari, and R. Malvano, Nuovo Cimento 29, 1262 (1963).
19. J.F. Chemin, S. Andriamonje, J. Roturier, J.P. Thibaud, S. Joly, and J. Uzureau, in Electronic and Atomic Collisions, N. Oda and K. Takayanagi, Eds. (North Holland, Amsterdam, 1980), p. 313.
20. J.F. Chemin, S. Andriamonje, J. Roturier, B. Saboya, J.P. Thibaud, S. Joly, S. Platard, J. Uzureau, H. Lorent, J.M. Maison, and J.P. Shapira, Nucl. Phys. A 331, 407 (1979).
21. S. Röhl, S. Hoppenau, and M. Dost, Phys. Rev. Lett. 43, 1300 (1979).
22. S. Röhl, S. Hoppenau, and M. Dost, Nucl. Phys. A 369, 301 (1981).
23. J.C. Hardy, J.A. Macdonald, H. Schmeing, H.R. Andrews, J.S. Geiger, R.L. Graham, T. Faestermann, E.T.H. Clifford, and K.P. Jackson, Phys. Rev. Lett. 37, 133 (1976).
24. J. Bang and J.M. Hansteen, K. Dan. Videns. Selsk., Mat. Fys. Med. 31, No. 13 (1959).
25. J.S. Blair and R. Anholt, Phys. Rev. A 25, 907 (1982).
26. J.M. Feagin and L. Kocbach, J. Phys. B 14, 4349 (1981).
27. K.W. McVoy and H.A. Weidenmüller, Phys. Rev. A 25, 1462 (1982).
28. N. Austern, Direct Nuclear Reaction Theories, (Wiley Interscience, New York, 1970).
29. J.S. Blair, Annual Report, Nuclear Physics Laboratory, University of Washington, Seattle (1980), p. 132.
30. J.F. Chemin, W.E. Meyerhof, R. Anholt, J.D. Molitoris, and Ch. Stoller, Phys. Rev. A 26, (1982) (in press).
31. J.F. Chemin, R. Anholt, Ch. Stoller, W.E. Meyerhof, and P.A. Amundsen, Phys. Rev. A 24, 1218 (1981).
32. W. Duinker, J. van Eck, and A. Niehaus, Phys. Rev. Lett. 45, 2102 (1980).
33. W.E. Meyerhof, G. Astner, D. Hofmann, K.O. Groeneveld, and J.F. Chemin, to be published.

34. R. Anholt, J.F. Chemin and P.A. Amundsen, Phys. Lett. B, to be published. G. Temmer first drew attention to the possible influence of IAR fine structure on $\bar{P}_K(0)$ (private communication).
35. C. Mahaux and H.A. Weidenmüller, Ann. Rev. Nucl. Sci. $\underline{29}$, 1 (1979).
36. J.F. Chemin, W.E. Meyerhof, R. Anholt, and Ch. Stoller, unpublished results.
37. W. Kretschmer and G. Graw, Phys. Rev. Lett. $\underline{27}$, 1294 (1971).
38. J.W. Harris, in Nuclear Physics with Heavy Ions and Mesons, R. Balian, M. Rho and G. Ripka, Eds. (North Holland, Amsterdam, 1978), p. 263.

DETERMINATION OF THE ELECTRON NEUTRINO MASS
FROM EXPERIMENTS ON ELECTRON-CAPTURE BETA-DECAY

B. Jonson[1], J.U. Andersen[2], G.J. Beyer[1]*, G. Charpak[1], A. De Rújula[1],
B. Elbek[3], H.Å. Gustafsson[4], P.G. Hansen[2], P. Knudsen[3],
E. Laegsgaard[2], J. Pedersen[3] and H.L. Ravn[1]
and
The ISOLDE Collaboration

1. CERN, 1211 Geneva 23, Switzerland.
2. Institute of Physics, University of Aarhus,
   8000 Aarhus, Denmark.
3. Tandem Accelerator, Niels Bohr Institute,
   Risø, 4000 Roskilde, Denmark.
4. Institute of Physics, Technical University of
   Lund, 22362 Lund, Sweden.

ABSTRACT

The IBEC theory has been tested for the case of ^{193}Pt. The Q_{EC} value is 56.3 ± 0.3 keV, and the data make it possible to give an upper limit on the electron neutrino mass of 500 eV (90% CL). The partial beta-decay half-life for M capture in ^{163}Ho is $(4.0 \pm 1.2) \times 10^4$ y and the ^{163}Ho - ^{163}Dy mass difference is 2.3 ± 1.0 keV. From these independent Ho results a half-life of $(7 \pm 2) \times 10^3$ y and an upper limit on m_{ν_e} of 1.3 keV can be deduced.

INTRODUCTION

The possibility that neutrinos can be massive is at present attracting considerable theoretical and experimental attention. Most of the modern particle theories could incorporate and even suggest non-zero masses ranging between some hundred electronvolts and downwards [see for example, De Rújula[1]]. Furthermore, the missing mass problems in cosmology could be solved if neutrinos had masses of the order of some tens of electronvolts[2,3]. In view of the fundamental importance of these problems, more and better experiments are clearly needed.

There are essentially two ways of performing measurements of neutrino masses : The neutrino oscillation experiments (see for example ref.3), and the studies of kinematics of weak interaction decay processes. Here we shall not concern ourselves with the former experiments which may be very sensitive but which depend on a combination of mass differences and mixing angles. We shall instead focus our attention on the kinematic experiments which provide a direct limit to the mass and in Table 1 we summarize the present status[4-10] for the different neutrino mass determinations.

*Visitor from Zentralinstitut für Kernforschung, 8051 Rossendorf, DDR.

0094-243X/82/940027-20$3.00 Copyright 1982 American Institute of Physics

Table I Experimental neutrino mass limits, August 1982

	Mass limit	Reference	Conf. level
$m_{\bar{\nu}_e}$	< 55 eV	Bergkvist[4]	90 %
	~ 35 eV[a)]	Lyubimov et al.[5]	99 %
	< 65 eV	Simpson[6]	95 %
m_{ν_e}	< 4.1 keV	Beck, Daniel[7]	67 %
	< 500 eV	This work	90 %
m_{ν_μ}	< 570 keV	Daum et al.[8]	90 %
	< 500 keV	Anderhub et al.[9]	90 %
m_{ν_τ}	< 250 MeV	Bacino et al.[10]	95 %

a) Limits 14 eV $< m_{\bar{\nu}_e} <$ 46 eV.

In all the experiments on the electron-antineutrino mass, where
the most stringent results are available, the determination of the
mass has been obtained from a study of one single, very favourable
decay process, namely the decay of tritium :

$$^3H \rightarrow {}^3He + e^- + \bar{\nu}_e \; ; \; Q = 18.6 \text{ keV} \qquad (1)$$
$$t_{1/2} = 12.3 \text{ y}$$

With its low Q-value, tritium is clearly ideal for studies of this type. The determination of the antineutrino mass is based on the expected modification of the high-energy end-point of the continuous
β^- spectrum[11]. The most striking result is, of course, the lower limit claimed by the ITEP group. This result has actuated many groups
all over the world to try to make a verification of it. One obstacle
to further progress seems, however, to be the corrections for atomic
and molecular excitations in the final state, which are specially pronounced in the triutium decay.

We shall now turn our attention to the corresponding antiparticle, the electron-neutrino, which is associated with β^+ and electron-
capture (EC) decays. A novel and intriguing method for determining
its mass was proposed by De Rújula[12] a little more than one year ago.
The suggestion is to study the shape near the end-point of the continuous internal bremsstrahlung spectrum (IB) in electron-capture beta
decay. The IBEC spectrum is modified by a finite neutrino mass in a
way that is completely analogous to that for the tritium electron
spectrum; it has a horizontal tangent if the mass of the neutrino is
zero and would be modified to have a vertical tangent if $m_\nu \neq 0$.

THE THEORY OF IBEC

The IBEC radiation is emitted in beta decay by the capture of bound atomic electrons according to the scheme

$$A_Z + e^-_{n,l,j} \rightarrow A_{Z-1} + \nu_e + \gamma_{IB}, \tag{2}$$

where n,l,j are the quantum numbers of the bound electrons. This process has a three-body final state, and the spectrum of emitted photons is continuous with a maximum energy

$$k^0_{max} = Q_{EC} - B[n,l,j]; \; m_\nu = 0, \tag{3}$$

where Q_{EC} is the energy available for EC decay and $B[n,l,j]$ is the binding energy of the captured electron. For Q_{EC} large, compared with atomic binding enrgies, the dominant contribution to the IBEC spectrum arises from capture from the S-states. The shape of the S-wave IBEC spectrum was first derived by Møller[13] and by Morrison and Schiff[14]. The shape is very simple and is found to be proportional to a universal three-body phase-space factor (daughter atom and two light particles) divided by k :

$$\frac{d\omega}{dk} \sim \phi(k) = k(k^0_{max}-k)[(k^0_{max}-k)^2-m_\nu^2]^{1/2}, \tag{4}$$

where k is the photon energy and k_{max} its maximum allowed value for vanishing neutrino mass. The ratio of the radiative-to-non-radiative S capture is of the order of 10^{-4} and is proportional to k^2_{max}. At a low value for k^0_{max} which is clearly most favourable for a neutrino-mass hunt, the S-wave IBEC intensities therefore become extremely low. The reason why an IBEC experiment may still compete successfully with the favourable tritium decay is due to the appearance of another mechanism : at low energies the IBEC spectrum arises mainly from P capture. This was first understood by Glauber and Martin[15]. In the P-wave IBEC, the capture takes place from a virtual S-state associated with a radiative P→S electromagnetic transition. The resonant nature of the process leads to important enhancements of the photon intensity at low energies and, in particular, close to the resonance energies corresponding to P→S transitions, i.e. the normal atomic x-rays. The continuous nP IBEC spectrum differs, in fact, from an ordinary nS non-radiative capture followed by the radiation of a characteristic x-ray only in a relaxation of the requirement of energy conservation in the intermediate state. In the immediate neighbourhood of the line, the process is identical to the real, resonance emission of x-ray. On the other hand, the P-state intensity remains high also for energies that are considerably different from the resonance energies. The intensity observed at the off-resonance energies arises in the nP IBEC process, and there, several virtual n'S-states contribute significantly to the photon spectrum.

The shape of the nP IBEC spectrum is, as demonstrated by Glauber and Martin[15], proportional to the product of $\phi(k)$ [eq.(4)] and a matrix element, $Q_{nP}^2(k)$:

$$\frac{d\omega}{dk} \sim \phi(k) \cdot Q_{nP}^2(k). \tag{5}$$

The Q_{nP} function was discussed in detail in ref.15 for energies above the 1S x-ray pole, $k > B_{1S} - B_{2P}$, and later extended by De Rújula[12] to the region below the pole. We show as examples the $Q_{2P_{3/2}}^2$ and $Q_{3P_{3/2}}^2$ functions for the case of ^{193}Pt in Fig. 1.

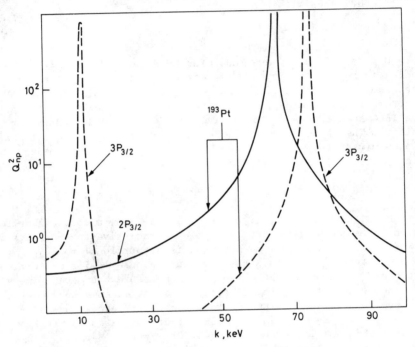

Fig. 1. The $Q_{2P_{3/2}}^2$ and $Q_{3P_{3/2}}^2$ functions for the element Pt calculated in the pole approximation [eqs. (10.5b) and (10.6b) in ref. 11]. The endpoints of the respective IBEC spectra are indicated.

The EC Q-value has nuclear origin and varies for different isotopes of the same element, whilst the atomic binding energies are essentially constant. The challenge is thus to find nuclei with Q-values giving the maximum profit for the resonance enhancement. The ideal case would be one where Q has a value just below the S binding energy of one of the shells. The two most interesting cases found so far are ^{163}Ho and ^{193}Pt. In the following sections we shall describe the first results from our studies of their decay.

DETAILED TEST OF THE THEORY FOR RADIATIVE P CAPTURE

The energy spectra of internal bremsstrahlung in electron capture have been measured for a large number of isotopes of elements, from Be (Z=4) to Tl (Z=81). The most recent review of the subject has been given by Bambynek et al.[16] in 1977. The 1S shape[13,14] of the spectra has been confirmed, and with the theory of Glauber and Martin[15] for the radiative P capture, the increase of the measured intensity at low energies could also be explained. The experimental results were, however, confined to the energy region above the 1S pole. As an example of the best data available we show in Fig. 2 the IB spectrum from ^{131}Cs[17] measured in singles and in coincidence with K x-ray.

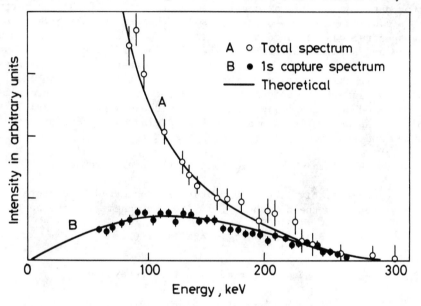

Fig. 2. Internal bremsstrahlung spectrum from ^{131}Cs measured by Biavati et al.[17]. The curves show the predictions from the theory[15].

With the extension of the Glauber-Martin theory by De Rújula, further tests of the theory were needed - especially data for energies below the 1S pole. An isotope with ideal parameters for this purpose is ^{193}Pt, which has a Q-value for electron capture some 20 keV below the K binding energy of the daughter atom, ^{193}Ir. Its decay can thus only take place via capture from the L, M and higher shells. In the following we shall give the main findings from our studies of this nuclide.

The preparation of the ^{193}Pt source and the experimental set-up

The source of ^{193}Pt used in our experiments contained 7×10^{13} atoms. The activity was extracted from about 1 kg of lead which had been used as target material at the ISOLDE facility at CERN. Only material that had had a cooling period of four years or more was chosen so that the only Pt isotope remaining was the long-lived isotope with mass 193 ($t_{1/2} = 50$ y). This activity could be extracted from the lead with a radio-chemical method. First the lead was removed through vacuum sublimation ; the remaining material, containing about 10^{16} atoms of ^{193}Pt, was then cleaned in a complicated sequence of liquid extractions and column separations. The decontamination factor obtained in this procedure was of the order of 10^{12} or better.

A cadmium plate was used as a source holder, and the bremsstrahlung spectrum was measured with an intrinsic Ge detector with 200 mm^2 surface and a thickness of 6 mm. The counting efficiency of this detector was 13% over the energy region of interest. A Si detector was employed for the detection of the L x-rays from the Pt decay, and the source was sandwiched between the two detectors. The IBEC spectrum was recorded both in singles and in coincidence with L x-rays. The whole set-up was built into a graded shielding consisting of lead, cadmium and copper.

Test of the dominance of P capture in the radiative decay of ^{193}Pt

The non-radiative electron capture is dominated by capture from the S states. For the n=2 shell in ^{193}Pt, the scheme for the capture is :

$$^{193}\text{Pt} \rightarrow {}^{193}\text{Ir} \; (2S_{1/2}) + \nu_e$$
$$\hookrightarrow L_I \text{ x-ray}$$

and the major L x-rays associated with this decay are

$$L_I \rightarrow M_{II}M_{III} \text{ at 10.51 and 10.86 keV,}$$
$$L_I \rightarrow N_{II}N_{III} \text{ at 12.84 and 12.92 keV,}$$
$$\text{and} \quad L_I \rightarrow O_{II}O_{III} \text{ at 13.35 and 13.36 keV.}$$

The shape of the Ir L x-ray spectrum is shown in Fig. 3.

The dominant contribution to the radiative capture comes from the 2P shell, with a predicted 2S/2P capture ratio[12] of about 1%.

$$^{193}\text{Pt} \rightarrow {}^{193}\text{Ir} \; (2P_{1/2,3/2}) + \nu_e + \gamma(\text{IBEC})$$
$$\hookrightarrow L_{II} \text{ or } L_{III} \text{ x-rays}$$

The L x-ray spectrum in coincidence with IBEC is also shown in Figure 3. One observes that the lines with L_I origin are strongly suppressed in the coincidence data. A limit of 1.5% for the $2S_{1/2}$ contribution is obtained from the absence of the L_I lines in the complex at 12-14 keV. This amounts to a direct experimental proof that the low energy internal bremsstrahlung originates in P-wave capture.

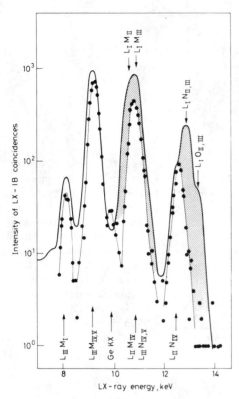

Fig. 3. Spectrum of Ir L x-rays
in coincidence with internal
bremsstrahlung. The curve shows
the shape of the singles L x-ray
spectrum from Ir. The components
originating in L_I ($2S_{1/2}$) capture
are indicated in the upper part
of the figure, whilst some of
the most prominent lines from
L_{II}, and L_{III} ($2P_{1/2,3/2}$) are
shown in the lower part. The L_I
lines are found to be suppressed
with about two orders of magnitude in the coincidence spectrum
(the hatched area). The weak
line at 9.9 keV in the coincidence spectrum is due to cross-
talk coincidences between radiation detected in the Ge detector
and escaping Ge K x-rays.

The ^{193}Pt Q-value

Hopke and Naumann[18] determined the Q_{EC} of ^{193}Pt from the end-
point of the bremsstrahlung spectrum. Their result is Q = 60.8 ±
3.0 keV. We have analyzed our coincidence data in order to try to
improve the precision of the Q-value.

The IBEC spectrum in coincidence with L x-rays has (disregarding the minute 2S contribution) two components originating in the
capture from the two spin-orbit partners $2P_{1/2}$ and $2P_{3/2}$. These differ
in energy by 1.6 keV which would complicate the Q-value analysis.
However, the high-resolution data allow the selection of a region
in the L spectrum with a relatively clean origin in one of the two
states. The best line for this purpose is the one at 9.14 keV (Fig.3)
which is of pure $L_{III}M_{IV}M_V$ origin. The presence of radiation less
transitions within the shell- Coster-Kronig transitions – give, however,
a contribution from the L_{II} sub-shell of about 10% in the coincidence
data. This gives only a small correction; thus we base our analysis on the 9.14 keV line. The Kurie plot of the data is shown in
Fig. 4. The end-point of the spectrum is at 45.3 keV and the
arrive at a Q-value for ^{193}Pt of :

$$Q_{EC} = 56.6 \pm 0.3 \text{ keV.}$$

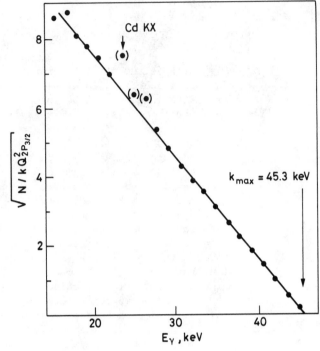

Fig. 4. Kurie plot of the ^{193}Pt IBEC spectrum. The points at 23–25 keV are influenced by the presence of Cd x-rays originating in fluorescence excitation of the Cd source holder. The Q-value obtained from this plot is 56.6 ± 0.3 keV.

Test of the predicted IBEC intensity and the spectral shapes

The IBEC spectra measured in singles and in coincidence with L x-rays are shown in figures 5 and 6. The experimental points are plotted on an absolute scale so that they are given as the intensity per keV. Both sets of data have been corrected for the detector response function. The curves show the spectra calculated according to the prescription given in ref. 12. Also these curves are given on an absolute scale. The comparison furnishes a very striking agreement with the theory. For the absolute intensities the agreement is quantitative !

It is also interesting to investigate the spectral shape in more detail by dividing out the trivial phase space factor from the data. The shape factor obtained in such an analysis is shown in Fig. 7 together with the calculated matrix element $\bar{Q}^2_{2p}$. Again we find excellent agreement with the theory. It should be noted that the predicted S spectrum, if plotted in Fig. 7, would be a straight line with a relative intensity of about 1%.

A limit on m_{ν_e} from the ^{193}Pt data

The case of ^{193}Pt with an endpoint for the 2P IBEC of 45.3 keV is not an ideal case for a precise determination of the electron-neutrino mass. The information about m_{ν_e} is contained essentially

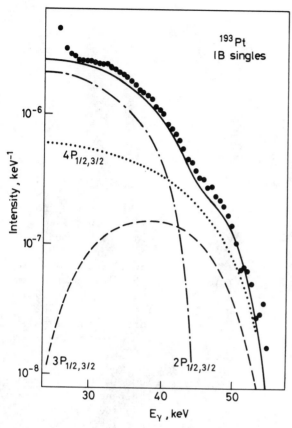

Fig. 5. IBEC spectrum from ^{193}Pt measured in singles. The curves show the calculated spectra from the capture in different subshells. The Q-value used in the calculation is 56.6 keV. The experimental points and the calculated curves are independently given on an absolute scale. The spectrum below 26 keV, which is not shown here, is dominated by L x-rays and pile-up from them.

within a distance of $m_{\nu_e}c^2$ from the endpoint and the useful fraction of the data is proportional to $(m_{\nu_e}c^2/k_{max})^3$. For a neutrino mass hunt, one should thus aim at a case with much lower Q-value. However, from our present data we may already improve on the limit on m_{ν_e} as compared to the best value given until now[7]. [Note : our limit is for m_{ν_e} as opposed to $m_{\bar{\nu}_e}$ (Table 1)]. In Fig. 8 we show the upper part of the Kurie plot of the data together with curves corresponding to different assumptions on m_ν. If we set a limit corresponding to a 90% confidence level, we arrive at $m_{\nu_e} < 500$ eV.

THE BEST CANDIDATE TO DATE : ^{163}Ho

The most promising case for future neutrino mass determinations from IBEC spectra is the long-lived radioactive isotope ^{163}Ho. This isotope was discovered by Naumann et al.[19] who detected[20] radiations of approximatively 1.3 keV which were interpreted as Auger electrons and x-rays following M capture. This assignment was recently confirmed through a measurement[21] with a high-resolution Si(Li) photon spectrometer. The result and the limit[20] set on the L capture branch brackets the Q-value between the limits $2.05 < Q < 9.1$ keV. These

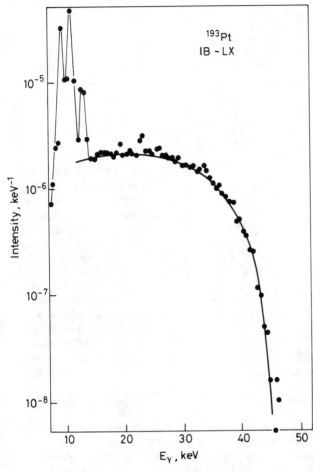

Fig. 6. Bremsstrah-
lung spectrum from
^{193}Pt obtained from a
measurement of coin-
cidences between IB
and L x-rays. The li-
nes 23 and 26 keV are
from fluorescence ex-
citation of the Cd
source holder. The
curve is the calcula-
ted $2P_{1/2}$ + $2P_{3/2}$
IBEC shape given on
an absolute scale.
Some left overs from
the random Lx-Lx
coincidences are
seen in the lower
part of the spectrum.

limits are, however, clearly inconsistent with the half-life of 33 ±
23 y (ref. 20) since the upper limit of the Q-value can be shown to
correspond to a lower half-life limit of 150 y. In the following we
shall report on a half-life determination of ^{163}Ho by counting M
shell radiations from carefully prepared sources containing a known
number of atoms. We have also determined the Q-value by a novel technique based on single-nucleon transfer reactions.

The preparation of the ^{163}Ho sources

The radiations from ^{163}Ho are of extremely low energy, and the
expected half-life is so long that it was felt necessary to aim at
having a source with a level of radioactive impurity well below
10^{-12}. The source would have to contain a known number of atoms,
preferably in the form of a mono-atomic layer in order to permit the
counting of M-shell radiations with energies just above 1 keV. We
have reached this objective by a combination of mass separation, ra-

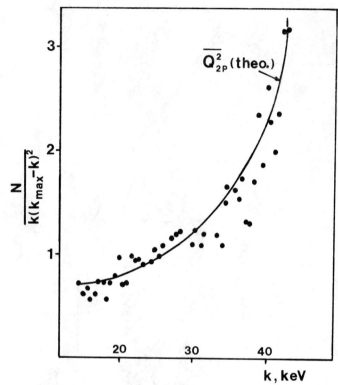

Fig. 7. Shape factor plot of the ^{193}Pt IBEC spectrum from 2P capture. The $\bar{Q}^2_{2P}$ function is given on the same arbitrary scale as the experimental points.

diochemistry and highly sensitive ion beam analysis.

The radioactivity was produced in the ISOLDE Facility connected to the CERN Synchro-cyclotron. A hot tantalum target, consisting of rolls of foil to give a target thickness of 122 g/cm^2, was bombarded with a 2.4 μA beam of 600 MeV protons; rare-earth elements evaporating from the target were ionized by surface ionization and mass separated. The sample collected at mass 163 consisted predominantly of the elements holmium to lutetium (the yield is increased by collecting also the isobars, which decay to ^{163}Ho). The collection took place in a Faraday cup; from the integrated current the number of atoms on the 10 μm tantalum collector foil was calculated to be 1.36 · 10^{15}. After some weeks of cooling, the main contaminants in the sample (determined by γ spectroscopy) were ^{147}Gd (1.4 × 10^{12} atoms collected as an oxide side band), and ^{167}Tm and ^{169}Yb (both about 4 × 10^9 atoms from "tails" in the separator). A further purification was obtained by a complex procedure based on wet chemistry. The collector foil was dissolved in the presence of 3 mg of lanthanum carrier and microgram amounts of other rare-earths, except holmium. For tracing purposes a small amount of short-lived ^{160}Ho was added. The separation was carried out in standard precipitation steps followed by ion-exchange chromatography on an Aminex A5 column with 0.11 M α-hydroxy-isobutyric acid as the eluant. The holmium fraction was cleaned twice,

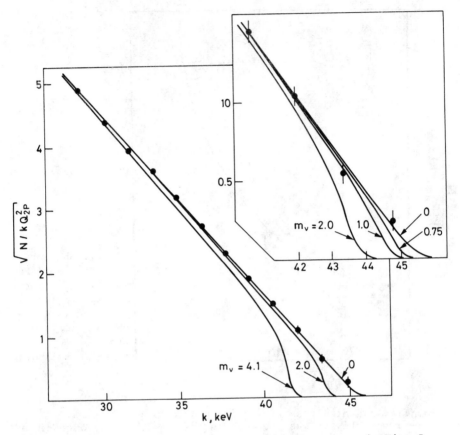

Fig. 8. The high energy part of the Kurie plot shown in Fig. 5
together with the shape of the theoretical IBEC spectrum. A detector resolution of 500 eV FWHM has been assumed. The curves
essentially display the expression

$$(k^0_{max} -k)^{1/2} \; [(k^0_{max} -k)^2 -m^2_\nu]^{1/4}$$

with different assumptions for the value of the electron neutrino mass.

again by the same procedure. From the shape of the elution curves
(see Fig. 9) it was estimated that each step gave a decontamination
of a factor 10^3 from the heavier rare earths (such as Tm, Yb) and
considerable more from the lighter ones. Thus the total decontamination offered by the combination of mass separation and chemistry was
of the order $10^{14} - 10^{15}$. The yield in the chemical procedure was
44%. Two thin sources were prepared from this stock of radioactivity by vacuum evaporation from a rhenium ribbon at 1800° C. The
first (source A) was made before the ^{160}Ho tracer had decayed, so

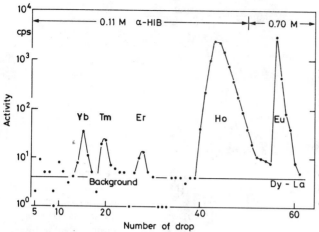

Fig. 9. Ion-exchange chromatogram for the first holmium separation. Note that the tailings, of which only Eu (from ^{147}Eu ^{16}O) is seen, have eluted at a higher concentration of the eluant.

that the yield could be determined : the result of 49% corresponds to 1.9×10^{13} atoms in the central part of a 64 cm^2 tantalum backing.

A second thin source (source B) was prepared by evaporating a sample of 14.6% of the original stock onto a $0.1 \times 20 \times 20$ mm^3 beryllium foil. In an experiment with a Sm tracer and the same geometry, the evaporation yield was estimated to be 35%, which would correspond to 3.1×10^{13} atoms. A second estimate came from counting sources A and B in a 10×10 cm^2 multiwire proportional chamber (MWPC) filled with an argon methane mixture; from the relative count rates, one finds 4.5×10^{13} atoms of ^{163}Ho. A third and entirely independent check on the contents of source B was obtained by means of a highly sensitive surface-analysis technique; elastic Coulomb scattering of particles in back angles at relatively low energy offers a sensitive probe for surface layers and a good mass resolving power arising from the kinematic energy loss in the collision. The experiment carried out with a beam of 3.5 MeV α particles from the Aarhus single-stage Van De Graaff accelerator, shows (Fig. 10) a well-resolved peak at the holmium mass position with an intensity (integrated over the source) of 5.0×10^{13} atoms. There is thus excellent agreement with the Faraday-cup and chemical-yield data. On the basis of these results, we stress that a consistent scale for the number of ^{163}Ho atoms on the sources has been obtained by two entirely independent techniques. We also note that the counting rates of 1 keV M-Auger electrons agree when measured from two different sources on backings with very different Z (4 and 73), so that back-scattering and self-absorption effects cannot be excessive.

The ^{163}Ho half-life

Knowing the number of atoms in the samples, the half-life may be determined from the decay rate of the source. The rate of M-Auger electrons in the MWPC gave a partial M-capture half-life of 51.000 y (see Table 2).

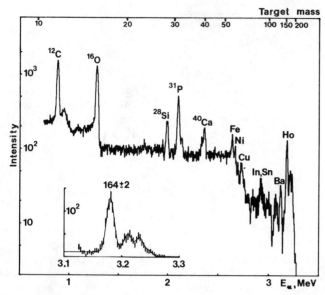

Fig. 10. Energy spectrum of 3.5 MeV α particles scattered from the centre of foil B through 170°. Surface impurities, such as holmium, appear as clearly resolved peaks, while bulk impurities give rise to a step function which reflects the energy loss of the α particle inside the target. The energy scale was calibrated by scattering the beam from a thin target of gold. The inset shows that the high-energy region is dominated by a peak at mass 164 ± 2 with weaker peaks around 183 and 203, presumably Ho, Ta/Re and Au/Pb, respectively. The intensity of the holmium peak integrated over the area of the foil, gives an independent check on the number of atoms in the source.

Table II Determinations of the partial M-capture half-life of ^{163}Ho

Source	No of atoms $(\times 10^{-13})$	M desintegration rate (s^{-1})	$t^M_{1/2}$ (y)
A	1.9	8.2[c]	5.1×10^4
B	4.5[a]	32.4[d]	
	5.0[b]	27.4[e]	3.0×10^4 [g]
		21.6[f]	

a) From the ratio of counting rates B/A = 2.34 measured with a MWPC.
b) From surface analysis by Rutherford scattering (see Fig. 10).
c) M-Auger electrons counted with a MWPC.
d) M-Auger electrons counted with a single-wire proportional counter.
e) M x-rays counted with an argon-filled proportional counter; 2 μm mylar absorber with calculated transmission for M x-rays of 48%. Assumed M fluorescence yield 0.98% (Fig. 11).
f) As (e) but 4 μm mylar absorber, calculated transmission 29%.
g) Based on (a) and (d).

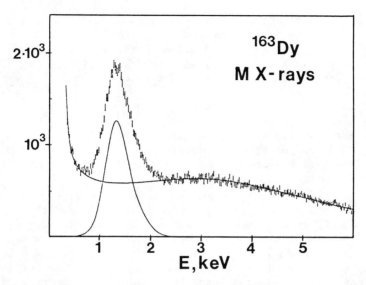

Fig. 11 . Spectrum of M x-rays from [163]Ho measured through a 2 μm mylar foil to absorb Auger electrons. After correction for background only a broad peak near 1.3 keV remains. The fit shown uses the experimental resolution determined with Al x-rays and a fine structure corresponding in energy and relative intensities expected theoretically.

Both the M-Auger electrons and the M x-rays (Fig. 11) were counted from source B. On the basis of the Auger electron decay rate of 32.5 s^{-1} and with the 4.5×10^{13} atoms in the source, we calculate a partial M half-life of 30.000 y.

The various measurements summarized above serve as a set of cross-checks for safeguarding against systematic errors. For this reason it would have little meaning to quote a best value, and we choose instead to represent the data by a single number $t_{1/2} = (4.0 \pm 1.2) \times 10^4$ y.

Estimate of the [163]Ho Q_{EC}-value

The transition probability for allowed electron-capture decay involving the atomic sub-shell x is given by the expression :

$$\lambda_x = \left(\frac{g^2}{4\pi^2}\right) \times [M_F^2 + M_{GT}^2] \times n_x \times \beta_x^2 B_x \phi_x , \qquad (6)$$

where g is the strength of the weak interaction, the M's denote Fermi and Gamow-Teller nuclear matrix elements, n_x is the occupation number (unity for a full shell); β_x^2 the squared electron wave function at the origin of the bound state x; and B_x is a correction for electronic exchange and overlap. The quantity ϕ_x represents the phase-space factor which, for a neutrino mass of zero, is simply the square of the neutrino energy $q_x = Q-E_x$, where Q is the total energy available and E_x the total excitation energy of the final atomic state, not counting the rest mass of the captured electron.

Consider now the [163]Ho experiment which determined

$$\lambda_{M_I} + \lambda_{M_{II}} = \lambda_M = \frac{\ln 2}{t_{1/2}^M} . \qquad (7)$$

It is clear that since the quantities $n_x \beta_x^2$ and B_x are known and tabulated[22], it is possible to determine the phase-space factor ϕ_{LI} *if* the nuclear matrix element can be calculated. (The capture from the M_{II} shell contributes only 10%.)

In the large majority of nuclear decays, the precise calculation of nuclear beta-decay rates is next to impossible. For the [163]Ho case, there is an accidental advantage that comes to our aid. In order to explain why this is so, we must briefly recall some of the properties of single-particle states of strongly deformed nuclei[23].

The single-particle states of strongly deformed nuclei may be characterized by a set of Nilsson (asymptotic) quantum numbers $\Omega^\pi[Nn_z\Lambda]$. Here Ω denotes the projection of total angular momentum on the nuclear symmetry axis, and the parity π is $=(-1)^N$, where N is the total number of quanta in an harmonic oscillator representation. The quantities n_z and Λ are the number of oscillator quanta along the nuclear symmetry axis and the projection of orbital angular momentum on the nuclear symmetry axis. A rare and well-studied set of allowed β transitions called "allowed, unhindered" (*au*) proceeds between states that obey the asymptotic selection rules $|\Delta\Omega| = 1$, (ΔN, Δn_z, $\Delta\Lambda$) = 0 and $\Delta\pi$ = no. These transitions have very large transition probabilities and have been the subject of several studies. From the general systematics of Nilsson states in the rare-earth region, it is clear that the [163]Ho β decay must be *au* and proceed from the state $\pi 7/2^-[523]$ (ground state of all deformed Ho isotopes) to $\nu 5/2^-[523]$. (Here π and ν denote "proton" and "neutron").

It turns out to be possible to write the reduced transiton probability for an *au* transition in a deformed odd-mass nucleus

$$|M_{GT}|^2 = \frac{C_A^2}{C_V^2} \times \gamma_1^2 \times {<I_i\Omega_i 1(\Omega_f-\Omega_i)|I_f\Omega_f>}^2 \times R_{pair} R_{\sigma\tau}, \qquad (8)$$

where the terms denote the ratio of GT/F coupling strength, the square of the single-particle matrix element γ_1 (which varies with nuclear deformation), a statistical factor, and corrections for pairing and spin-isospin polarization. The last term has a large absolute influence as it contributes with a factor of about one eighth[25], but it is expected to vary little between neighbouring nuclei as it is dominated by the large Coulomb energy shift of the Gamow-Teller Giant Resonance.

By taking only the statistical factor and the pairing correction into account, it is possible to calculate $|M_{GT}|^2$ for [163]Ho from the same β transition observed in neighbouring nuclei. It turns out that this conversion is most accurate if R_{pair} for the case chosen for comparison contains the same emptiness and fullness coefficients as ^{163}Ho[24]. This effect, illustrated in Table 3, arises because the errors in the assumed single-particle energies roughly cancel in this case.

Table III Calculations of pairing corrections for electron-capture β decays (EC) of the type $^7/_2{}^-[523] \to {}^5/_2{}^-[523]$. (The framed pair was used to calculate the ^{163}Ho nuclear matrix element.)

	R_{pair}	Ref. 25	Ref. 24
^{163}Ho	$u_p^2\ u_n^2$	0.21	0.354
^{161}Ho		0.36	0.508
^{165}Er	$v_p^2\ v_n^2$	0.52	0.436
^{163}Er		0.42	0.271

Using as our reference ^{161}Ho ($t_{1/2}$ = 2.48 h, Q = 853 ± 3 keV, $^7/_2 \to {}^5/_2$ branch 75%, log ft = 4.97 (Lederer et al.[26]), we find for ^{163}Ho a log ft value of 5.12.

We assume a 25% error in the semiempirical calculation and combine it with the observed ^{163}Ho M-shell capture rate to obtain

$$\phi^{1/2}(M_I) = 0.53 \pm 0.10 \text{ keV} \qquad (9)$$

after correction for M capture. With the assumption of m_{ν_e} = 0, this corresponds to

$$Q_{EC} = 2.58 \pm 0.10 \text{ keV}. \qquad (10)$$

The partial half-life and the theoretically known ratios for M/N/O capture (given by the electron wave functions at the origin) result in a ^{163}Ho half-life of $(7\pm2)\ 10^3$ y.

A m_{ν_e} limit from the combination of the phase space factor and the Q-value from nuclear reaction studies.

A measurement of Q without any neutrino mass dependence may be obtained from nuclear reaction experiments. Such a direct determination of the $^{163}Ho-^{163}Dy$ mass difference has been performed at the Niels Bohr Institute Tandem Accelerator and multigap magnetic spectrograph[27]. The reaction chain used was*)

$$^{163}Dy(d,t)\ ^{162}Dy(\tau,d)\ ^{163}Ho. \qquad (11)$$

*)We have used the conventional symbols t and τ for ^3H and ^3He.

Nearby reference lines were detected in the same spectrum from elastic scattering

$$^{163}Dy(d,d') \ ^{163}Dy \ \text{and} \ ^{162}Dy(\tau^{++},\tau^{+}) \ ^{162}Dy. \qquad (12)$$

The novel feature of this technique is that one detects the τ's as singly charged. These follow a path through the spectrograph close to the one of the tritons so that a nearby reference line is obtained. The Q-value can then be extracted from the sum of the energy differences of the two pairs (d,d') and (t,τ^{+}) corrected for the difference in recoil energy ΔE_R and the $(t,^{3}He)$ mass difference

$$Q = \Delta E(\tau^{+},t) + \Delta E(d,d') + \Delta E_R + \Delta m \ (t,^{3}He) \qquad (13)$$

The result from this measurement[27] is 2.3 ± 1.0 keV. The ultimate precision in the experiment is expected to be appreciably better than the one quoted here.

The reaction Q-value and the phase-space factor may be combined to give a limit on m_{ν_e}. This is graphically illustrated in Fig. 12. where $\phi^{1/2}$ is plotted against Q for different assumptions on m_{ν_e}. The hatched areas correspond to the present experimental limits on Q and $\phi^{1/2}$. The limit set on m_{ν_e} in this way is 1.3 keV. This limit can certainly be further improved with this technique, but the most promising prospects for future progress are to measure the IBEC spectrum from ^{163}Ho and to deduce the mass from its shape.

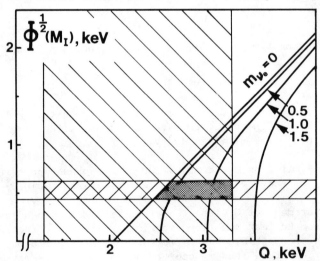

Fig. 12. Plot of $\phi^{1/2}(M_I)$ versus Q for different assumptions on the neutrino mass. The shadowed area corresponds to the limits set by the two present experiments and gives an upper limit of 1.3 keV.

FUTURE PROSPECTS FOR A DETERMINATION OF m_{ν_e}.

In this report we have demonstrated the existence of large enhancements of the IBEC intensities for the case of ^{193}Pt. In the case of ^{163}Ho the Q_{EC}-value is only about 500 eV from the 3S binding energy. The close presence of the 3S pole thus gives a very large enhancement effect in this case. The isotope ^{163}Ho would then be the best case known, although better ones might still be found.

The next step in our measurements will most likely be the detection of IB-Auger electron coincidences. The best rates are expected to come from the 4P capture:

$$^{163}\text{Ho} \rightarrow {}^{163}\text{Dy}^{N_2} + \gamma + \nu_e \quad (k_{max} \sim 2250 \text{ eV})$$

$$\hookrightarrow {}^{163}\text{Dy}^{N_4;N_{6,7}} + e^- (E_e \sim 175 \text{ eV}) \qquad (14a)$$

and

$$^{163}\text{Ho} \rightarrow {}^{163}\text{Dy}^{N_3} + \gamma + \nu_e \quad (k_{max} \sim 2290 \text{ eV})$$

$$\hookrightarrow {}^{163}\text{Dy}^{N_5;N_{6,7}} + e^- \quad (E_e \sim 135 \text{ eV}) \qquad (14b)$$

Calorimetric experiments may even be more promising since the process of electron ejection in electron capture[28] (EEEC) dominates over the IBEC spectrum by almost three orders of magnitude in the Ho case.

REFERENCES

1. A. de Rújula, Nucl. Phys. A374, 619 (1982) and references therein.
2. W.J. Marciano, Comments on Nuclear and Particle Physics IX, No 5, 169 (1981).
3. P.H. Frampton and P. Vogel, Phys. Reports Vol. 84, 342 (1982).
4. K.E. Bergqvist, Nucl. Phys. B39, 317 (1972); Phys. Scripta 4, 23 (1971).
5. V.A. Lyubimov, E.G. Novikov, V.Z. Nozik, E.F. Tretyakov and V.S. Kosik, Phys. Lett. 94B, 266 (1980).
6. J.J. Simpson, Phys. Rev. D23, 649 (1981).
7. E. Beck and H. Daniel, Z. Phys. 216, 229 (1968).
8. M. Daum, G.H. Eaton, R. Frosch, H. Hirschmann, J. McCulloch, R.C. Minehart and E. Steoner, Phys. Lett. 74B. 126 (1978).
9. H.B. Anderhub, J. Boecklin, H. Hofer, F. Kothermann, P. Le Coultre, D. Makowiecki, H.W. Reist, B. Sapp and P.G. Seiler, Phys. Lett. 114B, 76 (1982).
10. W. Bacino, T. Ferguson, L. Nodulman, W.E. Slater , H.K. Ticho, A. Diamant-Berger, G. Donaldson, M. Duro, A. Hall, G. Irwin, J. Kirkby, F. Merritt, S. Wojcicki, R. Burns, P. Condon and P. Cowell, Phys. Rev. Lett. 42, 749 (1979).
11. F. Perrin, Comptes Rendus 197, 1625 (1933). E. Fermi, Nuovo Cimento 11, 1 (1934); Z. Physik 488, 161 (1934).
12. A. De Rújula, Nucl. Phys. 188B, 414 (1981).

13. C. Møller, Phys. Z. Sowjetunion 11, 9 (1937) ;   Phys. Rev. 51, 84 (1937).
14. P. Morrison and L.I. Schiff, PHys. Rev. 58, 24 (1940).
15. H.J. Glauber and P.C. Martin, Phys. Rev. 104, 158 (1956); P.C. Martin and R.J. Glauber, Phys. Rev. 109, 1307 (1958).
16. W. Bambynek, H. Behrens, M.J. Chen and B. Crasemann, M.L. Fitz-patrick, K.W.D. Kedinghan, H. Genz, M. Mutterer and R.L. Intemann, Rev. Mod. Phys. 49, 77 (1977).
17. M.H. Biavati, S.J. Nassif and C.S. Wu, Phys. Rev. 125, 1364 (1962).
18. P.K. Hopke and R.A. Naumann, Phys. Rev. 185, 1565 (1969).
19. R.A. Naumann, M.C. Michel and J.C. Power, J. Inorg. Nucl. Chem. 15, 195 (1960).
20. P.K. Hopke, J.S. Evans and R.A. Naumann, Phys. Rev. 171, 1290 (1968).
21. C.L. Bennett, A.L. Hallin, R.A. Naumann, P.T. Springer, M.S. Witherell, R.E. Chrien, P.A. Baisden and D.H. Sisson, Phys. Lett. 107B, 19 (1981).
22. W. Bambynek, B. Crasemann, R.W. Fink, H.-U. Freund, H. Mark, C.D. Swift, R.E. Price and F.V. Rao, Revs. Mod. Phys. 44, 716 (1972).
23. Aa. Bohr and B.T. Mottelson, Nuclear Structure (W.A. Benjamin Inc Reading. Mass., 1975), Vol. II, pp. 245, 296 and 306.
24. R. Bengtsson and I. Ragnarsson, private communication (1981).
25. J. Żylicz, P.G. Hansen, H.L. Nielsen and K. Wilsky, Ark. Phys. 36, 643 (1967).
26. C.M. Lederer, V.S. Shirley, E. Browne, J.M. Dairiki, R.E. Doebler, A.A. Shibab-Eldin, L.J. Jardine, J.K. Tuli and A.B. Buyrn, Table of Isotopes (J. Wiley & Sons, Inc., New York, 1978).
27. J.U. Andersen, G.J. Beyer, G. Charpak, A. De Rújula, B. Elbek, H.Å. Gustafsson, P.G. Hansen, B. Jonson, P. Knudsen, E. Laegsgaard, J. Pedersen and H.L. Ravn, Phys. Lett. 113B, 72 (1982).
28. A. De Rújula and M. Lusignoli, CERN preprint TH. 3300, (1982); to be published in the proceedings of the Int. Conf. on Unified Theories and their Experimental Tests, Venice, 16-18 March 1982.

RELATIVISTIC THEORY OF SHAKEOFF ACCOMPANYING β-DECAY

J Law
Guelph Waterloo Program for Graduate Work in Physics,
Guelph Campus, University of Guelph,
Guelph, Ontario, Canada, N1G 2W1.

ABSTRACT

The relativistic theory of shakeoff of innershell electrons during β-decay is reexamined. In the case of negative β-decay a resolution of the discrepancy that existed between theory and experimental data has been achieved. This resolution comes about by considering the many-body aspects of the problem, and the wave functions involved in the initial and final states have to be evaluated more precisely. In the case of electron capture shakeoff the agreement still leaves room for probing electron correlation. The glaring fault seems to be in the case of positron decay shakeoff, where the theoretical values are about half of the measured ones. A plausible argument is given to resolve this discrepancy by considering the positron-ejected electron final state interaction.

INTRODUCTION

The understanding of the problem of shakeoff accompanying β-decay has gone through various phases. Beginning with the early calculations of Feinberg[1] and Migdal,[2] through the later reformulation and calculations of Law and Campbell (LC)[3] and Isozumi, Shimizu and Mukoyama (ISM),[4] our understanding of the processes have slowly crystallised owing in a large part to the excellent experiments performed. This interplay between theory and experiment is going on even now as will become evident later. I will not spend time in this talk tracing the history of the subject, as it is well covered in the literature,[5] but concentrate on more current progress.

Let me define the shakeoff process, before proceeding further. When an atom β-decays, there is a small probability that an innershell electron can be promoted to higher levels or to the continuum, leaving an innershell vacancy. The case where the electron is ejected is termed "shakeoff" and where the electron is simply promoted to higher bound levels "shakeup". Since shakeup is an even less probable process than shakeoff, I will simply refer to both processes as "shakeoff". Experimentally then, the signature of the process is a simultaneous detection of the X-ray from filling of the vacancy and the ejected electron (and/or the β-particle). The term β-decay covers both β⁻ decay and electron capture. In this talk I will restrict myself to K-vacancy creation in β± decay and K-vacancy creation in K-(electron) capture.

CURRENT STATUS

(a) β^- decay shakeoff

In the case of β^- decay, the calculations of LC and ISM using Dirac wave functions gave at most only ~50% of measured shakeoff probability in comparison with more than 33 measurements. The suggestion that this discrepancy could be accounted for by Direct Collision or Final State Interaction (FSI) was laid to rest by Intemann[6] in his recent work. Many body effects which were estimated by Cooper and Åberg[7] and by Mukoyama and Shimizu,[8] helped in lessening the discrepancies, but did not remove them. The approach of Law and Suzuki[9] calculates the many body effects in a relativistic self consistent field framework, and has resolved the discrepancy. The basis of the model lies in the definition of the final state and may be illustrated as follows:
Consider

$$Cu \rightarrow Zn^{++} + e^- + e^- + \bar{\nu},$$

$$\{[Ar]3d^{10}4s\}_{29} \rightarrow \{[Ar]1s^{-1}\ 3d^{10}4s\}_{30} + e_p + e_s + \bar{\nu},$$

where we have written down the configurations in the second form. The subscript is the nuclear charge and the argon core [Ar] in Zn^{++} has a K vacancy, hence the $1s^{-1}$; e_p, e_s are the electron continuum states. In all the calculations performed to date, the argon core and outer orbitals $3d^{10}$ and $4s$ basis states are solved for in the final state in the configuration

$$\Psi_g^f = \{[Ar]3d^{10}4s\}_{30} \quad \text{for } Zn^+,$$

with a K-shell occupation number of 2. However, the actual final state may approximate more to a K shell occupation number of 1. Thus we should solve for the configuration

$$\Psi_v^f = \{[Ar]1s^{-1}\ 3d^{10}\ 4s\}_{30}\ e_p \quad \text{for } Zn^{++} + e_p.$$

Note that the former configuration over-compensates for the interaction of the shakeoff electron by replacing it into the K shell. The LS approach is to solve for the Zn^{++} system first and then generate a SCF potential for the shakeoff electron e_p. The various ways of defining this potential are described in the paper by LS, within the framework of the Optimised-Dirac-Fock-Slater scheme as introduced by Lindgren and Rosén.[10] Example of a spectrum fit is given in fig. 1, and Table I shows the calculated values for P_K. The results also indicate that in this approach the shakeup contribution is at the most 1% of the shakeoff results, and therefore it has not been included in the table. There is still some room for final state interaction.

(b) K-shakeoff in K-capture

An extension of the model to K-shakeoff in K-electron capture was made by Suzuki and Law (SL).[11] The results are shown in Table II. The agreement is less spectacular than the β^- case. The tendency here is for theory to be slightly on the high side. A possible explanation is that initial state correlation is underestimated by the SCF model. Taking as an example the case of ^{207}Bi, fig. 2 shows the effect of taking a correlated wave function in the initial state of the form

$$\Phi_K(r_1, r_2) = A\{\phi_1(r_1)\phi_2(r_2) \exp(\gamma r_{12}/2)\}$$

The ϕ's are SCF single particle wave functions. The value of γ needed to fit Bi lies between 0.6 to 1.0, which is somewhat larger than expected since the SCF model already includes some mutual correlation, whereas the corresponding result for Cs gives a more reasonable range of γ from 0.37 to 0.43. Anyhow, this indicates the sensitivity to initial state correlation.

(c) β^+ decay shakeoff

The most interesting results have to do with β^+ decay shakeoff. Within the same model as used for the β^- decay calculations, the results for β^+ decay shakeoff show definite disagreement. The results are shown in Table I. The various other possible processes such as Internal K-Annihilation (IKA) and Charge Exchange (CE) have been estimated to be small. A process that has not been estimated is Final State Interaction (FSI). We will show that this may be the possible cause of the discrepancy.

THE IMPORTANCE OF FSI

Consider a simplified model atom of charge Ze consisting of only 2 electrons both in the K shell. Let us assume that the electrons' mutual interaction and correlation produces exactly a shielding of unit charge, so that we can write the exact initial state wave function as

$$\Phi_i = \psi_K(Z-1)\psi_K(Z-1)$$

(i) Suppose this system K-captures, then the final state is

$$\Psi_f^e = \psi_K(Z-1)\phi_\nu.$$

In this case, the non captured electron has to remain in the K-shell, since the final state K-orbital overlaps completely with the initial state K-orbital. Thus no shakeoff can occur.

(ii) Suppose we consider β^- decay occurring with shakeoff. Ignoring the β^- particle first, our method of calculation involves defining a potential for the ejected electron. The bound

50

electron sees a full charge of Z+1. The continuum electron is
shielded by the bound electron, so that the final state is

$$\Psi_f^- = \psi_K(Z+1)\psi_p(Z+1-\sigma)\phi_\nu$$

where σ is the appropriate shielding. Thus a measure of
shakeoff is the overlap integral matrix element
$|\langle\psi_p(Z+1-\sigma)|\psi_K(Z-1)\rangle|^2$, and indirectly the charge difference
$(Z+1-\sigma)-(Z-1) = 2-\sigma$ between initial and final states. If we
were to bring back the effect of the β^- particle and ascribe it
as an extra shielding factor δ, then the relative contribution
due to it is $|\delta/(2-\sigma)|$.

(iii) For β^+ decay, a similar argument gives a final state wave
function of

$$\Psi_f^+ = \psi_K(Z-1)\psi_p(Z-1-\sigma)\phi_\nu$$

hence a shakeoff matrix element of $|\langle\psi^P(Z-1-\sigma)|\psi^K(Z-1)\rangle|^2$ and a
corresponding charge difference of $(-\sigma)$. If we place the β^+ e^-
final state interaction as an effective charge δ, then we see
that the relative contribution is $|\delta/\sigma|$. Thus it is plausible
that the FSI of the β^+ with the ejected e^- could account for a
large part of the discrepancy.

CONCLUSION

We have presented indications that the theory of shakeoff
accompanying β-decay may be finally succeeding in confronting
experimental results. However, there are still problems to solve.
The use of Relativistic Hartree-Fock wavefunctions as opposed to
Relativistic Hatree-Fock-Slater wavefunctions and the effect of
correlations over and above the SCF correlations should be explored.
FSI should also be incorporated especially in the positron decay
cases. Finally, experiments on higher shell vacancies would be very
useful to extending the theory.

ACKNOWLEDGMENTS

I would like to thank the Natural Sciences and Engineering
Research Council of Canada for a grant in aid of this research and my
colleagues Drs. Akira Suzuki and J. L. Campbell for invaluable
discussions.

REFERENCES

1. E. L. Feinberg, J. Phys. (USSR) 4,423 (1941).   E. L. Feinberg,
    Yad. Fiz. 1, 612 (1965) [Sov. J. Nucl. Phys. 1, 438 (1965).
2. A. Migdal, J. Phys. (USSR) 4, 449 (1941).
3. J. Law and J. L. Campbell, Nucl. Phys. A185, 529 (1972). J. Law
    and J. L. Campbell, Phys. Rev. C 12, 984 (1975).

4. Y. Isozumi, S. Shimizu, and T. Mukoyama, Nuovo Cimento A 41, 359 (1977).
5. M. S. Freedman, Annu. Rev. Nucl. Sci. 24, 209 (1974). R. J. Walen and Chantal Briançon, Atomic Inner-Shell Processes, edited by B. Crasemann (Academic, New York, 1975), p. 233.
6. R. L. Intemann, Probability of Internal Ionization during Nuclear β-decay: Effect of Final-State Interaction, (sub to Phys. Rev. C, 1982).
7. J. W. Cooper and T. Åberg, Nucl. Phys. A298, 239 (1978).
8. T. Mukoyama and S. Shimizu, J. Phys. G 4, 1509 (1978).
9. J. Law and Akira Suzuki, Phys. Rev. C 25, 514 (1982).
10. I. Lindgren and A. Rosén, Case Stud. At. Phys. 4, 93 (1974).
11. Akira Suzuki and J. Law, Phys. Rev. C 25, 2722 (1982).

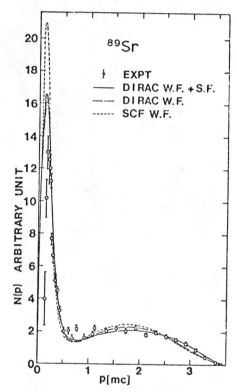

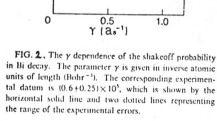

FIG. 2. The γ dependence of the shakeoff probability in Bi decay. The parameter γ is given in inverse atomic units of length (Bohr^{-1}). The corresponding experimental datum is $(0.6+0.25)\times10^5$, which is shown by the horizontal solid line and two dotted lines representing the range of the experimental errors.

FIG. 1. K-shell ejected electron spectrum measured in coincidence with K x rays in the β decay of ^{89}Sr. The curves ($--$) Dirac w.f., and (----) SCF w.f. do not include the shape factor corrections. The theoretical curves are normalized to the data point at $p=2.5$ mc.

TABLE I. Shakeoff probabilities in units of 10^{-4}.

	E_0 (keV)	B_K (keV)	P(exp)	P(ODFS)	P(LDA)	P(AKF)	k_f (mc)
beta minus decay							
Cl	710	3.21	22.1 ±3.8	46.09	43.77	39.81	0.10
Ca	252	4.49	24.3 ±3.9	28.57	26.82	24.44	0.11
Ni	65.9	8.98	4.6 ±0.4	5.54	5.25	4.81	0.13
Cu	573	9.66	11.8 ±0.8	14.27	13.43	11.95	0.14
¹⁹Sr	1463	17.04	8.32 ±0.63	8.97	8.39	7.54	0.17
⁹⁰Sr	546	17.04	6.0 ±0.9	7.30	6.85	6.12	0.17
Y	2270	18.00	7.2 ±1.2	8.89	8.31	7.52	0.17
Nb	160	20.00	3.4 ±0.4	2.88	2.70	2.46	0.18
Tc	292	22.12	3.65 ±0.11	3.88	3.65	3.34	0.18
In	1978	29.20	5.40 ±0.14	5.42	5.05	4.54	0.20
Pr	930	43.57	2.89 ±0.14	2.90	2.70	2.42	0.23
Pm	225	46.84	0.906 ±0.047	0.78	0.73	0.66	0.23
Sm	76	48.52	0.022 ±0.003	0.020	0.019	0.018	0.24
Er	335	59.39	1.0 ±0.2	0.81	0.77	0.70	0.26
W	429	71.68	1.00 ±0.25	0.78	0.74	0.67	0.27
Hg	214	85.53	0.13 ±0.04	0.13	0.12	0.11	0.29
Tl	765	88.01	1.12 ±0.06	1.05	0.99	0.88	0.29
Bi	1160	93.11	1.30 ±0.07	1.38	1.29	1.15	0.30
beta plus decay							
Cu	656	8.33	13.23 ±0.65	5.80	7.03	˙8.25	0.13
Co	474	7.11		6.70	7.96		

TABLE II. Shakeoff probabilities (in multiples of 10^5). See text for description of the various columns. $B_K(i)$ and $B_K(f)$ represent the K-shell binding energies of the parent and the daughter atoms, respectively.

	E (keV)	$B_K(i)$ $B_K(f)$ (keV)	P (exp)	P (ODFS)	P (LDA)	P (AKF)
Ar	813.8	3.21	44 ± 8	49.36	52.94	55.40
		2.82	37 ± 9			
Fe	231.4	7.11	38 ±17	18.80	20.06	21.14
		6.54	28			
			12 ± 4			
			10.1 ± 2.7			
Ge	235.7	11.10	13.3 ± 1.4	11.12	11.84	12.55
		10.37	13 ± 5			
			24			
Pd	507.2	24.35	3.13± 0.31	5.59	6.03	6.45
		23.22				
Cd	94.0	26.71	72 + 5	0.83	0.89	0.97
			− 15			
		25.51	15.2 ± 2.4			
			2.8 ± 0.7			
			1.02± 0.36			
Cs	355.6	35.99	5.0 ± 1.0	2.99	3.22	3.48
		34.57	2.5 ± 0.2			
			2.0 ± 1.3			
			1.33± 0.33			
			2.3 ± 0.3			
			1.4 ± 0.1			
Er	377.1	57.49	1.5 ± 0.4	1.58	1.71	1.88
		55.62	0.67± 0.39			
Bi	772.4	90.53	0.6 ± 0.25	1.81	1.97	2.20
		88.01				

ALIGNMENT IN IONIZATION BY ION, ELECTRON AND PHOTON IMPACT

Werner Mehlhorn
Fakultät für Physik, Universität Freiburg, D-7800 Freiburg, FRG

INTRODUCTION

Atoms are said to be aligned if the population probabilities of magnetic substates are unequal with the restriction that they are independent of the sign of the magnetic quantum number. The alignment in electron impact excitation of atoms and its measurement by the polarization of subsequent line radiation has been known since 1926[1]. The question of an alignment in the <u>ionization</u> process came up first in connection with the new technique of Auger electron spectrometry, i.e. the use of gaseous targets and the ionization by an external directed beam of X-rays or electrons[2]. It was only in 1968 that it was predicted[3] that ionization of an inner-shell electron with $j \geq 3/2$ by a directed unpolarized particle beam would lead to an alignment of the ionic state. Auger electrons or X-rays emitted in the subsequent decay manifest this alignment through their anisotropic angular distribution or through the polarization of the X-rays. After the first experimental evidence of an alignment in an inner-shell ionization by electron impact had been given (HgL$_3$ in 1970[4], ArL$_3$ in 1971[5]), it became clear that the measurement of the alignment would also offer the possibility of determining the ionization cross sections of magnetic subshells nℓjm and would thus give a more detailed information on the collision process. The first calculation of the alignment for electron impact ionization was done by McFarlane[6] and the theory of nonisotropic angular distribution of Auger electrons was given by Cleff and Mehlhorn[7] in 1974.

In the case of ionization by ion impact, Volz and Rudd[8] tried to measure the alignment of ArL$_3$ for 300 keV H$^+$ projectiles but did not get a clear result. (We now know that for the alignment of ArL$_3$ for 300 keV H$^+$ projectiles, the anisotropy of Auger electrons is of the order of 7 %.). The first successful measurements of an alignment in ion-atom collisions were performed for MgL$_3$ in Mg$^+$-Ar collisions[9] and for AlKL$_{2,3}$ ionization by H$^+$ and He$^+$ impact[10].

The alignment in inner-shell photoionization was predicted in 1972[11], the first experimental evidence was given by Caldwell and Zare[12] in 1977. Since 1977 the interest in investigations, experimentally as well as theoretically, on the alignment in inner-shell ionization has been rapidly increasing.

In the present contribution, the alignment of inner-shell ionized atoms is discussed in the light of our present theoretical understanding and of recent experimental results. In particular, we consider the following two-step processes:

Step I: Ionization

$$\left\{ \begin{array}{c} P \\ \\ \gamma \end{array} \right\} + A(J_o) \rightarrow A^+_{\text{inner shell}}(JM) + \left\{ \begin{array}{c} P + e^- \\ \\ e^-_{ph}(\ell j) \end{array} \right\} \qquad \begin{array}{c} (1a) \\ \\ (1b) \end{array}$$

Step II: Decay

$$A^+_{\text{inner shell}}(JM) \left< \begin{array}{l} A^{++}(J_f) + e^-_{\text{Auger}}(\ell'j') \qquad\qquad (2) \\ \\ A^+(J'_f) + \gamma' \qquad\qquad\qquad\quad (3) \end{array} \right.$$

As projectiles P either electrons or light ions (H^+, He^+) will be considered (the alignment in inner-shell ionization by heavy ion-atom collisions is discussed by Jitschin[13]). We shall assume that the target atoms are randomly oriented and that the beam of projectiles P is unpolarized. The photons γ are either unpolarized or linearly or circularly polarized. We shall further assume that the SLJM coupling is valid, i.e.

fine structure splitting $\gg$ level widths Γ $\gg$ hyperfine structure splitting,

which is certainly the case for inner-shell vacancy states.

In almost all experimental alignment studies the axial symmetry (non-coincidence experiments) has been used. Only very recently in coincidence experiments[14,15] has the symmetry of the experiment been determined by the plane of the incident and scattered particle P.

THEORY

The ionic state formed by impact ionization or by photoionization is in general anisotropic. Theoretically, the anisotropy of a state is described by a set of anisotropy parameters $\mathcal{A}_{kK}$ (with $\mathcal{A}_{oo} \equiv 1$) which are uniquely related to the matrix elements of the density matrix of the relevant ionization process (here and in the following we will use the notation of Berezhko at al.[16-18]). Due to the basic assumption that ionization and decay can be treated independently as a two-step process, the general angular correlation of emitted Auger electrons e^- or photons γ' can be expressed in terms of the parameters $\mathcal{A}_{kK}$ and of coefficients α_k, where the latter being dependent only on the decay process. The number of independent non-zero parameters $\mathcal{A}_{kK}$ of an ionic state is determined by the symmetry of the experiment. Furthermore, the parameters $\mathcal{A}_{kK}$ that could be deduced from experiment depend also on the mode of decay of the inner-shell vacancy. The Auger decay, where parity is conserved, restricts k to only even values, the electric dipole radiative decay restricts k to only 1 and 2.

In the following we shall outline the theoretical formulation only for the symmetries of those experiments which have already been performed: 1) axial symmetry in particle and photon ionization and

2) plane symmetry in coincidence experiments with incident
particles.

For simplicity we shall only consider closed shell atoms A
with $J_o = 0$ before the inner-shell ionization.

Particle impact experiments with axial symmetry

Here we assume an unpolarized beam of projectiles P, defining
the z-axis, and randomly oriented target atoms. We also assume that
neither the scattered projectile nor the ionized electron is detected. Consequently, the state of the ion is axially symmetric about z
and symmetric with respect to reflection in the plane perpendicular
to z. This restricts the anisotropy parameters $\mathcal{A}_{k\kappa}$ of the ionic
state $A^+(JM)$ to $\mathcal{A}_{k0}$, where k can take on only even values 2,4,...,
2J-1. From this it follows that only inner-shell vacancy states with
$J \geq 3/2$ are anisotropic. For a given value of J the $\mathcal{A}_{k0}$ can be ex-
pressed in terms of the occupation probabilities $P(J|M|)$, since
$P(JM) = P(J, -M)$. For $J_o = 0$ the $P(J|M|)$ are equal to the ionization
cross section $Q(j|m|)$ of electron jm with $j = J$ and $|m| = |M|$. For
example, for a $p_{3/2}$ vacancy we have[7]

$$\mathcal{A}_{20} = \frac{Q(3/2\ 3/2) - Q(3/2\ 1/2)}{Q(3/2\ 3/2) + Q(3/2\ 1/2)} = \frac{Q(p1) - Q(p0)}{Q(p0) + 2Q(p1)} , \qquad (4)$$

where the second equality is in terms of cross sections $Q(\ell m_\ell)$[7].
Through Eq. (4) the anisotropy (or alignment) parameter $\mathcal{A}_{20}$ is related to the spatial shape of charge distribution of a $p_{3/2}$ vacancy
(see Fig. 1). For $\mathcal{A}_{20} = 0$ the charge distribution is isotropic, for
$\mathcal{A}_{20} > 0$ the charge distribution becomes oblate along the z-axis,
and for $\mathcal{A}_{20} < 0$ the charge distribution becomes prolate along the
z-axis.

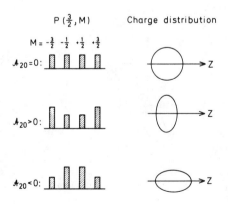

Fig. 1. Population probabilities P(3/2, M) and charge distributions
of a $p_{3/2}$ vacancy state for various values of the alignment
parameter $\mathcal{A}_{20}$.

The angular distribution of Auger electrons is then given by

$$I_{e^-}(\theta) = \frac{I_{o,e^-}}{4\pi} \left\{ 1 + \sum_{k=2}^{k_{max}} \alpha_k^{e^-} \mathcal{A}_{ko} \cdot P_k(\cos\theta) \right\}, \tag{5}$$

where $P_k(\cos\theta)$ are the Legendre polynomials, θ is the angle relative to the incident direction and k_{max} is determined by[7,16,19]

$$k_{max} = 2 \min \left\{ J, \ell', j' \right\}, \tag{6}$$

where J, ℓ' and j' are as given in Eq. (1) and (2). The coefficients $\alpha_k^{e^-}$ depend on the Auger decay amplitudes and relative phases between different Auger partial waves. Only in the special case with $J_F = 0$, where a single Auger wave is emitted, the $\alpha_k^{e^-}$ are independent of the decay amplitude. This special case is therefore particularly suited to an experimental determination of the parameters $\mathcal{A}_{ko}$. For example, Auger electrons of the transition

$$A^+(np_{3/2}^{-1} \, {}^2P_{3/2}) \rightarrow A^{++}(J_f = 0) + e^-_{Auger}$$

have[16]

$$\alpha_2^{e^-} = -1. \tag{7}$$

The angular distribution of X-rays (electric dipole transitions) is given by[16,17]

$$I_{\gamma'}(\theta) = \frac{I_{o,\gamma'}}{4\pi} \left\{ 1 + \alpha_2^\gamma \mathcal{A}_{20} P_2(\cos\theta) \right\} \tag{8a}$$

and the linear polarization by

$$P_{1in, \gamma'} = \frac{I_{||} - I_\perp}{I_{||} + I_\perp} = \frac{3\alpha_2^\gamma \mathcal{A}_{20}}{\alpha_2^\gamma \mathcal{A}_{20} - 2}, \tag{8b}$$

where the coefficient α_2^γ depends only on J and J_f' of the initial and final state of the decay[16].

In Eq. (8b) $I_{||}$ and $I_\perp$ are the X-ray intensities with electric vector parallel and perpendicular, respectively, to the incident direction and measured perpendicular to the incident direction. Note that due to the dipole character of the decay radiation, only the parameter $\mathcal{A}_{20}$ can be determined via Eqs. (8a) and (8b).

Photon impact experiments with axial symmetry[20,23]

For unpolarized photons the direction of incidence z defines the symmetry axis (see Fig. 2a). However, for linearly-polarized photons, incident also along z, the direction of the electric vector (in direction of x) defines the axis of symmetry (see Fig. 2b).

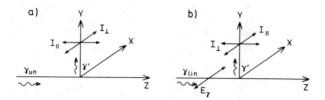

Fig. 2. Different definitions of $I_{\parallel}$ and $I_{\perp}$ of fluorescence radiation γ' for the incident radiation being unpolarized (a) or li-nearly polarized (b).

Then again, the axial symmetry of the ionization process allows only the $\mathcal{A}_{ko}$ of even k to be non-zero, but the additional assumption that the photoionization is by an electric dipole transition reduces k to only 2.

Then the angular distributions of Auger electrons and dipole-photons γ' are given by Eq. (5) (but with only $\mathcal{A}_{20} \neq 0$) and by Eq. (8a), respectively, where the angle θ is relative to the axis of symmetry (z for γ_{un}, x for γ_{1in}). The linear polarization P_{lin} of dipole photons γ' is given by Eq. (8b), where $I_{\parallel}$ and $I_{\perp}$ are the in-tensities of radiation γ' with electric vectors as indicated in Fig. 2a and 2b.

The alignment parameter $\mathcal{A}_{20}(\gamma_{1in})$ for linearly polarized in-cident radiation γ_{1in} is related to $\mathcal{A}_{20}(\gamma_{un})$ for unpolarized inci-dent radiation as follows[18,20-22].

$$\mathcal{A}_{20}(\gamma_{1in}) = -2\, \mathcal{A}_{20}(\gamma_{un}). \tag{9}$$

Relation (9) is exactly analogous to the relation in the angular distributions of photoelectrons for incident light γ_{1in} and γ_{un}.

For example, the degree of linear polarization P_{lin} of emitted photons γ' for unpolarized incident light γ_{un} is given by

$$P_{lin}(\gamma') = \frac{3\alpha_2^{\gamma}\, \mathcal{A}_{20}(\gamma_{un})}{\alpha_2^{\gamma}\, \mathcal{A}_{20}(\gamma_{un}) - 2} \tag{10a}$$

and for linearly polarized incident light γ_{1in} by

$$P_{lin}(\gamma') = \frac{3\alpha_2^{\gamma}\, \mathcal{A}_{20}(\gamma_{1in})}{\alpha_2^{\gamma}\, \mathcal{A}_{20}(\gamma_{1in}) - 2} = \frac{3\alpha_2^{\gamma}\, \mathcal{A}_{20}(\gamma_{un})}{\alpha_2^{\gamma}\, \mathcal{A}_{20}(\gamma_{un}) + 1}. \tag{10b}$$

If synchrotron radiation is used as light source then the incident radiation is in general elliptically polarized. This case has recently been treated by Klar[21].

In the case of circularly polarized light along the z-axis, the experimental situation is still axially symmetric about the z-axis but, due to the helivity of the photons, is not symmetric with respect to reflection in the plane perpendicular to the z-axis. From that, it follows that the anisotropy of the ionic state is given by an alignment ($A_{20} \neq 0$) and an orientation ($A_{10} \neq 0$), the latter being caused by the helicity of the incident photons. The alignment $A_{20}(\gamma_c)$ obtained with circularly polarized incident light γ_c is exactly the same as that obtained with unpolarized light, i.e.

$$A_{20}(\gamma_c) = A_{20}(\gamma_{un}). \tag{11}$$

From this, it follows that the angular distributions of Auger electrons and photons γ' (without polarization dection) and the linear polarization P_{lin} of photons γ' are given exactly by the same relations as for unpolarized incident radiation (Fig. 2a). In order to determine also the orientation parameter A_{10} one has to observe the circular polarization of fluorescence light γ' (see e.g. Klar[20] and Greene and Zare[22]).

Experiments with plane symmetry

The symmetry plane (zx) is defined by the incident and the scattered particle (see Fig. 3). Again we deal with unpolarized incident projectiles P and an unpolarized target. Then, the state of the ion formed is symmetric about reflection in the reaction plane (zx) and the anisotropy parameters become interrelated

$$A_{k\kappa} = (-1)^{k + \kappa} A_{k,-\kappa}. \tag{12}$$

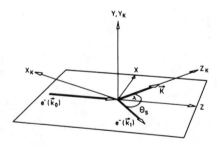

Fig. 3. Experiment with plane symmetry. The incident and scattered electrons define the reaction plane zx. $\vec{K} = \vec{k}_o - \vec{k}_1$ is the momentum transfer.

We consider the following special cases (for the general case see Berezhko et al.[17]):

1) For ionization in any subshell with $j \geq 3/2$ and for photon emission γ' the number of independent observable parameters is reduced to 4, namely, $\mathcal{A}_{11}$, $\mathcal{A}_{20}$, $\mathcal{A}_{21}$ and $\mathcal{A}_{22}$.

2) For ionization in a subshell with $j = 3/2$ and for electron emission e^- the number of independent and observable parameters is reduced to only 3, namely $\mathcal{A}_{20}$, $\mathcal{A}_{21}$ and $\mathcal{A}_{22}$.

The in-plane angular distributions $I_{\gamma'}(\theta)$ in case (1) and $I_{e^-}(\theta)$ in case (2) are given by the same relation[17]

$$I_{\gamma'(e^-)}(\theta) \sim 1 + \beta_{\shortparallel}\cos 2(\theta - \psi). \tag{13a}$$

Here $\beta_{\shortparallel}$ gives the amplitude and $\theta = \psi$ gives the minimum (or the maximum) of the distribution. It is to be noted that $\beta_{\shortparallel}$ and ψ are functions of $\mathcal{A}_{20}$, $\mathcal{A}_{21}$, $\mathcal{A}_{22}$ and of $\alpha_2^{\chi(e^-)}$.

The angular distribution of photons γ' or Auger electrons in a plane perpendicualr to the z-axis appears as[17]

$$I_{\gamma'(e^-)}(\phi) \sim 1 + \beta_{\perp}\cos 2\phi, \tag{13b}$$

where $\beta_{\perp}$ can be expressed in terms of $\mathcal{A}_{20}$, $\mathcal{A}_{22}$ and of $\alpha_2^{\chi(e^-)}$.

Furthermore, it has been shown[17], that in Born approximation due to the axial symmetry along the momentum transfer $\vec{K} = \vec{k}_0 - \vec{k}_1$ (see Fig. 3), the anisotropy parameters $\mathcal{A}_{K\kappa}^K$ defined with direction of $\vec{K}$ as quantization axis reduce to exactly $\mathcal{A}_{k0}^K$ with k taking only even values. From this it follows that the ion is aligned with respect to the direction $\vec{K}$. Then, the angular distribution should be symmetric with respect to a plane perpendicular to $\vec{K}$, i.e., the position ψ of the extremum of the angular distribtion (Eq. 13a) coincides with the direction λ of the momentum transfer $\vec{K}$ (see Fig. 3). Deviation of the experimental values of ψ from λ may be taken as a measure of violation of the Born approximation.

Again, in order to measure the orientation parameter $\mathcal{A}_{11}$, one has to observe the circular polarization of fluorescence light γ' [17]. Finally we note, that the anisotropy parameters $\mathcal{A}_{11}$, $\mathcal{A}_{20}$, $\mathcal{A}_{21}$ and $\mathcal{A}_{22}$ used here are fully equivalent to the parameters O_0^{coll} and $A_{0,1+,2+}^{coll}$ introduced by Fano and Macek[24].

GENERAL PROPERTIES OF THE ALIGNMENT

What kind of information can we obtain from a study of the alignment? Since up to now only the alignment parameter $\mathcal{A}_{20}$ has been determined in axial symmetric experiments, we will discuss only the properties of this parameter.

Particle impact ionization

The alignment parameter $\mathcal{A}_{20}$ for ion impact ionization for different subshells nl has been calculated recently by Sizov and Kabachnik[25] in the Born approximation (BA). The incoming ion was considered as a structureless particle of charge Z and for the target wave functions different approximations were employed (hydrogen-like with the same effective charge Z_{eff} for the bound and continuum state (HL), hydrogen-like with effective charge and outer screening (SHL), Hartree-Slater (HS)). It has been shown[6,25] that in the HL model and for a given projectile the alignment as function of the reduced velocity $V = v_{ion}/v_{n\ell j}$ of the projectile is a universal curve for each subshell $n\ell j$ (v_{ion} and $v_{n\ell j}$ are the velocities of the projectile and the $n\ell j$ electron of the target, respectively). For a discussion of the main properties the alignment parameter $\mathcal{A}_{20}$ for various subshells is plotted as function of the reduced velocity V for proton impact ionization in Fig. 4a. For comparison, the alignment parameter $\mathcal{A}_{20}$ for electron impact ionization of a $2p_{3/2}$ electron calculated in the HL model[16] is also given in Fig. 4b beside that for proton impact (note that, for electron impact, the only possible region is V > 1). In the following we will discuss separately the high region (V > 1) and the low region (V < 1) of reduced velocity.

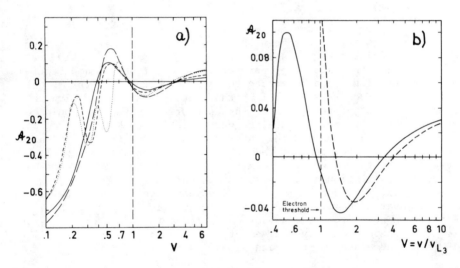

Fig. 4. a) Theoretical alignment of the $2p_{3/2}$(———), $3p_{3/2}$(— — — —), $4p_{3/2}$(....) and $3d_{5/2}$(—— ——) subshells due to proton impact ionization as function of the reduced velocity $V = v_{ion}/v_{n\ell j}$. The calculations have been performed in the HL model.
   b) Theoretical alignment of the $2p_{3/2}$ subshell for electron impact(— — — —) and proton impact(———) ionization using the HL model. Taken from Sizov et al.[25].

As can be seen, for V > 1 the alignment curves for different subshells are similar. The alignment passes through zero at a reduced velocity V between 3 and 4 and approaches in the limit $V \to \infty$ a positive value given simply by the squares of the dipole matrix elements (Bethe approximation, see Ref. 6, 26). The reason for the similar structure of alignment curves is that for large reduced velocities V the main contribution to the cross sections $Q(n\ell m_\ell)$ comes from small momentum transfers K and, therefore, the outer parts of the wave functions (large r) dominate the matrix element of the Born operator exp $(i\vec{K}r)$.

Also, for electron impact ionization the alignment curves exhibit, at least for $V = v_e-/v_{n\ell j} > 1.5$, the same structure independent of the subshell $n\ell j$[16]. The zero-crossing of the alignment curves at about V = 4 for both electron and proton impact ionization is due to a kinematic effect[26-28]: The angles λ between the various momentum transfers $\vec{K}$ and the incident beam direction are such as to make $\mathcal{A}_{20}$ vanish. E.g. for $p_{3/2}$ ionization, the parameter $\mathcal{A}_{20}$ given by

$$\mathcal{A}_{20} = \left\{ \frac{1}{2} Q(2p) \right\}^{-1} \cdot \iint P_2(\cos\lambda) \left\{ \frac{d^2Q(2p,\mu=1)}{dKdE} - \frac{d^2Q(2p,\mu=0)}{dKdE} \right\} dKdE, \tag{14}$$

has zero value. Here $d^2Q(2p\mu)/dKdE$ are the double differential cross sections for a collision with momentum transfer K and energy transfer E, μ is the magnetic quantum number in a frame with the direction of $\vec{K}$ as quantization axis. $P_2(\cos\lambda)$ is due to the rotational transformation from the frame with $\vec{K}$ as axis to the frame with the incident beam direction as axis.

We now turn to the low velocity region (V < 1). Here the alignment curves show a structure which strongly depends on the subshell $n\ell$ and which can be traced back to the nodes of the radial part of the wave functions. Qualitatively, for smaller reduced velocities of the projectile the main contribution to the cross sections comes from larger momentum transfers K and, consequently, the major contribution comes from the inner parts of the spatial wave functions. If there are several nodes in the radial distribution of the wave function $n\ell$, the different nodes at increasing r lead to structures in the alignment curve also at increasing values of V. It is important to realize from Fig. 4a that in non-coincidence experiments the inner part of the radial distribution of wave functions can only be probed effectively in collisions with heavy projectiles.

In Fig. 5 the influence of the various approximations of target wave functions (HL, SHL and HS) on the alignment parameter $\mathcal{A}_{20}$ for the $2p_{3/2}$ ionization of Ar by proton[25] and electron impact[25,16] is shown. The outer screening parameter in the SHL model takes into account the difference between the experimental binding energy E_B^{exp} and the binding energy in the HL model, $E_B^{HL} = Z_{eff}^2 R/n^2$, and is defined by[29]

$$\theta = E_B^{exp}/E_B^{HL} = (E_B^{exp} n^2)/(Z_{eff}^2 R). \tag{15}$$

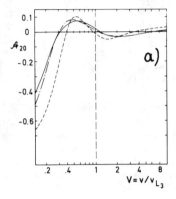
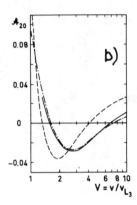

Fig. 5. The alignment of the $2p_{3/2}$ subshell of Ar due to proton impact (a) and electron impact (b) ionization as function of reduced velocity V. HS model (——), SHL model with $\Theta = 0.38$ (—— · ——), HL model (— — — —). Taken from Sizov et al.[25] and Kabachnik[57].

This yields $\theta = 0.38$ for ArL$_3$ electrons. It is interesting to note that the HS and SHL calculations give almost the same $\mathcal{A}_{20}$. The same situation occurs also for the $2p_{3/2}$ shell of heavier elements[25] but not for light target atoms, e.g. Mg[25].

Photoionization

It was shown above that in an axially symmetric photoionization experiment using unpolarized or linearly polarized incident radiation the only observable anisotropy parameter was $\mathcal{A}_{20}$. In the relativistic independent particle model the photoionization (electric dipole transition) is completely described by three radial matrix elements $R_{n\ell j, \epsilon l' j'}$ and two relative phases (in the non-relativistic limit this reduces to two radial matrix elements and one relative phase). It has been shown[18,23,32], that only the squares of the matrix elements $R_{n\ell j, \epsilon l' j'}$ enter into $\mathcal{A}_{20}$ but not the relative phases. Thus, the determination of $\mathcal{A}_{20}$ of photoions offers the possibility to get an additional independent information on the radial matrix elements of the photoprocess. This supplements the information on these 5 quantities which is obtained through the subshell cross section $\sigma_{n\ell j}$, the ß-parameter of the angular distribution of photoelectrons and, most recently, the spin-polarization of photoelectrons[30,31]. The alignment parameters $\mathcal{A}_{20}$ have been calculated for various subshells $n\ell j$ in the HS model by Berezhko et al.[18]. For example, the alignment $\mathcal{A}_{20}$ of the Xe $4d_{5/2}$ vacancy is displayed in Fig. 6 as function of photoelectron energy ϵ. As can be seen, $\mathcal{A}_{20}$ can reach large values. This is due to 1) the suppression of the $n\ell \rightarrow \epsilon l+1$ transition for samll ϵ caused by the centrifugal barrier in the effective potential for the electron with angular momentum $\ell + 1$ (Manson and Cooper[33]) and 2) the presence of a Cooper minimum ($R_{n\ell,\varepsilon\ell+1} = 0$)[34].

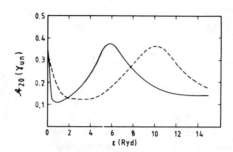

Fig. 6. The alignment $\mathcal{A}_{20}(\gamma_{un})$ of the $4d_{5/2}$ subshell of Xe due to photoionization by unpolarized radiation $\gamma_{un}V$ as function of the photoelectron energy ε. Hartree-Slater model (——), Hartree-Fock model (— — — —). Taken from Berezhko et al.[18].

EXPERIMENTAL INVESTIGATIONS

Experimentally, all three methods mentioned above have been used to measure the alignment in particle impact ionization. In the region of low transition energy, Auger electron spectrometry is best suited[26,27,35] because of the ease of electron energy analysis and of very small fluorescence yield. For high transition energies the alignment has been measured via the angular distribution of X-rays using in most cases Si(Li) detectors[36]. For medium transition energies (< 3 keV) the resolution of Si(Li) detectors is too poor and the partial superposition of unresolved lines with opposite angular distribution leads to rather small anisotropies. Here the linear polarisation of X-radiation using crystal spectrometers with high resolution has been measured[10,37-40]. For detailed references on experimental investigations of the alignment parameter $\mathcal{A}_{20}$ in H[+] and He[+] impact ionization of the L3-subshell see Mehlhorn[28] and Jitschin et al.[36]. Most recently the proton-induced alignment $\mathcal{A}_{20}$ in the M3, M4 and M5 subshells of Th have also been measured[40].

For electron impact ionization the alignment $\mathcal{A}_{20}$ has been measured for MgL3[35,41], ArL3[7,26,42], KrM4, M5[27], XeL3[43] and HgL3[4].

The only two coincidence experiments so far performed have been done by Schuch et al.[14] (α + Dy(L3)) and by Sewell et al.[15] (e[-] + Ar(L3)).

In the case of photoionization as incident radiation both unpolarized resonance radiation[12,44] (HeI/II, NeII) and monochromatized synchrotron radiation[45-47] have been used. The alignment of photoions was measured either via the polarization of fluorescence radiation[12,44,47] or via the angular distribution of Auger electrons[45,46].

COMPARISON BETWEEN EXPERIMENT AND THEORY

Ion impact ionization

In Fig. 7 experimental values of $\mathcal{A}_{20}$ for $2p_{3/2}$ ionization in silver[39], xenon[48], dysprosium[36,48], gold[36,38-50] and uranium[36,48] by proton impact are compared to the theoretical curves for PWBA in the SHL model with different θ for Ag ($\theta = 0.54$) and Au ($\theta = 0.63$). For reduced velocities $V > 0.2$ there is good agreement between experiment and theory, whereas for reduced velocities $V < 0.2$ the experimental values deviate systematically from theory and tend to zero. The origin for this dealignment is the Coulomb deflection of the projectile in the field of the target nucleus, i.e. breakdown of PWBA. In order to correct the alignment parameter for the Coulomb deflection, two somewhat different models have been proposed[51,49]. Both models account for the change in quantization axis caused by the Coulomb deflection. However, they differ in the assumption whether the new quantization axis is tangent to the projectile path at the point of closest approach[51] or coincides with the direction of Coulomb deflected particle[49]. In the former case the new quantization axis makes an angle $\Theta_s/2$ with the incident direction (z-axis) in the latter case the angle is Θ_s. Clearly, in both models the rotation of quantization axis leads to a dealignment, with a larger effect in the case of rotation through Θ_s (dotted curve in Fig. 7). Both models qualitatively

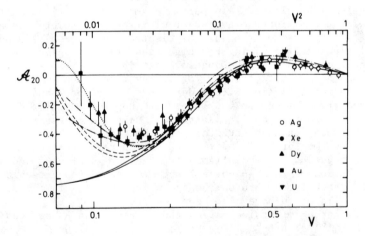

Fig. 7. L_3-subshell alignment for proton impact as function of the reduced projectile velocity V. Experimental data: ○ Ag[39], ● Xe[48,36], ▲ Dy[48,36], ■ Au[36,48-50], ▼ U[36,48]. Theory: ——PWBA[25], – – – –PWBA with correction for Coulomb deflection by $\Theta_s/2$[51], upper curves correspond to Ag, lower curves to Au; .... PWBA with correction for Coulomb deflection by Θ_s for Au[49]; relativistic SCA for Ag(– ·· –) and for Au[53] (– · –). Taken from Jitschin et al.[48].

describe the experimental data. They have been refined by correcting
for the velocity retardation of the projectile in the target field[52]
but with only a small effect. Recently, SCA calculations for Ag and
Au became available[53] which employ relativistic wave functions as
well as curved projectile trajectories. The results (dot-dashed and
double dot-dashed curves in Fig. 7) agree reasonably with the experimental data for all projectile energies.

In Fig. 8 the experimental A_{20} values[35] for $2p_{3/2}$ ionization
in Mg by impact of H^+ and He^+ are compared to theoretical alignment
curves calculated for the HS and the SHL model ($Z_{eff} = 7.85$,
$\theta = 0.27$) for proton impact[25]. It is interesting to note that the
theoretical curves for the two models are quite different in contrast
to the case of Ar and heavier elements, where the results of the two
models agreed well with each other (see Fig. 5). Furthermore, the
experimental values for different projectile masses show qualitatively the same structure but neither do they fall on an almost universal
curve, as is predicted in PWBA[54], nor do the values for proton impact
agree with the theoretical curves. It was shown by Kabachnik et
al.[55,56] that the reason for this discrepancy with theory in the case
of $Mg(2p_{3/2})$ is due to the neglect of the electron capture process,
which for small reduced velocities ($V < 0.7$) dominates the direct
Coulomb ionization. Since the direct ionization cross section as a
function of V is proportional to $1/Z_{eff}^4$ whereas the electron capture
cross section is proportional to $1/Z_{eff}^7$ (for $Z_{projectile} \ll Z_{eff}$),
for H^+ or He^+ impact on heavy atoms the direct ionization is by far
the dominant process and determines the alignment, as demonstrated by
Fig. 7.

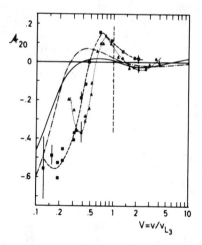

Fig. 8. Alignment of the L_3-subshell of Mg for impact ionization by
different projectiles. Experiment[35,41]: $\blacktriangle$ = H^+, $\blacksquare$ = He^+.
The curves through the experimental values are only meant as
a visual guide. Theory[25]: Values are for BA and HS model
(——) or SHL model with $\theta = 0.27$ (— · —).

66

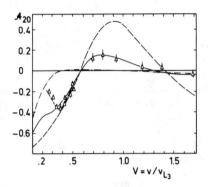

Fig. 9. Alignment of the L₃-subshell of Mg due to impact ionization or/and electron capture by protons. Experiment[35,41] = ▲. Theory[56]: Direct ionization in PWBA and HS model (− · −), electron capture in OBK and HS model (− − −), direct ionization and electron capture (——). Taken from Sizov and Kabachnik[56].

In Fig. 9 the experimental alignment data $\mathcal{A}_{20}$ of Mg(2p₃/₂) for proton impact[35,41] are compared to theoretical alignment values for direct ionization (HS-model and PWBA), electron capture (HS model and OBK approximation) and for both processes[56]. Besides the fact that the capture process itself causes large alignment values it can be seen that the theoretical values including both processes are in good agreement with experiment except for small reduced velocities V < 0.4. It is not yet clear whether this decrease of the magnitude of alignment is already caused by the dealignment via the Coulomb deflection. The observed shift of the zero-crossing of experimental alignment curves for He⁺ and H⁺ impact at small velocities could also be explained by the inclusion of electron capture[54-56].

For $\mathcal{A}_{20}$ values for 2p₃/₂ ionization in heavy ion-atom collisions see Jitschin[13]. In the case of 3p₃/₂, 3d₃/₂ and 3d₅/₂ ionization of Th by proton impact the experimental $\mathcal{A}_{20}$ values[40] agree well with the theoretical data[25] in the range of investigation (V = 0.2, ..., 0.8). In particular they confirm the additional minimum of $\mathcal{A}_{20}$ (3p₃/₂) at about V = 0.3.

Electron impact ionization

As an example, in Fig. 10a and b the experimental $\mathcal{A}_{20}$ values of ArL₃[7,26] and MgL₃[35,52] ionization by electron impact are compared to the theoretical results. As can be seen, the experimental $\mathcal{A}_{20}$ values agree well in magnitude and shape with the theoretical PWBA values for HS wave functions[16,57] in the high velocity range (V > 4). The deviations for ArL₃ for V > 10 are due to relativistic effects of the incident electrons[26]. For the near-threshold range (V < 2.0) the

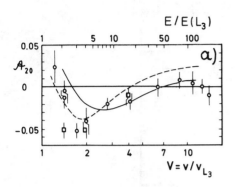
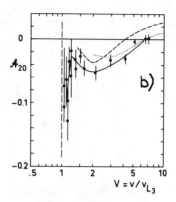

Fig. 10. Alignment of the L_3-subshell of Ar (a) and Mg (b) due to
electron impact ionization as function of the reduced velocity V of incident electrons.
a) Experiment[7,26]: $\square$, $\bigcirc$. Theory: HL model[16] (— — —), HS model[16,57] (——)
b) Experiment[35,52]: $\bullet$. The solid line through the experimental values are only meant as a visual guide. Theory[16]: HL model (— — —), HS model (...).

experimental values deviate systematically from the theoretical HS
values in PWBA indicating the invalidity of the plane-wave Born approximation in this region. Recently, Berezhko and Kabachnik[58] tried
to restore the agreement between experiment and theory for the case
of ArL$_3$ also in the low velocity region by using a distorted-wave
Born approximation (see Fig. 11). Indeed, they could improve the
agreement down to V $\cong$ 1.4. However, a most recent experiment by Du-
Bois and Rødbro[42] near to threshold seems to contradict the earlier
experimental values[26]. The strong increase of $\mathcal{A}_{20}$ to almost -0.1
for V close to threshold in the ArL$_3$ ionization[42] (Fig. 11) is similar to the near-threshold behaviour of $\mathcal{A}_{20}$ in the MgL$_3$ case (Fig.
10b) and has not yet been explained theoretically. Besides the general break-down of the PWBA, one possible explanation of the discrepancy between theory and experiment is the existence of the post-collision interaction (PCI). So far, only the shift and the distortion of
the line shapes of Auger lines due to the PCI for inner-shell ionization by electron impact have been studied (see e.g. Huster and
Mehlhorn[59]). In the case of electron <u>excitation,</u> however, the occurence of a strong dip in the experimental alignment curve near threshold, which is in contradiction to the theoretical prediction, has
been interpreted as a PCI effect (transfer of angular momentum)[60].

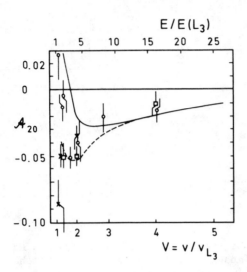

Fig. 11. Alignment of the L_3-subshell of Ar due to electron impact ionization. Experiment: $\square$, $\bigcirc$ = Ref. 7, 26; x = Ref. 42. Theory: PWBA and HS model[57] (– – –), DWBA and HS model[58] (——).

In the case of X-ray emission after inner-shell ionization, no post-collision interaction is possible[59]. It is therefore interesting to measure the alignment $\mathcal{A}_{20}$ near threshold via the angular aniso-tropy or the polarization of X-rays. In the case of HgL_3, the linear polarization of $L\alpha_1$ radiation has been measured[4] to be +0.14±0.04 for $V = 1.55$ (or $E_e-/E(L_3) = 2.4$), which in turn results in $\mathcal{A}_{20} = -0.98\pm0.25$ (note that Hrdy et al.[4] used a definition of the linear polarization which differs in sign from Eq. (8b)). In the case of XeL_3, the linear polarization of $L\alpha_{1,2}$ radiation for $V = 1.12$ ($E_e-/E(L_3) = 1.25$) was found[43] to be +0.066, which again yields a value for $\mathcal{A}_{20}$ close to -1.0. Both these results give support for $Q(2p0) \gg Q(2p1)$ at threshold (see Eq. 4), a threshold behaviour which has not yet been found for the Auger channel decay (see Fig. 10b and 11). The different behaviour of the X-ray and the Auger chan-nel may be attributed to the manifestation of the post-collision in-teraction and should be studied more systematically. It is worthwhile to note that a relation $Q(2p0) \gg Q(2p1)$ for threshold <u>ionization</u> does not[26] follow from the same general angular momentum considera-tion as for threshold <u>excitation</u>[24].

Photoionization

The alignment in inner-shell photoionization has been studied so far for $Cd^+(4d\bar{5}/2)$ using both unpolarized resonance radiation (He I/II, Ne II)[12,44] and monochromatized linearly polarized synchrotron radiation[47] and for $Xe^+(4d_{5/2}^{-1})$ using synchrotron radiation[45,46].

The results obtained[12,44,47] for the alignment of $Cd^+(4d_{5/2}^{-1})$ via the linear polarization of radiation γ' of the decay $Cd^+(4d^9 5s^2\ ^2D_{5/2}) \rightarrow Cd^+(4d^{10} 5p\ ^2P_{3/2}) + \gamma'(4416\ \text{Å})$ were consistently smaller in magnitude than the predicted theoretical values. The latter had been calculated either non-relativistically (with Hartree-Slater functions[18] or with a many-body perturbation theory[61]) or relativistically (with Dirac-Slater functions[32] or with a random-phase approximation theory[62]). Only quite recently, it has been found[47] that the too small magnitude of the alignment is due to a depolarization caused by a spurious magnetization of the stainless steel of the Cd-oven. Preliminary values[47] of the polarization of radiation γ' for He I/II radiation now agree with the theory within 30 %.

In the case of Xe $4d_{5/2}$ photoionization by linearly polarized synchrotron radiation the alignment was measured[45,46] via the angular distribution of Auger electrons for the decay

$$Xe^+(4d_{5/2}^{-1}) \rightarrow Xe^{++}(5s^{-2}\ ^1S_0) + e_{A1}^- \tag{16a}$$

$$\rightarrow Xe^{++}((5s5p)^{-1}\ ^1P_1) + e_{A2}^- \tag{16b}$$

For Auger electrons with final $^1S_0$ state the angular anisotropy (see Eq. (5) with only $\mathcal{A}_{20} \neq 0$ and Eq. (9)) is given by[16,46]

$$\beta_2 = \alpha_2^{e^-} \cdot \mathcal{A}_{20}(\gamma_{1in}) = 2\ \sqrt{8/7}\ \mathcal{A}_{20}(\gamma_{un}). \tag{17}$$

In Fig. 12 the experimental values β_2 of Auger electrons e_{A1}^- and e_{A2}^- are compared to the theoretical values of $2\ \sqrt{8/7}\ \mathcal{A}_{20}(\gamma_{un})$. The theoretical values of $\mathcal{A}_{20}(\gamma_{un})$ as function of photoelectron energy have been obtained[46] by using HF photoionization radial matrix elements[63]. The experimental anisotropy β_2 of Auger electrons e_{A1}^- agrees well with the general shape of the theoretical curve although it is somewhat smaller near threshold. The $\alpha_2^{e^-}$ of the Auger transitions of (16b) depends on three Auger radial matrix elements and their phases, nevertheless it has a constant value. Consequently, the anisotropies β_2 of electrons e_{A1}^- and e_{A2}^- should only differ by a constant factor which is qualitatively the case (see Fig. 12).

Coincidence experiments

In only two experiments so far, the scattered particles have been measured in coincidence with either the emitted X-rays[14] or the ejected Auger electrons[15]. Schuch et al. measured the alignment parameters $\mathcal{A}_{20}(b)$ and $\mathcal{A}_{22}(b)$ as functions of the impact parameter b for the system $He^{++} + Dy(L_3)$. Sewell and Crowe[15] measured the in-plane angular distribution of $L_3M_{2,3}M_{2,3}(^1S_0)$ Auger electrons of Ar in coincidence with the scattered incident electrons. The data obtained at an incident electron energy of 1 keV for electron scattering angles of 15 and 21° and for a scattering energy of 746.4 eV (energy of ionized electrons $\varepsilon = 5$ eV) is shown in Fig. 13.

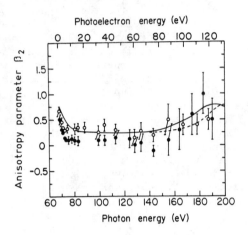

Fig. 12. Auger electron anisotropy parameters for the transitions
$N_5O_1O_1(^1S_0)$ (O) and $N_5O_1O_{2,3}(^1P_1)$ (●). The theoretical curves, given by $2\sqrt{8/7}\,A_{20}(\gamma_{un})$, are for a transition of the
general type $N_5OO(^1S_0)$. Dashed curve: HF length; full curve:
HF velocity (from Southworth[46]).

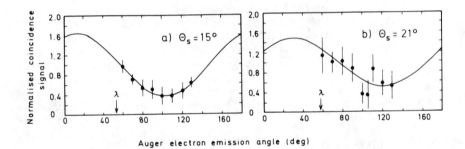

Fig. 13. Angular correlation between the scattered electron and the
$L_3M_{2,3}M_{2,3}(^1S_0)$ Auger electron of Ar for an incident electron
energy of 1000 eV, scattered electron angles Θ_s of (a) 15 and
(b) 21° and a mean ejected electron energy of 5 eV. The full
curve is a least-squares fit of Eq. (13a) to the data. The
arrows indicate the angle λ of momentum transfer $\vec{K}$ (from
Sewell and Crowe[15]).

The measured angular anisotropies are much larger than have been found
in experiments with axial symmetry (see Fig. 10a). Furthermore, assuming the validity of the Born approximation, the in-plane angular distribution of $L_3M_{2,3}M_{2,3}(^1S_0)$ Auger electrons (Eq. 13a) reduces to the

expression

$$I_{e^-}(\Theta_K) \sim 1 - \mathcal{A}_{20}^K \cdot P_2(\cos\Theta_K), \tag{18}$$

where Θ_K is the angle relative to the direction of $\vec{K}$. The two cases of scattered electrons of Fig. 13 (Θ_s = 15 and 21^0) have momentum transfers Ka_0 = 2.4 and 3.1 and theoretical values[17] of $\mathcal{A}_{20}^K$ in the range of 0.1 to 0.2. Then, with Eq. (18), the minima of the angular distributions of Fig. 13 should coincide with the angle λ of $\vec{K}$. In both cases the minimum positions strongly deviate from the directions λ (given by the arrows in Fig. 13) indicating the non-validity of the PWBA. Also the theoretical values of $\mathcal{A}_{20}^K$ are much smaller than the experimental ones of 0.85 and 0.65. More experimental investigations using the coincidence technique and better approximations than the PWBA in theoretical calculations of the relevant alignment parameters $\mathcal{A}_{20}$, $\mathcal{A}_{21}$ and $\mathcal{A}_{22}$ are highly desirable.

ACKNOWLEDGEMENTS

I thank Prof. W.N. Asaad and Dr. H. Klar for valuable comments. I am indebted to Drs. B. Cleff, N.M. Kabachnik, R. Schuch and S.H. Southworth for using their results prior to publication.

REFERENCES

1) H.W.B. Skinner, Proc.Roy.Soc.Lond. A 112, 642 (1926).
2) W. Mehlhorn, Z.Physik 160, 247 (1960).
3) W. Mehlhorn, Phys.Lett. 26 A, 166 (1968).
4) J. Hrdy, A. Henins, and J.A. Bearden, Phys.Rev. A 2, 1708 (1970).
5) B. Cleff and W. Mehlhorn, Phys.Lett. 37 A, 3 (1971).
6) S.C. McFarlane, J.Phys. B 5, 1906 (1972).
7) B. Cleff and W. Mehlhorn, J.Phys. B 7, 605 (1974).
8) D.J. Volz and M.E. Rudd, Phys.Rev. A 2, 1395 (1970).
9) P. Dahl, G. Hermann and M. Rødbro, Abstracts of Contributed Papers, IV. Int.Conf. Atomic Physics, Heidelberg 1974, p. 693.
10) K.A. Jamison and P. Richard, Phys.Rev. Letters, 38, 484 (1977).
11) S. Flügge, W. Mehlhorn, and V. Schmidt, Phys.Rev.Lett. 29, 7 (1972).
12) C.D. Caldwell and R.N. Zare, Phys.Rev. A 16, 255 (1977).
13) W. Jitschin, Invited paper at X-82 Conference, this book
14) R. Schuch, R. Hoffmann, J. Konrad, H. Schmidt-Böcking, H.J. Specht and F. Ziegler, Contr. Paper at X-82 Conference, Eugene (Oregon) 1982, p. 297.
15) E.C. Sewell and A. Crowe, J.Phys. B 15, L 357 (1982).
16) E.G. Berezhko and N.M. Kabachnik, J.Phys. B 10, 2467 (1977).
17) E.G. Berezhko, N.M. Kabachnik and V.V. Sizov, J.Phys. B 11, 1819 (1978).
18) E.G. Berezhko, N.M. Kabachnik and V.S. Rostovsky, J.Phys. B 11, 1749 (1978).
19) J. Eichler and W. Fritsch, J.Phys. B 9, 1477 (1976).
20) H. Klar, J.Phys. B 13, 2037 (1980).

72

21) H. Klar, J.Phys. B 15, (1982) to be published.
22) Ch.H. Greene and R.N. Zare,Ann.Rev.Phys.Chem. 33, (1982) to be published.
23) K.-N. Huang, Phys.Rev. A 25, 3438 (1982).
24) U. Fano and J. Macek, Rev.Mod.Phys. 45, 553 (1973).
25) V.V. Sizov and N.M. Kabachnik, J.Phys. B 13, 1601 (1980).
26) W. Sandner and W. Schmitt, J.Phys. B 11, 1833 (1978).
27) W. Sandner, M. Weber and W. Mehlhorn in Coherence and Correlation in Atomic Collisions (Ed. H. Kleinpoppen and J.F. Williams), Plenum Press 1980, p. 215.
28) W. Mehlhorn, Proc.Int.Sem. on High-Energy Ion-Atom Coll., Debrecen 1981, Publ. House of the Hungarian Academy of Sciences, 1982, p. 83.
29) E. Merzbacher and H.W. Lewis in Handbuch der Physik (Ed. S. Flügge), Springer-Verlag, Heidelberg, 1958, vol. 34, p. 166.
30) U. Heinzmann, J.Phys. B 13, 4353 and 4367 (1980).
31) K.-N. Huang, Phys.Rev. A 22, 223 (1980).
32) C.E. Theodosiou, A.F. Starace, B.R. Tambe, and S.T. Manson, Phys. Rev. A 24, 301 (1981).
33) S.T. Manson and J.W. Cooper, Phys.Rev. 165, 126 (1968).
34) J.W. Cooper, Phys.Rev. 128, 681 (1962).
35) R.D. DuBois, L. Mortenson, and M. Rødbro, J.Phys. B 14, 1613 (1981).
36) W. Jitschin, A. Kaschuba, H. Kleinpoppen and H.O. Lutz, Z.Phys. A 304, 69 (1982).
37) K.A. Jamison, P. Richard, F. Hopkins and D.L. Matthews, Phys.Rev. A 17, 1642 (1978).
38) N.M. Kabachnik, V.P. Petukhov, E.A. Romanovskii and V.V. Sizov, Zh.Eksp.Teor.Fiz. 78, 1733 (1980).
    Engl.Transl.: Sov.Phys. JETP 51, 869 (1980).
39) G. Richter, M. Brüssermann, S. Ost, J. Wigger, B. Cleff and R. Santo. Phys.Lett. 83 A, 412 (1981).
40) J. Wigger, S. Ost, M. Brüssermann, G. Richter and B. Cleff, to be published.
41) M. Rødbro, R. DuBois and V. Schmidt, J.Phys. B 11, L 551 (1978).
42) R. DuBois and M. Rødbro, J.Phys. B 13, 3739 (1980).
43) M. Aydinol, R. Hippler, I. McGregor and H. Kleinpoppen, J.Phys. B 13, 989 (1980).
44) W. Mauser and W. Mehlhorn, Extended Abstracts of the VI. Int.Conf. on VUV Rad.Phys., Charlottesville, Virginia (USA), II-7 (1980).
45) S.H. Southworth, P.H. Kobrin, C.M. Truesdale, D. Lindle, S. Owaki, and D. Shirley, Phys.Rev. A 24, 2257 (1981).
46) S.H. Southworth, Ph. D. Thesis, Lawrence Berkeley Laboratory, Univ. of California, 1982.
47) W. Kronast, Universität Freiburg, private communication, 1982.
48) W. Jitschin, H. Kleinpoppen, R. Hippler and H.O. Lutz, J.Phys. B 12, 4077 (1979).
49) J. Pálinkás, L. Sarkadi nad B. Schlenk, J.Phys. B 13, 3829 (1980).
50) J. Pálinkás, B. Schlenk, A. Valek, J.Phys. B 14, 1157 (1981).
51) W. Jitschin, H.O. Lutz and H. Kleinpoppen, X-80 Conference, Stirling, Book of Abstracts, p. 32 (1980) and in: Inner Shell Phy-

sics of Atoms and Solids (eds. D.J. Fabian, H. Kleinpoppen, L.M. Watson), New York, Plenum Press, p. 89 (1981).

52) C.V. Barros Leite, N.V. de Castro Faria, R.J. Horowicz, E.C. Montenegro and A.G. de Pinho, Phys.Rev. A $\underline{25}$, 1880 (1982).

53) F. Rösel, D. Trautmann, G. Baur, Z.Phys. A $\underline{304}$, 75 (1982).

54) E.G. Berezhko, N.M. Kabachnik and V.V. Sizov, J.Phys. B $\underline{11}$, L 421 (1978).

55) E.G. Berezhko, V.V. Sizov and N.M. Kabachnik, J.Phys. B $\underline{14}$, 2635 (1981).

56) V.V. Sizov and N.M. Kabachnik, J.Phys. B $\underline{15}$, (1982), in print.

57) N.M. Kabachnik, private communication 1978.

58) E.G. Berezhko and N.M. Kabachnik, J.Phys. B $\underline{13}$, L 445 (1980).

59) R. Huster and W. Mehlhorn, Z.Phys. A $\underline{307}$, 67 (1982).

60) H.G.M. Heidemann, W. van de Water and L.J.M. Moergestel, J.Phys. B $\underline{13}$, 2801 (1980).

61) S.L. Carter and H.P. Kelly, J.Phys. B $\underline{11}$, 2467 (1978).

62) W. Johnson and K.-N. Huang, private communication, 1982.

63) D.J. Kennedy and S.T. Manson, Phys.Rev. A $\underline{5}$, 225 (1972).

ALIGNMENT STUDIES IN INNER-SHELL PROCESSES

W. Jitschin

Fakultät für Physik, Universität Bielefeld, 4800 Bielefeld 1,
W. Germany

INTRODUCTION

Inner-shell ionisation by energetic projectiles has been subject of numerous investigations. Basic information about the collisional interaction can be obtained from total cross sections measurements [1]. The need for more detailed information has led to the development of angular correlation studies, as e.g. the impact parameter dependence of the ionisation probability [2] or of the energy distribution of the ejected electrons [3]. Recent developments procede towards experiments which are complete in a quantum mechanical sense. These are dealing with the determination of the spacial distribution of collisional induced vacancies and with amplitudes and phases of the corresponding states [4]. A convenient and frequently used description employs the irreducible alignment tensor components of the density matrix [5-8]. The simplest quantity related to the spacial distribution is the total alignment $\mathcal{A}_{20}$ which can be determined in an axially symmetric (non-coincident) experiment. Other components of the density matrix can be obtained from coincidence experiments which employ planar symmetry.

Alignment studies for electron and photon impact have already been performed several years ago and are discussed in another article of this volume [9]. Recently alignment studies were extended to the field of inner-shell ionisation by ion impact. Single K-, L_1-, L_2-vacancies cannot be aligned since the condition $j > 1/2$ is not fulfilled [10]. Measurements of the proton induced L_3-subshell alignment $\mathcal{A}_{20}$ have been performed for proton impact on Cu [11], and were later extended to heavier target atoms and a wide range of impact energies [12-18]. Also the alignment of the M-subshells by proton impact [19], of KL double vacancies [20-22], and the impact parameter dependence of $\mathcal{A}_{20}$ and $\mathcal{A}_{22}$ has recently been investigated [23]. A review of ion-induced inner-shell alignment has been prepared one year ago [22], but since several new experimental and theoretical results have been obtained. Also rather comprehensive data for heavy ion projectiles are now available [14,16,24,25]. The present paper gives a short account of the basic principles and discusses experimental data of the L_3-subshell alignment induced by heavy projectiles.

PRINCIPLE

The production of a vacancy and its decay are assumed to be independent processes. The atomic vacancy state after the collision can be uniquely characterized by a density matrix. Its diagonal elements correspond to the ionisation cross sections of the magnetic substates. In general, these are not equal; the resulting anisotropy of the ionic state can be described by alignment parameters [5,6,8].

For example, the $\mathcal{A}_{20}$ parameter for the L_3-subshell reads:

$$\mathcal{A}_{20} = \frac{\sigma(1)-\sigma(0)}{2\sigma(1)+\sigma(0)} = \frac{\sigma(3/2)-\sigma(1/2)}{\sigma(3/2)+\sigma(1/2)} \tag{1}$$

where the term in brackets denotes the magnetic quantum numbers m_ℓ = 0, 1 and m_j = 1/2, 3/2, respectively. In the derivation of eq. 1 it was assumed that the level width is small as compared to the spin-orbit splitting, i.e. the LSJM coupling scheme holds.

The induced vacancy decays by emission of an Auger electron or a x-ray. In investigations employing heavy target atoms, x-rays are preferrentially observed since high efficiency Si(Li)-detectors and foil targets can be used. The angular distribution of characteristic x-rays follows a dipole pattern; experiments with non-coincident detection of the x-rays have axial symmetry and the angular intensity distribution $I(\theta)$ is symmetric about the axis, i.e. the direction of incoming beam:

$$I(\theta) = (I_0/4\pi)\ (1+\alpha\kappa\mathcal{A}_{20}P_2(\cos\theta)) \tag{2}$$

θ denotes the angle between the direction of observation and ion beam axis, P_2 is the second Legendre polynomial, κ is a correction for transfer of isotropic vacancies from the L_1- and L_2-subshells to the L_3-subshell by Coster-Kronig transitions [12,24], and α depends on the quantum numbers j_i and j_f of the initial and final state of the x-ray transition [5,6,8]:

$$\alpha = (-)^{j_i+j_f+1}\ \sqrt{3j_i+3/2}\ \begin{Bmatrix} 1 & j_i & j_f \\ j_i & 1 & 2 \end{Bmatrix} \tag{3}$$

For example, the coefficient α is 0.05 for the strong Lα line and 0.5 for the weak Lℓ line.

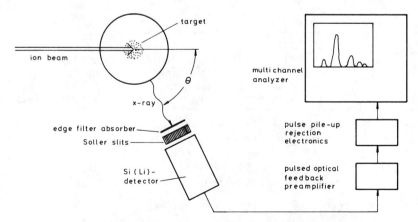

Fig. 1: Experimental setup for alignment studies (from ref. 17).

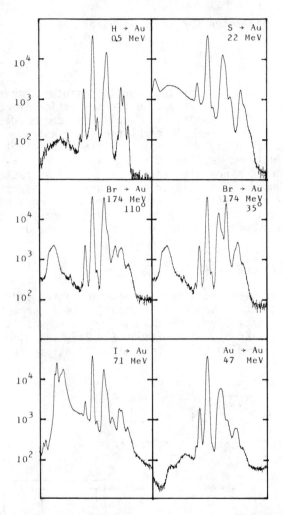

Fig. 2: Au L x-ray spectra induced by different projectiles. For Br+Au the Doppler-effect shifts the Br Kα line across the Au Lβ line.

Experimentally x-ray spectra are recorded at different observation angles (Fig. 1,2) and the angular distribution of a x-ray line (Lℓ) is conveniently determined by normalisation to a line which is predicted to be (practically) isotropic (Lα,Lγ). This procedure has been confirmed by an absolute measurement of the angular dependence [24]. By fitting the measured angular distribution $I(\theta)$ according to equation (2) the alignment $\mathcal{A}_{20}$ is obtained.

Anisotropy effects are small in general; even for the highest observed absolute alignment $|\mathcal{A}_{20}| \approx 0.5$, the intensities of the Lα and Lℓ lines change only by 2 % and 20 %, respectively, while changing the observation angle θ from 90° to 30°. This might explain why earlier investigations failed to observe any anisotropy.

PROTON IMPACT - FIRST-ORDER PERTURBATION APPROACH

The proton induced alignment $\mathcal{A}_{20}$ of the L_3-subshell has been carefully studied by several groups, using various target atoms (47 < Z < 92) and measuring the polarization [26,27] or angular distribution [12,13,15,17,18] of diagram lines. The data (Fig. 3)

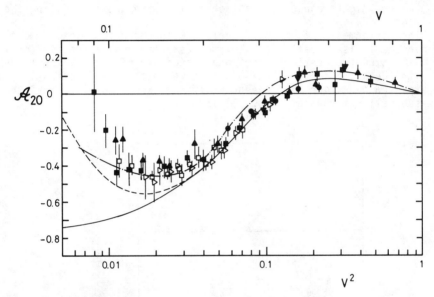

Fig. 3: Alignment of the L_3-subshell induced by proton impact. Targets ● Xe, ▲ Dy, ■□ Au, ▷ Pb, ▼ U. Curves for Au targets: ——— PWBA, - - - with correction for Coulomb deflection, -··-·· RSCA (from ref. 25).

show no pronounced dependence on the target if the projectile velocity v_1 is plotted in relative units $V=v_1/v_{L_3}$ (v_1 is the classical orbital velocity, as calculated from the experimental binding energy E_{L_3}). For a theoretical description first-order perturbation theories (plane-wave Born approximation, PWBA; semiclassical approach, SCA) have been employed. These theories assume stationary target electron wavefunctions which are fixed in the laboratory frame. Accounting of outer screening is essential for the description of alignment [12] as is already known from subshell cross sections. For V > 0.2 good agreement between experimental data and PWBA [8,13,28] as well as SCA [29,30] calculations is obtained. Discrepancies at smaller V are mainly due to the increasing disturbance of the projectile path by the field of the target nucleus which is not contained in the PWBA. The resulting change of the projectile direction has a strong influence on the alignment [13,17,31], whereas the velocity reduction in the target field is of much smaller importance [18]. This is in contrast to total cross sections which are sensitive to the retardation [32]. Relativistic SCA calculations (RSCA) employing a hyperbolic projecile path have been performed which account for both, deflection and retardation [30]. These calculations yield a good description of the experimental data for all V although a small discrepancy at V < 0.1 seems to remain. This might possibly be explained by the fact that in the low velocity regime the L-shell electrons will have time to relax adiabatically to the combined field of projectile and target [33] which might cause a rotation of the wavefunctions during the collision.

The alignment of higher shells with more complex radial wavefunctions is expected to show more structure as a function of V [8]. The structure reflects the nodal structure of the wavefunctions and its detailed shape is quite sensitive to the wavefunctions [8]. The expected behaviour has been verified experimentally for the $M_{3,4,5}$-subshells of thorium (Fig. 4) [19].

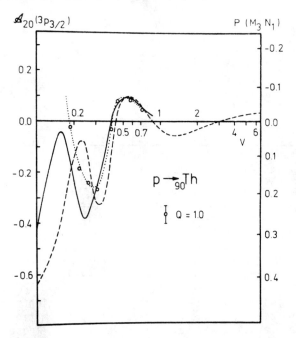

Fig. 4: Alignment of the M_3-subshell of Th induced by proton impact. Curves: ——— PWBA with HS wavefunctions, - - - PWBA with hydrogenlike wavefunctions, ... line through experimental points (from ref. 19).

A special case is the alignment of double vacancy states: by measuring the anisotropy of satellite lines the alignment of KL vacancies has been studied [20-22]. Due to simultaneous K-shell ionisation, the L-shell alignment is investigated at small impact parameters without employing coincidence techniques.

LIGHT ION IMPACT ($2 \leq Z_1 \leq 16$)

For He impact first-order perturbation theories are still expected to be valid. However, the trajectory of He (and heavier) projectiles is less influenced by the target nucleus field than the trajectory of H ions due to the smaller charge-to-mass ratio. Experimental results for the He induced total alignment A_{20} [13,14,25] are close to the proton data and, at low impact velocities V, exhibit deviations from RSCA calculations employing hyperbolic projectile trajectories [30]. Recently, difficult coincidence experiments have been performed in which the impact parameter dependence of the parameters $A_{20}(b)$ and $A_{22}(b)$ for 16 MeV He impact on Dy has been measured [23]. Whereas for $A_{22}(b)$ no theoretical predictions are available, $A_{20}(b)$ has been calculated in RSCA including trajectory effects [30]. The calculations reproduce the measured shape of the impact parameter dependence, but there are significant deviations. The observed discrepancies for the He induced alignment are probably not due to trajectory effects since the path of He ions is comparatively weakly influenced and trajectory effects are included in the calculations. It should be noted that for He impact on Au significant deviations also for the L_1-, L_2-, L_3-subshell cross sections at small impact velocity have been observed [34]: calculations predict $\sigma_{L_1} << \sigma_{L_2}$, σ_{L_3} , which is not found experimentally [34]. This has been explained by a vacancy transfer during the collision between the L_1-L_2-, L_1-L_3- [35] and L_2-L_3-subshells [33]. The transfer probability between L_1-L_2 and L_2-L_3 is estimated to amount to some 20 % for low velocity He impact on Au [33,35]. It can be expected that the transfer also influences the alignment, although such effects cannot be quantified yet.

For heavier ion projectiles ($Z_1 > 2$) transfer probabilities scale with Z_1^2 and become of the order of unity for $Z_1 \geq 4$ at small velocities indicating that the L-subshells are almost completely coupled. This conclusion is confirmed by experimental results (Fig. 5) where in the velocity range $0.15 < V < 0.3$ the measured alignment values $A_{20} = -0.2 \pm 0.1$ are almost independent of the projectile ($4 \leq Z_1 \leq 16$) [25]. At $V \sim 0.1$ the experimental alignment data exhibit an unexpected behaviour: with decreasing V the alignment rises steeply to large positive values [25]. Alignment values up to +0.47 have been observed for 2.4 MeV N on Au ($V \sim 0.09$) [24], which is close to the maximum attainable value of +0.5 (equation 1). The steep rise of the alignment is correlated to a surprisingly large L_2-subshell cross section σ_{L_2} which becomes larger than σ_{L_1} and σ_{L_2} . These experimental findings are not yet understood; possibly they are due to formation of molecular Stark-states during the collision.

Electron capture by the projectile may contribute significantly to the ionisation process. A comparison of calculated cross sections for direct ionsation (CPSSR) [36] and electron capture (Oppenheimer-Brinkman-Kramers, OBK) [37] reveals that for bare nuclei of $Z_1 = 10$ both processes are of the same order of magnitude, whereas for $Z_1 = 14,16$ electron capture is dominating. The alignment produced by electron capture has also been investigated theoretically [38]. The predicted alignment depends on the approximations and the

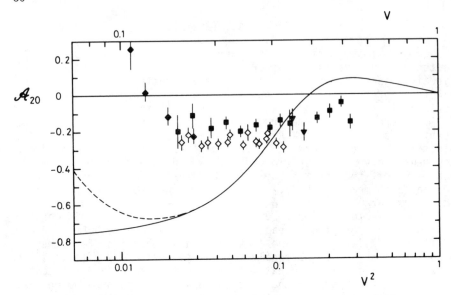

Fig. 5: Alignment of the L₃-subshell of Au, Bi, U targets induced by O impact. Curves ——— modified CPSSR, - - - with correction for Coulomb deflection (from ref. 25).

wavefunctions used, and on the charge state of the ion. At low impact velocity large negative alignment of the L_3-subshell is expected. Obviously, electron capture as described in OBK does not resolve the discrepancy between experiment and theory. For heavier projectile ions no detailed experimental studies of inner-shell alignment by electron capture have been performed so far, although these should be feasible by using thin gas targets and different projectile charge states.

HEAVY ION IMPACT ($Z_1 \geq 35$) - MOLECULAR EXCITATION

For heavy ion impact at low velocities quasi-molecular electron promotion is known to play a decisive role. Alignment studies have been performed for the collision systems Br+Au and Br+Pb (Fig. 6) [25] in which the K-shell binding energy of the projectile almost matches with the binding energy of the target L_2- or L_3-subshell, respectively. As has been derived from total cross-section measurements, target L-shell ionisation in these systems occurs via a two-step process: initially vacancies are formed in the $3d\sigma$ molecular orbital and then shared between the collision partners on the outgoing part of the trajectory [39]. Since in the molecular picture the magnetic quantum number is conserved, one expects only the creation of m=0 vacancies, i.e. large negative alignment. The experimental data, however, show only small negative values. This discrepancy is not yet understood. Presumably the Coulomb excitation is strong enough to thoroughly mix states and thus destroy the

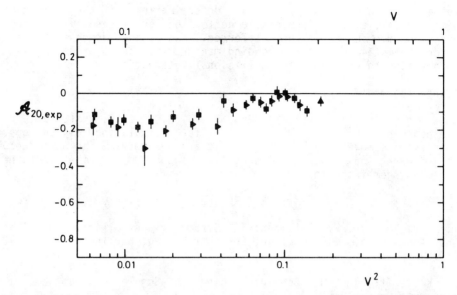

Fig. 6: Alignment of the L_3-subshell of ■ Au and ▶ Pb induced by Br impact (from ref. 25).

alignment. Furthermore, the molecular orbitals which are related to the rotating internuclear axis possibly do not completely follow the rotation (electron slippage) as is known for the anisotropy of molecular orbital radiation [40]. One should keep in mind that in close heavy-ion atom collisions the probability for additional creation of vacancies in higher shells is large; the existence of such vacancies has not been taken into account in the relationship which connects measured anisotropy and alignment. (equation 3).

The alignment induced by I impact on Au and Pb impact has also been studied for a broad range of impact energies and $\mathcal{A}_{20}$ values between -0.15 and 0 have been obtained [25]. For nearly symmetrical systems, alignment measurements are scarce. The available data exhibit no large alignment effects [16,41].

CONCLUSION

Inner-shell alignment measurements have provided new information about collisional processes which in part is complementary to total cross section measurements. In case of H and He impact the alignment $\mathcal{A}_{2,0}$ has proved to be a powerful tool for investigating projectile trajectory effects. Smaller discrepancies at low impact velocities and for the impact parameter dependence remain. For heavier projectile ions and impact velocities $V>0.1$ the experimentally observed alignment effects become monotoneously smaller. This could indicate that other mechanisms effectively reduce any alignment which might be produced in the primary excitation process. It is doubtful if total alignment measurements can fully resolve this problem.

Rather, studies involving the determination of the alignment differential, e.g., in projectile deflection or secondary electron emission may have to be performed. As has been demonstrated for outer shells [42,43], corresponding measurements can provide decisive information to elucidate the collisional interaction.

Acknowledgements

The author is indebted to Prof. Dr. H.O. Lutz, Prof. Dr. H. Kleinpoppen, and Dr. R. Hippler for stimulating discussions. Clarifying comments by Dr. R. Anholt are gratefully acknowledged. This work has been financially supported by the Deutsche Forschungsgemeinschaft.

REFERENCES

1. P.H. Mokler, F. Folkmann: in ´Structure and Collisions of Ions and Atoms´, ed. I.A. Sellin, Springer Verlag, Berlin, pp. 201-272 (1978)

2. K.E. Stiebing, H. Schmidt-Böcking, R. Schule, K. Bethge, I. Tserruya: Phys. Rev. A 14, 146-151 (1976)

3. T. Weiter, R. Schuch: Z. Phys. A, in press (1982)

4. H.O. Lutz: in ´Inner Shell and X-Ray Physics of Atoms and Solids´, ed D.J. Fabian, H. Kleinpoppen, L.M. Watson, Plenum Press, New York, pp 79-88 (1981)

5. E.G. Berezhko, N.M. Kabachnik: J. Phys. B 10, 2467-2477 (1977)

6. E.G. Berezhko, N.M. Kabachnik, V.V. Sizov: J. Phys. B 11, 1819-1832 (1978)

7. K. Blum, H. Kleinpoppen: Phys. Rep. 52, 205-261 (1979)

8. V.V. Sizov, N.M. Kabachnik: J. Phys. B 13, 1601-1610 (1980)

9. W. Mehlhorn: this volume (1982)

10. W. Mehlhorn: Phys. Lett. 26 A, 166-167 (1968)

11. A. Schöler, F. Bell: Z. Phys. A 286 163-168 (1978)

12. W. Jitschin, H. Kleinpoppen, R. Hippler, H.O. Lutz: J. Phys. B12, 4077-4084 (1979)

13. J. Palinkas, L. Sarkadi, B. Schlenk: J. Phys. B 13, 3829-3834 (1980)

14. V. Zoran, I. Piticu: private communication (1980)

15. J. Palinkas, B. Schlenk, A. Valek: J. Phys. B 14, 1157-1159 (1981)

16. Z. Stachura, F. Bosch, D. Maor, P.H. Mokler, W.A. Schönfeldt: in ´3. Arbeitsbericht, Arbeitsgruppe Energiereiche Atomare Stösse´ pp. 70-71 (1982)

17. W. Jitschin, A. Kaschuba, R. Hippler, H.O. Lutz: Z. Phys. A 304, 69-73 (1982)

18. C.V. Barros Leite, N.V. de Castro Faria, R.J. Horowicz, E.C. Montenegro, A.G. de Pinho: Phys. Rev. A 25, 1880-1886 (1982)

19. J. Wigger, S. Ost, M. Brüssermann, G. Richer, B. Cleff: Phys. Lett. 84 A, 110-114 (1981)

20. K.A. Jamison, P. Richard: Phys. Rev. Lett. 38, 484-487 (1977)

21. K.A. Jamison, J. Newcomb, J.M. Hall, C. Schmiedeskamp, P. Richard: Phys. Rev. Lett. 41, 1112-1115 (1978)

22. B. Cleff: Acta Physica Polonica A61, 285-319 (1982)

23. J. Konrad, R. Hoffmann, H. Schmidt-Böcking, R. Schuch: in ´3. Arbeitsbericht, Arbeitsgruppe Energiereiche Atomare Stösse´ pp. 72-73 (1982)

24. J. Palinkas, L. Sarkadi, B. Schlenk, I. Török, Gy Kalman: J. Phys. B 15, L451-L454 (1982)

25. W. Jitschin, R. Hippler, R. Shanker, H. Kleinpoppen, H.O. Lutz, R. Schuch: submitted to J. Phys. B (1982)

26. V.P. Petukhov, E.A. Romanovskii, S.V. Ermakov: JETP Lett. 29, 385-387 (1979)

27. G. Richter, M. Brüssermann, S. Ost, J. Wigger, B. Cleff, R. Santo: Phys. Lett. 82 A, 412-414 (1981)

28. M. Kamiya, Y. Kinefuchi, H. Endo, A. Kuwako, K. Ishii, S. Morita: Phys. Rev. A 20, 1820-1827 (1979)

29. L. Kocbach: Abstracts of the VIth Int. Sem Ion/Atom Collisions, Tokaimura, Ibaraki, Japan pp. 188-190 (1979)

30. F. Rösel, D. Trautmann, G. Baur: Z. Phys. A 304, 75-78 (1982)

31. W. Jitschin, H. Kleinpoppen, H.O. Lutz: in ´Inner Shell and X-Ray Physics of Atoms and Solids´, ed. D.J. Fabian, H. Kleinpoppen, L.M. Watson, Plenum Press, New York, pp. 89-92 (1981)

32. L. Kocbach, J.M. Hansteen, R. Gundersen: Nucl. Instr. Meth. 169, 281-291 (1980)

33. K. Aashamar, P.A. Amundsen: to be published (1982)

34. W. Jitschin, R. Hippler, R. Shanker, H.O. Lutz: Abstr. 8th Int. Conf. Atomic Physics, Göteborg, p. B63 (1982)

35. L. Sarkadi, T. Mukoyama: J. Phys. B 14, L255-L260 (1981)

36. W. Brandt, G. Lapicki: Phys. Rev. A 20, 465-480 (1979)

37. V.S. Nikolaev: Soviet Physics JETP 24, 847-857 (1967)

38. E.G. Berezhko, V.V. Sizov, N.M. Kabachnik: J. Phys. B 14, 2635-2646 (1981)

39. W.E. Meyerhof, R. Anholt, J. Eichler, A. Salop: Phys. Rev. A 17, 108-119 (1978)

40. R. Anholt: Z. Phys. A 288, 257-276 (1978)

41. W. Jitschin, R. Shanker, B. Wisotzki: unpublished (1982)

42. R. Hippler, G. Malunat, M. Faust, H. Kleinpoppen, H.O. Lutz: Z. Phys. A 304, 63-68 (1982)

43. L. Zehnle, Th. Hall, R. Schinke, V. Kempter: Z. Phys. A 304, 95-101 (1982)

INNER-SHELL IONIZATION BY RELATIVISTIC ELECTRON IMPACT*

Harald Genz
Institut für Kernphysik,Technische Hochschule Darmstadt
6100 Darmstadt, Germany

ABSTRACT

The present paper deals with the improvements that have been obtained in the calculation of total inner-shell ionization cross sections by relativistic electron impact and it contains a comparison with experimental data. Special attention is given to the observed discrepancy between experiment and theory for impact energies up to 2.0 GeV where, due to the expected polarization of the target medium, the total cross section is predicted to saturate, while the experimental data exhibit a still rising behaviour. The missing density effect is not yet understood and the suggestions to explain these findings are not satisfactory. Progress, however, has recently been achieved in the measurement of cross sections which are differential in energy and angle of the final-state electrons by observing the electrons either in a single arm experiment i.e. integrated over all shells or in coincidence with the corresponding K respectively L X ray. Finally, the greatly desired completely differential cross-section measurement, i.e. differential in energy and angle of both final-state electrons has been reported for the bombardment of Ag by 500-keV electrons. The comparison with theory exhibits agreement only for some regions of momentum transfer.

INTRODUCTION

Since the last conference on inner-shell ionization[1] several authors have published papers on the subject of electron induced total and differential cross-section measurements and theoretical predictions. Due to the limitation of space I will concentrate on four major subjects while giving reference to the other investigations only without claiming completeness.

(i) The first subject will contain a comparison of experimental data of total K-shell cross sections and theory. Here the elaborate relativistic Hartree-Fock calculations by Scofield[2] are considered as well as a method by Eschwey and Manakos[3] which applies the one-photon-exchange approximation to express the K-shell ionization cross section in terms of two Lorentz invariant structure functions for the atom. This method, widely used in nuclear and particle physics, has not been so popular in atomic physics but it provides a rather instructive method to perform cross-section calculations and is especially suitable for a model independent discussion of the cross section. It also reveals the connections with the K-shell photoabsorption cross section and pro –

*
Work supported by Deutsche Forschungsgemeinschaft

vides a natural way to examine the validity of the Weizsäcker-Williams approximation.

(ii) The second chapter deals with the fact that the atoms of a target are not isolated of each other and thus a polarization of the target medium occurs – caused by the electromagnetic field of the incident electron. The influence of this polarization on the cross section is negligible for low impact energies. For very large energies, typically of the order of several hundred MeV, however, one expects a saturation of the cross section[2-5]. This phenomenon is called the density effect. While its influence on the ionization energy loss of relativistic charged particles has well been established it has not[6] or at least not unequivocally[7] been detected for inner-shell ionization.

(iii) The third item treated in more detail covers the class of measurements of cross sections which are differential in energy and angle. This type of experiments has been demanded for since a long time[8] since they expose the most detailed informations of the electron impact ionization process.

(iv) At the end of the talk I shall report on recent investigations about the electron induced production rate of multiple vacancies which exhibits an unexpected energy dependence.

We would finally like to draw the attention to a new field that gives information about the behaviour of electron induced inner-shell vacancies as is reported on this conference by Müller et al.[9] who detected the charge state of recoil ions produced by the bombardment of rare gases by electrons with 20 and 50 MeV impact energy.

TOTAL CROSS SECTION

In the one-photon-exchange approximation – neglecting exchange effects – the cross section for inner-shell ionization of isolated atoms by incident electrons can be written in the laboratory in the form

$$\sigma = \sigma^{\ell} + \sigma^{t} \tag{1}$$

where the so-called longitudinal contribution is given by[3]

$$\sigma^{\ell} = \frac{4\pi \, \alpha^2}{mc^2 \vec{k}_0^2} \int_{\Delta E_{min}}^{\Delta E_{max}} \int_{\hat{q}^2_{min}(\Delta E)}^{\hat{q}^2_{max}(\Delta E)} \frac{1}{\vec{q}^4} \left[\frac{E_0(E_0 - \Delta E)}{\hbar^2 c^2} + \frac{1}{4} \hat{q}^2 \right] \cdot$$

$$\left(-\frac{\vec{q}^2}{\hat{q}^2} \right) w^{\ell}(\hat{q}^2, \Delta E) d(\hat{q}^2) d(\Delta E) \tag{2}$$

with

$$w^{\ell}(\hat{q}^2, \Delta E) = - W_1(\hat{q}^2, \Delta E) - (\vec{q}^2/\hat{q}^2) W_2(\hat{q}^2, \Delta E) \tag{3}$$

and the so-called transverse contribution by[3]

$$\sigma^t = \frac{4\pi\,\alpha^2}{mc^2\vec{k}_0^2} \int\limits_{\Delta E_{min}}^{\Delta E_{max}} \int\limits_{\hat{q}_{min}^2(\Delta E)}^{\hat{q}_{max}^2(\Delta E)} \frac{1}{\hat{q}^4} \cdot \left[-\left(\frac{m^2c^2}{\hbar^2} + \frac{1}{2}\,\hat{q}^2\right) - \right.$$

$$\left. - \frac{\hat{q}^2}{\vec{q}^2}\left(\frac{E_0(E_0-\Delta E)}{\hbar^2c^2} + \frac{1}{4}\,\hat{q}^2\right) \right] W_1(\hat{q}^2,\Delta E)\,d(\hat{q}^2)\,d(\Delta E). \tag{4}$$

Here

$$E_0 = (\hbar^2c^2\vec{k}_0^2 + m^2c^4)^{1/2} \tag{5}$$

is the total energy of the incident electron, m is the electron rest mass and α is the fine structure constant.
Furtheron

$$\hbar\hat{q} = \hbar(\hat{k}_0 - \hat{k}_1) = (\Delta E/c, \hbar\vec{q}) \tag{6}$$

is the four-momentum transfer and $\hat{q}^2$ is given by

$$\hat{q}^2 = [(\Delta E)^2/\hbar^2c^2] - \vec{q}^2 . \tag{7}$$

It should be noted that $\hat{q}$ is a space-like four-vector ($\hat{q}^2 < 0$). The integration limits follow from kinematics with

$$\Delta E_{min} = I \text{ (ionization energy)} \tag{8}$$

$$\Delta E_{max} = E_0 - mc^2$$

$$\hat{q}_{min}^2(\Delta E) = \frac{2}{\hbar^2c^2}\{m^2c^4 - E_0(E_0-\Delta E) - \hbar c|\vec{k}_0|[(E_0-\Delta E)^2 - m^2c^4]^{1/2}\} \tag{10}$$

$$\hat{q}_{max}^2(\Delta E) = \frac{2}{\hbar^2c^2}\{m^2c^4 - E_0(E_0-\Delta E) + \hbar c|\vec{k}_0|[(E_0-\Delta E)^2 - m^2c^4]^{1/2}\} \tag{11}$$

All relevant information about the structure of the target atom is contained in the two dimensionless structure functions $W_1(\hat{q}^2,\Delta E)$ and $W_2(\hat{q}^2,\Delta E)$. The method of introducing two structure functions to express the cross section is well known in nuclear and particles physics[10,11]. The longitudinal contribution to the cross section (2) is attributed to the Coulomb interaction since the corresponding structure function is related to the transition charge density[3]. Similarly, the structure function of the transverse part of the cross section is determined by the transverse component of the current density operator, i.e. the magnetic part of the electromagnetic interaction.

For the deduction of the total cross section (1) the structure functions (3) were calculated for K-shell ionization in a simple one particle model[3], where the bound and ejected electron are

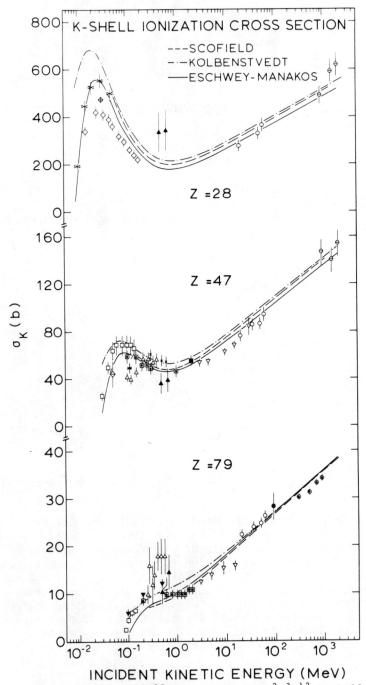

Fig.1. Experimental[5,6,13-27] and theoretical[2,3,12] K-shell cross sections for nickel, silver and gold.

described by one-particle Darwin-wavefunctions and where the structure functions are empirically corrected near the photon point ($\hat{q}^2=0$) to be consistent with K-shell photoionization data. The results are shown in Fig. 1 together with the relativistic Hartree-Fock calculations by Scofield[2] and with the semiclassical calculations[6,12] based on the Weizsäcker-Williams-method of equivalent photons. The agreement between the results of the semiclassical and quantum mechanical calculations is rather striking at first. However, this agreement can be understood since it was shown[3] that the semiclassical formulas can be obtained from the quantum mechanical formulas assuming some approximations.

The comparison with experiment[5,6,13-27] exhibits a satifactory agreement although certain distinctions can be drawn.

(i) In the energy region of $\beta<1$, i.e. up to a few MeV impact energy, the experimental data are widely scattered and new more accurate and consistent measurements are called for[28].

(ii) The region of $\beta=1$, e.g. up to a few hundred MeV, shows a rather good agreement between experiment and theory especially with the calculations of Ref.3.

(iii) In the ultra relativistic region, i.e. for energies exceeding several hundred MeV, the presented theoretical treatments should not hold since the atoms cannot anymore be considered to be isolated of each other. This subject is treated in detail in the following chapter.

It should be mentioned, finally, that an extension of the Kolbenstvedt method[12] based on the virtual photon concept has recently been published[29] as well as an addition[30,31] to Scofield's calculations which also yield rather good agreement with experiment especially for low electron energy impact. For this region an empirical expression for the production of K-shell vacancies by electrons has lately been derived[32].

THE DENSITY EFFECT

The atoms in a target have to be considered in general as non isolated of each other. Therefore in a quantum mechanical treatment of the inner-shell ionization cross section for non isolated atoms in the one-photon-exchange approximation, it has to be taken into account that the virtual photon exchanged between the incoming electron and a specific target atom may excite the other target atoms into virtual states. These virtual excitations are equivalent to the polarization of the target material and cause the modification of the photon propagator as shown in the Feynman diagram of Fig. 2. This modification of the photon propagator by the target medium is called density effect.

The modified photon propagator can be calculated approximately by perturbation theory, Fig.2. In second order perturbation theory Yura[33] has shown that the modification of the photon propagator can be described approximately by a dielectric function $\varepsilon(\vec{q}, \Delta E)$. With reference to Yura the cross section for inner-shell ionization

90

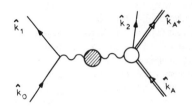

Fig.2. The Feynman diagram that takes into account the density effect. Here $h\hat{k}_o$, $h\hat{k}_1$ and $h\hat{k}_2$ denote the four-momenta of the ingoing, scattered and emitted electron, and $h\hat{k}_A$ and $h\hat{k}_A+$ the four-momenta of the atom before and of the ion after the scattering.

of non isolated atoms deviates from (1), (2) and (4) in such a way that one has to replace in the longitudinal contribution (2)

$$\frac{1}{\vec{q}^4} \rightarrow \frac{1}{\vec{q}^4 |\varepsilon(\vec{q},\Delta E)|^2} \tag{12}$$

and in the transverse part (4)

$$\frac{1}{\hat{q}^4} \rightarrow \frac{1}{\left| \left(\frac{\Delta E}{hc}\right)^2 \varepsilon(\vec{q},\Delta E) - \vec{q}^2 \right|^2} \tag{13}$$

To derive an expression of $\varepsilon(\vec{q},\Delta E)$ that enables the calculation of cross sections, it is assumed[3],[27] that the dielectric properties of the target medium are independent of the direction of $\vec{q}$. This assumption seems to be reasonable because the targets used in experiments are of an amorphous structure and not crystalline. Now it turns out that the cross section is mainly determined by the structure functions for small energy loss and $\hat{q}^2$ values near the photon point ($\hat{q}^2=0$). For ultra relativistic energy E of the incoming electron and small energy loss ΔE ($\Delta E \ll E$) the upper limit of integration $\hat{q}^2_{max}(\Delta E)$ lies very near the photon point. Therefore the main contribution to the cross section is due to those virtual photons which satisfy in good approximation the relation

$$hc|\vec{q}| = \Delta E \tag{14}$$

which holds for real photons. Hence it is assumed that the dielectric function, which is in general complex, depends only on one variable

$$\varepsilon(\vec{q},\Delta E) = \varepsilon(\Delta E) = \varepsilon'(\Delta E) + i\varepsilon''(\Delta E). \tag{15}$$

Further it is to be expected because of (14) that the dielectric function (15) has to be consistent with measurements of the index of refraction $n(\Delta E)$, that means

$$\varepsilon(\Delta E) = [n(\Delta E)]^2. \tag{16}$$

Yura[33] has been shown that from a general expression for $\varepsilon(\Delta E)$ one obtains the relation

$$\varepsilon(\Delta E) = \varepsilon'(\Delta E) = 1 - (\hbar\omega_p/\Delta E)^2 , \qquad (17)$$

where ω_p is the plasma frequency which is related to the electron density N (= number of electrons per volume) of the target medium via

$$\omega_p = (4\pi e^2 N/m)^{1/2} . \qquad (18)$$

This dielectric function (17) is the same as for a plasma of free electrons and does not depend on the specific structure of the target. It has to be expected from the derivation that the expression (17) is valid for energies ΔE above the K-shell ionization energy I_K. This has been verified in measurements[34] of the index of refraction $n(\Delta E)$ using real photons of energy $\hbar\omega = \Delta E \gg I_K$. However, near K threshold ($\Delta E \approx I_K$) the dielectric function (17) does not hold because of the strong absorption of photons of energy $\hbar\omega = \Delta E \approx I_K$. This implies that near threshold the imaginary part cannot be neglected. An imaginary part different from zero, however, does not change the results for K-shell ionization cross sections significantly.

In Fig. 3 the calculation of the cross section for K-shell ionization with and without the density effect are displayed for several Z values and electron incident energies, $0.07 \leq E < 2.0$ GeV, ($E = E_o - mc^2$). The dielectric function (17), using the plasma energies of Table I, has been applied to describe the density effect.

Table I. Plasma energies $\hbar\omega_p$ used for the calculation of the dielectric function(17)

Target	$_{16}$S	$_{20}$Ca	$_{25}$Mn	$_{28}$Ni	$_{29}$Cu	$_{32}$Ge	$_{47}$Ag	$_{79}$Au
$\hbar\omega_p$ (eV)	29.4	25.1	52.2	59.4	58.1	44.3	61.7	80.3

From Fig. 3 it becomes obvious that for large energies the cross sections including the density effect should reach a constant value and the onset of this saturation starts earlier for low Z elements. These results are in qualitative agreement with the predictions of Dangerfield[4], Dangerfield and Spicer[5] and of Scofield[2].

Three experiments[6,7,27] have been performed in the energy region where the cross section should saturate. While the data of Middleman et al.[6] are well described by the theory for isolated atoms and impact energies of 300<E<900 MeV, the authors of Ref.7 interpret their experimental findings to be consistent with the onset of the density effect in the region 70<E<300 MeV, although the effect found was much smaller than predicted.

To solve this discrepancy in a recent investigation[27] the electron impact energy was extended into the ultra relativistic region, $0.9 \leq E \leq 2.0$ GeV. Absolute measurements of K- and L-shell cross sections for Ni(K), Cu(K), Ag(K,L) and Au(L) are reported as well as the relative K-shell X-ray yields for the low Z elements S, Ca, Mn, Ni and Ge at 0.9 and 2.0 GeV. Finally, the K X-ray yield of a composite Ca-Mn target at these two energies was determined very precise-

92

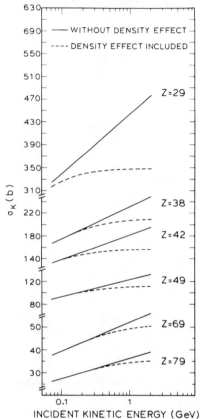

630
690 ── WITHOUT DENSITY EFFECT
550 ---- DENSITY EFFECT INCLUDED
510
470   Z=29
430
390
350
σ_K(b)
310
220   Z=38
180   Z=42
140
120   Z=49
80
50   Z=69
40
30   Z=79

0.1   1
INCIDENT KINETIC ENERGY (GeV)

Fig.3. Calculations[3] of the
K-shell ionization cross-
section assuming isolated
atoms (solid line) and including the density effect
(dotted line).

ly. The experimental method was standard, i.e. bombardment of thin solid
targets and the detection of characteristic X rays under 90° to the beam
axis by means of a Si(Li) detector.
The total cross sections are obtained
from the X-ray production cross section divided by the fluorescence
yield. The results are shown in
Fig. 4 and 5 together with other
data[6,7,16,27].

From Fig. 4 it becomes evident that the K-shell cross section
of Cu and the L-shell cross section
of Au at 900 MeV agree for the two
different experiments[6,27]. Furthermore, the cross sections at 2.0 GeV
exceed always the values at 900 MeV.
Finally, a comparison with theory
exhibits that experimental values
agree with the calculations without
the density effect. Thus the cross
sections do not saturate but seem
still to be rising as is also mani-
fested by the fact that the data measured in the ultra relativistic region follow the scaling behaviour[16]
that has been derived for impact
energies 20<E<60 MeV.

The displayed values of the
cross-section ratios in Fig.5 again
exhibit an increasing behaviour with
increasing energy which is clearly
indicated by plotting the ratios obtained for each element at 2.0 and
0.9 GeV as function Z (Fig.6). The
ratios for all five target atoms are
different from unity as is not expected from theory with density
effect due to the predicted saturation, and the average of these
ratios yields

$$\bar{R} = \frac{\sigma_K(Z;E=2.0 \text{ GeV})}{\sigma_{K(Z;E=0.9 \text{ GeV})}} = 1.08\pm0.01 \ .$$

This result is again in agreement with the calculation without the
density effect and with the low energy scaling behaviour. Finally,
the most crucial check is obtained from the rate of the K X-ray
yield ratio of the Ca-Mn target at 0.9 and 2.0 GeV which is

$$RR = \frac{\sigma_K(Ca;2.0 \text{ GeV})/\sigma_K(Mn; 2.0 \text{ GeV})}{\sigma_K(Ca;0.9 \text{ GeV})/\sigma_K(Mn; 0.9 \text{ GeV})} = 0.99\pm0.02$$

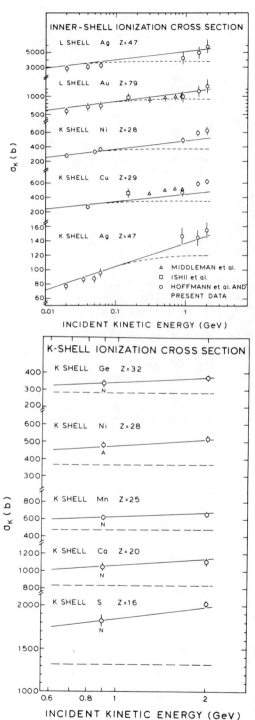

Fig.4. Absolute K and L-shell cross sections[6,7,27]. The solid lines represent the theory[3] for the K-shell assuming isolated atoms and the dashed lines the calculations including the density effect. For the L-shell the calculations of Scofield[2] are shown with (dashed line) and without (solid line) the density effect.

Fig.5. K-shell cross section obtained at 0.9 and 2.0 GeV for low Z elements. The cross sections are normalized at 0.9 GeV to the calculations[3] for isolated atoms. The value of σ_K for Ni is in agreement with the absolute measurement Fig.4.

94

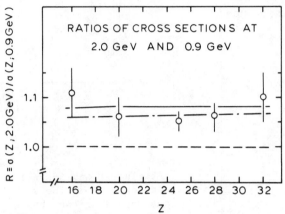

Fig.6. The K-shell cross section taken at 2.0 GeV is devided by the value at 0.9 GeV for various Z elements. The calculations[3] for isolated atoms (solid line) and including the density effect (dashed line) are shown together with the low energy scaling behaviour (dotted line)[16].

Assuming the density effect less pronounced than predicted and its onset to be observable only for Ca - being the lower Z element - in the investigated energy interval and no influence of the density effect yet for the heavier element Mn, then this ratio should be less than one. However, if both cross sections are still rising between 0.9 and 2.0 GeV, the ratio RR is expected to be close to one.

The reason for the missing density effect still remains to be revealed. It has been suggested[7,35] to introduce in (17) instead of ω_p an effective plasma frequency $\bar{\omega}_p < \omega_p$. However, in this case the dielectric function is not in agreement with measurements of the index of refraction and the experimental results of Ref.27 prove that these assumptions are incorrect. To solve the puzzle Amundsen[36] has proposed a "process-dependent" dielectric function without deriving such a quantity, however.

DIFFERENTIAL ANGLE AND ENERGY CROSS SECTIONS

Following the procedure including the use of structure functions as outlined in the two preceding chapters the cross section differential in angle and energy of both electrons can be expressed in the most general form as follows[37]:

$$\frac{d^3\sigma}{d\Omega_1 dE_1 d\Omega_2} = \Gamma(W^t + PW^\ell + \sqrt{P(P+1)}W^i\cos\phi + PW^P\cos 2\phi). \qquad (19)$$

Here Γ measures the flux and P the polarization of the virtual photons produced by the scattered electron and ϕ is the azimutal angle of the emitted electron in a coordinate system where the direction of the momentum transfer is parallel to the z axis. The experiment thus allows to determine four independent structure functions. The transverse polarization term, $\sim W^P$, and the longitudinal/transverse interference term, $\sim W^i$, are characterized by their dependence on the azimuthal angle, $\cos 2\phi$ and $\cos\phi$, respectively. The structure functions

$$W^{t,\ell,i,P} \sim (d\sigma^{t,\ell,i,P})/d\Omega_2 \qquad (20)$$

with
$$W = W(\hat{q}^2, \Delta E, \Theta, E) \qquad (21)$$

allow to probe the spatial distribution of the response for the bound electron. In (21) Θ stands for the angle between the momentum transfer and the z axis and E denotes the energy of the emitted electron.

The simplest form of measurements differential in angle and energy are those where only one of the two final-state electrons is detected, i.e. $d^2\sigma/d\Omega_1 dE_1$, while it is integrated over all angles of the other electron. With this single arm method it is not possible to distinguish the shell from which an electron is ejected.

Recently, double differential cross section have been determined by bombarding thin solid targets of 10-50 $\mu g/cm^2$ with 225-keV electrons at the Darmstadt superconducting pilot model c.w. accelerator[38] in order to examine the discrepancy of the measurements by Missoni et al.[39] and the theoretical predictions by Cooper and Kolbenstvedt[40]. The cross section, obtained by bombarding a thin carbon foil of 13 $\mu g/cm^2$ with electrons of 200 keV and by detecting one of the outgoing electrons, is shown in Fig.7. together with the theory[40]. These data were deduced by taking into account effects caused by multiple scattering and by the energy resolution of the detector. Contrary to the findings of Ref.39 the energy and the value of the cross section at the Møller peak are in good agreement with the calculations. However, similar as in the former work contributions to the cross section above the Møller energy have been found that can not be explained by theory and might be caused by the selection of plane waves[8].

A higher level of sophistication is obtained from measurements of cross sections which are differential in the energy and angle of one of the final-state electrons by observing the electron in coincidence with a K X ray[41]. Since the second electron is not observed this cross section is integrated over all angles. For the first time coincidence measurements have been reported for the L ionization of gold[42] by 300-keV electron impact.

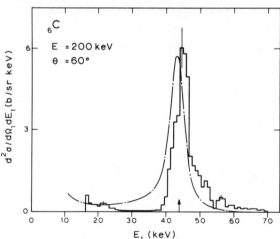

Fig.7. Differential cross section[38], summed over all shells, for ionization of carbon by 200-keV incident electrons as a function of the energy of one of the electrons emitted as a scattering angle of 60° and theoretical calculations[40]. The arrow is the Møller energy.

In Fig. 8 the energy dependent cross section at a scattering angle of 20° is shown, but unfortunately for the L shell there is no theoretical prediction the data could be compared with.

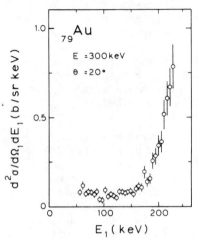

Fig.8. Differential cross section[42] for ionization of the L shell of gold by 300-keV incident electrons as a function of the energy of one of the electrons emitted at a scattering angle of 20°.

A completely differential cross section measurement, i.e. differential in energy and angle of both final-state electrons, has been reported by Schüle and Nakel[43] for the K-shell ionization of silver by 500-keV electron impact. Until now, these so-called (e,2e) experiments have been performed only at low energy ranges and not for inner-shell ionization, although they have been called for since a long time because the most detailed aspects of the electron ionization process can be obtained as well as informations about the momentum distribution of the atomic electron at least from those experimental conditions for which (19) factorizes.

The results of the experiment[43] are shown in Fig.9 for the angular combination -20°/40° and -25°/40°. It is plotted $d^3\sigma/d\Omega_1 dE_1 d\Omega_2$ versus the kinetic energy E_1 of the electrons detected at -20° and -25° respectively. The solid lines are calculations applying the formula given by Das and Konar[44], who used

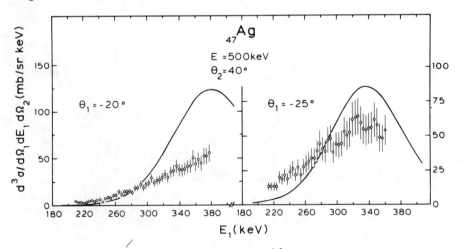

Fig.9. Triple differential cross section[43] for K-shell ionization of Ag by 500-keV electron impact as a function of the energy of the electrons scattered at -20° and -25° respectively. The solid line is calculated from the formula of Das and Konar[44].

a relativistic interaction Hamiltonian, relativistic plane waves for the incident and the high energy scattered electron, a non relativistic hydrogen wavefunction for K-shell electrons and a Sommerfeld-Maue wavefunction for the low energy ejected electron. Electron exchange is included in the Ochkur approximation, while electron correlations are neglected. The agreement with experiments is not always satisfactory.

MULTIPLE VACANCIES

In a recent investigation[45] the energy and Z dependence for electron induced multiple ionization has been systematically studied for $9 \leq Z \leq 29$ and $10 \leq E \leq 200$ keV. A typical spetrum - obtained with a crystal spectrometer - shows the satelite KL^1 resp. KL^2 and the diagram line KL^0 for the bombardment of Mg with 50-keV electrons (Fig.10). The ratio KL^1/KL^0 agrees for low bombarding energies with the results of other measurements [46,47] and the shake-off theory by Åberg[48] but exhibits for increasing energies a decreasing behaviour (Fig.11). The origin for this deviation is not yet understood. We are currently planing an experiment[49] to measure the angular distribution of the KL^1/KL^0 ratio in order to investigate the possibility if an unexpected strong energy dependent alignment could cause this behaviour.

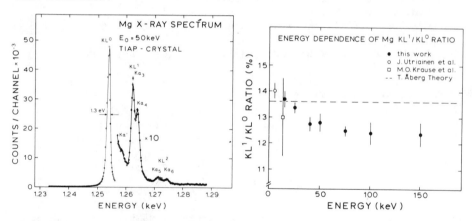

Fig. 10. Diagram and satelite lines.   Fig. 11. The KL^1/KL^0 ratio

CONCLUSIONS

Since the last reviews[8,35] progress has been achieved in the theoretical description of inner-shell ionization by relativistic electron impact and agreement with experiment is in general good, besides the region of $\beta < 1$ where consistent experimental values are needed. There is no convincing explanation for the severe discrepancy in the ultra relativistic region where the density effect has not been observed. The first investigations of cross sections differential in energy and angle of one or even both final-state electrons exhibit only partial agreement with theoretical predictions. More experiments especially also away from the photon point ($\hat{q}^2=0$) i.e. towards larger scattering angles, are desired as well as refined calculations. The reported energy dependent satelite production rate needs further investigations.

ACKNOWLEDGEMENT

The experimental and theoretical work at Darmstadt has been performed by C. Brendel, P. Eschwey, H.G., H.D. Gräf, C.Jakoby, M. Janke, U. Kuhn, W. Löw, P. Manakos, A. Richter, P. Seserko, E. Spamer, O. Titze. Furthermore I am grateful to Prof. Nakel, Dr. E.Schüle, Prof. Salzborn and Dr. A. Müller for providing their data prior to publication.

REFERENCES

1. Inner-shell and X-Ray Physics of Atoms and Solids,(Plenum Press, New York and London, 1981) eds. D.J.Fabian, H.Kleinpoppen and L.M. Watson.
2. J.H. Scofield, Phys. Rev. A18, 963 (1978).
3. P. Eschwey, PhD Thesis, Institut für Kernphysik, Technische Hoch-Hochschule Darmstadt (1981) D17 (unpublished); and P. Eschwey and P. Manakos, to be published.
4. G.R. Dangerfield, Phys. Lett 46A, 19 (1973).
5. G.R. Dangerfield and B.M. Spicer, J. Phys. B: At Mol. Phys. 8, 1744 (1975).
6. L.M. Middleman, R.L. Ford and R. Hofstadter, Phys. Rev. A2, 1429 (1970).
7. M. Kamiya, A. Kuwako, K. Ishii, S. Morita and M. Oyamada, Phys. Rev. A22, 413 (1980).
8. D.H. Madison and E. Merzbacher, Atomic Inner-Shell Processes (Academic Press, New York, 1975) ed. B.Crasemann, p.1
9. A. Müller, W. Groh, U. Kneißl, H. Ströher, R. Heil and E. Salzborn, this conference.
10. T. de Forest and J. D. Walecka, Adv. Phys. 15, 1 (1966).
11. G. B. West, Phys. Rev. 18C, 263 (1975).
12. H. Kolbenstvedt, J. Appl. Phys. 38, 4785 (1967); 46, 2771 (1975) compare also Ref. 6.
13. G. W. Green, Proc. 3rd Int.Symp. on X-ray Microscopy, Stanford (1962), Academic Press New York.
14. L. T. Pockman, D. L. Webster, P. Kirkpatrick and K. Harworth, Phys. Rev. 71, 330 (1947).
15. S. A. H. Seif el Nasr, D. Berényi and Gy. Bibok, Z. Phys. 267, 169 (1974).
16. D. H. H. Hoffmann, C. Brendel, H. Genz, W. Löw, S. Müller and A. Richter, Z. Phys. A293, 187 (1979).
17. D. V. Davis, V. D. Mistry and C. A. Quarles, Phys. Lett. 38A, 169 (1972).
18. D. H. Rester and W. E. Dance, Phys. Rev. 152, 1 (1966).
19. H. Hansen and A. Flammersfeld, Nucl. Phys. 79, 135 (1966).
20. A. Li-Scholz, R. Collé, I.L. Preiss and W. Scholz, Phys. Rev.A7, 1957 (1973).

21. K. Ishii, M. Kamiya, K. Sera, S. Morita, H. Tawara, M. Ayamada and T. C. Chu, Phys. Rev. A15, 906 (1977).

22. J. W. Motz and R. C. Placions, Phys. Rev. A136, 662 (1964).

23. J. Jessenberger and W. Hink, Z. Phys. A275, 331 (1975).

24. B. Fischer and K.-W. Hoffmann, Z.Phys. 204, 122 (1967).

25. S. Ricz, B. Schlenk, D. Berényi, G. Hock and A. Valek, Acta Phys. Hung. 42, 269 (1977).

26. H. Hübner, K. Ilgen and K.-W. Hoffmann, Z. Phys. 255, 269 (1972).

27. H. Genz, C. Brendel, P. Eschwey, U. Kuhn, W. Löw, A. Richter and P. Seserko, Z. Phys. A305, 9 (1982).

28. C. Jakoby, H. Genz and A. Richter, in progress.

29. K. K. Sud and D. K. Sharma, J. Phys. B: At. Mol. Phys. 15, 1905 (1982).

30. J. T. Ndefru, J. G. Wills and F. B. Malik, Phys. Rev. A21, 1049 (1980).

31. J. T. Ndefru and F. B. Malik, Phys. Rev. A25, 2407 (1982).

32. E. Casnati, A. Tartari and C. Baraldi, J. Phys. B: At. Mol. Phys. 15, 155 (1982).

33. H. T. Yura, The quantum electrodynamics of a medium. The RAND Corporation, Santa Monica, Ca Report No R-410-PR (1963).

34. B. K. Agarwal, X-Ray Spectroscopy, Berlin, Heidelberg, New York, Springer Verlag, 1979.

35. H. Tawara, Electronic and Atomic Collisions, X ICPEAC, ed. G. Watel, p.311, Amsterdam:North Holland Publishing Company 1978.

36. P. A. Amundsen, to be published.

37. D. Drechsel, Lecture Notes in Physics, Vol.137: From Collective States to Quarks in Nuclei, Springer Verlag, Berlin Heidelberg New York, 1981, ed. H. Ahrenhövel and A. M. Saruis, p.358.

38. P. Seserko, H. Genz, W. Löw and A. Richter, to be published.

39. G. Missoni, C.E. Dick, R.C. Placious and J. W. Motz, Phys. Rev. A2, 2309 (1970).

40. J. W. Cooper and H. Kolbenstvedt, Phys. Rev. A5, 677 (1972).

41. M. Komma and W. Nakel, J. Phys. B: At. Mol. Phys. 12, L587 (1979); 15 , 1433 (1982).

42. W. Bleier, E. Geisenhofer, A. Grammer and W. Nakel, Europhysics Conference Abstracts, European Physical Society, eds. J. Kowalski, G. zu Putlitz, H. G. Weber, C5A, 755 (1981)

43. E. Schüle and W. Nakel, J. Phys. B: At. Mol. Phys., accepted for publication.

44. J. N. Das and A. N. Konar, J. Phys. B: At. Molec. Phys. 8, 2427 (1974).

45. W. Löw, H. Genz and A. Richter, to be published.

46. M.O. Krause, T.A. Carlson and R.D. Dismukes, Phys.Rev.170,36 (1968).

47. J. Utriainen, M. Linkoaho, E. Rantavuori, T. Åberg, G. Graeffe, Z. Naturforschung 23a, 1178 (1968).

48. T.Åberg, Phys. Rev. 156,35 (1967); Phys. Lett. 26A, 515 (1968)

49. S. Reusch, H. Genz and A. Richter in progress.

MULTI-VACANCY EFFECTS IN ARGON K-SPECTRA

R.D. Deslattes, P.L. Cowan and R.E. LaVilla
National Bureau of Standards
Washington, D.C. 20234

K. Dyall
Department of Theoretical Chemistry
University of Oxford
Oxford, England

ABSTRACT

We have carried out coordinated measurements of K series
emission and absorption spectra in atomic argon. Specificially,
emission spectra (especially in the region of $K\beta_{1,3}$) were recorded
with photon excitation energies ranging from below the single-
vacancy threshold to energies above most important double-vacancy
thresholds. Satellite emission spectra were modelled using both
Dirac-Fock and Configuration Interaction calculations. Comparisons
with experiment show reasonable agreement of the CI calculations for
the first high energy satellite complex, β^V, but not for the second,
β''.

INTRODUCTION

Although elementary discussions of X-ray emission and absorption spectra emphasize generally dominant single-vacancy processes,
it is well-known that observed profiles contain substantial contributions from more complex, multiple-vacancy processes. In closed
shell atoms, monoatomic metals and alkali halides, it has been
possible to achieve partial discrimination between these effects on
a semi-empirical basis. For example, the long-understood regularities
represented by Mosely diagrams lead to fairly ready identification
of single-vacancy emission lines; historically this was important
for the identification of "unknown" elements.

In the case of multiple vacancy emission lines or satellites,
there are analogous regularities which can be summarized in semi-
Mosely diagrams. Other distinguishing features of the satellite
lines include their added complexity and, in general, energetic
thresholds greater than those for the associated single-vacancy
processes. Evidently it is possible to distinguish between emission
features originating from single-vacancy configurations and those
originating from multiple-vacancy ones by threshold measurements.
Such diagnostics have been practiced in the past with good success[1]
and will evidently be further needed in the future as suggested
below.

In addition to their manifestation through emission features,
there are also long-known extra structures in atomic absorption
spectra located near energies corresponding to double-vacancy
thresholds[2]. Such features are easily recognized in spectra of free

0094-243X/82/940100-05$3.00

atoms and are plausibly associated with certain emission features. In the case of more complex systems such as polyatomic molecules, elemental as well as compound solids and ionic solutions, the corresponding absorption spectra are already expected to have great complexity. To this rather rich array of structure, one must presume there is added much additional structural complexity by the action of multiple vacancy processes mentioned above.

The particular studies of Kβ emission and K absorption of argon on which this report focusses were undertaken in response to the above-described overall situation. We wanted, at least in this simple case, to show directly the association between satellite emission features and already known structural complexities in the absorption spectrum. Methods which have now been demonstrated for such satellite diagnostics in the simple free atom case should carry-over to the more complicated situations encountered in poly-atomic systems. In addition the free atom case is accessible to theory and provides, in the case of double vacancy processes, a rather stringent and, it turns out, informative testing place for multi-configuration relativistic self-consistent field calculations.

EXPERIMENTAL BACKGROUND

Our experimental results and procedures are reported elsewhere[3]. A brief outline is, however, included here for the sake of complete-ness. The procedure entailed production of an intense tuneable monochromatic photon beam to excite fluorescent spectra in gas phase argon. At each of many primary excitation energies through the region of single (K-shell) and multiple (K+M) thresholds the emission spectra were recorded with high resolution in the region of the $K \rightarrow M_{II,III}$ ($K\beta_{1,3}$) transitions and its principal satellite groups, β^V and β''.

Primary radiation originated as a synchrotron radiation continuum from a six-pole wiggler at an electron storage ring. The storage ring, SPEAR, operated in dedicated mode at 3 GeV electron energy and an average stored current of about 40 ma. Primary radiation was monochromatized to a bandwidth of approximately 1 eV by a separated function monochromator using principally Si 111 crystals[4].

Primary radiation entered a small (1 cm) sample cell containing argon at about 0.3 atm. Before and after the sample there were He filled ionization detectors which monitored incident and transmitted radiation in the manner customary for EXAFS measurements. Repeated scans of the absorption spectrum were required in order to keep track of the (drifting) energy output of the primary monochromator.

Fluorescent radiation in the Kβ region was analyzed by a 2 meter Johann type focussing spectrometer equipped with a linear, position-sensitive proportional counter of the backgammon variety[5]. The gain in data acquisition efficiency associated with simultaneous registration of the entire region of interest was important for the success of this rather low intensity experiment. A possibly less obvious, but also quite important, feature of the mode of operation used is that this parallel registration gives results which are independent of source instability.

THEORETICAL PROCEDURES

The methods used to calculate the details of the Kβ emission spectrum are reported in full elsewhere[6]. They follow in principle the model proposed by Dyall and Larkins[7]. It is assumed that the primary photoelectron is ejected sufficiently rapidly that it perturbs the remaining electrons little; these then relax in response to the core vacancy to produce a number of different states which decay by X-ray and Auger electron emission. The populations of the various initial states for the decay process are described here by shake theory[8,9]. The energies of the X-ray emission lines were obtained from relativistic CI calculations for initial and final states. These procedures included both the relativistic contributions from the full finite frequency transverse interaction and estimates of quantum electrodynamic effects. Correlation contributions from relaxation between initial and final states in the emission process were also included.

The configuration sets used in the initial state CI calculations contained only those configurations most likely to be populated by shake-up or shake-off, i.e., the primary hole states with single 3s and 3p excitations and double 3p excitations in the single ion and single 3p excitations in the double ion, and also those necessary to correlate any 3s hole states. The corresponding configuration sets were employed in the final state CI calculations, i.e., those with a 3p hole in place of a 1s hole. Transition rates were calculated in the dipole approximation using the same CI expansions for the wavefunctions but with a common set of radial functions for initial and final states calculated with a "transition orbitals" approach[10,11]. Transition rates calculated in the length and velocity forms of the dipole operator agreed within about 10% in absolute value, but relative rates differed by less than 1%. The remaining quantities required for comparison with experiment are the total decay rates for the initial states. To calculate them would have been an enormous task; instead it was assumed that they were all equal, though there is evidence that total decay rates decrease as valence electrons are removed[12].

COMPARISON OF THEORY AND EXPERIMENT

The end result of the theoretical calculations is a "stick" spectrum consisting of many lines. To compare with experiment the lines have been given a Lorentzian shape with a full width at half maximum of 1 eV, somewhat larger than the approximate natural line width. The resulting spectrum is shown in Fig. 1 along with the experimental spectrum and a spectrum based on simple Dirac-Fock calculations. The experimental spectrum is that corresponding to excitation at high photon energies where all satellite processes are energetically possible.

The DF calculations indicated in the figure used only the ground state of the minimum configurations needed for description of the single and double hole states. As described above, the CI calculations included substantially all configurations populated in

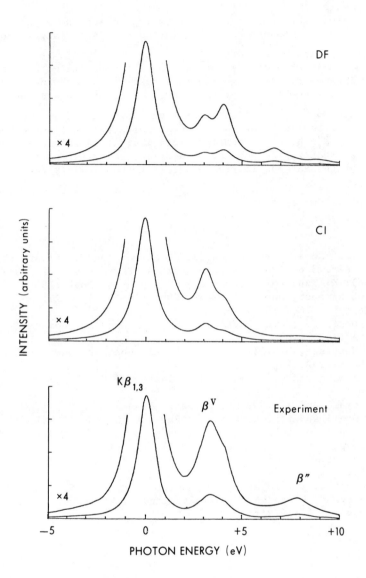

Fig. 1. Comparison of experimental emission profile with Dirac–Fock and configuration interaction models for atomic argon. The structure β^V was identified experimentally as arising from double vacancy configurations $KM_{II,III} \rightarrow M_{II,III}^2$ by threshold studies. This region is also fairly well reproduced by the CI calculations. Threshold measurements for β'' are consistent with a KM_I origin but not conclusive. Also neither calculation appears to satisfactorily account for both regions.

the shake model.

It is clear that the CI calculation reproduces the experimental β^V satellite better than the DF calculation. This is largely due to the contribution made by the $\underline{1s}\ \underline{3p}(^1P)4p\ ^2S - \underline{3p}^2(^1D)4p\ ^2P^o$ transition. Neither model reproduces the β'' satellite either in energy or intensity. Although the peak just below 7 eV in the DF spectrum could conceivably belong to β'', it corresponds to the $\underline{1s}\ \underline{3s}\ ^3S - \underline{3s}\ \underline{3p}\ ^3P$ transition which moves down in energy when correlation of the 3s holes is included in the calculations. The assignment of β'' therefore remains tentative; it is sufficient to comment here that in the absence of detailed calculations of the lifetimes of the various initial states, the possibility cannot be excluded that it is due to $\underline{1s}\ \underline{3p}^2 - \underline{3p}^3$ transitions with or without Rydberg spectator electrons.

REFERENCES

1. See for example, C.F. Hague, J.-M. Mariot and G. Dufour, Phys. Lett. A78, 328 (1980); R.D. Deslattes, Phys. Rev. 133, A399 (1964); K. Tsutsumi, J. Phys. Soc. Japan 14, 1696 (1959).
2. H.W. Schnopper, Phys. Rev. 131, 2558 (1963); F. Wuilleumier, J. de Phys. 26, 776 (1965).
3. R.D. Deslattes, R.E. LaVilla, P.L. Cowan and A. Henins, Phys. Rev. A (in press).
4. J.B. Hastings, B.M. Kincaid and P. Eisenberger, Nucl. Inst. and Meth. 152, 167 (1978).
5. R. Allemand and G. Thomas, Nucl. Inst. and Meth. 137, 141 (1976).
6. K.A. Dyall and I.P. Grant, to be published.
7. K.A. Dyall and F.P. Larkins, J. Phys. B 15, 203 (1982).
8. K.A. Dyall and F.P. Larkins, J. Phys. B 15, 219 (1982).
9. T. Åberg, Phys. Rev. 156, 35 (1967).
10. I.P. Grant, D.F. Mayers and N.C. Pyper, J. Phys. B 9, 2777 (1976).
11. O. Goscinski, G. Howat and T. Åberg, J. Phys. B 8, 11 (1975).
12. J. Nordgren, H. Ågren, C. Nordling and K. Siegbahn, Phys. Scripta 19, 5 (1979).

ATOMIC INNER-SHELL THRESHOLD EXCITATION
WITH SYNCHROTRON RADIATION

G. E. Ice
Metals and Ceramics Division, Oak Ridge National Laboratory
Oak Ridge, Tennessee 37830 USA

G. S. Brown
Stanford Synchrotron Radiation Laboratory
Stanford, California 94305 USA

G. B. Armen, M. H. Chen, B. Crasemann, J. Levin, and D. Mitchell
Department of Physics, University of Oregon
Eugene, Oregon 97403 USA

ABSTRACT

Monochromatized synchrotron radiation from a focused 8-pole wiggler beam line has been used to excite the L_2 and L_3 holes of atomic Xe and the K hole of atomic Ar near their photoionization thresholds. We have measured the Xe L_3-$M_4 M_5$ ($^1G_4$) Auger line, the Xe L_2-$L_3 N_{45}$ Coster-Kronig line, and the Ar K-$L_{23} L_{23}$ Auger line with incident x-ray energies from ∿20 eV below threshold to ∿200 eV above threshold. The spectra allow accurate determination of spectator-satellite energy shifts, post-collision interaction (PCI) shifts in the diagram-line energy, and spectral changes associated with the resonant Raman Auger (RRA) effect. The energies of satellites corresponding to photoelectrons promoted to bound states exhibit linear dispersion as a function of the initial photon energy. The measured post-collision interaction energy shifts of the diagram lines agree qualitatively with semiclassical PCI theory above threshold. In all cases, the energy of the diagram line is observed to decrease below threshold. The intensity of the 5d spectator line associated with the Xe L_3-$M_4 M_5$ line is larger than expected from absorption measurements.

INTRODUCTION

Near the photoionization threshold, the existence of a radiationless process analogous to x-ray resonant Raman scattering (RRS)[1,2] has recently been observed.[3] We refer to this process as the resonant Raman Auger effect (RRA). Both RRS and RRA are observed when x rays are used to excite an inner-shell hole below the photoionization threshold. The radiationless or radiative decay products which result from the initial hole are observed to be

resonantly enhanced as the x-ray energy approaches the
photoionization threshold. Furthermore, the energy of the decay
products is observed to be linearly dispersed with energy, below
threshold. If the photoejected electron is promoted to a narrow
unfilled bound state, linear dispersion continues above threshold.

Resonant Raman x-ray scattering and the companion resonant
Raman Auger effect are analogous processes in which energy is conserved in the excitation-deexcitation sequence devoid of
intermediate relaxation. Consider a narrow-bandwidth x ray used to
photoionize an inner-shell electron. In the photoionization
process, the x-ray quantum is absorbed by a single electron which is
ejected. In accordance with Einstein's law, the inner-shell hole
energy is thus determined by the initial photon energy $h\nu_1$ and the
energy of the photoelectron. The inner-shell hole energy can take
on a range of values due to its lifetime width, analogous to
photoabsorption by an energy band with a Lorentzian density-of-
states distribution. When the hole is filled, the atom "remembers"
the energy of the vacancy.

Because the matrix elements for photoionization change slowly
near the absorption edge, the RRA process can be estimated by
assuming the transition probability to be proportional to the
Lorentzian inner hole-state probability associated with a given
photon and photoelectron energy. For a given x-ray energy, the
resonant Raman process causes a distribution in the Auger energy.
As the initial hole state is excited below threshold, the Auger
energy increases linearly with photon energy up to the
photoionization threshold. Above the photoionization threshold, the
energy remains nearly fixed. The line width of the Auger-electron
peak is broad below threshold, reaches a minimum at threshold, and
then increases towards the diagram limit above threshold.

Post-collision interaction is another effect which has been
demonstrated to shift the Auger energy near threshold.[4,5] The
Coulomb field of the slowly receding photoelectron perturbs the
Auger transition energy. Energy shifts produced by PCI have been
predicted semiclassically by Niehaus[4] and are discussed in the paper
by V. Schmidt in these Proceedings. Unlike RRA, the theoretical PCI
shifts are always positive, and the slope is opposite to that of RRA
shifts below threshold. The FWHM due to PCI has a maximum near
threshold. In calculations of shifts from incident x rays below
threshold, the existence of Rydberg levels is assumed at suitable
energies. For the Auger lines studied here, however, unfilled
states are widely spaced or do not exist 0.5 eV or more below the
vacuum level.

EXPERIMENTS

The experiments here described were performed in the Stanford
Synchrotron Radiation Laboratory. X rays from an 8-pole wiggler
operating at 14 kG were focused onto a target-gas jet by a platinum-
coated doubly-curved toroidal mirror. To preserve the intrinsic
vertical collimation of the x-ray beam and to insure thermal stability of the mirror, slits upstream of the mirror were used to
limit the horizontal acceptance of ∿2 milliradians.

A narrow energy band was selected from the white synchrotron-radiation beam by a tunable Si 111 (1, -1) double-crystal mono-chromator. The monochromator bandpass was ∿2 eV at 5 keV and ∿0.5 eV at 3 keV. The storage-ring electron energy of 3 GeV and current of 60 ma yielded a flux of ∿5x10^{11} photons/sec at 5 keV through a 1x2-mm aperture upstream of the target.

The target consisted of a gas jet formed by a gold-plated glass capillary (0.2-mm i.d.); the jet intersected the x-ray beam at right angles in the horizontal plane. The gas nozzle was mounted on a micrometer-controlled vacuum manipulator which allowed the tip to be positioned within ∿0.2 mm of the electron-spectrometer source volume. The pressure in the intersection region was calculated to be ∿0.1 Torr, yielding a background pressure in the spectrometer of 2x10^{-4} Torr.

Electrons emitted from the gas-x-ray intersection region were analyzed by a commercial double-pass cylindrical-mirror analyzer, with its symmetry axis in the vertical plane (Fig. 1). The spectrometer energy resolution was ∿2.0 eV FWHM.

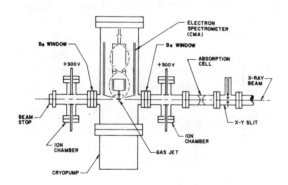

Fig. 1. Schematic drawing of gas-phase electron spectrometer.

The energy of the incident x rays was placed on an absolute scale using absorption edges calibrated to ∿0.5 eV by Breinig et al.[5] The edges were least-squares fitted with an estimated uncertainty of ∿0.5 eV.

Electron energy was calibrated by measuring photoelectrons emitted with x rays incident well above threshold. The electron-spectrometer voltage is

$$V_{pass} = h\nu - E_{binding} - V_{retarding}. \tag{1}$$

Since the photon energy was known and the binding energy is known to 0.2 eV, the spectrometer pass voltage could be easily measured. This measured value agreed within uncertainty with the nominal spectrometer pass voltage determined by the applied potentials.

The measured Auger energies of the diagram lines, assuming the calibrated pass voltages, were found to be ∿2 eV from the theoretical values when excited by x rays well above threshold. This discrepancy was attributed to nonlinearity in the computer-controlled retarding voltage over the nearly 3-kV range of operation. The Auger data were therefore corrected for this offset by normalizing to theoretical values.

System stability was monitored by measuring the photopeak before and after each Auger measurement, and by frequently measuring the Auger line far above threshold, where the measured Auger energy is insensitive to $h\nu$ but is sensitive to drifts in pass voltage.

Photopeaks and Auger peaks were least-squares fitted with Pearson 7 functions. A Pearson 7 function is a 4-parameter function which can be changed smoothly from a Gaussian to a Lorentzian, by varying a single parameter. This function allows excellent fitting of Voight profiles.

The Auger-peak fitting function was built from up to six Pearson 7 functions whose relative intensities and positions were fixed by theory. The average position, FWHM, shape, and intensity of the multiplets were then fitted as a single 4-parameter function. A typical fit of the theoretical multiplet structure of the Xe $L_3-M_4M_5$ transition to the measured Auger-electron spectrum is illustrated in Fig. 2. The multiplet structure of each Auger line served as a finger-print which greatly aided in the interpretation of the data and shortened fitting time.

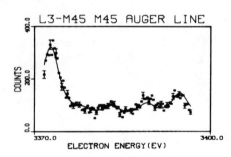

ELECTRON ENERGY (EV)

Fig. 2. The $L_3-M_{45}M_{45}$ Auger spectrum of Xe, excited near threshold. In order of increasing energy, the $^1G_4$, $^1D_2$, $^3F_3$, $^3F_2$, and $^3F_4$ lines can be seen, shifted ~+6 eV by RRA + PCI.

RESULTS

Xe $L_3-M_4M_5$ Measurements

The x-ray resonant Raman Auger effect was first observed in the xenon $L_3-M_4M_5$ transition.[3] We have repeated this measurement, taking advantage of the greater x-ray intensity and stability provided by the new wiggler beam line in the Stanford Synchrotron Radiation Laboratory.

The Xe L_3 hole state has an experimentally measured energy of 4786.3 (6) eV, a theoretical energy of 4786.54 eV, and a theoretical lifetime width of ~2.8 eV. Nearly 18% of all L_3 holes are filled by Auger decay through the $L_3-M_4M_5$ transitions. The width of the M_{45} holes is ~0.5 eV.[6]

Selected L_3-M_4M_5 ^1G Auger spectra are shown in Fig. 3. The spectra are seen to undergo dramatic change when excited near threshold. If the fitted Auger-peak positions are plotted as a function of x-ray energy, as in Fig. 4, two lines exhibit different photon-energy dependence. The higher-energy Auger line is a satellite that originates from the Xe L_3-M_4M_5 $^1G_4$ transition accompanied by promotion of the photoelectron to the open 5d shell. The measured energy shift of this "5d spectator satellite" with respect to the energy of the diagram line far above threshold is 8.3 eV, close to a value of 7.2 eV calculated with a relativistic relaxed-orbital Auger-energy code. Assignment of this satellite line to the 5d spectator transition is confirmed by plotting the intensity of the line as a function of x-ray energy (Fig. 5). We note that the intensity shows a resonance at 4783.8 eV, close to the

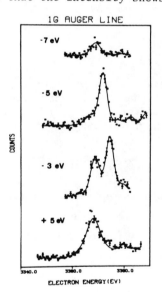

Fig. 3. Xenon L_3-M_4M_5 ^1G Auger electrons (normalized to photoelectron-peak intensity), excited with x rays from −7 eV to +5 eV with respect to the L_3 binding energy.

Fig. 4. Xenon L_3-M_4M_5 $^1G_4$ 5d spectator-satellite energy and diagram-line energy, as functions of exciting x-ray energy (in eV).

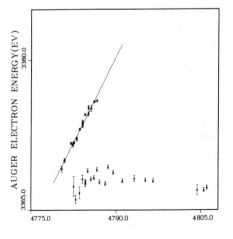

110

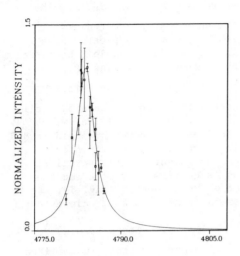

Fig. 5. Intensity of the Xe $L_3-M_4M_5$ ^1G 5d-spectator satellite line, as a function of exciting x-ray energy (in eV).

theoretical cross-section peak of 4784.2 eV.[5] The width of the resonance is ∿4 eV. Linear dispersion of the 5d spectator satellite Auger line is due to the resonant Raman effect.

The lower Auger-peak energies in Fig. 4 correspond to the normal Xe $L_3-M_4M_5$ $^1G_4$ diagram line.

Ar $K-L_2L_3$ Measurements

Measurements of the Ar $K-L_2L_3$ line shifts were made for the first time, with x-ray energies ∿10 eV below to ∿200 eV above the L_3 threshold. The Ar K hole has a theoretical lifetime width of 0.66 eV. Nearly 44% of all K holes decay via the $K-L_2L_3$ $^1D_2$ multiplet line, with about 5% decaying via the $K-L_2L_2$ $^1S_0$ (∿8 eV lower). Other Auger lines arising from the K hole are considerably weaker. The widths of the L_2 and L_3 holes are about 0.12 eV. Because of the small width of the Ar K hole state, more structure is seen in the absorption edge, but the PCI shifts are smaller. For example, evidence of excitation of the ^1S electron to 4p and 5p states is seen in high-resolution absorption measurements of the Argon K edge.[5]

The plotted positions of fitted Auger peaks in the vicinity of the argon threshold show a spectator-satellite line with linear dispersion as well as the diagram line. The measured shift of the 4p spectator satellite line is 4.5 eV. The intensity of the satellite line shows a resonance of 3203.6 eV, in agreement with the resonance assumed in the absorption spectrum[5] at 3203.54 eV. Energy shifts in the diagram line are small, except near threshold, where a large contribution can be expected from the unresolved 5p spectator satellite.

Xe L_2-L_3N_{45}

A particularly interesting case is the Xe L_2-L_3N_{45} Coster-Kronig line excited near threshold. The Xe L_2 hole state has an experimentally measured energy of 5107.0 (5) eV, a theoretical energy of 5106.73 eV,[7] and a width of $\sim$3.05 eV. About 8% of all L_2 holes decay via the L_2-L_3N_{45} line, with 7% in the L_3N_4 J=3 and J=2 multiplet components, separated by 0.2 eV. The width of the L_3 hole is $\sim$2.8 eV,[6] and that of the $N_{4,5}$ lines is less than 0.1 eV. What makes the L_2-L_3N_{45} line so interesting is the low energy (228.4 eV) of the Coster-Kronig electrons. In the semiclassical theory, the Auger electron is treated as "fast" compared with the photoelectron, permitting a Born-Oppenheimer separation. Interaction between the two continuum electrons is neglected. The validity of these assumptions dissolves in the Coster-Kronig case.

In Fig. 6 the near-threshold energy dependence of the observed Coster-Kronig lines are plotted as a function of x-ray energy. The measured 5d spectator-satellite shift of 7.02 eV, with a dispersion slope of 1.0.

The Niehaus semiclassical PCI theory[4] is fitted to the L_2-L_3N_4 data in Fig. 7. A good fit is seen to be attained, but only if a width of 4.7 eV is used in the Niehaus formula, well in excess of the actual Xe L_2-hole width[6] of 3.1 eV. It is possible that this discrepancy could be caused by the final-state energy distribution.

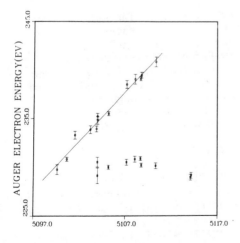

Fig. 6. Xenon L_2-L_3N_4 5d spectator-satellite energy and diagram-line energy, as functions of exciting x-ray energy (in eV).

112

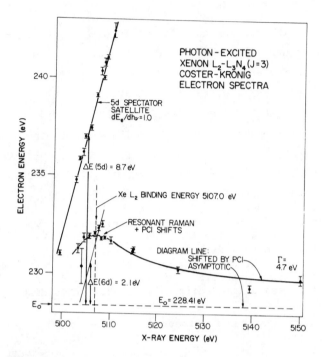

Fig. 7. Xenon $L_2-L_3N_4$ Coster-Kronig electron energies, as a function of exciting x-ray energy. The solid curve indicates the prediction of the semiclassical Niehaus theory, above threshold.

ACKNOWLEDGMENTS

We are indebted to Cullie Sparks for useful discussions. This research was sponsored in part by the Division of Materials Science, U.S. Department of Energy, under Contract W-7405-eng-26 with the Union Carbide Corporation, and by Air Force Office of Scientific Research Grant No. AFOSR-79-0026 to the University of Oregon. The experiments were done at SSRL which is supported by the NSF through the Division of Materials Research and the NIH through the Biotechnology Resource Program in the Division of Research Resources, in cooperation with the Department of Energy.

REFERENCES

1. C. J. Sparks, Phys. Rev. Lett. <u>33</u>, 262 (1974).
2. P. Eisenberger, P. M. Platzmann, and H. Winick, Phys. Rev. Lett. <u>36</u>, 623 (1976), and references therein; P. Suortti, Phys. Status Solidi(b) <u>91</u> 657 (1979).
3. G. S. Brown, M. H. Chen, B. Crasemann, and G. E. Ice, Phys. Rev. Lett. <u>45</u>, 1937 (1980).

4. A. Niehaus, J. Phys. B $\underline{10}$, 1845 (1977).

5. M. Breinig, M. H. Chen, G. E. Ice, F. Parente, B. Crasemann, and G. Brown, Phys. Rev. A $\underline{22}$, 520 (1980).

6. M. H. Chen, B. Crasemann, and H. Mark, Phys. Rev. A $\underline{24}$, 177 (1981).

7. K.-N. Huang, M. Aoyagi, M. H. Chen, B. Crasemann, and H. Mark, At. Data Nucl. Data Tables $\underline{18}$, 243 (1976).

EVIDENCE FOR INTENSITY SHIFTS OF K X-RAYS

G.L. Borchert, T. Rose and O.W.B. Schult
IKP, KFA Jülich, D-5170 Jülich, Germany

ABSTRACT

During the last years the study of minute energy shifts of K X-rays has revealed interesting information about nuclear and atomic properties. The corresponding changes in the wave functions of the atomic electrons should affect the transition probabilities as well. Therefore the investigation of the K X-ray intensities should yield complementary information. Accordingly, an experiment has been performed, the results of which show evidence for such intensity shifts.

INTRODUCTION

With increasing performance of experimental technique the study of K X-rays yields new information about atomic and nuclear properties. Changes in the electromagnetic quantities of the atomic system cause a characteristic rearrangement of the electrons. As a consequence, the atomic levels were somewhat displaced, which can be detected in favourable cases as a generally small shift or even a characteristic shift pattern of the involved X-ray transition energies. By a proper design of the experimental performance at a high precision level the energy shifts of K X-rays could be attributed to different physical origins. From the K X-ray shifts in different isotopes the changes of the mean square charge radius of various nuclei have been deduced[1,2,3]. By a comparison of the K X-ray energies of elements in different chemical compositions the influence of the bond on specific atomic orbitals has been demonstrated[4,5]. The shifts of K X-ray transitions due to the hyperfine splitting of the K level and in internal conversion allow in special radioactive decay processes to determine the nuclear magnetic moments[6,7,8]. Finally, shifts due to dynamic effects and due to atomic structure effects in electron capture decay yield information about atomic shell processes at a very short time scale[9,10].

On the other hand, a change of the wave functions of the atomic electrons should give rise to a change of the probability of electronic transitions as well. Therefore one could think of obtaining complementary information about atomic processes by the investigation of the intensities of K X-ray transitions. As to our knowledge no one had studied these properties for K X-ray transitions so far, we felt it worth trying an experiment to investigate the influence of the atomic configuration on the intensity of K X-ray transitions. When the configuration of an atom is changed due to the chemical bond especially the outer electron shells are affected. Therefore it is essential to study the complete set of K transitions including the weak transitions from the N and O shells.

GENERAL CONSIDERATIONS

It is obvious that the investigation of intensity shifts for
X-rays represents a much more difficult problem than the determination
of energy shifts. The X-ray intensity as it is produced by photoionization for instance, is usually changed by several processes before the
detection, which cannot be avoided by simple means.

First of all there is the self absorption in the source which
causes an intensity shift. Even for a thin source with sufficient
homogeneity as it is used for a DuMond type crystal spectrometer it
is very difficult to calculate this contribution reliably. Therefore
it has to be determined in an extra experiment in case it is not
possible to reduce source thickness or density any longer due to
count rate problems.

A special problem arises for the KO transition because of its
energy being close to the K absorption edge. As it is well-known
from X-ray absorption studies[11],[12], the chemical bond changes the
position and structure of the K absorption edge. Therefore the intensity of the KO line depends sensitively on the change of the absorption behaviour in the surroundings of the K edge. In turn, the KO line
can be used as a probe for absorption edge studies instead of specially prepared monochromatic photons.

Besides the self absorption in the source the reflectivity of the
Bragg crystal and the detector efficiency depend on the energy as
well. As these corrections are not known with sufficient accuracy,
an absolute measurement of the K X-ray intensities is very difficult.
Therefore we used one K X-ray line as a reference and normalized all
other transitions to its intensity. In this way only the change of
the self absorption in the source itself has to be corrected for. As
reference line we chose the $K\alpha_1$ transition assuming that analog to
the chemical and atomic structure energy shift[4],[8], the influence on
the intensity of the L shell transitions should be negligible. This
implies of course that a shift of these lines is not detectable with
the present method.

In this way, the strategy of the present experiment to investigate the intensity shifts of K X-rays due to changes in the chemical
configuration implies for each source an experiment to measure the
intensity of the individual K X-ray transitions and series of measurements to determine the self absorption behaviour of the source.

APPARATUS AND MEASUREMENT

The experiments have been performed with the DuMond bent crystal
spectrometer of the Institut für Kernphysik at the Kernforschungsanlage
Jülich. The apparatus is described in detail elsewhere[13]. It was
equipped with a twin source arrangement similar to the one used in
the instrument at CERN[14]. The main features are shown in fig. 1. The
inactive samples denoted by source 1 and source 2 are mounted very
close to each other on two plexiglas blocks, which can be adjusted
in the direction of dispersion by means of micrometer screws and be
rotated around a vertical axis. Opposite to the samples a radioactive
^{169}Yb source with an activity of about 40 Ci is inserted into a U

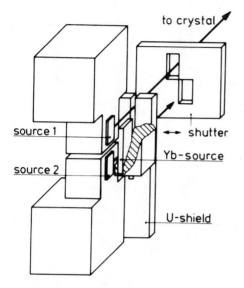

Fig. 1: Schematic view of the source arrangement from the DuMond spectrometer. The positions of the inactive samples at the face of two separate plexiglas blocks are denoted by source 1 and source 2. A very strong ^{169}Yb source is placed close to them exciting the fluorescence radiation in the samples. To reach the Bragg crystal of the spectrometer this radiation has to pass through slits in a moveable U plate acting as a shutter.

shield. Its radiation excites the samples to photofluorescence. These X rays have to pass through slits in a movable U plate acting as a shutter. Therefore the radiation of only one source at a time can pass to the Bragg crystal and reach the detector. By means of opto-electronical control, the intensity of the two sources, after being processed by the same electronical chain is stored in separate registers. In this way the influence of mechanical uncertainties and electronical drifts is sufficiently reduced. As detector an intrinsic Ge detector with an energy resolution of 2 keV is used which permits a drastic background reduction as compared to NaI scintillators. The sources have a typical size of 1 x 5 x 0.1 mm^3. They are adjusted so that the corresponding X-ray reflections coincide and rotated to optimize the geometrical widths to the natural line widths. For the study of the self absorption correction both sources were rotated by another small amount and the K X ray reflections were remeasured. Then this procedure was repeated with sources of different thickness or density. The X-ray reflections are scanned across several times at positive and negative Bragg angles within a range of about 3 to 4 times the FWHM. A typical example is given in fig. 2. A scan across the Kα_1, Kβ_1, Kβ_{24} and KO reflections of the Gd and Gd$_2$O$_3$ are shown. The intensity of Gd metal is denoted by X that of the Gd$_2$O$_3$ source is marked by dots. Note the different time scales T which are given for each reflection in the upper part of figure 2.

The measured data are submitted to a computer analysis. After substraction of the background, constant or from neighbouring lines for the Kβ_1 or KO from the Kβ_3 or K$\beta_{2,4}$ respectively, the intensity of each reflection is determined within a range of typically the full width at one tenth of the maximum. From these data the intensity ratios of the individual X-ray transitions were calculated. To simplify the comparison with other sources the intensity ratios then were normalized to one for the Kα_1 transitions.

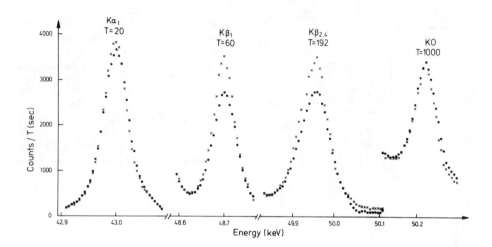

Fig. 2: Prominent K X-ray transitions from the Gd, Gd$_2$O$_3$ experiment. The intensity of the Gd metal is denoted by X, that of the Gd2O3 by dots. The measuring time T per point is given above the corresponding reflections.

RESULTS AND DISCUSSION

For a first experiment the rare earth region seemed advantageous. As the K binding energy is below 63 keV the ^{169}Yb source yields a high photofluorescence efficiency due to the intense 63 keV γ line. Because of the large effect of the 4f electrons on the KO line in the energy shift studies[4],[10], one would expect a significant influence on the intensity as well.

Therefore we chose the pairs Tb metal, Tb$_4$O$_7$ and Gd metal, Gd$_2$O$_3$. The data of these first experiments were submitted to the mentioned analysing procedure and normalized to the Kα$_1$ intensity ratio. These results, as measured without applying any absorption correction are listed in table 1. There is a qualitative and quantitative similarity of the intensity ratios of the corresponding X-ray transitions for both elements.

To study the effect of the self absorption in the sources, the experiments have been repeated with

1) the metallic source by a factor of 9 thinner
2) the oxide source by a factor of 5 diluted with Al$_2$O$_3$
3) the sources rotated by about 1° to change the effective absorption length

A preliminary evaluation of part of these data indicates that for the system where the self absorption is reduced the shift pattern essentially remains unchanged. A maximum absorption correction

compatible with the preliminary uncertainty would amount to about
5 %. At this state the intensity ratios suggest that changes of the
chemical structure in Gd and Tb atoms have a significant influence on
the transitions from the M, N and O shells. Especially the drastic
intensity shift of the KO transitions is remarkable. Because of our
normalization on the $K\alpha_1$ line intensity we have no means to study an
effect of the $2p_{3/2}$ shell.

Further experiments to study the self absorption correction in
more detail are on the way and we will continue to other elements with
interesting electronic structure in various compounds.

Table 1: The intensity ratios of corresponding K X-ray transitions in
% normalized to the $K\alpha_1$ intensity (uncorrected) Metal/Oxid

	$K\alpha_1$	$K\beta_1$	$K\beta_{2,4}$	KO
Gd_2O_3 – Gd	100(1)	122(2)	121(2)	86(4)
Tb_4O_7 – Tb	100(1)	119(2)	119(2)	83(4)

REFERENCES

1. F. Boehm and P.L. Lee, Atomic Data and Nuclear Data Table, 14,
   605 (1974)
   P.L. Lee and F. Boehm, Phys. Rev. C8, 819 (1973)
2. O.I. Sumbaev and A.F. Mezentsev, Soviet Phys. JETP 22, 323 (1966)
3. G.L. Borchert, O.W.B. Schult, J. Speth, P.G. Hansen, B. Jonson,
   H.L. Ravn, and J.B. McGrory, Nuovo Cimeno, in press, 1982
4. O.I. Sumbaev, Chemical effects in x-ray spectroscopy in Modern
   Physics and Chemistry, Vol. 1, 31, Academic Press, London, 1976
   O.I. Sumbaev, Usp. Fiz. Nauk 124, 281 (English translation
   Soviet Physics Usp. 21, 141) 1978
5. F. Boehm, in Atomic Inner-Shell Processes (B. Crasemann, ed.),
   Vol. I, 411, Academic Press, London and New York, 1975
6. G.L. Borchert, P.G. Hansen, B. Jonson, H.L. Ravn, O.W.B. Schult,
   and P. Tidemand-Petersson, Phys. Lett. 63A, 15 1977
7. A.I. Egorov, A.A. Rodinov, A.S. Ryl'nikov, A.E. Sovestnov,
   O.I. Sumbaev, and V.A. Shaburov, Pis'ma Zh. Eksp. Teor. 27,
   514 (English translation: JETP Lett. 27, 483), 1978
   A.I. Grushko, K.E. Kir'yanov, N.M. Miftakhov, A.S. Ryl'nikov,
   Yu.P. Smirnov, and V.V. Fedorov, Zh. Eksp. Teor. Fiz. 80, 120
   (English translation: Soviet Phys.-JETP 53, 59) 1981
8. G.L. Borchert, P.G. Hansen, B. Jonson, H.L. Ravn and O.W.B.
   Schult, Inner-Shell and X-ray Physics of Atoms and Solids, Eds.
   D.J. Fabian, G. Watson and H. Kleinpoppen, Plenum, N.Y., 163
   (1982)
9. G.L. Borchert, P.G. Hansen, B. Jonson, I. Lindgren, H.L. Ravn,
   O.W.B. Schult, and P. Tidemand-Petersson, Phys.Lett. 66A, 374,
   1978
10. G.L. Borchert, P.G. Hansen, B. Jonson, H.L. Ravn and O.W.B. Schult,
    Proceedings of the XVIII International Winter Meeting on Nuclear
    Physics, Bormio 1980, 579, Ed. I. Iori, University of Milan, 1980

11. P.A. Lee, P.H. Citrin, P. Eisenberger, and B.M. Kincaid, Rev. Mod. Phys. 53, 769 (1981)

12. R.A. Martin, J.B. Boyce, and J.W. Allen, Phys. Rev. Lett. 44, 1275 (1980)
     H. Launois, M. Rawiso, E. Holland-Moritz, R. Pott, and D. Wohlleben, Phys. Rev. Lett. 44, 1271 (1980)

13. G.L. Borchert, W. Scheck and O.W.B. Schult, Nucl. Instr. Meth. 124, 107 (1975)

14. G.L. Borchert, P.G. Hansen, B. Jonson, H.L. Ravn, O.W.B. Schult, and P. Tidemand-Petersson: Nucl. Instr. & Meth. 178, 209 (1980)

INFORMATION-THEORETICAL DESCRIPTION OF HEAVY-ION –ATOM COLLISIONS

T. Åberg
Helsinki University of Technology, 02150 Espoo 15, Finland

O. Goscinski
Uppsala University, 75120 Uppsala, Sweden

ABSTRACT

In the keV/u to MeV/u range most experimental studies of inelasticity have provided limited resolution of the accessible final states. In this work the information content regarding   excitation processes in such measurements has been analyzed from a stochastic point of view using the minimum cross entropy principle in the construction of electronic density matrices. The final nonstatistical distribution of the observed radiation or fragments is obtained by minimizing the cross entropy with respect to a prior distribution and subject to constraints imposed by the experimental arrangement and by the atomic structure of the projectile-target system. Application to equilibrium charge-state distributions is made. Observed asymmetries are related to a single energy-independent parameter characterizing the rate of charge formation. An information-theoretical interpretation of the frequently observed binomial KL^n x-ray and Auger electron distributions is given and contrasted with conventional derivations.

INTRODUCTION

Most experiments concerning heavy-ion –atom collisions

$$A^{q_o} + B \rightarrow A^{q_1} + B^{q_2} + (q_1 + q_2 - q_o)e \qquad (1)$$

have provided only very limited resolution of the large number of accessible final many-electron states which are associated with the residual ionic fragments. A typical example is the observation of the distribution of the projectile charge q_1 after collisions between A and B in a gas target. The situation often resembles circumstances encountered in statistical physics although atomic structure may induce simplifications in the low- and high-velocity limits and especially when $q_1 \simeq q_o \gtrsim Z_A$. However even in such cases our understanding of reactions (1) has remained fragmentary due to the complexity of a rigorous quantum-mechanical treatment. Hence in order to understand these collisions it is necessary to study cases in which details of the collision dynamics are gradually faded out to the extent statistical models become applicable.

A statistical model based on phase-space arguments has previously been developed by Russek and coworkers [1] for the analysis of the degree of ionization in ion-atom collisions under single scattering conditions. Brandt and coworkers [2] have applied a statistical random-walk approach by Mittleman and Wilets [3] to the K- and L-shell vacancy production with mixed success. [4] Baudinet-Robinet et al have considered equilibrium charge distributions using a modified chi-squared (χ^2) parametrization which is superior to the classical Bohr-Lindhard model. [5] However as far as we know there has been no previous attempt to develop a model-independent stochastic approach to reactions of type (1).

Our description is based on the maximum entropy principle (MEP) [6] or to be more precise on the minimum cross-entropy principle (MCEP) [7] which is a more general information-theoretical method of inference than MEP. These methods have their roots in the development of statistical mechanics by Boltzmann, Maxwell and Gibbs. In somewhat restricted forms they were introduced into quantum mechanics by von Neumann [8], Stratonovich [9] and Fano [10]. An early application to the distribution of nuclear energy levels was given by Bethe [11] but apparently Jaynes [12], inspired by the work of Shannon [13] in communication theory, was the first one who fully realized the generality of MEP as applied to statistical physics and to inference problems in general. Since then MEP and MCEP have found a remarkable variety of applications.[6,7] In collision physics MEP has been used by Levine et al. [14] for a systematization of data on various molecular and nuclear scattering processes. A maximal-entropy approach to the S matrix has been developed for nuclear reactions by Mello and Seligman. [15]

Below we briefly describe our formulation of collision problems. In the subsequent sections we discuss two examples, namely, the information-theoretical interpretation of equilibrium charge-state distributions and of inner-shell vacancy distributions. We conclude with remarks concerning future applications.

THEORY

The probability that a property x (e.g. charge of either fragment or both in coincidence) is measured in a scattering experiment is given by

$$P_x = Tr\rho_x = Tr[\rho_{out}F_x] \tag{2}$$

The detector (efficiency) operator[16]

$$\hat{F}_x = \sum_\gamma \varepsilon_\gamma(x) |\gamma><\gamma| \tag{3}$$

describes the response of the detector system in the experimental apparatus in terms of the eigenstates $|\gamma>$ of the final system and their relative efficiencies $\varepsilon_\gamma(x)$. The density matrix ρ_{out} is related to the density matrix $\rho_{in} = \rho_A\rho_B$ of the projectile-target system prior to the collision by $\rho_{out} = T\rho_{in}T^\dagger$. The transition matrix T describes the collision dynamics. Note that x may refer to separate fragments (ions) after the collision or it may refer to the collision complex (e.g. observation of quasimolecular MO x-rays) depending on the definition of $\hat{F}_x$. Since Eq. (2) refers to single-collision conditions ρ_x should in the case of multiple collisions be properly generalized.

In most cases an ab initio calculation of T and, hence the

unknown P_x, is impossible. Hence one may apply MCEP which is a general method of inference about an unknown probability distribution when there is a prior estimate of the distribution and new information in the form of constraints on expected values. Specific prior estimates p(x) of P_x are always possible on physical grounds, e.g., electronic structure and boundary conditions including phase space partitioning. In the simplest case p(x) is the uniform distribution. Furthermore one may know how the system behaves globally with respect to the observations. Mathematically this can be expressed in the form of integral constraints (generalized moments)

$$\int P_x <A_\nu>_x dx = a_\nu \qquad (\nu = 1, \ldots, N) \qquad (4)$$

where

$$<A_\nu>_x = \frac{Tr[\rho_{out} A_\nu F_x]}{Tr[\rho_{out} F_x]} \qquad (5)$$

are x-dependent expectation values of linearly independent operators $\hat{A}_\nu$ in the space defined by $\hat{F}_x$.

According to MCEP Eqs. (4) can now be used to improve the prior estimate p(x) of P_x in a consistent and unique way.[7] This results in a posterior probability distribution

$$P(x) = p(x)\exp[-\lambda_o - \sum_{\nu=1}^{N} \lambda_\nu <A_\nu>_x] \qquad (6)$$

This equation is obtained by minimizing the cross entropy

$$S = \int P(x) \ln \frac{P(x)}{p(x)} dx \qquad (7)$$

subject to the normalization condition $\int Tr\rho_x dx = 1$ and the constraints (4) which determine the Lagrange parameters λ_o and λ_ν. The generalization to the case, where there are also inequalities among the constraints is straightforward.[7]

Alhassid and Levine[17] have used MEP to obtain a most probable ρ_{out} for a given set of constraints. This implies a construction of T in terms of a finite operator basis by maximizing the entropy $S[\rho] = -Tr[\rho \ln \rho]$ which corresponds to a uniform and diagonal prior density matrix ρ_o. The crucial influence of F_x on the measured quantities is neglected in their analysis. It is simple to circumvent these problems within MCEP provided ρ_o is diagonal and all the operators $\hat{A}_\nu$ commute with $\hat{F}_x$ and each other. This is usually the case.

Both MEP and MCEP lead to unique probability distributions in the sense that maximizing (or minimizing) any other function than the entropy (or a function with identical maxima or minima) would violate fundamental consistency axioms concerning the procedure for obtaining posterior probability distributions from the priors.[7] It should be noted that these axioms are compatible with those of quantum statistics. The MEP distribution is equal to the frequency distribution that can be realized in the greatest number of ways subject to a given set of constraints.[18] With respect to reaction (1) it means that for uniform prior distributions the processes leading to the final distribution are treated as random events.

EQUILIBRIUM CHARGE-STATE DISTRIBUTIONS

We consider a recent information-theoretical description of charge-state distributions of heavy ions which penetrate through gas targets and thin foils. [19] It is well known that these distributions reach equilibrium, i.e. become independent of the initial projectile charge q_0, when the target is thick enough. They are generally asymmetric rather than Gaussian as predicted by the Bohr-Lindhard theory. [20] An analysis of this poorly understood feature, based on ab initio quantum mechanical calculations, is certainly unfeasible.

We shall use MCEP as formulated in Eqs. (4)-(7) for the analysis of the asymmetry. The prior probability is represented by a family of monotonously increasing functions $p(x) = x^{\mu-1}$, where $x = q + 2$ and $\mu > 1$. This choice only accounts for electron loss and for the fact that $q \geq -2$. In reality electron capture competes with electron loss which results in a dispersion of the charges and a lowering of the probability of high charge states. After a sufficient number of collisions this charge distribution becomes independent of q_0. Hence x approaches on the average a constant value as a function of the distance z travelled by the ions and $(dx/dz) \to 0$, i.e., $d(\log x)/dz \to 0$. According to Eqs. (4) one has to account for this behaviour in terms of a constraint.

For this purpose we assume that the charge $q = x - 2$ follows the rate equation

$$\frac{dx}{dz} = rxG(x) = \frac{rx}{\alpha}\left[1 - \left(\frac{x}{k}\right)^\alpha\right] \tag{8}$$

which is well known in the theory [21] of population growth and extinction. The right-hand side describes the net effect (loss mimus capture) on q in terms of three parameters: $r(>0)$ is the net electron loss ($x_0 < k$) or capture ($x_0 > k$) per unit charge, $k - 2$ the final charge, and α (≥ 0) controls the rate of charge formation (Fig. 1). Note that Eq. (8) is consistent with known facts [22] about the q dependence of electron loss and capture cross sections as long as $0 \leq \alpha \leq 3$ which is the range which we shall consider in the following. For $\alpha = 0$, $G(x) = -\ln(x/k)$. Equation (8) leads to the desired asymptotic property that on the average (with respect to the final distribution $P_\alpha(x)$) $d(\log x)/dz \to 0$ in form of the moment condition

$$\langle G(x) \rangle = \frac{1}{\alpha} \int_0^\infty P_\alpha(x)\left[1 - \left(\frac{x}{k}\right)^\alpha\right]dx = 0 \tag{9}$$

The minimization of Eq. (7) subject to the constraint (9) and normalization lead according to Eq. (6) to the probability distribution

$$P_\alpha(x) = \frac{\alpha(x/k)^{\lambda\alpha-1}\exp\left[-\lambda\left(\frac{x}{k}\right)^\alpha\right]}{k\lambda^{-\lambda}\Gamma(\lambda)} \tag{10}$$

where $\mu = \lambda\alpha$ according to Eq. (9). It also follows that

$$P_\beta = \lim_{\alpha \to 0} P_\alpha(x) = (\beta/2\pi x)\exp\left[-\frac{\beta}{2}(\ln\frac{x}{k})^2\right] \tag{11}$$

where $\beta = \lambda\alpha^2 \simeq \nu/2 = \langle x\rangle^2/\sigma^2$. Here $\langle x\rangle$ is the mean value $\langle q\rangle + 2$ and σ^2 the variance $\langle q^2\rangle - \langle q\rangle^2$. Using the maximum value of $P_\alpha(x)$ corresponding to $x_{max} = q_{max} + 2$ Eqs. (10) and (11) take the form

$$\Phi_\alpha(x) = -N_\alpha(\nu)\ln[P_\alpha(x)/P_\alpha(x_{max})] = \begin{cases} 2\alpha^{-2}[y^\alpha - \alpha\ln y - 1], & \alpha>0 \\ (\ln y)^2, & \alpha=0 \end{cases} \tag{12}$$

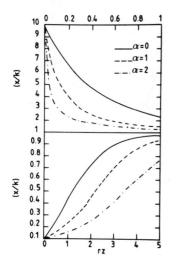

Fig. 1. The deterministic solution of Eq. (8) as a function of rz for $(x_0/k) = 10$ (capture) and $= 0.1$ (loss) and for $\alpha = 0, 1, 2$.

where $y = (x/x_{max})$ and $N_\alpha(\nu) \simeq 4\nu^{-1}$. Equation (12) is the testing ground of our theory since the functional forms of the right-hand side can be compared with the left-hand side, expressed itself in terms of experimental quantities $\langle q\rangle$, σ^2, $P_{max}(q)$, and $P(q)$. The expressions for $N_\alpha(\nu)$ and for $x_{max} = \langle x\rangle f(\nu)$ used in $\Phi_\alpha(x)$ are exact ($\alpha = 0, 1$) or accurate to the second order in ν^{-1}.[19] The test consists of finding an α value such that the right-hand side of Eq. (12) agrees with the experimental $\Phi_\alpha(x)$ points. Figure 2 shows two typical $\Phi_\alpha(x)$ plots for gas targets. The $\alpha=0$ curve represents the most asymmetric distribution. For $\alpha>2$ the $\Phi_\alpha(x)$ curves become more like the Gaussian $(y-1)^2$ as indicated by the skewness $\gamma \simeq (3-\alpha)/(\nu/2)^{1/2}$ which vanishes for $\alpha = 3$.

Similar trends like those shown in Fig. 2 have been found for each projectile-target combination that has been analyzed so far. The optimum α values are listed in Table I. It is remarkable that α is independent of the projectile velocity which ranges from 0.4 to 4 a.u. in the experiments considered. It also implies that the asymmetry is independent of $\langle q\rangle$ for each collision system studied. Note that it has not been necessary to use non-integer α values in our analysis although $\Phi_\alpha(x)$ is a sensitive function of α for both small and large y.

According to Table I, for light targets including the carbon foil, there is a minimum in α as a function of the projectile atomic number Z. Table I also shows that heavy targets have systematically smaller α values than light targets. According to Fig. 1 this trend is a manifestation of more frequent losses than captures. This leads to an excess of higher charge states and consequently to a larger asymmetry independently of the velocity. The opposite trend exhibited by the carbon foils is to be noted. We have thus related the asymmetry of the observed distributions to the rate of charge formation by an energy-independent parameter.

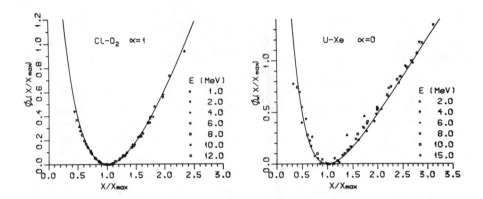

Fig. 2. The Φ_α function, defined by Eq. (12), as a function of x/x_{max} (x = q + 2) for (a) Cl-O_2 and (b) U-Xe collisions. For the experimental data see footnote a of Table I.

Table I. The asymmetry parameter α for various targets as a function of projectile atomic number Z.[a]

Target \ Z	H_2	He	C[b]	N_2	O_2	Ar	Kr	Xe
17	2	2	3	1	1	0	0	-[c]
35	1	1	3	0	0	0	0	-[c]
53	1	1	1	-[c]	0	0	0	-[c]
73	2	2	2	1	1	0	0	0
92	2	2	2	1	1	0	0	0

[a] The experimental values are from the following references: G. Ryding, A. B. Wittkower, and P. H. Rose, Phys. Rev. <u>185</u>, 129 (1969); A. B. Wittkower and G. Ryding, Phys. Rev. <u>A4</u>, 226 (1971); A. B. Wittkower and H.-D. Betz, Phys. Rev. <u>A7</u>, 159 (1973).

[b] Carbon foils

[c] No experimental data available

MULTIPLE VACANCY PRODUCTION

Observations of x-rays and Auger electrons in the MeV/u range have shown that even under single-collision conditions $n = q_1 + q_2 - q_0$ can be very large in processes of type (1). [23, 24] In the following we shall consider the vacancy distribution of the target atom B (or A) that is created <u>before</u> the x-rays or Auger electrons are emitted.

Previous attempts [25] to understand the multiple ionization problem can be summarized as follows. One assumes a partition

$$\hat{H}(\bar{R}(t),X) = \sum_{\nu=1}^{M} \hat{H}_o(\bar{R}(t),x_\nu) \tag{13}$$

of the semiclassical many-electron Hamiltonian $\hat{H}$ into effective molecular one-electron Hamiltonians $\hat{H}_o$. In Eq. (13) $\bar{R}(t)$ is the time-dependent internuclear vector and $X = x_1, \ldots, x_M$ represents the electronic coordinates including spin. Suppose the electrons occupy the spin orbitals $|\mu'\rangle$ ($\mu' = \nu_1', \nu_2', \ldots, \nu_M'$) when A and B are far apart at $t = -\infty$. In analogy with shake theory[26] of multiple photon-excitation processes it follows that

$$P_{\mu\mu'} = |\det\{\langle \nu_i|\nu_j'\rangle\}|^2 = \begin{vmatrix} \langle \nu_1|\nu_1'\rangle, & \ldots, & \langle \nu_M|\nu_1'\rangle \\ \langle \nu_1|\nu_2'\rangle, & \ldots, & \langle \nu_M|\nu_2'\rangle \\ & \ldots & \\ \langle \nu_1|\nu_M'\rangle, & \ldots, & \langle \nu_M|\nu_M'\rangle \end{vmatrix}^2 \tag{14}$$

represents the probability that the electrons occupy the spin orbitals $|\mu\rangle$ ($\mu = \nu_1, \nu_2, \ldots, \nu_M$) properly distributed among the two nuclei after the collision. In Eq. (14)

$$\langle \mu|\mu'\rangle = \lim_{t=+\infty} \langle \nu_\mu|\chi_{\mu'}^\dagger,(\bar{R}(t))\rangle \tag{15}$$

where the scattering wave function $\chi_{\mu'}^\dagger$ satisfies the time-dependent Schrödinger equation pertaining to $\hat{H}_o$. The proper outgoing-wave boundary condition requires that ν_μ (for $t = +\infty$) and $\nu_{\mu'}$ (for $t = -\infty$) include the electron translational factors. [27] Note that electron capture is implicitly included insofar as the molecular one-electron Hamiltonian $\hat{H}_o$ is defined. Reading and Ford [25] have emphasized that $\langle \nu_i|\nu_j'\rangle \neq 0$ for $i \neq j$. Nevertheless if one only keeps the diagonal elements in each $P_{\mu\mu'}$ one may easily sum over restricted sets $\{\mu\}$. Suppose $\{\nu_1', \nu_2', \ldots, \nu_N'\}$ refers to a subshell nl ($N = 2(2l + 1)$) of B and consider all possible $\{\nu_1, \nu_2, \ldots, \nu_N\}$ that leave N' spin orbitals occupied and N-N' unoccupied regardless what happens to the remaining M-N electrons. Using completeness relations gives the corresponding probability as

$$P_{N'N} = \binom{N}{N'}(1-p)^{N'} p^{N-N'} \tag{16}$$

if it is assumed that $p = \sum_{\nu_1 \neq \nu_k} |\langle \nu_1|\nu_k'\rangle|^2 = 1 - |\langle \nu_k|\nu_k'\rangle|^2$ is independent of k, i.e., the magnetic quantum numbers. Since p can be interpreted as an impact-parameter ($\bar{b}$) dependent probability in the semiclassical approach the total probability is given by $\mathcal{P} = \int P_{N'N}(\bar{b})d\bar{b}$ and this is the formula most calculations are based on. If h subshells are involved, then $P_h = P_{N_1'N_1} P_{N_2'N_2^x}, \cdots, P_{N_h'N_h}$.

Usually the probabilities p have been calculated using the semiclassical [28] or related first-order perturbation theory (Born) approximations in which the key quantity is the structure factor $\langle v_1 | \exp(i\bar{K}\cdot\bar{r}) | v_k \rangle$, where $\bar{K}$ is the momentum transfer vector. If one from the very beginning were to assume first-order perturbation theory and the independent-electron approximation in the resulting many-electron structure factor, $P_{\mu\mu}$, would vanish whenever there are two or more initial and final spin orbitals that differ. Hence this approach does not predict multiple ionization at all whereas Eq. (16) does. The difference between these two approaches lies merely in the order of the application of the approximations. Furthermore, the description of multiple ionization using higher-order semiclassical perturbation theory does not lead to Eq. (16) and can easily be shown to break down as compared with experiments.[29] The question then arises whether Eq. (16) is valid on more general grounds than anticipated by the semiclassical derivation given above, since it has had a remarkable semiempirical value.[23,24] Under such circumstances a model-independent approach is desirable.

The only assumption that can be made in the information-theoretical description concerning the x-ray and Auger electron measurements is that the distribution of the excited states in B (or A) can be classified in terms of the occupancy of one-electron states such that the maximum occupancy is one according to the Pauli principle. There is no need to specify the one-electron basis nor the associated many-electron states. For N accessible one-electron states the number of final many-electron states $|\kappa\rangle$ is $N_0 = 2^N$. The electrons and holes are distributed among these states such that for each κ, $N_\kappa' + N_\kappa'' = N$, where N_κ' is the number of electrons and N_κ'' the number of holes. If p_κ is the probability that the state $|\kappa\rangle$ occurs in B, then

$$\sum_{\kappa=1}^{N_0} p_\kappa N_\kappa' = \langle N' \rangle \qquad \sum_{\kappa=1}^{N_0} p_\kappa N_\kappa'' = \langle N'' \rangle \qquad (17)$$

represent the average number of electrons and holes in the initial states of the x-ray or Auger emission. The condition $\langle N' \rangle + \langle N'' \rangle = N$ corresponds to the normalization

$$\sum_{\kappa=1}^{N_0} p_\kappa = 1 \qquad (18)$$

The minimization of the discrete version

$$S = \sum_{\kappa=1}^{N_0} p_\kappa \ln \frac{p_\kappa}{p_{0\kappa}} \qquad (19)$$

of cross entropy (7) subject to the constraints (17) and (18) yields a factorized representation

$$P_\kappa = \frac{1}{Z} x^{N'_\kappa} y^{N''_\kappa} \tag{20}$$

for the probabilities provided p_o is chosen to be the uniform distribution. In Eq. (20) $x = \exp(-\lambda_e)$ and $y = \exp(-\lambda_h)$, where λ_e and λ_h are Lagrange parameters associated with electrons respectively holes. The normalization is accounted for by the partition function

$$Z = \sum_{\kappa=1}^{N_0} x^{N'_\kappa} y^{N''_\kappa} \tag{21}$$

Note that according to MCEP Eq. (20) is not only the most probable but also the only possible distribution that is consistent with the constraints (17). A given pair (N'_κ, N''_κ) occurs $\binom{N}{N'_\kappa} = \binom{N}{N''_\kappa}$ times in Eq. (21) since x and y are independent of κ. Hence one has the binomial form $Z = (x + y)^N$ and $p_\kappa = x^{N'_\kappa} y^{N''_\kappa}/(x + y)^N$. Consequently the probability $P_{N'N}$ which describes how frequently a given pair (N', N'') appears is given by Eq. (16), but where

$$p = y/(x + y) = [1 + \exp(\lambda_h - \lambda_e)]^{-1} \tag{22}$$

is now the information-theoretical probability that a hole has been created in the collision. The difference $\lambda_h - \lambda_e$ is in principle determined by the constraints (17) which give $<N'> = N(1 - p)$ and $<N''> = Np$. Since the experimental results are usually classified in terms of subshells for which $N = 2(2l + 1)$ there is a need for a generalization to several subshells. Each subshell is represented by a pair of constraints, given by Eqs. (17). Consequently, if there are h subshells the probability that the pairs (N'_1, N''_1), ..., (N'_h, N''_h) occur in each subshell is given by the product probability $P_h = P_{N'_1 N_1} P_{N'_2 N_2} x, \ldots, P_{N'_h N_h}$, where each $P_{N'N}$ is defined by Eq. (16).

The considerations given above demonstrate that the frequently observed binomial x-ray satellite and Auger electron distributions follow from much more general arguments than anticipated by the semiclassical derivation of Eq. (16). It is merely a consequence of the classification of the observed transitions in terms of the number of initial holes and of the Pauli principle. In fact, Eq. (22) would reduce to the Fermi-distribution law for the holes provided a temperature T and chemical potential μ could be defined such that $\lambda_h - \lambda_e = (\epsilon - \mu)/kT$, where ϵ is the average excitation energy. An analogous situation occurs in the case of the representation of level populations in beam-foil excitation.[30]

Figure 3 shows a fit of Eq. (22) to experimental $<N''_L>/8 = p_L$ values as a function of the projectile charge q_o. The data have been obtained from binomial representations of neon KL^n x-ray satellite data after fluorescence-yield corrections.[23] The various projectiles had energies between 1.4 to 1.5 MeV/u. Our fit is obtained with $\lambda_h = 6/q_o + 1/8$, corresponding to the choice $\lambda_h = 1$

130

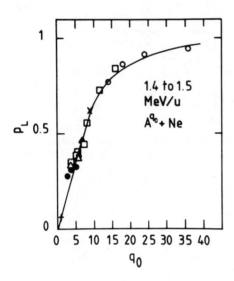

1.4 to 1.5 MeV/u

A^{q_o} + Ne

Fig. 3. The average number p_L of L holes as a function of the projectile charge q_o. The solid line follows from Eq. (22) and $\lambda_h = \lambda_e^{-1} = 6/q_o + 1/8$. The experimental data are from Ref. 23 (measurements by Kauffman et al. and Beyer).

for $p = 0.5$, which leads to $\lambda_h = \lambda_e^{-1}$. A study of the velocity-dependence [31] of λ_h would give a better idea of the significance of this analytical result.

CONCLUSIONS

Total ionization and charge transfer cross sections are often succesfully described by models, based on crude assumptions of ill-defined range of validity. Hence MCEP may be useful for constructing total cross sections from a stochastic point of view. This requires constraints which reflect dominant global features of pseudocrossings and correctly describe the boundary conditions under single-collision conditions. Such a procedure, if succesful, would lead to the identification of dynamical parameters which have a simple dependence on the external variables in analogy to the projectile-velocity dependence of the asymmetry parameter in our analysis of the equilibrium charge-state distributions. Since our formalism is suitable for comparing single and multiple collision regimes as well as the approach to equilibrium the ultimate goal will be relations between relevant dynamical parameters similar to "equations of state". More acquisition of experimental data on inelastic single and multiple collisions between many-electron ions and atoms is thus highly desirable.

ACKNOWLEDGEMENTS

We would like to thank Eugen Merzbacher for pointing out to us Ref. 11 and Maija Peltonen for her valuable assistance in the charge-state analysis.

REFERENCES

1. A. Russek and J. Meli, Physica $\underline{46}$, 222 (1970) and references therein.
2. W. Brandt and K. W. Jones, Phys. Lett. $\underline{57A}$, 35 (1976); B. M. Johnson, K. W. Jones, W. Brandt, F. C. Jundt, G. Guillaume and T. H. Kruse, Phys. Rev. A $\underline{19}$, 81 (1979).
3. M.H. Mittleman and L. Wilets, Phys. Rev. $\underline{154}$, 12 (1967).
4. J. Stähler and G. Presser, in Europhys. Conf. Abstr., At. Phys., eds. J. Kowalski, G. zu Putlitz, and H. G. Weber, (Heidelberg 1981) Vol. 5A Part II, p. 795; P. H. Woerlee, R. J. Fortner and F. W. Saris, J. Phys. B $\underline{14}$, 3173 (1981).
5. Y. Baudinet-Robinet and D. Lamotte, J. Phys. B $\underline{12}$, 3329 (1979); Y. Baudinet-Robinet, Phys. Rev. A $\underline{26}$, 62 (1982).
6. For a recent review, see The Maximum Entropy Formalism, eds. R. D. Levine and M. Tribus (MIT Press, Cambridge, 1979).
7. J. E. Shore and R. W. Johnson, IEE Trans. Inform. Theory $\underline{26}$, 26 (1980) and $\underline{27}$, 472 (1981).
8. J. v. Neumann, Göttinger Nachr. 273 (1927).
9. R.L. Stratonovich, J. Exper. Theoret. Phys. USSR $\underline{28}$, 409 (1955) (Sov. Phys. JETP $\underline{1}$, 254, 1955).
10. U. Fano, Rev. Mod. Phys. $\underline{29}$, 74 (1957).
11. H. Bethe, Rev. Mod. Phys. $\underline{9}$, 69 (1937).
12. E. T. Jaynes, Phys. Rev. $\underline{106}$, 620 (1957) and Phys. Rev. $\underline{108}$, 171 (1957).
13. C. E. Shannon, Bell Syst. Tech. J. $\underline{27}$, 379 and 623 (1948).
14. R. D. Levine in Ref. 6; see also Y. Alhassid, R. D. Levine, J. S. Karp, and S. G. Steadman, Phys. Rev. C $\underline{20}$, 1789 (1979).
15. P. A. Mello and T. H. Seligman, Nucl. Phys. A$\underline{344}$, 489 (1980; T. A. Brody, J. Flores, J. Bruce French, P. A. Mello, A. Pandey, and S. S. M. Wong, Rev. Mod. Phys. $\underline{53}$, 385 (1981).
16. see e.g. K. Blum Density Matrix Theory and Applications (Plenum Press, New York 1981).
17. Y. Alhassid and R. D. Levine, Phys. Rev. A $\underline{18}$, 89 (1978).
18. E. T. Jaynes in Statistical Physics, Vol. 3, p. 182 (Benjamin 1963).
19. O. Goscinski, T. Åberg and M. Peltonen, Phys. Lett. A, in press; T. Åberg, M. Peltonen and O. Goscinski, in Abstr. of Eight Int. Conf. At. Phys., eds. I. Lindgren, A. Rosén, S. Svanberg (Gothenburg 1982), B59.
20. H.-D. Betz, Rev. Mod. Phys. $\underline{44}$, 465 (1972).
21. N. S. Goel, S. C. Maitra and E. W. Montroll, Rev. Mod. Phys. $\underline{43}$, 231 (1971); N. S. Goel and N. Richter-Dyn, Stochastic Models in Biology (Academic Press, New York 1974).
22. R. K. Janev and P. Hvelplund, Comm. At. Mol. Phys. $\underline{11}$, 75 (1981).
23. P. Richard in Electronic and Atomic Collisions, eds. N. Oda and K. Takayanagi (north-Holland 1980) p. 125.
24. D. Schneider, M. Prost, B. DuBois, and N. Stolterfoht, Phys. Rev. A $\underline{25}$, 3102 (1982).

132

25. J. H. McGuire and L. Weaver, Phys. Rev. A 16, 41 (1977); J. F. Reading and A. L. Ford, Phys. Rev. A 21, 124 (1980).
26. T. Åberg, Ann. Acad. Sci. Fenn. AVI, 308, 1 (1969).
27. J. S. Briggs, Rep. Progr. Phys. 39, 217 (1976).
28. J. M. Hansteen, O. M. Johnsen, L. C. Kochbach, At. Data and Nucl. Tables 15, 305 (1975).
29. D. H. Madison and E. Merzbacher in Atomic Inner Shell Processes I, ed. B. Crasemann (Academic Press, New York 1975) p. 1.
30. T. Åberg and O. Goscinski, Phys. Rev. A 24, 801 (1981).
31. H. F. Beyer, R. Mann, and F. Folkmann, in Phys. Electronic and Atomic Collisions, Abstr. Contr. Papers, Vol. 2, ed. S. Datz (Gatlinburg 1981), p. 850.

IMPACT PARAMETER DEPENDENCE OF L AND K-SHELL EXCITATION MECHANISMS IN NEAR SYMMETRIC HEAVY-ION COLLISIONS

E. Morenzoni,[*] M. Nessi, P. Bürgy, Ch. Stoller, and W. Wölfli
Laboratorium für Kernphysik, Eidg. Technische Hochschule, Hönggerberg
CH-8093, Zürich

ABSTRACT

Results, on the impact parameter dependence of L-shell vacancy production during heavy ion collisions are discussed in the framework of molecular orbital (MO) excitation mechanisms. It appears that excitation in the $4f\sigma$ MO and vacancy transfer in the $3p\sigma$, $3d\pi$, and $3d\sigma$ MO through $3p\pi - 3p\sigma$ and $3d\delta - 3d\pi - 3d\sigma$ rotational couplings reproduce fairly well the experimental results for near-symmetric systems.

Using the experimental results on the impact parameter dependence of the projectile L-vacancy production probability, the multiple collision process of Meyerhof et al. of $2p\sigma$-vacancy production is considered in the case of I - Ag collisions. It is found that this process dominates K-vacancy production at large impact parameters, whereas at small impact parameter the experimental results can be explained by a single-collision direct excitation process from the coupled $2p\pi$ and $2p\sigma$ MO's.

INTRODUCTION

In the last decade a large number of experimental and theoretical investigations have been made to understand inner-shell vacancy production mechanisms during heavy ion collisions.[1] It appears that for not too asymmetric systems ($Z_p \approx Z_t$, Z_p, Z_t projectile and target atomic number, respectively) and quasiadiabatic conditions ($v_p << v_e$, v_p projectile velocity, v_e electron orbital velocity) quasimolecular orbitals (MO's) are transiently formed around projectile and target nuclei during the collision.[2] In this case, the experimental results can be explained in terms of electronic excitation out of these orbitals, due to couplings (rotational or radial) between few selected orbitals. As the couplings act at different intermolecular distances the most detailed information on the time dependent MO excitation process has been obtained by measuring the inner-shell x-ray or Auger emission probability as a function of the projectile impact parameter b.

So far most of these investigations have been concentrated on K-shell excitation mechanisms. In symmetric and near-symmetric collision systems (with Z_p, $Z_t \stackrel{<}{\sim} 60$, where relativistic effects do not considerably affect the MO binding energies) the commonly accepted mechanism for the production of K-vacancies in target and projectile is rotational coupling of $2p\pi$ vacancies to the $2p\sigma$ orbital and sharing of these vacancies between target and projectile in the outgoing part of the trajectory.[1]

---

* Present address: Department of Physics, Stanford University, Stanford, CA 94305, USA.

Provided that 1sσ excitation is negligible the summed projectile and target K-vacancy production represents the 2pσ excitation probability $P_{2p\sigma}(b)$. Theoretical calculations of the 2pπ - 2pσ rotational coupling[3] predict a characteristic b-dependence for $P_{2p\sigma}(b)$, with a broad adiabatic maximum at large impact parameters and a narrow kinematic maximum at impact parameters corresponding to a c.m. scattering angle of 90°.

In experiments with light gas targets (Z_p, Z_t < 20) good agreement between the rotational coupling theory and experiment has been found. In solid and heavier target experiments (Z_p, Z_t > 30) significant discrepancies have been reported.[4] In particular, the observed broad maximum in $P_{2p\sigma}$ is shifted towards smaller impact parameters and strongly suppressed compared to the theoretical 2pπ - 2pσ adiabatic maximum. The reasons for the observed discrepancies are not well understood.

Contrary to K-vacancy production, the impact parameter dependence of L-vacancy production $P_L(b)$ has been studied in few cases only. Nevertheless, this problem is of relevance for several reasons. First, experimental data on $P_L(b)$ provide us with an additional field where theories on inner-shell excitation mechanisms can be investigated. Second, measurements of the impact parameter dependence of L and K vacancy production can give us information on the influence and interplay between outer shell and K-shell excitation. For instance, in using the 2pπ - 2pσ rotational coupling model it is important to understand the mechanism producing 2pπ vacancies, that is vacancies in the L-shell of the heavier collision partner.

In this paper recent measurements of the impact parameter dependence of L and K-shell excitation are discussed with special regard to the two mentioned points.

EXPERIMENTAL RESULTS

After the pioneering work of Stein et al.,[5] who measured the impact parameter dependence of the total L x-ray emission in I - Te collisions, mainly strongly asymmetric systems have been investigated to study K-L level matching effects.[6-8] Recently, L-shell excitation in symmetric and near symmetric collisions in the quasimolecular regime has been measured in different experiments.[3,10,11,12] Results from these investigations are shown in Fig. 1, 2, 3 and 5. In the I - Ag case at 63 MeV, we also measured the target and projectile K-vacancy production to study the connections between L and K-vacancy production (see Fig. 6).

MO L-SHELL EXCITATION MECHANISMS

The essential quasimolecular excitation channels contributing to the target and projectile L-vacancy production can be easily recognized by considering a correlation diagram. In Fig. 4, the adiabatic correlation diagram for the near symmetric system I(Z_p = 53) + Ag(Z_t = 47) is shown. The diagram represents a non-relativistic calculation for a neutral system using the variable screening model of Eichler and Wille.[14] The diabatic orbitals relevant for the L-shell

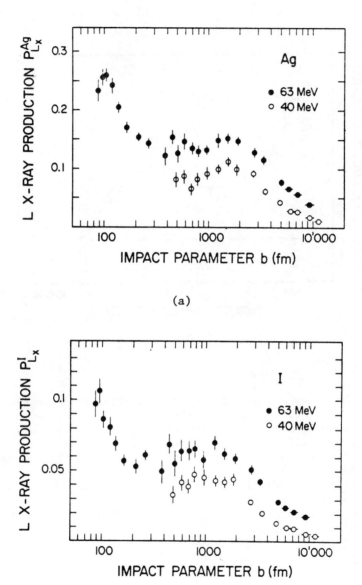

(a)

(b)

Fig. 1. (a),(b), L-shell x-ray emission probability in 40 and 63
MeV I - Ag collision for target and projectile as a function
of the impact parameter.

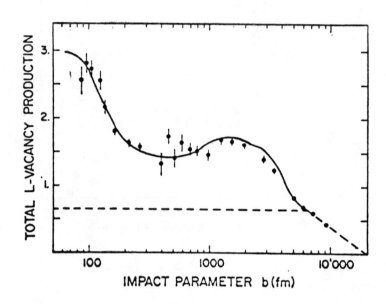

Fig. 2. Total L-shell vacancy production probability in 63 MeV
I - Ag collisions. The probability was obtained by
assuming a fluorescence yields w_L = 0.13 for I and
w_L = 0.118 for Ag.[13] The solid line indicates the
predicted L-shell excitation arising from the $3p\pi$-$3p\sigma$
($N_{3p\pi}$ = 2.5) and $3d\delta$-$3d\sigma$($N_{3d\delta}$ = 1.25) rotational
couplings superimposed to the $4f\sigma$ excitation (dashed
line).

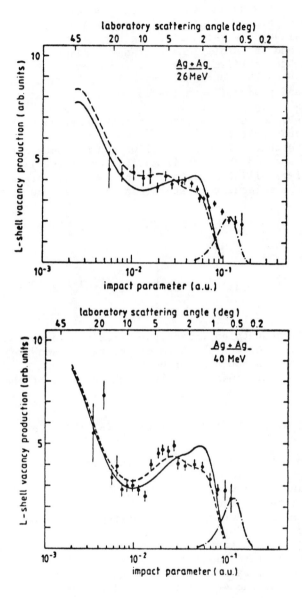

Fig. 3. Impact parameter dependence of Ag L-vacancy production in
26 and 40 MeV Ag - Ag collisions. The full curve shows
calculations for the 3pπ-3pσ and 3dδ-3dπ-3dσ rotational
couplings by assuming $N_{3p\pi} = N_{3d\delta}$. The dashed curve
assumes $N_{3p\pi} = 2N_{3d\delta}$. The broken   curve gives the
theoretical shape for 4fφ-4fδ-4fπ-4fσ rotational coupling.
The experimental points are normalized to the theoretical
curves (from Ref. 10).

excitation are the 4fσ-MO and the orbitals correlating to the M-shell of the united atom e.g. the 3dπ, 3dσ, and the 3pσ orbitals. These orbitals are indicated by thick lines in Fig. 4. Vacancies created in the 4fσ, 3dπ and 3pσ orbitals are transferred primarly to the L-shell of the lighter collision partner, those created in the 3dσ orbital to the L-shell of the heavier partner. Vacancy sharing between target and projectile on the outgoing part of the trajectory, before the atoms reseparate, by means of a long range radial coupling of the Meyerhof-Denlov type can then redistribute the vacancies between target and projectile.[15]

According to this molecular model we can decompose the impact parameter dependence of the L-shell excitation in the lighter $P_L^\ell(b)$ and in the heavier collision partner $P_L^h(b)$ into the vacancy production probability of the molecular orbitals 4fσ, 3dπ, 3dσ and 3pσ:

$$P_L^\ell(b) = [P_{4f\sigma}(b) + P_{3d\pi}(b) + P_{3p\sigma}(b)][1 - w_{LL}(b)] + P_{3d\sigma}(b)w_{LL}(b),$$

$$(1)$$

$$P_L^h(b) = [P_{4f\sigma}(b) + P_{3d\pi}(b) + P_{3p\sigma}(b)] \, w_{LL}(b) + P_{3d\sigma}(b)[1 - w_{LL}(b)].$$

$$(2)$$

Here, w_{LL} represents an impact parameter dependent mean vacancy sharing probability between the L levels of target and projectile.[15] The summed L-vacancy production probability is the sum of the different excitation channels independently of the details of the excitation mechanism.

$$P_L^\ell(b) + P_L^h(b) = P_{4f\sigma}(b) + P_{3d\pi}(b) + P_{3p\sigma}(b) + P_{3d\sigma}(b) . \qquad (3)$$

Vacancy production in inner quasimolecular levels may occur by three main physical processes:
(a) Vacancy transfer between approaching MO's via radial coupling at large internuclear distances,
(b) Vacancy transfer via rotational coupling near the united-atom limit (e.g. 3dδ → 3dπ,σ, 3pπ → 3pσ),
(c) Direct electron excitation to continuum or vacant quasicontinuum states at small internuclear distances.

In the following, we discuss the importance of these processes for the excitation of the 4fσ, 3dπ, 3dσ and 3pσ orbitals. From the magnitude of the binding energy and the behavior of the MO's as a function of the internuclear distance, it can be expected that the 4fσ excitation channel is dominant in collision at large impact parameters, whereas 3d and 3p orbitals are effective mainly at small impact parameters.

4fσ Excitation

The peculiar role played by the highly promoted 4fσ orbital in L-vacancy production has been recognized early from cross-section and inelastic energy loss measurements.[16] In light collision systems

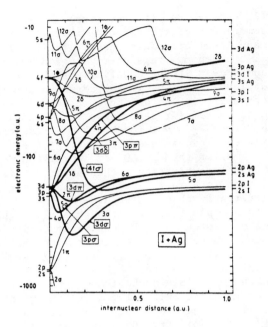

Fig. 4. Correlation diagram for the I – Ag system.[19] The diabatic molecular orbitals relevant for the L-vacancy production are shown as heavy lines.

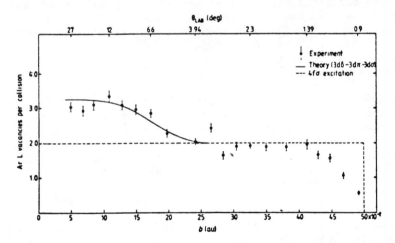

Fig. 5. Impact parameter dependence of the Ar L-vacancy production probability in .3 MeV Ar⁺- Ar collisions. The broken curve shows the prediction from the Kessel model. The full curve is a calculation for the 3dδ-3dπ-3dσ rotational coupling. (From Ref. 9.)

like Ar + Ar ($Z_p = Z_t = 18$), if a critical internuclear distance r_c is reached (promotion radius), the 4fσ orbital rises into an energy region where many level crossings occur. Taking into consideration the collisional broadening, this energy region belongs effectively to the continuum. Even if the coupling probability is small, the large number of level crossing will lead to an almost unity probability to empty the 4fσ orbital. According to this picture (Kessel model)[15] a step-like impact parameter dependence for $P_{4fσ}(b)$ is expected: $P_{4fσ} = 2$ for $b < r_c$, $P_{4fσ} = 0$ for $b > r_c$. This behavior has been found in Ar - Ar collisions[9] (see Fig. 5). The second rise shown in Fig. 5 for b smaller than 13,000 fm is related to the onset of an additional L-shell excitation mechanism as we discuss below.

4fσ excitation in higher Z-collision systems (such as I - Ag) may proceed differently than in the Ar - Ar case, because the 4fσ level rises less steeply. Also, although still highly promoted, the 4fσ MO does not merge into the continuum region at small internuclear distances. Moreover, in contrast with the situation in the Ar - Ar system, the 4f and other closely lying levels are filled in the static united-atom system. This has led Shanker et al. to consider another possible mechanism of 4fσ excitation, namely vacancy transfer from the initially empty 4fϕ MO into the 4fσ via rotational coupling among the diabatic 4fϕ (1ϕ in adiabatic notation of Fig. 4), 4fδ (3δ - 2δ), 4fπ (6π - 5π - 4π - 3π) and 4fσ orbitals.[12] Qualitative evidence for this mechanism has been found by measuring the Kr L - x-ray emission probability in 1.4 MeV Kr-Xe collisions.[12] The very low energy guarantees that the adiabatic criterion is fulfilled for the united atom 4f level ($v_p/v_e = 0.023$). Presser et al.[10] to explain their experimental data at large impact parameters (Fig. 3) assumed that this rotational coupling has still a dominating influence for the 4fσ excitation even at higher energies (broken curve in Fig. 3). In 63-MeV I-Ag collisions, the v_p/v_e ratio for 4f electrons is 0.6, which barely fulfills the adiabatic criterion. In such a situation it may be questionable to consider a well defined isolated MO coupling.

Another possibility to excite 4fσ electrons is direct coupling to higher lying empty or continuum states at small internuclear distances.[18] We expect this process to be enhanced in this case because the quantity $dE/dr \cdot v_p/E$, which is a measure of the coupling strength, is large. Since calculations of direct 4fσ excitation are lacking we have used some qualitative arguments to estimate its contribution to the total L-vacancy production probability. It has been shown experimentally[17] and theoretically[18] that if MO's have an almost constant binding energy for a large region of internuclear distances, direct excitation is largely impact parameter independent for impact parameters lower than an adiabatic radius defined by $r_{ad} = \hbar v_p/E$, v_p is the projectile velocity and E is the binding energy of the relevant orbital (in our case the 4fσ) in the UA system. For 63 MeV I + Ag, we obtain $r_{ad} \cong 6,500$ fm. At such large impact parameters L-excitation channels involving the 3d and 3p orbitals are negligible, (see below). Therefore, as a first guess, in Fig. 2, we have taken the experimental value at b = 6,500 fm as probability for (direct) 4fσ excitation.

3pσ, 3dπ, 3dσ Excitation

At smaller impact parameters, the 3dπ , 3dσ and 3pσ orbitals can also be excited and carry additional vacancies into the L-shell of target and projectile. Among the three molecular excitation mechanisms mentioned before, vacancy transfer via rotational coupling seems to play a leading role. For instance, vacancies created in the 3dδ or 3pπ MO in the ingoing part of the collision or in a previous collision (in a solid target) can be transferred by two rotational couplings, 3dδ-3dπ-3dσ and 3pπ-3pσ into the 3dπ, 3dσ and 3pσ orbitals. Qualitatively, the impact parameter dependence for each excitation process shows the characteristic behavior of a rotational coupling, with a broad adiabatic maximum at large impact parameters and a peak, due to kinematic effects at small impact parameters.[10,19] The position of the adiabatic maximum is shifted towards larger impact parameters as one goes from the 3pσ to the 3dπ and 3dσ excitation.

In the Ar-Ar system only the 3dδ-3dπ-3dσ coupling appears to be effective (see Fig. 5). In fact in the neutral Ar atom the 3d level (from which the 3dδ-MO originates) is completely empty, whereas the 3p level (which generates the 3pπ MO) is completely filled. Moreover the 3p level is much more bound compared to the 3d. The theoretical curve shown in Fig. 5 presupposes a half empty 3dδ MO when the 3d rotational coupling begins to act.[9]

In a heavier system like I-Ag (Fig. 1,2), the 3dδ and 3pπ orbitals both correlate to the 3d level of the heavier partner, according to Ref. 14. One would expect, therefore, the number of incoming 3dδ and 3pπ vacancies ($N_{3d\delta}$, $N_{3p\pi}$) to be nearly equal. On the other hand as can be seen from correlation diagram, the 3pπ level crosses the 3dδ level at 30,000 fm and becomes less bound below this internuclear distance. Coupling to continuum or quasicontinuum levels on the way into the collision, therefore, would preferably ionize the 3pπ level. Consistent with this expectation better agreement with the experimental data is found if one assumes $N_{3p\pi} > N_{3d\delta}$. Fig. 2 shows that the total L-vacancy production probability in 63-MeV I - Ag collisions is qualitatively and quantitatively well reproduced over the entire impact parameter range by a combination of a constant contribution from 4fσ excitation and a contribution from the 3dδ-3dπ-3dσ and 3pπ-3pσ rotational couplings at small impact parameters (with $N_{3p\pi}$ = 2.5 and $N_{3d\delta}$ = 1.25). In particular, the pronounced increase of the excitation probability at small impact parameters can be attributed to the onset of the kinematic peaks of both rotational couplings. The contributions from the rotational couplings also fit qualitatively fairly well with the results on L-vacancy production probabilities in Ag - Ag and Ag - Cs collisions between 17 and 40 MeV. Better agreement with the data is found again by assuming a larger vacancy occupation number in the 3pπ level than in the 3dδ.[10] In principle, direct excitation can also ionize the 3dπ, 3dσ and 3pσ orbitals. Indeed, results on L-shell excitation in strongly asymmetric collision systems close to the K - L level matching condition seems to indicate that if the projectile is the lighter collision partner, 3dσ direct excitation is predominant,[18] whereas when the projectile is the heavier partner some evidence for the 3dδ-π-σ rotational coupling has been found.[8] In the near symmetric systems discussed before we estimated the contribution of the direct excitation would be obscured by the previously-discussed excitation processes.[11]

Information on $w_{LL}(b)$ [Eqs. (1),(2)] has been obtained from the separate target and projectile L-shell excitation probabilities. For instance from $P_L^I(b)/P_L^{Ag}(b) \cong w_{LL}(b)/1 - w_{LL}(b)$ (for b $\gtrsim$ 4000 fm) we obtained $w_{LL}(b)$ nearly constant and in good absolute agreement with the the prediction of the Meyerhof formula.[15]

CONNECTIONS BETWEEN L AND K-VACANCY PRODUCTION

In spite of discrepancies in experiments with solid targets, $2p\pi$-$2p\sigma$ rotational coupling is believed to play an important role in the K-vacancy production of target and projectile in near symmetric collisions. The proposed mechanisms to provide for $2p\pi$ vacancies in a solid target, when the incoming projectile has all its L-orbitals filled, can be classified as single and multiple collisions processes.[20] In a multiple collision process projectile 2p vacancies made in a previous collision can live long enough to be carried into the $2p\pi$ MO in a subsequent collision and can be transferred through $2p\pi$-$2p\sigma$ rotational coupling into the $2p\sigma$ MO. K-total cross sections on solid targets have been successfully interpreted, by assuming this mechanism.[20]

From the measurement of the impact parameter dependence of the I-L x-ray emission probability, the contribution of the muliple-collision process to the impact parameter dependence of the $2p\sigma$ excitation in I-Ag collisions can be determined. The contribution is calculated according to the equation[21]

$$P_{2p\sigma}^{MC}(b) \frac{d\vec{b}}{d\Omega} = C(v_p) \int \frac{d\vec{b}_1}{d\Omega_1} d\Omega_1 P_{L_x}^I(b_1) \frac{d\vec{b}_2}{d\Omega_2} P_{2p\pi-2p\sigma}(b_2). \quad (4)$$

$C(v_p)$ is the probability to transfer a L-projectile vacancy into the $2p\pi$ orbital.[20] The other quantities which enter in Eq. (4) are the experimental data for $P_{L_x}^I(b)$ (Fig. 1b) and $P_{2p\pi-2p\sigma}$, the probability to transfer a $2p\pi$ vacancy into the $2p\sigma$ orbital.[22]

The integration in Eq. (4) takes into account the possibility that the projectile may suffer a large deflection in the first collision (with impact parameter b_1), where the 2p vacancy is created, and then a smaller deflection in the second collision (with impact parameter b_2), in which the vacancy is transferred into the $2p\sigma$ orbital (b indicates the impact parameter corresponding to the total deflection angle). In Fig. 6, the calculated multiple-collision contribution for 63 MeV-I-Ag (curve 1) collisions is compared with our experimental results and data of Anholt et al.[23] at 60 MeV. The integration in Eq. (4) was performed by a Monte Carlo method.[24] Curve 2 shows the unfolded multiple collision contribution given by $P_{2p\sigma}(b) = Co_{L_x}^I P_{2p\pi-2p\sigma}(b)$. The comparison of curves 1 and 2 shows that the folding affects the $2p\sigma$ dependence on b, slightly shifting the position of the $2p\pi$-$2p\sigma$ adiabatic maximum to smaller impact parameters and improving the agreement at large impact parameters. In the region of the $2p\pi$-$2p\sigma$ kinematic peak, the multiple collision probabilities accounts only for about 30% of the observed total K-vacancy production probability.

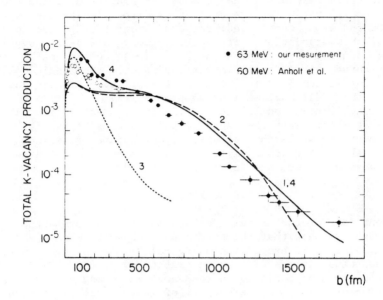

Fig. 6. Total K-vacancy production probability for I and Ag
$P_K^I(b) + P_K^{Ag}(b) = P_{2p\sigma}(b)$. Closed symbols from our
experiment (63 MeV) open symbols from Ref. 23 (60 MeV).
The solid (curve 1) and the broken lines (2) are the
multiple collision contributions with and without folding
respectively, The dotted line (3) shows the single-
collision direct excitation probability.[24] Curve 4
is the sum of 1 and 3.

As a possible explanation for the enhanced probability at small impact parameters Jakubassa[24] has proposed a single-collision process, where $2p\sigma$ and $2p\pi$ vacancies are created by direct ionization of the united atom and subsequently redistributed among the $2p\sigma$ and $2p\pi$ orbitals in the outgoing part of the trajectory by $2p\pi$-$2p\sigma$ coupling. The calculation for this process in 63-MeV I + Ag collisions is shown in Fig. 6 (curve 3). At small impact parameters, the single-collision process dominates compared to the multiple-collision process and can explain the additional excitation probability.

By adding single and multiple collision process (curve 4) a reasonable agreement with the experimental features of $P_{2p\sigma}(b)$ is reached. over the whole investigated impact parameter region. In particular, the peaked probability around $\theta_{cm}= 90°$ and the disappearance of the adiabatic maximum are fairly well reproduced.

CONCLUDING REMARKS

Present data of the L-shell excitation in near symmetric collision systems show that the experimental results can be understood within the model of MO excitation using relatively few couplings. On the other hand, it has been shown that study of the L- and K-shell vacancy production probability in the same system may provide a better understanding of the complex processes of inner shell excitation in a solid target. Further investigations may include measurements of the impact parameter dependence of L-vacancy production for a complete identification of the kinematic peaks, and measurements with gas targets to study L-shell excitation under single collision conditions. Although not easily feasible, similar investigations for the K-shell excitation in systems comparable to I - Ag, but with gas targets would provide more stringent tests of the various possible K-shell excitation mechanisms.

ACKNOWLEDGEMENTS

We thank U. Wille and D. Jakubassa for performing the calculations discussed in the text and sending us results prior to the publication. We are also indebted to W. E. Meyerhof for many useful comments. This work was partially supported by the Swiss National Science Foundation.

REFERENCES

1. For a recent review see R. Schuch, in Physics of Electronic and Atomic Collisions, S. Datz editor, North Holland, 1982, p. 151.
2. U. Fano and W. Lichten, Phys. Rev. Lett. 14, 627 (1965).
3. K. Taulbjerg, J. S. Briggs and J. Vaaben, J. Phys. B: At. Mol. Phys. 9, 1351 (1976).
4. R. Schuch, G. Nolte, and H. Schmidt-Böcking, Phys. Rev. A22, 1447 (1980), and references therein.
5. H. J. Stein, H. O. Lutz, P. H. Mokler, and P. Armbruster, Phys. Rev. A5, 2126 (1972).
6. P. Fintz, G. Guillaume, F. C. Jundt, K. W. Jones, and B. M. Johnson, Phys. Lett. 71A, 432 (1979).

7. B. M. Johnson, K. W. Jones, W. Brandt, F. C. Jundt, G. Guillaume, and T. H. Kruse, Phys. Rev. A19, 81 (1979).

8. A. Warczak, D. Liesen, P. H. Mokler, and W. A. Schönfeldt, Phys. Lett. 84A, 239 (1981).

9. R. Shanker, R. Bilau, R. Hippler, U. Wille, H. O. Lutz, J. Phys. B: Atom. Molec. Phys. 14, 997 (1981); and G. M. Thomson, Phys. Rev. A 15, 965 (1977).

10. G. Presser, J. Stähler, R. Werner and U. Wille, Arbeitsbericht EAS 1982, p. 38, Fricke et al. editors.

11. E. Morenzoni, M. Nessi, P. Bürgy, Ch. Stoller, and W. Wölfli, Arbeitsbericht EAS 1982, p. 33, Fricke et al. editors.

12. R. Shanker, R. Bilau, R. Hippler, U. Wille, and H. O. Lutz, Arbeitsbericht EAS 1982, p. 36, Fricke et al. editors.

13. S. Hagmann, P. Armbruster, G. Kraft, P. H. Mokler and H. J. Stein, Z. Physik A 288, 353 (1978).

14. J. Eichler, U. Wille, B. Fastrup and K. Taulbjerg, Phys. Rev. A 14, 707 (1976).

15. W. E. Meyerhof, A. Rüetschi, Ch. Stoller, M. Stöckli and W. Wölfli, Phys. Rev. A 20, 154 (1979).

16. Q. C. Kessel and B. Fastrup, Case Studies in Atomic Physics, Vol. 3, Ed. E. W. Mc Daniel and M. R. C. McDowell (North Holland), p. 137.

17. D. Maor, D. Liesen, P. H. Mokler, B. Rosner, H. Schmidt-Böcking, and R. Schuch, Physics of Atoms and Molecules, Ed. Fabian et al. (Plenum, 1980), p. 67.

18. W. E. Meyerhof, Phys. Rev. A 20, 2235 (1979).

19. U. Wille, private communication.

20. W. E. Meyerhof, R. Anholt, T. K. Saylor, Phys. Rev. A 16, 169 (1977).

21. R. Anholt, Stanford University, preprint 1981.

22. D. H. Jakubassa, K. Taulbjerg, J. Phys. B: At. Mol. Phys. 13, 757 (1980).

23. R. Anholt, Ch. Stoller and W. E. Meyerhof, J. Phys. B: Atom. Mol. Phys. 13, 3807 (1980).

24. D. H. Jakubassa-Amundsen, private communication.

RESONANT BEHAVIOR IN THE PROJECTILE K X-RAY YIELD ASSOCIATED WITH ELECTRON CAPTURE IN ION-ATOM COLLISIONS

J. A. Tanis* and E. M. Bernstein*
Western Michigan University, Kalamazoo, MI. 49008

W. G. Graham+
The New University of Ulster, Coleraine, BT521SA, N. Ireland

M. Clark† and S. M. Shafroth†
University of North Carolina, Chapel Hill, N. C. 27514

B. M. Johnson, K. Jones, and M. Meron
Brookhaven National Laboratory, Upton, N. Y. 11973

ABSTRACT

A review of recent efforts to observe simultaneous electron cap-ture-and-K-shell excitation in ion-atom collisions is presented. This process is qualitatively analogous to dielectronic recombina-tion (inverse Auger transition) in free-electron-ion collisions, and, hence, is expected to be resonant. Experimentally, events having the correct signature for simultaneous capture-and-excitation are iso-lated by detecting projectile K x rays in coincidence with ions which capture a single electron. In a recent experiment involving 70-160 MeV S^{13+} ions incident on Ar, resonant behavior was observed in the yield of projectile K x rays associated with electron cap-ture. This resonance is attributed to simultaneous capture-and-ex-citation. The position (120 MeV) and width (60 MeV) of the observed resonance are in good agreement with theoretical calculations. The data indicate that this resonant process is an important mechanism in inner-shell vacancy production in the energy range studied.

*Supported in part by grants from the Precision Kawasaki Company (Japan), Shinotest Shoji K. K. (Japan), and Nakano Transportation Company (Japan).

+Supported in part by a grant from the NATO Research Grants Pro-gramme (Grant No. 1910).

†Supported in part by the U. S. Department of Energy, Division of Chemical Sciences.

INTRODUCTION

In MeV/amu heavy ion-atom collisions several fundamental inner-shell processes such as excitation, ionization, and charge transfer occur. Recently it was suggested that projectile K-shell excitation may occur <u>simultaneously</u> with electron capture in ion-atom collisions.[1] Such a process, which is due to the Coulomb interaction of the projectile with the target electrons, is qualitatively analogous to an inverse Auger transition and is expected to be resonant for projectile velocities corresponding to the energy of an exiting electron in the Auger process. Since the captured electron is initially bound in the target, the width of the resonance should be reflective of the distribution of electron momenta in the target, i.e., the Compton profile of the target electrons. The <u>simultaneous</u> electron-capture-and-excitation process has recently been referred to as resonant-transfer-and-excitation (RTE).[2]

In the case of free electron recombination, this process is called <u>dielectronic recombination</u> which occurs when a highly stripped ion captures a continuum electron and simultaneously excites an electron from the ground-state configuration of the ion. Since radiation can be emitted following the formation of this excited state, dielectronic recombination is believed to be an important energy-loss mechanism in high temperature fusion plasmas.[3] Dielectronic recombination has been identified in plasmas but has never been successfully investigated in laboratory experiments.

The process of dielectronic recombination for a H-like atom is shown schematically in Fig. 1. An electron from the continuum is captured simultaneously with the excitation of an electron from the K shell to the L shell. In the case of RTE in an ion-atom collision, the captured electron is supplied by the neutral target atom. The signature for either dielectronic recombination or RTE is the emission of a K x ray in coincidence with a single capture event.

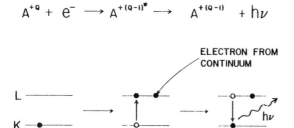

$$A^{+Q} + e^- \longrightarrow A^{+(Q-1)*} \longrightarrow A^{+(Q-1)} + h\nu$$

ELECTRON FROM CONTINUUM

Fig. 1 Schematic of dielectronic recombination process for a H-like ion.

For a free electron incident on a highly stripped ion, the resonant width is very narrow (< 1 eV). However, for ion-atom collisions, calculations by Brandt[4], which include the effect of the Compton profile of the target electrons, indicate that the width of the resonance is sufficiently wide (tens of MeV) so as to be easily observable.

In preliminary experiments[1,5] investigating K-shell excitation associated with electron capture in ion-atom collisions, it has been found that as much as 25% of the total K x-ray yield results from events associated with electron capture for Li-like projectiles. A process which can compete with RTE is one-electron capture following excitation (by the target nucleus or target electrons) in the same collision with one target atom. Since RTE is a resonant process, observation of a resonant behavior in the x-ray yield associated with capture would identify the mechanism and distinguish it from competing channels.

To our knowledge, three investigations of RTE involving the coincidence technique have been performed to date. These works are described in Refs. 1, 5 and 6.

In another type of experiment, the group at Kansas State has recently tried to observe the RTE process using a high resolution Bragg crystal spectrometer[7]. In this work, a search was made for x-ray transitions associated with the formation of specific intermediate RTE states for F^{8+} ions incident on several target gases. No definite conclusion concerning RTE was obtained from these measurements. Only those investigations involving the coincidence measurements will be considered here.

In Ref. 1, 70 MeV S^{q+} ions (q=13-16) were incident on an argon gas target. The beam energy in this work was below the expected resonance energy ($\sim$ 100 MeV) for sulfur ions, but still within the calculated resonance profile.[4] While coincidence events were observed, the data obtained was not sufficient to identify the mechanism involved.

In the experiment of Ref. 5, 160 and 180 MeV Ar^{q+} (q=14, 15 and 17) were incident on Xe. The results for Ar^{15+} are shown in Fig. 2. The expected resonance energy for Ar^{15+} is 160 MeV. As seen, the cross section for K x-ray emission associated with capture σ_{COINC}^{q-1} shows a slightly higher ($\sim$ 20%) value for 160 MeV than for 180 MeV. Measurements at additional energies would be needed, however, before any conclusions can be drawn concerning the existence of resonance behavior. Also shown in the figure is the cross section for total K x-ray emission $\sigma_{K\alpha}$ and the ratio of coincidence K x rays to singles K x rays $\sigma_{COINC}^{q-1}{}_{K\alpha}/\sigma_{K\alpha}$.

Furthermore, in this work projectile x-ray coincidences with other, outgoing charge states of the projectile, e.g., q+1, q+2, etc., were also measured. This was done for 180 MeV Ar^{14+} ions and Ar^{15+} ions as shown in Fig. 3. From these results it is apparent that most of the emitted x-rays are associated with the outgoing charge state q of the projectile.

In the most recent investigation of RTE[6], which will be described in detail, strong experimental evidence is obtained for the

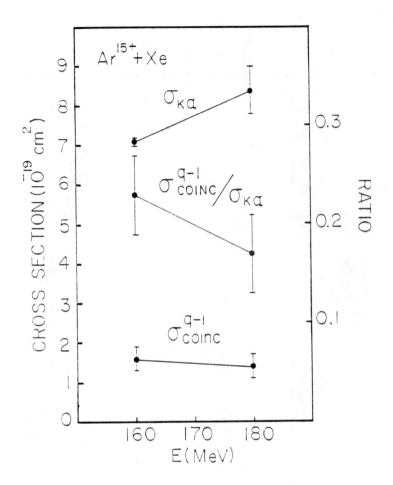

Fig. 2. Total cross section $\sigma_{K\alpha}$ for argon K x-ray emission and the cross section σ_{COINC}^{q-1} for argon K x-ray emission following electron capture for Ar^{15+} + Xe collisions. Also shown is the fraction $\sigma_{COINC}^{q-1}/\sigma_{K\alpha}$ of x rays associated with capture events.

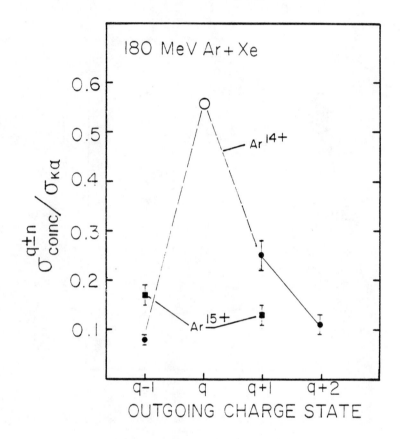

Fig. 3. Fraction of K x rays associated with outgoing charge states of projectile for 180 MeV Ar[14+] and Ar[15+] ions incident on Xe.

existence of this process. In this work, resonant behavior is observed in the yield of projectile K x rays in coincidence with single electron capture for 70-160 MeV S+ Ar collisions. The results indicate that this resonant process is an important mechanism for inner-shell vacancy production in the energy range studied.

EXPERIMENTAL

In the work of Ref. 6, sulfur K x rays coincident with single electron capture and   loss were measured for 70-160 MeV S^{13+} ions incident on argon under single collision conditions. Li-like S is used since it has no long lived excited states. This experiment was performed at Brookhaven National Laboratory using the MP tandem Van de Graaff accelerator. The beam of S^{13+} ions was collimated by two 0.5 mm square apertures separated by about 2.8 m prior to entering a differentially pumped gas cell. X rays from the interaction region were detected with a 200 mm^2 Si(Li) detector located about 1 cm from the beam line and positioned at 90° to the incident beam direction. Ions emerging from the gas cell passed between a pair of vertically mounted electrostatic deflection plates which separated the various charge state components of the beam in a horizontal plane. Emerging ions undergoing single electron capture (q-1) and single electron loss (q+1) were detected in surface barrier detectors. The non-charge-changed component of the emerging beam (q=13+) was collected in a Faraday cup located about 40 cm behind the surface barrier detectors. A capacitance manometer was used to determine the pressure in the target gas cell which was less than 7 mTorr for all measurements. All cross sections of interest were found to be independent of gas pressure in the range studied.

Coincidences between S K x rays and single electron capture and loss events were recorded with a time-to-amplitude converter (TAC). The singles x-ray spectrum and the coincidence spectra from the q-1 and q+1 TACs were stored in a two-dimensional array as x-ray energy vs. time. This allowed energy "cuts" to be made in the TAC spectra to give the charge-changing events which were coincident with the projectile x rays.

RESULTS

Figure 4 shows the various projectile cross sections obtained from our measurements. Relative errors for a given cross section are generally about 5%. Absolute errors in the values shown are estimated to be about 25%.

From the figure it is seen that the cross section for sulfur K x-ray emission associated with single electron capture σ_{COINC}^{q-1} exhibits a maximum in the region of 120 MeV. We attribute this peak to resonant transfer and excitation (RTE). RTE is expected to occur as a result of the formation of intermediate resonant states such as $1s^22s \rightarrow 1s2s^22p$, $1s2s2p^2$, $1s2s^23p$, $1s2s2p3d$, etc. Qualitatively, a maximum in the σ_{COINC}^{q-1} cross section would occur when the energy of

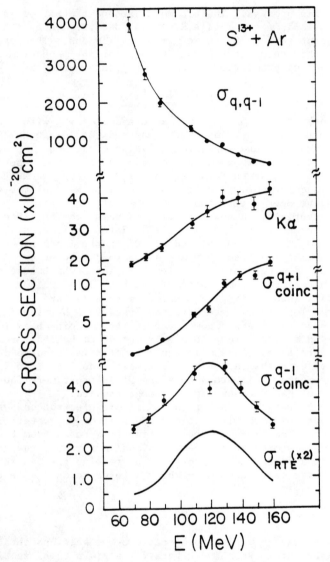

Fig. 4 Cross sections for 70-160 MeV S^{13+} Ar collisions. σ^{q-1}_{COINC} and σ^{q+1}_{COINC} are cross sections for projectile K X rays associated with single electron capture and loss, respectively, and $\sigma_{K\alpha}$ is the cross section for S singles Kα emission. The solid lines are fits to the data using Eq. (1). The curve labeled σ_{RTE} is a theoretical calculation (Ref. 4) of the resonant part of σ^{q-1}_{COINC}. $\sigma_{q,q-1}$ is the total capture cross section. In this latter case the curve is drawn to guide the eye.

the captured electron in the projectile rest frame matches the Auger electron energy for a given transition. Since the velocity component of the target electrons along the beam axis contributes to the relative velocity, the shape of σ_{COINC}^{q-1} should reflect the target electron momenta distribution, i. e., their Compton profile.

It is also seen that there is considerable background under the σ_{COINC}^{q-1} peak giving a signal-to-noise ratio of ∿2:1. The background may be due to <u>uncorrelated</u> excitation and single capture events taking place during the same collision with a single target atom. Since the single capture cross section is large (∿10^{-17}cm^2) it is possible that an ion which undergoes K-shell excitation due to the Coulomb interaction <u>with the target nucleus</u> will also capture during the same collision.

Also shown in Fig. 4 is the cross section for sulfur K x-ray emission associated with single electron loss σ_{COINC}^{q+1} and the cross section for singles Kα x-ray emission $\sigma_{Kα}$. In both σ_{COINC}^{q+1} and $\sigma_{Kα}$ an anomalous variation is noted in the region of the resonance. A possible explanation for the origin of these anomalies will be given later.

DISCUSSION

As mentioned previously, the existence of a resonance in the σ_{COINC}^{q-1} cross section would provide a signature for the RTE process. Hence, it is essential to determine whether the observed maximum in σ_{COINC}^{q-1} is a real resonance or whether it results some how from multi-step processes. One such two step process is K-shell excitation followed by electron capture in the same collision with a single target atom. In order to establish the origin of the observed anomalies, the data has been analyzed using general resonance theory as described by Soga and Bernstein.[8]

According to resonance theory, the energy dependence of a cross section, σ, near an isolated resonance can be expressed as

$$\sigma = B + \frac{R}{1 + x^2} + \frac{Ax}{1 + x^2} \tag{1}$$

where $x = \frac{2(E - E_R)}{\Gamma}$, E_R is the resonance energy and Γ is the resonance width. The first term in Eq. (1) is the background, the second term is the Breit-Wigner resonance function, and the last term is an asymmetric interference term. If the same resonance is observed in the excitation functions for several channels, the data should be consistent with the same values of E_R and Γ.

The solid curves in Fig. 4 for σ_{COINC}^{q-1}, $\sigma_{Kα}$, and σ_{COINC}^{q+1} show fits of Eq. (1) to the data using the procedures of Soga and Bernstein.[8] In these fits the resonance parameters were assumed to be energy independent and the background was assumed to vary linearly over the energy range of interest. Fits using resonance parameters in the vicinity of $E_R \sim 120$ MeV and $\Gamma \sim 60$ MeV were found to reproduce the data very well. The specific values used for the curves shown in Fig. 4, $E_R = 120$ MeV and $\Gamma = 60$ MeV, were chosen because of their theoretical significance which is discussed below. The fitting parameters obtained for these values of E_R and Γ are listed in Table I.

The fact that the <u>same resonance parameters</u> can be used to produce such good fits to the observed anomalies in all three channels provides strong evidence that these effects are due to a real resonance and negates the possibility that the peak in σ^{q-1}_{COINC} is produced by uncorrelated events as an accidental consequence of a decreasing capture cross section and an increasing excitation cross section in a multi-step process.

Further strong evidence that the observed anomalies are due to a real resonance and, in fact, are produced by RTE is provided by a comparison of the data with theoretical calculations of the RTE process. A calculation of the resonant part (RTE) of σ^{q-1}_{COINC} for $S^{13+} + Ar$ has been carried out by Brandt[4] using the dielectronic recombination cross sections of McLaughlin and Hahn[9] and the Compton profile of the target electrons[10]. It should be noted that this calculation requires the projectile velocity to be much greater than the bound electron velocity. Hence, only those target electrons which satisfy this condition were included. Furthermore, all intermediate states which can result in $K\alpha$ emission were included in Brandt's computations since the x-ray detector used in the present experiment could not resolve the different $K\alpha$ transitions.

The theoretical curve for σ^{q-1}_{COINC} based on this calculation is labeled σ_{RTE} in Fig. 4. The relatively large width of the resonance is partly due to the distribution of target electron momenta. An additional contribution to the width as well as a slight asymmetry in σ_{RTE} occur because this curve is not a single resonance but is instead the envelope of the contributions from the various intermediate states. It is important to note, however, that Eq. (1) with the background term set to zero and with a small asymmetry term provides an excellent approximation to the actual shape of the theoretical RTE curve. The maximum in the theoretical RTE "resonance" (E_R) is at 120 MeV and the full width at half maximum (Γ) is 60 MeV which are the values used in the fits to the data shown in Fig. 4. Thus, the position and shape of the theoretical curve are in excellent agreement with the observed peak in σ^{q-1}_{COINC}.

If the apparent background is subtracted from the experimental values, then the absolute magnitudes of the experimental and theoretical results agree to within a factor of two. Reasons for the discrepancy could be twofold: (1) the calculated dielectronic recombination cross sections[9] are too small and/or (2) important effects arise from the initial binding energies of the target electrons which are not accounted for in the calculations. These points need further study.

As previously noted the resonance interpretation of the peak in σ_{COINC}^{q-1} is strongly supported by the observation of the resonance in $\sigma_{K\alpha}$ and σ_{COINC}^{q+1}. The origin of the resonance effects in these latter cross sections is not obvious and requires further consideration. Since $\sigma_{K\alpha}$ is the total x-ray yield, a true resonance in σ_{COINC}^{q-1} would produce an equal strength resonance in $\sigma_{K\alpha}$. However, the magnitude of the resonance term obtained from the fit of Eq. (1) to $\sigma_{K\alpha}$ is about a factor of two larger than that for the σ_{COINC}^{q-1} fit as seen from Table I. This implies that other charge states in addition to q-1 must be contributing substantially to the resonance in $\sigma_{K\alpha}$. The fit of Eq. (1) to σ_{COINC}^{q+1} gives a very small resonance amplitude; in fact, only the interference term (in addition to the background) is significant in this case (see Table I). Thus, since all other out-going charge states (q-2, q+2, etc.) except q-1, q, and q+1 have negligible intensities, the additional intensity in the $\sigma_{K\alpha}$ reso-nance is attributed almost entirely to charge state q.

Table I. Resonant parameters obtained from fitting Eq. (1) to the data of Fig. 4 using E_R = 120 MeV and Γ = 60 MeV. The values of R and A are given in units of $10^{-20}cm^2$.

Cross Section	R	A
σ_{COINC}^{q-1}	2.8	0.02
σ_{COINC}^{q+1}	0.0	3.0
$\sigma_{K\alpha}$	6.2	1.5

A possible explanation for the appearance of resonant effects in the q and q+1 channels is the existence of strong coupling be-tween the three intense charge states (q-1, q, q+1). Such strong coupling between these various charge states is reasonable in view of the very high probability for charge exchange in the relatively close collisions expected for RTE. Since charge state q is related to q-1 by a single electron exchange, a resonance in q-1 could have a large effect in the outgoing q channel. Moreover, the fact that only a resonance interference term (term 3 in Eq. (1)) is observed for σ_{COINC}^{q+1} is consistent with this explanation since less probable two electron exchange is involved in the coupling of the resonant q-1 channel to the q+1 channel. Channel coupling effects such as these are well known in nuclear reactions. Further experimental and theoretical work is needed in order to confirm or reject this channel coupling possibility.

If the discussion above concerning channel coupling is assumed to be valid the calculation of Brandt for σ_{RTE} should be compared with the resonance strength in $\sigma_{K\alpha}$ instead of σ_{COINC}^{q-1}. In this case the theoretical intensity is about a factor of four lower than the experimental resonant intensity obtained for $\sigma_{K\alpha}$ as seen from Table I.

CONCLUSION

In conclusion, strong experimental evidence is presented for a new resonant process (RTE) in high energy ion-atom collisions. Additional support for the interpretation of this resonant process as the analog of dielectronic recombination is provided by comparison of the results with theoretical calculations.[4] The data indicates that RTE is very important in the production of inner-shell vacancies in the velocity regime investigated. To our knowledge, RTE has not been included in any theoretical description of inner-shell excitation/ionization in ion-atom collisions.

Furthermore, if the RTE interaction occurs primarily with the weakly bound target electrons for which the projectile velocity is much greater than the electron velocity then RTE may, in fact, approximate dielectronic recombination. Although the captured electron in RTE cannot be represented as a free electron in a rigorous physical description of the process, it may be that the initial binding of this electron does not significantly alter calculated values of dielectronic recombination cross sections such as those of Ref. 9. This point needs further theoretical study. The free electron treatment has been found to be highly successful in the case of radiative electron capture[11] in ion-atom collisions. Another well-known example is that of the Compton effect in which the target electrons are considered "free".

If the free electron approximation is valid for RTE, then this process can be used as a benchmark for theoretical calculations of dielectronic recombination cross sections. We emphasize, however, that such a close association between RTE and dielectronic recombination remains to be established.

It is planned to extend the present measurements to other target atoms, ionic species, and charge states. A particularly interesting target to investigate is He due to its simple electronic structure.

We gratefully acknowledge the assistance of Mr. Marc Soller in making these measurements. The BNL portion of this research was supported by the U. S. Department of Energy, Division of Basic Energy Sciences under Contract No. DE-AC02-76CH00016.

REFERENCES

1. J. A. Tanis, S. M. Shafroth, J. E. Willis, M. Clark, J. Swenson, E. N. Strait, and J. R. Mowat, Phys. Rev. Lett. 47, 828 (1981)
2. D. Brandt, M. Clark, T. McAbee, S. Shafroth, and J. S. Swenson, Bull. Am. Phys. Soc. 26, 1302 (1981).
3. A. L. Merts, R. D. Cowan, and N. H. Magee, Jr., Los Alamos Scientific Laboratory Report No. LA-6220-MS, 1976 (unpublished).
4. D. Brandt (to be published in Phys. Rev. A).
5. J. A. Tanis, E. M. Bernstein, M. P. Stöckli, W. G. Graham, K. H. Berkner, D. J. Markevich, R. H. McFarland, R. V. Pyle, J. W. Stearns, and J. E. Willis, Bull. Am. Phys. Soc. 26, 1304 (1981).
6. J. A. Tanis, E. M. Bernstein, W. G. Graham, M. Clark, S. M. Shafroth, B. M. Johnson, K. Jones, and M. Meron (submitted to Phys. Rev. Lett.).
7. P. Richard, P. Pepmiller, J. Newcomb, R. Dillingham, J. M. Hall, T. J. Gray, and M. Stöckli, Bull. Am. Phys. Soc. 27, 513 (1982)
8. M. Soga and E. M. Bernstein, Phys. Rev. C 13, 473 (1976).
9. D. J. McLaughlin and Y. Hahn, Phys. Lett. 88A, 394 (1982).
10. F. Biggs, L. B. Mendelsohn, and J. B. Mann, At. Data Nucl. Data Tables 16, 201 (1975).
11. J. A. Tanis, S. M. Shafroth, J. E. Willis, and J. R. Mowat, Phys. Rev. A23, 366 (1981)

158

ANGULAR DISTRIBUTION OF ELECTRONS AND X-RAYS FROM HIGH ENERGY ION-ATOM COLLISIONS

D. BERÉNYI

Institute of Nuclear Research of the Hungarian Acad. of Sci. (ATOMKI),
H-4001 Debrecen, Hungary

Recent, partly preliminary results on angular dependence of the broad peak near $v_e = v_i$ and Auger electrons in the spectrum of electrons emitted in high energy ion-atom collisions are given for different projectiles (H^+, H_2^+, He^+, Ne^{3+}, Ne^{10+}, with impact energies from 0.8 to 112 MeV) and targets (Ar, Ne), respectively. The angular distribution of L_ℓ X rays from H^+, He^+, C^+, N^+ and Ne^+ impact (ion energies from 0.2 to 18.2 MeV, respectively) on Au was studied and the alignment parameter determined.

INTRODUCTION

It is well-known that the study of the energy and the angular distribution, i.e. doubly differential cross-sections for electron and X-ray emission in high energy ion-atom collisions are very informative from the point of view of the research of the mechanism of the collision processes concerned. To continue the earlier investigations on ion-atom collisions in our institute (e.g. ref. [1-4]), various measurements were carried out recently to study the angular distributions of electrons and X-rays.

In the present report some of our new, yet unpublished, partly preliminary results will be given on the angular distribution of electrons and X-rays emitted in high energy ion-atom collisions.

STUDY OF ELECTRONS

Light ion projectiles

    To study the angular distribution of electrons from ion–atom collisions, a special triple-pass electrostatic electron spectrometer, ESA-21 (a combination of a spherical and a double-pass cylindrical mirror type spectrometer) was constructed in this institute (a detailed description will be published later[5]; preliminary publications on ESA-21 under ref. 6-9). By using this equipment, the electron spectra can be measured in thirteen angles at the same time from 0° to 180° with a good resolution (the best being 0.1 % in orders of magnitude).

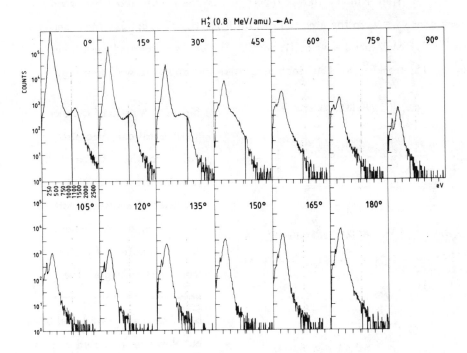

H_2^+ (0.8 MeV/amu) → Ar

Fig. 1.   A series of the spectra of electrons emitted in impact of H_2^+ on Ar taken by the triple-pass electrostatic electron spectrometer ESA-21, simultaneously. The scale is changed at 1100 eV on the X-axis.

The above equipment was used at first at the 5 MV Van de Graaff of our institute where ion–atom collisions with light ion impact (H^+, H_2^+, He^+) were studied. A series of experimental electron spectra (without any correction) at 13 emission angles relative to the direction of the ion beam from the accelerator is shown in Fig. 1. These spectra were measured by ESA-21 simultaneously.

Some results on Z^2 scaling and effective Z of the projectile in H_2^+, He^+-Ar collision have already been published[10]. This study was expanded for low energy range (down to 25 eV) of the electron spectra at a wider choice of impact energies of the ions at the 7 MV Van de Graaff of the Institute of Nuclear Physics of J.W. Goethe University, in collaboration[11].

Here I should like to report first on the results about the broad peak near $v_e = v_i$ in the electron spectra from H_2^+ (0.8; 2.0 MeV/amu), He^+ (0.8 MeV/amu) - Ar collisions[12]. The peaks concerned were taken at every $15°$ from $0°$ to $180°$ including these last angles as well (see Fig. 1).

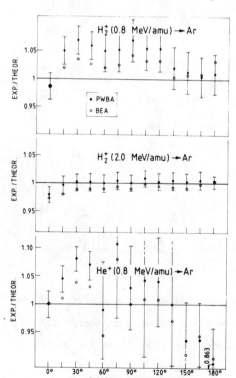

Fig. 2. A ratio of the experimental and theoretical values for the position of the peak near $v_e = v_i$ in the spectrum of emitted electrons at H_2^+ and He^+ impact on Ar as a function of the electron emission angle. The errors are about the same for both theories (PWBA and BEA) but they are indicated only at the values from PWBA.

After the necessary corrections (subtraction of the background and the continuous component, correction on the spectrometer efficiency, etc.) the peak position, the width (FWHM) and the shape as well as the intensity (summed the number of electrons in the peak - SDCS) was studied as a function of the emission angle.

As regards the position of the peaks in the case of the H_2^+ projectiles, the range of the variation is 30-40 eV at both energies in the whole angular region while for He^+ this range is nearly 100 eV. Fig. 2 shows the comparison of these values with BEA and PWBA calculations. As it can be seen, the agreement is better with BEA than with PWBA theory within the limit of the errors at 0.8 meV/amu impact energies (for both H_2^+ and He^+) but for the 2 MeV H_2^+ impact the PWBA seems to be somewhat better approach.

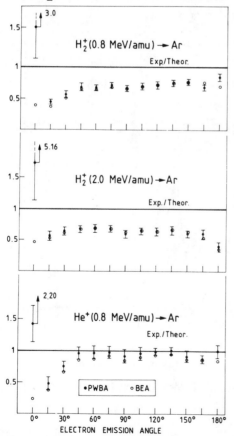

Fig. 3. A ratio of the experimental and theoretical (PWBA) values for the width of the peak near $v_e = v_i$ in the spectrum of emitted electrons at H_2^+ and He^+ impact on Ar as a function of the electron emission angle. (On the errors see the figure caption at Fig. 2).

However the agreement both with BEA and with PWBA, on the width (FWHM) of the peak is very bad for H_2^+ at both impact energies, while for He^+ it is rather good (PWBA is somewhat better), except the lower angles. A comparison of the experimental data and calculations is given in Fig. 3. The disagreements here for H_2^+ projectiles are probably due to the insufficient approach on the molecular ion in the calculations. The relatively great deviation at 0° claims further analysis and corrections.

Various comparisons on the shape of the peak were carried out in the case of different projectiles at the same angle and at different angles and projectiles with theory. Shortly, one can say that both the BEA and the PWBA give the experimental shape approximately well (the difference here between the two approaches is not substantial in general). However, the agreement between experiment and theory is better at larger angles.

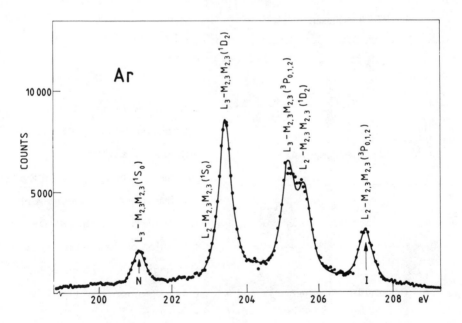

Fig. 4. The $L_{2,3}M_{2,3}$ Auger spectrum of Ar taken in the present measurements by ESA-21 spectrometer at 3 MeV proton impact on Ar gas target

By summing up the electrons in the peak, the single differential cross-section (SDCS) is determined. The experimental angular distribution pattern seems to show a more intensively increasing tendency at larger angles than predicted by the theories.

Similar studies were carried out by Menendez and Duncan first of all for H and H^- and partly for H_2^+ and He^+ projectiles[13-16] but a systematic study in the whole angular region was missing up till now.

It is well known that the number of measurements for the angular

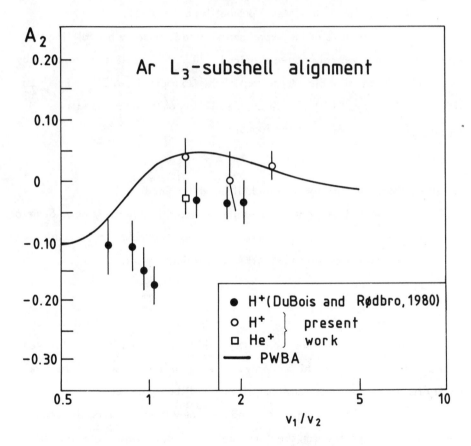

Fig. 5. The A_2 alignment parameter from the present study and that according to DuBois and Rødbro[18] as a function of v_1/v_2 where v_1 is the impact ion velocity and v_2 is the velocity of the 2p electron.

distribution of Auger electrons especially from ion-atom collisions is rather few. The non-isotropic angular distribution of Auger electrons following inner – shell ionization by electron impact was first proved by Cleff and Mehlhorn[17] for the $L_3 - M_{2,3}^2 (^1S_o)$ state of Ar. Recently DuBois and Rødbro received similar results at 250 – 2000 keV proton impact on argon[18].

The $L_{2,3}M_{2,3}$ Auger spectrum of Ar taken in our measurement[19] with 3 MeV proton projectiles is shown in Fig. 4. Spectra were recorded at 13 angles from $0°$ to $180°$ for H^+ (0.8; 0.9; 1.0; 1.6; 3.0 Mev), H_2^+ (1.6 MeV) and He^+ (3.2 MeV) projectiles by ESA-21.

The analysis of the experimental data is made by normalizing the intensity of the expectedly non-isotropic $L_3-M_{2,3}^2 (^1S_o)$ line (indicated by "N" in the spectrum in Fig. 4) to that of the isotropic $L_2-M_{2,3}^2 (^3P_{0,1,2})$ line (indicated by "I"). By chi-squared fitting of the normalized $L_3-M_{2,3}^2 (^1S_o)$ intensities to the formula

$$w (\vartheta) = A_o \left[1 + A_2 P_2 (\cos \vartheta) \right] ,$$

A_2 alignment parameters for L_3-subshell are determined.

The evaluation of the experimental data is under way now. Some of our preliminary data together with those of DuBois and Rødbro for A_2 are given in Fig. 5 in comparison with the PWBA theoretical values.

Heavy ion projectiles

The ESA-21 electron spectrometer, after having used it at the Van de Graaff of our institute, was transported to the Dubna Institute last year to study the electrons from high energy ion-atom collisions. Some of the preliminary results of the measurements in Dubna will be given here.

First of all we have obtained some new information on the peak near $v_e = v_i$ in the electron spectrum. Sellin and coworkers[20] carried out a series of fine work on $v_e = v_i$ peak at $0°$ in the case of various high energy heavy ion projectiles in the last years. Stolterfoht and coworkers[21-23], on the other hand, studied this peak from $20°$ to $150°$.

In Fig. 6 the shape of the peak in question is compared for 112 MeV Ne^{10+} projectiles (Ne target)[24] at $0°$ of the electron emission to the direction of the impact ion beam. The shape of the " cusp" (for Ne^{10+}) and that of the electron loss peak (for Ne^{3+}) is similar to those by Sellin and coworkers [25] obtained by 40 MeV 0^{8+} and 0^{3+} impact on Ar.

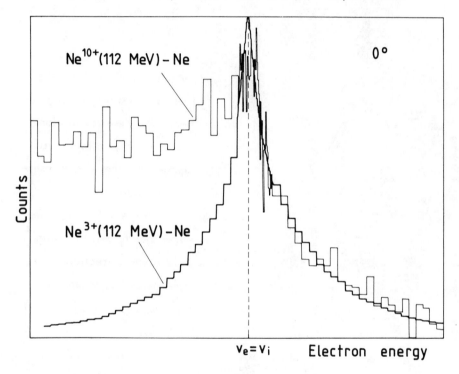

Ne^{10+}(112 MeV) – Ne

$0°$

Ne^{3+}(112 MeV) – Ne

Counts

$v_e = v_i$ Electron energy

Fig. 6. A comparison of the shape of the peak near $v_e = v_i$ in the forward direction for 112 MeV Ne^{10+} and Ne^{3+} impact on Ne. (Preliminary results[24]).

As regards the angular dependence of the EL peak, i.e. the peak in the case of N^{3+} impact (here the contribution of " cusp " mechanism is negligible even at $0°$ [26]) at the lowest ($0°$, $15°$, $45°$) and the highest ($165°$, $180°$) angles, the position of the peak is in good agreement with PWBA (scaled hydrogen wavefunctions) calculations except $165°$ and to a lesser extent $0°$. In these preliminary measurements, however, the width of

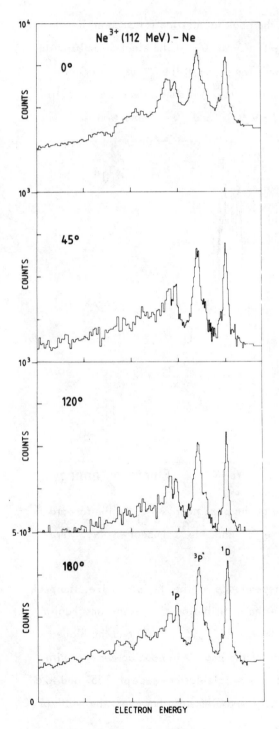

Fig. 7. Auger spectra (row data) from Ne^{3+} (112 MeV) – Ne collisions[24].

the peak (FWHM) and the shape shows a larger disagreement with these calculations but we have good reason to suppose that they are of instrumental origin.

In the following, some Auger spectra will be shown[24] without any correction from high energy heavy ion-atom collision measured by ESA-21 in Dubna.

In Fig. 7 the Auger spectrum from 112 MeV Ne^{3+} on Ne collision is shown at four different angles of elec - tron emission including 0° and 180° . Corresponding da- ta are available at further 9 angles. Some tentative assign- ments are indicated. One can see some differences in the spectra at the various angles

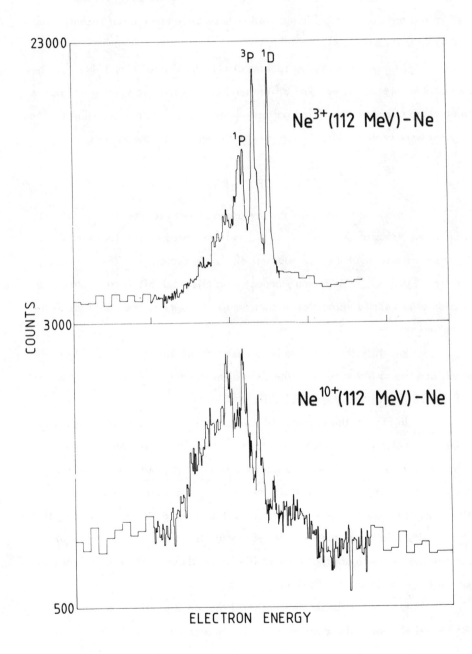

Fig. 8. A comparison of the row Auger spectra from Ne^{3+} (112 MeV)-Ne, respectively.

partly certainly of instrumental origin. The evaluation is going on and the investigations are continued. Similar studies have been carried out recently by Stolterfoht and coworkers[21, 23, 27].

In Fig. 8 the Auger spectra at 112 MeV Ne^{3+} and Ne^{10+} impact on Ne are compared. The difference between the two spectra and so the strong charge dependence of the pattern of the Auger spectrum is striking. These diagrams were made by summing up all the thirteen angular channels.

STUDY OF X-RAYS

In our institute there are some experiences on the study of the angular distributions of X-rays[28-30]. It has been stated earlier that the $\mathcal{A}_2$ alignment parameter of the L_3 subshell of Au in proton (0.25-1.5 MeV) and He^+ (0.2 - 1.1 MeV/amu) impact ionization exhibits a minimum as a function of projectile velocities in contrast to the formerly predicted theoretical behaviour[29].

Recently these studies have been continuing at our 5 MV Van de Graaff and the 5 MV tandem of the Zentralinstitut für Kernforschung, Rossendorf (GDR) with heavy ion projectiles.

In Fig. 9 the recent, till now unpublished data taken at our Van de Graaff[31] and at the Rossendorf tandem[32] in collaboration are shown together with the similar values of Jitschin et al.[33]. It should be emphasized here that the isotropy of the intensity of the L_γ line used as a reference line in the earlier measurements was checked in this investigation by measuring the L_γ intensity relative to the intensity of the scattered ions at a given angle. One can see in the figure that none of the known theoretical calculations can explain the experimental behaviour.

Finally, it should be mentioned that cross-sections for the different L-subshells were also determined and compared with calculations in this study.

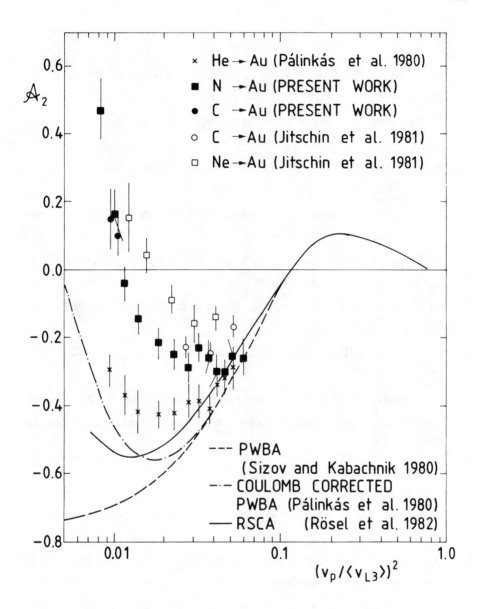

169

Fig. 9. The A_2 alignment parameter for L_3-subshell of gold at heavy ion impact ionization. Here v_p is the ion velocity and v_{L3} is the average velocity of the L_3 electrons. The experimental values partly from the present study, partly from Pálinkás et al.[29], Jitschin et al.[33] and the theoretical curves from Sizov and Kabachnik[34], Pálinkás et al.[29], Rösel et al[35].

REFERENCES

1. E. Koltay, D. Berényi, I. Kiss, S. Ricz, G. Hock and J. Bacsó, Z. Physik A278 , 299 (1976)

2. T. Mukoyama, L. Sarkadi, D. Berényi and E. Koltay, Phys. Letters 67A, 180 (1978)

3. Á. Kövér, S. Ricz, Gy. Szabó, D. Berényi, E. Koltay and J. Végh, Phys. Letters 79A, 305 (1980)

4. T. Mukoyama, L. Sarkadi, D. Berényi and E. Koltay, J. Phys. B : Atom. Molec. Phys. 13, 2773 (1980)

5. D. Varga, J. Kádár, S. Ricz, J. Végh, Á. Kövér, to be published

6. D. Varga and J. Végh, ATOMKI Közl. 22, 15 (1980)

7. D. Varga, J. Végh, Á. Kövér, S. Ricz and A. Domonyi, ATOMKI Közl. 23, 40 (1981)

8. I. Kádár, B. Sulik, I. Cserny, T. Lakatos and J. Végh, ATOMKI Közl. 23, 42 (1981)

9. D. Berényi in " High Energy Ion-Atom Collisions " (ed. D. Berényi, G. Hock). Akadémiai Kiadó, Budapest - Elsevier Sci. Publ. Co., Amsterdam, 1982. p. 131)

10. A. Kövér, Gy. Szabó, D. Berényi, D. Varga, I. Kádár, S. Ricz and J. Végh, Phys. Lett. 89A, 71 (1982)

11. J. Schader, R. Lotz, D. Berényi, H. J. Frischkorn, K. O. Groeneveld, P. Koschar, A. Kövér and Gy. Szabó, to be published

12. A. Kövér, D. Varga, Gy. Szabó, D. Berényi, I. Kádár, S. Ricz and J. Végh, in course of publication : J. of Phys. B : Atom. Molec. Phys.

13. M. G. Menendez and M. M. Duncan, Phys. Rev. Lett. 40, 1642 (1978)

14. M. M. Duncan and M. G. Menendez, Phys. Rev. A19, 49 (1979)

15. M. G. Menendez and M. M. Duncan, Phys. Rev. A20, 2327 (1979)

16. M. M. Duncan and M. G. Menendez, Phys. Rev. A23, 1085 (1981)

17. B. Cleff and W. Mehlhorn, J. Phys. B : Atom. Molec. Phys. 7, 605 (1974)

18. R. DuBois and M. Rødbro, J. Phys. B : Atom. Molec. Phys. 13, 3739(1980)

19. D. Varga, I. Kádár, D. Berényi, S. Ricz, J. Végh, L. Sarkadi and Á. Kövér, to be published

20. M. Breinig, S.B. Elston, S. Huldt, L. Liljeby, C.R. Vane, S.D. Berry, G.A. Glass, M. Schauer, I.A. Sellin, G.D. Alton, S. Datz, S. Overbury, R. Laubert and M. Suter, Phys. Rev. A 25, 3015 (1982), and references contained therein

21. M. Prost, Doctoral dissertation, Die Freie Universität, Berlin, 1980

22. D. Schneider, M. Prost and N. Stolterfoht, Book of Abstracts, XII. Int. Conf. on Phys. Electronic and Atomic Coll. (IDPEAC). Gatlinburg, 1981. p. 806

23. M. Prost, D. Schneider, R. DuBois, P. Ziem, H.C. Werner and N. Stolterfoht, in " High Energy Ion-Atom Collisions " (ed. D. Berényi, G. Hock). Akadémiai Kiadó, Budapest - Elsevier Sci. Publ. Co., Amsterdam, 1982. p. 99

24. D. Varga, I. Kádár, S. Ricz, J. Végh, B. Sulik and D. Berényi, to be published

25. M. Breinig, M.M. Schauer, I.A. Sellin, S.B. Elston, C.R. Vane, R. S. Thoe and M. Suter, J. Phys. B : Atom. Mol. Phys. 14, L291 (1981)

26. C.R. Vane, IEEE Trans. on Nucl. Sci. NS-26, 1078 (1979)

27. D. Schneider, M. Prost, B. DuBois and N. Stolterfoht, Phys. Rev. A25, 3102 (1982)

28. J. Pálinkás, B. Schlenk and A. Valek, J. Phys. B : Atom. Molec. Phys. 12, 3273 (1979)

29. J. Pálinkás, L. Sarkadi and B. Schlenk, J. Phys. B : Atom. Molec. Phys. 13, 3829 (1980)

30. J. Pálinkás, B. Schlenk and A. Valek, J. Phys. B : Atom. Molec. Phys. 14, 1157 (1981)

31. J. Pálinkás, L. Sarkadi, B. Schlenk, I. Török and Gy. Kálmán, J. Phys. B : Atom. Molec. Phys., in course of publication

32. J. Pálinkás, L. Sarkadi, B. Schlenk, I. Török, Gy. Kálmán, C. Bauer,

172

K. Brankhoff, D. Gramhole, C. Heiser, W. Rudolph and H.J. Thomas, to be published

33. W. Jitschin, A. Kaschuba, R. Hippler, L. Sarkadi, H. Kleinhoppen and H. Lutz, Book of Abstracts, Int. Conf. on Phys. Electronic and Atomic Coll. (ICPEAC). Gatlinburg, 1981. p. 843

34. V.V. Sisov and N.M. Kabachnik, J. Phys. B : Atom. Molec. Phys. 13, 1601 (1980)

35. F. Rösel, D. Trautmann and G. Baur, Z. Phys. A, Atoms and Nuclei 304, 75 (1982)

IN SEARCH OF SPONTANEOUS POSITRON CREATION †

Jack S. Greenberg *
Yale University, Wright Nuclear Structure Laboratory
New Haven, Connecticut 06511 USA

ABSTRACT

We review the recent experiments investigating the behavior of the electron-positron field in strong external electromagnetic fields. In particular, the search for spontaneous positron creation in superheavy collision systems is discussed. A striking new result emerging from this search is the observation of narrow peaked structures in the positron spectra whose origin cannot be associated with the established dynamic mechanisms of positron production involving Coulomb trajectories only. The possibility that these peaks may be signatures for the spontaneous decay of the vacuum and for the formation of metastable giant nuclear complexes is considered in the context of recent experimental results.

INTRODUCTION

On the scale usually associated with the energy and momentum transfer encountered in classical atomic phenomena, the collision of two massive Uranium atoms with GeV kinetic energy has to be considered a very violent event. There is a deep interpenetration of the electronic structure accompanied by extensive excitations, the emission of a multitude of electron and X-rays, and even the nuclear surfaces may overlap leading to nuclear reactions. Such encounters, therefore, are apparently quite complex. Naively, they would not be expected to provide the most accommodating conditions to obtain the detailed information needed for exploring fundamental aspects of atomic structure and quantum electrodynamics. Yet recent theoretical and experimental studies have demonstrated that the opposite situation may prevail.

We find that for specific phenomena involving the inner shells and the spacial environment in the immediate vicinity of the two nuclear charges when they are close together, these collision systems provide a unique source of information. In fact, the violence of the collision and the extreme conditions imposed on the electron, or more precisely the electron-positron field, can be exploited as sources of interesting new physics which will be the subject of my talk.

*The recent experiments on positron production discussed in this report were carried out in collaboration with H. Bokemeyer, K. Bethge, H. Folger, H. Grein, A. Gruppe, S. Ito, R. Schulé, D. Schwalm, J. Schweppe, N. Trautmann, P. Vincent and M. Waldschmidt.

Our main focus is on the extreme fields that penetrate the space around the nuclei and on the giant quasi-atoms formed transiently in the collision that can simulate the large nuclear charge needed to produce the strong fields. Our special interest in these strong fields is connected with the opportunity to observe for the first time a general property of the vacuum associated with quantum field theory; the theory allows for the spontaneous creation of real particles in strong <u>static</u> external fields. Thus when conditions are suitable for this process to occur, the normal vacuum state is unstable and decays into a new state that contains real particles. This process may occur for any elementary particle field, but it is difficult to envision achieving the necessary physical situations except for the electron-positron field where the formation of superheavy quasiatoms in heavy-ion collisions may offer the only possibility presently. In the latter case, as we shall see, the instability of the vacuum appears as the spontaneous emission of positrons which, in principle, can be readily detected.

My discussion today will center on the experimental effort to observe this fundamental process. In more specific terms it will be structured around the following questions:

1) Can the binding energy of an electron exceed the supercritical value of twice the electron rest mass, and does such supercritical binding exist in atoms?

2) Can spectroscopic information on the structure of superheavy atomic species be extracted from studying superheavy collision systems?

3) In particular, is it possible to utilize collision systems such as U + U and U + Cm, where supercritical binding is expected to occur, to observe the spontaneous decay of the supercritically bound state?

4) An intriguing question which has emerged from the most recent data concerns the possibility of exploiting atomic processes, such as positron creation, as probes of exotic nuclear properties such as the formation of superheavy nuclear species. (This interplay between atomic and nuclear physics is becoming a recurrent theme in heavy-ion physics and is discussed at this conference by E. Merzbacher and W. E. Meyerhof.)

The experimental picture pertinent to answering these questions is yet to be completed. But a sufficient level of achievement has been reached to indicate that some exciting results may not only be emerging on the electrodynamic aspects of the questions, but also on nuclear physics. Among these is the possibility that the positron spectrum may indeed provide evidence for the formation of superheavy metastable nuclear complexes with $Z > 180$ and that it may be utilized to probe the properties of such systems.

THE UNSTABLE VACUUM

Let us first consider the conditions that lead to an unstable vacuum state and see how these conditions can be simulated in quasiatoms.

If we venture well beyond the atomic species, which are ordinarily accessible to study, to a part of the periodic table where $Z\alpha > 1$, some very unusual properties are predicted by theory [1] which are quite foreign to ordinary atomic species. The superheavy atom is an interesting object.

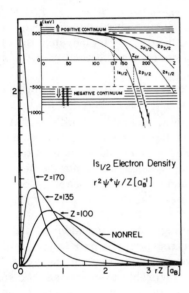

Fig. 1. The radial electron density of the $1s_{\frac{1}{2}}$ state for Z = 100, 135, and 170 together with the non-relativistic Schrodinger wavefunction. [3] The axes are scaled with powers of Z. The insert shows energy levels vs. Z.

For illustration, Figure 1 shows that for a nuclear charge center with finite dimensions, the K-electron binding energy increases very rapidly with Z as we pass the sacred $Z\alpha = 1$ boundary. The binding energy of the electron soon exceeds its own rest mass at Z ∼ 150 and plunges even further towards $-mc^2$ as the charge is increased. The relativisitc effects which are responsible for this property also produce some drastic further modifications of the energy level structure: The fine structure splitting of the $2p_{3/2} - 2p_{1/2}$ states becomes very pronounced achieving a value of ∼ mc^2, while the $2p_{1/2}$ and $2s_{1/2}$ energy levels interchange and the close degeneracy observed in light atoms is severely violated. We see that the energy level structure of a superheavy atom looks very unlike that of a hydrogen-like atom.

The relativistic effects that dominate the energy level structure of superheavy atoms also profoundly modify their wavefunctions. Figure 1 [3] illustrates that the $1s_{\frac{1}{2}}$ wavefunction shrinks to small dimensions, well within the Compton wavelength, for $Z\alpha \gtrsim 1$. As a consequence, it becomes very sensitive to the nuclear charge dimensions. Figure 1 also compares this severe contraction to the non-relativistic density. For orientation it is interesting to note that $< r >_{1s}$ for a Z = 170 is ∼ 120fm while for Pb $< r >_{1s}$ is ∼750fm.

More dramatic is the prediction [3] that for Z = 170 the relativistically calculated density at the nucleus exceeds the corresponding non-relativistic value by more than 3 orders of magnitude. The latter property clearly illustrates the unusual nature of the electronic structure of superheavy atoms, and it is this behavior of the wavefunctions that is responsible for many of the effects that are encountered in superheavy quasiatoms.

But probably the most novel feature of superheavy atoms occurs when the binding energy reaches and exceeds $2mc^2$. The required "critical" charge, Z_{cr}, is predicted to be ~ 173 for 1s states. [4] Although a complete field theoretic treatment of the many-body problem is required to describe the physics as we cross this boundary, [5-8] the physical significance of reaching and passing the critical charge can be readily envisioned qualitatively.

With finite size nuclei there are no undue difficulties in tracing the energy levels, as Z increases, to the value Z_{cr} where they encounter the negative energy continuum. At this threshold, theory [9-11] tells us that the bound 1s state ceases to exist as it joins the negative energy continuum, but instead develops into a resonance shared over the continuum states. The width of this resonance grows with deeper penetration into the negative energy continuum as Z increases beyond Z_{cr}. The appearance of this resonance is the crucial aspect of the physics since it marks a decaying state and thus an instability.

The physical process that follows is suggested by considering the smallest energy required to create an electron-positron pair in the vicinity of such an overcritically charged nucleus. As the binding energy of $2mc^2$ is reached, it becomes energetically favorable to create an electron-positron pair if the 1s state is unoccupied by an electron. The spontaneous emission of such a pair is not forbidden by any conservation law. A positron is emitted and an electron remains bound under supercritical conditions.

Thus, as depicted on Figure 2, after K-shell ionization an atom with Z > 173 always will spontaneously shield itself with two electrons (real vacuum polarization charge) while emitting positrons with a rather well defined kinetic energy centered at the binding energy less $2mc^2$. This two-electron state becomes the lowest energetically stable state for this atom. Calculations show that it forms on a time scale of $\sim 10^{-19}$ sec, corresponding to the width of the resonance state referred to above. [9-11] As the central charge is increased arbitrarily--so that successive bound states $1s_{1/2}$, $2p_{1/2}$, $2s_{1/2}$ join the negative energy continuum--successive phase transitions occur in which the vacuum increases its negative charge. The vacuum thus sparks in supercritical fields and becomes a "charged vacuum." [10-13] Clearly the "charged vacuum" is a new ground state in supercritical potentials.

Calculations of the interaction of the electron with the radiation field

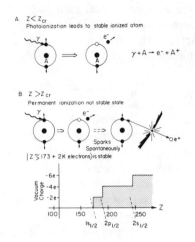

A $Z < Z_{cr}$
Photoionization leads to stable ionized atom

$\gamma + A \rightarrow e^- + A^+$

B $Z > Z_{cr}$
Permanent ionization not stable state

Sparks
Spontaneously

$|Z \gtrsim 173 + 2K$ electrons$)$ is stable

Fig. 2. For $Z > Z_{cr}$ an ionized $1s_{\frac{1}{2}}$ state is unstable to spontaneous positron emission. As successive bound states join the negative energy continuum, the vacuum increases its negative charge.

do not seem to alter these predictions. B. Müller at this conference will discuss recent calculations which show that the self energy decreases the binding by only ~ 11 keV [20] for $Z = 170$, while vacuum polarization enhances the binding of the 1s electron.

Spontaneous positron emission is, therefore, a unique signature for establishing the change in the ground state of QED as the supercritical field threshold is passed. As previously mentioned, it is the search for this process and its theoretical implications that has motivated much of the experimental interest which I am discussing today.

It bears mentioning that, in principle, the atomic domain is not necessarily the only environment where we can seek this effect. Potentials greater than $2mc^2/e$ extended over space can continuously produce pairs. [14, 15] However, with the requirement that this potential difference has to be achieved over approximately a Compton wavelength in order for the process not to be suppressed exponentially, [16] (a critical electric field of ~ 1.3 keV/fm is needed) "sparking" is not very likely to be observed at macroscopic potentials generated in the laboratory.

Of course, the atom with a supercritical charge > 173 is not readily available either. (The promised islands of stability for superheavy systems have turned to shifting sands.) Fortunately, the existence of modern heavy-ion accelerators may supply us with a suitable vehicle to pursue the question of supercritical binding. Although a semi-stable supercritical charge has been denied us so far, nature can be circumvented to some extent by probing into the electron structure of heavy-ion collision systems. Super-heavy atoms seem to be more accessible than their nuclear counterparts, but only for a fleeting moment. The duration of this moment turns out to be an important consideration.

OVERCRITICAL FIELDS IN HEAVY-ION COLLISIONS

The essential idea that has presented a serious experimental challenge to look for spontaneous positron emission is the suggestion [17, 18] that heavy-ion collisions can be used to assemble a supercritical charge for a short interval of time. A quasi-atom is formed with an effective charge equal to the combined Z of the colliding nuclei. The basis of this idea is in the disparity between the nuclear collision velocity and the velocities of the orbiting electrons.

As an example, let us consider $^{238}U_{92} + {}^{238}U_{92}$ collisions near the Coulomb barrier. The relative velocity required to bring the two nuclei into contact is approximately $0.1c$ corresponding to a bombarding energy of ~ 1.4 GeV. The distance of closest approach for a head-on collision $2a = 17.5$ fm, $< r_k >_u \simeq 700$ fm and $< r_k >_{Z = 184} \simeq 100$ fm. The collision trajectories can clearly be treated classically since the Sommerfeld parameter $\eta = a/\lambda \simeq 500$. Thus, for the inner-shell fast moving electrons, when velocities are close to c, the Coulomb potential of the colliding nuclei varies sufficiently slowly so that the electrons can adjust adiabatically. At small internuclear separations, well within the orbiting radii, the electrons cannot distinguish between the two nuclear centers and act as if they were bound by the combined $Z = 184$ of the two nuclear charges. Asymptotically the quantum numbers are, therefore, those of this superheavy quasiatom. In the intermediate region, between the very distant and very close internuclear separations, the electrons evolve through a series of quasimolecular states in the two-center field as the internuclear distance decreases and subsequently increases again in the collision.

Figure 3 [19] illustrates more specifically how the most bound $1s\sigma$ orbital, of particular interest to us, varies in binding energy and mean square radius as a function of united atom Z and internuclear separation. In evidence are the two prominent extreme-relativistic properties which are characteristic of the superheavy systems. [5-8] The binding energies increase rapidly with decreasing internuclear separation; the rate of increase grows with Z, and for the largest Z values the binding energies exceed $2mc^2$. This rapid energy variation is accompanied by the typical relativistic collapse of the wavefunction about the two nuclear centers, as we observed is the case of stable atoms. As we shall see, it is these two properties that focus all affects at the turning points of the heavy-ion trajectory so that they especially probe the properties of the quasiatom in the united atom limit and effectively exclude complications that could be introduced by the outer electronic structure.

It can be readily visualized how the formation of superheavy quasi-molecules can simulate the conditions required to observe spontaneous positron emission. The process is sketched schematically on Figure 4, which provides a representation of the time evolution of a collision, the associated energy level structure and the atomic excitation and deexcitation processes. If a vacancy is produced in the $1s\sigma$ molecular orbital by ionization at a suitable prior time, spontaneous positron emission can occur during the fraction of the collision time when the $1s\sigma$ binding is supercritical.

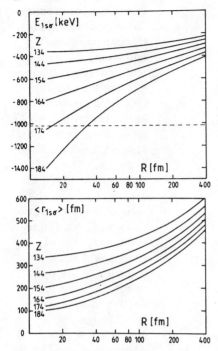

Fig. 3. Binding energies and radial expectation values $<r_{1s\sigma}>$ vs. internuclear separation R. [19]

It would seem that if this were the only process involved and if the concept of superheavy quasiatoms had some validity, demonstrating spontaneous positron emission should not be too difficult a task. A simple Gedanken-experiment would involve the selection of collision trajectories and the total Z of the collision system so as to proceed systematically from a situation of subcritical to supercritical binding. Some sort of threshold behavior would be expected for the positron production. However, contributing to this very simple scenerio are several important ingredients which play a central role in affecting the detection of spontaneously emitted positrons.

One of these concerns the crucial first step of vacancy formation which triggers the vacuum breakdown. Not only does the total probability for $1s\sigma$ vacancy formation have to be large, but it has to be concentrated at the close collisions which lead to supercritical binding. Another important factor is the short interval of time spent when the internuclear separation is less than R_{cr}. A typical time of $\sim 2 \times 10^{-21}$ sec noted earlier has to be compared with the much slower process of spontaneous decay of the $1s\sigma$ resonance ($> 10^{-19}$ sec). We see that there is little time in a Coulomb collision for the positron to be emitted from the filling of the $1s\sigma$ vacancy.

But equally significant is the consideration that the short collision time implies high Fourier frequencies associated with the collision which can lift deeply bound electrons into empty bound states and into the continuum. The nuclear motion that provides the $1s\sigma$ vacancies (ionization) therefore also leads to additional electron-positron pair production processes by the time-dependence of the electromagnetic potentials in the collision. Moreover, these processes are highly enhanced in the high Z systems.

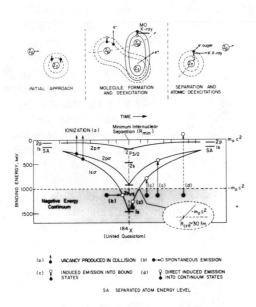

Fig. 4. Schematic presentation of the time evolution of quasimolecular orbitals in U + U collisions showing the atomic excitation and deexcitation processes. Positron production mechanisms in heavy-ion collisions are denoted by processes "b", "c", and "d".

As depicted schematically in Figure 4, in addition to process "b" representing spontaneous positron emission, the dynamical process "c" and "d" can also contribute to the positron spectrum from both subcritical as well as supercritical collision systems. The matrix elements involved are similar to those responsible for the ionization of bound electrons and, indeed, we shall see that the behavior of the dynamic positron production much resembles the ionization process. Process "c" occurs in second order and is very similar to spontaneous positron emission, but can contribute with or without supercritical binding. Process "d" is a direct ionization to the continuum. Although not our central focus, both processes are inherently interesting since they provide an opportunity to study both bound and highly distorted continuum electronic states in very-strong electromagnetic fields when perturbation theory is not applicable. They can be studied in the absence of spontaneous emission in subcritical collision systems. It bears emphasis that "b", "c", and "d" are all coherent so that on a given Rutherford trajectory it is not possible to isolate them individually.

Without having to consider details, which we defer to the later discussion, it is clear that dynamic effects introduce major complications into the search for spontaneous positron emission from quasiatoms. The large $1s\sigma$ vacancy probabilities which are needed also simultaneously imply large dynamic amplitudes for positron production which can overwhelm the spontaneous process. Any threshold effect for the latter process, therefore, can be obscured. In fact, in a more general sense, we can conclude that short collision times obviate any sudden threshold detection and exclude the appearance of sharp structures in the positron spectra; several hundred keV distributions are associated with collision times of $\sim 2 \times 10^{-21}$ sec.

In addition to these complicating effects from atomic processes, nuclear excitations in the collision from multiple Coulomb excitation and nuclear reactions can produce positrons from the internal pair conversion of γ-ray transitions above 1.02 MeV. This constitutes an important background and, of course, must be dealt with before confronting experiment with theory.

The quasiatom, therefore, may be the only vehicle available for our search, but it is certainly far from an ideal one if the time evolution cannot be altered from that associated with a pure Coulomb trajectory. We will return to this important point again.

But, from our present discussion, it is evident that understanding the dynamic effects in detail constitutes a critical aspect of utilizing heavy-ion collisions in study of overcritical binding phenomena. They enter in the excitation processes of ionization and delta electron emission, and they are also a prominent feature of the deexcitation modes, such as positron emission and molecular orbital x-ray emission.

ELECTRONIC EXCITATIONS IN QUASIATOMS

The experimental approaches to studying superheavy quasiatoms has revolved about studying the dynamic effects involving the excitation and deexcitation mechanisms mentioned. Since April 1976 when the first Uranium beams were accelerated at the UNILAC at Darmstadt to energies sufficient to reach the Coulomb barrier for the heaviest nuclear species, a large amount of data has been accumulated which provides evidence for the formation of quasiatoms in heavy-ion collisions and information on the behavior of the electronic structure.[5] Because of their importance to the positron search experiments, let us consider briefly some aspects of the ionization process. They provide a good illustration of the essential physics which governs the behavior of not only the ionization process but all of the dynamic effects associated with the Coulomb excitation process, such as "c" and "d" in Figure 4.

In the molecular description of the excitation processes, the time variation of the initial quasimolecular states due to the radial variation and rotation of the internuclear axis induces the excitation and ionization of the electrons. The two types of excitations display different radial dependence and different selection rules. For superheavy systems radial coupling becomes the dominant term for the $s_{\frac{1}{2}}$ and $p_{\frac{1}{2}}$ states in the last 100fm of approach (the vicinity of penetration of the levels into the negative energy continuum). In fact, it turns out that this is the region where electrons are kicked out and where the positrons are dynamically produced,

182

reflecting the new features of superheavy systems that are not found in light systems.

We can anticipate this behavior by considering the extreme relativistic features of superheavy systems I referred to earlier. The rapid increase of the binding energy at small internuclear separations, as well as the severe shrinkage of the electron distribution about the nuclear center of the combined system, lead to several important consequences. As we illustrate in Figure 5 schematically, taken from the calculations of the Frankfurt group, [23, 24] the radial matrix elements become large and dominate at small internuclear separations. The ionization is concentrated at the distance of minimum approach since it is here the states are changing so rapidly that the electron cannot keep up. The main results that emerge from such considerations are large cross sections for $1s\sigma_{\frac{1}{2}}$ and $2p\sigma_{\frac{1}{2}}$ excitation and a rapid falloff with impact parameter b. [23-25]

The large cross sections reflect the very large enhancement of the inner-shell electron wavefunction at small distances about the two nuclear centers. We saw previously in Figure 1 that the electron density for a combined $Z = 170$ could be as much as a factor of 10^3 larger than expected for an equivalent system with the relativistic contraction absent.

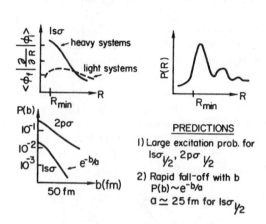

PREDICTIONS

1) Large excitation prob. for $1s\sigma_{\frac{1}{2}}$, $2p\sigma_{\frac{1}{2}}$

2) Rapid fall-off with b
$P(b) \sim e^{-b/a}$
$a \simeq 25\,\text{fm}$ for $1s\sigma_{\frac{1}{2}}$

Fig. 5. Sketch of the behavior of the radial matrix elements and the excitation probability, P(b), with internuclear separation R and impact parameter b for the ionization of inner-shells of superheavy collision systems. [23,24]

The physical origin of the strong localization of the excitation probability is readily apparent from considering the kinematics of the collision. To ionize the strongly bound states ($E_B \sim 1$ MeV) of the quasimolecule (or quasiatom in the limit) requires a large energy transfer, ΔE, and therefore a large minimum momentum transfer $q_{min} = \Delta E/\hbar v$ in a collision with a velocity v. This sets the scale for spacial confinement of the process to an average impact parameter $\bar{b}$ determined by $\bar{b}q_{min} \sim 1$, as suggested, for example, by the Bang-Hansteen scaling rule for Coulomb ionization. [26] Taking $\Delta E \sim 1$ MeV and $v/c = .1$, yields $\bar{b} \simeq 20\,\text{fm}$.

The important point to note is that the observation of such a rapid falloff with b (or equivalently R_{min}) can only be a signature for the participation of states in the ionization process that are associated with the quasiatom and not with the individual collision partners. High momentum transfer implies a large initial momentum for the ejected electron. Such large momenta can only be found in the high momentum components of a wavefunction which is strongly confined in space. A sufficient degree of confinement is possible for the electrons around the combined supercharge of the quasiatom for $\sim 10^{-20}$ sec, but not for the electrons in the inner-shells of the colliding atoms. As a specific example, consider that for a Pb + U quasiatom ($Z_1 + Z_2 = 174$), $< r_{1s\sigma} > \simeq 120$fm, the momentum transfer required for ionization is ~ 5 MeV/c, while the mean momentum for a 1s electron in Pb is ~ 0.5 MeV/c.

These features of the inner-shell ionization in superheavy systems are explicitly displayed in a scaling law derived from an analytic model by Müller et al.[27] and developed further by Bosch et al.[28] (The Bang-Hansteen scaling law is closely related.[26]) It is based on first order perturbation theory and makes use of a number of approximations relying on the concentration of the ionization near the turning point of the collision. Although we know that multi-step processes[29,30] play an essential role in the ionization and cannot be neglected, the scaling law, nevertheless, provides a very useful guide for anticipating the behavior of Coulomb ionization on the two important quantities--the energy transfer and the electron density at the nucleus. In the scaling law written as

$$P_{1s\sigma} = F(Z)e^{-\gamma R_{min} \Delta E(R_{min}, Z)/\hbar v}$$

$F(Z)$ reflects primarily the effects of the concentration of the electron density and increases with Z as the coupling to the continuum increases, while the energy transfer depends on the binding energy at R_{min}, where it is assumed the ionization takes place. The competition between an increasing $F(Z)$ and the decreasing exponential determines the behavior of $P_{1s\sigma}$ with Z. The interesting point is that the scaling law seems to be equally applicable to the bound electron states and states of the negative energy continuum. We, therefore, find that it also describes essential features of positron emission by the changing Coulomb field produced by the combined nuclear charges.

Many of the anticipated characteristics of inner-shell ionization outlined above have been corroborated in the numerous experimental studies of superheavy collision systems.[31] Let us consider some of the most relevant to the positron experiments.

Greenberg et al.[32] established in one of the early measurements that the ionization of the $1s\sigma$ state in superheavy quasimolecules is predominantly concentrated at impact parameter < 50fm. This observation

184

together with a concurrent determination by Behncke et al. [33] and Anholt et al. [34] that the cross section of $1s\sigma$ ionization was of the order of several barns for collision systems such as Pb + Pb and U + U are, of course, of central importance to the spontaneous positron search experiments. The confinement of the $1s\sigma$ ionization probability for heavy systems at small impact parameters is now well documented, as evidenced by data such as that displayed in Figure 6. [35] The scaling law we referred to earlier unifies the data in a remarkable fashion (see Figure 7). [36] If we empirically determine F(Z), it seems to correctly describe some central points of the excitation mechanism, although it is not able to predict absolute values. For the latter we have to go to full coupled channel calculations and include screening. [29, 30]

Carrying this analysis further, [28] ΔE, identified with $E_{1s\sigma}$ at R_{min}, has been used to extract binding energies by measuring at a fixed R_{min} the excitation probability at different projectile velocities. Some results are shown in Figure 8. [31, 37] This type of "$1s\sigma$ - spectroscopy" can at best be indirect (we are not considering transitions between well-defined states in the usual sense) and there are obvious limitations to the scaling law, as well as ambiguities in defining the relevant quantities where averages are involved. [28] But from the consistency found in many data it becomes clear that we can at least conclude that electrons in the superheavy collision complex are bound by several hundred keV.

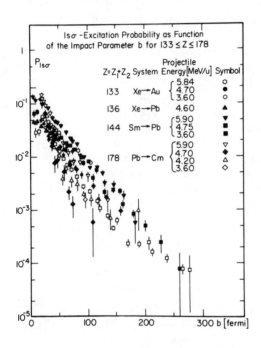

Fig. 6. The $1s\sigma$ - excitation probability, $P_{1s\sigma}$, as a function of impact parameter, b. [35] The maxima of cross sections are confined well within 50fm.

A similar conclusion emerges from investigating delta-electron emission in superheavy collision systems. It is obvious that in detecting the delta electrons we are acquiring one more differential piece of information not available in the total vacancy production studies via x-ray deexcitation;

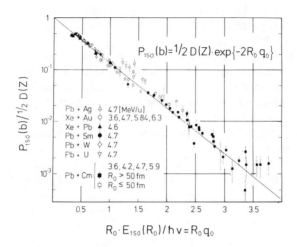

Fig. 7. Unification of the $1s\sigma$ excitation data [28] for superheavy collision systems by the scaling law. [27, 28] Here R_o is the minimum internuclear separation and q_o is the minimum momentum transfer.

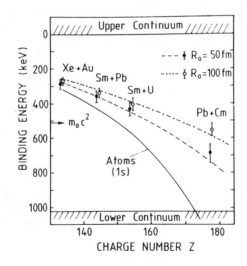

Fig. 8. $1s\sigma$-binding energies at two internuclear separations for $Z_u = 133, 178$ extracted from $P_{1s\sigma}$ according to scaling law. [28] The dashed and dash-dotted lines are calculations by G. Soff, and the full line is taken from reference 2.

i.e., we define the energy transferred to the ejected electron and thus the total energy transferred if the binding energy is known. The energy transfer, ΔE, in the scaling law for $P_{1s\sigma}$ is now defined more explicitly and consistency can be sought in determining $E_{1s\sigma}$ as a function of electron energy.

But even a qualitative examination of the delta-electron spectrum yields significant information. Simple kinematics show that the high energy components of this spectrum must reflect high momentum components of the quasiatomic wavefunction which, in turn as we saw, provide the information regarding close encounters. (The classically allowed maximum energy transfer to an electron at rest by a heavy nucleus is in the 10 keV region.)

The detection of such a high energy electron component in Pb + Pb collisions is illustrated in Figure 9 from data by Kozhuharov et al.[38]

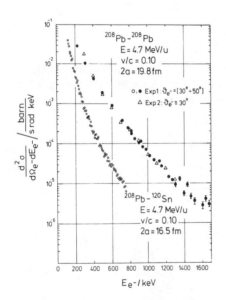

Furthermore, Figure 10 shows how the cross section at a particular electron energy grows with the united atom Z_u, exposing in a dramatic fashion the effects of the contraction of the wavefunction and the growth of the electron density around the nuclear centers.

We have only sampled a small fraction of the information available from studies of vacancy production, delta-electron emission and also molecular orbital x-ray emission, which has not been mentioned. A great deal more data is available which I have not discussed.[39] But even this short overview is sufficient to provide convincing evidence on two important questions: 1) We seem to be forming superheavy quasi-molecules (quasiatoms) in collisions near the Coulomb barrier with very deeply bound $1s\sigma$ states. The quasiatom as a source of information on superheavy atomic systems appears to be a meaningful concept; 2) The characteristics of $1s\sigma$ vacancy formation are favorable to spontaneous positron emission. The $1s\sigma$ ionization probability is large and concentrated at small internuclear separations where the vacancies are needed

Fig. 9. Delta-electron spectra from Pb-Sn ($Z_u = 132$) and Pb-Pb ($Z_u = 164$) collisions.[38] Excited states not specified.

during the time of supercritical binding.

With these essential ingredients available, the obvious question to be addressed is how spontaneous positron emission can be identified in the presence of the overabundant dynamic positron production processes referred to earlier.

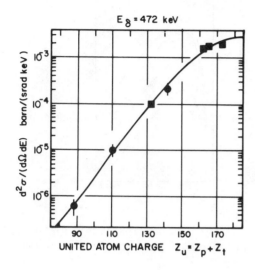

Fig. 10. Double differential cross section for emission of delta-rays of 472 keV kinetic energy for the collision systems with total charge Z_u. [38]

POSITRON DETECTION EXPERIMENTS

If we confine ourselves to Coulomb trajectories, theory does not provide encouraging answers to this question. For a sub-Coulomb barrier scattering experiment, theory predicts that the features of positron emission evolve from subcritical to supercritical conditions without any distinguishing signatures to identify the transition. The absence of any prominent effect is evident, for example, in the calculated spectra and emission probabilities as the critical charge and internuclear separation are passed (see Figure 11 [40, 41]). As noted earlier, the lack of any threshold merely reflects the dynamical widths associated not only with induced emission but also with the spontaneous amplitude due to the short collision times. Peaked structures are not expected, and production rates continue smoothly from the subcritical to the supercritical situation.

Therefore, such calculations indicate that a contribution to the positron spectrum can only be sought in a detailed quantitative comparison with theory rather than appearing in a clear qualitative signature. Fortunately, it now appears that Nature, in fact, may be more magnanimous in providing conditions which lead to a more dramatic demonstration of spontaneous positron emission.

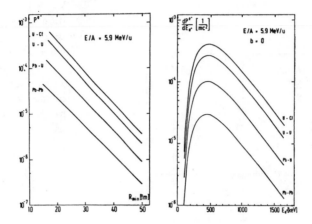

Fig. 11. Calculated [40] energy spectra of positrons created in 5.9 MeV/amu head-on collisions and the emission probability of positrons as a function of R_{min}.

(a) Positron Detectors

The experimental approach has been dictated by the need to detect a small cross section--of the order of 100 μb--in an overwhelming radiation field of γ-rays, delta electrons and nuclear constituents such as neutrons. To meet this challenge, detection systems [42, 43, 46, 47] have been developed with a large sensitivity for primary positrons while suppressing secondary positron production processes such as external pair creation by electrons and γ-rays. Three detectors have been employed in experiments at the GSI laboratory. [42] For orientation to much of the data I will be presenting, let me discuss some of the essential features of the instrument used to obtain these data.

A schematic presentation of this spectrometer is shown in Figure 12. [43] It employs a solenoidal transport system to focus positrons from the target onto a cylindrical Si(Li) detector which provides the positron energy determination. The spectrometer exploits the two basic features of a solenoidal transport system, namely the inverted spiralling directions of oppositely charged electrons and positrons, and the property that the particles return to the field lines passing through their point of creation. Using these properties and a helical baffle system, with the spiralling direction of positrons, positrons are very effectively selected from the copious flux of delta electrons that exceed the positron intensity by several orders of magnitude in the energy region of interest. The axial geometry

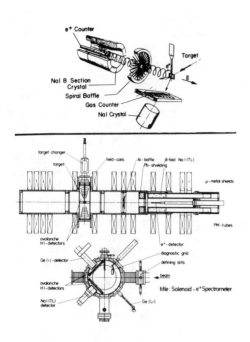

Fig. 12. Schematic and plane-view of solenoidal spectrometer EPOS for positron detection in heavy-ion collisions. [43]

of the positron detector combines maximum efficiency for target-produced positrons with minimum sensitivity for positrons and electrons which have been scattered or do not originate from the target. The transport efficiency is maximized by using a magnetic minor on the side of the target opposite to the positron detector and a field depression at the position of the spiral baffle.

A positron event it defined as a coincidence between the Si(Li) detector and at least one of the two 511 keV positron annihilation γ-quanta. The annihilation radiation is detected in an eight-segment cylindrical NaI-detector array surrounding the positron counter. By summing up the signals of the eight separate NaI crystals and requiring a total energy sum ≥ 440 keV, the NaI efficiency attains a value of $\sim 58\%$ per positron detected in the Si(Li) counter, while the total peak efficiency of the system is $\sim 13\%$. The energy resolution of the positron detector is ~ 10 keV for 662 keV conversion electrons.

The ability to closely define the reaction kinematics constitutes a significant part of the measurements. Although we remain close to the Coulomb barrier, we will see that nuclear reactions may play a substantial role, and the selection of events close to elastic scattering is an important requirement.

For this reason, both the scattered projectile and the recoiling target nucleus are detected in two symmetrically arranged parallel plate avalanche detectors with continuous delay-line read-out for scattering-angle determination. [44] The angular acceptance of each of the two particle detectors is $20^{\circ} < \theta < 70^{\circ}$ at constant $\Delta\varphi = 60^{\circ}$. In addition to the particle scattering-angle information, the time of flight difference between both particles as well as rough Z-information is obtained via the energy loss of the particles in the detector gas.

The coincident detection of both particles has the advantage that for asymmetric systems, such as U + Pb, the impact parameter (or equivalently the distance of closest approach, R_{min}) can be determined uniquely for Coulomb scattering from the angle-angle correlation. Figure 13 illustrates the excellent separation that can be obtained between the forward and backward scattering events in U + Pb collisions. Some information on inelasticity is obtained from the sum of the two scattering angles $(\theta_1 + \theta_2)$. Moreover, the kinematic coincidence can be exploited together with the rough Z-identification to select desired reaction channels, and to discriminate against backgrounds from the target-backing and/or covering foils.

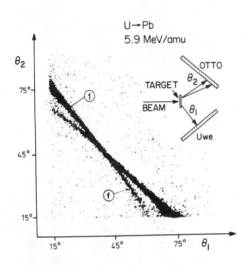

Fig. 13. Intensity of scattered particles as a function of two laboratory angles, θ_1 and θ_2, for the system U + Pb at 5.9 MeV/amu. An event in branch 1 corresponds to U being observed in lower counter.[43]

An important component of the apparatus is gamma-ray detectors to monitor gamma-rays from nuclear reactions. We have already mentioned that internal pair conversion of gamma-ray transitions can be an important background. In U + U collisions the contribution of positrons from this source is approximately 25%-30% of the observed positron yield. The gamma-ray spectra, measured in coincidence with the scattered ions, are utilized to construct these background positron spectra by using theoretically[45] or experimentally determined internal pair conversion coefficients. Since the multipolarities of the gamma-rays are unknown, the mix has been determined semi-empirically by systematically studying positron production in a number of nuclear systems with a variety of nuclear properties where quasiatomic positron production can be considered to be negligible (e. g., ^{238}U + ^{154}Sm, ^{238}U + ^{165}Ho). The apparent similarity that has been found both in the measured gamma-ray spectra and in the subsequently constructed positron spectra for a large variety of systems greatly simplifies and provides confidence in this procedure. Of course, EO transitions have to be taken into account separately. It should be noted that in general the almost continuous gamma-ray spectra lead to smooth positron spectra. Although they are a consideration in determining absolute positron yields, they have little effect on prejudicing the search for any sharp peak structures in the positron spectrum. We will return to this point.

(b) Gross Features of Positron Production

For orientation to the most recent results, let us first consider briefly the gross features of positron production obtained in some of the first measurements carried out.[46, 47, 5]

One of the first experimental goals in the search for spontaneous positron emission was to establish the level of positron production from the atomic processes relative to that from nuclear effects, and to confirm our theoretical understanding of the dynamic positron production processes in heavy-ion collisions. In this regard, the first measurements on the ^{208}Pb + ^{208}Pb collision system played a particularly important role. The nuclear background is especially simple to evaluate in this case because it is associated with a well-known excitation mechanism and known energy levels so that it can be simulated exactly.

The results from measurements on the energy-integrated positron production in the Pb – Pb system,[46] shown in Figure 14, provided the first convincing demonstration of positron production in heavy-ion collisions which originates with the electronic quasimolecular complex and not with the nuclear structure alone. In fact, the nuclear component is only a fraction of the total positron yield, particularly at large values of R_{min}. The good agreement of theory[40, 41] with experiment and, in particular, the observed rapid exponential fall-off of the cross section with R_{min} within distances of a few nuclear diameters stresses that the positron production mechanisms are closely associated with the relativistic properties of the superheavy quasiatom already

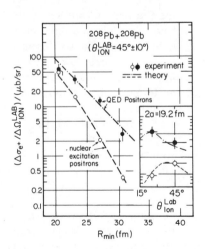

Fig. 14. Differential positron cross sections as a function of R_{min} for ^{208}Pb + ^{208}Pb collisions. The inset shows an angular distribution for $2a = 19.2$fm.[47]

described, and with the time-changing Coulomb field of this complex. Being a subcritical system, only the dynamic processes are expected to contribute with the direct excitation amplitude (processes "d" in Figure 4) supplying the largest fraction. [40, 41]

Measurement with Pb + U and U + U collisions [47] have carried these investigations into heavier systems, but under different and more complex background conditions. In order to investigate the consequences arising from this nuclear background in more detail, a systematic investigation was carried out on the ratio of positron to gamma-ray intensity over a broad range in Z. The interesting qualitative observation emerging from these studies is illustrated in Figure 15. [42, 47] Above $Z_u \simeq 160$, there is a spectacular increase in the total positron yield, with increasing Z_u over that expected from nuclear internal pair conversion as it is extrapolated from the positron to gamma-ray ratios measured for $Z_u < 160$. More precisely, for constant R_{min} and v, positron production in superheavy collision systems is found to increase as $(Z_1 + Z_2)^{21}$. In this striking feature, which seems to have no other analogue in nature, the theory of dynamic positron creation in heavy-ion collision again correctly anticipates the experimental results.

The data shown in Figures 16 and 17 [47, 48] illustrate that the positron emission probability behaves very similarly with R_{min} and with the relative velocity for all three systems. In fact, the first experiments did not reveal any surprises which could be considered as anomalously deviating from the theoretical positron production rates dominated by the dynamic

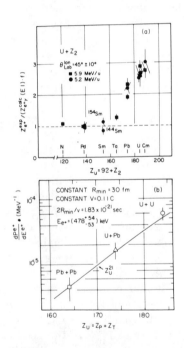

Fig. 15. (a) The ratio of measured positron yields to those calculated from the γ-ray yields assuming E1 multipolarity. The factor f is close to unity. The measurements were carried out with the spectrometer described by H. Backe et al. [46]
(b) The dependence of the positron emission probability on the total charge of the collision system. [47]

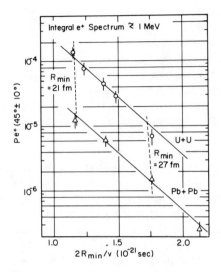

Fig. 16. Positron emission probability vs. R_{min}/v for U + U and U + Pb collisions. Measurements carried out with spectrometer in reference 46. Scattering angle was fixed at $45 \pm 10°$ and relative velocity v was varied to determine R_{min}.

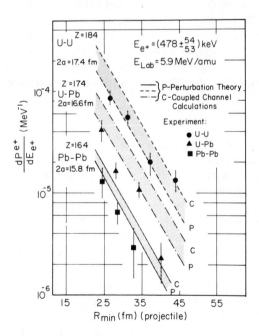

Fig. 17. Positron production probabilities at 5.9 MeV/amu as a function of R_{min}. [47]

mechanisms, although the U + U system could be expected to produce super-critical binding for some of the bombarding conditions used. The results can be summarized as follows:

(i) Atomic QED processes account for the major fraction of the total positron creation cross section in superheavy collision systems.

(ii) The positron creation probability associated with quasiatomic processes exhibits a rapid exponential fall-off with R_{min} at constant collision velocity. This behavior parallels that found for the ionization probability of inner-shell electrons and reflects similar mechanisms. In this case, the minimum energy transfer is given by the energy gap $2mc^2$ for pair creation which sets the scale for the spacial confinement of this process as it did for the ionization process through $e^{-\alpha R_{min} \Delta E / \hbar v}$.

(iii) The almost exponential increase of the positron creation probability with Z_u cannot be accounted for without invoking a non-perturbative treatment of positron production in the quasiatomic picture of $(Z_1 + Z_2)$ acting in unison to generate very strong electric fields.

But, although these experiments represented very significant steps towards providing confidence in our understanding of the dynamic positron production processes, they also indicated that a signature for spontaneous positron emission was not going to be readily forthcoming, at least not in the gross features that had been studied. To search for such a signature, the experiments, therefore, have evolved naturally into examining some finer details of the positron emission process. Particularly, the spectra had not been explored with any detail, and little was known about the correlation between the spectral distribution and the kinematic variables of the scattered ions. A focussing of such a correlation to selected parts of the spectrum could have readily been missed in the first experiments.

(c) Positron Spectra

The most recent investigations,[42] therefore, have concentrated on studying the positron spectra and on extending the investigations to collision systems with higher total charge. I will consider primarily the data[49] obtained with the spectrometer described above, which has developed some of the compelling evidence for some interesting new features of positron production, which may be connected with the effects being sought. The data I present is preliminary, and should be considered indicative of the kind of information being accumulated, rather than a final result.

The first indication for us that a previously undetected phenomenon could be occurring was from a measurement that compares the R_{min} dependence for the low energy and high energy parts of the spectrum from U + Cm collisions at 5.8 MeV/amu.[50] The positron detector utilized for this measurement was a plastic scintillator with modest resolution. For this bombarding energy, the U + Cm system, whose combined Z = 188 is expected to become supercritical within the range subtended by the two particle detectors.

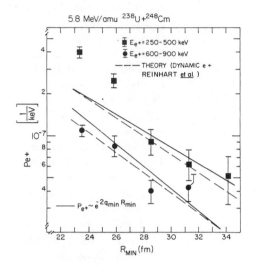

As shown in Figure 18, it was observed that the slope of dP_{e+}/dE_{e+} versus R_{min} is greatest for the low energy component of the positron spectrum. The significance of this observation is that it contrasts with the expectations for dynamically induced positrons. In fact, theory predicts that the higher energy positron band has the larger slope in the ratio $\sim e^{-Const(\Delta E)}$, where ΔE is the energy transferred to the positron, $\Delta E = 2mc^2 + E_{e+}$. Indeed, as indicated in Figure 18, the average slope of dP_{e+}/dE_{e+} for the low energy positrons is considerably steeper than that found for all the Coulomb ionization processes we have examined in the superheavy collision systems. The implication, of course, is one of structure in the positron spectrum which could indicate the

Fig. 18. The dependence of the positron production probability for selected kinetic energy regions of the positron spectrum on the distance of minimum separation in $^{238}U + ^{248}Cm$ collisions.[50] Nuclear background has been subtracted. The dashed lines show the slopes theory predicts (Reinhardt et al.).

presence of an additional positron production process in the U + Cm collision system which contributes particularly at low positron kinetic energies.

The question of structure was pursued with further measurements using the higher resolution of a Si(Li) detector discussed in the description of the spectrometer. Both the gross features and the finer details of these measurements have exhibited interesting aspects.

For example, Figure 19 compares total positron spectra from U + Pb, U + U and U + Cm collisions. They have been accumulated in coincidence with scattered ions detected between 20° and 70° in the laboratory and include all elastic and quasielastic two-body events. As shown, the spectra are not corrected for background contributions from internal pair conversion of gamma-ray transitions in excited nuclei and they have not been adjusted for the solenoid transport efficiency. The latter factor partly determines the shape of the spectra at the lowest and highest energies.

There is little to distinguish the bell-shape form of the U + Pb

196

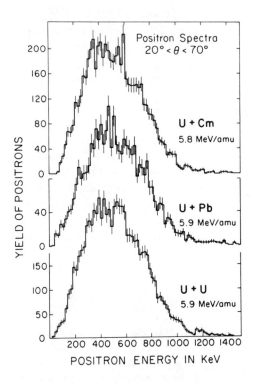

YIELD OF POSITRONS

Positron Spectra
20° < θ < 70°

U + Cm
5.8 MeV/amu

U + Pb
5.9 MeV/amu

U + U
5.9 MeV/amu

POSITRON ENERGY IN KeV

Fig. 19. Positron energy spectra for
the system ^{238}U + ^{208}Pb, ^{238}U + ^{238}U
and ^{238}U + ^{248}Cm integrated over particle scattering angles 20° to 70°. [49]
The spectra are not corrected for the
solenoid transport efficiency, for the
line-shape of the Si(Li) detector and
for the contribution from the nuclear
background.

spectrum, peaking near
~ 500 keV, from the dy-
namically induced posi-
tron energy distributions
theoretically predicted
for Rutherford trajec-
tories, and shown in
Figure 11. However,
although this overall
general shape is also
followed in the spectra
from U + U and U + Cm
collisions, there seem to
exist statistically significant structures super-
imposed on this envelope
which grow more promi-
nent with increasing
total nuclear charge.
Moreover, it is observed
that imposing kinematic
constraints developes
this structure even fur-
ther. Let us consider
some interesting exam-
ples and comparisons.

The U + Cm colli-
sion system represents
the largest combined
nuclear charge, $Z_u = 188$,
investigated to date.
Positron spectra emitted
from U + Cm colliding at
an energy close to the
Coulomb barrier are dis-
played in Figure 20. It is particularly striking to observe in Figure 20(a)
the well-defined peak centered at an energy of ~ 320 keV. The intensity of
this peak above the smoother continuum is correlated with choosing a sample of scattering events which attempts to preferentially select the backward scattering angles and suppress the forward scattering angles. In
this near symmetric collision system a clear separation is not achieved
between the close and distant collision as is the case in Figure 13 for
U + Pb. By comparison, Figure 20(b) shows that singling out scattering
angles, preferentially forward in the center of mass system, largely excludes this peak and leads to a spectrum which mirrors the general shape

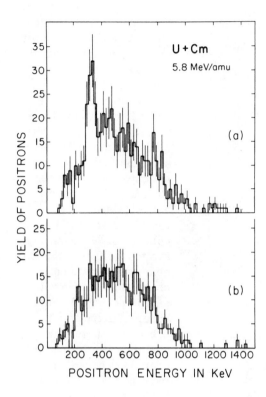

Fig. 20. Positron energy spectra from 5.8 MeV/amu ^{238}U + ^{248}Cm collisions selected in coincidence with kinematic conditions to emphasize (a) backward scattering and (b) forward scattering of projectiles.[49] The spectra are not corrected for the solenoidal transport efficiency, for Doppler broadening, for the Si(Li) detector line shape and for the nuclear background.

of the continuum underlying the peak in Figure 20(a). It was also found that scattering angles around 45° in the laboratory are correlated with such a smooth spectrum.

It is important to point out that the prominent dependence of the peak intensity on the choice of ion scattering angle excludes the possibility that the appearance of a peak is produced by instrumental effects associated with the positron detection method used. For reasons to be discussed, it is also significant to find that the measured width of the peak in Figure 20(a) is ~ 80 keV. Moreover, this width is consistent with the Doppler broadening expected for a positron line spectrum emitted from a system moving with the velocity of the quasimolecular system. Therefore, the intrinsic width of the peak is surely less than 80 keV and, indeed, it could be <u>very much smaller</u> than this value.

U + U collisions produce different but equally interesting features for scattering conditions that parallel those found in the U + Cm measurements. In this case, Figure 21 suggests that a peak pattern with several peaks of almost equal intensity appears in the spectrum associated predominately with larger angle scatterings, while the more forward scattering angles correlate again to the smooth continua expected from dynamic processes. There is also a suggestion in the data not displayed that there is multiple peak structure in the U + Cm collision system, although in this case the

198

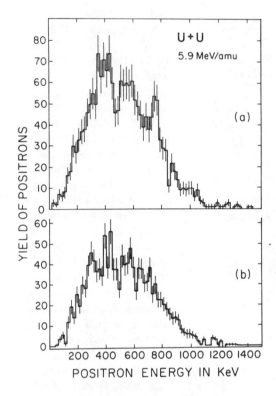

Fig. 21. Positron energy spectra for
5.9 MeV/amu ^{238}U + ^{238}U collisions
under the conditions specified in
Figure 20. [49]

spectrum is dominated by the 320 keV peak and it is difficult to draw any conclusion presently with the limited statistical accuracy on hand.

Of course, it is of great interest also to examine a collision system such as U + Pb where supercritical binding is not expected to be present. Figure 22 shows some data obtained for 5.9 MeV/amu U + Pb collisions. As demonstrated in Figure 13, excellent separation of close and distance collision is achievable in this case. If structure is present in these spectra, certainly it is much less prominent here than in the other two collision systems. One of the difficulties encountered in studying the U + Pb system is that ~ 50% of the total positron yield is contributed by the smooth background from pair conversion of nuclear gamma-ray transitions. This fraction, for instance, is to be compared to < 30% for U + U and < 20% for U + Cm. It, therefore, becomes important in the case of U + Pb to evaluate this background with some care before a conclusive judgment can be made regarding the presence of any structure in its positron spectrum, although presently there does not seem to be much evidence that any exists.

(d) Origin of Peak Structure

Whatever the sources of the peaks observed in U + U and U + Cm, it is apparent that the explanation must be sought outside the scope of the theory for dynamic positron creation based on Rutherford scattering alone since this theory does not allow for narrow peak structures in the positron

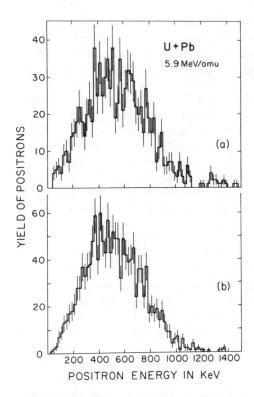

Fig. 22. Positron energy spectra for 5.9 MeV/amu ^{238}U + ^{208}Pb collisions under the conditions specified on Figure 20. [49]

spectrum. Deviations from this theory in U + U collisions have also been demonstrated in other experiments carried out by Kozhuharov, Kienle et al. and Backe, Kankeleit et al. at the GSI laboratory. [42] Kozhuharov, Kienle et al. have observed peak structures in U + U collisions at several bombarding energies between 5.7 and 5.9 MeV/amu. Backe, Kankeleit et al. have explored, in addition, bombarding energies much above the Coulomb barrier and found that the positron spectra are smooth but narrower than can be explained [51] by considering scatterings confined to Coulomb trajectories alone. Details of the effects observed differ among experiments due to differences in the detection schemes and in the measurement conditions utilized, but the broad conclusions are similar in the following sense: All experiments carried out to date indicate that there are positron producing mechanisms in superheavy collision systems which do not originate with the known dynamic mechanisms associated in a simple way with the time-varying electric field produced in Coulomb trajectories.

It is also difficult to connect these deviation from smooth positron spectra with pure nuclear effects. There are two prominent candidates: One is the internal pair conversion of a nuclear transition which leads to a positron energy distribution which may be peaked in character, [45] and the other is the internal pair conversion process followed by the capture of the electron into empty atomic orbits which leads to positron line spectra. We have found that the gamma-ray spectra observed for the same kinematic conditions as the positron peaks are smooth and do not show any prominent single lines at energies relevant to the positron structure.

However, strong EO transitions, undetected in the gamma-ray spectrum, can be a potential source of positrons. This possibility seems to be precluded particularly for the peak observed in the U + Cm system since its width is considerably narrower than the triangular shapes calculated [44] for the positron spectra from EO transitions at this energy. More generally, the smooth angular distributions expected for the scattered ions after Coulomb excitation of the colliding nuclei contrast with the observed behavior for production of the positron peaks. We also find that intensity considerations exclude the second mechanism by orders of magnitude. But more direct studies of these and other effects are still required, and are being pursued, to exclude any connections with nuclear transitions.

Of course, if these less interesting explanations for the structure are excluded, the observation of the line-like positron emission from the U + Cm system together with its apparent selective appearance only under particular scattering conditions opens up to serious consideration the possibility that we may be observing spontaneous positron emission. We are encouraged circumstantially by the fact that this peak happens to occur at an energy consistent with a calculation [51] of the $1s\sigma$ resonance in this quasiatom. Obvious systematic confirmation is required to follow up on these very suggestive data, but already these new developments suggest the possibility of another new observation. For if indeed the narrow positron peak reflects spontaneous positron emission, the parent nuclear supercritical charge must exist for a time long compared to sub-Coulomb scattering collision times.

This possibility recalls a proposal put forward by Rafelski, Müller and Greiner [52] which at the time of its suggestion had only academic interest. We have seen that the subordinate role played by the spontaneous component in the sub-Coulomb collisions is, clearly, an outgrowth of the short collision time which does not allow the spontaneous amplitude to develop. Therefore, they proposed that the effect of the spontaneous amplitude would be enhanced if advantage could be taken of time delays which may be associated with nuclear reactions. The situation then becomes as depicted in Figure 23. If two colliding ions stick together for an extended period Δt before separating again, the time spent by the quasimolecular levels in an overcritical binding is correspondingly prolonged. During the sticking time, the energies of the electronic states do not change in the limit that a static nuclear shape is maintained. Two important consequences follow: [52] (a) if vacant supercritically-bound states are present, a clear peak is predicted to develop for spontaneous positron emission as the spontaneous amplitude grows with time; (b) even without supercritical binding occurring the positron spectrum also is expected to exhibit an oscillating structure as a function of positron energy which reflects the delayed interference between incoming and outgoing induced positron production amplitudes along the trajectory of the colliding ions. Figure 23 [52] attempts to depict these two situations schematically with some arbitrary

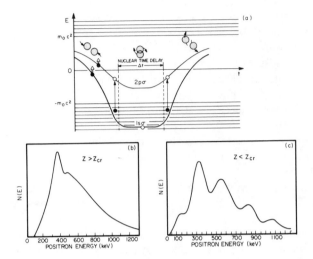

Fig. 23. Schematic representation of the time evolution of the inner electronic shells in inelastic U - U collisions involving the formation of a metastable nuclear complex for a time Δt. (b) Example of positron energy spectrum for a fixed scattering angle if a supercritically-bound electronic state, such as the 1sσ state shown is vacant during Δt. (c) Example of positron energy spectrum from induced emission if a vacant state, such as the 2 pσ state shown is subcritically bound during the entire collision time.

choice for the ratio of nuclear reactions to Rutherford scattering.

Adopting this mechanism, J. Reinhardt, et al.[51] have suggested that the observation of spontaneous positron emission as a sharp line must simultaneously imply the formation of a metastable superheavy nuclear composite system at bombarding energies close to the Coulomb barrier with a lifetime sufficiently long to account for the relatively narrow peak observed. Widths of 100 keV or narrower correspond to lifetimes for the dinuclear system longer than ~ 40 times the Rutherford scattering collision time during which the 1sσ state is overcritically bound. As mentioned above, the intrinsic peak width must be much narrower than this, implying a very long-lived system. Indeed, it is difficult to invent any mechanism, associated with atomic positron emission, which would explain either the narrow peak width found in the U + Cm spectrum or the multiple peaks in the positron distribution emitted from U + U collisions without introducing a time delay such as could be supplied by the formation of a rather cold intermediate superheavy nuclear complex as the nuclei barely touch in overcoming the Coulomb barrier. Of course, in considering the effect found in U + U as well as in explaining the U + Cm spectrum it is necessary to account in detail for the relative intensity of peaks to the underlying continuum produced by Rutherford scatterings, as well as explaining the time-delay structure required to reproduce the data.

It bears mentioning in connection with the spectra shown in Figures 20 and 21 that a rapidly falling efficiency for the detection of low energy positrons in our spectrometer would have inhibited the observation of a peak below 200 keV. This energy region in the U + U spectra is potentially of some interest to us since it could include positrons related to the peak observed in the U + Cm collisions, but lower in energy and intensity. Theory calculates [51] that the ratio of the widths for spontaneous emission in U + Cm and U + U is approximately 4 for similar nuclear configurations.

Thus, we presently have the situation where several independent measurements confront us with evidence that peak structures exist in positron spectra emitted from collision systems where the quasiatom can have overcritically bound electrons. There remains the task to identify unambiguously the sources for these structures among the possibilities we have discussed or yet to be considered.

The directions for future experiments are self-evident:

(i) It has to be determined whether the positrons of interest originate with separated atoms, implying nuclear sources of positrons, or with quasimolecules. For this purpose the Doppler broadening and shift that otherwise plagues the resolution of the spectrometers can be utilized to advantage. In this connection, it would be very useful to track down nuclear EO transitions via internal electron conversion decays.

(ii) Questions associated with the nuclear physics aspects of the problem have to be addressed in more detail. Are we forming metastable superheavy nuclear species? Here the techniques of nuclear physics have to be mobilized with careful excitation functions and many-body break-up studies. It would be preferable to carry out these studies independent of the positron probe if the level of sensitivity allows. Most likely, this will not be the case, and the heavy-ion detection in the positron spectrometers will have to be developed into more elaborate configurations. It is also of particular importance to compare measurements near the Coulomb barrier with measurements using projectile energies that ensure the absence of nuclear reactions. Unfortunately, the latter consume excessive accelerator time, which explains the absence of such data to date.

(iii) The electrodynamic aspects have to be explored with studies of the Z dependence. Here the possibilities for supercritical systems is limited, but the obvious tracking of positron peak energies with $(Z_1 + Z_2)$ and nuclear shapes is a major aspect to be studied.

The future for these experiments is both challenging and interesting.

ACKNOWLEDGEMENTS

The author is grateful for the stimulation and collaboration of the many co-workers which are listed in the references associated with our work. In particular, I am especially obliged to H. Bokemeyer, A. Gruppe, J. Schweppe and P. Vincent, who analyzed and prepared much of the data discussed in this report. I thank W. Greiner, B. Müller, U. Müller, J. Reinhardt and G. Soff for many helpful discussions on theoretical aspects of this work. We are obliged to the Trans-plutonium Program of the USDOE for the loan of the ^{248}Cm isotope material used for target fabrication.

† This work has been supported in part by the U.S. Department of Energy Contract No. DE-AC02-76ER03074, and the Bundesministrium für Forschung and Technologie.

REFERENCES

1. W. Pieper and W. Greiner, Z. Phys. <u>218</u>, 327 (1969).
2. B. Fricke and G. Soff, At. and Nucl. Data Tabl. <u>19</u>, 83 (1977).
3. G. Soff, J. Reinhardt, B. Müller and W. Greiner, private communication.
4. W. Pieper and W. Greiner, Z. Physik <u>218</u>, 126 (1969).
   Earlier studies of this problem are: I. Pomeranchuck and I. Smordinsky, J. Phys. USSR <u>9</u>, 97 (1945); F. G. Werner and J. A. Wheeler, Phys. Rev. <u>109</u>, 126 (1958); V. V. Voronkov and H. N. Koleznikov, Sov. Phys. JETP <u>12</u>, 136 (1961); V. S. Popov, Sov. Phys. JETP <u>12</u>, 235 (1971); G. Soff, B. Müller and J. Rafelski, Z. Natur-forsch <u>29A</u>, 1267 (1974).
5. For a general review of the present status of the field see "Quantum Electrodynamics of Strong Field," Proceedings of the NATO Advanced Study Institute at Lahnstein/Germany, June 1981, Plenum Press, and also review in references 6, 7, and 8, which provide many of the references to the original literature.
6. J. Reinhardt and W. Greiner, Rep. Prof. Physics <u>40</u>, 219 (1977).
7. J. Rafelski, L. P. Fulcher and A. Klein, Phys. Reports <u>38C(5)</u>, 227 (1978).
8. S. J. Brodsky and P. J. Mohr, Structure and Collisions of Ions and Atoms, I. A. Sellin, (ed), Springer, New York (1978).
9. B. Müller, H. Peitz, J. Rafelski and W. Greiner, Phys. Rev. Letters <u>28</u>, 1235 (1972).
10. B. Müller, J. Rafelski and W. Greiner, Z. Physik <u>257</u>, 62 (1972).
11. Ya. B. Zel'dovitch and V. S. Popov, Soviet Phys. Uspekhi <u>14</u>, 673 (1972).
12. J. Rafelski, B. Müller and W. Greiner, Nucl. Phys. <u>B68</u>, 585 (1974).
13. L. P. Fulcher and A. Klein, Phys. Rev. <u>D8</u>, 2455 (1973).
14. J. Schwinger, Phys. Rev. <u>82</u>, 664 (1951).

204

15. O. Klein, Z. Physik 53, 157 (1929); F. Hund, Z. Physik 117, 1 (1941).
16. F. Sauter, Z. Physik 69, 742 (1931) and Z. Physik 73, 547 (1932).
17. S. S. Gershstein and Ya. B. Zel'dovitch, Sov. Phys. JETP 30, 358 (1970); Lettre Nuovo Cimento 1, 835 (1969).
18. J. Rafelski, L. P. Fulcher and W. Greiner, Phys. Rev. Letters 27, 958 (1971).
19. T. de Reus, U. Müller, J. Reinhardt, P. Schlüter, K. H. Wietschorke, B. Müller, W. Greiner and G. Soff, Proceedings of the NATO Advanced Study Institute on "Quantum Electrodynamics of Strong Fields," Lahnstein, June 1981.
20. G. Soff, P. Schlüter, B. Müller and W. Greiner, Phys. Rev. Letters 48, 1465 (1982).
21. K. Smith, H. Peitz, B. Müller and W. Greiner, Phys. Rev. Letters 32, 554 (1974).
22. G. Soff, J. Reinhardt, B. Müller and W. Greiner, Phys. Rev. Letters 38, 592 (1977); D. H. Jakubassa and M. Kleber, Z. Physik A227, 41 (1976).
23. W. Betz, G. Soff, B. Müller and W. Greiner, Phys. Rev. Letters 37, 1046 (1976).
24. G. Soff, W. Greiner, W. Betz and B. Müller, Phys. Rev. A20, 169 (1979).
25. D. H. Jakubassa and M. Kleber, Z. Physik A227, 41 (1976).
26. See discussion in P. Armbruster, H. H. Behncke, S. Hagmann, D. Folkmann and P. H. Mokler, Z. Physik A228, 277 (1978). J. Bang, J. M. Hansteen, Kgl Danske Vid. Selsk. Mat.-Fys. Medd. 31, 13 (1959).
27. B. Müller, G. Soff, W. Greiner and V. Ceaucescu, Z. Phys. A285, 27 (1978).
28. F. Bosch, D. Liesen, P. Armbruster, D. Maor, P. H. Mokler, H. Schmidt-Böcking and R. Schuch, Z. Physik A296, 11 (1980).
29. J. Reinhardt, B. Müller, W. Greiner and G. Soff, Phys. Rev. Letters 43, 1307 (1979).
30. G. Soff, B. Müller and W. Greiner, Z. Phys. A299, 189 (1981).
31. See lectures by P. Armbruster and F. Bosch in reference 5.
32. J. S. Greenberg, H. Bokemeyer, H. Emling, E. Grosse, D. Schwalm and F. Bosch, Phys. Rev. Letters 39, 1404 (1977).
33. H. H. Behncke, P. Armbruster, F. Folkman, S. Hagmann, J. R. Macdonald and P. H. Mokler, Z. Physik A289, 333 (1979).
34. R. Anholt, H. H. Behncke, S. Hagmann, P. Armbruster and P. H. Mokler, Z. Physik A289, 359 (1979).
35. From work by P. Armbruster, F. Bosch, D. Liesen, P. H. Mokler et al., private communication.

36. D. Liesen, P. Armbruster, F. Bosch, S. Hagmann, P. H. Mokler, H. J. Wollersheim, H. Schmidt-Böcking, R. Schuch and J. B. Wilhelmy, Phys. Rev. Letters <u>44</u>, 983 (1980).

37. P. H. Mokler and D. Liesen, "X-Rays from Superheavy Collision Systems," preprint.

38. C. Kozhuharov, Physics of Electronic and Atomic Collisions, S. Datz, (ed), North Holland (1982), p. 179.

39. J. S. Greenberg, Electronic and Atomic Collisions, M. Oda and K. Takayanagi, (eds), North Holland (1980), p. 351.
    J. S. Greenberg, Conference Summary, NATO Advanced Study Institute at Lahnstein/Germany, June 1981, Plenum Press.

40. J. Reinhardt, B. Müller and W. Greiner, Phys. Rev. <u>A24</u>, 103 (1981).

41. T. Tomoda and H. A. Weidenmüller, Phys. Rev. <u>A26</u>, 162 (1982); T. Tomoda, Phys. Rev. <u>A26</u>, 174 (1982).

42. See lectures by H. Backe, H. Bokemeyer and P. Kienle in reference 5 for details.

43. A. Balanda, K. Bethge, H. Bokemeyer, H. Folger, J. S. Greenberg, H. Grein, A. Gruppe, S. Ito, S. Matsuki, R. Schule, D. Schwalm, J. Schweppe, P. Vincent and M. Waldschmidt, GSI, Darmstadt, GSI-preprint 80-16 (1980).

44. P. Fuchs, H. Emling, E. Grosse, D. Schwalm, H. J. Woltersheim, R. Schulze, GSI, Darmstadt, Jahresbericht 1977, GSI-J-1-78 (1978).

45. P. Schlüter, G. Soff and W. Greiner, Z. Physik <u>A286</u>, 149 (1978).

46. H. Backe, L. Handschug, F. Hessberger, E. Kankeleit, L. Richter, F. Weik, R. Willwater, H. Bokemeyer, P. Vincent, Y. Nakayama and J. S. Greenberg, Phys. Rev. Letters <u>40</u>, 1443 (1978).

47. C. Kozhuharov, P. Kienle, E. Berdermann, H. Bokemeyer, J. S. Greenberg, Y. Nakayama, P. Vincent, H. Back, L. Handschug and E. Kankeleit, Phys. Rev. Letters <u>42</u>, 376 (1979).

48. E. Kankeleit, Nukleonika <u>25</u> – No. 2/80, p. 253.

49. H. Bokemeyer, K. Bethge, H. Folger, J. S. Greenberg, H. Grein, A. Gruppe, S. Ito, R. Schule, D. Schwalm, J. Schweppe, N. Trautmann, P. Vincent and M. Waldschmidt, described partly in reference 5 by H. Bokemeyer and to be published.

50. H. Bokemeyer, H. Folger, H. Grein, S. Ito, D. Schwalm, P. Vincent, K. Bethge, A. Gruppe, R. Schule, M. Waldschmidt, J. S. Greenberg, J. Schweppe and N. Trautmann, GSI, Darmstadt, Scientific Report 1980, GSI 81-2 (1981), p. 127.

51. J. Reinhart, U. Müller, B. Müller and W. Greiner, Z. Physik <u>A303</u>, 73 (1981).

52. J. Rafelski, B. Müller and W. Greiner, Z. Physik <u>A285</u>, 173 (1978).

THEORY OF POSITRON CREATION IN HEAVY ION COLLISIONS

U. Müller, J. Reinhardt, T. de Reus,
P. Schlüter, G. Soff, W. Greiner, and B. Müller[*]
Institut für Theoretische Physik
der Johann Wolfgang Goethe-Universität,
Robert-Mayer-Straße 8-10, Postfach 111 932,
D-6000 Frankfurt am Main, West Germany

INTRODUCTION

Collisions of very heavy ions at energies close to the Coulomb barrier are a unique tool to study non-perturbative aspects of quantum electrodynamics, in particular spontaneous pair-creation in strong electric fields[+,1]. It is well established by now[2] that such collisions offer the possibility to perform a spectroscopy of electronic states in transient superheavy systems within a charge range of $100 < Z \leq 188$ ($_{92}U +_{96}Cm$). The central question here is whether the binding energy of an electronic state can exceed twice the electron rest mass, as it is predicted by theory.

An important part in this endeavour is the investigation of electron excitation processes in superheavy systems[3]. In contrast to the physics of light ion collisions the multi-step excitation processes are crucial for a quantitative understanding of inner-shell vacancy formation in scattering processes involving very heavy ions where they actually dominate the vacancy production probabilities[4]. This fact makes it necessary to integrate the coupled equations for the electron occupation amplitudes in the basis of quasi-molecular adiabatic states rigorously and, in particular, to include couplings between continuum states. That large ionization probabilities are found despite the tremendously increased binding energies is the consequence of the relativistic shrinkage of the inner-shell wavefunctions. This is best visible if the electron density at the origin, $|\psi(0)|^2$, is plotted as a function of Z (Fig. 1). Apart from the Z^3-dependence known from the nonrelativistic theory the density increases by three additional orders of magnitude going from Z=1 to Z=180. The figure also displays the electron density at the origin for the $2p_{1/2}$-wavefunction. For a nonrelativistic p-wave, characterized by the sharp angular momentum $\ell = 1$, this value is exactly zero. The $2p_{1/2}$-Dirac spinor, however, carries a mixture of $\ell = 1$ (upper component) and $\ell = 0$ (lower component) orbital angular momentum. In atoms with $Z\alpha > 1$ the 'small' and 'large' components of the wavefunction become of the same magnitude. As a consequence the behaviour of electrons in $s_{1/2}$- and $p_{1/2}$-states becomes increasingly similar. This explains the steep rise of the $2p_{1/2}$-density as well as the strong increase of its binding energy as a function of Z.

The semi-classical treatment of electron-positron excitation processes is based on the time-dependent two-centre Dirac equation

---

[*] Invited speaker at the International Conference on X-Ray and Atomic Inner-Shell Physics, Eugene/Oregon, August 1982.

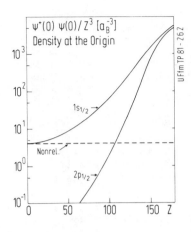

Fig. 1. The drastic increase of the
electron density at the origin over its
nonrelativistic value is shown. The
densities of the $1s_{1/2}$ and $2p_{1/2}$-states
become of comparable size in the region
$Z\alpha > 1$. (Note that $\Psi_{2p_{1/2}}$ (0) vanishes
in the nonrelativistic limit!)

$$i\partial/\partial t \; \Phi_i \; (\vec{R}(t)) = H_{TCD} \; (\vec{R}(t)) \; \Phi_i \; (\vec{R}(t)) \; , \tag{1}$$

where H_{TCD} is the relativistic two-centre Hamiltonian depending on
the internuclear separation $\vec{R}(t)$. At non-relativistic bombarding
energies it is useful to expand the wavefunction Φ_i into Born-Oppenheimer states ϕ_j given by the instantaneous molecular eigenstates
of the Hamiltonian:

$$\Phi_i (\vec{R}(t)) = \sum_j a_{ij}(t) \; \phi_j(\vec{R}(t)) \; \exp \{-i\chi_j(t)\} \; . \tag{2}$$

The sum includes an integration over continuum states of positive and
negative energy. The phase factors χ_j are conveniently chosen as

$$\chi_j(t) = \int^t dt' \; <\phi_j(\vec{R}(t')) | H_{TCD} (\vec{R}(t')) | \phi_j(\vec{R}(t'))> \; . \tag{3}$$

Inserting the expansion of eq. (2) into eq. (1) and projecting with
stationary eigenfunctions we obtain a set of coupled differential
equations for the amplitudes $a_{ij}(t)$

$$\dot{a}_{ij}(t) = -\sum_{k \neq j} a_{ik}(t) \; <\phi_j | \partial/\partial t | \phi_k> \; \exp \{i(\chi_j - \chi_k)\} \; , \tag{4}$$

with the initial condition $a_{ij}(-\infty) = \delta_{ij}$.
After splitting the time derivative operator in terms of a radial and a rotational coupling and neglecting the latter one, the
coupled equations (4) may be solved by numerical integration. In the
independent-particle approximation excitations of the many-electron
system are described by incoherent summation over one-electron transition probabilities. After the collision the number of particles
occupying a state above the Fermi level, up to which the quasimolecular levels are initially filled, is

$$N_p = 2 \sum_{k<F} |a_{kp}(\infty)|^2 \qquad (p > F) \; , \tag{5}$$

while the number of holes in a state below the Fermi level is

$$N_q = 2 \sum_{k>F} \left| a_{kq}(\infty) \right|^2 \qquad (q<F) \; . \tag{6}$$

A hole in the negative energy continuum after the collision is a positron. For the number of correlated particle-hole pairs $N_{p,q}$ one has to calculate

$$N_{p,q} = 4N_p \cdot N_q + 2 \left| \sum_{k<F} a^{\ast}_{kp} \, a_{kq} \right|^2 \; . \tag{7}$$

This expression is relevant for instance if positrons and electrons would be measured in coincidence. Eqs. (5)-(7) contain a summation over spin orientations (for possible polarization effects see ref. 5).

THE SELF-ENERGY OF ELECTRONS IN CRITICAL FIELDS

The K-electron binding energy increases strongly as a function of the nuclear charge Z (see Fig. 2). For Z = 150 it amounts to

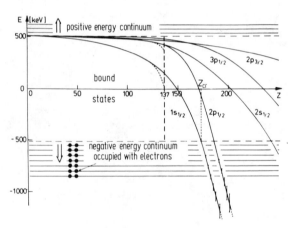

Fig. 2. Lowest bound states of the Dirac equation for nuclei with charge Z. While the Sommerfeld fine-structure energies (broken lines) for κ=-1 end at Z≈137, the solutions with extended Coulomb potential (full curves) can be traced down to the negative-energy continuum which is reached at critical charge Z_{cr}. The states entering the continuum obtain a spreading width.

about the electron rest mass and hence one enters the truly relativistic domain. For Z $\gtrsim$ 170 the binding energy exceeds twice the electron rest mass and the K-shell electron becomes a resonance imbedded in the negative energy continuum, which opens the possibility of spontaneous positron production[6-8]. It has long been a heavily debated question whether these assertions that hold in the independent particle approximation (external field approximation) continue to be valid if quantum field-theoretical corrections, such as vacuum-polarization and self-energy (see Fig. 3) are taken into account. In short: can such corrections prevent the occurrence of critical binding?

The dominant vacuum-polarization contribution is provided by the attractive Uehling potential. Its influence on electronic binding energies for superheavy systems has been calculated by various

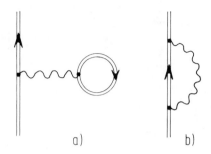

a)   b)

Fig. 3. Feynman diagrams for the lowest order vacuum-polarization (a) and self-energy (b). The double lines indicate the exact propagators and wavefunctions in the Coulomb field of a nucleus.

authors[6,9,10]. For the critical nuclear charge Z_{cr} the Uehling potential leads to an energy shift $\Delta E_{VP}^{(n=1)}=-11.8\text{keV}$, thus Z_{cr} will decrease by 1/3 of a unit. The remaining vacuum-polarization effects in lowest order of the fine structure constant α but in all orders of $(Z\alpha)^n$ were evaluated by M. Gyulassy[11] and by Rinker and Wilets[12] following the method of Wichmann and Kroll[13]. They obtained an additional energy shift of $\Delta E_{VP}^{(n>1)}=+1.15\text{keV}$ that is very small compared with the total K-shell binding energy of 1 MeV.

Electronic self-energy corrections for high-Z systems have been first studied in the pioneering work of G.E. Brown et al.[14], who showed how the traditional expansion in powers of the coupling constant $(Z\alpha)$ of the external field was avoided. This method was further refined and successfully applied in computations of electron energy shifts in high-Z elements by Desiderio and Johnson [15], who allowed for a realistic nuclear charge distribution as well as the electron-electron interaction in the Hartree-Fock approximation. Cheng and Johnson[16] continued these calculations up to Z = 160, where they found a repulsive energy shift for K-shell electrons of $\Delta E_{SE}=+7.3\text{keV}$. With respect to the critical charge Z_{cr} their calculations remained inconclusive.

Our calculations are based on the same method, which may be slightly simplified for the special case of K-shell electrons in hydrogen-like systems. The external potential energy V(x) is determined by the nuclear charge distribution, for which a homogeneously charged sphere with a radius $R = 1.2\ A^{1/3}$ fm has been assumed.

To check our computer code, we computed the self-energy contribution to the K-shell binding energy for various nuclear charges from Z=80 (mercury) up to Z=160 and found good agreement with the published values of the previous calculations (Fig. 4). Our calculation for Z=169 yielded ΔE_{SE} = 10.839keV. For the critical nuclear charge Z=170 we adjusted the nuclear mass number and hence the nuclear radius such that the K-electron energy eigenvalue differed only by 10^{-3} eV from $(-mc^2)$, the border line of the negative energy continuum. In this case we found an energy shift of ΔE_{SE} = 10.989 keV $\pm$ .3 keV, which still represents only a 1% correction to the total K-electron binding energy[18]. We conclude that self-energy effects may be safely neglected at the present status of precision in the investigations of ionization probabilities in superheavy quasimolecular systems. Self-

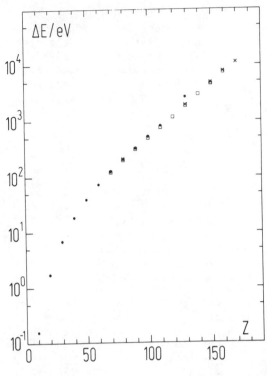

Fig. 4. The self-energy shift ΔE of K-shell electrons as a function of the nuclear charge Z. The calculations are performed to first order in α but to all orders in the coupling constant $(Z\alpha)$ of the external field. The dots denote the numerical results of P. Mohr[17] for 1s-electrons in the Coulomb field of point-like nuclei. The squares represent the values obtained by Cheng and Johnson[16] for a Hartree-Fock potential and extended nuclei. The results of the present calculations[18] for extended nuclei are indicated by crosses.

energy corrections certainly cannot be responsible for any deviations from simple scaling laws as was suggested in ref. 19.

If one adds the shifts due to vacuum-polarization and self-energy for critical external potentials, the total energy shift caused by field-theoretical corrections of order α amounts only to 300 eV. Whether this almost total cancellation is accidental or a systematical effect is presently unknown. We thus conclude that vacuum-polarization and self-energy may not prevent the K-shell binding energy from exceeding $2mc^2$ in superheavy systems with $Z > Z_{cr} \sim 170$. It would be desirable to be able to extend the self-energy calculation into the supercritical region. Here one would need to compute the shift of the peak of the bound state resonance. Due to the necessary renormalization of the divergent integrals this extension is beset with various difficulties and remains a task for the future.

POSITRON CREATION

An adequate description of positron production in supercritical collision systems, where Z_T+Z_p exceeds 173, requires a slight modification of the formalism set forth in the Introduction. In a supercritical system the 1s-state is represented as a resonance in the positron s-wave continuum and not by a single eigenstate of the Hamiltonian H_{TCD}. A formalism that avoids those difficulties and moreover has heuristic value for the interpretation of the positron creation process was developed by Reinhardt et al.[1] and later also discussed by Tomoda and Weidenmüller[20]. The method is based on the observation that the continuum wavefunction of the supercritical system at resonance energy $E_p = E_{res}$ is quite similar to the discrete 1s-state in the subcritical case except for an oscillating tail, small in amplitude but reaching out to infinity. This structure reflects the occurrence of a tunneling process through the barrier separating the particle- and antiparticle solutions of the Dirac equation in a semi-classical picture. Apart from the asymptotic behaviour the 1s-wave-function retains many of its properties, e.g., the strong localization and the radial matrix elements which may be continued smoothly to the supercritical region if the tail of the wavefunction is neglected.

This idea can be used to develop a general method to treat resonance scattering. In this context Wang and Shakin[21] introduced a projection formalism for resonances in the nuclear continuum shell model: After having defined a normalizable quasibound wavefunction ϕ_R, a new continuum $\tilde{\varphi}_{Ep}$ is constructed which spans a subspace orthogonal to ϕ_R and replaces the old continuum ϕ_{Ep}. The modified continuum states satisfy the original Dirac equation supplemented by an inhomogeneous term that ensures orthogonality with respect to the resonance wavefunction ϕ_R:

$$(H_{TCD} - E_P) \, |\tilde{\varphi}_{Ep}> = <\phi_R|H_{TCD}|\tilde{\varphi}_{Ep}> \, |\phi_R> . \tag{8}$$

If the states ϕ_R and $\tilde{\varphi}_{Ep}$ are used as part of the basis in eq. (2) the 1s-state ϕ_R couples to the new positron continuum by two separate coupling operators

$$\dot{R} \, <\tilde{\varphi}_{Ep}|\partial/\partial R|\phi_R> + i/\hbar \, <\tilde{\varphi}_{Ep}|H_{TCD}|\phi_R> . \tag{9}$$

The second matrix element arises since ϕ_R and $\tilde{\varphi}_{Ep}$ are not exact eigenstates of the two-centre Hamiltonian H_{TCD}. It does not depend on the nuclear motion and leads, in the static limit $R(t) = const < R_{cr}$, to an exponential decay of a hole prepared in ϕ_R. The decay width

$$\Gamma = 2\pi \, |<\tilde{\varphi}_{Eres}|H_{TCD}|\phi_R>|^2 \tag{10}$$

is identical to the width of the resonance in the unmodified positron continuum.

The formalism thus leads naturally to the emergence of 'induced' and 'spontaneous' positron creation, the latter resulting from the presence of an unstable state ϕ_R in the expansion basis. In practice, however, this does not result in a marked threshold behaviour at the border of the supercritical region for two reasons. Firstly, both couplings enter via their Fourier transforms depending on the time development of the heavy ion collision. Their contributions have to be added coherently so that in a given collision there is no physical way to distinguish between them. Secondly, in collisions below the Coulomb barrier the rapid variation of the quasimolecular potential, especially in the supercritical region, causes significant contributions from the dynamical coupling, whereas the period of time for which the internuclear distance $R(t)$ is less than R_{cr} is usually very short ($\sim 10^{-21}$ sec) as compared with the decay time of the 1s-resonance ($\sim 10^{-19}$ sec).

Therefore, the predicted production rates and energy spectra of positrons continue smoothly from the subcritical to the supercritical region. Qualitative deviations of the positron production rate in supercritical collision systems are expected only under favourable conditions: Since the 'spontaneous' and 'dynamical' couplings exhibit a different functional dependence on the nuclear motion, an increase in collision time can be expected to provide a clear signature for supercritical collisions. Therefore Rafelski, Müller, and Greiner[22] suggested the study of positron emission in heavy ion reactions at bombarding energies above the Coulomb barrier, where the formation of a di-nuclear system or of a compound nucleus would eventually lead to a time delay within the bounds of the critical distance R_{cr}. During this sticking time T the spontaneous decay of the $1s\sigma$-resonance, by filling dynamically created K-shell holes under emission of positrons, might be strongly enhanced.

A variety of experiments concerning positron creation have been performed at the Gesellschaft für Schwerionenforschung (GSI) in Darmstadt during the past four years. They are subject of J.S. Greenberg's review talk at this conference to which we refer the reader. Here we wish to concentrate on theoretical results and comparisons with selected experiments concerning non-Coulombic collisions. First of all some general remarks.

We have integrated the modified system of differential equations (4), (9) in the framework of the monopole approximation including up to 8 bound states and ~ 17 states in the upper continuum for each angular momentum channel ($s_{1/2}$ and $p_{1/2}$-waves, i.e. $\kappa = -1$, $+1$, respectively). Positron emission rates increase very fast with total nuclear charge, flattening somewhat for the highest Z-values. If parametrized by a power law $(Z_T + Z_p)^n$, the exponent takes values of 20 down to 13, if an initial Fermi-level above $3s\sigma$, $4p_{1/2}\sigma$ is assumed, or even $n \simeq 29$ for bare nuclei (F=0). This highly nonlinear behaviour clearly expresses the non-perturbative nature of the mechanism of positron production in such superheavy systems. Mainly responsible for the enhancement for fully stripped nuclei is the contribution of the 1s-state which in normal collisions (F>0) is suppressed by the small K-vacancy probability. If the K-shell is empty it becomes the dominant final state for pair production due to the strong coupling between the 1s-state and the antiparticle continuum which it approaches and even enters in the supercritical region. In sub-Coulomb barrier collisions $s_{1/2}$ and $p_{1/2}$-waves contribute about equally to the total result.

At this point we must address the major problem in analysing the experimental data. Already for bombarding energies well below the Coulomb barrier E_c (E/E$_c$ $\sim$ .8) the nuclei can be excited by Coulomb excitation, and the emitted photons with energy above 1022 keV can undergo internal pair conversion. Thus one has to measure the γ-spectrum simultaneously and to fold it with the conversion coefficients. Here one has to know - or to assume - the γ-ray multipolarity. Monopole conversion cannot be handled by this method. Up to now, all conclusions on positron production in heavy-ion collisions had to rely on the described procedure for background subtraction. For further details we refer to J.S. Greenberg's contribution.

The first generation of experiments[23,24] established the dependence of positron excitation rates on the kinematic conditions as well as on the combined charge Z. The Z-dependent increase, which spans an order of a magnitude while $\Delta Z/Z$ is only 12%, is well described by theory. Also the shape of the theoretical curves is in good agreement with the experimental data. In the Pb+Pb system and, for smaller distances of closest approach, even in Pb+U and U+U collisions the data agree also in absolute values. In the heaviest accessible system U+Cm (Z_u=188) and for larger distances R_{min} the theory has a tendency to overestimate the measured data by up to 40%.

From these data no qualitative signature for the 'diving' of the 1sσ-state in U+U, U+Cm collisions could be extracted, in agreement with theoretical predictions. More sensitive information can be obtained by the measurement of energy spectra of positrons detected in coincidence with the scattered ions. Their knowledge is most useful if one wants to find deviations hinting to the positron creation mechanism. Fig. 5 shows the first published positron spectra of Backe et al.[25,26] for three collision systems, U+Pd, U+Pb, and U+U, at 5.9 MeV/u bombarding energy; the ions are detected in an

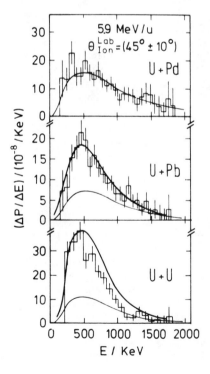

Fig. 5. Spectra of emitted positrons in 5.9 MeV/u collisions measured by Backe et al.[25,26] in coincidence with ions scattered in the angular window Θ_{lab}=45°±10°. The spectrum in the lightest system, U+Pd, is explained by nuclear pair conversion alone (light line). In the U+Pb and U+U systems the sum (full lines) of nuclear and calculated atomic positron production rates is displayed.

angular window $\theta_{lab}=45^O\pm10^O$. For U+Pd (Z=138) no atomic positrons are expected, the data can be fully accounted for by nuclear conversion (light curve). Extrapolating this procedure to the U+Pb system (light curve) the sum of background and calculated QED positron rates (full curve) is in excellent agreement with the observed emission rates. In the spectrum of the supercritical U+U system ($E_{1s} \simeq$ 1200 keV for $R_{min} \simeq 21$ fm) some deviations are seen.

A possible source which could cause deviations in the shape of the positron spectra from the results presented so far will be discussed in the following. To obtain theoretical predictions for positron production it is essential to include the 'spontaneous' coupling for collisions where the 1sσ-state joins the lower continuum. If it is left out of the calculation the resulting positron spectra would be strongly altered: The 'induced' radial coupling is changed at the same time as the spontaneous coupling becomes important. Both contributions add up coherently and cannot be observed separately. A promising strategy to get a clear qualitative signature for the diving process is to try to modify the time structure and to select heavy-ion collisions with prolonged nuclear contact time. Such nuclear reactions are expected to occur at energies close to or above the Coulomb barrier. The nuclear delay time T should provide a handle to distinguish supercritical systems.

Using a schematic model for the trajectory we have performed coupled channel calculations for the four heavy-ion collision systems Pb+Pb, Pb+U, U+U, and U+Cm, corresponding to Z_{united} = 164, 174, 184, and 188, respectively[27]. Independently of assumptions on the incoming and outgoing path, of dissipation of nuclear kinetic energy or angular momentum during the reaction, and of the position of the initial Fermi level, all positron spectra exhibit the following features:

In subcritical collision systems ($Z_u+Z_p \lesssim 173$) a delay time T causes modulations in the positron spectrum with a width $\Delta E=h/T$. In Fig. 6a positron spectra are displayed for a Pb+Pb collision ($E_{lab}=$ 8.73 MeV/u, b=7.11 fm, F=3sσ, $4p_{1/2}$σ) with delay times T=0 (pure Rutherford scattering), 3·, 6·, and $10\cdot10^{-21}$ sec. The modulations are due to interference effects in much the same way as predicted for the δ-electron spectra in deep inelastic heavy-ion collisions[28].

In addition to the interference patterns an enhancement of positron production in time-delayed supercritical collisions is observed, where the binding energy of the lowest bound states exceeds the value $2mc^2$. For long delay times a distinct peak in the positron spectrum is found at the location of the supercritical bound state resonance (binding energy minus $2mc^2$) due to the spontaneous pair-creation mechanism. A detailed analysis of the spectra reveals that this peak emerges gradually as $Z_u=Z_u+Z_p$ exceeds Z_{cr}. Positron spectra for the supercritical system U+U ($E_{lab}=7.35$ MeV/u, b=3.72 fm, F=3) are shown in Fig. 6b. With increasing delay time the position of the maximum drifts slowly from the kinematic maximum to the 'resonance energy', which depends on the combined charge, the separation of the two nuclei and on the nuclear charge distribution[27].

However, for any chosen set of experimental parameters, the nuclear reaction time T may not (and will not) be sharp but distributed

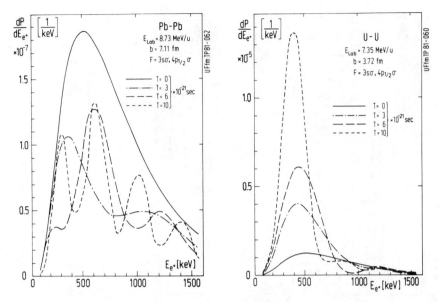

Fig. 6. Spectra of positrons created in subcritical (part a) and supercritical (part b) heavy-ion collisions assuming grazing Coulomb trajectory (full lines) and nuclear reactions leading to delay times T=3·, 6·, and $10 \cdot 10^{-21}$ sec, resp., using a schematic model for the trajectory[7]. Whereas for the lighter collision systems modulations in the positron spectra are present, a distinct peak at the 'resonance' energy $E_{1s\sigma}(R_{min})$ builds up for systems with $Z_u > 173$.

over a certain range with a time distribution function $f(T)$. As an assumption we took a Gaussian centered at $\bar{T}$

$$f(T) = \frac{1}{\sqrt{2\pi}\tau} \exp\left(-\frac{(T-\bar{T})^2}{2\tau^2}\right). \tag{11}$$

The resulting positron spectra for parameters $\bar{T} = 16 \cdot 10^{-21}$ sec and $\tau = 0$ or $\tau = 2 \cdot 10^{-21}$ sec are displayed in Fig. 7 for a head-on U+U collision at $E_{lab} = 5.9$ MeV/u. If we consider the 'subcritical' p-states only (part a) we observe that the oscillations disappear already for $\tau = 2 \cdot 10^{-21}$ sec. Thus we conclude that the appearance of several oscillations in the spectrum can be expected only for sufficiently sharp nuclear reaction times if the system is subcritical. Also in the total spectrum (part b) the oscillations are damped out for increasing τ. But most striking is the invariance of the dominant nonvanishing first peak, which originates from the spontaneous part of the positron production mechanisms.

A similar effect is expected in the continuum spectrum of quasi-molecular X-rays. Calculations for the Pb+Pb system (Fig. 8) show that a line emerges at the united atom K transition energy if the delay time becomes sufficiently long[29]. The results for the U+U system will be similar, but there is a formidable background from nuclear

216

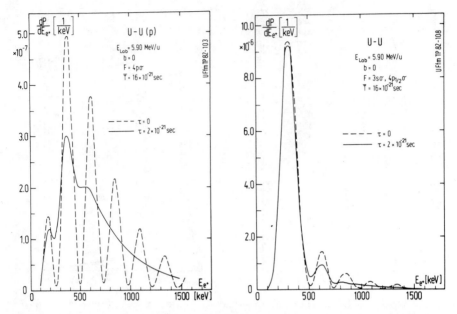

Fig. 7. Differential positron production probability versus kinetic positron energy E_{e^+} in a central U+U collision at E_{lab} = 5.9 MeV/u, assuming a Gaussian nuclear reaction time distribution centered at $\overline{T}$ with a width τ. (a) p-states only representing a subcritical system, (b) including also the contribution of the s-states. Most striking is the appearance of the first pronounced peak which originates from spontaneous positron production.

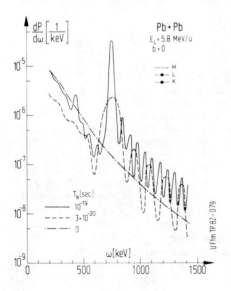

Fig. 8. Quasimolecular X-rays of the Pb+Pb system for various sticking times.

γ-rays which probably makes an experimental observation very difficult.

The results described so far were obtained within a schematic model for the nuclear motion, which facilitates a systematic study of the time delay effect and allows for an investigation of the conceptually interesting limit of large sticking times. To analyse a given experiment, however, the employed nuclear trajectories should be consistent with the elastic and inelastic heavy-ion scattering data. Many reaction models with different degrees of refinement have been discussed in the literature. We have calculated trajectories with the macroscopic friction model of Schmidt et al.[30], which includes nuclear neck formation. Strong deviations from Coulomb trajectories are found and an energy loss up to ~30% (for b~0) can be obtained. The change of the positron spectrum for collisions with varying degree of nuclear contact is demonstrated in Fig. 9. Part (a) shows the modified U+U-trajectories R(t) for several orbital angular momenta from $\ell=0$ head-on to $\ell=400\hbar$ near grazing collisions. The corresponding positron spectra, part (b), show a gradual enhancement at $E_e+ = 500$ keV. As expected a longer delay time ΔT leads to increased positron production in the s-channel. On the other hand the change in kinematics causes a drift to lower kinetic energies in the $p_{1/2}$-partial wave spectra due to destructive interference. Both effects taken together lead to an enhancement of the maximum and a drift towards lower energies also in the total spectrum.

Measurements by Backe et al.[31] seem to indicate such tendencies: In U+U and U+Cm collisions at energies above the Coulomb barrier positron spectra have been measured in coincidence with fission fragments in order to get a signature for close nuclear contact. The analysis shows an enhancement of dP/dE_e+ at lower kinetic energies in qualitative agreement with Fig. 9b. For a quantitative comparison one has to integrate the impact parameter-dependent positron spectra over all values of b which lead to a nuclear reaction, weighted by the corresponding probability w(b) to induce nuclear fission. Performing the integration with a weight factor w(b) = 1 for $b<b_{grazing}$ and w(b)=0 elsewhere, there remains an energy shift of ~50 keV in the experimental data in comparison with the theoretical curves, which might be due to electron screening effects. Furthermore, as mentioned above, the theoretical values have to be reduced by an overall factor ~ 2/3. Fig. 9c shows the experimental data for U+U collisions at E_{lab} = 5.9 MeV/u, 7.5 MeV/u, and 8.4 MeV/u, in comparison with theoretical results excluding electron screening, but reduced by the factor mentioned above. Dashed lines indicate pure Rutherford scattering trajectories, whereas the solid lines display spectra calculated with the modified trajectories of Fig. 9a. For an even better agreement longer delay times ΔT (~ $2 \cdot 10^{-21}$ sec) may be needed. Further investigations along these lines seem to be very promising, both to establish the mechanism of positron production and to deduce the nuclear reaction time scale.

As another interesting theoretical problem one might speculate about the presence of nuclear collisions with very long reaction times. What would positron spectra look like if, at a given scattering angle, a superposition of Rutherford scattering and long-lasting

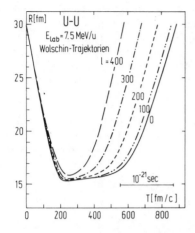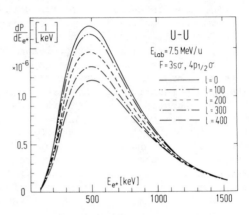

nuclear reactions is assumed? If, for the sake of simplicity, nuclear scattering with definite delay time T is assumed, a ratio q<<1 can be used as a measure of the relative cross section of reactions leading to long contact times, as compared with the Rutherford cross section. As positron production is very strongly enhanced for reaction times larger than 10^{-20} sec, a peak superimposed on the smooth spectrum of positrons emitted in the much more frequent 'distant' Coulomb collisions could emerge. A rough estimate shows that for long delay times a peak may be prominent even if the differential reaction cross section is less than 1% of the Rutherford cross section.

To obtain the full shape of the positron spectrum, an assumption

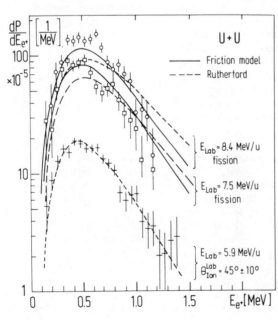

Fig. 9. (a) Nuclear trajectories calculated in the friction model[30] for 7.5 MeV/u U+U collisions at various values of the orbital angular momentum ℓ between 0 and 400ħ. (b) Energy spectra of positrons emitted in the collisions shown in (a). The results for the angular momentum channels s and $p_{1/2}$ have been added. (c) Comparison of theoretical predictions for U+U at E_{lab}=5.9 MeV/u, 7.5 MeV/u, and 8.4 MeV/u, with experimental data of Backe et al.[31], as described in the text.

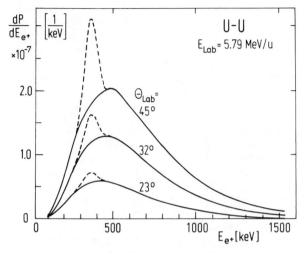

Fig. 10. Spectra of positrons emitted in 5.79 MeV/u U+U collisions in coincidence with a scattered nucleus for three selected lab ion angles. The fully drawn curves are calculated assuming Rutherford scattering only. The dashed lines show the effect of an additional nuclear reaction with a lifetime $T = 4 \cdot 10^{-20}$ sec. A relative fraction of $q = 2.4 \cdot 10^{-3}$ reactions per elastically scattered ion (at 45°) has been assumed.

about the angular distribution $d\sigma^N/d\theta$ of the nuclear reaction component is needed. We shall discuss two extreme simplified cases: (i) isotropic break-up of the compound system, (ii) focussing of the reaction fragments into a narrow angular window in the c.m. system. If the reaction products were emitted isotropically, a line should, in principle, be observable in the positron spectrum at all ion scattering angles. However, it would be most pronounced at $\theta_{c.m.} = 90^{\circ}$, being suppressed at other angles relative to the elastic scattering cross section. Fig. 10 shows spectra of positrons, reduced by a factor $\sim 2/3$, for 5.79 MeV/u U+U collisions for several fixed scattering angles. A long lived nuclear reaction component (dashed lines) with $T = 4 \cdot 10^{-20}$ sec and admixture $q = 2.4 \cdot 10^{-3}$ (at $\theta_{c.m.} = 90^{\circ}$) has been added under the assumption of an isotropic distribution in the reaction plane.

If the emitted reaction fragments are focussed under a certain scattering angle, a detailed quantitative analysis depends strongly on the strength of the focussing effect. Such an effect could be produced, if the life-time T of the nuclear molecular or compound system depended strongly on the angular momentum ℓ carried into the reaction. This is not an unreasonable possibility since it is evident that T must vanish for very large values of ℓ. A detailed investigation of this scenario would require a quantum mechanical treatment of the nuclear motion together with a partial wave analysis of the scattering cross-section.

Two experimental groups[32,33] recently have performed experiments with U+Th, U+U, and U+Cm at energies close to the Coulomb barrier. Contrary to the results of Backe et al.[31] their positron spectra seem to show remarkable structures. C. Kozhuharov, P. Kienle et al.[32] measured at 5.8 MeV/u beam energy positron spectra in coincidence with ions scattered into various narrow angular windows. A preliminary analysis shows sharp maxima in the spectra that are most pronounced under laboratory scattering angles around $\Theta_{lab} \sim 45^O$. In the U+U measurement the position of the peak was found to be located at $\sim$ 370 keV with a width of about $\sim$ 90 keV, for U+Th the effect is less well established. After subtraction of a smooth background the number of positrons per detected ion emitted in the peak is roughly 10^{-5} at $\Theta \sim 45^O$. If the observed structure is of quasimolecular origin, it must be produced in very long-lasting nuclear reactions because of the small width. If one compares the experimental width with spectra from coupled channel calculations based on the schematic sticking model a minimum value $T \sim 4 \cdot 10^{-20}$ sec is required. Should the observed line width have instrumental origins or be due to additional broadening effects (such as Doppler broadening) the reaction time T would have to be even longer. According to our calculations the probability for positron production in a delayed ($T=4 \cdot 10^{-20}$ sec) collision should be $\sim 4.7 \cdot 10^{-3}$, which must be compared with the observed probability $\sim 10^{-5}$. Thus a fraction of $q = P_{e^+}(exp)/P_{e^+}(theor) \simeq 2 \cdot 10^{-3}$ delayed collisions per elastically scattered ion is sufficient to produce the observed effect (at 45°). This number should serve only for a general orientation, since it depends on the details of the model.

Similar structures were detected in the experiment of H. Bokemeyer, J.S. Greenberg et al.[33] in U+U and U+Cm collisions. For details we refer to J.S. Greenberg's contribution to this conference.

One might suppose that those peaks are caused by nuclear background processes. However, nuclear transitions of, e.g., multipolarity E1 or E2 should also be observable in the emitted photon spectra, provided that proper Doppler shift corrections are performed. Whether the peaks can be caused by EO-processes will be discussed later on. On the other hand, if a sharply focussed nuclear reaction takes place it is not surprising that an experiment not triggering for the optimal kinematic conditions might smear out any evidence for structure. Further investigations are needed to settle this question. Should the observed phenomena indeed be caused by reactions with a very long time scale, this would have far reaching consequences for the physics of nuclear systems in the superheavy region. More about those aspects can be found in ref. 34.

The superheavy nuclear compound system very likely is not a static object. Its internal dynamics may influence the spectrum of emitted positrons. Clearly, a rigorous treatment of such effects cannot be based on the semiclassical approximation, but requires a fully quantum mechanical reaction theory for the nuclear scattering. While we are working on such an extension of the work of Anholt and Blair[35] and McVoy and Weidenmüller[36], we have attempted to understand possible consequences in the framework of very simple (and probably oversimplified) models.

Let us imagine a classical picture for a nuclear excitation of the supercritical compound system in the spirit of the correspondence principle. The 'internuclear separation' R(t), or better: the quadrupole deformation of the dinuclear system, is supposed to oscillate around the distance of closest approach R_o of the Coulomb trajectory for a fixed duration T

$$R = R_o \ (1- \alpha_o \ \sin(2\pi\nu t)) \ , \qquad\qquad (12)$$

where $h\nu$ is the quantum mechanical oscillator energy.

In Fig. 11 positron spectra are plotted for a central U+U collision at $E_{lab} = 6.2$ MeV/u. In order to demonstrate the qualitative effect we have restricted the calculations to couplings between s-states only. The parameters in eq. (12) are fixed by $\alpha_o = .25$ and $\nu = .125 \cdot 10^{21} sec^{-1}$ corresponding to an oscillator energy $h\nu = 517$ keV. Various sticking periods between T=0 and $T=108 \cdot 10^{-21}$ sec are considered in Fig. 11a. Fig. 11b shows on a linear scale the computed positron spectrum for $T=36 \cdot 10^{-21}$ sec. As in Fig. 6b we find the dominant 'spontaneous peak'. But in addition a second pronounced peak appears at about $E_{e^+} = 800$ keV. This reflects the fact that part of the vibrational energy of the dinuclear system is transferred to the emitted positron.

This is nothing but the classical analogue of a pair conversion process in the supercritical nuclear compound system. If the nuclear system is vibrationally excited, it may decay to the ground state transferring its excitation energy to an electron-positron pair. The electron occupies the vacant 1s-state and the positron carries the energy balance. Without the nuclear de-excitation this would be the spontaneous positron creation process, but with it the positron line is shifted by the nuclear excitation energy. We shall discuss this

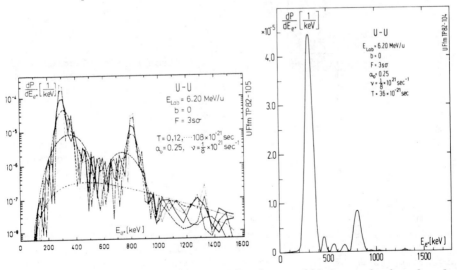

Fig. 11. Positron spectra in a U+U head-on collision calculated under the assumption that the distance between the nuclear centres oscillates around the distance of closest approach R_o of the Coulomb trajectory (cf. eq. (12)). Nuclear reaction times T = 0, 12, 36, 60, 84, and $108 \cdot 10^{-21}$ sec are considered. The longest duration T corresponds to the most pronounced 'spontaneous peak', etc. Note the logarithmic scale in part a).

process further within the context of pair-conversion processes in the next chapter.

CONVERSION PROCESSES IN SINGLE ATOMS AND IN SUPER-CRITICAL COMPOUND SYSTEMS

In collisions of very heavy ions with E_{lab} > 3 MeV/u both nuclei are Coulomb excited. For bombarding energies at about the Coulomb barrier transfer reactions or even deep inelastic nuclear reactions can take place which lead to additional excitations of the nuclei. This internal excitation energy may be carried away by a photon or may be transferred to a bound electron or to an electron of the negative energy continuum, which leads to ionization and electron-positron pair creation, respectively. The latter process requires nuclear transition energies ω larger than twice the electron rest mass. Nuclear E0-transitions are characterized by the absence of single photon emissions, because a photon must carry at least one unit of angular momentum. As mentioned before such processes form the main source of non-atomic positrons, and they have to be well understood[37,38], if one wants to draw firm conclusions about the presence or absence of spontaneous pair creation.

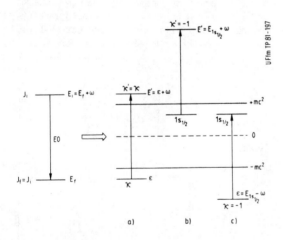

Fig. 12. Schematic representation of electron conversion processes accompanying nuclear E0-transition from a state $\{E_i, J_i\}$ to a state $\{E_f = E_i - \omega, J_f = J_i\}$. a) Electron-positron pair production leading to a continuous energy distribution of positrons and electrons. b) Conversion of K-shell electrons - a monoenergetic electron-production mechanism. c) Monoenergetic positron-production - a negligible process.

The basic processes under investigation are depicted schematically in Fig. 12. The nucleus which makes an E0-transition is labelled by its initial and final state angular momenta J_i, $J_f = J_i$ and eigenenergies E_i, $E_f = E_i - \omega$. Process a) describes the electron-positron pair creation. An electron of the negative energy continuum ($\varepsilon = -E < -mc^2$) with Dirac quantum number κ is lifted to the positive energy continuum. The final state energy obviously amounts to $E' = \varepsilon + \omega$ whereas the angular momentum quantum number remains unchanged. Since neither the initial state energy nor the final state energy is fixed one expects a continuous energy distribution for the emitted positrons. Process b) indicates the conversion of a K-shell

electron $(n=1, l=0, j=\frac{1}{2}, \kappa=-1)$ with energy eigenvalue $E_{1s_{1/2}}$. Thus
bound states with definite energies are involved. Energy conservation then simply causes monoenergetic lepton emission for a fixed nuclear transition energy ω. Process c) symbolizes monoenergetic positron production. Here an electron of the negative energy continuum is excited to a bound state, e.g. to the $1s_{1/2}$-state. This represents a rather rare process[37], which can be neglected.

   Thus we focus our attention on

i)   the pair conversion coefficient β, defined as the ratio of the pair production probability (process a)) compared with that of photon emission for a specific nuclear transition with energy ω. Since the energy of the electron and the positron takes continuous values we may express β also as integral of the differential pair conversion coefficient $d\beta/dE$. The lower bound of the integral is determined by the rest mass of the electron, which corresponds to vanishing kinetic energy, while the upper bound is given by the nuclear transition energy ω minus m_e.

ii)   the conversion coefficient α, defined as the ratio of the probabilities of inner-shell vacancy formation (process b)) and photon emission. In particular this mechanism is important for low energy nuclear transitions.

iii)   the ratio η of the two conversion probabilities for electron-positron pair creation and for the ionization of bound state electrons. This ratio is completely determined by the density of the electron wavefunctions at the nuclear origin, thus being independent of the nuclear wavefunction.

   We computed the differential conversion coefficient $d\beta/dE$ for nuclear E1 and E2 transitions. For the nucleus $_{92}$U the energy distributions of emitted positrons is shown in Fig. 13a. Nuclear transition energies of 1323 keV, 1423 keV, 1523 keV, and 1623 keV are considered. Fig. 13b shows the equivalent differential conversion coefficient $d\eta/dE$ for nuclear E0-transitions. As bound state only the atomic K-shell has been taken into account. The conversion probability of higher bound states is at least one order of magnitude smaller.

   Now we can discuss the possibility whether the observed structures in positron spectra may originate from nuclear E0-transitions. One convincing argument against this interpretation is related to the shape of the e^+-energy distribution. According to Fig. 13b the halfwidth of the spectra should be at least 150 keV. However, the observed structure is much narrower. The second argument is connected with the energy distribution of the emitted δ-electrons. It was shown in ref. 37 that, if the observed structures are caused by nuclear E0-transitions, one should also observe a distinct peak in the δ-electron distribution. Such a peak does not seem to exist[39].

   We now turn to the discussion of electron-positron pair conversion in supercritical compound systems. A supercritical nucleus $Z=184$ which undergoes a transition with $\omega > 2mc^2$ during the nuclear reaction period T may transfer this excitation energy to one of the electrons in the negative energy continuum. The remaining hole is defined as positron. But also the K-shell electron can be lifted

224

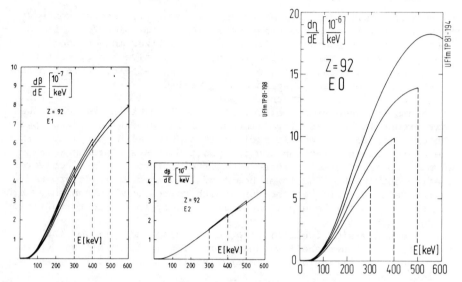

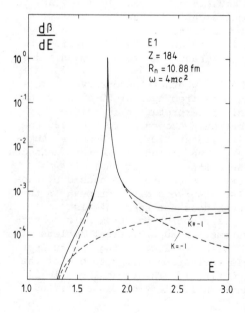

Fig. 13. (a) Differential conversion coefficient dβ/dE with respect to the kinetic positron energy E for nuclear E1- and E2- transitions in $_{92}$U. Nuclear transition energies ω = 1323 keV, 1423 keV, 1523 keV, and 1623 keV are considered, corresponding to maximum kinetic positron energies of E_{max} = 300 keV, 400 keV, 500 keV, and 600 keV, respectively. (b) Differential conversion probability ratio dη/dE with respect to the kinetic positron energy E for nuclear E0-transitions in U. The same transition energies as in part (a).

Fig. 14. Differential pair conversion coefficient dβ/dE as function of the positron energy E. Nuclear transitions with ω=4 mc^2 and of multipolarity E1 in a supercritical nucleus Z=184 with a radius R_n = 10.88 fm are considered. The dashed curves denote the various contributions of the electron angular momentum states.

to the upper continuum. If T is longer than the spontaneous decay
width of the K-shell resonance the K-vacancy will be filled again
leading to spontaneous positron emission.

For the nuclear charge distribution a homogeneously charged
sphere with a radius R_n = 10.88 fm has been assumed. In Fig. 14 we
show the differential pair conversion coefficient $d\beta/dE$ as function
of the positron energy E. Nuclear transitions with $\omega = 4mc^2$ and of
multipolarity E1 are considered. The appearance of the pronounced
peak at $E = E_{res}$ is striking. The dashed curves represent the various contributions of electron states with κ = -1 and $\kappa \neq$ -1. As expected the resonance shows up only in the $(\kappa = -1)$-channel. Similar
ratios $d\beta/dE$ are obtained for other multipolarities. Further details
can be found in ref. 34.

We now come back to the conversion process in the nuclear compound system in the presence of a K-vacancy, which was discussed at
the end of the previous chapter in the framework of a classical model.

In ordinary stable nuclei this process is rather slow, having
transition times in the order of $>10^{-12}$ sec (corresponding to partial
decay widths $\Gamma_{conv} <10^{-3}$ eV). The contraction of bound and continuum
wavefunctions in superheavy atoms strongly enhances this process.
This is most prominent in electric monopole (E0-) conversion, where
the width to a good approximation is given by the simple expression

$$\Gamma_{conv} (i \rightarrow f, E0) = 2\pi \left| \frac{1}{6} e^2 \phi_f (0) \phi_i (0) R^2 \rho \right|^2 \tag{13}$$

with the nuclear E0-transition matrix element

$$\rho = \sum_p \int \Psi_f (r_p/R)^2 \Psi_i dV. \tag{14}$$

Γ_{conv} is proportional to the electron (positron) densities in the
initial and final state at the origin. A simple inspection of these
densities assuming ϕ_i to be a state in the positron continuum and
ϕ_f the 1s-state leads to the observation that Γ_{conv} (E0) increases
by more than five orders of magnitude when going from a single U nucleus to the combined Z=184-system, assuming constant ρ.

Depending on the strength of the nuclear matrix element this
means that the width of the conversion process can approach the spontaneous decay width within one or two orders of magnitude.

In the case of a subcritical charge, $Z<Z_{cr}$, positron lines could
be produced by monoenergetic conversion of excited nuclear states
filling a hole in, e.g., the 1s-, $2p_{1/2}$-, etc. inner shell levels.
The width of the line again would be inversely proportional to the
lifetime T of the nuclear system, provided this is shorter than the
decay time of the excited state. In the same way for supercritical
Z the spontaneous decay of the 1s-resonance could be accompanied by
a conversion process leading to one (or more) additional weaker lines,
the relative intensity being determined by the ratio $\Gamma_{spont}/\Gamma_{conv}$.
Here also a new physical phenomenon is possible, as indicated in the
right part of Fig. 15: In an inverse conversion process the nucleus
may take up energy released by the filling of a 1s-hole. The emitted positron has an energy reduced by the absorbed amount E_N.

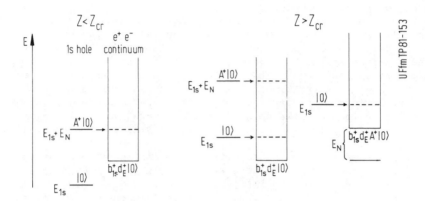

Fig. 15. Monoenergetic pair conversion filling a hole in the 1s-state, induced by a nuclear transition with energy E_N, in a subcritical (left) and supercritical (centre) system. In the latter case also the inverse process is possible, where the nucleus becomes excited while a positron with reduced energy is emitted (right).

Under favourable circumstances, therefore, 'sidebands' might appear in the positron spectrum, allowing for a spectroscopy of the compound system. To gain a full understanding of these processes it will be necessary to develop a theory accounting for both the classical and quantal aspects of the combined electron and nucleus system in a heavy ion collision.

ACKNOWLEDGEMENTS

We acknowledge very fruitful discussions with P. Armbruster, H. Backe, K. Bethge, H. Bokemeyer, F. Bosch, J.S. Greenberg, A. Gruppe, P. Kienle, W. Koenig, Ch. Kozhuharov, M. Krämer, D. Liesen, P. Mokler, D. Schwalm, J. Schweppe, and P. Senger concerning their experiments.

This work was supported by the Bundesministerium für Forschung und Technologie (BMFT) and the Deutsche Forschungsgemeinschaft (DFG). One of us (G.S.) acknowledges the support of the DFG-Heisenberg Programm.

REFERENCES

+   For a general survey of the present status of the field we refer to the book "Quantum Electrodynamics of Strong Fields", Proceedings of the NATO Advanced Study Institute at Lahnstein/Germany, June 1981, to be published by Plenum Press.

1. J. Reinhardt, B. Müller, and W. Greiner, Phys. Rev. A24, 103 (1981).

2. G. Soff, B. Müller, and W. Greiner, Phys. Rev. Lett. 40, 540 (1978).

227

3. G. Soff, W. Greiner, W. Betz, and B. Müller, Phys. Rev. A20, 169 (1979).

    G. Soff, J. Reinhardt, B. Müller, and W. Greiner, Z. Physik A294, 137 (1980).

    G. Soff, B. Müller, and W. Greiner, Z. Physik A299, 189 (1981).
4. J. Reinhardt, B. Müller, W. Greiner, and G. Soff, Phys. Rev. Lett. 43, 1307 (1979).
5. G. Soff, J. Reinhardt, and W. Greiner, Phys. Rev. A23, 701 (1981).
6. W. Pieper and W. Greiner, Z. Physik 218, 327 (1969).
7. Ya. B. Zeldovich and V.S. Popov, Sov. Phys.-Usp. 14, 673 (1972).
8. B. Müller, H. Peitz, J. Rafelski, and W. Greiner, Phys. Rev. Lett. 28, 1235 (1972).

    B. Müller, J. Rafelski, and W. Greiner, Z. Physik 257, 62 and 183 (1972).

    J. Rafelski, B. Müller, and W. Greiner, Nucl. Phys. B68, 585 (1974).
9. F.G. Werner and J.A. Wheeler, Phys. Rev. 109, 126 (1958).
10. G. Soff, B. Müller, and J. Rafelski, Z. Naturforsch. 29a, 1267 (1974).
11. M. Gyulassy, Phys. Rev. Lett. 33, 921 (1974).
12. G.A. Rinker and L. Wilets, Phys. Rev. A12, 748 (1975).
13. E.H. Wichmann and N.M. Kroll, Phys. Rev. 101, 843 (1956).
14. G.E. Brown and G.W. Schaefer, Proc. Roy. Soc. (London) A233, 527 (1956).

    G.E. Brown, J.S. Langer, and G.W. Schaefer, Proc. Roy. Soc. (London) A251, 92 (1959).

    G.E. Brown and D.F. Mayers, Proc. Roy. Soc. (London) A251, 105 (1959).
15. A.M. Desiderio and W.R. Johnson, Phys. Rev. A3, 1267 (1971).
16. K.T. Cheng and W.R. Johnson, Phys. Rev. A14, 1943 (1976).
17. P.J. Mohr, Ann. Phys. 88, 26 and 52 (1974).
18. G. Soff, P. Schlüter, B. Müller, and W. Greiner, Phys. Rev. Lett. 48, 1465 (1982).
19. D. Liesen, P. Armbruster, F. Bosch, S. Hagmann, P.H. Mokler, H.J. Wollersheim, H. Schmidt-Böcking, R. Schuch, and J.B. Wilhelmy, Phys. Rev. Lett. 44, 983 (1980).
20. T. Tomoda and H.A. Weidenmüller, Phys. Rev. A26, 162 (1982).

    T. Tomoda, Phys. Rev. A26, 174 (1982).
21. W.L. Wang and C.M. Shakin, Phys. Lett. 32B, 421 (1970).
22. J. Rafelski, B. Müller, and W. Greiner, Z. Physik A285, 49 (1978).
23. H. Backe, L. Handschug, F. Hessberger, E. Kankeleit, L. Richter, F. Weik, R. Willwater, H. Bokemeyer, P. Vincent, Y. Nakayama, and J.S. Greenberg, Phys. Rev. Lett. 40, 1443 (1978).
24. C. Kozhuharov, P. Kienle, E. Berdermann, H. Bokemeyer, J.S. Greenberg, Y. Nakayama, P. Vincent, H. Backe, L. Handschug, and E. Kankeleit, Phys. Rev. Lett. 42, 376 (1979).
25. H. Backe, W. Bonin, W. Engelhardt, E. Kankeleit, M. Mutterer, P. Senger, F. Weik, R. Willwater, V. Metag, and J.B. Wilhelmy, GSI Scientific Report 1979, GSI 80-3, 101 (1980).
26. H. Backe, W. Bonin, W. Engelhardt, E. Kankeleit, M. Mutterer, P. Senger, F. Weik, R. Willwater, V. Metag, and J.B. Wilhelmy, 'Positron Production in Heavy Ion Collisions', preprint (1979).

228

27. J. Reinhardt, U. Müller, B. Müller, and W. Greiner, Z. Physik A303, 173 (1981).
28. G. Soff, J. Reinhardt, B. Müller, and W. Greiner, Phys. Rev. Lett. 43, 1981 (1979).
29. J. Kirsch, Frankfurt, to be published.
30. R. Schmidt, V.D. Toneev, and G. Wolschin, Nucl. Phys. A311, 247 (1978).
31. H. Backe, W. Bonin, E. Kankeleit, M. Krämer, R. Krieg, V. Metag, P. Senger, and J.B. Wilhelmy, GSI Scientific Report 1981, in press.
32. E. Berdermann, F. Bosch, M. Clemente, F. Güttner, P. Kienle, W. Koenig, C. Kozhuharov, B. Martin, B. Povh, H. Tsertos, W. Wagner, and Th. Walcher, GSI-Scientific Report 1980, GSI 81-2, 128 (1981).
33. H. Bokemeyer, H. Folger, H. Grein, S. Ito, D. Schwalm, P. Vincent, K. Bethge, A. Gruppe, R. Schulé, M. Waldschmidt, J.S. Greenberg, J. Schweppe, and N. Trautmann, GSI Scientific Report 1980, GSI 81-2, 127 (1981).
34. U. Müller et al., 'Spectroscopy of superheavy quasimolecules and quasiatoms', in: Proc. Conference on Fundamental Aspects in Heavy Ion Physics, La Rabida/Spain, June 1982.
35. J.S. Blair and R. Anholt, Phys. Rev. A25, 907 (1982).
36. K.W. McVoy, X.T. Tang, and H.A. Weidenmüller, Z. Physik A299, 195 (1981).
37. G. Soff, P. Schlüter, and W. Greiner, Z. Physik A303, 189 (1981).
38. P. Schlüter, G. Soff, and W. Greiner, Phys. Rep. 75, 327 (1981).
39. Ch. Kozhuharov, GSI, private communication.

Chapter 4   Collisionally Excited Few-Electron Systems

COLLISIONALLY EXCITED FEW-ELECTRON SYSTEMS: THEORETICAL
INTRODUCTION AND SURVEY

A. L. Ford and J. F. Reading
Texas A&M University*, College Station, Tx. 77843

R. L. Becker
Oak Ridge National Laboratory†, Oak Ridge, Tn. 37830

ABSTRACT

We consider excitation, ionization, and charge transfer in collisions of protons (and antiprotons) with the single-electron targets H, He^+, and Li^{2+}. These collisions are first compared to other types of ion-atom collisions. A brief review of our own theoretical method is given; in particular we describe how we allow for both large charge transfer and ionization probabilities while retaining the computational efficiency that allows us to consider a variety of collision partners and collision energies. We comment on the comparison of our results to other theoretical work and to experiment. The qualitative features of the various inelastic cross sections are discussed, in particular how they scale with collision energy, target nuclear charge, and the sign of the projectile charge.

INTRODUCTION

There are many reasons why theoretical and experimental studies of atomic collisions are important. Aside from the technological applications of atomic collisions and the role they play in other branches of physics, atomic collisions provide a means for learning about collision physics itself. In this regard atomic collisions have a distinct advantage over nuclear and elementary particle collisions, in that for the atomic case the fundamental two-body (nonrelativistic) force is known. Furthermore, due to the strong forces between the electrons and atomic nuclei the independent particle model (IPM) is always a reasonable starting point and is often (particularly for inner-shell processes) an excellent approximation. For bare ion projectiles, the projectile is clearly distinguishable from the target electrons whose excitation, ionization, and capture are being studied. For all but very low impact velocities the bare ion projectile accurately can be assumed to be moving on a classical path. An accurate description of the electronic structures of the separated projectile and target is usually not difficult to obtain. With all this the dynamical collision

*Research sponsored at Texas A&M by the NSF under Grant No. PHY-7909146 and the center for Energy and Mineral Resources.
†Research sponsored at ORNL by the Division of Basic Energy Sciences, US DOE, under Contract W-7405-eng-26 with Union Carbide.

problem can be reduced to one for the individual electrons.

Despite these simplifying features, there is a great richness of phenomena in atomic collisions. As this paper will attempt to emphasize, the collision parameters (collision velocity, target Z, and projectile Z) can be, and have been, varied over wide ranges. There are several quite distinct types of collisions, each with its own dominant processes and important physics.

This is a conference on atomic inner shell physics. One subclass of collisions involving inner shell electrons is K- and L-shell excitation, ionization, and charge transfer in collisions of small bare ions (nuclear charge $Z_p \leq 3$ or so) with neutral target atoms of nuclear charge Z_T greater than 10 or so. Inner-shell inelastic processes are important for collision energies that give velocity matching between the projectile and the target electrons in the shells in question. For inner shell electrons the electron-nucleus force dominates, and the IPM is an excellent approximation. The description of the target electronic structure is thus relatively simple, and one can concentrate on the collision physics. For these collisions ionization is by far the dominant inelastic process; excitation and charge capture probabilities are quite small. The most widely used theoretical model, and one which is quite successful in describing many features of the collisions, is the increased binding correction of the plane wave Born approximation (PWBA) that was developed some 10 years ago by Brandt and co-workers.[1] Our own contributions involve coupled-channel calculations with a single-center basis.[2] Discrete pseudostates are used to span the ionization continuum. Projectile centered functions are not needed in the expansion of the time-dependent single electron orbitals of the collision system because the flux in the charge transfer channels is very small. We have shown that the small charge transfer probability can be computed perturbatively from our single-centered expansion.[3] In asymmetric collisions $(Z_p \ll Z_T)$ an important electron capture mechanism is a two-step process in which the electron to be captured is first excited to a target continuum state which is resonant in energy with the state on the moving projectile into which the capture is to occur. The width of the energy resonance of this two-step process is very narrow when $Z_p \ll Z_T$, so care must be exercised in the manner in which the exact continuum spectrum of the target hamiltonian is replaced by a discrete pseudostate representation.[4,5]

This paper is however part of a symposium on collisional excitation of few-electron systems, and the remainder of the paper will deal specifically with single-electron collision systems rather than inner shells of neutral atoms. The inner shell case has the complications (and therefore interesting physics!) of a many-fermion system. Even though the independent particle model may be accurate, the Pauli exclusion principle imposes a correlation among the electrons.[6] Consider K-shell vacancy production as a simple example. One mechanism is direct ionization of a K-shell electron. But there are others. For example, in a single collision an

L-shell electron could be ionized and then the K-shell vacancy made
by excitation of a K-shell electron into the L-shell hole that was
just made. On the other hand, a K-shell vacancy is not made when a
K-shell electron is ionized but then in the same collision a L- or
M-shell electron is deexcited into the K-hole. All such processes
must at least formally be considered,[6] and in some cases are important in practice.[3,7] For multi-electron targets there is also
the possibility of breakdown of the IPM. Rearrangement of a number
of the electrons during the course of a collision can cause the effective single electron potentials to vary in time and to be different in each multi-electron channel.

   Single-electron collision systems avoid these complications.
There has been to our knowledge no experimental work on single-
electron collision systems in the $Z_p \ll Z_T$ asymmetric region in which
our inner-shell studies have been carried out. Such experiments
would be very useful in testing our understanding of this region of
collision parameters. There has been a lot of effort, both experi-
imental and theoretical, in the inverse region where $Z_p \gg Z_N$, i.e.
highly stripped or bare heavy ions in collision with atomic hydrogen. One characterizing feature of these collisions is the importance, particularly at lower collision energies and for large ionic
charge q of the projectile, of capture into a large number of highly excited orbitals on the heavy projectile. Due to the large number of discrete states involved, standard coupled-channel calculations with all important channels included can be prohibitively ex-
pensive. We have done some computational work on $He^{2+}+H(1s)$ and
$Li^{2+,3+}+H(1s)$ collisions,[8,9] but will have nothing further to say
here about such collisions.

   The rest of this paper deals with single-electron collisions
for which $Z_p \lesssim Z_N$, specifically p+H, p+He$^+$, and p+Li^{2+}. As we will
show, the ratio of capture to ionization cross sections varies over
a wide range as the collision velocity is changed. Excitation
cross sections, for example to the 2s and 2p states, are the same
order of magnitude as the ionization, and are included in the discussion. Our method, particularly how it accounts for the large
charge transfer flux that can occur in these collisions, is very
briefly reviewed in the next section. We then discuss our results.
We briefly outline the comparison between our results and those of
experiment and other theories. But the emphasis is on the qualitative features of the cross section results, and in particular how
they scale among the various sets of collision partners.

THEORETICAL METHOD

   The method we used for the calculations being discussed here
has been extensively described in the literature,[10] and here we
only briefly review some of the most important features. We are
considering collisions for which the projectile velocity is approximately (to within a factor of 10) equal to the velocity of the

target electron. For heavy projectiles (specifically protons) the projectile motion can be treated as classical. In all our calculations discussed here the projectile is taken to move with constant speed and in a straightline.

Coupled-channel methods use a basis set expansion of the time-dependent wave function $\Psi_\lambda(\vec{r},t)$ of the electron in the collision system. The target-centered expansion functions we use are obtained by diagonalizing the target hamiltonian H_T in underlying finite basis sets of square integrable functions. The pseudostates for which the eigenvalues of the projected hamiltonian lie above the ionization threshold provide a discrete representation of the ionization continuum. Examination of the spectrum of eigenvalues from the diagonalization of H_T has proven to be a very useful assessment of the adequacy of the basis sets being used. For example, certain basis functions can lead to pseudostates of such high energy that they do not participate in the dynamics of the collision. Adding such functions to the basis without prediagonalization of H_T would lead to a false sense of convergence.

The conventional two-center expansion (TCE) used by others employs an expansion of the form

$$\Psi_\lambda \ (TCE) = \sum_{n=0}^{N} a_{n\lambda}(t)\chi_n(\vec{r},t) + \sum_{m=0}^{M} b_{m\lambda}(t)\phi_m(\vec{r},t). \tag{1}$$

The functions χ_n and ϕ_m are target and projectile centered, respectively, and diagonalize the respective hamiltonians. This expansion has the defect that with it solving the coupled equations for the expansion coefficients is very time consuming. The problem arises from the fact that for finite times the χ_n and ϕ_m are not orthogonal, and that for large N and M the expansion leads to a large number of the difficult exchange matrix elements. The TCE also has the at least formal defect that for finite t the expansion is overcomplete. For finite t and large N the ϕ_m have unit projection onto the set χ_n, and the equations for determining $a_{n\lambda}$ and $b_{m\lambda}$ must become ill-conditioned. This overcompleteness also points up the lack of economy in the TCE: why include the ϕ_m for those times where they add nothing to the expansion?

Our first coupled-channel calculations were for asymmetric $Z_p \ll Z_N$ collisions, in which the charge transfer flux comprises a very small part of the wave function Ψ_λ. For calculations of ionization the projectile centered part of the wavefunction is hence unnecessary, and the charge transfer probability can be calculated with a t-matrix expression.[4] We thus used a single-center expansion

$$\Psi_\lambda(SCE) = \sum_{n=0}^{N} a_{n\lambda}(t) \; \chi_n(\vec{r},t) \tag{2}$$

and found it to be not only very efficient but also accurate.

For the collisions we are considering in the present paper though, for which $Z_p \sim Z_N$, the charge transfer probability can be large and the single center expansion is not an efficient way to proceed. To retain the computational speed of the SCE but still allow for the charge transfer flux we invented[10] what we call the 'one and a half centered expansion', in which

$$\Psi_\lambda(OHCE) = \sum_{n=0}^{N} a_{n\lambda}(t)\chi_n + \sum_{m=0}^{M} b_{m\lambda}(\infty)\beta_m(t)\phi_m. \tag{3}$$

The OHCE differs from the TCE in that the OHCE prechooses $\beta_m(t)$ as a fixed function. The choice made for $\beta_m(t)$ is constrained in principle only by the boundary conditions that $\beta_m(-\infty)=0$ and $\beta_m(\infty)=1$, so that the $b_{m\lambda}(\infty)$ are charge transfer amplitudes. The time dependent expansion coefficients $a_{n\lambda}(t)$ and time independent ones $b_{m\lambda}(\infty)$ are determined from applying the conditions

$$\left\langle \chi_n \left| i\hbar \frac{\partial}{\partial t} -H \right| \Psi_\lambda(OHCE) \right\rangle =0 \tag{4}$$

and the auxiliary constraint

$$0= \int_{-\infty}^{\infty} \langle \phi_m | i\hbar \frac{\partial}{\partial t} -H | \Psi_\lambda(OHCE) \rangle dt. \tag{5}$$

With this specific choice of auxiliary constraint we have what we have called the perturbative version of the OHCE (POHCE), and it is the only version of the method that we consider here. If $\Psi_\lambda(OHCE)$ is replaced by $\Psi_\lambda(SCE)$ in eq. (4) we recover the method we used previously for asymmetric collisions where charge transfer is small; hence the name perturbative. All calculations described here take $\beta_m(t)$ to be a unit step function at t=0 and retain only a single state, the 1s, in the projectile-centered part of the expansion in eq. (3).

Our expansion then is characterized by a large number of target-centered functions (which we are able to use because our method is comparatively so much more efficient than the TCE), but also with allowance for flux loss in the dominant charge transfer channel. The large target-centered expansion allows the ionization of the target to be well represented. A necessary, but not sufficient, test of the adequacy of our representation of the ionization

234

continuum is that in the limit of small Z_p, where our calculation reduces to the semiclassical first Born approximation for ionization, we accurately reproduce the plane wave Born ionization calculated with exact target continuum wavefunctions.

RESULTS AND DISCUSSION

We have applied the POHCE to collisions of protons with ground state H, He[+], and Li[2+]. Comparison of our cross sections to experiment and other calculations has been given elsewhere,[8-10] so we only make a few brief remarks here. For p+H and collision energy E≥15 keV the POHCE was found to give very good agreement with experiment and with extensive TCE calculations of Shakeshaft[11] for n=2 and n=3 excitation, ionization, and charge transfer. An example is given in fig. 1, where our ionization cross section is compared to that of Shakeshaft and to the very recent experiment of Shah and Gilbody.[12] The agreement among the three results, while not perfect, is overall quite good. We note that we used only s, p, and d angular momentum states in our target-centered expansion,

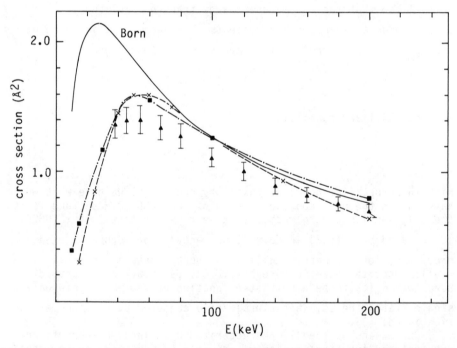

Fig. 1. Ionization cross section for p+H collisions. The experimental points are from Shah and Gilbody[12]. The 3.8% uncertainty in overall normalization has been included in the error bars. The squares, connected by a chain curve as a guide to the eye, are our POHCE calculation, and the crosses, connected by the dashed curve, are from the TCE calculation of Shakeshaft[11].

so from the coupled equations calculation we have only the ionization in these partial waves. To approximately include the higher
partial waves, we added their contribution as given in first Born.
It is difficult to see how to make a similar correction in the TCE,
and that may partly account for why our cross section at 200 keV
lies above that of Shakeshaft.

For p+He[+] the situation is not as good.[8] An example is
given in fig. 2, where our POHCE capture cross section is compared
to experiment[13] and to the calculations of Winter,[14] the most
extensive TCE calculations published so far for this system. We

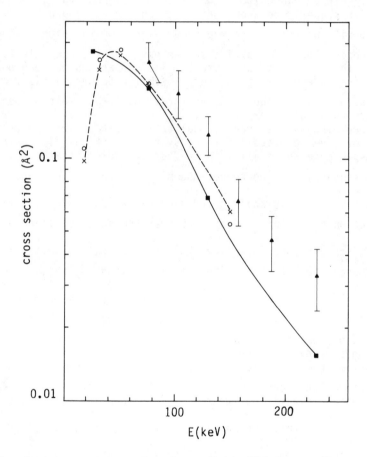

Fig. 2. Capture cross section for p+He[+] collisions. The experimental points are from Angel et al[13]. The 13.6% uncertainty in
absolute magnitude has been included in the error bars. The
squares connected by a solid line to guide the eye are our POHCE
calculations, where we calculate only capture into H(1s) and assume
excited state capture is 20% of this. The crosses connected by a
dashed curve, are from the TCE calculation of Winter[14]. The open
circles are 1.2 times Winter's ground state capture.

have included the 13.6% uncertainty in the absolute magnitude of the experimental results in the error bars. We calculate only capture into H(1s) and assume that excited state capture is the 20% of this that one gets by assuming n^{-3} scaling. The agreement between our calculation and that of Winter is only fair. Part of the discrepancy is that his excited state capture differs from our assumption of 20% of ground state capture. We are significantly below the experiment, particularly at the higher energies. It is interesting that our total electron loss cross section (capture plus ionization) is in rather good agreement with experiment. Apparently either our calculation or the experiment fails to distinguish properly between ionization and charge capture.

For p+Li^{2+} we know of no experimental data with which to compare. There is data for p+Li$^+$, and we have carried out calculations for this two-electron system.[9] The comparison between our calculation and experiment is overall similar to what it is for p+He$^+$, except that in addition the total loss cross section we compute is somewhat below experiment.

We now discuss the overall qualitative features of the cross sections we have computed for these systems, particularly their energy dependence and scaling in the target nuclear charge Z_T. The 1s to 1s capture, σ_{KK}, decreases sharply with collision energy E, and at a given E/Z_T^2 it scales approximately as Z_T^{-5}. The ionization cross section σ_I is fairly flat in E over the same collision energy range and scales approximately as Z_T^{-4} (the Z_T scaling in the first Born). Hence the ratio of capture to ionization, σ_{KK}/σ_I, scales roughly as Z_T^{-1} and falls sharply with E in the energy range we are considering. This is shown in fig. 3. As Z_T and E are varied in the collisions we have considered, the ratio of capture to ionization changes by over two orders of magnitude.

For asymmetric collisions ($Z_p \ll Z_T$) the excitation and ionization cross sections are accurately given by the first two terms in the Born expansion, $\sigma = \sigma_B(1+f\rho)$, where $f = Z_p/Z_T$ and ρ is independent of Z_T and Z_p. The Born cross section σ_B is proportional to Z_p^2, so the correction to the Born scales like Z_p^3. For $Z_p > 0$, at low energies ρ is negative and has been associated physically with the increased binding of the target electron when the projectile is well inside the electron's orbital radius[1]. A novel way we have illustrated this effect is by performing coupled-channel calculations with negative Z_p[15]. For low energy p$^\pm$+Cu collisions then the cor-

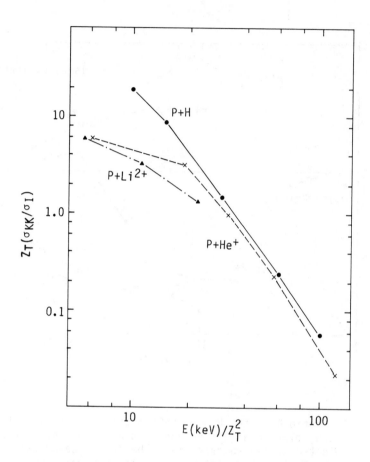

Fig. 3. The ratio of capture to ionization cross sections, multiplied times Z_T and plotted as a function of scaled collision
energy for the indicated collision partners.

rection $Z_p\rho$ is negative for $Z_p=+1$ (increased binding) and positive for $Z_p=-1$ (decreased binding). In figs. 4-6 the behavior of
the ratio $R=\sigma/\sigma_B$ is followed into the higher scaled collision
energy and more symmetric collision regime. The $Z_p=-1$ results for
Cu connect smoothly, in a qualitative sense, to the results for
smaller Z_T. There is a change in the magnitude of the Z_p^3 correc-

238

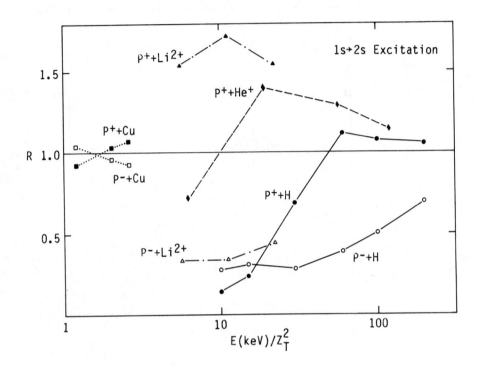

Fig. 4. The ratio R of the POHCE 1s→2s excitation cross section to the first Born, plotted as a function of scaled collision energy for the indicated collision partners. The copper calculations were done using the Hartree-Fock potential of the neutral atom.

tion as Z_T is decreased by over a factor of ten in going from the highly asymmetric collision to the more symmetric ones. In each case (excitation to 2s and 2p and ionization) there is 'decreased binding' (R>1) at low energies and a transition to 'antipolariza-tion' (R<1) as E is increased. The Born correction in R does not scale as f for the small Z_T cases, but at least does decrease as Z_T increases.

The behavior for Z_p=+1 is on the other hand quite different.

For p$^\pm$+Cu the p$^+$ and p$^-$ cross sections are symmetrically placed on opposite sides of the Born, but for $Z_p \sim Z_T$ that is the case only

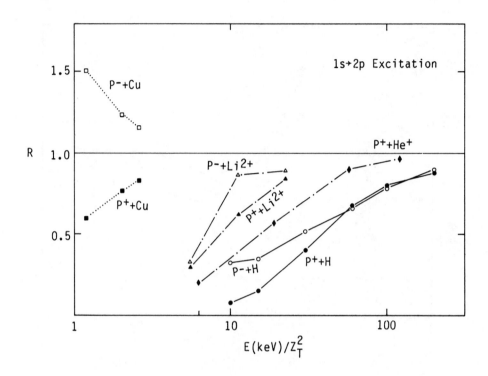

Fig. 5. As in fig. 4, but for 1s→2p excitation.

at high energy. We ascribe the qualitative difference in the behavior of R for Z_p=+1 to be due to the existence of a charge transfer channel for Z_p=+1 but its complete absence for Z_p=-1. It appears that when charge transfer is large, it robs flux from both
the excitation and ionization channels. For example, at E/Z_p^2~10
keV charge transfer becomes very large as Z_T is decreased. Correspondingly, R is sharply depressed below the value one would extra-
polate from p^++Cu when one goes from Li^{2+} to He^+ to H targets.
    Another striking feature of the results shown in figs. 4-6 is
the very close agreement above E~50 keV between the 2p excitation
cross sections for p^++H and p^-+ H, and to a lesser extent for the

240

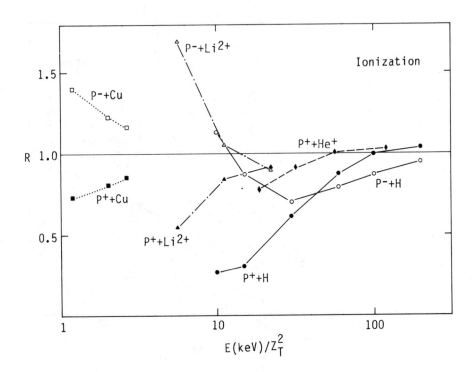

Fig. 6. As in fig. 5, but for ionization.

ionization. We do not presently understand the physical origin of this effect, and feel in fact that there is much still to be learned from results of the sort presented in these figures. We note that it would be very useful to have experimental results for 2s and 2p excitation in the $p+He^+$ and Li^{2+} cases, so that our Z_T scaling of these excitation cross sections could be tested.

In conclusion, collisions of bare ions of Z =±1 with one-electron targets of small nuclear charge Z_T constitue one distinct type of ion-atom collision. Such collisions are characterized by large collision strength $f=Z_p/Z_T$ and large charge transfer. Our UHCE method allows us to consider a variety of values for the collision parameters. We find qualitative differences in the collision physics here compared to what obtains in the $Z_p \ll Z_T$ asymmetric region that is most often explored in inner-shell collision studies. We plan further study of these Z_p~Z_T collisions by extending our OHCE methods to system of several electrons where IPM breakdown is extensive.

REFERENCES

1. G. Basbas, W. Brandt, R. Laubert, A. Ratkowski, and A. Schwarzchild, Phys. Rev. Lett. 27, 171 (1971); G. Basbas, W. Brandt, and R. Laubert, Phys. Rev. A 7, 983 (1973).
2. A. L. Ford, E. Fitchard, and J. F. Reading, Phys. Rev. A 16, 133 (1977).
3. A. L. Ford, J. F. Reading, and R. L. Becker, Phys. Rev. A 23, 510 (1981).
4. J. F. Reading, A. L. Ford, G. L. Swafford, and A. Fitchard, Phys. Rev. A 20, 130 (1979).
5. J. F. Reading and A. L. Ford, J. Phys. B: Atom. Molec. Phys. 12, 1367 (1979).
6. J. F. Reading, Phys. Rev. A 8, 3262 (1973); J. F. Reading and A. L. Ford, Phys. Rev. A 21, 124 (1980).
7. R. L. Becker, A. L. Ford, and J. F. Reading, J. Phys. B: Atom. Molec. Phys. 13, 4059 (1980).
8. J. F. Reading, A. L. Ford, and R. L. Becker, J. Phys. B: Atom. Molec. Phys. 15, 625 (1982).
9. A. L. Ford, J. F. Reading, and R. L. Becker, J. Phys. B: Atom. Molec. Phys., in press (1982).
10. J. F. Reading, A. L. Ford, and R. L. Becker, J. Phys. B: Atom. Molec. Phys. 14, 1995 (1981).
11. R. Shakeshaft, Phys. Rev. A 18, 1930 (1978).
12. M. B. Shah and H. B. Gilbody, J. Phys. B: Atom. Molec. Phys. 14, 2361 (1981).
13. G. C. Angel, E. C. Sewell, K. F. Dunn, and H. B. Gilbody, J. Phys. B: Atom. Molec. Phys. 11, L297 (1978).
14. T. G. Winter, Phys. Rev. A 25, 697 (1982).
15. M. H. Martir, A. L. Ford, J. F. Reading, and R. L. Becker, J. Phys. B: Atom. Molec. Phys. 15, 1729 (1982).

PRODUCTION OF COLLISIONALLY EXCITED FEW-ELECTRON IONS

C. L. Cocke
Kansas State University, Manhattan, KS 66506

ABSTRACT

This paper reviews some characteristics of the production of slow, highly ionized recoils in the bombardment of gaseous targets by fast heavy ion beams. One example of the use of an ion source based on this idea is discussed.

INTRODUCTION

Conventional ion sources produce highly charged ions by a succession of single ionizations generated by electron-ion collisions. An alternative approach is to try to remove many electrons in a single encounter by using a highly stripped heavy ion whose Coulomb field is at least as strong as that binding the electrons to the target. Over the past decade a great deal of accelerator-based atomic-collisions work has been done on systems whose collision incurs multiple-ionization of both the projectile and the target. For example, in Fig. 1 is shown a high-resolution K-X-ray spectrum of Ne, taken by a KSU – Univ. Tenn. collaboration,[1] which shows very high ionization states of neon are created in single collisions with fast Cl projectiles. This work I believe to be the first to reveal Ly-α K target radiation resulting from the loss in a single collision of 9 of the 10 neon electrons.

It is important to distinguish between collisionally produced highly-stripped <u>projectiles</u> and <u>target</u> <u>ions</u>. Highly-stripped <u>fast</u> projectiles can easily be generated by passage through thin foils[3] or in single collisions with gases.[4] If their electronic "excitation" energy is to be of order E_e, however, they must have a CM energy $E \sim (M/m) E_e$ to achieve this, where m and M are the electron and projectile mass, respectfully. In the single collision case, E_e will be about the same for projectile and recoil, but the recoil $E \sim (m/M) E_e$, a reversal of the same mass ratio (See Fig. 2). Thus the recoil can be a <u>slow</u> highly stripped ion.

There are at least two uses for <u>slow</u> highly-stripped ions:

(1) For spectroscopic purposes, they are essentially free from the Doppler

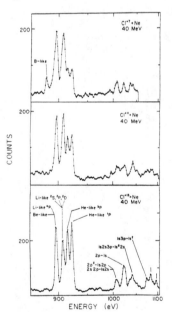

NeK-X-ray spectrum from 40 MeV Cl^{+13} bombardment (Brown et.al., ref. 1).

shifts and broadening which characterize radiation from a fast foil- or single-collision excited projectile beam. This spectroscopic advantage was suggested and used[2] by Sellin in the ultraviolet. Several groups today at laboratories including GSI,[4,5] Berkeley, Aarhus, and Hahn-Meitner Institute[6] are doing X-ray or Auger-electron spectroscopy using a fast-beam-pumped source.

(2) As a source of ions for studying collisions between low-energy highly-charged (LEHQ) ions and neutral targets, they lie in a velocity regime even lower (v ~ 10^6cm/sec) than that so far accessed by EBIS and ECR ion sources. The electronic excitations are in the range relevant to MFE plasmas (E_e ~ 1-5keV) and below.

In the following discussion I will review some of the experimental systematics and theoretical models proposed for the production of the ions and will outline some of the uses to which they are now being put. As an example, I will finally summarize one recent experiment we have carried out using this source of LEHQ ions.

LEHQ PRODUCTION

This section reviews, in order of increasing "differentiality", some data which cast light on the LEHQ production mechanism.

A. Total charge production:

An integral measurement of the cross sections for LEHQ recoil production is provided by passing a fast ion beam between the appropriately shielded plates of a parallel-plate capacitor in the presence of a tenuous target gas and measuring the net charge production. In terms of the cross sections σ_q^Q for producing a slow recoil of charge q by a projectile of charge Q, the net charge-prodcution cross section σ_+^Q is defined as $\sum_q \sigma_q^Q$. Fig. 3 shows results[8] for heavy ion bombardment of Ar, Ne, and He.

These data are compared with results from a theoretical model by Olson.[9] In this model, the probability p for removal of a single electron from a target of appropriately chosen effective charge is calculated as a function of impact-parameter (b) using a Classical-Trajectory Monte-Carlo (CTMC) method. Multiple ionization at each b is

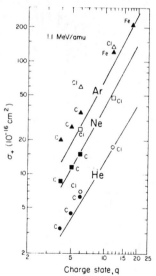

Fig. 3 - Cross sections for total ionization. (Schlachter, et al., ref. 8).

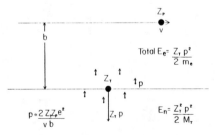

Fig. 2 - Schematic of fast, large-b collision. E_e and E_n are energies imparted to electrons and target nucleus, respectively.

treated using binomial statistics whereby $P_q(b) = \binom{M}{q} p^q(1-p)^{M-q}$, where the target shell has M electrons and P_q is the probability for removing q electrons. The cross section given by $\sigma_q^Q = 2\pi \int P_q(b) \, b \, db$. Agreement with the data is seen to be quite good.

B. <u>Total cross sections σ_q^Q</u>:

Individual charge state cross sections have been measured recently by several groups,[11-14] often using a simple time-of-flight spectrometer to separate different charge states of the recoils. In Fig. 4 are shown charge spectra from ~ 1MeV/amu Cl^{+12} and U^{+44} on neon. The higher Q is clearly important to generate Ne^{+9} and Ne^{+10}. In Fig. 5 the dependence of σ_q^Q on Q, q and E are shown. The decreasing energy dependence occurs because one is far above velocity matching for the L-shell. The steep Q-dependence for high q is probably due partially to the role of K-hole production for high q, and partially to the high power of p which enters for this case.

The Olson model gives a good representation for low q, but overestimates σ_q^Q for high q. This is probably due at least in part to the fact that it is necessary to impart much more energy per electron than the first ionization potential (Ip) in order to remove several electrons. Only Ip enters the CTMC calculation. An alternative theoretical model[11] which avoids this problem invokes a two-step picture of the ionization process. First, the fast projectile passes through the atom at b, imparting an energy E_e to the electrons. Second, after the projectile has disappeared down the beam pipe, the excess electronic energy is dissipated by electron emission. Population of the final charge

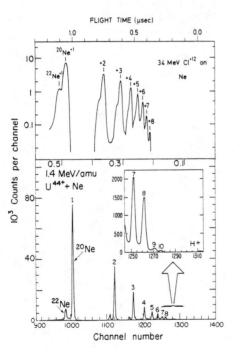

Fig. 4 - Recoil time-of-flight spectra. Cl^{+12}, Cocke, ref. 11; U^{+44}, Schlachter et al., ref. 13.

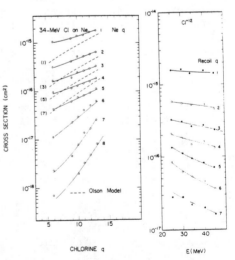

Fig. 5 - Projectile-Q and bombarding energy dependence of σ_q^Q for Cl on Ne. (Cocke, ref. 11); Olson model, ref. 9.

state distribution is calculated according to the expressions of Russek et al.,[14] which prescription essentially weights each final state proportional to the phase space available to the corresponding number of electrons emitted. The empirical ionization potentials for each ionization stage thus enter calculation. Results from this so-called Energy-Deposition (ED) model are seen to be in slightly better agreement with the data for high-q in Fig. 6. This model generates too high a value for σ_1^Q, probably because, at very large b, it allows unphysical accumulations of very small energy transfers per electron to add up sufficiently to produce single ionization.

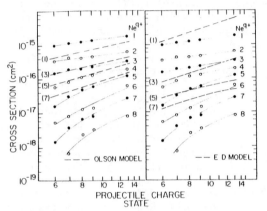

Fig. 6 - Comparison of σ_q^Q at 1MeV/amu with Olson model (ref. 9) and ED model (ref. 11). Data are for bare nuclei (ref. 12) and Cl^{+13} (ref. 11).

C. $\underline{P_L(b=0)}$ from spectroscopic data

From the relative satellite intensities of a K-X-ray or K-Auger electron spectrum, the relative probabilities of exciting various L-vacancy configurations can be obtained, provided the average fluorsecence yield appropriate to each satellite is known. The parameterization of the resulting charge state distribution in terms of a single L-ionization probability $p_L(0)$ has been frequently used in the interpretation of such spectra.[16] The assumption that the relevant probability is at b ≃ 0 relies on the K-vacancy production probability being sharply peaked near b = 0 on scale characterizing L-vacancy production.

Although the use of binomial statistics is not justifiable on first principles for large $p_L(0)$, this parameter is useful in comparing the relative ionizing powers of different beams over a wide range of velocity and charge state. Schneider, et al.[17] showed Fig. 7 that an empirical relationship between $p_L(0)$ and v/Q exists for a wide range of data from both K-X-ray and K-Auger spectra.

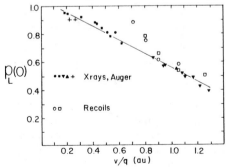

Fig. 7 - Universal $p_L(0)$ vs projectile velocity to charge ratio. Solid points from spectroscopic data (refs. 17 - 20); open points from data of Fig. 10.

D.   Direct Measurements of $P_L(0)$:

   Although both the Olson and ED models are formulated as a function of b, it is experimentally very difficult to test their results directly due to the small scattering angles involved.   For example, a 1 MeV/amp F^{+9} scattered in a collision with a bare nucleus of neon at b = 0.2 a.u. will scatter at a laboratory angle of 0.68 mrad.   In Fig. 8 are shown theoretical values of $P_n(b)$ for 1 MeV/amu projectiles of Q = +10 (Olson model) and Q = +9 (ED model) onto neon, from which it can be seen that even such a small scattering angle samples only $P_q(b)$ for b $\simeq$ 0.

   Experimental coincidence measurements of $P_q(0)$ carried out at KSU are shown in Figs. 9 and 10 for 1 MeV/amu projectiles of 0 and F in various charge states incident on Ne.   The mean scattering angle was approximately 0.49 mrad, corresponding to b = 0.26 a.u. for the case of F^{+9}.   Since every collision at this "large" angle generates a charged recoil, the P_n can be expressed as charge state fractions.

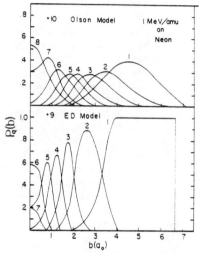

Fig. 8 - Theoretical $P_q(b)$ from Olson (ref. 9) and ED (ref. 11) models for +10 and +9 point projectiles.   Numbers above curves give q.

These data can be used to elucidate the role of <u>inner</u> shell (here K-shell) vacancy production in the generation of LEHQ recoils. It is immediately clear that a qualitative shift to high q in the mean recoil charge state occurs for high projectile q, becoming most marked for projectile charges for which K-to-K vacancy transfer

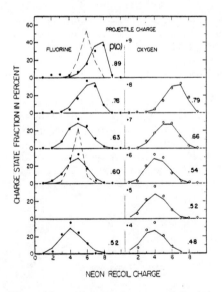

Fig. 10 - Measurements of $P_q(0)$ for F^{+Q} and O^{+Q} on neon.   Solid lines are binomial statistics fits, with $p(0)$ indicated. Dashed curves are ED model.

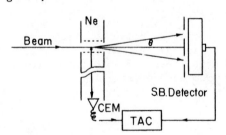

Fig. 9 - Schematic of apparatus for $P_q(0)$ measurements.

is possible. In this case, the probability per collision for the creation of neon K-vacancies $P_K(b)$ by electron capture by F becomes of order unity for b ~ 0.2 a.u.. To illustrate this point, we show in Fig. 11 the experimental $P_K(b)$ of Hagmann, et al.[21] for the creation of single (satellite) neon K-vacancies for F^{+9} on neon. Also shown in that figure are $P_K2(b)$, the probability for creating double K-vacancies (hypersatellite). The shift seen in Fig. 10 may thus interpreted as follows for the F case (the O case is clearly similar): For $Q \leq 7$, K-vacancy production per collision is weak, and the recoil charge state distribution is nearly equal to the L-shell charge state distribution deduced spectroscopically. Indeed, for this case the ED prediction does rather well at predicting the charge state distribution. For $Q = 8, 9$, single and double K-vacancy production is very probable for $b \approx 0.26$ a.u. Each collision removes ≈ 5 L-electrons, by direct ionization, but a K-electron is removed as well, usually succeeded by K-Auger electron emission. Thus a shift of one or two charge states is created, as observed. Such an interpretation would require that the p (O) deduced from binomial fits to the curves of Fig. 10 would agree with the spectroscopic values of p_L (O) in Fig. 7 so long as K-vacancy production is small, but would rise above the universal curve for $F^{+8,9}$ and $O^{+7,+8}$ on neon. Such is seen to be the case in Fig. 7.

An interesting, if indirect, support of this interpretation is provided by data of Hagmann[21] which show that delayed Ne K-Auger electrons from 10 MeV F^{+9} on Ne have a P(b) reflecting that for producing neon recoils with double K-vacancies, not single K-vacancies (See Fig. 11). The time-delayed neon K-Auger electrons follow the time-consuming capture of a third electron in the target gas by $(1s2s)^3S$ cores. The production of these cores appears to be through Auger cascades from $(1s)^0(2s,p)^3$ states, not directly from the $(1s)^1$ $(2s)^1$. In other words, it is easier to remove two K-electrons than it is to ionize the L-shell beyond $(2s,p)^3$.

To summarize, and generalize, from this case study of neon, the collisional production of LEHQ recoils probably involves at least two important physical processes. The outer shells are directly ionized with a probability weakly decreasing with v/Q. To reach really high charge states requires removing inner-shell electrons as well, however, which in turn is very much enhanced if inner shell capture is enabled. The latter requires that the projectile bear vacancies whose binding energy is comparable to or greater that that of the target inner shell.

It is interesting for the case discussed above to estimate the center of mass energies which one might expect for, e.g., a $(1s2s)^3S_1$ Ne^{+8} recoil created by a 10 MeV/amu F^{+9} Beam. Double K-vacancies are created at $b \approx 0.1$ Å, for which L-vacancy production has already saturated. Rutherford scattering at small angles gives $E_r = (Q^2q^2e^4M_p)/(E\,b^2M_t) = 15.8$ eV. This energy goes down rapidly with increasing E and b^2. The above expression is probably an overestimate, since some momentum exchange occurs between the neon nucleus and its electrons during the collision. Schlacter et al.[13] argue that the above expression should be divided by two.

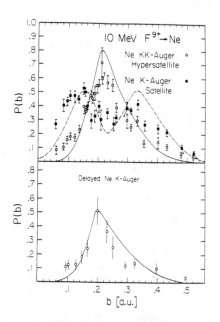

USES FOR COLLISION WORK

Fig. 11 – P(b) for K-Auger-electron production in Ne by F^{+9} projectiles (Hagmann, et al., ref. 21).

As a source of ions for studying low-energy electron capture, LEHQ recoils have been used in situ,[23] in ion traps,[24] and extracted to form a low energy beam.[25,27] Our own application has been of the last variety. We have been able to extract useable quantities of such species as Ne^{+10}, C^{+6} and Ar^{+15} and have performed cross-section and energy-gain-spectroscopy studies for these and other rare gas projectiles on various neutral targets for $v \sim 10^6 - 10^7$ cm/sec, $2 < q < 12$.

As an example, I describe here a recent electron capture experiment we[28] carried out at Berkeley Super HILAC using a secondary ion recoil source. We have not yet been able to produce useful quantities of neon above q = 8 at the KSU Tandem, and thus need the more highly charged pump beam to produce sufficient quantities of bare and hydrogenlike neon. The purpose of the experiment was to measure capture cross sections for Ne^{+10} and Ne^{+9} (and other) projectiles on rare gases and to try to determine, via energy-gain spectroscopy on the post-collision projectiles, the final n-distribution for the electrons.

The apparatus is shown schematically in Fig. 12. An electric field of 2V/cm extracts the recoils at right angles to the beam. They are then accelerated to 20-50 eV/q and charge-analyzed by a double-focussing magnetic spectrometer. Charge state spectra obtained with a 4.9 MeV/amu Au beam as the pump are shown in Fig. 13. For total cross section measurements a retarding-grid analyzer (apparatus(B)) with a wide angular acceptance is used. For energy-gain spectroscopy, we use a hemispherical electrostatic analyzer (apparatus(A)). An overall resolution of about 0.8 eV/q is achieved.

Fig. 14 shows the energy spectra of 492 eV Ne^{+9} ions from bombardment of various rare-gas targets by bare neon nuclei. Even at this low energy, the reaction proceeds dominantly at scattering angles less than 3°, and the energy released when a loosely bound target electron is transferred to the highly charged projectile goes almost entirely to the projectile center-of-mass motion. Thus the projectile energy-gain spectrum directly reflects the energy-release spectrum. In each spectrum, an n-scale shows the energy at which the projectile should appear if a state of principal quantum number n is populated on the projectile. The data show that the reactions populate very selectively excited states on the projectile, with n ranging from 5 to 7, depending on the target ionization potential. This result is also known from spectroscopic work at GSI for selected recoil projectiles.[5,6] Population inversion on the final ion is readily obtained.

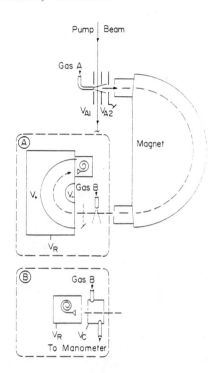

Fig. 12 - An apparatus used LEHQ capture studies.

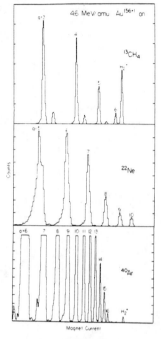

Fig. 13 - Charge State spectra from recoil source.

The n-values populated are in excellent agreement with those calculated from the classical barrier model.[5,6,29] This model ascribes the transfer to the first crossing between entrance and exit molecular orbitals for which the target electron can classically surmount the potential barrier between the two collision centers (see Fig. 15). The resulting $n \leq n_c = (2I_p[1+(q-1)/2\sqrt{q}+1)])^{-\frac{1}{2}}$, where n_c is a "critical" n-value parameter defined here to be non-integral.

The usefulness of being able to combine total capture cross sectiondata with energy-spectra can be illustrated by a final example. For each reaction one (or at most, two) n-value(s) is apparently populated, which more or less pins down the crossing radius R_e as $(q-1)/\Delta E$, where ΔE is the measured energy defect. In the single crossing model, the cross section is given by $\sigma = \pi R_e^2 <P>$ where $<P>$ is the b-averaged probability for electron capture. If both σ and R_e are known experimentally, one can deduce $<P>$. For the cases of Fig. 14, this procedure gives $<P>$ values of 0.26, 0.16, 0.15, and 0.28 for targets of He, Ne, Ar and Xe. The frequently assumed[29] classical model value of 0.5 is not adequate. It appears that those systems which in some sense have optimized crossing radii, i.e., a curve crossing just at that internuclear distance where the barrier goes down, are those for which transfer occurs most easily, This conclusion still lies in the realm of experimental systematics, not yet having received detailed theoretical examination.

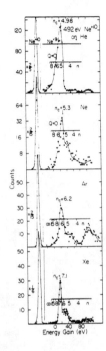

Fig. 14 – Energy-gain spectra from rare gases.

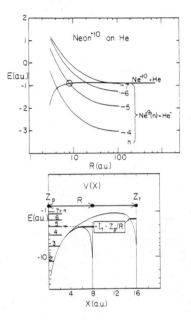

Fig. 15 – Upper figure shows curve crossings for Ne^{+10} on He. Lower figure gives electron potential V as a function of X from Ne^{+10} for internuclear distances R of 8 and 16 a.u.

ACKNOWLEDGMENTS

I am grateful especially to R. Mann and T. Gray for numerous discussions and continuing collaboration on this subject. This work was supported in part by the U. S. Department of Energy, Chemical Sciences Division.

REFERENCES

1. M. D. Brown, J. R. Macdonald, P. Richard, J. R. Mowat and I. A. Sellin, Phys. Rev. A9, 1470 (1974); R. L. Kauffman, C. W. Woods, K. A. Jamison and P. Richard, ICPEACIX, contr, abst. p. 939 (1975); see also D. L. Matthews, B. M. Johnson, G. W. Hoffman and C. F. Moore, Phys. Lett. 49A, 195 (1974).

2. I. A. Sellin, Advances in Atomic and Molecular Physics, Vol. 12, eds. D. R. Bates and B. Bederson, (Acad. Press, N. Y., 1976), p. 215; I. A. Sellin et al., Z. Physik A283, 329 (1977).

3. H. D. Betz, Rev. Mod. Phys. 44, 465 (1972).

4. Q. C. Kessel, M. D. McCaughey and E. Everhart, Phys, Rev. 153, 57 (1967).

5. H. F. Beyer, K.-H. Schartner and F. Folkmann, J. Phys. B13, 2459 (1980).

6. R. Mann, R. Folkmann and H. F. Beyer, J. Phys. B14, 1161 (1981); R. Mann, H. F. Beyer and F. Folkmann, Phys. Rev. Lett. 46, 646 (1981).

7. D. Schneider, M. Prost, B. Dubois and N. Stolterfoht, Phys. Rev. A25, 3102 (1982).

8. A. S. Schlachter, et al., Phys. Rev. A23. 2331 (1981).

9. R. Olson, J. Phys. B12, 1843 (1979).

10. J. H. McGuire and O. L. Weaver, Phys. Rev. A16, 41 (1977).

11. C. L. Cocke, Phys. Rev. A20, 749 (1979).

12. T. J. Gray, C. L. Cocke and E. Justiniano, Phys. Rev. A22, 849 (1980).

13. A. S. Schlachter, W. Groh, A. Müller, H. F. Beyer, R. Mann and R. Olson, Phys. Rev. A, in press (1982).

14. S. Kelbch, H. Ingwersen, H. Schmidt-Böcking, R. Schuch and J. Ullrich, U. Frankfurt Inst. für Kernphysik progress report, p. 36 (1981).

15. A. Russek and J. Meli, Physica 46, 222 (1970).

16. For a review, see F. Hopkins, "Ion-Induced X-ray Spectroscopy" (p. 355); D. Matthews, "Ion-Induced Auger Electron Spectroscopy" (p. 433) in Methods of Experimental Physics, Vol. 17, ed. P. Richard (Academic Press, 1980).

17. D. Schneider, M. Prost and N. Stolterfoht, in abstracts of Contributed Papers, 7th Int. Conf. on Atomic Physics, p. 102 (1980).

18. R. L. Kauffman, C. W. Woods, K. A. Jamison and P. Richard, Phys. Rev. A11, 872 (1975).

19. R. Mann, GSl-Report PA-6-77 (1977); H. F. Beyer, K.-H. Schartner and F. Folkmann, J. Phys. B13, 2459 (1980).

20. D. Schneider, C. F. Moore and B. M. Johnson, J. Phys. B9, L153 (1976); N. Stolterfoht, D. Schneider, R. Mann and F. Folkmann, J. Phys. B10, L281 (1977); N. Stolterfoht, IEEE Trans, Nucl. Sci. 76CH1175-9, NPS311 (1976).

21. S. Hagmann, C. L. Cocke, P. Richard, H. Schmidt,-Böcking and R. Schuch, private comm. (1982).

22. W. Groh, A. Miller, C. Achenbach, A. S. Schlachter and E. Salzborn, Phys. Lett. 84A, 77 (1981).

23. H. F. Beyer, R. Mann and F. Folkmann, J. Phys. B (to be published, 1982).

24. R. Vane, M. H. Prior and R. Marrus, Phys. Rev. Lett. 46, 107 (1981).

25. C. L. Cocke, R. Dubois, T. J. Gray, E. Justiniano and C. Can, Phys. Rev. Lett. 46, 1671 (1981).

26. E. Justiniano, C. L. Cocke, T. J. Gray, R. D. Dubois and C. Can, Phys. Rev. A24, 2953 (1981).

27. C. L. Cocke, R. Dubois, T. J. Gray and E. Justiniano, IEEE Trans, Nucl. Sci NS-28, 1032 (1981).
28. R. Mann, C. L. Cocke, A. S. Schlachter, M. Prior and R. Marrus.
29. H. Ryufuku, K. Sasaki and F. Watanabe, Phys. Rev. A21, 745 (1980).

X RAYS AND AUGER ELECTRONS FROM FEW-ELECTRON RECOIL IONS

Heinrich F. Beyer, Rido Mann
GSI Darmstadt, Bundesrepublik Deutschland

Finn Folkmann
Det Fysiske Institut, Århus Universitet, Århus, Danmark

ABSTRACT

A summary is given of recent experiments in which highly ionized atoms of low laboratory velocities are created by distant collisions with highly-charged very heavy projectiles in the MeV/amu specific-energy range. Emphasis is put on very-high recoil-ion charges and on excited states carrying K vacancies. The potential of x-ray and Auger-electron emission studies is illustrated in view of both the primary multiple ionization process and the selective electron capture observed in a secondary collision. The relevance of the latter process for the population of specific few-electron states is outlined. Briefly discussed are also a few aspects of electron-rearrangement processes following molecular-target ionization.

INTRODUCTION

There is presently a great interest in collision and radiation processes involving highly ionized atoms. X-ray transitions of such ions are observed in solar spectra [1] and in fusion plasmas [2] where they play an important role as they significantly contribute to plasma cooling and serve as plasma diagnostics. Similar states of excitation can be observed in ions of several MeV/amu specific energy traversing thin foils [3].

Another type of accelerator-based experiments [4,5] makes use of the observation of target atoms ionized in a single collision with fast heavy projectiles. Even though the total energy corresponding to the electronic excitation can easily reach several keV the kinetic energy of the recoil ion may be as low as a few eV.

Whereas other experiments [6] use such ions after they have returned to their ground states or do not discriminate against electronic states, we concentrated on *excited* states and the related radiation processes. In our studies slow recoil ions of atomic number $6 \leq Z_2 \leq 18$ were produced by bombarding monatomic and molecular target gases with fast heavy ions from the UNILAC. A wide range of projectiles with atomic number of $18 \leq Z_1 \leq 92$, ionic charges of $12 \leq q_1 \leq 70$ and specific energy between 1.4 and 9.0 MeV/amu was covered. The strong field suddenly introduced by the high charge of the projectile leads to electron removal already at impact parameters which are much larger than the electron-shell radii of the targets. For the collision regimes under discussion collision velocities approximately match orbital velocities of inner-shell electrons whereas they are large compared to those of outer-shell electrons. This mismatching can be compensated for by a very high projectile charge which then leads to depletion of *outer* shells whenever the collision is close enough to remove an inner-shell electron of the target. The K x-ray and Auger-electron emission spectra, sensitive only for K-hole states, therefore, sample an impact-parameter range where high recoil charges are obtained. In the present article we will discuss the projectile dependence of high-charge state recoil-ion production in the *primary* single collision. Because the excited recoil ions move so slowly (typically 0.01 mm/ns) they can easily be observed until their electronic states are changed by *secondary* charge changing collisions with surrounding gas atoms. The importance of such processes for the population of few-electron excited states will be outlined. When *molecular* targets are used, a fast electronic relaxation limits the number of few-electron ions emerging from the dissociating molecule.

PROJECTILE DEPENDENCE OF MULTIPLE TARGET IONIZATION

The degree of multiple ionization can be obtained from both x-ray and Auger-electron satellite structure [7],[8]. In this section we are discussing the results obtained on the two monatomic targets studied i.e. neon and argon. Figures 1 and 2 demonstrate the projectile velocity and charge-state dependence for a given projectile species. The Auger-electron spectra of figure 1 are taken at the fixed charge state 16+ (18+ in case of 1.4 MeV/amu). At high bombarding energy they reveal a complicated line structure due to simultaneous excitation of 3- to 5-electron states. When the bombarding energy decreases the structure more and more changes into a simplified pattern which in case of the lowest energy of 1.4 MeV/amu reflects the auto-ionization of a pure three-electron system. A simple relation between the number of L-shell vacancies and the centroid of the Auger-electron satellite distribution was found by Stolterfoht et al. [9] allowing to extract the degree of ionization from the measured data. The procedure starts to get inaccurate, however, when the spectra saturate at Li-like states as does the bottom spectrum of figure 1. A further increase of the multiple ioni-

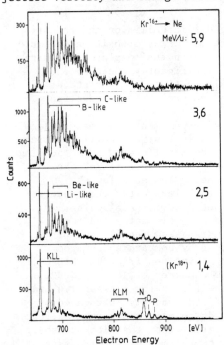

Fig. 1. Neon-K Auger-electron spectra induced by krypton impact at various bombarding energies.

zation does not change the spectra appreciably. The Kα satellite structure shown in figure 2 reveals a strong charge-state effect as observed through the change in relative intensity between He-like and Li-like lines going from $q_1 = 30$ to $q_1 = 16$ at 5.9 MeV/amu. For comparison the 1.4 MeV/amu Kr^{18+} spectrum is also displayed showing the increase of ionization when the velocity is lowered.

For characterizing the degree of multiple ionization we use the probability $p_L(0)$ for L-shell electron removal in small impact parameter i.e. K-shell ionizing collisions. An empirical scaling of p_L is obtained when p_L is plotted as a function of the projectile velocity devided by the incoming charge state v_1/q_1. The results are shown in figure 3 together with data for bare projectiles reported by Kauffman et al. [10]. Both, the bare and partially stripped ion data fit a common straight line. Taking into account the ionic rather than the nuclear charge seems to be justified - as we will see below - from the argument that the ionic radius of the highly charged heavy projectile is smaller than the mean impact parameter at which the recoil ion is created. Therefore, the projectile's inner-shell electrons will shield its nuclear charge during the collision i.e. the highly charged heavy ions behave like point charges. That can be quite different for less highly charged lighter ions where the inner [11] but not the outer [12] electrons shield the nuclear charge. Under the extreme condition of a very high projectile charge as in case of 5 MeV/amu U^{65+} the neon K x-ray spectrum is dominated by the hydrogen-like line series as demonstrated in figure 4. Part of the observed intensity is already caused by first stripping the target atom completely to Ne^{10+} and afterwards redressing it with one electron from the surrounding gas atmosphere. Those secondary-capture processes will be discussed in detail further below.

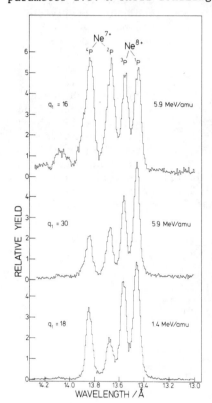

Fig. 2. Neon-Kα x-ray satellites induced by Kr^{q_1+} impact.

Similar results as for neon were obtained in case of the argon [13] target for which we show representative spectra in figures 5 and 6. A strong increase in the degree of ionization can be observed in the Kα satellite distribution of figure 5 when going from Kr^{30+} to Pb^{60+} impact. In case of the high projectile charge the spectrum reduces to essentially He- and Li-like states. Comparison with

256

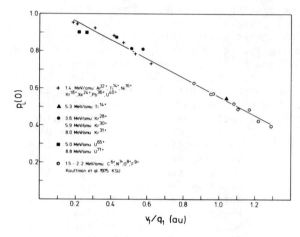

Fig. 3. L-shell ionization probability $p_L(0)$ of Ne in K-vacancy producing collisions.

theoretical wavelength positions of the individual satellites indicate that for the high degree of L-shell ionization, as observed in the present experiment, the M-shell is totally depleted. Figure 6 demonstrates the charge-state effect for 5.9 MeV/amu uranium ions. Only with the high charge state individual lines are discernable which then represent the three-electron Auger decay. The argon satellite structure was studied for projectiles ranging from Kr to U in the specific-energy range of 2.5 - 9.0 MeV/amu. In Table I are listed the L-shell ionization probabilities p_L extracted from the Kα satellite distribution.

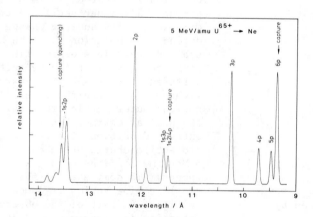

Fig. 4. Neon x-ray lines produced by 5 MeV/amu U^{65+}.

For lighter projectiles ($Z_1 \leq 17$) Schmiedekamp et al.[14] found a scaling of p_L as a function of the collision velocity v_1 and the nuclear charge Z_1 employing the BEA model[15] on the basis of an increased binding effect[16]. Not discussed so far has been the dependence on the ionic charge q_1. This, however, plays an important role in the heavy-ion regime as demonstrated above. Our data suggest

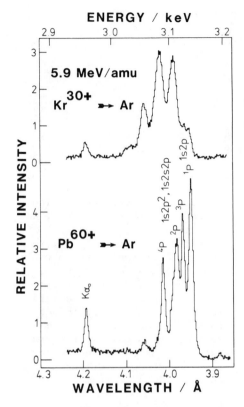

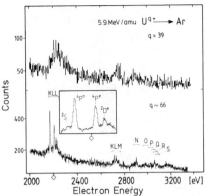

Fig. 6. Argon Auger-electron spectra showing the charge-state dependence for uranium projectiles.

Fig. 5. Argon Kα satellite struc-ture reveiling the strong increase of the degree of ionization with projectile charge.

to use q_1 as the effective charge of the projectile. This is supported by the small ionic radii r_{proj} listed in Table I. The latter were estimated via the relation

$$r_{proj} = n_i(-2E)^{-\frac{1}{2}} \tag{1}$$

using the potential energies E of Carlson et al. [17]. n_i denotes the principal quantum number of the outermost shell of the ion. We applied the BEA model and a binding correction using q_1 instead of Z_1 to scale our data. In the BEA approximation the cross section for L-shell ionization then reads

$$\sigma_L = \pi N q_1^2 u_L^{-2} G(V) \tag{2}$$

where N is the number of electrons and u_L the binding energy of the L-shell. G(V) denotes the BEA ionization curve as a function of the scaled velocity V which is the ratio of projectile and L-shell orbital velocity.

258

Assuming the ionization probability to have the form

$$p_L(b) = \sigma \cdot (\pi NR^2)^{-1} \quad b \leq R \qquad (3)$$
$$ = 0 \qquad\qquad\quad b > R$$

yields

$$p_L = q_1^2 R^{-2} u_L^{-2} G(V) \qquad (4)$$

The binding energy correction is taken into account by replacing u_L by $u_L \cdot \varepsilon$ and V by $V \cdot \varepsilon^{-\frac{1}{2}}$ arriving finally at

$$p_L = q_1^2 R^{-2} u_L^{-2} \varepsilon^{-2} G(V \varepsilon^{-\frac{1}{2}}) \qquad (5)$$

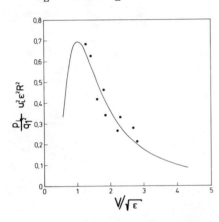

The binding-energy correction factor ε, listed in Table I, was determined by equation (17) of Brandt and Lapicki [16]. Figure 7 shows the data plotted in the BEA scaling according to equation (5). The only fit parameter used is the cut-off radius R which is determined to $R = 1.1$ au. Besides some scattering of the experimental data the BEA curve reproduces the dependence on q_1 and v_1 quite well.

Fig. 7. Scaled L-hole probability of argon according to equation (5) and Table I.

Table I Experimental L-hole probabilities p_L and L-shell binding-energy correction factors ε after Brandt and Lapicki [16] for various heavy ions colliding with argon.

ION	SPEZ. ENERGIE (MeV/amu)	V	q_1	r_{proj}/au	p_L	ε
Kr	2.5	2.35	26	0.14	0.53	2.82
	5.9	3.61	30	0.13	0.51	2.42
	8.0	4.20	31	0.12	0.40	2.27
Xe	2.6	2.39	36	0.22	0.73	3.48
	4.7	3.21	41	0.20	0.78	3.16
	9.0	4.46	45	0.09	0.75	2.74
Pb	5.9	3.61	60	0.14	0.83	3.83
	8.0	4.20	64	0.13	0.81	3.62
U	5.0	3.31	64	0.13	0.92	4.28

RECOIL-ENERGY DISTRIBUTION

Kinematical broadening of Auger-electron lines allows access to the velocity of the emitting ion provided other sources of broadening (natural width, instrumental) are small. This is the case for the lines observed in figure 6 where the instrumental width ΔE_i = 3.5 eV due to the limited resolution of the electron-energy analyzer is substantially smaller than the one observed in the spectrum. In our case the recoil ions are produced by large-impact parameter collisions and, thus, are recoiling nearly perpendicular to the direction of the projectile. For a fixed kinetic energy E_R of the recoil ions the width ΔE_D of the Doppler profile, which can be approximated by a rectangular shape, is given by [18]

$$\Delta E_D^{\ 2} = 16 \ \frac{m_e}{m} \ E_A \cdot E_R \cdot \sin^2\alpha \tag{6}$$

where m_e and m denote the electron mass and the recoil-ion mass, and E_A and α denote the Auger-transition energy and the observation angle relative to the primary beam, respectively.

Because the recoil ions have a continuous energy distribution the experimental line profile U(E) will be a convolution of the Doppler profile D with the transmission function G of the energy analyzer and with the recoil-energy distribution F :

$$U(E) = \iint D(E',E_R) \cdot G(E-E') \cdot F(E_R)dE_R dE' \tag{7}$$

The transmission function G can be approximated by a Gaussian. If the observed line U(E) is Gaussian as well F(E_R) can be easily de-convoluted analytically which results in a mean recoil energy of

$$\bar{E}_R = \frac{3}{8\ell n2} \ \frac{\Delta E^2 - \Delta E_i^{\ 2}}{16m_e/m \ E_A \ \sin^2\alpha} \tag{8}$$

where ΔE is the width of the observed line. If U(E) has some arbi-trary shape F(E_R) can be deconvoluted numerically. In our case, the experimental line shapes follow quite well a superposition of two Gaussians separated by only 1 eV. This is demonstrated in figure 8 in case of the 1s2s2p $^4P^0$ line [13]. From the smooth curve of figure 8a follows the recoil-energy distribution displayed in figure 8b. The distribution maximizes already at 7.5 eV whereas the centroid is at $\bar{E}_R$ = 21±3 eV. These small recoil energies show that the recoil ions are created in very large-impact-parameter collisions. Provided the projectile-target interaction potential is known, the recoil energy distribution can be converted into an impact-parameter dependent ionization probability. In a simplified treatment one can assume the repulsion of the target to occur after the shielding target electrons are removed, i.e. in the second half of the collision one assumes Rutherford scattering of two point charges with the in-coming and the final recoil charge state, respectively. In such a treatment, using the charge states 66 and 15, the curve of figure 8b would translate into an ionization probability which maximizes near an impact parameter of 0.3 au and which levels off at impact parameters larger than 6 au.

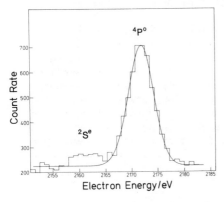

Fig. 8. (a) Experimental profile   (b) The deconvoluted recoil-
        for the 1s2s2p $^4P^0$ energy distribution.
        Auger-electron line
        of Ar^{15+}.

EXCITATION BY SECONDARY ELECTRON CAPTURE

The possibility of observing secondary electron capture by
emission spectra of K-vacancy bearing states enters through the
high fraction of metastable states present when the degree of target
ionization is high. A recoil ion in a long-living K-hole state produced in the primary heavy-ion encounter survives until its electronic state is changed in a second collision with a neutral target
gas atom. At practical gas pressures the second collision is observed typically after a few nanoseconds which corresponds to a
travel distance of a few 10^{-2} mm. In that collision the electron-
capture reaction

$$A^{q+} + B \rightarrow A^{(q-1)+}(n) + B^+ + E_{exo} \tag{9}$$

takes place, where an electron is transferred from a neutral atom B
to an outer shell n of the highly stripped recoil ion. A simple one-
electron model [7,19,20] assuming an over-barrier transition predicts
the principal quantum number n as a function of the ionization
potential J_B according to

$$n^2 \leq \frac{q^2}{2J_B[1+(q-1)/(1+2q^{1/2})]} \tag{10}$$

whereas the crossing distance is given by

$$R = \frac{q-1}{q^2/2n^2 - J_B} \tag{11}$$

In our experiments [7,20,21] we have studied

(i) excited states distribution of the product ion $A^{(q-1)+}(n)$ by
      its Auger or x-ray decay,

(ii)   the exothermic energy defect E_{exo} by kinematic line
        broadening,
(iii)   capture rates by measuring the production rate of $A^{(q-1)+}(n)$
        or the quenching rate of A^{q+}.
        Because the transfer of the electron occurs selectively into
mainly one shell of principal quatum number n specific 'capture
lines' can be observed which sometimes dominate the spectra. In
figure 4, for instance, the extraordinarily high intensity of the
$6p \rightarrow 1s$ transition in NeX is due to the reaction $Ne^{10+} + Ne \rightarrow$
$Ne^{9+}(n=6) + Ne^{+}$. A variety of gas mixtures was employed for investigating reaction (9) for a large number of different collector ions
and donor atoms (of different ionization potentials). Furthermore,
delayed Auger-electron spectra were measured making use of the
pulsed primary beam. There the capture events occuring with several

Table II   Predicted (equation (10)) and experimental principal
            quantum number for electron capture in various systems.

Ion	Neutral	Principal Quantum Number		Observed Radiation
		Calculated	Experimental	
C^{4+}	CH_4	3.2	3	Auger
N^{5+}	NH_3	4.3	4	Auger
"	CH_4	4.0	4	Auger
"	H_2	3.6	3(4)	Auger
"	N_2	3.4	3(4)	Auger
"	Ne	3.0	3	Auger
"	He	2.8	3	Auger
N^{6+}	N_2	4.0	4	Auger
O^{6+}	H_2O	4.6	4(5)	Auger
"	CO_2	4.4	4	Auger
"	O_2	4.7	5(4)	x ray
"	Ne	3.5	3(4)	x ray
"	He	3.3	3	Auger
O^{7+}	O_2	5.3	5	x ray
"	Ne	4.0	4	x ray
"	He	3.7	4	x ray
Ne^{8+}	NH_3	6.4	6	Auger, x ray
"	Xe	5.9	6	Auger
"	CH_4	5.8	6	Auger, x ray
"	H_2	5.2	5	Auger, x ray
"	Ar	5.2	5	x ray
"	Ne	4.4	4(5)	Auger, x ray
"	He	4.2	4	Auger, x ray
Ne^{10+}	CH_4	7.0	7	x ray
"	H_2	6.3	6	x ray
"	Ar	6.2	6	x ray
"	Ne	5.3	6(5)	x ray
"	He	5.0	5	x ray

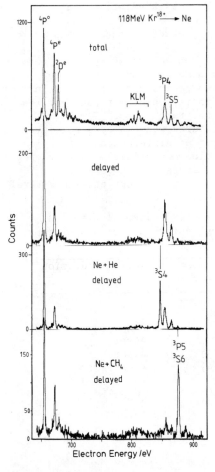

ns delay time are separated from the prompt events. This is demonstrated in figure 9 showing the Li-like Auger-electron-spectra of Ne excited by 1.4 MeV/amu Kr^{18+} ions. The corresponding collector ions are in the metastable $1s2s$ ^3S or $1s2p$ $^3P_{0,2}$ states to which an electron is added into an outer shell producing the pronounced peaks observed at electron energies larger than about 840 eV. The selectivity and the dependence on the admixed gas and its ionization potential is clearly observed in agreement with equation (10). Part of the produced 3-electron product ions are in quartet states which do not directly decay to the K-shell but via cascades feed the low-lying $^4P^0$ and $^4P^e$ states found in the low-energy part of the spectra.

In Table II experimental results on the principal quantum numbers n are compared with the predictions of equation (10).

Fig. 9. Total and delayed Ne-K Auger-electron spectra from different gas mixtures. The lines labelled ^{2S+1}Ln arise from states with 2-electron cores $(1s2\ell)$ ^{2S+1}Ln and a captured electron in the outer shells n.

When monatomic gases are used the exothermic energy defect E_{exo} is substantially larger [20] than the initial recoil energy. As the energy defect is shared as kinetic energy among the two repulsive product ions according to their mass ratio the energy defect is observed as an Auger-line broadening which is obtained analog to equation (6). The effect as observed on the ^4P lines is demonstrated in Table III. In case of delayed observation where the lines are produced by capture they are systematically broader than for prompt observation where they are due to direct population. The crossing distances R derived via the relation

$$E_{exo} = \frac{q-1}{R} \qquad (12)$$

are, within the experimental accuracy, in agreement with the prediction of equation (11).

Table III Experimental line widths in eV for Ne (KLL)^{4}P lines in the prompt and in the delayed spectra for bombardment with 1.4 MeV/u Kr^{18+}. The detector resolution was around 1.4 eV and the uncertainty for the widths ± 0.2 eV.

$$\Delta E_{diss.} = (\Delta E^2_{del.} - \Delta E^2_{prompt})^{1/2}$$

Target Pressure (mtorr)	Ne + Xe 25 + 25	Ne 20	Ne + CH$_4$ 20 + 20	Ne + He 40 + 40	Ne + H$_2$ 25 + 25
^{4}P^0 prompt	1.1	1.3	1.4	1.2	1.6
^{4}P^0 delayed	2.7	2.1	2.0	1.9	2.2
^{4}Pe prompt	1.9	1.9	1.9	1.9	2.3
^{4}Pe delayed	3.9	2.8	2.4	2.4	2.5
^{4}Pe $\Delta E_{diss.}$	3.4	2.0	1.5	1.4	1.0

For the same Ne^{8+} recoil ions capture rates were measured employing the photon-quenching method which makes use of the fact that metastable states of proper radiative lifetime get collisionally depopulated by the capture process whereas prompt states keep uneffected [21]. Figure 10 shows the neon Kα satellites measured at target pressures of 12 mbar and 93 mbar, respectively. The spectra are normalized to the ^{1}P intensity. At high pressure one notices a strong quenching of the ^{3}P$_1$ state and of the Li-like ^{4}P states whereas the intensity ratio of the prompt lines ^{1}P/^{2}P remains practically constant. The ^{1}P/^{3}P intensity ratio as a function of target density and target composition can be shown to be

$$I(^1P)/I(^3P) = R[1 + \tau(^3P) (k_{c1}N_1 + k_{c2} N_2)] \qquad (13)$$

where R denotes the intensity ratio at zero density. N_1 and N_2 denote the density of the gas from which the recoil ions are studied and the density of a second admixed gas, respectively. k_{c1} and k_{c2} are the corresponding capture rates and $\tau(^3P)$ is the lifetime of the ^{3}P$_1$ state which amounts to [22] 0.18 ns. Our experimental results are consistent with the linear pressure dependence predicted by equation (13). This is demonstrated in figure 11 for the pure Ne, the Ne+He, and the Ne+Xe mixture. From the slopes of the straight lines fitted to the data capture rates k_c are extracted which are in the range of $1 - 23 \cdot 10^{-9}$ cm^3 s^{-1}. The corresponding cross sections $\sigma_c = k_c/\bar{v}_{rec}$ are obtained using the mean recoil velocity $\bar{v}_{rec} = 0.8 \cdot 10^6$ cm s^{-1} ($\bar{E}_{rec} = 6.7$ eV). In figure 12 the cross

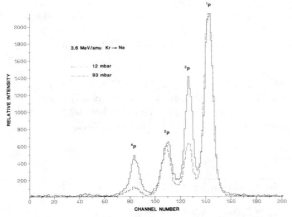

sections are compared to the geometrical cross section πR^2 predicted by equation (11). The oscillations are due to capture into different outer shells according to equation (10), where n has to be an integral value.

Fig. 10. Neon Kα satellite spectra induced by 3.6 MeV/amu krypton impact and taken at two different target pressures.

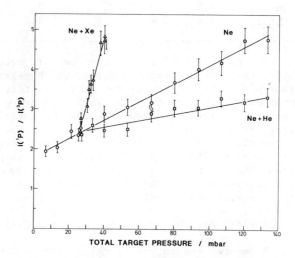

Fig. 11. Pressure dependence of the 1s2p $^1P/^3P$ intensity ratio in NeIX for different target compositions. In case of the mixed gases the neon partial pressure amounts to 27 mbar.

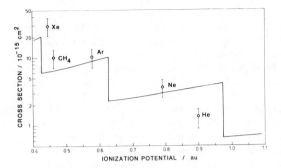

Fig. 12. Experimental charge-exchange cross sections for Ne^{8+} recoil ions on various neutrals. The full line represents the geometrical cross section determined from equation (11).

SPECTROSCOPY OF RYDBERG LINES

The high degree of target ionization achieved with the highly charged heavy projectiles results in a considerable reduction of the number of lines present in the spectra and single lines of the few-electron systems dominate (cf. figure 4). As we have seen above, the intensities of those lines may be manipulated by choosing a set of experimental parameters which determine primary and secondary excitation. Because relativistic and quantum-electrodynamic (QED) contributions to both term energies and decay rates scale as high powers of the nuclear charge Z it is interesting to investigate a high-Z target. Until now argon is the heaviest target studied for which we can present a sample x-ray spectrum in figure 13. There

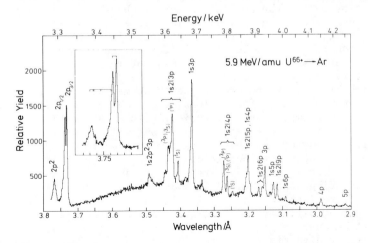

Fig. 13. Wavelength scan showing lines of the argon few-electron systems.

the line series extending approximately from 3.8 Å down to 2.9 Å
are shown in a case of favourable conditions for a high degree of
ionization. The H-like line series starting with the Lyman-α doublet
extends up to principal quantum number n = 5, whereas the He-like
series is observed up to n = 6. For the Li-like series the multiplet
splitting of the He-like 1s2ℓ core is observed and is indicated in
brackets. The line labelled 1s2ℓ9p is due to secondary electron
capture by Ar^{16+} ions into n = 9. There is, however, no indication
of capture by bare or one-electron ions. The broad background observed in figure 13 is caused by uranium-M radiation of the projectile. As the experiments presented were not aimed at providing
high wavelength precision the accuracy is limited to about ± 0.0008 Å.
Experimental wavelengths of the line positions together with a comparison with theoretical transition wavelengths are given elsewhere [13].

MOLECULAR RELAXATION

The study of secondary electron capture as discussed above represents the first step towards tracing the slowly moving ion on
its way back to charge neutrality. Once the K vacancy is refilled,
however, further charge-exchange processes can not be detected in
the spectral ranges under discussion. The situation gets much more
complicated when heavy molecules are used. *Light*-ligand molecules
reveal a similar behaviour [18] as monatomic targets, whereas the
presence of *heavy* ligands with their many electrons results in fast
charge-quenching processes which readily compete with the radiative
deexcitation observed. Despite of the complexity of the process a
few aspects may be delineated experimentally.

Because of the long-range nature of the target ionization by
very highly charged projectiles the ligands get multiply ionized at
least in their outer shells, in addition to the atom whose Auger or
x-ray decay is observed. This molecular ionization leads to the
'Coulomb explosion' of the molecule which is observed by means of
the kinematical broadening of metastable Auger-electron lines [18].
The respective line width is, via the dissociation velocity, a
measure for the degree of molecular ionization. Analogous to the
monatomic-target results a drastic increase of molecular ionization
with projectile charge was found.

The time history of the charge quenching may be diagnosed with
the aid of x-ray and Auger transitions of various atomic lifetimes.
Transitions from prompt states of typically 10^{-13} s lifetime occur
when the electronic state is close to the one initially produced by
the heavy ion encounter. At that instant many electrons (the number
depending on the degree of ionization) are excited to the continuum
or very high n states either by direct excitation or by very fast
electron transitions from excited ligands. As time proceeds those
electrons will redistribute and will fill the initial vacancies.
Metastable states will experience this relaxation and the radiation
emitted will be characteristic of the changed electronic state.
Examples are given in figures 14 and 15 for fluorine ions emerging
from SF_6 or CF_3CL. Those molecules contain 60 or 40 electrons,

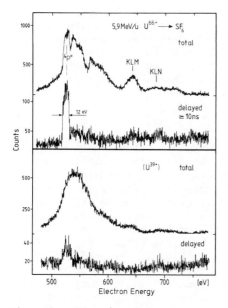

Fig. 14. Fluorine-K Auger-electron spectra of SF_6 bombarded with U^{66+} and U^{39+}.

respectively. Figure 14 shows the total and delayed ($\geq$ 10 ns) fluorine Auger-electron spectra for SF_6 bombarded by U^{66+} and U^{39+} projectiles, respectively. No single lines can be observed in the total spectra because of the many-electron states produced. The delayed spectra show the fraction of metastable three-electron states directly produced which have survived electron refilling. In case of U^{39+} impact that fraction is close to zero, whereas for the high charge of U^{66+} it is much higher. This might be explained by the circumstances that in the latter case both the number of initially produced three-electron states and the number of ionized molecular electrons are higher. The large line width of 12 eV corresponds to a fragment energy of 560 eV.

As opposed to the total Auger-electron spectra of figure 14 a clear line structure is observed in the K x-ray spectra of figure 15. Even with the less violent projectile Kr^{18+} the structure is re-

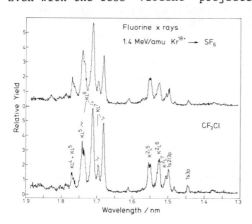

Fig. 15. Fluorine-K x-ray produced by bombardment of SF_6 and CF_3CL, respectively.

markably close to that of the monatomic Ne target. The extent of fluorine-fragment ionization represented by the K x-ray spectra is only marginally higher in case of CF_3Cl as compared to the heavier SF_6 molecule. Because most of the lines originate from states with lifetimes in the 10^{-13} s range the molecular electrons might not be fully redistributed at the time of x-ray emission. One might speculate about electrons occupying high-n states which lead to small satellite-wavelength shifts, even though existing data have not yet been analyzed with respect to that. An interesting finding is the quenching of the 1s2p ^{3}P intensity relative to the ^{1}P. Because of the low target pressure (~ 0.05 mbar) this effect is not explained by electron capture of a released F^{7+} ion colliding with a surrounding molecule. Due to the greater lifetime of the ^{3}P$_1$ level of 1.6 ns the x-ray transition occurs at times when the relaxation is more advanced. The ^{3}P/^{1}P intensity ratio, therefore, is a diagnostic for the evolution of the molecular relaxation.

CONCLUSION

In summary we have tried to delineate the potential of x-ray and Auger-electron emission spectra as an effective probe for some of the interesting features of the recoil-ion light source. With the use of very heavy highly charged projectiles very high recoil charges can be produced. The corresponding x-ray and Auger-electron spectra of the few-electron systems reveal single-lines structure. For collision velocities much higher than the orbital velocity of target K-shell electrons the extent of multiple target ionization decreases. The simple BEA scaling used to explain the projectile dependence should not obscure the fact that further effort is necessary to ilucidate the complex process of ionizing many electrons in a single collision. In applications where an outmost degree of ionization is required it might be advantageous to accelerate the projectiles, to strip them to a high equilibrium charge, and afterwards to decelerate them again. From the kinematical analysis a recoil-energy distribution follows which covers the few-eV range. From such measurements as well as from more direct determinations of recoil energies a deeper insight into excitation mechanisms is expected.

The analysis presented has also aimed to show the relevance of selective electron capture for the population of high-n states. Spectroscopic investigation of the capture process allows to use the recoil ions without acceleration. Thereby the collision regime of high charge states and ultimately low velocities is explored. The information extracted from the x-ray and Auger-electron measurements comprises final-state distribution, as well as capture rates and capture kinematics. Extension to other spectral ranges would allow to investigate those processes also for core ions without K-vacancies.

It should be emphasized that the recoil-ion light source can be regarded as stationary with regard to precision wavelength measurements and wavelength-calibration difficulties as known from the fast beam-foil source are missing. The study of molecular relaxation after introducing a fast disruption by a heavy-ion encounter may be relevant for tackling the even more complex situation of a very heavy ion interacting with matter of high density.

ACKNOWLEDGEMENT

It is a pleasure to thank the many people who have contributed to the results presented, either by assistance in the data acquisition, analysis, or preparation of the manuscript.

REFERENCES

1. A. Gabriel and K. Phillips, Mon. Not. R. Astron. Soc. 189, 319 (1979);
   U. Feldman, Physica Scripta 24, 681 (1981).
2. M. Bitter, S. von Goeler, N. Santhoff, K. Hill, K. Brau, D. Eames, M. Goldmann, E. Silver, and W. Stodiek, Inner-Shell and X-Ray Physics of Atoms and Solids, D.J. Fabian, H. Klein-poppen, and L.M. Watson eds. (Plenum, N.Y., 1981) p. 861;
   V.A. Boiko, A.Ya. Faenov, S.A. Pikuz, and U.I. Safronova, Mon. Not. R. Astron. Soc. 187, 107 (1977);
   D.J. Nagel, P.G. Burkhalter, C.M. Dozier, J.F. Holzrichter, B.M. Klein, J.M. McMahon, J.A. Stamper, and R.R. Whitlock, Phys. Rev. Lett. 33, 743 (1974).
3. P. Richard, R.L. Kauffman, F.F. Hopkins, C.W. Woods, and K.A. Jamison, Phys. Rev. Lett. 30, 888 (1973);
   H.-D. Dohmann and R. Mann, Z. Physik A291, 15 (1979).
4. J.R. Mowat, R. Laubert, I.A. Sellin, R.L. Kauffman, M.D. Brown, J.R. Macdonald, and P. Richard, Phys. Rev. A10, 1446 (1974).
5. H.F. Beyer, F. Folkmann, and K.-H. Schartner, J. Physique 40, CL-17 (1979);
   R. Mann and F. Folkmann, J. Physique 40 CL-236 (1979).
6. C.L. Cocke, Phys. Rev. A20, 749 (1979);
   C.R. Vane, M.H. Prior, and R. Marrus, Phys. Rev. Lett. 46, 107 (1981).
7. H.F. Beyer, K.-H. Schartner, and F. Folkmann, J. Phys. B13, 2459 (1980);
   H.F. Beyer, K.-H. Schartner, F. Folkmann, and P.H. Mokler, J. Phys. B12, L363 (1978).
8. N. Stolterfoht, D. Schneider, R. Mann, and F. Folkmann, J. Phys. B10, L281 (1977);
   R. Mann, GSI Report PA-6-77 (1977).
9. N. Stolterfoht, in Fourth Conference on the Application of Small Accelerators (Publication No. 76 CH-117-9 NTS, Denton, Texas, Oct. 1976).
10. R.L. Kauffman, C.W. Woods, K.A. Jamison, and P. Richard, Phys. Rev. A11, 872 (1975).
11. R.L. Kauffman, C.W. Woods, K.A. Jamison, and P. Richard, J. Phys. B7, 1335 (1974).
12. C.F. Moore, J. Bolger, K. Roberts, D.K. Olsen, B.M. Johnson, J.J. Mackey, L.E. Smith, and D.L. Matthews, J. Phys. B7, L451 (1974).
13. H.F. Beyer, R. Mann, F. Folkmann, and P.H. Mokler, to be published.
14. C. Schmiedekamp, B.L. Doyle, T.J. Gray, R.K. Gardner, K.A. Jamison, and P. Richard, Phys. Rev. A18, 1892 (1978).
15. J.H. McGuire and P. Richard, Phys. Rev. A8, 1374 (1973).
16. W. Brandt and G. Lapicki, Phys. Rev. A10, 474 (1974).
17. T.A. Carlson, C.W. Nestor, N. Wassermann, and J.D. McDowell, At. Data 2, 63 (1970).
18. R. Mann, F. Folkmann, R.S. Peterson, Gy Szabó, and K.O. Groene-veld, J. Phys. B11, 3045 (1978).

19. H. Ryufuku, K. Sasaki, and T. Watanabe, Phys. Rev. $\underline{A21}$, 745 (1980).

20. R. Mann, F. Folkmann, and H.F. Beyer, J. Phys. $\underline{B14}$, 1161 (1981); R. Mann, H.F. Beyer, and F. Folkmann, Phys. Rev. Lett. $\underline{46}$, 646 (1981).

21. H.F. Beyer, R. Mann, and F. Folkmann, J. Phys. $\underline{B14}$, L377 (1981); J. Phys. $\underline{B15}$, 1083 (1982).

22. C.D. Lin, W.R. Johnson, and A. Dalgarno, Phys. Rev. $\underline{A15}$, 154 (1977); T.W. Tunnell, C.P. Bhalla, and C. Can, Phys. Lett. 75A, 195 (1980).

HIGH PRECISION SPECTROSCOPIC STUDIES OF FEW ELECTRON IONS

Jean Pierre Briand

Institut Curie, Section de Physique et Chimie *
and Université Pierre et Marie Curie
11, rue Pierre et Marie Curie
75231 Paris Cedex 05

The spectroscopy of few electron heavy ions which can provide elementary information on atomic structure and on some of the fundamental processes has recently received a great deal of interest. The one-electron heavy ions for which exact calculations can be carried out certainly constitute the best candidates for studying quantum electrodynamics in very intense fields or the relativistic corrections in the high velocity stationary systems. On the other hand the two-electron ions provide the most elementary systems to study all the processes involved in the relativistic electron-electron interaction in atoms (Breit interaction i.e. mostly spin-spin interaction, retardation effects...).

There are various ways in which few electron heavy ions can be obtained
   . from gas target bombardment by energetic heavy ions
     (recoil ions of very low energy range : 1 up to few eV),
   . from the new sources Cryebis or ECR (medium energy
     range : few keV up to hundreds of keV),
   . from stripping through thin carbon foils of the beams
     delivered by the most powerful heavy ion accelerators
     (high energy range : 100MeV up to GeV).

The heaviest ions which we shall mainly deal with have only been produced for the moment by the third method (in table 1 are presented for instance the charge state distributions for Argon and Iron ions of 6-8MeV/amu energy).

* laboratoire associé au CNRS n°198.

Table 1. Equilibrium charge state distribution for

Fe ions at 8.5MeV/amu

Fe^{26+}	Fe^{25+}	Fe^{24+}
(bare)	(H)	(He)
5%	32%	46%

Ar ions at 6MeV/amu

Ar^{18+}	Ar^{17+}	Ar^{16+}
(bare)	(H)	(He)
20%	32%	29%

We shall describe in this report some experiments in which the Lyman α lines of hydrogenlike and heliumlike Argon and Iron ions have been measured. The principle of all these experiments was to study with a crystal spectrometer the X rays emitted in flight by the ions, at 90° with respect to the direction of the beam.

THE HYDROGENLIKE IONS

The main goal of the experiments on hydrogenlike ions was the measurement of the (1s) Lamb shift and the study of the higher order relativistic corrections. The principle of the measurement of the (1s) Lamb shift was to compare the energy of the Lyman α lines obtained from the eigenvalues of the Dirac equation and the experimental values. All the considered (initial and final) levels in the transition are moved with respect to the Dirac eigenvalues when considering quantum electrodynamics corrections, and the accuracy with which the (1s) Lamb shift can be measured strongly depends on the relative values of the 2p level shifts and the experimental error bar. We present in table 2 the expected values of all the radiative corrections which clearly show that in experiments with an absolute precision of the order of few tenths of eV, only the (1s) Lamb shift is measured.

The (1s) Lamb shift is only known for the moment for neutral hydrogen. Its value is a very small fraction of the energy of the Lyman α line (3ppm). Its values have been measured at Stanford by Wieman et al[1] in comparing in the first and fourth orders of reflection the Balmer β line and the Lyman α one.

In the case of Iron for instance, the (1s) Lamb shift becomes a very substantial fraction of the energy of the Lyman α line : ∿4eV for a line of ∿7keV energy. Taking into account the typical precision in the energy measurement of X rays, this then makes possible the measurement of these radiative corrections.

The principle of the experiment is to compare the energy of the Lyman α lines emitted by the ions to a well-known X-ray line emitted by a conventional X-ray tube (the Co Kα lines for hydrogenlike

Iron and the K Kα line for hydrogenlike Argon).

Table 2. 1s Lamb shift measurement

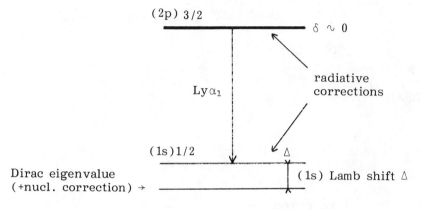

$$\Delta = E_{Dirac} - E_{experiment}$$

	δ	Δ
Ar	$<8 \ 10^{-3} \text{eV}$	$+ 1.1 \text{eV}$
Fe	$- 0.03 \text{eV}$	$+ 3.9 \text{eV}$

     A schematics of the experimental device is presented in Fig.1. The Bragg spectrometer consisted of a flat Si (111 or 220) crystal mounted on a very accurate goniometer controlled by a stepping motor. The crystal was typically located at 3.6m apart from an entrance slit fixed as close as possible of the axis of the ion beam. The emitted X rays after passing through the slit were reflected by the crystal to a position sensitive detector of the Backgammon type mounted on a circular track fixed around the center of the crystal goniometer. The position of the detector along its track was controlled by a precision translator (1μm) fixing the 2θ angle of the spectrometer. The whole system except the position sensitive detector was under vacuum. A curved beryllium foil fixed all around the goniometer housing allowed the transmission of the X rays after being reflected by the crystal, to the detector.

     In a first experiment (Fig.2) the reference (Kα₁ and Kα₂) X rays emitted by the X-ray tube were reflected on the axis of the crystal goniometer at the theoretical Bragg angle and detected in a given channel of the position sensitive detector.

274

Fig.1. Experimental set-up

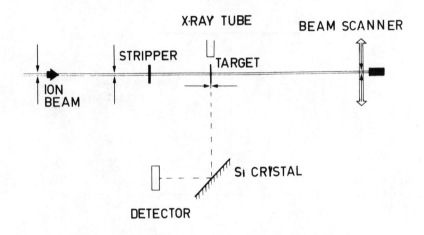

Fig.2. Principle of the experiments

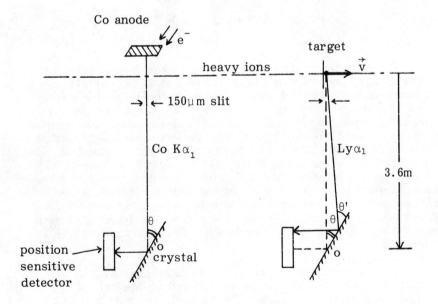

Without moving anything a thin carbon foil was then introduced in front of the slit and the   ion beam switched on. The
characteristic Lyman α line emitted by the ion nearby the target
(Δx < 1μm) was reflected through the slit to another channel of the
position sensitive detector and the energy of the Lyα line obtained
by comparison with the reference lines.

In order to achieve an absolute energy measurement the
experiment was repeated at two angles symmetric to the normal of the
line of flight of the ions (classical θ, -θ independent measurement,
Fig.3).

Fig.3. 2 independent (θ, -θ) measurements

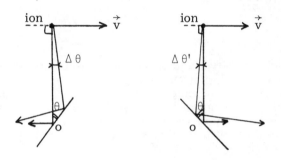

To get rid of the waving of the carbon foil after being
heated by the beam, the target was mounted on a very precise stepping
translator and on a goniometer in order to continuously optimize the
transmission of the X ray through the slit.

The main difficulty in such experiments comes from that
the reference X rays are emitted by an X-ray tube linked to the ground
while the X rays under study come from ions travelling at a very large
velocity (1/10 of the velocity of the light). This introduced very strong
relativistic Doppler aberrations which have to be controlled to achieve
an absolute energy measurement. There are in such an experiment
four independent degrees of freedom (Fig.4) :

    1)   the value of the velocity of the ion
    and for a given position of the crystal, three angles
    2) θ Bragg for the Co Kα line
    3) θ Bragg for the Lyα line
    4) the angle between the trajectory of the ion and the
      detected X rays.

Fig.4. The degrees of freedom in the measurement

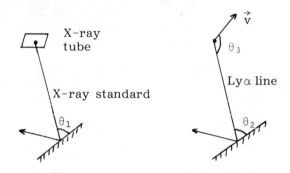

3 "angular" degrees of freedom + the absolute value of
the velocity of the ion.

At such velocities and for 7keV photons the Doppler shift
is about ±1keV when looking at the X rays emitted in the forward or
backward direction, and a change in the emission angle of 0.1° provided a change in the energy of the X ray of 1.6eV. The sensitivity
at the value of the velocity of the ions is also very large : an error
in the energy measurement of the ions of $4 \ 10^{-3}$ providing an uncertainty in the X-ray energy of 0.25eV.

The way in which these problems were solved was to use
a plane crystal spectrometer. It means a set-up whose angular
acceptance is very small. In the case of the Iron experiment the
acceptance angle of the crystal (a Si 220 crystal) which is equal to
its diffraction pattern was only of $10^{-3}d°$(or 0.22eV in the energy
scale). The typical transmission of this spectrometer was then very
small : 10^{-7} for Ar and 10^{-9} for Iron.

The exact measurement of all the considered angles was
achieved in using a complex optical system merging the trajectories
of the X rays and the ions with a laser beam. The angles of these
trajectories were measured with a precision better than $10^{-2}d°$
(Doppler effects) while the Bragg angles were determined with a precision of $3 \ 10^{-4}d°$.

The overall contribution to the X-ray measurement of all
these angles measurements were, in these θ, $-\theta$ experiments, found
to be less than 1/10 of eV.

The quality of the ion beam also plays a crucial role in
such experiments.

The line of flight of the ions was known with a precision better than 10^{-2} degree and the angular divergence of the beam was better than 1/100 of degree. These two important characteristics of the beam were accurately measured with a precision scanner located in front of the Faraday cup.

In order to reach such an angular divergence the beam intensity was reduced, taking into account the emittance of the beam, by a factor of 16. The intensity on the target was of the order of 1μA in the case of Iron.

The measurement of the energy of the ions constituted one of the main difficulties of the experiment. It was determined in three independent measurements carried out in crossed ways using either magnetic deflection, recoil protons, time of flight, or crystal charge accumulation. All these measurements provided an energy value known within a 0.4% confidence.

In Fig. 5 we present the Kα spectrum of Co which was used to calibrate the hydrogenlike Iron experiment. As one can see at such an energy resolution the Kα₁ line clearly exhibits its complex multiplet structure.

Fig.5. Co Kα spectrum

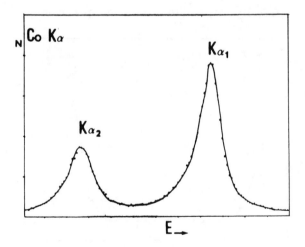

Following Bearden[2] we chose to use for calibration the top of this line but we must take into consideration that in this energy region it does not seem possible to obtain precise energy standards. This contributes, in a non negligible part, to the error bar in the measurement. In the next two figures (Fig. 6 and 7) are shown the Lyα spectra of Argon

278

and Iron.

Fig.6. Hydrogenlike Argon Lyα spectrum

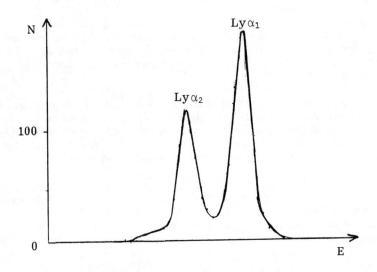

Fig.7. Hydrogenlike Iron Lyα spectrum

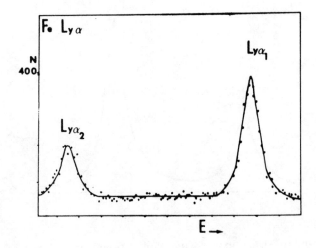

The observed lines are very symmetric purely gaussian peaks which are well resolved. In Fig.8 is presented the Cl Lyman α lines very recently observed at Brookhaven by P. Richard et al[3].

Fig.8. Hydrogenlike Chlorine Lyα spectrum

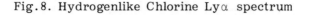

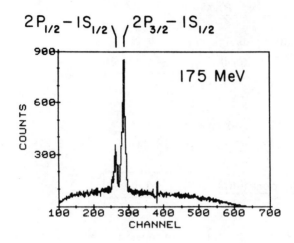

DISCUSSION OF THE RESULTS

In tables 3 and 4 are summarized all the theoretical predictions (P. Mohr[4]) and the experimental results.

In the case of Argon a very good agreement between the theory and the experiment is found within 0.1eV confidence.

In the case of Iron a quite good agreement is also observed between the predicted values and the experimental data when using for the Lamb shift the main value that one can extract from the energies of the Lyman α_1 and Lyman α_2. One must however point out that these two values strongly differ by an amount which is much too big to be explained for the moment : the Lyman α_1 being 0.5eV higher than expected (this having been confirmed in all the experiments carried out) while the Lyman α_2 was always found to be much closer to the theoretical value.

In conclusion to this discussion on the hydrogenlike spectra one may point out how such high Z ions can exhibit some effects which cannot be observed in hydrogen, even when doing experiments at the highest level of precision.

In table 5 are reported the two formulae that may be used to calculate for instance the (1s) energy of hydrogenlike ions, the approximate formula : Schrödinger eigenvalue + relativistic corrections

Table 3. (1s) radiative corrections for Argon (Z=18)

1s Lamb shift (theory) *P. Mohr*		
Self energy	+ 1.221 eV	
Vacuum polarization	− 0.085 eV	
total	+ 1.136 eV	
	theory	experiment
$Ly\alpha_1$	3 323.05 eV	3 323.2±0.5
$Ly\alpha_2$	3 318.23 eV	3 318.1±0.5
1s Lamb shift (experiment)	+ 1.0 eV	

Table 4. (1s) radiative corrections for Iron (Z=26)

1s Lamb shift (theory) *P. Mohr*		
Self energy	+ 4.263 eV	
Vacuum polarization	− 0.33 eV	
total	+ 3.93 eV	
	theory	experiment
$Ly\alpha_1$	6 973.29 eV	6 973.8±0.6
$Ly\alpha_2$	6 952.08 eV	6 951.9±0.7
1s Lamb shift (mean value)(experiment)	+ 3.55 eV	

(mass, spin orbit and Darwin) at the first order of perturbation theory, and the exact Dirac eigenvalue.

Using the best value of the Rydberg constant (table 6), and both formulae, it is possible to calculate the energy of the 1s energy level of hydrogen. As one can see these two values only differ at the last tenth digit. This means that even with the precision at which the Rydberg constant is known, it is not possible to be sensitive to the second order relativistic effects.

Table 5. The second order relativistic corrections

First order perturbation theory

$$E(n,\ell) = R\,\frac{Z^2}{n^2} + \Delta E_{mass} + \Delta E_{so} + \Delta E_{Darwin}$$

$$E(n,k) = R\,\frac{Z^2}{n^2}\left[1 + \frac{(\alpha Z)^2}{n}\left(\frac{1}{k} - \frac{3}{4n}\right)\right]$$

The Dirac eigenvalue

$$E(n,k) = \frac{2R}{\alpha^2}\left[1 - \frac{1}{\sqrt{1 + \left(\dfrac{\alpha Z}{(n-k) + \sqrt{k^2 - \alpha^2 Z^2}}\right)^2}}\right]$$

Table 6. (1s) energy values of hydrogen

Experimental Rydberg constant $R\infty$ = 109,737.314 <u>76</u> (32) cm^{-1}

{0.27ppm}

For hydrogen:

$$E(1s)_{Dirac} - E(1s)_{(Schrödinger + rel.\ corr.)} = 0.000,04\ cm^{-1}$$

      The same calculation for the hydrogenlike Iron ions is presented in table 7 where it can be shown that the difference in energy given by the two formulae just appear at the fourth decimal... Expressed in eV this energy difference is few times larger than the error bar in the experiment (0.6eV).

      This clearly shows how precise energy measurements for hydrogenlike ions can allow us to check fundamental theories and open the way to the direct observation of a lot of unknown phenomena, like
. the diamagnetic effect in the spin orbit
. the influence of the finite size of the nucleus, and
. the coulombic and magnetic homogeneity of the nucleus...

Table 7. (1s) energy level of hydrogenlike Iron (Z=26)

Energy of (1s) ground state (cm^{-1})	
Schrödinger + first order relativistic corrections	74,850,027.546
	↑
Dirac	74,862,321.042
δ = 160 ppm	

THE HELIUMLIKE IONS

The physical problems which can be studied with heavy heliumlike ions when everything is clear in the one electron systems is the three-body problem in the relativistic case.

In the case of helium the energy levels are known with a precision of the order of 1ppm, it means of the same order with which the fine structure constant is known (the last measurement of the fine structure of the helium atom, at Harvard[5], has been used to re-measure the fine structure constant).

But like in the case of hydrogen, this precision is not sufficient to measure some relativistic or QED effects which in fact are of a great importance for the heavy atoms.

Among these effects we shall mainly discuss two of them which seem now to be studied :
. the multiboby QED effects , and
. the magnetic correlation effects.

In heavy atoms these effects may represent quite large changes in the energy of the levels(10 to few tens of eV), but they cannot be specifically measured when taking into account the large number of electrons which are present in heavy neutral atoms.

Heliumlike heavy ions certainly constitute the best systems to specifically study these effects. There are unfortunately only known in very few cases and at very low precision (200 to 450ppm), a precision which is by far not sufficient.

In Fig.9 a typical energy diagram of a heliumlike ion is shown. Three of the four possible transitions between n=2→n=1 level can be observed in such a kind of experiment: the 2 E1 transitions decaying the $^1P_1$ and $^3P_1$ (1s) (2p) levels. The M2 line ($^3P_2$ level) whose lifetime is about 100 times larger than for the E1 can also be partially observed but the M1 line whose lifetime is still much longer has not been observed.

Fig. 9. Typical energy diagram of a heliumlike ion

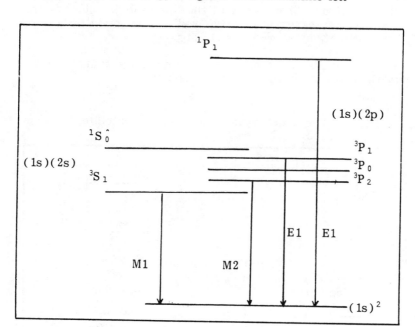

Before describing the experiments we must discuss an important problem which appears in the energy measurements and can drastically change the position of the lines : the contamination of the lines by the so-called dielectronic satellites.

The heliumlike ion spectra have very often been observed (but not precisely measured) in hot plasmas
. fusion plasmas
. solar corona plasmas.
In these plasmas Iron constitutes usually the most abundant heavy contaminent and at the considered temperatures (millions of degrees) it strongly appears as heliumlike ions.

The excitation of the heliumlike ions is however very differ-
rent than that appearing by foil excitation, the hot electron gas producing strong dielectronic recombination processes.

The dielectronic recombination process may be described
as the simultaneous capture of one electron and the excitation of the
ion. The excitation of a heliumlike ion via dielectronic recombination
then produced a heliumlike excited core plus an extra (outershell)
"spectator" electron. The energy of the line (satellite line) is then
slightly shifted in energy when compared to that given by a pure singly
excited heliumlike ion (diagram line) (Fig.10).

In hot plasmas these satellite lines have a much larger
intensity than the so-called "diagram line".

Fig.10. The dielectronic recombination satellites

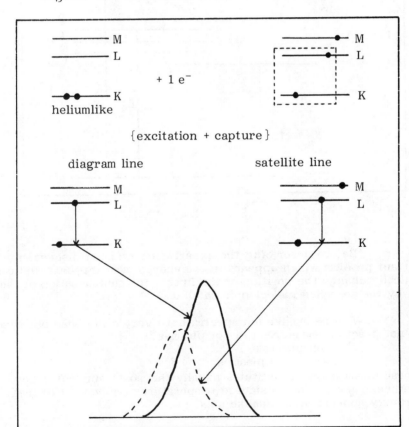

When using foil excitation of ion beams, it is possible to control the number of additional outermost shell electrons of the heliumlike ions. The thicker the foil is the more of outershell electrons we can get.

In Fig. 11, we can see how the contamination of the spectra by satellites (far satellites as well as contamination satellites) can be modified in changing the nature or thickness of the foils. It is then possible to vary the contamination of the lines in decreasing the target thickness until pure diagram lines can be obtained.

In Fig. 12, the $^1P_1$ line which is usually the more contaminated one has been drawn at a scale such as it is easy to evaluate how important the errors in the energy measurement can be when we do not take into account the satellite contamination.

In Fig. 13 and 14, are presented the best heliumlike spectra we obtained in cases where satellite contamination is negligible.

There has been up to now a lot of different calculations of the energy levels of heliumlike ions done by various authors and various techniques (Johnson, Drake, Sofranova, Dubau and Bely Dubau, Chen and Crasemann) and lots of others but unfortunately no accurate experimental values have been so far carried out.

We shall specially compare the experimental data with the J.P. Desclaux multiconfiguration Dirac-Fock calculations. In table 8, a comparison is made between these calculations and the experimental values, where a very good agreement (better than 1/10 of eV) is found. In order to have a better physical insight of the problem we have shown in table 9 the details of the calculations of J.P. Desclaux on the $^1P_1$ line of heliumlike Iron (table 9 A and B).

Two important points have to be emphasized when considering the very good agreement between experiment and theory and the degree of accuracy which is reached in such experiments. ·

First we can see that the Lamb shift contribution to the energy of the transition 3.5eV is a little smaller than in the case of hydrogenlike ion where we found 3.9eV. This is due to the screening effects on the self-energy value and to the three-body QED effects. In the case of heliumlike Iron, the energy difference found is larger than the error bar in the measurement (A).

Another interesting point appears when looking at the various contributions to the correlation energy. In the first column(B) we see the nature of the configurations introduced by Desclaux. In the third one the additional energy introduced by the correlation to the 1s energy value. In the last column, the value of the Breit term

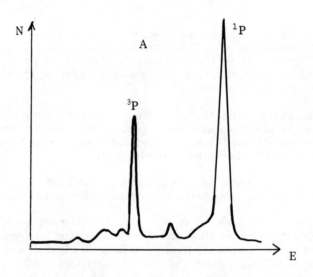

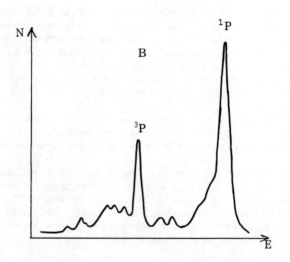

Fig. 11. Heliumlike Iron spectrum
A. 100μg/cm² Carbon foil
B. 100μg/cm² Aluminium foil
see also Fig. 12: 50μg/cm².

Fig. 12. Shape of the $^1P_1$ contaminated by
"dielectronic recombination satellites"

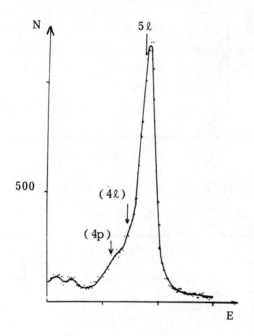

Fig. 13. Heliumlike Iron spectrum

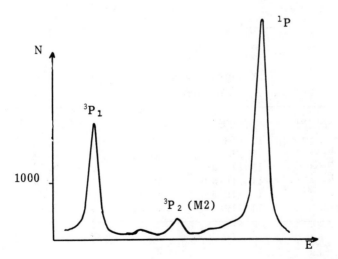

Table 8. Heliumlike Argon

	n=2 ⟶ n=1	
transition	experiment (our values)	theory (J.P. Desclaux)
2 $^1P_1$ (E1)	3 139.6±0.3	3 139.7
2 $^3P_1$ (E1*)	3 123.6±0.3	3 123.6
2 $^3P_2$ (M2)	3 126.4±0.4	3 126.3

Table 9. Heliumlike Iron

A

	$^1P_1 \rightarrow {}^1S_0$	$^3P_1 \rightarrow {}^1S_0$	$^3P_2 \rightarrow {}^1S_0$
Fe^{24+} Hartree-Fock	6638.81	6616.77	6616.77
Dirac correction	70.22	59.00	74.06
Breit interaction	-6.10	-5.70	-6.08
QED corrections	-3.52	-3.58	-3.55
Correlation	1.29	1.30	1.28
Total	6700.70	6667.79	6682.48
Our exp. values	6700.9(±.25)	6667.8(±.25)	6682.7 (±.4)

B

The magnetic correlation
1s energy

configuration	energy	ΔE	Breit term
$1s^2$	11 118.48	0	-6.2
$2s^2$	11 118.85	0.37	-6.15
$3s^2$	11 118.88	0.40	-6.14
$2p^2$	11 119.50	1.02	-6.02
$3p^2$	11 119.57	1.09	-6.00
$3d^2$	11 119.64	1.16	-5.98

Fig.14. Heliumlike Argon spectrum

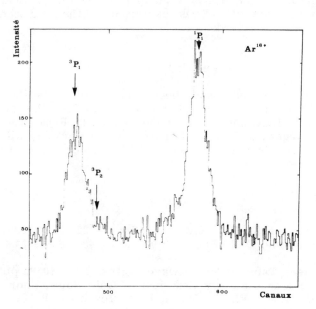

when calculated with all these configurations is shown. As one can see the contribution of various terms clearly varies when considering more and more configurations. This is due to the effect of magnetic correlation which means the instantaneous correlated movement between the two electrons induced mainly by the spin-spin interaction.

The contribution of this effect is now of the order of magnitude of the error bar in the measurement but there is no doubt that this effect which fastly increases with Z and is of a great importance in heavy atoms will be soon very sensitively observed in experiments.

CONCLUSION

In conclusion we can say that the few electron ions can provide interesting information on some effects which are of importance for heavy many-electron atoms.

The interest is obviously increasing with the atomic number of the ion but there is no doubt that very soon, making use of the new heavy ion accelerators such as GSI(new version),Bevalac, Ganil and so on,the heaviest ion will be obtained (hydrogenlike Uranium...).

But however it has also to be pointed out that the quality of the available heavy ion beams is going to be also dramatically improved, allowing very likely remeasurements of the fundamental constants or X-ray standards.

Acknowledgments

These experiments have been carried out at the Heavy Ion Cyclotron (CEVIL) in Orsay in collaboration with J.P. Mossé, P. Chevallier, P. Indelicato, D. Vernhet and M.T. Ramos for the experiment of Argon, and at the SuperHILAC of Berkeley with the collaboration of R. Marrus, H. Gould, M. Tavernier and P. Indelicato for the experiment of Iron.

REFERENCES

1. C. Wieman and T.W. Hansch, Phys. Rev. A 22 192 (1980).
2. J.A. Bearden and C.H. Shaw, Phys. Rev. 48 18 (1935)
   and
   J.A. Bearden, Table of X-ray wavelengths, NYO 10586 published by the U.S. Atomic Energy Commission, Oak Ridge, Tenn. (1964).
3. P. Richard, M. Stockli, R. Mann, R.D. Deslattes, P. Cowan, B. Johnson, K. Jones, M. Meson and K. Schartner, this conference.
4. P.J. Mohr, Phys. Rev. Letters 34 1050 (1975)
   and
   P.J. Mohr, Annals of Physics 88 26 and 52 (1974).
5. W. Frieze, E.A. Hinds, V.W. Hughes and F.M. Pichanict, Phys. Rev. A 24 279 (1981).

REVIEW OF SINGLE AND DOUBLE K ELECTRON TRANSFER
IN MeV/u HEAVY ION ATOM COLLISIONS

A. Chetioui
Institut Curie and Université Pierre et Marie Curie
11, rue Pierre et Marie Curie F-75231 Paris Cedex 05

ABSTRACT

The process of single and double electron transfer from target K to projectile K or excited state is discussed for collisions in the high and intermediate velocity range. A description is given for the concepts underlying different theoretical models and recent detailed experimental results are presented.

INTRODUCTION

We consider in this review the process in which a swift bare ion captures a K electron of a target atom in one of its K or excited states. Indeed such a process keeps approximately the simplicity of the $p \rightarrow H$ collision, a pure three-body problem, since non captured electrons can generally be considered as passive and their effect can be reduced to a screening effect. However by varying the ratio of projectile to target nuclear charges, one can explore different mechanisms in which the three-body effects appear in different ways. We will reduce the scope of this review to MeV/uma collisions ; that is to collisions in which the velocity of the incoming particle, v, is generally greater or equal to the mean orbital velocity of the active electron before (v_i) or after (v_f) capture. One usually divides this field into "high velocity" and "intermediate velocity" collisions depending if $v \gg \sup \{v_i, v_f\}$ or $v \sim v_i, v_f$. When levels i and f are regarded as hydrogenic, their energies can be expressed as $\varepsilon_{i,f} = -1/2 \, m v_{i,f}^2$ and the high velocity condition becomes $E_{lab} \, (MeV/uma) \gg 0.05 \sup \{\varepsilon_i, \varepsilon_f\}$ with $\varepsilon_{i,f}$ in atomic units.

The main result of comparison between experiments and theory is that, whereas for a wide range of high relative velocities or low projectile to target nuclear charge ratios, the excitation and ionization processes may be evaluated in the first Born approximation, this happens to be never the case for electron capture. That means that one has to deal with a problem which can never be reduced to a two body problem.

We discuss in this review some of the approximations which have been developed to treat the three-body effects and we make some comparisons between theoretical predictions and recent detailed experimental results.

0094-243X/82/940291-12$3.00 Copyright 1982 Aemrican Institute of Physics

I. K ELECTRON CAPTURE IN HIGH VELOCITY COLLISIONS

It is well known that excitation and ionization processes dominate over charge exchange at high impact velocity. It is thus expected that the capture process is then strongly coupled to the above mentioned channels.

Classical theory

The classical model of Thomas (1) exhibits the main features of the high velocity capture mechanism ; this capture appears as :
- a three-body process
- a two symmetric step process.

The classical mechanism is a double scattering process. The electron first scatters off the projectile ion through 60° with respect to the beam direction, getting a velocity with a magnitude equal to the projectile one. Then the electron scatters through 60° off the target nucleus and is left with almost zero momentum with respect to the projectile nucleus.

In the first scattering the transverse momentum transfer is $mv_1 \sin 60° = \sqrt{3}/2\, mv$ and the projectile B is deflected from an angle θ such that $tg\,\theta = \sqrt{3}/2\, mv/M_B v$ i.e. $\theta \sim \sqrt{3}/2\, m/M_B$. This well defined non zero projectile deflection angle is characteristic of the double scattering process. $\theta = 1.6'$ for the p→H collision and θ decreases for heavier projectiles.

The Thomas cross section, $\sigma_{Thomas} = \dfrac{2^{13/2}}{3} \pi Z_A^2 (\dfrac{e^2}{mv^2})^{11/2}$ has the well known v^{-11} dependence.

General theories

We examine now the physical meaning of quantum theories and their connection to the classical picture. The quantal cross section is expressed as

$$\sigma = \int |\frac{\mu T}{2\pi}|^2 \, d\Omega$$

where the general form of the transition amplitude T is (in the past form) :

$$T^+ = <\Phi_j\, \beta\, |v\beta|\, \Psi_i^{\alpha+}>$$

$$v_\beta = \frac{Z_A Z_B}{R} - \frac{Z_A}{x} \quad ; \quad v_\alpha = \frac{Z_A Z_B}{R} - \frac{Z_B}{s}$$

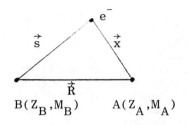

Φ_j^β (channel β, state j) is the wave function for the electron bound to the B nucleus in the j state,

Φ_i^α is the wave function for the electron bound to the A nucleus in the i state,

$\Psi_i^{\alpha+}$ is the collision wave function.

The Born series

The transition amplitude may be expanded into a two-body perturbation series, the Born series.

At first order $\quad T_{BI}^{+} = \langle\, \Phi_j^\beta \,|v_\beta|\Phi_i^\alpha\, \rangle$

At second order $\quad T_{BII}^{+} = \langle\, \Phi_j^\beta \,|v_\beta(1+G_0^{+}\, v_\alpha)|\Phi_i^\alpha\, \rangle .$

The Born I amplitude represents a one-step process in which capture occurs for electrons with a high orbital velocity, close to the projectile velocity

Born I terms :

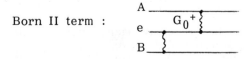

OBK

Jackson-Schiff

The full amplitude has been calculated by Jackson and Schiff (2) but the version currently used is the Oppenheimer, Brinkman Kramers one (OBK) (3) in which the nucleus nucleus interaction is dropped out. The predicted cross section for the 1s→1s charge exchange varies as v^{-12} instead of v^{-11} in the classical picture. The cross section for electron exchange between arbitrary (n,ℓ) hydrogenic states has been given by Omidvar (4) and has been often used as a reference. Because the OBK calculations overestimate strongly cross sections, a scaling factor is generally used (5). A first order method taking into account residual attraction from the target nucleus (6) has given a tentative interpretation of this scaling factor.

The Born II amplitude represents a double scattering process.

Born II term :

The equivalence of Born II with the classical double scattering process has been demonstrated by Shakeshaft (7). As shown by Spruch (8), for transfer between states of large n and ℓ quantum numbers total cross sections become identical. One can thus expect the second order term to become very important at high velocity. Drisko (9) first showed that the first Born term does not provide the leading contribution at high velocity. Dettman and Leibfried (10) found that for forward electron transfer, the second order Born term provides the leading v^{-11} dependence of the asymptotic cross section. Finally Shakeshaft (11) showed that the third order term is also important. The problem of the convergence of the Born series in the case of rearrangement collisions is not yet solved. Aaron (12) has shown that, due to the existence of disconnected diagrams, the series of transitions operators diverges. However this does not necessarily mean that the series of amplitudes diverges.

Because of computational difficulties, the second Born approximation has not yet been evaluated, except in the simplest $1s \rightarrow 1s$ case (13). An approximate analytical form has recently been given by Briggs and Dubé (14) for charge transfer from initial $n'\ell'm'$ to final $n\ell m$ hydrogenic one-electron state.

The continuum distorted wave approximation

Another approach to charge transfer consists in using the distorted wave method in order to avoid all disconnected diagrams contained in the Born series. A three-body series is then derived.

The continuum distorted wave (CDW) method has been introduced by Cheshire (15) in the impact parameter method and by Gayet (16) in the wave version.

At first order, the transition amplitude is equal to :

$$T_{CDW} = < \chi_j^{\beta^-} |\sigma| \chi_i^{\alpha+} >,$$

an expression in which the distorted wave function $\chi_j^{\beta^-}$ ($\chi_i^{\alpha+}$) is the product of the wave function for an electron in a $j(i)$ state bound around nucleus B(A), by a Coulomb wave for a free electron around nucleus A(B) and by the Coulomb wave describing the relative motion of nuclei ; σ, an operator proportional to $\vec{\nabla}x.\vec{\nabla}s$, represents the transition of the electron from nucleus A to nucleus B.

By its inclusion of a virtual intermediate channel, the CDW method allows for coupling to the continuum states already at first order of perturbations. It thus describes a process related to the Thomas process and can be compared to second order theories (Born II). For $p \rightarrow H$ collision, the CDW and Born II asymptotic forms of $1s \rightarrow 1s$ cross section are identical. However they are different for capture into excited states (17).

The CDW approximation has recently been generalized to complex systems (17) for which the initial and final orbitals of captured electrons may be represented as a superposition of Slater type orbitals.

First comparisons between theoretical and experimental K capture cross sections at high velocities

Two main questions must be anwered :

- What is the relative importance of first and second order processes at high velocity ?
- Are three-body second order theories (CDW) better than two-body ones (Born II) ?

No answer yet exists for the first question. Some experiments have been suggested :

- to look at the Thomas peak in the differential capture cross section (fig. 1) but this peak occurs at a very small angle (1.6' for the p→H collision and less for collisions of heavier projectiles)
- to study the velocity dependence of the 1s→nℓ capture cross sections or of the 1s→nℓ/1s→nℓ' ratios.

Fig. 2 shows indeed that first order (OBK) and second order (CDW) predictions exhibit markedly different trends.

For asymptotically high velocities for instance, first order theories show a predominance of capture in low angular momentum states since target electrons which possess a high momentum with respect to the projectile must be captured at very small distances from the nucleus. On the other hand, in the double scattering process the velocity of the electron relative to the projectile is reduced and higher angular momentum states may be reached.

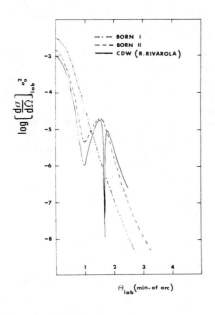

Fig. 1. Differential cross sections for p→H collision 8 MeV.

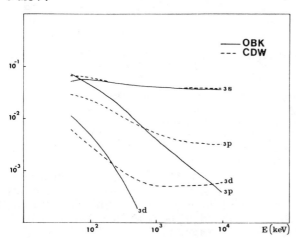

Fig. 2. 1s→nℓ cross sections for p→H collision
---- CDW
—— OBK

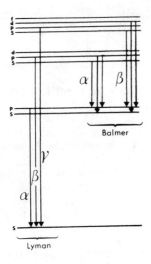

Capture probabilities in np states can be deduced from measurements of Lyman X-ray intensities. For low n quantum numbers, these rays are strongly affected by cascades which must be substracted. Their intensities may be deduced from the Balmer X-ray intensities and branching ratios in hydrogenlike atoms. Capture probabilities in ns, nd states can be derived from Balmer X-ray intensities under certain approximations.

The importance of second order effects at high relative velocities can be studied either at high impact velocities or for capture in high n levels. Such experiments have been performed by Tawara and al (18) and Chevallier and al (19) by looking at the high n Lyman by means of high resolution spectroscopy (see fig.3).

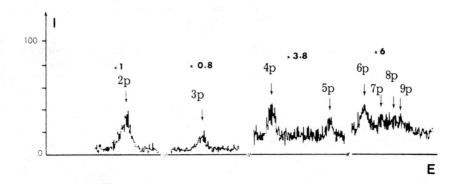

Fig.3. Lyman spectrum observed in the collision $^{18+}$Ar (250MeV)$\rightarrow$N$_2$(19).

A discrimination between the two-second order theories, Born II and CDW can probably be done when considering collisions of highly charged projectiles, since CDW theory allows for the distortion of the electron wave in the continuum.To this end, experiments have been performed with $^{26+}$Fe projectiles (20).

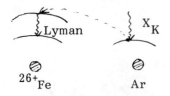

Doubly differential 1s$\rightarrow$nℓ cross sections have been measured by looking at the projectile Lyman spectrum emitted in coincidence with target K X rays.

Results are very well reproduced by CDW calculations (see figures
4 and 5). Born II calculations (14) are under performance

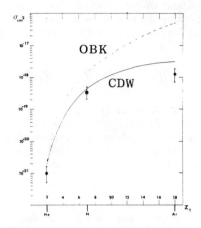

Fig. 4. 1s→2p capture cross
sections, 400MeV $^{26+}$Fe collisions.

Fig. 5. 1s→3p capture cross sections, 400MeV $^{26+}$Fe collisions.

II. K ELECTRON CAPTURE IN INTERMEDIATE VELOCITY COLLISIONS

A picture of capture process in intermediate velocity collisions must
lie between the high velocity image of capture in the field of target
nucleus'residual attraction and the molecular picture in which the
electron does not belong to any particular nucleus during the collision.

Starting from the high velocity case and considering decreasing
velocities one can imagine that when the velocity is slow enough, the
captured electron has time to jump back to the stationary target nucleus.
If the velocity decreases further the electron can even make multiple
transitions between the two nuclei.

high velocity case   intermediate velocity case   low velocity case

The hypothesis of the electron going back and forth has been formulated by Cheshire (21) to explain the oscillations in the p→H charge
exchange cross section at a fixed scattering angle as a function of the
projectile energy.

Thus the capture process at intermediate velocity appears as
strongly coupled to the excitation process.

Theories

The methods used in the intermediate energy field are usually based

298

on the semi-classical approximation : the two nuclei are treated as classical particles moving with a constant velocity and then the full three-body problem is reduced to the problem of an electron moving in a time-dependent field. The electronic wave function is expanded on a set of wave functions centered about both nuclei with time-dependent coefficients :

$$i\hbar \frac{\delta \Psi}{\delta t} = H\Psi \; ; \; \Psi = \sum_i a_i(t)\phi_i \; ; \; P_{if}(\rho) = |a_{if}(\rho)|^2 \; ; \; \sigma = 2\pi \int_0^\infty P_{if}(\rho)\rho d\rho$$

Capture and excitation processes are thus treated on the same footing. Most calculations are restricted to the p→H collision. Cheshire and al (22) used an expansion in hydrogenic states and added pseudo-states in order to simulate the molecular features at small inter-nuclei separations. Shakeshaft used Sturmian basis vectors (23) or a scaled hydrogenic function expansion (24).

The 1s→1s charge exchange has been calculated by the two-state atomic expansion introduced by Mc Carroll (25). While for the symmetric $H^+ + H$ system the resonant K-K transfer can be described properly with the TSAE over a broad range of velocities, calculations for one electron asymmetric systems (26) indicate an increased sensitivity of cross sections to the character and size of the basis set. The TSAE formalism for 1s→1s charge exchange has recently been extended by Lin (27) to collisions of multielectron atoms. It is usually assumed (28) that at very low velocity the electronic system is described more appropriately as a completely relaxed system by choosing a relaxed molecular hamiltonian and molecular orbitals to represent the evolution of passive and active electrons. However in the intermediate velocity range Fritsch and al (29) investigated the possible use, for active orbitals, of MO orbitals calculated from a frozen two-center hamiltonian in which all outer electrons are fixed in their respective atomic orbitals during the collision.

Experiments

K-K capture cross sections have been measured by many authors (30). The technique consists in comparing target K X-ray emission for collisions of bare or two-electron ions.

Systematical studies have been performed as a function of velocity (29) (Fig.6) or as a function of projectile to target nuclear charge ratios (31) (Fig.7).

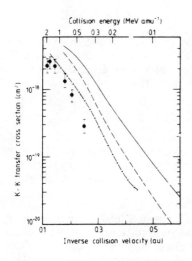

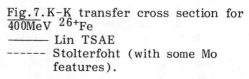

Fig.6. K-K transfer cross section in $^{9+}$F +Si collisions
-·-··-·- TSAE
------- MO (frozen two-center hamiltonian)
———— MO (relaxed hamiltonian).

Fig.7. K-K transfer cross section for 400MeV $^{26+}$Fe
———— Lin TSAE
------ Stolterfoht (with some Mo features).

For $v > v_e$ a relatively good agreement with TSAE calculations has been observed. For $v < v_e$ it has proved to be very important to introduce a realistic potential (such as a Herman-Skillman one) to represent the target atom. Still some unexplained discrepancies persist at low energy. Calculations including relaxation (29) lead to an increase of the discrepancy.

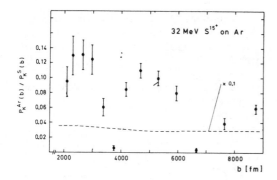

Fig.8. Vacancy sharing ratio dependence on impact parameter

Studies of the impact parameter dependence of charge transfer are much more sensitive to model's assumptions than total cross sections and would provide useful information. Only few studies (32) include such measurements (Fig.8). They show the characteristic oscillations of the two passage case involving a localized coupling zone. A very distinctive feature of the non relaxed and fully relaxed case is the position of the peak at small impact parameter (29).

III. DOUBLE K ELECTRON TRANSFER

The simplest way to deal with this many-body problem is to use the independent electron approximation. This requires that correlations between electrons are small. Then the many-electron amplitude can be expressed as a product of single electron amplitudes :

$$\text{ST} \qquad\qquad\qquad\qquad\qquad \text{DT}$$
$$P_{if}(\rho) = |a_{if}(\rho)|^2 \qquad\qquad P_{if}(\rho) = P_{if}(\rho)^2.$$

The comparison of such calculations with experiment tests at the same time the independent electron approximation (that one can expect to be valid for not too light atoms) and the method to calculate single electron amplitudes.

Most studies have been performed at intermediate energies where cross sections are not too low. Theoretical calculations are relatively scarce. The total double capture cross section has been measured for $He^{+}{\to}He$ collision at MeV energies (33) and compared to the $1s^2{\to}1s^2$ electron transfer. Two states (34) and three states (35) expansions within the independent model show good agreement with experiment. On the contrary, in the case of CDW calculations (36), cross sections have been found to be very sensitive to the choice of the final $1s^2$ wave function (with or without electron interaction) ; this feature disappears when heaviest systems are considered ($^{9+}F{\to}Ar$) and this would lead to the conclusion that correlation effects in Helium are large in contradiction with other results (34) (35). It is very important to test further this point since both amplitude and energy dependence of double capture cross sections in the CDW treatment of the Helium case appear to be very sensitive to the inclusion of electron interaction.

For heaviest systems the double K-K transfer has been obtained experimentally from the target hypersatellite K X-ray intensity (37).

For collisions of N to Cl projectiles on Ti target in the energy range of 2 to 6MeV/u the cross sections have been well reproduced by three state calculations and one center multi-state calculation (37) in the independent electron picture.

First measurements of double K K transfer to excited projectile states have been recently performed (38) by measuring coincidences between the two Lyman X rays emitted by a bare projectile after a double capture. In the case of triplet final states the transition amplitude appears as a linear combination of products of amplitudes ; if interference effects are not too small, such measurements could constitute sensitive tests of theories.

CONCLUSION

The field of electron capture will probably develop quickly in the near future. First exact Born II calculations for capture in excited states will appear as well as CDW calculations in high n and ℓ quantum number levels. If computational difficulties are solved, coupled state calculations will be also developped for excited states.

Detailed experimental studies are required such as measurements of substate to substate transitions, differential cross sections as a function of impact parameter or scattering angle and multiple capture process.

The use of high energy accelerators will finally raise the problem of relativistic effects in non radiative as well as in radiative electron capture.

302 REFERENCES

1. L.H. Thomas, Proc. Roy. Soc. 114 561 (1927).
2. J.D. Jackson and H. Schiff, Phys. Rev. 89 359 (1953).
3. H.C. Brinkmann and H.A. Kramers, Proc. K. Ned. Akad. Wet.
   33 973 (1930) ; J.R. Oppenheimer, Phys. Rev. 31 349 (1928).
4. K. Omidvar, Phys. Rev. 153 121 (1967).
5. A.M. Halpern and J. Law, Phys. Rev. Lett. 31 4 (1973).
6. F.T. Chan and J. Eichler, Phys. Rev. A20 1841 (1979).
7. R. Shakeshaft, J. Phys. B Atom. Molec. Phys. 7 1059 (1974).
9. R.M. Drisko, Thesis Carnegie Institute of Technology (1955).
10. K. Dettman and G. Liebfried, Z. Phys. 210 43 (1968), Z. Phys.
    218 1 (1969).
11. R. Shakeshaft, Phys. Rev. A18 1011 (1978).
12. R. Aaron, R.D. Amado and B.W. Lee, Phys. Rev. 121 319 (1961).
13. P.J. Kramers, Phys. Rev. A6 2125 (1972).
14. J.S. Briggs, L. Dubé, J. Phys. B. Atom. Molec. Phys. 13 771
    (1980).
15. I.M. Cheshire, Proc. Phys. Soc. 84 89 (³964).
16. R. Gayet, J. Phys. B. 5 483 (1972).
17. Dz Belkic, R. Gayet R., A. Salin, Physics Reports, 56 280 (1979).
18. H. Tawara, P. Richard, K.A. Jamison, T.J. Gray, J. Newcomb
    and C. Schmiedekamp, Phys. Rev. A19 1960 (1979).
19. P. Chevallier, J.P. Rozet, A. Chetioui, A. Jolly, K. Wohrer,
    F. Fernandez and C. Stephan, to be published.
20. A. Chetioui and al, to be published.
21. I.M. Cheshire, J. Phys. B1 428 (1968).
22. I.M. Cheshire, D.F. Gallagher, and A. Joanna Taylor, J. Phys.
    B3 813 (1970).
23. R. Shakeshaft, Phys. Rev. A14 1626 (1976).
24. R. Shakeshaft, Phys. Rev. A18 1930 (1978).
25. R. Mc Carroll, Proc. R. Soc. A264 547 (1961).
26. T.G. Winter, G.J. Hatton, and N.F. Lane, Phys. Rev. A22 930
    (1980).
27. C.D. Lin, and L.N. Tunnel, Phys. Rev.A22 76 (1980).
28. J.S. Briggs, Rep. Prog. Phys. 39 217 (1976).
29. W. Fritsch, C.D. Lin, and L.N. Tunnel, J. Phys. B. Atom.
    Molec. Phys. 14 2861 (1981).
30. P. Richard, Electronic and Atomic Collisions ed.NODA and
    K. Takayanagi (Amsterdam : North Holland) 125 (1980).
31. K. Wohrer and al, to be published.
32. R.R. Randall, J.A. Bednar, B. Curnutte and C.L. Cocke,
    Phys. Rev. A13 204 (1976).
    R. Schuch, G. Nolter, H. Schmidt-Böcking and H. Lichtenberg,
    Phys. Rev. Lett. 43 1104 (1979).
33. E.W. Mc Daniel, M.R. Flannery, H.W. Ellis, F.L. Eisele, and
    W. Pope, US Army Missile Research and Development Command
    Technical Report, H78-1 (1977).
34. T.C. Thessen and J.H. Mc Guire, Phys. Rev. A20 1406 (1979).
35. C.D. Lin, Phys. Rev. A19 1510 (1979).
36. R. Gayet, R.D. Rivarola and A. Salin, J. Phys. B. Atom. Molec.
    Phys. 14 2421 (1981).
37. J. Hall, P. Richard, T.J. Gray, and C.D. Lin, Phys. Rev. A24
    2416 (1981).
38. A. Chetioui and al, to be published.

SOURCES OF LOW ENERGY, VERY HIGHLY IONIZED ATOMS -
PRESENT STATUS AND POSSIBLE FUTURE DEVELOPMENTS

Vaclav O. Kostroun
Nuclear Science and Engineering Program and School of
Applied Physics, Cornell University, Ithaca, NY 14853

ABSTRACT

The states of highly charged ions and their interactions with
electrons, other ions and atoms are difficult to investigate experimentally at low ion kinetic energies. Investigations in this area
have long been hampered by the lack of suitable multiply charged ion
sources capable of producing ion beams of a few keV or less kinetic
energy and intense enough to permit the type and variety of spectroscopic and collisional experiments available to study neutral or
singly charged ions.

Over the past decade, a number of sources of multiply charged
ions have been developed for nuclear particle accelerators. Two
of these, the electron cyclotron resonance ion source (ECRIS) and
electron beam ion source (EBIS) show particular promise for use in
atomic physics experiments involving low energy, highly ionized
atoms. While neither of these sources is ideal, each possesses
features which make it suited for a particular class of atomic
physics experiments. For example, the EBIS can be used to generate
ions of elements throughout the periodic table with the highest
extractable charge state of any source, in some case Ar^{+18} and Xe^{+52}
have been obtained, while the ECRIS can generate intense beams of
highly charged, lower Z elements.

In this paper the present status of existing sources of low
energy, very highly ionized atoms is reviewed, the potential of
these sources for atomic physics experiments is evaluated, and possible future developments in source technology are discussed.

INTRODUCTION

Multiply charged ions and their interactions with electrons,
other ions and atoms are of ever increasing theoretical and experimental interest. In plasma physics for example, properties and
behavior of plasmas can be deduced from a spectroscopic analysis of
electromagnetic radiations emitted, and the qualitative and quantitative interpretations of these radiations require a detailed understanding of the atomic interactions involved. In particular, efforts
to produce high-temperature plasmas in the laboratory with the objective of developing thermonuclear fusion power sources have shown an
even greater need for more accurate and detailed atomic structure
and collision data in order to understand the behavior of present
and proposed fusion plasma devices.[1] In atomic physics, multiply
charged ions and their interactions with other particles are recognized as interesting in their own right, exhibiting experimentally

observed features and theoretical difficulties not present when dealing with neutral atoms.

The structure of multiply charged ions and the details of their
interactions with electrons, other ions, atoms or electromagnetic
radiation can be deduced from spectroscopic and collisional studies.
With presently available multiply charged ion sources, discussed
more fully below, it is possible to investigate multiply charged
ion-atom collision and electron-ion ionization (excitation) processes.
Multiply charged ion-atom collisions, whether with simple systems
such as atomic, molecular hydrogen or helium, or atoms of the same
or different atomic number, are the easiest to carry out. Total
electron capture cross section measurements,[2,3] or energy gain spectra measurements in small-angle capture collisions,[4] require very
small currents of multiply charged ions. On the other hand, any
moderate to high resolution analysis $(E/\Delta E > 100)$ of radiations
emitted in single and double electron capture (photons and/or
autoinizing electrons) requires more intense ion beams, but probably
within the output range (depending on the charge state) of presently
available multiply charged ion sources.

Multiply charged ion beam fluxes required for multiply charged-
ion-electron ionization (excitation) studies and ion-ion collision
studies involving crossed beams, are greater than those needed for
spectroscopic studies, and possibly out of the intensity range of
presently available sources (except for highly charged lower Z
elements). Ion-photon interactions with other than laser beams
are at present more speculative and will require special considerations.[5]

The general types of experiments mentioned, require low energy
~ 1 keV/nucleon or less, collimated beams of particles with as small
as possible energy spread. These requirements generally eliminate
laser induced plasmas and theta pinch devices in which high densities of multiply charged ions for very short lengths of time are
available, but which cannot be extracted and converted into intense
enough beams.

This leaves essentially two ways of producing multiply charged
ions in appreciable amounts in the laboratory: by electron bombardment of ions confined in a plasma for a sufficient length of
time, and by accelerating slightly ionized atoms to fairly high
(MeV) energies and stripping off electrons by passing the ions
through matter (possibly reaccelerating the emergent highly charged
particles). The latter method has been, and continues to be used
in accelerator based atomic physics to create multiply charged
ion beams with kinetic energies in the range of several hundred keV
to tens, even hundreds of MeV which are used in beam foil spectroscopic studies,[6] atomic collisions studies,[7] Lamb shift measurements,[8] etc.

Beams of singly, and in some instances multiply charged ions
can be obtained from classical gaseous arc discharge sources such as
the duoplasmatron[9] or Penning discharge source (PIG)[10] and their

variants which will not be discussed. Instead, we consider only
sources of multiply charged ions which have been developed over the
past 15 years for nuclear particle accelerators. These sources can
provide 10^8 to 10^{14} particles/cm^2-s. with kinetic energies of the
order of a keV per nucleon or less, and their availability coupled
with present day readily accessible ultra-high vacuum technology
open up completely new experimental possibilities in atomic physics.
In particular, it is now possible to generate sufficiently intense
beams of low energy multiply charged ions which can be used in both
spectroscopic and collision studies to investigate fundamental processes in a systematic and thorough manner. Of the various sources
developed, two, the electron cyclotron resonance ion source (ECRIS)
and electron beam ion source (EBIS) show particular promise for use
in atomic physics experiments involving low energy, multiply charged
ions. While neither of these sources may be ideal, each possesses
features, often complementary, which make it suitable for a particular class of atomic physics experiments. After a brief discussion
of the characteristics of multiply charged ion sources and a general
description of ECRIS and EBIS sources, the present status of existing sources of low energy, multiply charged ions is reviewed, and
possible future developments in source technology are discussed.

GENERAL CHARACTERISTICS OF MULTIPLY CHARGED ION SOURCES

A given ion source, whether duoplasmatron, hot or cold cathode
Penning discharge source, electron cyclotron resonance or electron
beam source, can be characterized by the following plasma parameters: E, the bombarding electron energy, n, the electron density
and τ, the ion confinement time. For efficient ion production, E
must be a few times the ionization potential of the last electron
removed from the ion, and nτ must be large enough to make high
charge states. The ranges of parameters are limited by practical
consideration and fix the ion yield, charge distribution and the
highest degree of ionization attainable, all characteristic of a
given source. Both the ECRIS and EBIS sources have E and nτ
greater than the other source and can therefore produce higher
charge state ions.

In ECRIS and EBIS sources it has been established that ion production results from sequential removal of electrons rather than
from multiple ionization following the removal of a deep core electron.[11] Since the multiply charged ion distribution obtainable from
an EBIS is more amenable to analysis than that from an ECRIS, a
simplified model of the latter will be presented, followed by an outline of modifications necessary to describe the output from an ECRIS.

In an EBIS, Figure 1, a high current density, energetic electron beam is launched in or into a solenoidal magnetic field which
prevents the beam from spreading due to its space charge. The
electron beam ionizes atoms introduced into the ionization region
of the source, and multiply charged ions are formed by successive
ionization by electron impact. The ions formed are trapped in the

radial direction by the attractive potential formed by the electron beam's space charge, and axially by a suitable potential distribution impressed on a number of cylindrical electrodes concentric with the beam, Figure 1 (a). After a predetermined ion containment time, the axial trap potential distribution is changed, and the ions are expelled from the ionization region of the source. At the electron collector-ion extractor, the electrons are separated from the ions and collected, while the latter are refocused and continue on.

Ion production in the source can be described by a simple balance model in which the net rate of production of ions of charge i is equal to the rate at which ions of charge i are produced from ions of charge i-1, less the rate at which they are depleted to form ions of charge i+1. In terms of the density of ions of charge i, $n_i(t)$, at time t and the production rate coefficient Q_i,

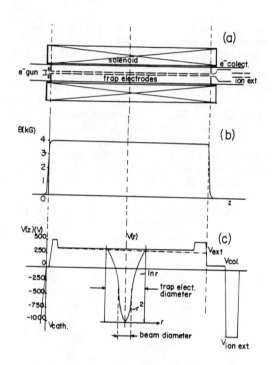

Figure 1. (a) Schematic showing main components of an EBIS. Electron beam confining solenoidal magnetic field profile and potential distributions applied to axial trap electrodes are shown in (b) and (c) respectively. In (c), the ———— curve represents the confining potential and the .___.___ curve represents the potential applied during extraction. The central portion of (c) shows the radial distribution of the space charge potential ($\sim r^2$) and trap potential ($\sim \ln r$).

$$\frac{dn_i(t)}{dt} = (1-\delta_{oi})Q_{i-1}N_{i-1} - (1-\delta_{\xi i})Q_i N_i \quad i=0,1,..\xi \tag{1}$$

where δ_{ij} is the Kronecker delta symbol and ξ is the highest charge state attainable under given conditions.

In this model, multiple ionization by electron impact, electron-ion recombination and ion losses in general are neglected,[11]

assumptions shown to be consistent with experimental observations.[11,12] The system of equations (1) can be solved by the
method of Laplace transforms[13] to give the number density $n_i(t)$ at
time t in terms of the neutral atom density N_o present at $t = 0$:

$$n_i(t) = \frac{N_o}{\sigma_i} \sum_{j=o}^{i} \sigma_j e^{-Q_j t} \left(\prod_{k=o}^{i}{}' \frac{\sigma_k}{\sigma_k - \sigma_j} \right), \qquad (2)$$

where $Q_j = (J/e)\sigma_j$, J is the bombarding electron current density, e
the electronic charge, and $\sigma_j \equiv \sigma_{j \to j+1}$ the ionization cross section
for going from charge state j to j+1. The prime on the product
sign excludes the term k = j.

The expression for $n_i(t)$, Eq. (2), consists of a series of ex-
potentials, each with a characteristic time $\tau_j = 1/Q_j = e/J\sigma_j$. The
confinement time τ necessary to obtain charge state ξ can be ap-
appximated by $\tau = \frac{e}{j} \sum_{j=o}^{\xi} \frac{1}{\sigma_j}$. The highest charge state ξ is thus

determined by the electron bombarding energy E, the current density
J and the confinement time $t = \tau$. From (1) it follows that:

$$n_\xi(t) = N_o \left(1 - \sum_{i=o}^{\xi-1} n_i(t) \right) \qquad (3)$$

To investigate the time variation of the $n_i(t)$'s, the following approximate cross sections can be used:[14]

$$\sigma_j(\varepsilon) = \frac{1.6 \times 10^{-13}}{[I_{j \to j+1}(eV)]^2} \frac{\ln \varepsilon}{\varepsilon} \quad cm^2 \qquad (4)$$

where $\varepsilon \equiv E/I_{j \to j+1}$ is the bombarding electron's energy, measured in
units of the ionization potential $I_{j \to j+1}$ of the most easily removed
electron in going from charge state j→j+1. (See reference 15 for
tabulated theoretical values of the ionization potentials.) It is
customary to express the charge state development (Eqs. (2) and (3))
as a function of the "ionization factor" $j\tau$ (Coulombs/cm^2), rather
than the containment time τ, and Figure 2 taken from reference 16,
shows the evolution of charge-state populations for argon bombarded
by 10 keV electrons. In this case to produce Ar^{18+}, a $J\tau$ of $\sim$
300 Coul/cm^2 is required. For a containment time of 30 msec this
requires an electron beam current density of 10^4 A/cm^2.

$J\tau$ is related to $n\tau$ by $n\tau = J\tau/e<v>$ where $<v>$ is the mean

electron velocity.
In order to pro-
duce fully strip-
ped ions of the
light elements,
$J\tau$ values
greater than
10 Coul/cm^2 are
needed.[16] At
5 keV bombard-
ing electron
kinetic energy,
this corresponds
to an $n\tau$ of $\sim$
10^{10} cm^{-3} s.
This simple
model neglects
the neutraliza-
tion of the

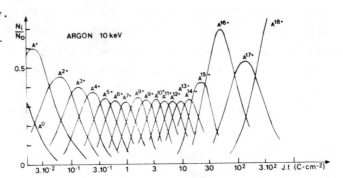

Figure 2. Charge state evolution in an EBIS as function of ionization factor $J\tau$ (Coulomb/cm^2) for argon bombarded by 10 keV electrons. Taken from reference 16.

electron beam's space charge by ionized background gas, which in turn
reduces the radial trapping ability of the beam. Neglecting this
neutralization, the maximum number N_q of ions of charge q created
after containment time τ can be estimated by equating the multiply
charged ion density to the electron density in the beam, since at
this point the electron beam is space charge neutralized and can
hold no more charge. One gets.

$$N_q = \frac{3.333 \times 10^9 \ I(A) \ \ell(cm)}{\sqrt{E(keV)}} \ \frac{f_q}{\sum\limits_{i=o}^{\xi} i f_i} \ \text{ions/pulse} \qquad (5)$$

where I is the electron beam current in amperes, ℓ the length of
the ionization region in cm, f_i is the fraction of ions of charge i
and ξ is the maximum charge attainable. Since in general the extraction time τ for ions is short to the confinement time, the number
of emitted particles per second is N_q/τ. For example, for a 100 cm
long ionization region, a 1 Ampere, 10 kV beam can contain up to 10^{11}
ions per confinement cycle.
    In reality residual gas is of course always present in any
vacuum system, and the pressure in the ionization limits the confinement time and charge states attainable. An upper limit on the
tolerable pressure P in the ionization region can be obtained by
again equating the electron beam ionized background gas ion density
N_i to some fraction of the bombarding electrons' density, i.e.,
$N_i = fJe/v$ where v is the electron velocity. Rewritten in a
slightly different form, one finds that:

$$P < \frac{1.92 \times 10^{-10} \, f}{\sigma(\pi a_o^2) \, \tau(\text{sec}) \, \sqrt{V_e} \, (\text{kV})} \quad \text{torr} \tag{6}$$

where σ is the residual gas ionization cross section for electrons of kinetic energy eV_e in units of π (Bohr radius),^{2} and τ is the confinement time. For typical values of τ, σ and V_e, assuming that 10% neutralization of the electron beam by background gas (mostly hydrogen) ions is tolerable, P is of the order of 10^{-9} torr.

The depth of the space charge potential along the radial direction for an axial Brillouin flow[17] electron beam is:[18]

$$V(R_a) - V(0) = \frac{1}{4\pi\varepsilon_o} \frac{1}{\sqrt{2\eta}} \frac{I_b}{\sqrt{V_z}} \tag{7}$$

where $V(R_a)$ is the potential at the outer surface of the beam (at radius R_a), $V(0)$ is the potential along the axis (r=0), V_z is associated with the forward kinetic energy of the electrons, I_b is the beam's direct current and η is the ratio of the electron's charge to mass. Numerically,

$$V(R_a) - V(0) = 15154 \frac{I_b (A)}{\sqrt{V_z (eV)}} \quad \text{Volts} \tag{8}$$

The above outlined simple theory does not take into account the time variation of space charge neutralization of the beam by multiply charged ions and ionized background gas ions during confinement, or the possibility of non-uniform electron beam density. In actuality, a complete description of an EBIS is much more complicated, and the behavior of these sources is presently not completely understood.

A similar situation exists for ECRIS sources. Conceptually, the basic principle of operation of an ECRIS is also quite simple. The source, however, cannot be subjected to such a simple analysis as the EBIS since one is dealing with motion of charged particles in combined dc magnetic and rf electromagnetic fields, and a complete solution of the problem involves a numerical integration of the relativistically correct equations of motion.[19-21]

Figure 3 shows a simplified sketch of an ECRIS source adapted from reference 16. The source consists of two cavities, C_1 and C_2 placed inside solenoidal magnetic mirrors, M_1 and M_2, which form magnetic bottles. Bottle M_2 has a minimum B structure obtained by superimposing solenoidal and hexapolar magnetic fields. The cavities C_1 and C_2 are fed microwave power at a frequency around 10 GHz (corresponding to an electron cyclotron resonance (ECR) frequency in a magnetic field of $\sim$ 3600 Gauss.) The cavity structure is such that the microwave wavelenth is small compared to cavity dimensions and the cavities can be considered multimode. A plasma of cold ions is formed in cavity C_1 and allowed to drift into C_2, located in the minimum B field. C_2 contains a hot

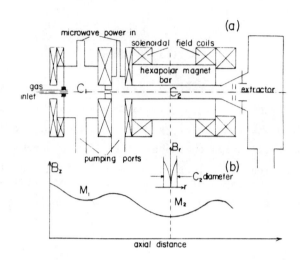

Figure 3. (a) Schematic showing main components of an ECRIS. (b) Axial and radial confining magnetic field profiles.

electron plasma in which energetic electrons are produced by ECR heating. Since minimum B structures generally suppress plasma turbulence and improve both electron and ion lifetimes through quiescent ambipolar diffusion, the ions take an appreciable time to pass through C_2 ($\sim 10^{-2}$ sec). While diffusing through the hot plasma, the ions are subjected to electron bombardment which results in creation of multiply charged ions. The multiply charged ions are then extracted from the source.

Thus far there appears to be no detailed theoretical analysis of the possible yield and charge state distribution from an ECRIS, though the problem is similar to the calculation of ionization balance commonly done in corona equilibrium[22] and non-stationary corona equilibrium. Rate equations similar to Eq. (1) are involved except that the Q_j's involve energy and time varying electron densities and energy varying cross sections.

PRESENT STATUS OF EXISTING SOURCES OF LOW ENERGY, MULTIPLY IONIZED SOURCES

Both the ECRIS and EBIS have undergone continuous development over the past 13 to 18 years by various groups around the world. The ECRIS was developed in France, while the EBIS in the Soviet Union, France and Federal Republic of Germany. More recently groups in the U.S. and Japan have constructed EBIS sources. At the time of writing, there is no ECRIS source in the U.S.

The latest version of the ECRIS source is the MICROMAFIOS

source constructed by R. Geller, et al. at Grenoble.[16,23] This source is a smaller scale version of earlier sources of a similar type. The minimum B field configuration is obtained by superimposing a solenoidal magnetic field upon a hexapolar field created by permanent $SmCo_5$ magnets. This results in a compact ion source with an interaction length of $\sim$ 50 cm, which consumes less than 100 kW of electric power. About 2 kW of 10 GHz microwave power are required to create and heat the plasma. The steady state operating pressure in cavity C_1, into which gas in injected, is 10^{-3} torr, while in cavity C_2, the pressure is 10^{-6} torr. Inside the closed magnetic surface within C_2 in which ECR heating occurs, the pressure is about a factor of 10 better. (In an earlier version of the source, the gas was injected directly into cavity C_2. However, in this case, the relatively high pressure in the cavity caused charge exchange between ions and gas, and high charge state could not be produced. The present arrangement reduces charge exchange in C_2 significantly, and high charge states can be obtained.) Ions are extracted through an 8 mm diameter hole at extraction voltages between 1 and 20 kV, and the ion yield from MICROMAFIOS for various gases at an extraction voltage of 8 kV is shown in Table 1 (taken from reference 16).

Table I   Ion yields (in electric microamperes) from ECRIS source MICROMAFIOS at 8 kV extraction (from reference 16).

Element	Charge state												
	1+	2+	3+	4+	5+	6+	7+	8+	9+	10+	11+	12+	13+
C	>30	>30	>30	>30	5	0.2							
N	>30	>30	>30	>30	20	3	0.1						
O	>30	>30	>30	>30	25	15	3	0.1					
Ne	>30	>30	>30	>30	>30	20	10	2	<0.1				
A	>30	>30	>30	>30	>30	>30	>30	>30	10	3	>1	<0.3	<0.1
Kr	20	20	20	12	12	12	11	11	11	10	6	6	2

Figure 4 shows ion current vs. time plots for 2+ and 8+ charge states of oxygen, obtained from MICROMAFIOS. The microwave heating pulse is shown on the same time scale. The O^{8+} disappears as soon as the rf power is turned off due to the rapid decrease in electron energy when cyclotron resonance ceases.

As mentioned above, there are a number of EBIS sources in existence. Two of the earliest, and currently in operation, are

KRION[24] and KRION II[25], constructed by E.D. Donets at
JINR, Dubna, USSR. Both
utilize a superconducting
solenoid to generate the electron beam confining magnetic
field, and the ionization
regions are cryopumped by
the solenoid inner bore
surfaces which are at liquid
helium temperature. Both
sources have electron guns
with cathodes immersed in
the magnetic field. KRION's
magnetic field is 1.2 Tesla,
and the electron gun operates
at 8 keV. The vacuum inside
the ionization region is
5×10^{-11} torr. KRION II's
magnetic field is 2.25 Tesla,
the electron gun operates at
20 keV and the vacuum is
better than 10^{-12} torr.
KRION is used as a source
of bare C, N, O and Ne
nuclei for the Dubna
synchrotron, while

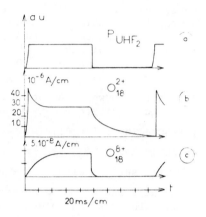

Figure 4. Current vs.
time plot for charge
states 2+ and 8+ of
oxygen obtained from an
ECRIS type source (taken
from reference 16). The
shape of the applied rf
pulse is shown in (a),
and the oxygen ion yield
in (b) and (c)

KRION II is used for source development and various experimental
studies involving low energy, multiply charged ions.[26] In KRION II,
an ionization factor of 2.4×10^3 Coulomb-cm^2 has been obtained with
20 keV electrons. This rather high ionization factor can generate
bare argon (Ar^{18+}), and helium-like krypton (Kr^{34+}) and xenon
(Xe^{52+}). The charge capacity of the electron beam trap is $\sim 10^{10}$
charges.

In KRION and KRION II, the electron beam is generated by a gun
immersed in the fringe field of the focusing solenoid. One of the
characteristics of immersed guns is that the obtainable electron
beam current density is limited by cathode size and magnetic flux
threading the cathode. Current densities greater than 500 A/cm^2
(in KRION sources 600 A/cm^2 has been achieved) are possible if electrons emitted from a large cathode are focused first electrostati-
cally, and then magnetically compressed.[27,28] In this case the
gun's cathode has to be located in a magnetic field free region,
and launching of the beam into the focusing solenoidal field is not
as simple as in the immersed gun case. Such external guns are
capable, at least in theory, of generating current densities of the
order of 10^4 A/cm^2.[28] Apart from the KRION sources, existing EBIS

sources use an external electron gun.

The most advanced and sophisticated of the external gun versions of an EBIS is CRYEBIS, built by Arianer, et al. at Orsay,[20] and intended as an ion source for the Saclay synchrotron. The main parameters of CRYEBIS are as follows: main magnetic field up to 3 Tesla, provided by a 1.5 m long, 0.2 m inner bore superconducting solenoid, electron beam of up to 2A, 10 kV with current densities $\sim 10^3$ A/cm^2, vacuum in the ionization region better than 10^{-11} torr, provided by a liquid helium cryopanel, and an electron beam trapping capacity of $\sim 10^{10}$ ions per pulse. CRYEBIS has generated bare argon, Xe^{44+} and Kr^{34+} in a confinement time of 5 ms.[31] The abnormally short confinement time required to produce such species implies an electron beam current density of 10^5 A/cm^2, approximately two orders of magnitude greater than that of the un-neutralized beam. The mechanism responsible for such "supercompression" of the electron beam is not fully understood, though it is believed that space charge compensation may cause a Brillouin focused beam to collapse to high current densities.[32] After the source was moved from Orsay to Saclay these results have not been reproduced, and recent performance of CRYEBIS is more modest.[33]

In addition to the above, there are, at the time of writing, four other laboratories around the world with EBIS sources. In Frankfurt, R. Becker and co-workers[34] have constructed an EBIS in which the focusing solenoidal field is provided by a 5 Tesla, 1 m long, 50 mm bore, superconducting solenoid. An external electron gun, capable of delivering up to 3A at 10 kV would generate an electron beam with a current density of the order of several thousand amperes/cm^2. This source has a very interesting feature in that the electrons are to oscillate axially instead of being continuously collected, thereby avoiding the need for a high power dissipating collector. Preliminary results obtained with this source appear very encouraging.[34]

In Japan, two EBIS sources have been built and are in operation. The first, called PROTO-NICE,[35] is a medium size EBIS with an ionization region 40 cm long. An external gun, capable of delivering 10 A/cm^2 generates an electron beam focused by a conventional 2 kG solenoid. The vacuum in the ionization region is 3×10^{-8} torr, and at this pressure, the source is capable of producing hydrogen and helium-like ions of carbon, nitrogen and oxygen in a confinement time of 10 ms ($J\tau = 0.1$ coul/cm^2). Even with this low ionization factor bare carbon has been been observed. The second source, NICE, which uses a superconducting solenoid, is very similar in design to the Dubna KRION source. With an electron beam current density about a factor of 10 less than the early KRION source, NICE has produced bare nuclei of carbon, nitrogen and oxygen. NICE is devoted to atomic physics studies and results on one-electron transfer processes to bare C,N and O were reported at the XII ICPEAC[36] and more recently at Stockholm.[4]

In the U.S., two EBIS sources are in existence. One is at the Lawrence Berkeley Laboratory, and was intended to increase the energy of the LBL 88 inch cyclotron,[37] the other, at Cornell University, is devoted to source development and atomic physics.[38] The LBL EBIS has an external gun, a conventional solenoid, and drift tubes at room temperature. Working material is injected into the electron beam from a metal vapor oven loaded with alkalis, and ultra high vacuum in the source ionization region is obtained with the help of unionized material which plates out and forms getter surfaces which pump background gasses. Ions of charge state as high as Ca^{15+} have been extracted.[39]

The Cornell EBIS is a medium sized, relatively simple mechanically low cost source. It also uses an external gun and a conventional 4.5 kG solenoid. The drift tubes, constructed from stainless steel mesh to improve their vacuum conductance, are at room temperature, and are surrounded by a distributed sputter-ion pump which maintains a pressure of $2x10^{-9}$ torr in the ionization region during operation. With a 1.5 kV, 20 mA electron beam and a confinement time of 5 ms, Ti^{6+} ions with a kinetic energy of 320 eV/q have been obtained.[38] The source is designed for an ionization factor $J\tau = 10$ Coul./cm^2.

FUTURE DEVELOPMENTS IN SOURCE TECHNOLOGY

The ECRIS and EBIS sources presently in existence are still to a large extent in a developmental stage, and neither type of source has approached its potential level of performance. The basic physical phenomena relevant to these sources are understood, but their role, importance and interplay are not always clear. This makes scaling and extrapolation of source design and performance somewhat difficult. For example, in an ECRIS, the electron density n, the confinement time τ and the electron energy E all have to be increased to generate higher charge states of higher Z elements and to produce more intense beams. How these quantities can be increased is relatively straightforward. The electron energy can be increased by putting more rf power , P_{rf}, into the plasma, and stochastic electron heating predicts that $E \sim \sqrt{P_{rf}}$. In actuality, ECRIS results are more consistent with $E \sim P_{rf}^{40}$. The electron density can be increased by increasing the microwave frequency since density is proportional to frequency squared.[16] Finally, though there appears to be no way of increasing the confinement time significantly, it can be increased by lengthening the drift region. With $n\tau = 10^{11}$ cm^{-3}s, completely stripped xenon can be produced and extrapolation to bare uranium nuclei is possible.[41]

Whether or not such highly charged species can be produced, and more importantly extracted, is another question. Thus far, only bare nuclei of low Z elements have been produced with an ECRIS.[16] Increasing the microwave frequency is the simplest way to increase the $n\tau$ product. However, higher frequencies require stronger confining fields (proportional to the frequency) and at frequencies above 10GHz, superconducting magnets are required. Two superconducting solenoid ECRIS sources are currently under construction, one at Julich, FRG, and the other at Louvain-la-Neuve, Belgium. It will be interesting to see the performance of these sources, since thus far, all ECRIS results are based on the performance of MICROMAFIOS,[16] which though impressive, have yet to be scaled up.

In an EBIS, which has proven that it is possible to generate highly charged species, (bare xenon has been obtained with the KRION sources) the parameters n, E and τ can be increased without many difficulties. The technology for producing high voltage, high current density electron beams is readily available. At electron energies and current densities greater than 10 kV and a few hundred ampere/cm^2, superconducting solenoids are required to generate the confining magnetic field. The confinement time is limited only by the vacuum in the ionization region of the source. Since in any vacuum system, excluding seal leaks, the limiting sources of contamination are desorption of molecules from surfaces, hydrogen diffusion through walls and helium diffusion through glass ports, the problem reduces to minimizing sources of desorbed and diffused gasses. Hydrogen and helium diffusion through wall material can be eliminated by cooling the walls far below the the diffusion activation energy, usually 77°K is sufficient. Outgassing from materials and surfaces within the vacuum chamber can be eliminated by cooling all materials below the desorption activation energy for each species.[42] In a number of sources this has been accomplished by cooling the ionization region and surrounding walls to liquid helium temperatures, resulting in pressures in the 10^{-12} torr range.

At present, there appears to be no problem in designing a source capable of generating fully stripped ions up to praseodymium and helium like ions up to uranium. A source capable of generating such species is currently under construction at Orsay, France.[43] Whether or not this source will live up to expectations is another matter. On one hand, one is faced with the very encouraging performance of KRION II, while on the other, the tantalizing, but thus far not reproducible performance of CRYEBIS. In any case, as more of these sources become operational around the world over the next few years, there will be a flurry of activity in both source development and in atomic physics involving low energy very highly charged ions.

ACKNOWLEDGEMENTS

This work was supported by the United States Department of Energy, Division of Chemical Sciences.

REFERENCES

1. M.R.C. McDowell and A.M. Ferendici, Eds., Atomic and Molecular Processes in Controlled Thermonuclear Fusion (NATO Advanced Study Institutes, Series B: Physics, Vol. 53, Plenum Press, New York 1980).
2. R.A. Phaneuf, Phys. Rev. 24A, 1138 (1981).
3. I. Iwai, Y. Kaneko. M. Kimura, N. Kobayashi, S. Ohtani, K. Okuno, S. Takagi, H. Tawara and S. Tsurubashi, Phys. Rev. 26A, 105 (1982).
4. S. Ohtani, Proc. Symposium on Production and Physics of Highly Charged Ions, Stockholm, Sweden, June 1982. In press.
5. V.O. Kostroun, Workshop on Atomic Physics at the National Synchrotron Light Source, BNL-28832, 79 (1982).
6. S. Bashkin in Beam Foil Spectroscopy, S. Bashkin, Ed. (Springer-Verlag, Berlin, Heidelberg 1976).
7. I.A. Sellin in Adv. Atom. and Mol. Phys 12, 215 (1976).
8. H.W. Kugel and D.E. Murnick, Rep. Prog. Phys. 40, 297 (1977).
9. J. Illgrew, R. Kirchner and J. Schulte, IEEE Trans. On Nucl. Sci. NS-19, 35 (1972).
10. J.R.J. Bennett, IEEE Trans. on Nucl. Sci. NS-19, 48 (1972).
11. A. Muller, H. Klinger and E. Salzborn, Nucl. Instr. and Methods 140, 181 (1977).
12. E.D. Donets and A.I. Pikin, Sov. Phys. JETP 40, 1057 (1976).
13. H. Bateman, Proc. Comb. Phil. Soc. XV, 423 (1910).
14. R. Becker and H. Klein, IEEE Trans. on Nucl. Sci. NS-23, 1017 (1970).
15. T.A. Carlson, C.W. Nestor, N. Wasserman and J.D. McDowell, Atomic Date 2, 63 (1970).
16. J. Arianer and R. Geller, Ann. Rev. Nucl. Part. Sci. 31, 19 (1981).
17. L. Brillouin, Phys. Rev. 67, 260 (1945).
18. W.G. Dow, Adv. in Electron and Elect. Physics X, 1 (1958).
19. E. Cannobio, Nuclear Fusion 9, 27 (1969).
20. F. Jaeger, A.J. Lichtenberg and M.A. Lieberman, Plasma Physics 14, 1073 (1972).
21. M.A. Lieberman and A.J. Lichtenberg, Plasma Physics 15, 125 (1973).
22. C. Jordan, Monthly Nat. Roy. Astron. Soc. 142, 501 (1969).
23. F. Bourg, R. Geller, B. Jacquot, T. Lamy, M. Pontonnier and J.C. Rocco, Nucl. Instr. and Methods 196, 325 (1982).
24. E.D. Donets, IEEE Trans. on Nucl. Sci. NS-23, 897 (1976).
25. E.D. Donets and V.P. Ovsyannikov, Preprint JINR PF-80-515, Dubna, USSR (1980).
26. E.D. Donets, Proc. Symposium on Production and Physics of Highly Charged Ions, Stockholm, Sweden, June 1982. In press.
27. G.R. Brewer, in Focusing of Charged Particles, A. Septier, Ed. (Academic Press, New York (1967)), Vol. II, p. 73.
28. B. Fogen, Proc. II EBIS Workshop, J. Arianer and M. Olivier, Eds., May 1981, Saclay/Orsay, France, p. 119.
29. J.F. Gittins, Power Traveling-Wave Tubes (American Elsevier

Publishing Co., Inc., New York (1965).

30. J. Arianer, S. Buhler, A. Cabrespine, C. Goldstein, M. Ulrich, G. Deschamps, F. Kircher and M. Abd El Baki, Orsay report IPNO-79-01.

31. J. Arianer, A. Cabrespine, C. Goldstein and G. Deschamps, IEEE Trans. on Nucl. Sci. NS-26, 3712 (1979).

32. M.C. Vella, Nucl. Instr. and Methods 187, 313 (1981).

33. J. Faure, Proc. II EBIS Workshop, J. Arianer and M. Olivier, Eds., May 1981, Saclay/Orsay, France, p. 13.

34. R. Becker, M. Kleinod and H. Klein, Proc. II EBIS Workshop, J. Arianer and M. Olivier, Eds., May 1981, Saclay/Orsay, France, p. 48.

35. H. Imamura, Y. Kaneko, T. Iwai, S. Ohtani, K. Okuno, N. Kobayashi, S. Tsurubashi, M. Kimura and H. Tawara, Nucl. Instr. and Methods 188, 233 (1981).

36. Y. Yaneko, T. Iwai, S. Ohtani, K. Okuno, N. Kobayashi, S. Tsurbushi, M. Kimura, H. Tawara and S. Takagi, Contributed Papers, XII ICPEAC, S. Datz, Ed., Gatlinburg, TN, 1981, p. 696.

37. B. Feinberg and I.G. Brown, Proc. II EBIS Workshop, J. Arianer and M. Olivier, Eds., May 1981, Saclay/Orsay, France, pp. 1 and 212.

38. V.O. Kostroun, E. Ghanbari, E.N. Beebe and S.W. Janson, Proc. Symposium on Production and Physics of Highly Charged Ions, June 1982, Stockholm, Sweden. In press.

39. I.G. Brown, private communication.

40. R. Geller, Proc. Symposium on Production and Physics of Highly Charged Ions, June 1982, Stockholm, Sweden. In press.

41. R. Geller, IEEE Trans. on Nucl. Sci. NS-26, 2120 (1979).

42. W. Thompson and S. Hanrahan, J. Vac. Sci. Technol. 14, 643 (1977).

43. J. Arianer, M. Brient, C. Goldstein, J. MacFarlane, M. Mallard P. Nicol, A. Serafini and S. Steinegger, Proc. II EBIS Workshop, J. Arianer and M. Olivier Eds., May 1981, Saclay/Orsay, France, p.240.

WIGGLERS AND UNDULATORS AS INTENSE X-RAY SOURCES

G. S. Brown
Stanford Synchrotron Radiation Laboratory
Stanford, Calif. 94305

INTRODUCTION

It is now evident, from both theoretical and operational experience, that the synchrotron radiation emitted by special purpose insertion devices (wigglers and undulators), is several orders of magnitude more intense than the radiation from conventional bending magnets. The radiation is highly polarized, and, in the case of undulators, quasi-monochromatic and highly polarized. These properties have motivated the construction of such devices on a number of electron storage rings, many of which have now fully instrumented beam lines. In what follows, we will briefly discuss the spectral properties of the radiation emitted by such devices, and summarize the facilities that are available for atomic physics research.

SPECTRAL PROPERTIES

The spectrum of synchrotron radiation emitted by electrons passing through a constant magnetic field is well known. The spectrum is peaked at a frequency $\omega = 0.3\omega_c$, where $\omega_c = (3/2) \gamma^3(c/\rho)$, ρ is the radius of curvature of the particle orbit and γ is the ratio of the electron energy to its rest mass. The spectrum falls off roughly exponentially at higher energies, approximately as $\exp(-\omega/\omega_c)$. The angular divergence of the photon beam with respect to the tangent to the particle orbit is roughly $(1/\gamma)$ in both directions, although in practice this divergence may be surpassed by the natural divergence of the electron beam, and certainly by the "effective" horizontal divergence of 2π. For the case of wiggler magnets, the spectrum is essentially the same, except that the radius of curvature is determined by the wiggler field, which can usually be freely adjusted through wide limits. More importantly, however, is the fact that the horizontal emission angles can be reduced, in principle, to an arbitrarily small value by reducing the spatial periodicity of the magnet. Stated in another way, the power emitted per unit horizontal angle is directly proportional to the number of poles of the wiggler, which can be up to 25 per meter while still maintaining a 13 kG magnetic field across a 1 cm gap (1).

An undulator is a wiggler magnet whose period has been reduced to the point that the amplitude of the angular oscillations of the electron beam are comparable to $1/\gamma$, the natural photon emission angle. In this limit, the photon emission amplitudes add coherently, and one observes a strict correlation (in the limit of a large number of poles) between the polar observation angle and the frequency of emission. This correlation is given by

$$\lambda = (\lambda_u/2\gamma^2)(1+ K^2/2 + \theta^2\gamma^2) \tag{1}$$

where λ_u is the period of the undulator and K is the fundamental undulator parameter, given by

$$K = \gamma\lambda_u/2\pi\rho = 0.0934B(kG)\lambda u \text{ (cm)}. \tag{2}$$

In the undulator limit, the photon brightness scales as N^2, where brightness is defined as the number of photons emitted per constant fractional bandwidth, per unit solid angle, at a given frequency. This scaling rule applies, of course, only when the betatron angles of the electron beam are small in comparison with the natural emission angles of the photon beam.

EXISTING FACILITIES

Wiggler and undulator facilities have been constructed or proposed in a number of laboratories, and the details have been reviewed in a paper by Brown, Halbach, Harris, and Winick (2), and references therein. The parameters of these devices are summarized in Table 1.

LABORATORY	B(kG)	λ_u(cm)	Periods	K	STATUS
Cornell	18.	35.	3	59.	yes
Frascati	19.	65.4	3	116.	yes
Frascati	4.8	11.6	20	5.2	yes
Novosibirsk	35.	9.	10	29.	yes
Novosibirsk	3.	10.	6	2.8	yes
Novosibirsk	7.	6.5		4.2	yes
Novosibirsk	8.	54.		40.	yes
Novosibirsk	23.		1		yes
NSLS	6.	17.4	3	97.	const.
NSLS	15.	13.6	12	19.	const.
NSLS	39.	6.5	38	2.4	const.
Orsay	4.6	4.	23	1.7	yes
Orsay		7.		2.5	yes
Photon Fact.	60.		1		yes
Photon Fact.	1.25	4.	10	0.47	yes
SRS	50.		1		yes
SSRL (2)	18.	45.	4	76.	yes
SSRL	13.	7.	27	8.5	const.
SSRL	2.4	6.1	30	1.4	yes

Table 1

ACKNOWLEDGEMENTS

The author gratefully acknowledges many helpful discussions with P. Eisenberger, K. Halbach, J. Harris, H. Wiedemann, and H. Winick. SSRL is supported by the National Science Foundation through the Division of Materials Research, and by the National Institutes of Health through the Biotechnology Resources Program in the Division of Research, Resources, and in cooperation with the Department of Energy.

REFERENCES

1. E. Hoyer et. al., Nuclear Instruments and Methods, (to be published).
2. G. Brown, K. Halbach, J. Harris, and. Winick, Nuclear Instruments and Methods (to be published).

Chapter 6   Theoretical Aspects of Inner-Shell Vacancy Production
and Decay-Fundamental Concepts and Effects of Relativistic Interactions

THE THEORETICAL BASIS FOR CALCULATIONS OF THE PRODUCTION
AND DECAY OF INNER SHELL VACANCIES*

Steven T. Manson
Department of Physics and Astronomy, Georgia State
University, Atlanta, Georgia 30303, U.S.A.

ABSTRACT

The similarities and differences between outer shell and inner
shell ionization are explored. It is shown that the outer shell
ionization process is well-defined, while the analogous inner shell
process is not owing to the decay of the hole. Under some conditions, however, it is shown that it is possible to separate the
inner shell vacancy production and decay to an excellent approximation. Examples of processes where the separation cannot be made
and where it can are presented and discussed.

INTRODUCTION

The creation and decay of inner shell vacancies have been
studied for a long time both theoretically and experimentally;[1]
both excitaion and decay have been studied including cross sections
for inner shell excitation and ionization induced by various
collision processes, inner shell binding energies, and decay rates
of inner shell vacancies via Auger and radiative processes.

From an experimental point of view a projectile impinges on the
target and either the energy loss cross section of the projectile
is measured or some emission of the target is scrutinized; in this
sense inner and outer shells are philosophically similar. From a
theoretical point of view there is, however, a crucial difference.
In an outer shell ionization process, the target makes a transition
from the initial eigenstate of the Hamiltonian (usually the ground
state) to a final continuum eigenstate of the Hamiltonian and the
process is well-defined. In an inner shell ionization process, the
target is initially in the same eigenstate of the Hamiltonian, but
the state with an inner shell vacancy is simply not an eigenstate.
Roughly speaking, this is because an inner shell vacancy state is
degenerate with continuum states with the vacancy in higher shells
so that the eigenstate is an admixture. In fact, it is just this
degeneracy which allows the decay, and only after the decay is the
target again in an eigenstate.

* Work supported by the U. S. Army Research Office.

In the following section, these ideas shall be discussed in greater detail using examples, and the problem of the separation of vacancy creation and delay is explored. The final section gives some examples of both cases.

A GUIDE TO THE THEORY OF INNER SHELL PHENOMENA

To focus clearly on the differences between inner and outer shells, let us consider the example of the photoionization of neon. The outer shell process is

$$h\nu + Ne(1s^2, 2s^2, 2p^6) \rightarrow Ne^+(1s^2, 2s^2, 2p^5) + e^- \tag{1}$$

In Eq. (1), the initial state of Ne, the final continuum state of $Ne^+ + e^-$, and the Ne^+ itself are both eigenstates of the Hamiltonian. Looking at the corresponding inner shell process,

$$h\nu + Ne(1s^2, 2s^2, 2p^6) \rightarrow Ne^+(1s, 2s^2, 2p^6) + e^-, \tag{2}$$

neither the final state of Ne^+ nor the $Ne^+ + e^-$ final system is an eigenstate; the Ne^+ in Eq. (2) is only a transient and will decay via Auger or radiation (or some mixture). The main decay possibilities of this core hole state are

$$Ne^+(1s, 2s^2, 2p^6) \rightarrow Ne^{+2}(1s^2, 2s^2, 2p^4) + e^- \tag{3a}$$

$$\rightarrow Ne^{+2}(1s^2, 2s, 2p^5) + e^- \tag{3b}$$

$$\rightarrow Ne^{+2}(1s^2, 2p^6) + e^- \tag{3c}$$

$$\rightarrow Ne^{+2}(1s^2, 2s^2, 2p^5) + h\nu \tag{3d}$$

Note that many other decays are possible. Note further that in processes (3b) and (3c), the Ne^{+2} decays further. Thus, if process (3a) occurs, the total reaction can be written as

$$h\nu + Ne(1s^2, 2s^2, 2p^6) \rightarrow [Ne^+(1s, 2s^2, 2p^6) + e^-] \rightarrow Ne^{2+}(1s^2, 2s^2, 2p^4) + 2e^- \tag{4}$$

or a process with one photon in and two electrons out. The total final state of Eq. (4) is an eigenstate, as is the Ne^{2+}; the "intermediate" state, in brackets, is not and only serves as an often useful way of thinking about the process. It is important to point out that the process shown in Eq. (4) is degenerate with the double photoionization process and has exactly the same initial and final states. Thus Eq. (4) represents a resonance in the double photoionization process. The conservation of energy constraint is satisfied by the sum of the energies of the two ejected electrons, not by either of them individually.

Does this mean that all Auger spectroscopy is a sham? Of course not! But it does mean that it must be treated with caution. In particular, if one thinks of Eq. (4) as a two step process, it is necessary that the first electron be fast enough to leave the scene of the atom _before_ the Auger decay occurs. In that way the two electrons are well separated spatially and there is virtually no (or very little) exchange of energy between them. On the other hand, if the photoelectron is slow, it can still interact with the Auger electron causing energy exchanges.

In principle, of course, these energy exchanges will occur at all energies, no matter how high, but they can be ignored to a very good approximation at the higher energies. The situation is very analogous to the ionization of atoms by electrons. The final state in this process is an ion plus two electrons in the continuum. These electrons, of course, can exchange energy. If, however, the incident electron is sufficiently fast compared to the secondary electrons, this final state interaction can be ignored. This is the underpinning of the (first) Born approximation. As an example, consider the electron impact ionization of He,

$$e^- + He \rightarrow e^- + He^+ + e^-. \tag{5}$$

Results have been reported for 2 keV electron impact[2] and a comparison between theory and experiment is shown in Fig. 1. Note that it is the double differential cross section (differential in secondary electron energy and emission angle) which is shown and not

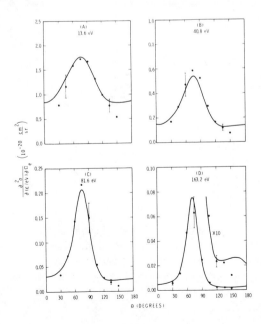

Figure 1. Angular and energy distributions of electrons (double differential cross sections), ejected from He by 2 keV electron impact ionization; comparison of Born approximation theory (solid line) and experiment (points) at fixed ejection energies taken from Ref. 2.

324

merely a total cross section. This double differential cross section (DDCS) is far more sensitive to any final state interaction effects than the total cross section but still, the agreement between theory and experiment is excellent. The fact that the experimental points are below the theory at the smallest and largest angles in all cases is due to an error in experimental analysis.[3]

In closing this section then, it is to be emphasized that the basic difficulty in inner shell vacancy phenomena is by no means in the experiments themselves, but rather in the theoretical framework used in interpreting these experiments.

EXAMPLES

As a first example of the phenomena discussed above, let us consider the $K-L_{2,3}L_{2,3}$ Auger process induced by electron impact ionization of neon.[4] In Fig. 2, the experimental Auger spectra of Hink et al. are shown for incident electron energies from 3 keV down to about 1 eV above the K-shell ionization threshold. These

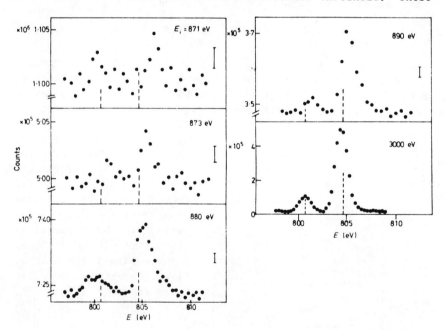

Figure 2. Measured $K-L_{2,3}L_{2,3}(^1D_2)$ and $^1S_0$ Auger lines of neon following electron impact ionisation. The nominal line energies are indicated by broken lines. PCI shift and a high energy tail are seen. E_i, electron impact energy; I, statistical error. This figure was taken from Ref. 4.

results illustrate the sharing of energy between ejected electrons and have come to be known as the post-collision interaction (PCI). It is seen that the Auger energies do not remain constant as a function of incident electron energy and that, closer to threshold, where the incident electron moves off more slowly, the effect is greater.

To investigate this more fully, the shift of one of the Auger lines from its asymptotic value is shown[4] in Fig. 3 vs. the energy of the incident electron after the collision, i.e., the incident energy minus the K-shell ionization energy. Near threshold, the shift is 2 eV or more. Furthermore, although this shift decreases with increasing incident electron energy, it never quite goes away. This is a very significant point since it means that at virtually any energy, PCI is there and can only be ignored to a certain level of accuracy. As measurements, and calculations, get more and more accurate, PCI will have to be taken into account further and further from threshold.

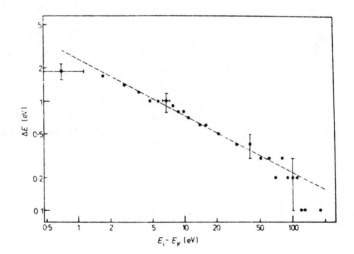

Figure 3. Measured energy shift ΔE of the $K-L_{2,3}L_{2,3}(^1D_2)$ Auger line of neon. E_i, electron impact energy; E_K, K-ionisation energy. This figure was taken from Ref. 4.

Lest the above example give rise to the notion that inner shells cannot be treated, we now turn to some examples where there is considerable agreement between theory and experiment.

Inner shell photoabsorption is one such area. As a first example, consider the photoelectron angular distribution of the 4d shell of xenon. The form of the angular distribution for low photon energies, where the electric dipole approximation is excellent, is given by

$$\frac{d\sigma}{d\Omega} = \frac{\sigma}{4\pi} \left[1 + \beta P_2(\cos\theta)\right] \tag{6}$$

where σ is the total subshell cross section, θ is the angle between photon polarization and photoelectron propagation directions, P_2 is a Legendre polynomial, and β, the asymmetry parameter, determines

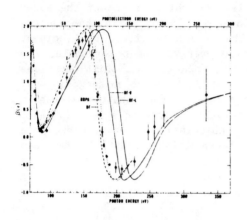

Figure 4. Energy dependence of the Xe 4d photoelectron angular distribution asymmetry, β, parameter. The points are the experimental results of Ref. 5, the theoretical HF curves from Ref. 6, the DF from Ref. 7, and the RRPA from Ref. 8. This figure was taken from Ref. 5.

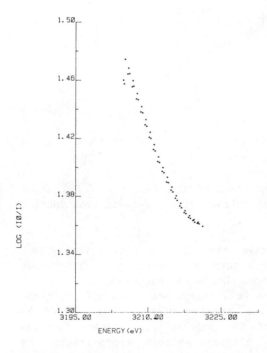

Figure 5. Experimental photoabsorption cross section at the K-edge of Argon from Ref. 9.

the shape of the angular distribution. A comparison of experimental[5] and theoretical values for   is given in Fig. 4 where it is seen that the agreement between experiment and the most sophisticated calculations is excellent. Note that the   parameter is essentially the differential cross section and, thus, more sensitive to the details of the calculation than the total cross section.

Another example in inner shell photoabsorption is the photoabsorption of Ar at the K-shell edge. Recent (unpublished) measurements of LaVilla and Deslattes[9] have found a structure in the relative photoabsorption cross section edge which is shown in Fig. 5. Our theoretical work is given in Fig. 6 where it is seen that the hydrogenlike calculation provides no hint of any structure, the Hartree-Slater (HS) calculation[10,11] gives some structure but in just the opposite direction from experiment, and a sophisticated relativistic random phase approximation (RRPA) calculation[12] gives rather good qualitative agreement with experiment. In previous theoretical work[10,11,13], it has been found that the K-shells of all atoms heavier than oxygen exhibit stucture of various kinds above the K-shell ionization threshold, and the photoabsorption cross sections are <u>not</u> hydrogenic. Although the initial state is rather hydrogen-like, the final continuum state "diffracts" through the whole atomic potential.[14] This "diffraction" near the outer edge of the atom is germane to the process in that it strongly effects the amplitude of the continuum wave function near the nucleus, i.e., in the vicinity of the K-shell. Studies of this amplitude and its variation with energy and Z are progress.[15]

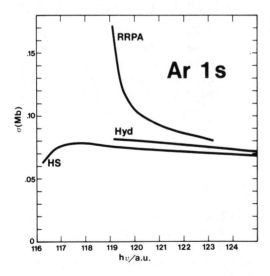

Figure 6. Theoretical K-shell photoabsorption cross section of Argon showing the hydrogenic, HS, and RRPA results.

This example points out that theory can do a reasonable job on K-shell vacancy production. In addition, it emphasizes the fact that there are still interesting unexplored phenomena in K-shells

328

that need to be scrutinized.

As a final example of inner shell vacancy production, we consider the energy and angular distribution of electrons ejected in 1.0 MeV proton impact ionization of Kr.[16] This double differential cross section (DDCS) is much more sensitive to the theoretical details than total or single differential cross sections are. The measurements are of the total electron distribution coming from both inner and outer shells, and are shown along with theory, in Fig. 7. Roughly speaking, the ejected electrons of low energy come from the outer 4p shell, while around 100 eV and higher, the emission is predominantly from inner shells. As seen in Fig. 7, for electrons emitted at 90° to the incident proton direction, agreement is most excellent. At 15°, a near forward direction, however, agreement is not so good, particularly in the range 100-500 eV. This is not due to any deficiency in treating inner shells, but rather an omission in theory. In addition to the direct ionization process, a charge transfer process can also occur. The charge transfer to the proton can go to the ground state of hydrogen, and excited discrete state, or a continuum state; this last possibility results obviously in a continuum electron at low energy with respect to the proton. In the laboratory system then, this leads to an electron distribution centered in the forward direction and centered at a velocity equal to the proton.[17] In this case, this "continuum charge transfer" process should maximize for 540 eV electrons. From Fig. 7 it is seen that this is where the discrepency between theory and experiment is greatest.

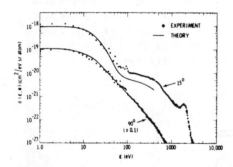

Figure 7. Energy distributions (double differential cross sections) for ejection of electrons at 15° and 90° by 1.0 MeV proton ionization of krypton. Taken from Ref. 14.

Up to this point we have talked about vacancy creation. The last example is of vacancy decay. Recent measurements[18] of x-ray energies and widths are compared with theory[19,20] in Table I. Although there are still some discrepencies larger than experimental error, the agreement is, over all, quite good. Certainly both x-ray energy and width are affected by the electron ejected in the creation of the inner shell hole, but at the present level of theory and experiment, this is not an important factor.

Z	Line	Energy (eV)		Width (eV)	
		Experimental	Theoretical	Experimental	Theoretical
47	$K_{\alpha 1}$	22163.06	22161.816	8.6 ± 0.4	9.33 ± .5
	$K_{\alpha 2}$	21990.44	21988.933	8.9 ± 0.4	9.16 ± .5
79	$K_{\alpha 1}$	68804.94	68802.663	57.5 ± 0.7	57.4 ± 1.7
	$K_{\alpha 2}$	66991.16	66988.486	59.2 ± 3.7	58.0 ± 1.7
92	$K_{\alpha 1}$	98432.21	98434.224	103.8 ± 2.4	103.5 ± 4.2
	$K_{\alpha 2}$	94651.45	94653.331	106.3 ± 6.1	105.4 ± 4.2

Table I: Experimental x-ray energies and widths from Ref. 16 compared to theoretical energies from Ref. 17 and semi-empirical widths from Ref. 18.

The small discrepencies are very likely due to some residual effects, not accounted for in the theory, of correlation between the bound electrons.[21] Eventually, when theory and experiment achieve a higher degree of accuracy, the ejected electron will have to be taken into account.

REFERENCES

1. B. Crasemann, ed., Atomic Inner-Shell Process (Academic Press, N.Y., 1975) V. I and II.
2. S. T. Manson, L. H. Toburen, D. H. Madison, and N. Stolterfoht, Phys. Rev. A 12, 60 (1975).
3. M. E. Rudd (private communication).
4. W. Hink, H. P. Schmitt, and T. Ebding, J. Phys. B 12, L257 (1979).
5. S. H. Southworth, P. H. Kobrin, C. M. Truesdale, D. Lindle, S. Owaki, and D. A. Shirley, Phys. Rev. A 24, 2257 (1981).
6. D. J. Kennedy and S. T. Manson, Phys. Rev. A 5, 227 (1972).
7. W. Ong and S. T. Manson (unpublished).
8. K. T. Cheng and W. Johnson (unpublished).

330

9. R. E. LaVilla and R. D. Deslattes (private communication).

10. S. T. Manson and M. Inokuti, J. Phys. B 13, L323 (1980).

11. S. T. Manson and M. Inokuti, in Inner-Shell and X-Ray Physics of Atoms and Solids, D. J. Fabian, H. Kleinpopen and L. M. Watson, eds. (Plenum Press, N.Y. 1981), p. 273-276.

12. P. Deshmukh (private communication).

13. B. W. Holland, J. B. Penory, R. F. Pettifer, and E. Boroas, J. Phys. C 11, 633 (1978).

14. U. Fano, C. Theodosiou, and J. L. Dehmer, Rev. Mod. Phys. 48, 49 (1976).

15. M. A. Dillon and M. Inokuti, J. Chem. Phys. 74, 6271 (1981) and private communication.

16. L. H. Toburen and S. T. Manson, X ICPEAC Abstracts of Papers (Commissariat a l'Energie Atomique, Paris, 1977), p. 990.

17. M. E. Rudd and J. H. Macek, Case Studies in Atomic Collision Physics 3, 47 (1972).

18. E. G. Kessler, R. D. Deslattes, D. Girard, W. Schwitz, L. Jacobs, and O. Renner (to be published).

19. M. H. Chen, B. Craseman, M. Aoyagi, K.-N. Huang, and H. Mark, Atomic Data and Nuclear Data Tables (accepted for publication).

20. M. O. Krause and J. H. Oliver, J. Phys. Chem. Ref. Data 8, 329 (1979).

21. M. H. Chen, B. Crasemann, and H. Mark, Phys. Rev. A 24, 1158 (1981).

RELATIVISTIC EFFECTS IN ATOMIC INNER-SHELL TRANSITIONS

Mau Hsiung Chen
Department of Physics and Chemical Physics Institute
University of Oregon, Eugene, Oregon   97403   USA

ABSTRACT

Theoretical calculations of atomic inner-shell transition rates based on independent-particle models are reviewed. Factors affecting inner-shell transition rates are examined, particularly the effects of relativity.

I.   INTRODUCTION

In this talk, I want to review the theoretical calculations of atomic inner-shell transition rates from independent-particle models. I will explore some factors affecting inner-shell transition rates, with special emphasis on the effects of relativity.

The effects of relativity on atomic structure have been thoroughly studied by Desclaux.[1] The main feature of the structure effects due to relativity is that the inner electrons are drawn closer to the nucleus as a result of the relativistic electron-mass increase. The relativistic contraction of the inner shells produces a considerable indirect relativistic effect on the outer electrons. The outer s electrons are drawn in because of the orthogonality constraint to the inner s orbitals in the SCF procedure. The outer d or f electrons move out farther from the nucleus than in the nonrelativistic approximation, due to the increased screening caused by the inner orbital contraction.

The effects of relativity on atomic transition probabilities can arise from several different factors: (i) changes in energies; (ii) relativistic orbital effects caused by the inclusion of the mass-velocity correction, the Darwin term and spin-orbit interaction in the Dirac equations; (iii) relativistic aspects of the pertinent operators, viz., the magnetic interaction and retardation correction in the two-electron operator, due to the electron-photon coupling. The net effect depends on the relative strengths and phases of these factors.

The effects of relativity on the transition rates of atoms with a single inner-shell vacancy can be seen more vividly in heavy elements. However, the effects of relativity can sometimes constitute a dominant mechanism in the decay of few-electron ions in the low-Z region, especially for the high-spin metastable states.

In this report, we will discuss the cases of atoms with a few inner-shell holes and of few-electron ions. Through extensive comparison between relativistic and nonrelativistic HF-type calculations, we will establish a few guidelines as to whether and when full relativistic calculations are required and under what circumstances a nonrelativistic calculation including certain relativistic corrections would be appropriate. From the comparisons between the theoretical results from the independent-particle model and experiment, we will try to determine the reliability of the independent-particle model in the inner-shell transition-rate calculations.

II. RELATIVISTIC TRANSITION ENERGIES

The importance of the relativistic effects on binding energies has long been established.[1-4] Recently, we have performed a systematic study of the neutral-atom binding energies using a relaxed Dirac-Hartree-Slater model with quantum-electrodynamic (QED) corrections and the finite nuclear size effect.[5-6] A first-order correction to the local approximation was made by computing the expectation value of the total Hamiltonian. The QED corrections are very small for light atoms but grow to a few hundred electron volts for heavy elements.

A systematic comparison between theoretical and experimental results[7] reveals the following observations: (i) for K, $L_{2,3}$, $M_{4,5}$ and $N_{6,7}$ levels, the agreement between theory and experiment is quite satisfactory (within 2 eV for Z < 60 and $\sim$6 eV for Z > 60 for the K shell and $\sim$2 eV for $L_{2,3}$, $M_{4,5}$, and $N_{6,7}$ shells). (ii) For L_1, M_1, $M_{2,3}$, N_1, and $N_{2,3}$ levels which have strong Coster-Kronig fluctuations, a systematic error has been found. For the L_1 levels, the discrepancy is $\sim$5 eV[8] and for M_1, $M_{2,3}$ levels, the discrepancies are $\sim$10 eV and 4 eV respectively.

The binding and x-ray energies for free atoms from the relativistic independent-particle model can be improved by including the following corrections: (i) accurate screening and relaxation corrections on K- and L-shell self-energies for heavy elements and an estimate of the self-energy correction for M_1 ($\sim$10 eV at Z = 90) and $M_{2,3}$ shells ($\sim$2 eV at Z = 90) for heavy elements;[7] (ii) the energy shift due to the interaction with the Coster-Kronig continua ($\sim$-5 eV for the L_1 shell for Z < 90,[8] $\sim$-3 eV for the M_3 shell for Z $\approx$92 [7]); (iii) the energy shift due to the virtual Coster-Kronig excitation (e.g., -4.3 eV for the 3s shell of $_{18}$Ar, -10 eV for the 4p shell of $_{55}$Cs [9]).

For nonradiative transition energies, several semi-empirical treatments have been developed.[10-12] Ab initio calculations with SCF methods have also been applied to calculate the K-LL,[13] L-LX,[14] and K Auger energies for Li-like ions.[15-16]

Auger transition energies calculated from the relativistic SCF model can also be improved by including similar corrections as discussed above, as well as the final two-hole configuration interaction such as $(2s)^{-2}$ and $(2p)^{-2}$ $^1S_0$ for K-LL transitions.

III. GENERAL APPROXIMATIONS

For most of the inner-shell transition-rate calculations, the creation of the hole states is assumed to be separate from the decay process, viz., a two-step process. In this approximation, the post-collision interaction between primary electron and Auger electron is ignored. This is a reasonable approximation for hole states created by excitation mechanisms with energy far above the ionization threshold.

The frozen-core approximation is invoked. The initial one-electron wave functions are assumed to be orthonormal to the final one-electron wave functions. The equivalence between few-hole and conjugate few-electron configurations can be called upon to simplify the derivation of the transition matrix elements which depend only on the wave functions of the active electrons.

The independent-particle models with central field, such as HF or HS, are used to generate one-electron wave functions.

The improvements of the frozen-core independent-particle model of inner-shell transitions are now discussed briefly. These include the effects of relaxation, intermediate coupling and configuration interaction.

IV. RADIATIVE TRANSITION

A. Brief History

In most of the early calculations of inner-shell x-ray transition rates,[17-19] hydrogenic wave functions were used. For El transitions, the nonrelativistic f values for the Coulomb potential are independent of atomic number.

With the advance of computers, several systematic studies with Hartree-Fock or Hartree-Slater wave functions have been performed recently. Manson and Kennedy[20] calculated the x-ray emission rates for all shells using nonrelativistic Hartree-Slater wave functions. Relativistic HS frozen-core calculations have been carried out by Scofield,[21] Rosner and Bhalla,[22] and Bhalla.[23] The only relativistic Hartree-Fock calculations including relaxation effects have been done by Scofield.[24] All these relativistic calculations were carried out in the Coulomb gauge.

B. Relativistic Theory

In the theory of quantum electrodynamics, the interaction potential between the electromagnetic field and the electron-positron field depends on the choice of gauge.[25] The transition matrix in turn also becomes gauge-dependent. The spontaneous emission rate for the transition $i \rightarrow f$ is[26]

334

$$A_{i \to f} = 2\alpha\omega\sum_L \frac{(2j_f+1)}{(2L+1)} \begin{pmatrix} j_i & L & j_f \\ 1/2 & 0 & -1/2 \end{pmatrix}^2 \left(|\overline{M}_{fi}^{(e)}(G_L)|^2 + |\overline{M}_{fi}^{(m)}|^2 \right), \tag{1}$$

where

$$\overline{M}_{fi}^{(e)}(G_L) = \overline{M}_{fi}^{(e)}(0) + G_L \overline{M}_{fi}^{(\ell)} \tag{2}$$

All the multipole transition matrices $\overline{M}_{fi}$ are defined in Grant's work.[26]

The Coulomb gauge and the length gauge correspond to the choices of $G_L = 0$ and $G_L = [(L + 1)/L]^{\frac{1}{2}}$, respectively. The necessary and sufficient condition for the transition matrices for all multipoles to be gauge-invariant is that the transition matrix of electric multipoles for longitudinal photons vanishes identically. This condition is automatically satisfied for a single-particle atomic model, but it may not necessarily hold for a Dirac-Fock model.

C. Survey of Theoretical Results

a. Gauge Invariance

We have calculated relativistic radiative transition rates to K- and L-shell vacancies,[27] using Dirac-Fock wave functions and employing both the Coulomb gauge and the length gauge. The difference between these two sets of results may shed some light on the uncertainty of independent-particle DF calculations.

In general, the results from the length gauge are larger than those from the Coulomb gauge. For K-shell radiative widths, the difference is $\sim20\%$ at $Z = 10$ and reduces to $\sim4\%$ at $Z = 30$. For the L_2 radiative widths, the difference is approximately a factor of 2 at $Z = 18$ and reduces to $\sim7\%$ at $Z = 48$ (Fig. 1). For individual transitions with $\Delta n = 0$, the discrepancy is large and persists to medium-heavy elements (e.g. 20% at $Z = 48$ for 2s-2p transitions).

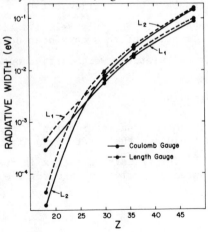

Fig. 1. Radiative widths of L subshells, computed with DF wave functions, vs. atomic number. Solid curves: Coulomb gauge; broken curves; length gauge. Dots represent calculated values.

b. Relativistic Effects

By comparing the electric-dipole line strength and transition rates from nonrelativistis HS[20] and relativistic DHS[21] calculations, one can study the relativistic orbital effects and energy effects on the transition rates. The line strength is independent of transition energy in the nonrelativistic electric-dipole approximation. In general, the line strength from relativistic theory is smaller than that from nonrelativistic theory, due to the relativistic orbital contraction which reduces the dipole matrix elements.[28] These relativistic orbital effects are very small for light atoms and reach about 10% at Z = 55 for Kα transitions; they are smaller for L-shell transitions (Fig. 2).

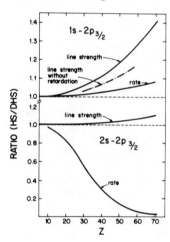

Fig. 2. Ratios of radiative $1s-2p_{3/2}$ and $2s-2p_{3/2}$ line strengths and transition rates from nonrelativistic HS calculations (Ref. 20) and from relativistic DHS calculations (Ref. 21), as functions of atomic number.

For the $K\alpha_1$ transition rates, the retardation effects are of the same order of magnitude as the relativistic orbital effects, and the effect due to the increase in $K\alpha_1$ energy is ∿15% at Z = 55. The difference between nonrelativistic and relativistic $K\alpha_1$ rates is only ∿8% at Z = 70, because of the partial cancellation between orbital effects, retardation correction, and change in transition energy.

The relativistic effects on the $2s-2p_{2/3}$ transition rates is dominated by the change of transition energy. The effect due to shifts in wave functions is only 10% at Z = 70, while the rate is increased by a factor of 20 due to the increase in transition energy. The relativistic line strength of $2p_{3/2}-3s$ transitions, instead of decreasing slowly as a function of Z, increases quite rapidly. The reason for this strange behavior is related to the fact that these transitions involve strong cancellations in the nonrelativistic calculations, and the differential orbital contraction tends to remove the cancellations.[28] The energy effects on these transitions are quite small due to the near-cancellation in the binding-energy increase.

c. Relaxation

To go beyond the frozen-core approximation, separate relativistic Hartree-Fock solutions were found for the initial and final states.[24] The transition rates were then calculated including exchange and overlap corrections. The systematic discrepancy between theoretical and experimental intensity ratios, especially for $K\beta/K\alpha$, has been removed with the inclusion of the relaxation effects.[24]

d. K X-Ray Hypersatellites

The $K\alpha$ hypersatellites $K\alpha_1^h$ and $K\alpha_2^h$ arise from the double-hole-state transitions $[1s^2] \rightarrow [1s2p_{3/2}]$ and $[1s^2] \rightarrow [1s2p_{1/2}]$, respectively, in the high-Z region. The energy shift of these $K\alpha$ hypersatellites with respect to the diagram lines has been calculated using a relativistic SCF approach in intermediate coupling including relaxation and QED corrections.[29-31] Although the contribution of the Breit interaction to the $K\alpha$ x-ray energy is only $\sim 0.4\%$ at $Z = 80$, the contribution to the energy shift of the $K\alpha$ hypersatellite is 16%. Therefore, precision measurements of the energy shift will provide a sensitive test of the QED corrections. The major deficiency in the theory comes from the lack of screening and relaxation correction to the self-energy and the neglect of electron-electron correlation. As of today, the agreement between experimental and theoretical energy shifts is good within the uncertainties, except for one precision measurement at $Z = 80$. More accurate measurements are needed at high Z in order to determine whether the discrepancy is real.

The $K\alpha_{1,2}$ hypersatellite intensity ratio has been analyzed in terms of the intermediate-coupling scheme.[30,31] For low-Z elements, the $K\alpha_1^h$ x-ray transition $[1s^2] \rightarrow [1s2p]$ $^3P_1$ is dipole-forbidden in LS coupling. X-ray emission then arises from mixing between $^3P_1$ and $^1P_1$ through the spin-orbit interaction. Including the Breit interaction in the intermediate-coupling calculations causes the $K\alpha_1^h$-to-$K\alpha_2^h$ intensity ratio to decrease at low Z ($\sim 25\%$ at $Z = 18$) and to increase for medium and heavy elements ($\sim 7\%$ at $Z = 60$).[31] The existing experimental results are not accurate enough to differentiate between the calculations.

V. RADIATIONLESS TRANSITIONS

Most of the early work on Auger transitions has been nonrelativistic, based on Wentzel's ansatz.[32] Howat, Åberg and Goscinski[33] have recently formulated the nonrelativistic radiationless transitions using a multichannel scattering approach and performed nonrelativistic calculations for K-LL rates of Ne including relaxation and channel-coupling effects. Massey and Burhop[34] attempted the first relativistic calculations using screened hydrogenic wave functions to deduce the K-LL rates of gold. Since then, several relativistic calculations[35-37] have been performed, with varied degrees of sophistication, for a few selected elements and transitions.

Recently, we have conducted rather extensive, systematic Dirac-Hartree-Slater calculations for the K, L, and M shells of atoms with an inner vacancy to study the effects of relativity on Auger rates, level widths, and fluorescence yields.[38-40] We have also calculated the relativistic Auger and x-ray emission rates for highly ionized atoms,[15,16,41] for the purpose of studying the relativistic effects on the decay of high-spin states.

A. Theory

In this report, the Auger transitions are treated as a two-step process. The Auger decay probabilities are calculated from perturbation theory. The transition rate is

$$T = \frac{2\pi}{\hbar} |\langle \psi_f | H - E | \psi_i \rangle|^2 \rho(\varepsilon). \tag{3}$$

Here, ψ_i and ψ_f are the antisymmetrized many-electron wave functions of the initial and final states of the ion, respectively, and $\rho(\varepsilon)$ is the energy density of final states.

In the frozen-core approximation, only the two-electron operator in Eq. (3) contributes to the transition rate, and the many-electron wave functions ψ_f, ψ_i can be replaced by the two-hole coupled wave functions with the appropriate phase factor for an atom with an initial inner-shell vacancy. In the nonrelativistic theory, the two-electron operator represents just the Coulomb repulsion between the two electrons. In the relativistic theory, the photon-electron coupling is automatically included in the two-electron operator. From quantum-electrodynamics, the electron-electron interaction operator is gauge-dependent.

In the currently available relativistic Auger calculations,[34-41] the two-electron operator is chosen to be the Møller operator, which is based on the Lorentz gauge:

$$V_{12} = (1 - \vec{\alpha}_1 \cdot \vec{\alpha}_2) e^{i\omega r_{12}} / r_{12}. \tag{4}$$

In the Coulomb gauge, the two-electron operator is

$$V_{12} = \frac{1}{r_{12}} - (\vec{\alpha}_1 \cdot \vec{\alpha}_2)\frac{e^{i\omega r_{12}}}{r_{12}} + (\vec{\alpha}_1 \cdot \vec{\nabla}_1)(\vec{\alpha}_2 \cdot \vec{\nabla}_2) \frac{e^{i\omega r_{12}} - 1}{\omega^2 r_{12}} \tag{5}$$

Here, the $\vec{\alpha}_i$ are Dirac matrices, and ω is the energy of the virtual photon.

In relativistic theory, the Auger matrix elements are calculated in j-j coupling and can be separated into angular and radial parts with the aid of Racah algebra.[39] The Auger matrix elements were evaluated numerically with Dirac-Fock or Dirac-Hartree-Slater wave functions corresponding to the initial hole-state configuration. In the frozen-core approximation, the same set of one-electron wave functions is used for initial and final states.

To make the full-scale calculation economically feasible, we use the j-j configuration average energies in the calculations. For Coster-Kronig transitions with small transition energies, we use transition energies computed from relativistic relaxed-orbital DHS calculations including QED corrections.[42] For Auger transitions that involve large transition energies, the empirical "Z + 1" rule is used. The effect on the Auger matrix elements caused by this error in energy ($\sim$30 eV out of a few keV) is quite negligible.

For few-electron ions, the Auger matrix elements are derived in the electron picture instead of the hole picture, and Auger transition energies are calculated in relativistic intermediate coupling including limited configuration interaction.[15-16]

To include the effects of relaxation, initial- and final-state wave functions are calculated separately, and the full perturbation Hamiltonian including the one-electron operator is used.[33]

B. Survey of Results for a Single Inner-Shell Vacancy

a. Gauge Invariance

We have performed relativistic Auger calculations with Dirac-Fock wave functions, using both the Lorentz gauge and Coulomb gauge, for selected elements in different ranges of atomic numbers, for K and L shells.[43] The agreement between the results from these two different gauges is better than 1% in all cases. Thus, in first-order perturbation theory the Auger rate is seen to be practically gauge invariant.

b. DF vs. DHS

The total Auger transition rates with DF wave functions are larger than those calculated with DHS wave functions, by $\sim$10% at Z = 18 and by 3% at Z = 80 for K shells and by 14% of Z = 30 and 3% at Z = 80 for L_3 shells.[43] A similar finding has been reported for x-ray emission rates.[24]

c. Effects of Relativity

The effects of relativity on radiationless transitions can come from the changes in transition energy, shifts in wave functions, and photon-electron coupling included in the two-electron operator. Energy effects obviously are very important for Coster-Kronig transitions. Without relativity, most of the super Coster-Kronig and Coster-Kronig transition would be energetically impossible.

In $L_1-L_{23}M_{45}$ Coster-Kronig transitions, the $L_1-L_3M_5$ transition energy is increased by a factor of 2 due to relativity. The large discrepancy between the results from HS and DHS for $25 \leqslant Z \leqslant 45$ is almost entirely due to the increase in transition energy. The energy effects on ordinary Auger transitions are much smaller, ranging from 2-4% at low to medium Z to 10-15% for high Z.

The relativistic orbital effects on K-LL transitions are typically ∿5-10% at medium Z and 25-30% at high Z. For $K-L_1L_1$ transitions, the relativistic orbital effects cause the transition rates to increase by 45% at Z = 50 and by a factor of 2.5 at Z = 80. The large effect on $K-L_1L_1$ is probably due to the fact that $K-L_1L_1$ transitions involve substantial cancellations in the nonrelativistic case. The relativistic orbital contraction removes the cancellation and causes the transition rate to increase as a function of Z. Similar relativistic orbital effects have been found for L- and M-shell transitions.

The contributions from the magnetic interaction and retardation correction to the Auger transition rates is very small at low Z. It can become as large as that from relativistic orbital effects for some K- and L-shell transitions (e.g., $K-L_1L_2$ and $L_2-M_4M_5$) (Fig. 3). In general, the contribution from the magnetic interaction and retardation correction is smaller for L-shell than for K-shell transitions and much smaller for M-shell transitions.

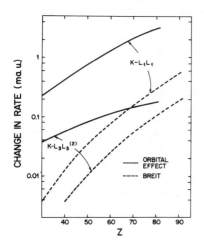

Fig. 3. Contributions (in milliatomic units) to Auger transition rates from the relativistic orbital effect (solid curves) and the Breit interaction (broken curves).

The interplay among the energy effects, relativistic orbital effects and Breit interaction determines the net relativistic effects on Auger transition rates. Constructive interference can lead to a large increase in transition rates (e.g., $K-L_1L_1$, $K-L_1L_2$). A near-cancellation can leave the transition rates with little change (e.g., $K-L_2L_3$).

The effects of relativity increase the total K Auger rate by a factor of 2, reduce the total L_2 Auger rate by 25%, and increase the total L_3 Auger rate by only 5% at Z = 80.

Good agreement between theory and experiment has been found for K- and L_{23}-shell fluorescence yields, Coster-Kronig yields f_{23} and group-intensity ratios (K - LX)/(K - LL).[38,39] A slight discrepancy (10-20%) for (K - XY)/(K - LL) ratios might be due to the neglect of the relaxation effect, similar to the case of Kβ/Kα ratios in x-ray transitions.[24]

d. Relativistic Intermediate Coupling with Configuration Interaction

Relativistic intermediate coupling with configuration interaction has been used successfully to analyze the K-LL Auger spectra.[44,45] We recently extended this same technique to treat the K-MM Auger spectra for medium and heavy elements.[46] Configuration interaction among all the possible final double MM hole states is included in the calculations. For these calculations, j-j coupled basis states are used. Coulomb as well as Breit interactions are included in the energy matrix. The eigenfunctions and eigenvalues are obtained by diagonalizing the energy matrix.

The relativistic effects are quite large on most of the K-MM transitions (e.g., approximately a factor of 2 at Z = 80 for $K-M_1M_1$). Intermediate coupling drastically improves the agreement between the theoretical and experimental $K-M_iM_j$ intensity ratios. The effects of configuration interaction among $M_1M_1(J = 0)$, $M_2M_2(J = 0)$, and $M_3M_3(J = 0)$ states persist to the heavy elements, similar to the case of $L_1L_1(J = 0)$, $L_2L_2(J = 0)$, and $L_3L_3(J = 0)$.[44,45] Very good agreement between theory and experiment[57] for the $K-M_1M_1$ intensity is obtained after including this configuration interaction (Fig. 4).

The $K-M_1M_{45}$ intensity has a strong peak at Z $\approx$ 63. This peculiar behavior is caused by level crossing. For Z < 60, the $M_3M_3(J = 2)$ level lies above the $M_1M_4(J = 2)$ and $M_1M_5(J = 2)$ levels. As Z increases, $M_3M_4(J = 2)$ first comes down to cross $M_1M_4(J = 2)$ at Z $\approx$ 62, then crosses $M_1M_5(J = 2)$ at Z $\approx$ 65. The $K-M_1M_{4,5}$ transitions pick up some intensity from $K-M_3M_3$ through the strong configuration interaction between $M_1M_{4,5}(J = 2)$ and $M_3M_3(J = 2)$ states. Level crossing is a very common phenomenon in multiply ionized atoms.[28] Great care has to be taken in treating these cases.

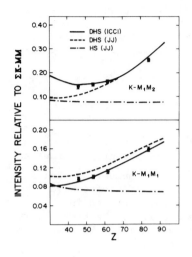

Fig. 4. Auger $K-M_iM_j$ intensities relative to the total K-MM intensity, as a function of atomic number. Theoretical relativistic results in intermediate coupling with configuration interaction [DHS (ICCI)] are compared with relativistic results in j-j coupling [DHS (JJ)], with nonrelativistic calculations in j-j coupling [HS (JJ)], and with experimental results with Ref. 47.

C. Few-Electron Ions

We have calculated the Auger and x-ray transition rates of the Li-like $1s2s^2$, $1s2p^2$, and $1s2s2p$ isoelectronic sequences, using relativistic intermediate coupling and configuration interaction. Breit interaction is included both in the Auger matrix elements and in the energy matrix of intermediate-coupling calculations.

In relativistic intermediate-coupling calculations, the full Breit interaction (magnetic and retardation) has to be included in the correctly coupled state description in order to accurately predict the fine structure of ^4P states of the $1s2p^2$ and $1s2s2p$ configurations.

For highly ionized atoms with an inner-shell vacancy in an otherwise closed shell (e.g., the $1s2s^22p^6$ F-like isoelectronic sequence), similar relativistic effects have been found with a slight increase over that of normal atoms.[41]

For open shells in intermediate coupling (e.g., $1s2p^2$ configurations), the effects of relativity can become important very early in the isoelectronic sequence. The Auger rate of the $^2D_{3/2}$ state of the $1s2p^2$ configuration is reduced by 12% already at Z = 20 due to the effects of relativity.

The ^2P and ^4P states of the $1s2p^2$ configuration and ^4P states of the $1s2s2p$ configuration are all Auger-forbidden in the nonrelativistic limit.[15,16] The Auger transition rates of these states are thus completely determined by the effects of relativity.

The $^2P_{1/2}$ and $^2P_{3/2}$ states of the $1s2p^2$ configuration acquire their Auger strength from $^2S_{1/2}$ and $^2D_{3/2}$ states, respectively, through the spin-orbit interaction. The Auger rates of ^2P states of $1s2p^2$ increase over four orders of magnitude from Z = 6 to Z = 30 (Fig. 5).[16]

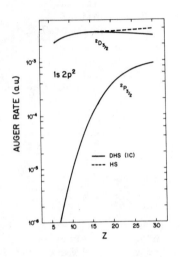

Fig. 5. Auger rates for $^2D_{3/2}$ and $^2P_{3/2}$ states of 1s2p configurations as functions of atomic number Z. Solid curves are from relativistic Dirac-Hartree-Slater calculations in intermediate coupling; the broken curve is from a nonrelativistic Hartree-Slater calculation.

The $^4P_{5/2}$ state of the 1s2s2p configuration can make Auger transitions by magnetic spin-spin interaction, or decay radiatively by magnetic-quadrupole emission.[15,16,48] The Auger decay of ^4P_J states of the $1s2p^2$ configuration and $^4P_{1/2,3/2}$ states of the 1s2s2p configuration can occur through mixing with doublet states or by the magnetic interaction. These two mechanisms have been found to be equally important for $^4P_{1/2,3/2}$ in the low-Z region.[15,16] The Auger decay of $^4P_{1/2,3/2}$ states of $1s2p^2$ and 1s2s2p configurations are very sensitive to the details of the atomic model because of the strong cancellation between these two effects. A nonrelativistic intermediate-coupling calculation neglecting the contribution from magnetic interaction would be in serious error.[15,16] Similar relativistic effects can be expected for other metastable states, such as $1s2s2p^2$ ^5P and $1s2p^3$ ^5S states.

There exists strong configuration interaction between the $1s2p^2$ ^2S and $1s2s^2$ ^2S states: the Auger decay of the former is reduced by a factor of 3 and the fluorescence yield of the latter is increased by six orders of magnitude at low and medium Z due to this effect.[16,41]

Comparing the lifetime of the ^4P_J states from the present relativistic calculations with the results from nonrelativistic theory and experimental data, we find that the relativistic results agree much better with experiment.[15,16]

VI.   GENERAL CONCLUSIONS

The relativistic independent-particle relaxed-orbital calculations of radiative transitions are in quite good agreement with experiments. The gauge dependence of total radiative rates is small ($\lesssim 3\%$ for K shells for Z $\gtrsim 30$, and $\lesssim 7\%$ for L shells for Z $\gtrsim 48$). Some general conclusions concerning the effects of relativity on radiative transitions can be drawn:

1.   Relativistic orbital contraction usually leads to line-strength reduction.

2.   For strong transitions (e.g., 1s-2p, $2p_{1/2}$-$3d_{3/2}$), the effect of relativity is small due to the near cancellation among the energy effect, relativistic orbital effect, and retardation correction.

3.   For weak transitions with $\Delta n = 0$ (e.g., 2s-2p), the effect due to the change in energy predominates.

4.   The relativistic orbital effects can be very large for transitions with large cancellations in nonrelativistic matrix elements (e.g., $2p_{3/2}$-3s).

Relativistic DHS calculations of Auger transitions in intermediate coupling with limited configuration interaction have been found to be in reasonable agreement with experiments. However, the independent-particle model is inadequate to treat low-energy Coster-Kronig and super-Coster-Kronig transitions. The effect of relativity on radiationless transitions has been studied in a systematic manner for atoms with an inner-shell vacancy and for few-electron ions. Some general conclusions can be drawn:

1.   For Coster-Kronig and super-Coster-Kronig transitions, the effect due to the change in transition energy is the dominant factor.

2.   For normal medium-Z and heavy atoms, the magnetic interaction and retardation corrections are as important as the relativistic orbital effect in inner shells, while at low Z, the orbital effect is more important.

3.   For transitions involving an outer shell, the wave-function effect is the most important.

4.   For highly ionized atoms, the effect of relativity can be very severe in the intermediate-coupling situation.

5.   The effect of relativity is very critical for the decay of metastable states (e.g., ^4P of $1s2p^2$ configurations), a consistent treatment including all aspects of relativistic effects is necessary.

Accurate measurements of lifetimes and fluorescence yields of the metastable states of highly ionized atoms are much needed.

344

ACKNOWLEDGMENTS

The work reported here was done in close cooperation with Bernd Crasemann and Hans Mark. Support was provided by the Air Force Office of Scientific Research.

REFERENCES

1. J. P. Desclaux, At. Data Nucl. Data Tables 12, 311 (1973).
2. I. Lindgren and Arne Rosén, Case Studies in Atomic Phys. 4, 93 (1974).
3. I. P. Grant, Adv. Phys. 19, 747 (1970).
4. J. B. Mann and W. R. Johnson, Phys. Rev. A 4, 41 (1971).
5. K. N. Huang, M. Aoyagi, M. H. Chen, B. Crasemann, and H. Mark, At. Data. Nucl. Data Tables 18, 243 (1976).
6. M. H. Chen, B. Crasemann, M. Aoyagi, K. N. Huang, and H. Mark, At. Data. Nucl. Data Tables 26, 561 (1981).
7. N. Mårtensson, M. H. Chen, B. Crasemann, and B. Johansson, unpublished.
8. M. H. Chen, B. Crasemann, and H. Mark, Phys. Rev. A 24, 1158 (1981).
9. D. R. Beck and G. A. Nicolaides, in Excited States in Quantum Chemistry edited by C. A. Nicolaides and D. R. Beck (Reidel, Holland, 1979).
10. W. N. Asaad and E. H. S. Burhop, Proc. Phys. Soc. Lond. 71, 369 (1958).
11. D. A. Shirley, Phys. Rev. A 7, 1520 (1973).
12. F. P. Larkins, At. Data Nucl. Data Tables 20, 311 (1977).
13. C. Briançon and J. P. Desclaux, Phys. Rev. A 13, 2157 (1976).
14. M. H. Chen, B. Crasemann, K. N. Huang, M. Aoyagi, and H. Mark, At. Data Nucl. Data Tables 19, 97 (1977).
15. M. H. Chen, B. Crasemann, and H. Mark, Phys. Rev. A 24, 1852 (1981).
16. M. H. Chen, B. Crasemann, and H. Mark, Phys. Rev. A (in press).
17. H. A. Bethe and E. E. Salpeter, Quantum Mechanics of One- and Two-Electron Atoms (Springer-Verlag, Berlin, 1957).
18. F. A. Babushkin, Acta. Phys. Pol. 25, 749 (1964); 31 459 (1967).
19. W. B. Payne and J. S. Levinger, Phys. Rev. 101, 1020 (1956).
20. S. T. Manson and D. J. Kennedy, At. Data Nucl. Data Tables 14, 111 (1974).
21. J. H. Scofield, Phys. Rev. 179, 9 (1969); At. Data Nucl. Data Tables 14, 121 (1974).
22. H. R. Rosner and C. P. Bhalla, Z. Phys. 231, 347 (1970).
23. C. P. Bhalla, J. Phys. B 3, 916 (1970); Phys. Rev. A 2, 2575 (1970).
24. J. H. Scofield, Phys. Rev. A 9, 1041 (1974).
25. A. I. Akhiezer and V. B. Brestetskii, Quantum Electrodynamics (Interscience, New York, 1965).
26. I. P. Grant, J. Phys. B 7, 1458 (1974).
27. M. H. Chen, B. Crasemann, and H. Mark, unpublished.
28. A. W. Weiss, in Beam-Foil Spectroscopy, edited by I. A. Sellin and D. J. Pegg (Plenum, New York, 1976).

29. J. P. Desclaux, Ch. Briançon, J. P. Thibaud, and R. J. Walen, Phys. Rev. Lett. 32, 447 (1974).
30. T. Åberg and M. Suvanen, in Advances in X-Ray Spectroscopy, edited by G. Bonnelle and C. Mandé (Pergamon, New York, 1980).
31. M. H. Chen, B. Crasemann, and H. Mark, Phys. Rev. A 25, 391 (1982).
32. G. Wentzel, Z. Physik 43, 524 (1927).
33. G. Howat, T. Åberg, and O. Goscinski, J. Phys. B 11, 1575 (1978).
34. H. S. W. Massey and E. H. S. Burhop, Proc. R. Soc. A 153, 661 (1936).
35. W. N. Asaad, Proc. R. Soc. A 249, 555 (1959).
36. M. A. Listengarten, Izv. Akad. Nauk SSR, Ser. Fiz. 25, 803 (1961) [Bull. Acad. Sci. USSR, Phys. Ser. 25, 803]; 26, 182 (1962) [Bull. Acad. Sci. USSR, Phys. Sev. 25, 182].
37. C. P. Bhalla, J. Phys. B 3, L9 (1970); 3, 916 (1970); Phys. Rev. A 2, 722 (1970); C. P. Bhalla and D. J. Ramsdale, J. Phys. B 3, L14 (1970).
38. M. H. Chen, B. Crasemann, and H. Mark, Phys. Rev. A 21, 436 (1980).
39. M. H. Chen, B. Crasemann, and H. Mark, Phys. Rev. A 24, 177 (1981); M. H. Chen, E. Laiman, B. Crasemann, M. Aoyagi, and H. Mark, Phys. Rev. A 19, 2253 (1979).
40. M. H. Chen, B. Crasemann, and H. Mark, Phys. Rev. A 21, 449 (1980).
41. M. H. Chen, B. Crasemann, K. R. Karim, and H. Mark, Phys. Rev. A 24, 1845 (1981).
42. M. H. Chen, B. Crasemann, K. N. Huang, M. Aoyagi, and H. Mark, At. Data Nucl. Data Tables 19, 97 (1977).
43. M. H. Chen, B. Crasemann, and H. Mark, unpublished.
44. W. N. Asaad and D. Petrini, Proc. R. Soc. London A 350, 381 (1976).
45. M. H. Chen, B. Crasemann and H. Mark, Phys. Rev. A 21, 442 (1980).
46. M. H. Chen, B. Crasemann and H. Mark, unpublished.
47. M. I. Babenkov, B. V. Bobykin, V. S. Zhdanov, and V. K. Petukhov, Phys. Lett. 56A, 363 (1976); J. Phys. B 15, 35 (1982); J. Phys. B 15, 927 (1982).
48. K. T. Cheng, C. P. Lin, and W. R. Johnson, Phys. Lett. 48A, 437 (1974).

INNER SHELL RELATIVISTIC FEATURES IN PHOTOEFFECT

R. H. Pratt
Department of Physics and Astronomy
University of Pittsburgh, Pittsburgh, Pennsylvania 15260 U.S.A.

ABSTRACT

We discuss relativistic features of radiation processes, with emphasis on those features produced in the interior of the atom, using photoeffect as our example. This is to be contrasted with the rather extensive discussion of relativistic effects on atomic structure and the recent studies of relativistic modifications of nonrelativistic exchange and correlation effects. We conclude that relativistic dipole and first higher multipole corrections suffice to characterize these processes through the x-ray regime, while at higher energies alternative characterization schemes need to be developed.

For our purposes it is sufficient to consider a radiation process as a single electron transition in a screened central potential (even though quantitative predictions may require a more sophisticated treatment of other aspects of the process). What we wish to understand are the differences from nonrelativistic dipole calculations in a nonrelativistic potential. These include differences in wave functions due to relativity, such as inner contraction and spin-orbit splitting, and resulting differences in screened potentials (contracted interior, extended exterior). Retardation effects enter in the same order as relativistic effects, and often the two effects tend to cancel. Higher multipoles must also be considered and can remain important even for low energy processes. Additional polarization correlations are possible, so that a "complete" experiment becomes more complex.

In photoeffect several such striking relativistic features have been observed. In total cross sections relativistic, retardation, and higher multipole effects can cancel throughout the x-ray regime, while in angular distributions retardation effects are important throughout and higher multipole effects persist to threshold. Once spin-orbit multiplets split, certain Cooper minima move to increasingly high energies above threshold (100's of eV) with increasing Z, leading to dramatic effects on β parameters and branching ratios. High energy behavior is totally different, and since with relativity the important region for the process stabilizes (rather than continuing to shrink), screening effects beyond normalization persist even in the high energy limit.

I. INTRODUCTION

Let us begin with several illustrations of relativistic effects in photoionization on different energy scales, as predicted in recent calculations:

- At high energy (> 1 MeV) subshell cross section branching ratios $B \equiv \sigma(P_{3/2})/\sigma(P_{1/2})$ are not the statistical value $B = 2$,

0094-243X/82/940346-11$3.00 Copyright 1982 American Institute of Physics but B(Z) ranges from >3 in low Z elements to <0.4 in high Z elements.[1]

- In high Z elements (Pb → Fm) the $6p_{1/2}$ (but not $6p_{3/2}$) asymmetry parameter β passes through 0 at 200 → 500 eV photoelectron energies and remains small over several hundred eV, the signature of a Cooper minimum.[2]

- The angular asymmetry ratio

$$R \equiv \frac{\frac{d\sigma}{d\Omega}(45°) - \frac{d\sigma}{d\Omega}(135°)}{\frac{d\sigma}{d\Omega}(45°) + \frac{d\sigma}{d\Omega}(135°)}$$

for the $5s_{1/2}$ subshell of Sn oscillates about one as a function of photoelectron energy,[3] as shown in Table I:

Table I   R-1 as a function of photoelectron kinetic energy T for the $5s_{1/2}$ subshell of Sn, where R is the angular asymmetry ratio.

T	4 eV	10 eV	400 eV	4 keV	10 keV
R-1	+0.07	–0.01	+0.07	–0.03	+0.12

We see in these examples that relativistic effects, largely inner shell in origin, lead to observable consequences in all energy regimes, affecting many different aspects of the photoelectric process, visible in both inner and outer shell phenomena.

There is a general theoretical interest in understanding these transitions in basic formalism, the transition from classical to quantum mechanics as well as the transition from nonrelativistic to relativistic quantum radiation theory discussed here. Our interest in understanding (as opposed to simply having the numerical predictions) is not just academic: understanding is needed so we can judge what to expect as we change the circumstances (as in introducing external fields, or a hot dense plasma environment) and it is needed to know how sophisticated a calculation will be needed in various circumstances.

Our focus here is on relativistic, retardation, and multipole effects within a single electron picture, while recognizing that in many circumstances quantitative predictions will also require going beyond the single electron picture. Beginning with single electron transitions in nonrelativistic dipole approximation, two main approaches have been (1) to include many electron effects or (2), as here, to include relativistic, retardation and higher multipole effects. We shall discuss briefly at the end what can be said about the need and the possibility for combining these two approaches.

Our focus here is on photoionization, a radiative vacancy creating process. We are not talking about structure. We are not talking about vacancy decay processes (radiative and Auger transitions). We are not talking about other vacancy production processes (Compton or electron scattering on bound electrons, photoexcitation). And we are not discussing radiation processes in which atomic state remains unchanged (bremsstrahlung, Rayleigh scattering).

II.   COMPARISON OF THE ELEMENTS OF
RELATIVISTIC AND NONRELATIVISTIC DESCRIPTION

We wish to calculate a differential cross section

$$\frac{d\sigma}{d\Omega} \sim |M|^2$$

from a matrix element

$$M \sim \int \psi^*_{final} \left[\begin{array}{c} \alpha \cdot \varepsilon \; e^{ik \cdot r} \\ \nabla \cdot \varepsilon \end{array} \right] \psi_{initial} \; d^3r$$

in relativistic and nonrelativistic cases respectively. This form implies a specific choice of gauge. The photon polarization vector $\bar{\varepsilon}$ appears in both cases and may be characterized in terms of the Stokes parameters. However the treatment of the photon in dipole approximation or in multipoles with retardation differs in the two cases. (We may distinguish the multipole expansion in angular momentum from the small k retardation expansion in each multipole -- there are situations in which the dipole term $e^{ik \cdot r} \sim j_o(kr)$ may suffice but the retardation expansion $j_o(kr) \sim 1$ is not satisfactory.) For the electron wave function ψ a nonrelativistic characterization of the bound state by ℓ becomes a relativistic characterization by $\underline{j}$, $\underline{s}$; the continuum state is characterized by its asymptotic linear momentum $\underline{p}$ in both cases but also by the spin direction $\underline{\sigma}$ in the relativistic case.

There is a relativistic contraction of Coulomb wave functions, as at small distances a behavior in r^ℓ is replaced by a behavior in $r^{\gamma-1}$, with

$$\gamma^2 = \kappa^2 - (Z\alpha)^2, \qquad \kappa = \mp(j + \tfrac{1}{2}) \text{ as } j = \ell \pm \tfrac{1}{2}.$$

The difference is greatest for large Z, small ℓ, and the $j = \ell - \tfrac{1}{2}$ case. The consequence of this interior contraction of atomic electrons is a greater shielding of the nucleus. This leads to a weakening of the potential at larger distances and so a greater extension of electron wave functions at larger distances. The balance of these two effects, as seen in bound state energies or low energy continuum phase shifts, favors expansion for large ℓ (not penetrating the inner region), contraction for small ℓ.

For consistency one should use a relativistic potential in the relativistic matrix element, a nonrelativistic potential in the non-relativistic matrix, as the mismatch can give large effects.[4] Screening <u>decreases</u> bound state wave functions at small distances, <u>increases</u> them at large distances. At high energies and small distances the ratio $R_{n\ell}/N_{n\ell}$ of a bound or continuum state radial wave function R to its normalization at the origin N is independent of n for given ℓ; the region of common shape is larger for larger n (or lower continuum energy T).

The photoeffect matrix elements M(T) may be characterized by radial matrix elements corresponding to transitions to continuum states of definite angular momentum through a definite photon multipole. At high energies such matrix elements, when divided by bound state normalization $N_{n\ell}$, will be independent of principal quantum number n, due to the corresponding property of the bound state wave function $R_{n\ell}$. In Fig. 1 we show as an example the (ns - εp) matrix

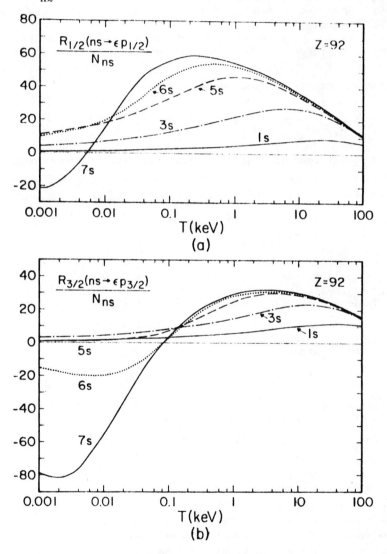

Fig. 1. Reduced dipole transition matrix elements in U as a function of photoelectron kinetic energy T, from Ref. 5.

elements of U, reduced by the bound state normalizations. We see
that these matrix elements indeed merge to a common curve at high
energy, and that the lower n, the sooner the deviation from the common
curve begins. We also observe that for large n the matrix element
changes sign above threshold, the so-called Cooper minimum phenomenon.
We see that this transition occurs for a substantially higher energy
for a continuum p_3 state than for p_1. This is a relativistic spin
   _2 _2
orbit splitting effect, resulting from the greater contraction of
the $j = \ell - \frac{1}{2}$ state, matched in $j = \ell + \frac{1}{2}$ only by going to substantially higher energy.

Cross sections are determined from radial matrix elements and
phase shift differences. The polarization information of photon and
continuum electron through the polarization correlations[6] $C_{ij}(\theta)$,
giving the differential cross section

$$\frac{d\sigma}{d\Omega} = \frac{d\sigma}{d\Omega}\bigg|_{unpol} \frac{1}{2} \sum_{ij} [1 + \xi_i \zeta_j \, C_{ij}(\theta)],$$

or an appropriate generalization if bound electron magnetic substates
are observed.[7] The unpolarized differential cross section may be
characterized by a series of coefficients B_n in a Legendre expansion

$$\frac{d\sigma}{d\Omega}\bigg|_{unpol} = \frac{\sigma}{4\pi} \sum_n B_n P_n(\cos\theta),$$

where σ is the total cross section, $B_o \equiv 1$, and $B_2 \to -\frac{1}{2}\beta$, β the
asymmetry parameter in dipole approximation. Similar expansions
are available for the C_{ij}. The usefulness of these expansions
(related to the multipole expansions) depends on their rapid convergence, and this is only possible through the x-ray regime; at higher
energies other approaches must be sought.

III. CONSEQUENCES FOR PHOTOEFFECT

The photoelectron kinetic energy T and the binding energy ε
determine the important distances for the matrix element. We may
distinguish three regimes: (1) ultra-relativistic regime, for $T > mc^2$,
(2) intermediate regime, becoming Coulombic for $T > (Z\alpha)^2 mc^2$, and
(3) low energy regime, with $T < 2\varepsilon$. We will note some relativistic
effects in each of these regimes.

Experiments on photoeffect have now[8] been pushed to 6.76 MeV
(for the K shell of U), more than twice as high in energy as any
previous measurements. Agreement between theory and experiment
appears good in this situation of substantial forward peaking, with a
cross section dropping two orders of magnitude by 30°, nearly three
orders of magnitude by 45°.

Theoretical studies[1] of subshell branching ratios, shown in Fig. 2,

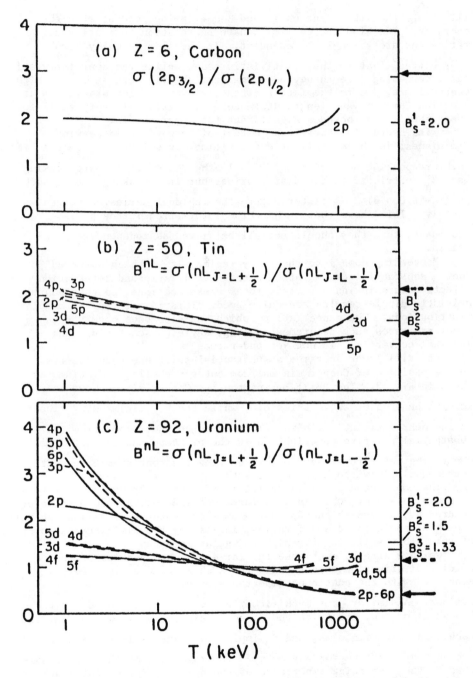

Fig. 2. Subshell branching ratios in C, Sn, and U as a function of photoelectron kinetic energy. Arrows indicate high energy limit values in the point Coulomb case. Statistical ratios B_s are also shown.

illustrate that these ratios indeed become n-independent at high energies. The high energy limits have no relation to statistical ratios and are strongly Z dependent. It has also recently been appreciated[9] that in the relativistic case, unlike the nonrelativistic case, screening effects beyond normalization will persist in the high energy limit, due to the fact that the relevant region stabilizes at electron Compton wave length distances. The effect is very small for inner shells, but becomes significant for outer shells.

With decreasing energy the extent of forward angle peaking diminishes; in the K shell of U one ultimately switches to substantial backward peaking near threshold.[10] In the region of forward peaking many B_n contribute to the distribution, but if a peaking factor such as $(1 - \beta \cos \theta)^{-4}$ is factored out, the residual series converges rapidly. This shows that much of the information in the B_n is redundant, that they should not all be viewed as independent parameters.

Proceeding down into the x-ray regime, it has been observed[11] that a substantial cancellation among relativistic and retardation effects occurs in inner s state total cross sections, so that nonrelativistic dipole results are quite good. This is not true in angular distributions. The cancellation, which has also been observed in other processes such as bremsstrahlung, Rayleigh scattering, and internal conversion, is not yet understood.

At still lower energies a substantial split has been observed[2] in the position of Cooper minima from outer p shells of Uranium, as shown in Fig. 3. The positions of the $p_{1/2}$ Cooper minima move to several hundred eV above threshold, while the $p_{3/2}$ minima stay closer to the nonrelativistic minima. This is to be understood from the substantial relative contribution in the $p_{1/2}$ case and the difficulty, due to the centrifugal barrier, in getting the continuum d to contract correspondingly without giving it substantially more energy.

The consequence of this splitting of minima for branching ratios[12] is shown in Fig. 4, where a very large oscillation, spread over several hundred eV, occurs in the 6p case. Smaller features at similar energies are observed in the d and f ratios, indicating the diminished importance of spin orbit splitting for higher angular momentum.

We have already mentioned the large backward peaking of the angular distribution expected for the K shell of U at low energy, connected with the continued importance of higher multipole contributions. A similar, but much richer effect, has been found[3] in the 5 s shell of tin, for which the B_n coefficients are shown in Fig. 5. Each change in sign of B_1 and B_3 implies a reversal in the sign of the angular distribution asymmetry. The magnitude of the term testifies to the continuing importance of the next multipole contributions. The sign changes are understood in terms of one sign change (Cooper minimum) of the dominant dipole matrix element, two sign changes in the quadrupole matrix element, and two sign changes of the cosine of the phase shift difference.

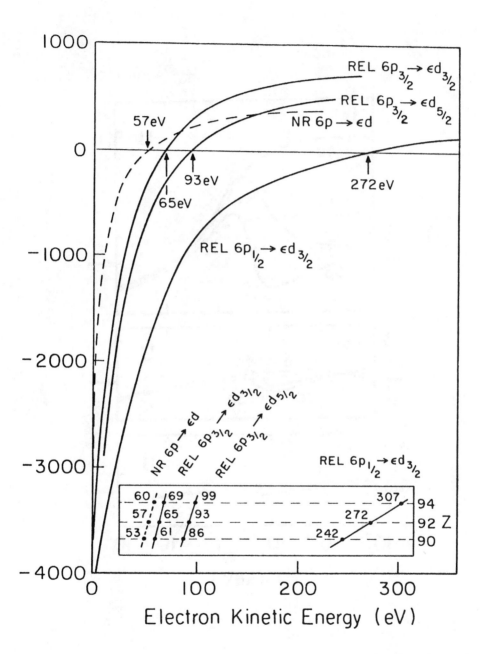

Fig. 3. 6p — εd matrix elements in U as a function of photoelectron kinetic energy, shown sign change, and (inset) energy positions of the Cooper minimum as a function of Z, from Ref. 2.

354

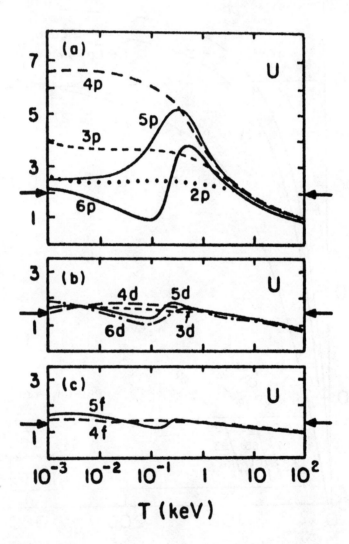

Fig. 4. Subshell branching ratios for U from Ref. 12. Arrows indicate the statistical values.

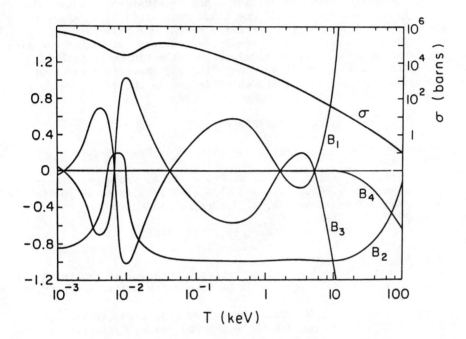

Fig. 5. Photoeffect cross section and angular distribution coefficients B_n for the 5s shell of tin as a function of photoelectron kinetic energy from Ref. 3, B_1 and B_3 have been multiplied by ten for greater visibility.

CONCLUSION

Our work suggests that if one goes beyond nonrelativistic dipole approximation and includes relativistic and retardation effects and the first higher multipoles, (1) one finds some significant changes in predictions and (2) through the x-ray regime one will have an adequate account of multipole effects. By contrast, at higher energies too many multipoles and radial matrix elements begin to contribute and such expansions no longer give an efficient characterization of the independent physical features which determine the processes; alternative characterization schemes are needed.

REFERENCES

1. Young Soon Kim, R. H. Pratt, and Akiva Ron, Phys. Rev. A 24, 1889 (1981).
2. Young Soon Kim, Akiva Ron, R. H. Pratt, B. R. Tambe and Steven T. Manson, Phys. Rev. Lett. 46, 1326 (1981); also to be published.
3. M. S. Wang, Young Soon Kim, R. H. Pratt, and Akiva Ron, Phys. Rev. A 25, 857 (1982).
4. Larry Spruch, Phys. Rev. Lett. 16, 1137 (1966).
5. Young Soon Kim, R. H. Pratt, Akiva Ron, and H. K. Tseng, Phys. Rev. A 22, 567 (1980).
6. R. H. Pratt, Richard D. Levee, Robert L. Pexton, and Walter Aron, Phys. Rev. 134, A916 (1964); R. D. Schmickley and R. H. Pratt, Phys. Rev. 164, 104 (1967).
7. Sung Dahm Oh and R. H. Pratt, Phys. Rev. 10, 1198 (1974).
8. S. Blakeway, K. Schreckenbach, H. Faust, and W. Gelletly, in Inner Shell and X-Ray Physics of Atoms and Solids, edited by Derek Fabian, Hans Kleinpoppen, and Lewis M. Watson, Plenum Press (New York and London 1981), p. 93.
9. Oscar V. Gabriel, Ph.D. thesis, University of Pittsburgh (1981).
10. H. K. Tseng, R. H. Pratt, Simon Yu, and Akiva Ron, Phys. Rev. A 17, 1061 (1978).
11. Sung Dahm Oh, James McEnnan, and R. H. Pratt, Phys. Rev. A 14, 1428 (1976).
12. Akiva Ron, Young Soon Kim, and R. H. Pratt, Phys. Rev. A 24, 1260 (1981).

PHOTON-ATOM ELASTIC SCATTERING

M. Gavrila

FOM-Institute for Atomic and Molecular Physics,

Kruislaan 407, 1098 SJ Amsterdam, The Netherlands

This review describes the status of elastic scattering of X-rays and γ-rays (up to about 1 MeV)
from atoms and molecules, free or bonded in solids.
At the energies considered Rayleigh scattering
from the atomic electrons entirely dominates the
process. We first present the basic matrix elements at low and high energies and the results obtained in their calculation. Then we review the
scattering experiments carried out in this energy
range.

I. INTRODUCTION

The object of this review is to present the theoretical and experimental status of the elastic scattering of photons by atoms and molecules.
For *free atoms* the process can be represented by

$$A_i + \gamma_i \rightarrow A_f + \gamma_f \ ,$$

where A and γ stand for the atom and the photon, and subscripts i and f
refer to the initial and final states. During the collision the particles will change of energy and momentum within limits imposed by the
conservation laws. The case of *elastic scattering* is the one in which
the *internal* energy of the atom does not change in the process.

New physics appears when dealing with the scattering from macroscopic
assemblages of atoms or molecules, such as gases, liquids or solids,
because the photons interact in principle with the system as a whole.
However, each atomic constituent acts as a scattering center and the
instantaneous total scattering amplitude can be expressed coherently
in terms of the amplitudes for the individual atoms (to the extent that
these remain unperturbed by their environment).

The coherence of the atomic or molecular contributions to the total-scattering depends critically on the degree of correlation of the atomic and molecular motions. In gases where the correlation is weak, diffraction phenomena are practically absent, whereas in liquids where correlation is important they become observable. They culminate in densly packed and ordered solid structures like crystals. Moreover, in this case, for a more accurate description, one has to allow for the distortion the atom (molecule, ion) undergoes when placed in the crystal.

Subsequently, we shall be interested in the scattering process on the atomic level, i.e. in the way in which the characteristics of free or bonded atoms (molecules) show up in the scattering.

The atom is a composite system and all its constituents, nucleus and bound electrons, contribute to the scattering of the photon. The atomic scattering matrix element M will contain a contribution from the nucleus, denoted by M_N and from the electrons, denoted by M_R:

$$M = M_R + M_N \qquad (1)$$

The bound-electron contribution will be referred to as *atomic Rayleigh scattering*. Moreover, there are several mechanisms contributing to the nuclear scattering matrix element, M_N: Thomson scattering from the nucleus viewed as a point charge, M_{NT}; nuclear resonance scattering from individual levels or from the giant dipole resonance, M_{NR}; and Delbrück scattering, which is the scattering of the photon by the electrostatic field of the nucleus, M_{ND}. Thus

$$M_N = M_{NT} + M_{NR} + M_{ND} \qquad (2)$$

The relative importance of Rayleigh and nuclear elastic scattering depends critically on the energy of the photon. For energies up to about 1 MeV Rayleigh scattering entirely dominates, and it is only at the upper end of this range and under special conditions that the nuclear contribution starts showing up (high Z elements, large scattering angles). Above 1 MeV nuclear scattering becomes increasingly important and eventually obscures the Rayleigh component, except at very small forward angles. Above 10 MeV, Rayleigh scattering can be entirely neglected under the present experimental conditions.

In the following we shall cover the elastic scattering of X-rays (energies below 100 keV) and of the softer γ-rays (energies below 1 MeV, the threshold of pair production). Under these circumstances the scattering is entirely dominated by the Rayleigh component and we shall not discuss nuclear scattering below, except briefly in § 4. For the theoretical aspects of the latter we refer to the review articles of Hayward (Ha; 1977) on nuclear resonance scattering, and of Papatzacos and Mork (PM; 1975) on Delbrück scattering. For the experimental side see Moreh (M; 1978) and Schumacher et al. (S; 1981).

In Sec. II we describe the theoretical status of the problem. Thus, in § 1, we recall some basic facts concerning the matrix elements of Rayleigh scattering, and then, in § 2, we review the results obtained in their calculation. Next, in Sec. III we consider the experimental situation: in § 3 we describe the experiments with X-rays and in § 4 those with γ-rays. The references given at the end have been grouped into: general and tabulations (designed by capital letters) theoretical, and experimental (X-rays and γ-rays).

II. THEORY

1. Rayleigh scattering matrix elements

The wave-vectors of the photons in the initial and final states will be denoted by $\vec{k}_\alpha$, and the corresponding (possibly complex) polarization vectors by $\vec{e}_\alpha$ ($\alpha = i, f$); $\vec{e}_\alpha \vec{k}_\alpha = 0$. For elastic scattering $k_i = k_f = k = \omega/c$. Further, let $\vec{\nu}_\alpha$ be the unit vector of $\vec{k}_\alpha$, and $\vec{\Delta} = \vec{k}_i - \vec{k}_f$, the photon momentum transfer. The photon scattering angle will be denoted by θ (and *not* by 2θ).

For a rotation invariant *free atom* (complete electronic n,l subshells) the Rayleigh matrix element necessarily has the form

$$M = A(\vec{e}_i \vec{e}_f^*) + B(\vec{e}_i \vec{\nu}_f)(\vec{e}_f^* \vec{\nu}_i), \tag{3}$$

where the amplitudes A and B depend on the frequency ω and on θ (or Δ). (Note that this form also applies to the nuclear scattering matrix element Eq. (2) and, therefore, also to Eq. (1)).

In the case of linear polarizations, by choosing $\vec{e}_i$ and $\vec{e}_f$ perpendicular to the scattering plane $\vec{k}_i$, $\vec{k}_f$, Eq. (3) yields the matrix element:

$$M_\perp = A \; . \tag{4}$$

For $\vec{e}_i$ and $\vec{e}_f$ parallel to the scattering plane $\vec{k}_i$, $\vec{k}_f$, one finds:

$$M_\parallel = A \cos\theta - B \sin^2\theta \; . \tag{5}$$

The atomic differential scattering cross section, summed up over the final photon polarizations and averaged over the initial ones, can be expressed simply as:

$$\frac{d\sigma}{d\Omega} = \frac{r_o^2}{2} \left(|M_\perp|^2 + |M_\parallel|^2 \right), \tag{6}$$

where $r_o = e^2/mc^2$ is the classical electron radius. Thus, $M_\perp$ and $M_\parallel$ entirely characterize the scattering; all recent computations have been made in terms of these quantities.

We shall now present and discuss the explicit forms of the matrix elements M for the process.

We first consider *nonrelativistic theory*, and denote by H_{NR} the Schrödinger Hamiltonian of the N-electron atom. (Because the recoil energy of the atom is quite small in the process, the nucleus can be in most cases considered to be at rest in the origin.) When coupling the atom to a radiation field of vector potential $\vec{A} = \vec{A}(\vec{r},t)$, by assuming that div $\vec{A} = 0$ and that the scalar potential vanishes, the perturbed Hamiltonian becomes:

$$\mathcal{H}_{NR} = H_{NR} - \frac{e}{mc} \sum_\sigma \vec{A}_\sigma \vec{P}_\sigma + \frac{e^2}{2mc^2} \sum_\sigma \vec{A}_\sigma^2 \; , \tag{7}$$

where $\vec{A}_\sigma = \vec{A}(\vec{r}_\sigma,t)$, $\vec{P}_\sigma = -ih\,\nabla_\sigma$, and the sums run over all atomic electrons ($\sigma = 1, \ldots N$).

The matrix element for Rayleigh scattering from an atomic state i is obtained by applying second order perturbation theory to the interaction Hamiltonian of Eq. (7). The Kramers-Heisenberg-Waller matrix element thus derived, can be written as (AB § 35):

$$M = (\vec{e}_i . \vec{e}_f^*) \; f_o(\Delta)$$

$$+ \frac{1}{m} <i| \; \sum_\sigma \vec{e}_f^* \; \vec{P}_\sigma e^{-i\vec{k}_f . \vec{r}_\sigma} \; \frac{1}{\Omega_1 - H_{NR}} \; \sum_\tau \vec{e}_i . \vec{P}_\tau \; e^{i\vec{k}_i . \vec{r}_\tau} \; |i>$$

$$+ \frac{1}{m} <i| \; \sum_\tau \vec{e}_i . \vec{P}_\tau \; e^{i\vec{k}_i . \vec{r}_\tau} \; \frac{1}{\Omega_2 - H_{NR}} \; \sum_\sigma \vec{e}_f^* . \vec{P}_\sigma \; e^{-i\vec{k}_f . \vec{r}_\sigma} \; |i> \; . \qquad (8)$$

Here $|i>$ denotes the atomic eigenstate, $H_{NR} \; |i> = E_i \; |i>$, and $(\Omega - H_{NR})^{-1}$ is the Green's operator associated to the Hamiltonian H_{NR}, the two values of Ω been given by

$$\Omega_1 = E_i + \hbar\omega + i\varepsilon$$
$$\Omega_2 = E_i - \hbar\omega - i\varepsilon, \qquad (9)$$

where ε is an infinitesimal positive quantity. $f_o(\Delta)$ is the "atomic form-factor":

$$f_o(\Delta) = <i| \; \sum_\sigma e^{i\vec{\Delta}.\vec{r}_\sigma} |i> \; . \qquad (10)$$

For spherically symmetric states this can be expressed in terms of the radial charge density of the atom, $\rho(r)$:

$$f_o(\Delta) = \int \rho(r) \; \frac{\sin \Delta r}{\Delta r} \; r^2 dr \; . \qquad (11)$$

Note that f_o does not depend on ω, and that in forward scattering each electron contributes one unit to f_o, since $f_o(0) = N$. The frequency dependence of M ("dispersion") is contained in the second and third terms of Eq. (8), the "dispersive terms".

The form-factor $f_o(\Delta)$ is obtained by treating the $\vec{A}^2$ terms in Eq. (7) to first order in perturbation theory, whereas the other two terms in Eq. (8) result from a second-order treatment of the $\vec{A}\vec{P}$ terms. A Feynman-type diagramatic representation of the three terms is given in Fig. 1.

When the energy of the photon $\hbar\omega$ is close to that of the excitation energy of a discrete (metastable) atomic state, the matrix element

362

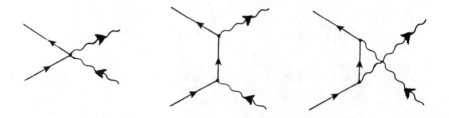

Fig. 1. Diagramatic representation of the three terms of the nonrelativistic Rayleigh matrix element Eq. (8) or (12). The full continuous line describes the atom propagating in the process, the
wavy lines the initial and final photons. The first graph represents the form-factor, the other two the second term (photon
absorption followed by emission) and third term (emission followed by absorption) of Eq. (8), respectively.

Eq. (8) has a resonant behaviour due to its second and third terms.
This is the phenomenon of *anomalous scattering*. However, Eq. (8) contains only the contribution to resonances from the configuration interaction of the metastable states with background ionization continua
(autoionization), but not that due to radiation damping. To include
the latter, general unified treatments for the radiative and nonradiative decay of metastable states are required (e.g. GW Chap. 8, Shore
1967, Åberg 1980). Thus the following matrix element can be derived:

$$
M = (\vec{e}_i \cdot \vec{e}_f^{\,*}) \, f_o(\Delta)
$$

$$
+ \frac{1}{m} \, \mathop{S}_{n} \, \frac{\langle f | \sum_\sigma \vec{e}_f^{\,*} \vec{P}_\sigma \, e^{-i\vec{k}_f \cdot \vec{r}_\sigma} | n \rangle \, \langle n | \sum_\tau \vec{e}_i \vec{P}_\tau \, e^{i\vec{k}_i \cdot \vec{r}_\tau} | i \rangle}{(\varepsilon_i + \hbar\omega) - \varepsilon_n}
$$

$$
+ \frac{1}{m} \, \mathop{S}_{n} \, \frac{\langle f | \sum_\tau \vec{e}_i \vec{P}_\tau \, e^{i\vec{k}_i \cdot \vec{r}_\tau} | n \rangle \, \langle n | \sum_\sigma \vec{e}_f^{\,*} \vec{P}_\sigma \, e^{-i\vec{k}_f \cdot \vec{r}_\sigma} | i \rangle}{(\varepsilon_i - \hbar\omega) - \varepsilon_n} \qquad . \qquad (12)
$$

Here the summation has to be carried out over a complete set of approximate eigenstates $|n>$ of H_{NR}, $H_{NR}|n> \simeq E_n|n>$, having stable (square integrable) excited states. We have denoted $\varepsilon_n = E_n - i\Gamma_n/2$, where Γ_n is the total width of the state $|n>$, radiative plus non-radiative (Auger-type). When the radiative widths are neglected in Eq. (12), this becomes essentially equivalent to Eq. (8). The best calculations have used as intermediate states $|n>$, states of the central self-consistent field approximation (antisymmetrized one-electron orbitals). However, low-energy calculations have been made also with electron-correlated wave functions.

It is customary to write the matrix element $M^{(o)}(\omega)$ for forward scattering with no change of polarization ($\vec{k}_i = \vec{k}_f$, $\vec{e}_i = \vec{e}_f$) as

$$M^{(o)}(\omega) \equiv f(\omega) = N + f'(\omega) + if''(\omega), \tag{13}$$

where $f'(\omega)$ and $f''(\omega)$ are the real and imaginary parts of the "dispersive terms" in Eq. (12), i.e. of the sums over intermediate states. It can be shown that in order to be consistent with the nonrelativistic framework of the theory adopted, $f'(\omega)$ and $f''(\omega)$ should be evaluated in the dipole approximation ($\vec{k}_i$ and $\vec{k}_f$ set equal to zero in Eqs. (8) and (12)).

By taking into account the analyticity properties of the matrix element $M^{(o)}(\omega)$ with respect to ω considered as a complex variable, dispersion relations of the Kramers-Kronig type can be derived (e.g. see G.W., Chap.10). Thus, for example

$$f' = \frac{2}{\pi} P \int_0^\infty \frac{\omega' f''(\omega')}{\omega'^2 - \omega^2} d\omega' . \tag{14}$$

Further the "optical theorem", written for vanishing Γ_n, gives:

$$\text{Im}M^{(o)}(\omega) = f''(\omega)$$

$$= -\frac{\pi}{2} \sum_n \omega_{ni} f_{ni} \delta(\omega_{ni} - \omega) - \frac{\omega}{4\pi r_o c} \sigma_{ph}(\omega) , \tag{15}$$

Here f_{ni} is the retarded oscillator strength for the transition $i \rightarrow n$ ($\hbar\omega_{ni} \equiv E_n - E_i$) and $\sigma_{ph}(\omega)$ is the atomic photoeffect cross section. (In the independent electron model, for each electron α the corresponding

$\sigma_{ph}^{(\alpha)}(\omega)$ is zero below its ionization threshold.) Equations (15) and (14) show that a knowledge of the f_{ni} and of the photoeffect cross section allows the determination of both anomalous terms $f''(\omega)$, and $f'(\omega)$. On the other hand, when these have been determined independently, the dispersion relation Eq. (14) allows a consistency check.

For photon energies in the hundreds of keV range ($\hbar\omega/mc^2 \gtrsim 1$) and high Z atoms (when $(\alpha Z)^2$ is no longer negligible with respect to 1) a *relativistic theory* is needed for the calculation of the matrix element. An approximate relativistic Hamiltonian for the N-electron atom H_R can be obtained by summing the individual one-electron Dirac Hamiltonians and including the electron-electron interactions in the form of the Coulomb repulsions plus corrective terms of the Breit-type. This Hamiltonian has been treated by Hartree-Fock procedures, but in photon scattering problems only the central field Hartree-Fock-Slater approximation was used. (The limitations of these approaches were discussed by Sucher, 1980).

By coupling H_R to the radiation field one gets the perturbed Hamiltonian (some higher-order corrections have been dropped):

$$\mathcal{H}_R = H_R - e \sum_\sigma \vec{\alpha}_\sigma \vec{A}_\sigma. \tag{16}$$

The relativistic matrix element for Rayleigh scattering can be obtained by treating the interaction term in eq. (16) to second order in perturbation theory (Waller), and can be written as:

$$M = m \langle i | \sum_\sigma \vec{e}_f^* \vec{\alpha}_\sigma \, e^{-i\vec{k}_f \vec{r}_\sigma} \frac{1}{\Omega_1 - H_R} \sum_\tau \vec{e}_i \vec{\alpha}_\tau \, e^{i\vec{k}_i \vec{r}_\tau} | i \rangle$$
$$+ m \langle i | \sum_\tau \vec{e}_i \vec{\alpha}_\tau \, e^{i\vec{k}_i \vec{r}_\tau} \frac{1}{\Omega_2 - H_R} \sum_\sigma \vec{e}_f^* \vec{\alpha}_\sigma \, e^{-i\vec{k}_f \vec{r}_\sigma} | i \rangle, \tag{17}$$

where Ω_1 and Ω_2 are again given by Eq. (9). Similarly to the nonrelativistic case this can be expressed as a sum over intermediate states, with decay widths included:

$$M = m \, \underset{n}{S} \, \frac{\langle i | \sum_\sigma \vec{e}_f^{*} \vec{\alpha}_\sigma \, e^{-i\vec{k}_f \vec{r}_\sigma} \, | n \rangle \, \langle n | \sum_\tau \vec{e}_i \vec{\alpha}_\tau \, e^{i\vec{k}_i \vec{r}_\tau} | i \rangle}{(\varepsilon_i + \hbar\omega) - \varepsilon_n}$$

$$+ \, m \, \underset{n}{S} \, \frac{\langle i | \sum_\tau \vec{e}_i \vec{\alpha}_\tau \, e^{i\vec{k}_i \vec{r}_\tau} \, | n \rangle \, \langle n | \sum_\sigma \vec{e}_f^{*} \vec{\alpha}_\sigma \, e^{-i\vec{k}_f \vec{r}_\sigma} | i \rangle}{(\varepsilon_i - \hbar\omega) - \varepsilon_n} \, . \quad (18)$$

The sums should be extended over all positive and negative energy states of the individual electrons. (This rises no difficulties in the independent electron approximation; for the more general Hartree-Fock case see Sucher (1980).) It can be shown that in the nonrelativistic limit the sums over the negative energy states in Eq. (18) yield the form-factor term in Eq. (12), while the sums over positive energy states yield the dispersive terms of Eq. (12), see AB § 35.

From the matrix element M, Eq. (18), one can derive via a relativistic dispersion relation the equation

$$M^{(o)}(\infty) = \sum_n f_{ni} + \frac{1}{2\pi^2 r_o c} \int_0^\infty [\sigma_{ph}(\omega') - \sigma_{pp}(\omega')] \, d\omega' \, , \quad (19)$$

where $M^{(o)}(\infty)$ represents the value for $\omega \to \infty$ of the forward scattering matrix element with no change of polarization, f_{ni} is the relativistic oscillator strength for the transition $i \to n$, $\sigma_{ph}(\omega)$ is the atomic photoeffect cross section, and $\sigma_{pp}(\omega)$ is the cross section for pair-production with the electron created in any *hole* one may make in the atom. (This is not to be confused with the usual pair production cross section in the field of the atom, describing an electron created in a *continuum* state.) In the independent electron model, for each hole α, $\sigma_{ph}^{(\alpha)}(\omega)$ is zero below the corresponding pair production threshold $mc^2 + E_\alpha$, where E_α is the relativistic binding energy. Eq. (19) is the relativistic generalization of the Thomas-Reiche-Kuhn sum rule (Gell-Mann et al., 1954; Levinger and Rustgi, 1956). Note that in the nonrelativistic version of Eq. (22) the sum represents the number of atomic electrons N, whereas in the relativistic case it represents the matrix element for forward scattering at infinite energy (only approximately equal to N, see § 2).

The discussion above refers to the scattering amplitudes of free atoms or molecules. We now briefly consider the case of the scattering from *crystal structures*, which gives rise to the well known diffraction patterns. In principle the general formula Eq. (12) applies also to the (instantaneous) scattering from a crystal, with the appropriate reinterpretation of the symbols. However, approximations are needed in order to make such a formula tractable.

If one assumes that there is little distortion of the individual atomic scatterers when they are incorporated in the crystal, the amplitude obtained for the Bragg peak associated with the reciprocal lattice vector $\vec{K}$ is proportional to the "structure factor" (e.g. J;W):

$$F(\vec{K},\omega) = \sum_j f_j(\vec{K},\omega)\, e^{i\vec{K}\vec{d}_j} . \tag{20}$$

Here $\vec{d}_j$ is the position of atom (ion) j in the unit cell (j = 1, ...n), and $f_j(\Delta,\omega)$ denotes the atomic matrix element Eq. (12). Eq. (20) applies to the case of a static lattice; to allow approximately for the thermal vibrations, it may be multiplied by the Debye-Waller factor e^{-w}, where

$$w = 8\pi^2\, \overline{u_\perp^2}\, (\sin^2 \frac{\theta}{2})/\lambda^2 ,$$

and $\overline{u_\perp^2}$ is the mean square displacement normal to the reflecting plane of the atoms from their mean position.

A more accurate approach for calculating the structure factor was developed by Dawson (1967,1969) by taking into account the distortions undergone by the atomic charge distributions in a crystal (see also Kurkki-Suonio and Ruuskanen, 1971). These are determined by the existing symmetry and the lattice vibrations. The deviations from the spherical free-atom symmetry were analyzed in terms of Kubic Harmonics. Anomalous atomic scattering can also be incorporated in the theory.

At sufficiently high energies the scattering amplitude of the crystal, Eq. (12), can be approximated by the form-factor term Eq. (10), which is the Fourier transform of the charge density. This contains contributions from the tightly bound, localized atomic electrons and from the electron bands. The former can be handled as in the free-atom case, but massive numerical calculations are needed to obtain the charge density of the bands from first principles, and then the crystal form-factor. Now, in order to account for the lattice vibrations, a canonical ensemble average of $\left|F(\vec{K},\omega)\right|^2$ is still needed. An approximate procedure for this was developed by Born (1942) in which the total charge density is decomposed into a sum of atomic densities centered on the different nuclei, and each of these densities is assumed to follow the motion of the corresponding nucleus. The procedure can be applied also to the scattering from molecules in gases (see Stewart et al. 1975, and Bentley and Stewart, 1975).

Finally, we recall the connection of the index of refraction with the forward scattering factors $f_j(\omega)$, Eq. (13), of the atomic constituents j of a medium:

$$ n = 1 - \frac{\lambda^2}{2\pi}\, r_o\, \sum_j \rho_j f_j(\omega), \tag{21} $$

where ρ_j is the number of atoms of species j per unit volume.

2. Results

We shall consider first the case of the *nonrelativistic matrix element* Eqs. (8), (12). For atomic hydrogen (or a Coulomb atomic model) it is possible to evaluate the matrix element of Eq. (8) analytically exactly in terms of hypergeometric functions which can be easily computed (K-shell: Gavrila, 1967, dipole approximation; Gavrila and Costescu, 1970, Gavrila and Pratt, 1982, retardation included; L-shell of a Coulomb atomic model: Costescu, 1976). This case has served as a guideline and a check for subsequent relativistic calculations (e.g. Kissel et al., 1980).

For many-electron atoms it is still possible to calculate the atomic form-factor Eq. (11) rather accurately, and increasingly improved charge densities have been used to this end over the years (screened-hydrogenic, Thomas-Fermi, Hartree-Fock, configuration-interaction). References to these and an extensive tabulation of results have been given by Hubbell et al. 1975, and Hubbell and Øverbø (1979). (For a critical assessment of the approximation of M by the form-factor see Roy et al., 1982.) Form-factors for ions in crystals were calculated by Schmidt and Weiss (1979), and for atoms in excited states (for use in plasma diagnostics) by Kuplyauskis et al. (1976), and by Ionushauskas et al. (1979).

It is much harder to deal accurately with the dispersive terms of Eq. (12). For many-electron atoms the most accurate atomic model used was that of a central Hartree-Fock self-consistent field of the Slater type (i.e. with electron-exchange effects incorporated approximately into the potential), denoted in the following by HFS. In these circumstances the electron orbitals are orthogonal and the atomic matrix element M reduces to the coherent sum of the individual electronic contributions:

$$M \simeq \sum_{\sigma = 1}^{N} M_{\sigma\sigma} = M_K + M_L + \ldots, \tag{22}$$

where $M_{\sigma\sigma}$ is now a one-electron matrix element of the form of Eq. (12), and M_K, etc. represent the contributions of the occupied shells. For an atom in a solid, however, the outer shells may be strongly distorted by the environment and an atomic HFS calculation will not suffice.

Thus, in an early calculation Hönl (1933) (see also J, Chap. IV, §1) essentially expanded the dispersive terms in Legendre polynomials of the scattering angle θ, expressed their coefficients in terms of integrals over the multipole contributions to the retarded photoeffect cross section (only continuum contributions were retained), which were then calculated with screened hydrogenic wave functions. The method was applied to the case of forward scattering (refractive indices), see Wagenfeld (1975) and the references therein to tabulations made by this method.

An alternative, semiempirical method to deal with the calculation of the dispersive terms for forward scattering assumed simple inverse power-law expressions for the photoeffect cross section $\sigma_{ph}(\omega)$ (with some parameters fitted from experiment or computed), obtain f"(ω) from Eq. (15) (only the continuum contributions are retained), and then calculate f'(ω) from the dispersion relation Eq. (14). For calculations

and tabulations along this line see J,Chap.IV, § 1, Parratt and Hem-
stead (1954), Dauben and Templeton (1955), the International Tables
for Crystallography (ITC; 1962), Savaria and Caticha-Ellis (1966), Cromer
(1965); see also Wang (1982).

In order to improve on these calculations Cromer and Liberman
(1970a,b) analyzed the nonrelativistic limit of the relativistic ma-
trix element Eq. (18), by assuming $\hbar\omega \ll mc^2$ and $(\alpha Z)^2 \ll 1$, and obtained for $f'(\omega)$ the usual dispersion relation Eq. (15) (with discrete
contributions neglected) plus a corrective term of order E_o/mc^2, where
E_o is the total binding energy of the atom (of order $(\alpha Z)^2$). The main
progress, over earlier calculations, however, was that they used for
$\sigma_{ph}(\omega)$ in Eq. (15) values computed from relativistic HFS-type wave
functions. The values tabulated for $f'(\omega)$, $f''(\omega)$ have been widely
used. Some numerical inaccuracies in the vicinity of absorption
thresholds were corrected in a later paper (Cromer and Liberman, 1981).
Jensen (1980) pointed out, however, that these authors had omitted in
their calculation corrective terms to the nonrelativistic dispersion
relation. By estimating them he concluded that values for $f'(\omega)$ tabulated by Cromer and Liberman could not claim an accuracy better than
of order E_o/mc^2.

By ignoring in Eq.(12) the damping effects, and the excited states
of free atoms or the valence bands, etc., of atoms embedded in
solids,none of the previous calculations of the anomalous terms is valid in the vicinity of the absorption edges, which is a case of considerable experimental interest. A discussion of dispersion near the
K-shell threshold of free (noble gas) atoms was given by Bremer (1979),
who evaluated Eq. (12) with damping included for an essentially
screened hydrogenic model.

The structure factors for *crystals* have been calculated in the
form-factor approximation. Form-factors have been derived either from
models (e.g. Dawson 1967, Dawson and Willis 1967, Ascarelli and Raccah
1970, Gray 1972) or from ab initio calculations. The latter are extremely tedious and time consuming, and quite a number of approximations are needed. Most band structure calculations contain also evalu-
ations of structure factors because of the relatively simple Fourier-

relationship Eq. (10). We shall not attempt to list them all here, but will merely quote those related to the experiments mentioned in § 3. Thus, the linear combination of atomic orbitals (LCAO) method was used by Kahane et al. (1973), and by Grosso and Parravicini (1978) to describe LiH; by Wang and Callaway (1977) for the case of ferromagnetic Ni, and by Laurent et al. (1978) for the case of V. The self-consistent augmented plane wave method (SCAPW) was used by Snow (1967) to calculate the band structure of Al. The alternative selfconsistent orthogonalized plane wave (SCOPW) approach was applied by Stukel and Euwema (1970), and by Raccah et al. (1970) to the study of C, Si, Ge, etc. (see also the references therein). Wakoh and Yamashita (1971) have calculated the structure factors of 3d transition metals (V, Cr, Fe, Ni, Cu) using the Green's function method with a selfconsistent potential. Full Hartree-Fock calculations have been carried out for C (diamond) by Euwema et al. (1973), and for crystalline Ne and LiF by Euwema et al. (1974).

Major progress has been achieved recently in the calculation of the atomic *relativistic Rayleigh matrix element* (Eq. 17). This development originates in the pioneering work of G.E. Brown and collaborators in the fifties (Brown et al., 1954; Brenner et al., 1954; Brown and Mayers, 1955, 1957; see also Cornille and Chapdelaine, 1959). The case of a K-shell electron bound by a Coulomb field was considered. The method used required tedious algebra and is based on multipole expansion of the photon wave functions $e^{i\vec{k}\vec{r}}$ appearing in Eq. (17), combined with a partial wave expansion of the Green's operators $(\Omega - H_R)^{-1}$ involved. The result for M is essentially an expansion in (associated) Legendre polynomials of the scattering angle θ. (The direct evaluation of radial Green's function was circumvented by integrating appropriate inhomogeneous equations.) The calculations were performed for Hg(Z = 80) at five photon energies ranging from 164 to 2620 keV and all scattering angles.

After a lapse of time of a decade, W.R. Johnson and collaborators reactivated the method by applying modern computational techniques. A relativistic HFS potential was used and the contributions of several atomic shells were included in M. The first calculations were performed for the noble gases He to Xe in the 1 to 10 eV range and M was expressed in terms of dynamical electric and magnetic susceptibilities (Johnson and Feiock, 1968). Afterwards, Lin et al. (1975) studied the electron-correlation corrections to the HFS one electron model. In this work they went beyond the second order matrix element Eq. (17) and calculated fourth-order S-matrix terms which include one-virtual photon interactions between electrons. The case they considered was that of He where correlation effects should be largest. The result has shown that, whereas for low energies up to several times the ionization potential I, electron correlation yields a sizable correction to the cross section (more than 10%), and the form-factor is a bad approximation, for energies of tens of times I the effect of electron-correlation entirely disappears and the form-factor becomes a very good approximation at all angles. In a subsequent publication Johnson and Cheng (1976) extended their calculation to high Z atoms ($30 \leq Z \leq 82$) and energies from 100 to 900 keV. The 5% to 20% disagreement they found with experiment was attributed to the neglect of some outer shell contributions.

An extensive computational effort was undertaken by R.H. Pratt and collaborators with the objective of obtaining total free atom Rayleigh amplitudes at the 1% level of accuracy or better, for energies ranging from 0.1 keV to 10 MeV. This is based on a code developed by L. Kissel (1976) from the earlier work of W. Johnson and collaborators. These "state of the art" computations are lengthy and require several hours on fast computers for a high-Z, and high-energy angular distribution. Presently, a substantial body of data is available for the amplitudes, the differential and total cross sections, for a sequence of energies ranging from 59 keV to 1.33 MeV for Zn, Sn, Pb and V, and data for Pb and U at higher energies, data on Al down to 0.1 keV, data on various ions of C, etc. Many of these results were presented and discussed in detail by Kissel et al. (1980), see also Pratt (1982), Roy and Pratt (1982), and Roy et al. (1982). Besides, these results yield a good understanding of the limitations of simpler theories, such as the non-relativistic approximation (in the dipole approximation or with retardation included), various type of form factor approximations, high-energy limit calculations, etc.

The behaviour of the Rayleigh matrix element at high energies (which is inaccessible to numerical partial-wave methods) was studied by Goldberger and Low (1968). They derived a formula for the high-energy limit of M at finite momentum transfers, i.e. $mc^2/\hbar\omega \ll 1$ and $c\Delta \ll \hbar\omega$, to be denoted here by $\bar{M}(\Delta,Z)$. (This formula had been actually obtained much earlier by Franz (1936), using less convincing mathematics, however.) Florescu and Gavrila (1976) computed $\bar{M}(\Delta,Z)$ for the K-shell of a hydrogen-like atomic model, for any Z and Δ. Further, a long standing theoretical debate was settled as to the value of $M^{(o)}(\infty)$ to be entered in the generalized Thomas-Reiche-Kuhn sum rule Eq. (19), $\bar{M}(\Delta = 0,Z)$ in our present notation. Contrary to earlier beliefs (Gell-Mann et al., 1954) this is not equal to N but is Z-dependent.

Finally, let us mention the fact that the recent progress achieved in the quantitative description of Rayleigh scattering has made it the best understood of the coherent processes contributing to the elastic scattering of γ-rays by atoms (see Sec.I). At the present time the least understood of these is Delbrück scattering, for which the Coulomb corrections to the lowest order matrix element represent the main source of uncertainty.

III. EXPERIMENTS

3. X-rays

3.1. *The angular distribution of the scattered intensity*

The angular distribution of the atomic scattered intensity is given by Eq. (6). Two amplitudes enter this equation, $M_\perp$ and $M_\parallel$ or, alternatively, A and B, see Eqs. (4), (5). However, for X-ray energies the amplitude B is quite small and can be usually neglected. Thus:

$$M_\perp \simeq A, \quad M_\parallel \simeq A \cos\theta . \tag{23}$$

The amplitude A can be decomposed further as

$$A(\omega,\Delta) \equiv f(\omega,\Delta) = f_o(\Delta) + f'(\omega,\Delta) + i \, f''(\omega,\Delta), \tag{24}$$

where $f_o(\Delta)$ is the atomic form-factor Eq. (11), and $f'(\omega,\Delta)$, $f''(\omega,\Delta)$ are real. (This parallels the decomposition of the *exact* matrix element $M^{(o)}$ for *forward scattering* given in Eq. (13)). Consequently, the angular distribution of the scattered intensity becomes

$$\frac{d\sigma}{d\Omega} = r_o^2 \frac{1}{2} (1 + \cos^2\theta) |f|^2, \tag{25}$$

where

$$|f(\omega,\Delta)|^2 = (f_o(\Delta) + f'(\omega,\Delta))^2 + f''^2(\omega,\Delta). \tag{26}$$

The object of earlier experiments was to obtain information on the atomic charge distribution, via the atomic form-factor, see Eq. (11). The scattered intensity was interpreted in terms of Eq. (26), and an "experimental" $f_o(\Delta)$ was derived by making an assumption on the angular dependence of $f'(\omega,\Delta)$, $f''(\omega,\Delta)$ (these had been calculated only for $\Delta = 0$, e.g. Cromer and Liberman, 1970), or neglect them altogether. For example, at low energies they were taken to be angle independent (which was approximately confirmed by subsequent calculations), but at higher energies the contribution of each shell to f', f'' was sometimes incorrectly taken to have the same angular dependence as its contribution to f_o. Finally, the "experimental" values for $f_o(\Delta)$ were compared to the numerous calculations available, thus trying to decide on the accuracy of the various atomic wave functions they were based on.

At higher energies, $B(\omega,\Delta)$ becomes important and the approximation Eq. (23) breaks down. Consequently, the angular distributions must be evaluated from the full Eq. (6) as was done in the computations of Kissel et al. (1980), see § 2.

Although the experimental study of atomic scattering factors was started in the thirties (e.g. Wollan, 1931), satisfactory accuracy has been achieved only in the last twenty years or so. Thus, one of the most accurate angular distribution measurements (to better than 0.5%) for the noble gases Ne to Xe was that of Chipman and Jennings in 1963, with MoK radiation (17.5 KeV). The small discrepancies with the then existing theory can be attributed to the uncertainties of the latter.

A recent experiment on He and H_2 was carried out by Ice et al.

(1978) using synchrotron radiation from the storage ring SPEAR at the Stanford Synchrotron Radiation Laboratory (SSRL). The energies ranged from 5 to 12 KeV, and the angles from 60° to 135° (large momentum transfers). Under these conditions the scattering can be interpreted solely in terms of the form-factor. The accuracy of the experiment was high enough to permit in both cases to favour the electron-correlated calculations of the form-factor as opposed to the Hartree-Fock ones.

Numerous experiments were carried out on various elements (C, Al, Si, V, Cr, Fe, Cu, Ge, W, etc.) and compounds by X-ray diffraction analysis with characteristic lines up to about 20 keV, see for example: Calder et al. (1962), Merisalo and Inkinen (1966), Paakkari and Suortti (1968), Raccah and Henrich (1969), Miller and Black (1970), Linkoaho (1972), Diana and Mazzone (1972), Temkin et al. (1972), Aldred and Hart (1973), Matsushita and Kohra (1974), Bilderback and Colella (1975,1976), Rantavuori and Tanninen (1977), Bonse and Terworte (1980), Takama and Sato (1981), Ohba et al. (1982), and the references therein (earlier experiments have been described in detail by Weiss, see We). The "experimental" $f_{o}(\Delta)$ derived differed from the best theoretical calculations available for free atoms and this was understood as due to the distortion of the atomic charge density by the solid state environment. More recently the interpretation of the results was made in terms of elaborate theories of crystal structure, such as the ones mentioned in § 2. The agreement with theory varies between good and fair. Thus, for example, the band calculations of Snow (1967) did not compare well with the results of Raccah and Henrich (1969), whereas the simple model used by Ascarelli and Raccah (1970) did. The structure factor measurements for V by Korhonen et al. (1971), see also Linkoaho (1972), are in rather good agreement with the band structure calculations of Wakoh and Yamashita (1971), but not so with those of Laurent et al. (1978). For Cu, Temkin et al. (1973) found good agreement with the calculations of Wakoh and Yamashita (1971), and Snow (see W 1967). The results for Si by Aldred and Hart (1973), and for α - Sn and InSb by Bilderback and Colella (1975,1976) could well be interpreted in the framework of the theory of Dawson (1967).

At higher energies Tirsell et al. (1975) have carried out an experiment with X-ray fluorescence lines in the 25 to 75 keV range on high Z metals, and angles from 45° to 135°. The agreement with the calculations of Kissel et al. (1980) was in general satisfactory.

3.2. *Studies of the anomalous dispersive corrections f' and/or f''*

A variety of methods have been used for the determination of the dispersion corrections f' and f". Most of the experiments (but not all) have been carried out in the vicinity of atomic absorption edges (K, L). In this case the energy variation of f' and f" can be quite sharp and show considerable structure due to solid state effects. Anomalous scattering is larger at the L edges than at the K edge. Thus, at the L edges values of $|f'|$ nearing 30 have been recorded, attaining about 50% of the form-factor value, $f_o = N$ (e.g. see Fuoss (1980), Table 1). This is clearly illustrated in Figs. 2 and 3 where we reproduce the variation of f' and f" in the vicinity of the K edge of Ni, as obtained by Bonse and Materlik (1976), and in the vicinity of the L edges of Cs, as obtained by Templeton et al. (1980). Note that the structure displayed by $f''(\omega)$ will have to show up (in a somewhat modified form) also in $f'(\omega)$, in view of the dispersion relation Eq. (14), and vice-versa.

The radiation sources used were either X-ray laboratory sources (characteristic lines, bremsstrahlung) or synchrotron radiation. The latter has the obvious advantage of continuous wavelengths, which is essential for studies near absorption edges, and of high intensity extending up to about 15 keV (gains in the range $10^3 - 10^6$ can be achieved at 1 Å over laboratory sources).

In the following we shall review the experiments grouped into three categories:

a. Direct measurements of f' by X-ray interferometry. The invention of the single-crystal Bragg reflection interferometer (Bonse and Hart, 1965) has allowed very accurate determinations of the refractive indices, Eq. (21), which, for X-rays, differ from 1 by less than 10^{-5}. These then give the forward scattering amplitude $\text{Re}f(\omega) = N + f'(\omega)$, see

Eq. (13). Typically, such measurements give $N + f'(\omega)$ with errors at the 0.2% level but the relative error on $f'(\omega)$ is markedly larger, because $f'(\omega)$ is usually small with respect to N. Nevertheless, in some cases $f'(\omega)$ could be derived to about 1% (Hart and Siddons, 1981). Reviews of the method have been given by Hart (1975,1980).

Some of the experiments were carried out with X-ray characteristic lines testing the dispersion far from edges (Bonse and Hellkötter, 1969; Creagh and Hart, 1970; Creagh, 1975, 1977). A number of others were made with bremsstrahlung radiation from laboratory sources. Anomalous dispersion near the K-edges of Zr, Nb and Mo were studied by Cusatis and Hart (1977), Hart and Siddons (1981), see also Hart (1980). Finally, synchrotron radiation (from the storage ring DORIS at the Deutches Elektronen-Synchrotron (DESY)) was used to study anomalous dispersion close to the K edges of Ni by Bonse and Materlik (1976), and Se by Bonse et al. (1980), revealing abundant structure in f', as shown in Fig. 2. The maxima and minima of f' on the short wavelength side of the edge correspond quite well to the ones observed in absorption by Cauchois and Manescu (1950), marked by vertical lines.

The merits of synchrotron radiation for X-ray interferometry were discussed by Hart (1980) and by Hart et al. (1980).

b. *Combined measurements of f'and f"* have been carried out in various ways. Most of the time the dispersion relation Eq. (14) was used as a consistency check on the results.

Fukamachi and Hosoya (1975), and Fukamachi et al. (1979) combined measurements of the intensity ratio between Friedel-pair reflextions in a polar crystal of known simple structure (GaP, GaAs) with measurements of the linear absorption coefficient to derive the variation of f' and f" at the K edges of the Ga and As atoms.

Templeton et al. (1980) determined f' and f" by a least squares adjustment from diffraction data measured with a crystal whose structure was known from work at other wavelengths. The method determines the values of f', f" at finite (nonzero) scattering angles (the angular dependence was not investigated, however). The experiment was carried out at the SPEAR storage ring of SSRL on Cs in the vicinity of the L

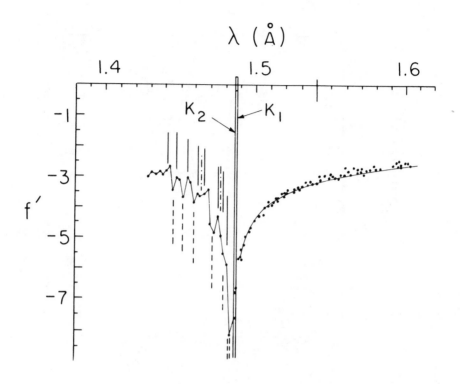

Fig. 2. Wavelength dependence of dispersion correction f' on either
side of the K-edge of nickel. K_1 onset, K_2 inflection point
of absorption edge. Full lines indicate positions of maxima,
dashed lines those of minima of the absorption spectrum given
by Cauchois and Manescu (1950). From U. Bonse and G. Materlik,
Z.Phys. B24, 189 (1976).

edges, with the result shown in Fig. 3 (for a theoretical discussion
see Wendin, 1980). The strong enhancements in f" appearing near the
edges are "white lines", familiar from absorption measurements (e.g.
Cauchois and Mott, 1949; Wey and Lytle, 1979, and the theoretical discussion by Brown et al., 1977).

Freund (1975) determined f' and f" from independent measurements
of the integrated intensity from a Cu crystal in reflection (Bragg
case), and from the absorption coefficient close to the K-edge. Finally, the total reflection approach has also been applied; for a description and earlier references see J, Chap. IV, § 2, for recent references see Fuoss (1981).

378

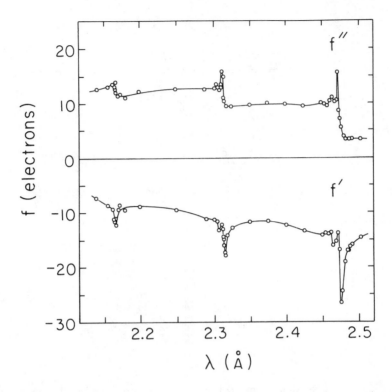

Fig. 3. Anomalous dispersion corrections f' and f" for cesium determined from diffraction data in the region of the L_1 absorption
edge (left), L_2 edge (center) and L_3, edge (right). From D.H.
Templeton et al., Acta Cryst. A<u>36</u>, 436 (1980).

    c. Direct measurements of f" have been made either from scattered
intensity or from absorption coefficients measurements. The first alternative was adopted by Engel and Sturm (1975) for heavier elements
(comparison of intensities of related reflections - Bijvoet pairs -
of an accentric crystal), and by Marezio et al. (1975) for light elements (determination of intensities from accentric inorganic compounds). A slight angular dependence was detected in f". (This is implicitly contained also in the earlier measurements of Borrmann, 1950.)

    Attenuation coefficients measurements in the vicinity of K edges
have been made by Fukamachi et al. (1977), Sasamachi et al. (1980),
Fuoss et al. (1980), and at other frequencies by Creagh (1977), Gerward et al. (1979) and by others. The value thus derived (combined
with supplementary information on f' at other frequencies) were used

to evaluate f' for forward scattering from the dispersion relation Eq. (14).

The most systematic and extensive effort to derive f' for forward scattering from *experimental* values of f" (total photo-ionization cross sections) was undertaken by Henke et al. (1982). These authors have made a "state of the art" compilation of photo-ionization cross sections of 94 elements in the energy range from 30 eV to 10 keV. The values of f' were then computed between 100 eV and 2 keV.

4. γ-rays

Numerous experimental studies on the angular distributions have been made over the years. Up to about 900 keV the nuclear scattering contribution cannot be detected with the present experimental means so that only Rayleigh scattering is tested. The experiments in this energy range are a natural extension of the ones in the X-ray regime considered in § 3.1. However, from about 900 keV upwards nuclear scattering can be detected under favorable experimental conditions and at still higher energies becomes dominant, obscuring the Rayleigh scattering at all angles (except $\theta = 0$).

Of all nuclear contributions Dellbrück scattering is the most spectacular, since it is a nonlinear effect of quantum-electrodynamics occurring via the virtual production and annihilation of electron-position pairs. Considerable effort has been invested in its search but it was only in the late sixties and early seventies that conclusive evidence for its existence was obtained by several groups of researchers (see PM; 1975). Also the experiments at the upper end of our energy range (i.e. around 1 MeV) were motivated by the desire to detect the weak Delbrück component.

The γ-ray scattering experiments were carried out with radioactive sources. There is a variety of them emitting at fixed frequencies, the energies of those which have been used being: 59.6; 84.3; 145; 245; 279; 317; 344; 444; 662; 779; 889; 964; 1086; 1112; 1121; 1173; 1274; 1333 MeV and higher. Typically, intensities ranged from tens to hundreds of millicuries.

A decisive role in the evolution of γ-ray experiments has been played by the detectors. In more recent years these used to be of the NaI (Tl) type, but by now they have been nearly entirely replaced by solid state Ge(Li) detectors. This has lead to a marked progress, since the resolution of the latter is considerably higher, e.g. 3 keV full width at half maximum for 1.3 MeV photons, as compared to the 100 keV of the former. (Their lower efficiency can be compensated by increasing their volume.) High resolution is especially required for small angle scattering in order to separate the elastic and inelastic (Compton) contributions.

A rather extensive survey of the experimental work on γ-ray elastic scattering was given by Roy (1980). We refer to this paper for more details and for earlier references. In the following we shall quote only the more recent experiments done with improved detection methods (mostly Ge(Li) detectors).

Usually experimentalists have adopted in their work one of the two alternatives, either (a) taking a fixed frequency source and scattering the photons on a series of targets, or (b) taking one element (sometimes several) and exposing it to various sources (or to different frequencies of the same source).

Experiments of kind (a) have been carried out at 59.5 keV by Schumacher and Stoffregen (1977); at 84.3 keV by Sen Gupta et al. (1979a, 1982); at 145 keV by Roy et al. (1975), by Prasad et al. (1978), by de Barros et al. (1981), and by Sen Gupta et al. (1982); at 279 keV by Smend et al. (1973); at 317 keV by de Barros et al. (1981); at 662 keV by Smend et al. (1973) and by Sen Gupta et al. (1979b).

Experiments of the kind (b) include those by Chittwattanagorn et al. (1980) on Pb, by Taylor et al. (1981) on W, and by Ramanathan et al. (1979) on Cu, Cd, Ta, Pb, all three using the multiple-frequency source ^{152}Eu; those by Schumacher (1969), by Schumacher et al. (1973), and by Sen Gupta (1979) on Pb; and those by Mückenheim and Schumacher (1980) on U.

More attention was focused on energies around 1 MeV because there Delbrück scattering starts showing up. Experiments at these energies (i.e. 0.889, 1.121, 1.173 and 1.333 MeV), besides the ones quoted

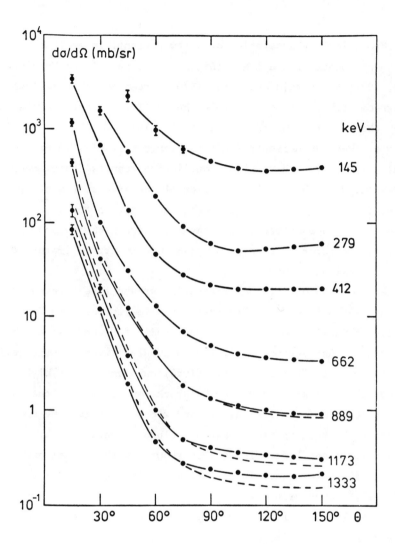

Fig. 4. Experimental differential cross section for uranium compared
with theoretical prediction at various energies. Full curves represent calculated cross sections, including Rayleigh, nuclear
Thomson and Delbrück contributions; broken curves, with Delbrück
omitted. From W. Mückenheim and M. Schumacher, J.Phys.G. <u>6</u>, 1237
(1980) and private communication.

above, were done by Dixon and Storey (1968), by Hardie et al. (1970,
1971), by Basavaraju and Kane (1970), and by Kane et al. (1978).

The agreement of recent selected experiments (Schumacher and Stoffregen, 1977; Mückenheim, 1981) with theory (Johnson and Cheng, 1975;
Kissel et al. 1980) is remarcably good, within the stated experimental

and theoretical uncertainties of a few percent. In another case, earlier experiments around 1 MeV (Dixon and Storey, 1968; Hardie et al., 1970, 1971; Basavaraju and Kane, 1970) agreed among themselves but disagreed with the then available theory. This discrepancy was cleared by the calculations of Kissel et al. (1980).

Some recent experiments (Chittwattanagorn et al., 1980; Ramanathan et al., 1979; etc.) have given conflicting results, both among themselves and with theory. A discussion of the situation was given by Roy et al. (1982). These authors conclude that, on the average, there is no clearly demonstrated disagreement between theory and experiment, and that, therefore, it would be premature to try to improve the numerical calculations.

The existence of Delbrück scattering around 1 MeV has now been definitely established. To illustrate this we reproduce in Fig. 4 the results of Mückenheim and Schumacher (1981) for the scattering of photons of various energies by U. The full curve gives the theoretical Rayleigh result of Kissel et al. (1980) supplemented at higher energies by the nuclear Thomson and Delbrück contributions, the latter being taken from calculations by Papatzacos and Mork. The broken lines represent the results *excluding* the Delbrück contribution. The need to include it is apparent. Fig. 4 also shows how the Rayleigh cross section peaks progressively in forward direction as the energy increases.

Finally, we mention that Bragg scattering of 412 keV radiation was studied by γ-ray diffractometry by Schneider (1976) on Cu, and by Schneider et al. (1978) on Au.

REFERENCES

General.Tabulations

AB   - Akhiezer A.I. and Berestetskii V.B.,Quantum Electrodynamics (Interscience Publishers, N.Y. 1965).

AS   - Anomalous Scattering, Eds. S. Ramaseshan and S.C. Abrahams (Munksgaard, Copenhagen, 1975).

E   - Egelstaff P.A., An Introduction to the Liquid State (Academic Press, London, 1967).

GW   - Goldberger M.L. and Watson K.L., Collision Theory (Wiley, N.Y. 1965).

Ha   - Hayward E., Photon Scattering in the Energy Range 5 - 30 MeV, in Photonuclear Reactions I (Lecture Notes in Physics, vol. 61 - Springer, N.Y., 1977) p. 340.

He   - Henke B., Lee P., Tanaka T., Shimambukuro R. and Fujikawa B., Low-energy X-ray Interaction Coefficients, Atom.Nucl.Data Tables 27, 1, 1982.

Hu   - Hubbell J.H., Veigele W.J., Briggs E.A., Brown R.T., Cromer D.T. and Howerton R.J., Atomic Form Factors, Incoherent Scattering Functions and Photon Cross Sections, J.Phys.Chem. Ref.Data 4, 471, 1975.

HØ   - Hubbell J.H. and Øverbø I., Relativistic Atomic Form Factors and Photon Coherent Cross Sections, J.Phys.Chem.Ref.Data 8, 69, 1979.

ITC   - International Tables for X-ray Crystalography (Kynoch Press, Birmingham, 1962), vol.III.

J   - James R.W., The Optical Principles of the Diffraction of X-rays (Bell, London, 1962).

K   - Kohler F, The Liquid State (Verlag Chemie, Weinheim, 1972).

M   - Moreh R., Studies of Fundamental Photon Scattering Processes Using n-capture γ-rays, Nucl.Instr.Meth. 166, 91, 1979.

PM   - Papatzacos P. and Mork K., Delbrück Scattering, Phys.Reports 21C, 81, 1975.

S   - Schumacher M., Smend F., Rullhusen P., Mückenheim W. and Börner H.G., Recent Developments in the Elastic Photon Scatter-

ing in the Energy Range 0.1 - 10 MeV, in <u>Proc.IV Int.Symp.</u>
<u>Neutron Capture γ-ray Spectroscopy</u>, Grenoble 1981 (Eds. von
Egidy, F.,Gönnenwein and B. Maier), p.598.

W   - Warren B.E., <u>X-ray Diffraction</u> (Addison Wesley, N.Y. 1969).

We   - Weiss R.J., <u>X-ray Determination of Electron Distributions</u>
       (North Holland, Amsterdam, 1966).

Theory

Åberg T., 1980, Phys.Scripta <u>21</u>, 495.

Bentley J. and Stewart R., 1975, J.Chem.Phys. <u>63</u>, 3794.

Born M., 1942, Rep.Progr.Phys. <u>9</u>, 294.

Bremer J., 1979, J.Phys. B<u>12</u>, 2797.

Brenner S., Brown G.E., and Woodward J.B., 1954, Proc.Roy.Soc.(Lond.)
       A <u>227</u>, 59.

Brown G.E., Peierls R.E., and Woodward J.B., 1954, Proc.Roy.Soc.(Lond.)
       A <u>227</u>, 51.

Brown G.E., and Mayers D.F., 1955, Proc.Roy.Soc.(Lond.) A <u>234</u>, 387.

Brown G.E., and Mayers D.F., 1957, Proc.Roy.Soc.(Lond.) A <u>242</u>, 89.

Brown M., Peierls R.E., and Stern E.A., 1977, Phys.Rev. B <u>15</u>, 738.

Cauchois Y., and Mott N.F., 1949, Philos.Mag. <u>40</u>, 1260.

Cornille H., and Chapdelaine M., 1959, Nuovo Cim., <u>14</u>, 1386.

Costescu A., 1976, Rev.Roumaine, Phys. <u>21</u>, 3.

Cromer D.T., 1965, Acta Cryst. <u>18</u>, 17.

Cromer D.T., and Liberman D., 1970a, J.Chem.Phys. <u>53</u>, 1891.

Cromer D.T., and Liberman D., 1970b, Los Alamos Scientific Report,
       LA-4403.

Cromer D.T., and Liberman D., 1981, Acta Cryst. A<u>37</u>, 267.

Dauben C.H., and Templeton D.H., 1955, Acta Cryst. <u>8</u>, 841.

Dawson B., 1967, Proc.Roy.Soc.(Lond.) A <u>298</u>, 255, 264, 379.

Dawson B., 1969, Acta Cryst. A <u>25</u>, 12.

Dawson B. and Willis B., 1967, Proc.Roy.Soc.(Lond.) A <u>298</u>, 307.

Euwema R.N., Wilhite D.L. and Surrat G.T., 1973, Phys.Rev. B <u>7</u>, 818.

Euwema R.N., Wepfer G.G., Surratt G.T. and Wilhite D.L., 1974, Phys.
       Rev. B <u>9</u>, 5249.

Florescu V. and Gavrila M., 1976, Phys.Rev. A <u>14</u>, 211.

Franz W., 1936, Zeits.Phys. <u>98</u>, 314.

Gavrila M., 1967, Phys.Rev. <u>163</u>, 147.

Gavrila M. and Costescu A., 1970, Phys.Rev. A <u>2</u>, 1752, and ibid. 1971, A <u>4</u>, 1688.

Gavrila M. and Pratt R.H., 1982, contributed paper to the X82 Conference.

Gell-Mann M., Goldberger M.L., and Thirring W., 1956, Phys.Rev. <u>95</u>, 1612.

Goldberger M.L. and Low F., 1968, Phys.Rev. <u>176</u>, 1778.

Gray M., 1972, Phys.Rev. B <u>5</u>, 253.

Grosso G. and Pastori Parravicini G., 1978, Phys.Rev. B <u>17</u>, 3421.

Hönl H., 1933, Ann.Phys.(Leipzig) <u>18</u>, 42.

Ionushauskas S.L., Kuplyauskene A.V. and Kuplyauskis Z.I., 1979, Opt. Spektrosk. <u>47</u>, 447 [Opt.Spectrosc. <u>47</u>, 248].

Jensen M.S., 1980, J.Phys. B <u>13</u>, 4337.

Johnson W.R. and Cheng K.T., 1976, Phys.Rev. A <u>13</u>, 692.

Johnson W.R. and Feiock F.D., 1968, Phys.Rev. <u>168</u>, 1.

Kahane S., Felsteiner J. and Opher R., 1973, Phys.Rev. B <u>8</u>, 4875.

Kissel L., Pratt R.H. and Roy S.C., 1980, Phys.Rev. A <u>22</u>, 1970.

Kuplyauskis Z.I. and Kuplyauskene A.V., 1976, Opt.Spektrosk. <u>41</u>, 677 [Opt.Spectrosc. <u>41</u>, 399].

Kurki-Suonio K. and Ruuskanen A., 1971, Ann.Acad.Sci.Fenn. A VI <u>358</u>, 1.

Laurent D.G., Wang C.S. and Callaway J., 1978, Phys.Rev. B <u>17</u>, 455.

Levinger J.S. and Rustgi M.L., 1956, Phys.Rev. A <u>103</u>, 439.

Lin C.P., Cheng K.T. and Johnson W.R., 1975, Phys.Rev. A <u>11</u>, 1946.

Parrat L.G., and Hempstead C.F., 1954, Phys.Rev. <u>94</u>, 1593.

Pratt R.H., 1982, to be published in the Proceedings of the Second Int.Symposium on Radiation Physics, Penang, Malaysia; also Internal report PITT-282.

Raccah P.M., Euwema R.N., Stukel D.J. and Collins T.C., 1970, Phys. Rev. B <u>1</u>, 756.

Roy S.C., and Pratt R.H., 1982, Phys.Rev. A <u>26</u>, 651.

Roy S.C., Kissel L., and Pratt R.H., 1982, PITT-276 preprint, to be published in Phys.Rev.A.

Savaria L.R. and Caticha-Ellis S., 1966, Acta Cryst. <u>20</u>, 927.

Schmidt P.C. and Weiss A., 1979, Z.Naturforsch. 34a, 1471.

Shore B.W., 1967, Revs.Mod.Phys. 39, 439.

Snow E.C., 1967, Phys.Rev. 158, 683.

Stewart R., Bentley J. and Goodman B., 1975, J.Chem.Phys. 63, 3786.

Stukel D.J. and Euwema R.N., 1970, Phys.Rev. B 1, 1635.

Sucher J., 1980, Phys.Rev. A 22, 348.

Wagenfeld H., 1975, see General References AS, p.13.

Wakoh S. and Yamashita J., 1971, J.Phys.Soc.Jap., 30, 422.

Wang C.S. and Callaway J., 1977, Phys.Rev. B 15, 298.

Wang M.S., 1982, Internal report PITT-281.

Wood J.H., 1967, in Energy Bands in Metals and Alloys, Eds. L. Bennett and J. Waber (Gordon and Breach, N.Y.).

Experiment

X-rays

Aldred, P.J. and Hart M., 1973, Proc.Roy.Soc.(Lond.) A 332, 223; 239 (1973).

Bilderback D.H. and Colella R., 1975, Phys.Rev. B 11, 793;
——————— , 1976, Phys.Rev. B 13, 2479.

Bonse U. and Hart M., 1965, Zeits.Phys. 188, 154.

Bonse U. and Hellkötter H., 1969, Zeits.Physik 223, 345.

Bonse U. and Materlik G., 1972, Zeits.Physik 253, 232;
——————— , 1976, Zeits.Physik B 24, 189.

Bonse U., Spieker P., Hein J.T. and Materlik G., 1980, Nucl.Instr. Meth. 172, 223.

Bonse U., and Terworte R., 1980 , J.Appl.Cryst. 13, 410.

Borrmann G., 1950, Zeits.Physik 127, 297.

Calder R., Cochran W., Griffiths D. and Lowde R., 1962, J.Phys.Chem. Solids 23, 631.

Cauchois Y., and Manesco I., 1950, J.Chim.Physique 47, 892.

Chipman D.R. and Jennings L.D., 1963, Phys.Rev. 132, 728.

Creagh D.C., 1975, Austr.J.Phys. 28, 543;
——— 1977, Phys.Stat.Sol.(a) 39, 705.

Creagh D.C. and Hart M., 1970, Phys.Stat.Sol. 37, 753.

Cusatis C. and Hart M., 1977, Proc.Roy.Soc.(Lond.) A 354. 291.

Diana M., and Mazzone G., 1972, Phys.Rev.B 5, 3832.

Engel D.W. and Sturm M., 1975, see General References AS, p.93.

Freund A., 1975, see General References AS, p.69.

Fukamachi T. and Hosoya S., 1975, Acta Cryst. A 31, 215.

Fukamachi T, Hosoya S,Kawamura T.and Okunuki M,1977,Acta Cryst.A 33, 54.

———————— 1979, Acta Cryst. A 35, 828.

Fuoss P.H., 1981, in Inner-Shell and X-ray Physics of Atoms and Solids,
    Ed. D. Fabian et al., (Plenum Press, 1981), p.875.

Fuoss P.H., Warburton W.K. and Bienenstock A., 1980, J.Non-Cryst.Sol.
    35 & 36, 1233.

Hart M., 1975, Proc.Roy.Soc.(Lond.) A 346, 1.

————— 1980, Nucl.Instr.Meth. 172, 209.

Hart M., Sauvage M. and Siddons D.P., 1980, Acta Cryst. A 36, 947.

Hart M. and Siddons D.P., 1981, Proc.Roy.Soc.(Lond.) A 376, 465.

Gerward L., Thuesen G., Jensen N.S. and Alstrup I., 1979, Acta Cryst.
    A 35, 852.

Ice G.E., Chen M.H. and Crasemann B., 1978, Phys.Rev. A 17, 650.

Linkoaho M.V., 1972, Phys.Scripta 5, 271.

Marezio M., Tranqui D. and Capponi J., 1975, see General References
    AS, p.263.

Matsushita T., and Kohra K., 1974, Phys.Stat.Sol. 24, 531.

Merisalo M. and Inkinen O., 1966, Ann.Acad.Sci.Fenn. A6 207, 3.

Miller G.G. and Black P.J., 1970, Acta Cryst. A 26, 527.

Ohba S., Saito Y., and Wakoh S., 1982, Acta Cryst. A 38, 103.

Raccah P.M. and Heinrich V.E., 1969, Phys.Rev. 184, 607.

Rantavuori E.. and Tanninen V.P., 1977, Phys.Scripta 15, 273.

Sasamachi T., Hosoya S. and Fukamachi T., 1980, Acta Cryst. A 36, 183.

Takama T., and Sato S., 1981, Jap.J.Appl.Phys., 20, 1183.

Temkin R.J., Henrich V.E. and Raccah P.M., 1972, Phys.Rev. B 6, 3572.

Templeton D.H., Templeton L.K., Phillips J.C. and Hodgson K.O., 1980,
    Acta Cryst. A 36. 436.

Tirsell K.G., Slivinsky V.W. and Ebert P.J., 1975, Phys.Rev. A 12,
    2426.

Wei P.S.P. and Lytle F.W., 1979, Phys.Rev.B 19, 679.

Wollan E.O., 1931, Phys.Rev. 38, 15.

388

γ-*rays*

Basavaraju G., and Kane P., 1970, Nucl.Phys. A 179, 49.

de Barros S., Eichler J., Gonçalves O., and Gaspar M., 1981, Z.Natur-
    forsch. 36a, 595.

Chittwattanagorn W., Taylor R.B., Teansomprasong P., and Whittingham
    I.B., 1980, J.Phys.G 6, 1147.

Dixon W. and Storey R., 1968, Can.J.Phys. 46, 1153.

Hardie G., Merrow W., and Schwandt D., 1970, Phys.Rev. C 1, 714.

Hardie G., de Vries J., and Chiang Ch.K., 1971, Phys.Rev.C 3, 1287.

Kane P., Basavaraju G., Mahajani J., and Priyadarsini A., 1978, Nucl.
    Instr.Meth. 155, 467.

Mückenheim W., and Schumacher M., 1980, Nucl.Phys. 6, 1237.

Prasad M.S., Raju G.S., Murty K.N., Murty V.A.N., and Lakshminarayana
    V., 1978, J.Phys.B 11, 3969.

Ramanathan N., Kenett, T.J., and Prestwich W.V., 1979, Can.J.Phys.
    57, 343.

Roy S.C., 1980, Internal Report PITT-233.

Roy S.C., Nath A., and Ghose A.M., 1975, Nucl.Instr.Meth. 131, 163.

Schneider J.R., 1976, J.Appl.Cryst. 9, 394.

Schneider J.R., Pattison P., and Graf H.A., 1978, Phil.Mag. 38, 141.

Schumacher M., 1969, Phys.Rev. 182, 7.

Schumacher M., Smend F., and Borchert I., 1973, Nucl.Phys. A 206, 531.

Schumacher M., and Stoffregen A., 1977, Zeits.Physik A 283, 15.

Sen Gupta S.K., Paul N.C., Basu J., and Chauduri N., 1979a, Phys.Rev.A.
    20, 948.

Sen Gupta S.K., Paul N.C., Roy S.C., and Chauduri N., 1979b, J.Phys.B.
    12, 1211.

Sen Gupta S.K., Paul N.C., Bose J., Goswami G.C., Das S.C., and Chauduri
    N., 1982, J.Phys.B 15, 595.

Taylor R.B., Teansomprasong P. and Whittingham I.B., 1981, Austr.J.
    Phys. 34, 125.

ELECTRON BREMSSTRAHLUNG EXPERIMENTS

Werner Nakel
Physikalisches Institut der Universität Tübingen
D - 7400 Tübingen, West Germany

ABSTRACT

The current status of experiments on electron bremsstrahlung emission from neutral atoms in thin foils or gases is discussed and a comparison with theory is given.

The application of the electron–photon coincidence technique is beginning to provide a deeper understanding of the elementary radiation process. Beyond measurements of the triply differential cross section it was possible to include the detection of the linear polarization of the emitted photons.

On the other hand, stimulated by recent calculations using the partial waves method, several noncoincidence experiments have been performed, measuring absolute doubly differential cross sections, angular distributions, energy spectra, and the dependence on the atomic number of the target material.

The production of bremsstrahlung by transversely polarized electrons has been measured, showing an azimuthal asymmetry in the spatial distribution of the photons emitted.

So far very few experimental results have been available for electron-electron bremsstrahlung generated by collisions with orbital electrons of the atoms.

1. INTRODUCTION

This survey is concerned with experimental investigations of the emission of a photon in the scattering of an electron by a neutral atom placed in a thin solid or gaseous target.

The main force acting on the incident electron, leading to bremsstrahlung emission, is from the nuclear charge. The effect of the atomic electrons is twofold. On the one hand, the atomic electrons screen the Coulomb field of the nucleus as a static charge distribution and thus reduce the cross section for bremsstrahlung emission. The recoil momentum is taken up by the atom as a whole, which remains in its ground state. The emission of bremsstrahlung in the screened nuclear Coulomb field is called atomic-field bremsstrahlung or electron–nucleus bremsstrahlung. On the other hand, the atomic electrons act as individual particles and the bremsstrahlung process may also take place in a collision with an atomic electron which then absorbs the recoil momentum and is ejected. This process is called electron-electron bremsstrahlung.

In addition, there is a radiation of the atom itself, dynamically polarized by the incident electron. The electron induces a dipole moment in the atom, which leads to the radiation. This radiation is determined by the atomic polarizability and therefore it is called "atomic" bremsstrahlung[1].

In this paper, we are dealing with the processes of atomic-field bremsstrahlung and electron-electron (e-e) bremsstrahlung. Atomic-field bremsstrahlung has been investigated mostly by the detection of the emitted photons without regard to the decelerated electrons. Many experiments had been performed over the years to measure the dependence of the radiation process on a variety of parameters as the initial electron energy, the atomic number of the target material, the photon emission angle, and the photon energy. However, for the comparison with theory, until recently, only approximations with various assumptions had been available. Since 1971 there has been extensive theoretical work done largely by Pratt and his co-workers[2,3] who calculate the atomic-field bremsstrahlung process numerically by means of a partial-wave expansion using screened potentials. These calculations have initiated a renaissance of measurements of singly and doubly differential cross sections, which we discuss in Sect. 3.1.

In bremsstrahlung measurements made without regard to the decelerated electrons, the results are necessarily averaged over all electron scattering angles and important details are lost in this averaging process. When, on the other hand, the bremsstrahlung photons are detected in coincidence with decelerated electrons scattered in a fixed direction, information on the elementary radiation process can be obtained and a more stringent check of the theoretical work becomes possible. The quantity measured in the coincidence experiments is the triply differential cross section, differential in photon energy and angle and in the scattering angle of the outgoing electron. This cross section, which is briefly discussed in Sect. 3.2, is the most detailed independently observable quantity if one does not regard the polarization of photons and the spin of electrons.

However, a perfect experimental study of the bremsstrahlung process would require the measurement of the triply differential cross section for transversely and longitudinally polarized electron beams and of the state of polarization of both the scattered electrons and the photons produced. Although the techniques are available for carrying out such a perfect experiment it has not yet been done. Until now, only the photon linear polarization had been measured in the triply differential cross section, which is discussed in Sect. 3.3.1(b).

For the doubly differential cross section, i.e., observing only the photons and not the scattered electrons, there are a variety of experiments which include the polarization of incident electrons and/or produced photons. These polarization effects are discussed in Sect. 3.3.1(a) and 3.3.2.

Electron-electron (e-e) bremsstrahlung generated in collisions of incident electrons with atomic electrons in general is very weak compared to atomic-field bremsstrahlung. Several experimental investigations show that it is difficult to isolate e-e bremsstrahlung from atomic-field bremsstrahlung in measurements in which the final electrons are not observed. In electron-photon coincidence experiments, however, it is possible to differentiate completely between the two bremsstrahlung components. Especially for low-Z targets where the target electrons can be considered essentially free and at rest, measurements of the triply differential cross section, the angular distribution, the spectral distribution, and the photon polarization have been possible and are discussed in Sect. 4.

An important topic not discussed here is the production of coherent bremsstrahlung radiation from high-energy electrons on crystals [4,5]. Under proper conditions roughly monochromatic, highly polarized photon beams can be produced.

2. EXPERIMENTAL DEVICES

In the schematic diagram of Fig. 1 the main components are shown employed in coincidence and noncoincidence bremsstrahlung experiments.

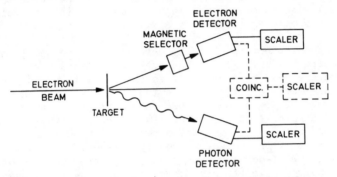

Fig. 1. Schematic diagram of experimental arrangements.

(a) Electron beam sources. In the low and medium energy range the usual electrostatic accelerators are used. For experiments with spin-polarized electron beams the following polarization methods have been employed until now: use of radioactive β-emitters, Mott scattering of fast electrons by heavy nuclei, and the Fano effect (see Sect. 3.3.2).

(b) Targets. To study the processes of bremsstrahlung generation mostly thin solid foils are used as targets, in some recent experiments gaseous targets were employed. In the ideal case the target should be so thin that the incident electron undergoes only one single collision on the way through it so that the bremsstrahlung process is not disturbed by additional (elastic or inelastic) scattering processes. The advantage in the use of thin targets is that the effect of one single interaction is much easier to calculate theoretically than the corresponding multiple electron scattering problem. Hence, such experiments can be compared much more directly with theory. However, in measurements of small cross sections the problems associated with getting sufficient counting rates from thin target foils without destroying the foil are considerable.

(c) Photon detectors. The detection of bremsstrahlung radiation is mostly done by photon counting techniques. In recent years, semi-conductor detectors have been improved considerably with regard to resolution and so have found increasing use in x-ray spectroscopy. Lithium-drifted germanium detectors, intrinsic germanium detectors, and lithium-drifted silicon detectors are employed, the latter down to energies in the region of 1 keV. In addition scintillation detectors are still in use in bremsstrahlung experiments.

Besides analysing the bremsstrahlung radiation in terms of intensity and photon energy, the polarization of the photons is a property of interest and one needs polarization sensitive detectors. All recent polarimeters for linear polarization depend on the polarization sensitivity of Compton scattering. The basic arrangement consists of a central scatter and a photon detector as an analyzer measuring the spatial distribution of the scattered photons. In some polarimeters the scatterer consists of a detector measuring the Compton recoil electrons and the analyzer operates than in coincidence with the scatterer (see Sect. 3.3.1).

(d) Electron detectors. For the detection of the decelerated outgoing electrons scintillation detectors as well as surface barrier detectors have been used. It is essential to avoid detecting elastically scattered electrons since the count rate for these in general is orders of magnitude higher than that of electrons after radiative scattering. In order to eliminate the elastically scattered electrons before reaching the detector a magnetic selector is inserted in front of the detector. Mostly, dispersive magnets are used, focusing a well-defined energy onto the detector. Recently, however, a nondispersive (triply focusing) magnet was employed in order to eliminate the elastically scattered electrons but to transmit the inelastically scattered electrons within a broad energy range. With this device a two-parameter coincidence experiment was performed (see Sect. 4.2.).

3. ATOMIC-FIELD BREMSSTRAHLUNG

3.1 SINGLY AND DOUBLY DIFFERENTIAL CROSS SECTIONS

The process of atomic-field bremsstrahlung has been investigated experimentally for decades as a function of many parameters. These investigations have been the subject of several reviews [6,7,8]. Until recently, however, for the comparison with theory only approximations of different accuracy had been available. Since 1971 Pratt and his coworkers [2,3] calculated the atomic-field bremsstrahlung process numerically by means of a partial-wave expansion using screened self-consistent relativistic central potentials. These calculations provide what is expected to be the best prediction over the entire range of atomic numbers and radiated photon energies of both the photon energy spectrum, or singly differential cross section $d\sigma/dk$, and the energy and angular distributions of the photons or doubly differential cross section $d^2\sigma/d\Omega dk$. Now Pratt et al [9,10] have examined the regions of validity of various approximations by comparison with the fully relativistic numerical calculations.

However, this best current calculation of bremsstrahlung also involves a number of model assumptions and has to be checked experimentally. Stimulated by these calculations several experimental investigations were performed recently in low and medium energy region.

Quarles and Heroy[11] measured the absolute cross section for atomic-field bremsstrahlung produced by 50-, 100-, and 140-keV electron bombardment of thin targets of aluminum, copper, silver, and gold at several photon emission angles using a planar Ge(Li) detector. Theory and experiment are generally found to be in very good agreement, although there are some small systematic discrepancies. An example for the angular dependence of the bremsstrahlung process is shown in Fig. 2. In the past, the usual way of presenting the angular distribution data was to plot the doubly differential cross section versus the photon emission angle for a fixed value of photon energy. Recently, Tseng et al[10] have defined a shape function $S(Z, T, k/T, \theta)$ as the ratio of the angular distribution to the energy spectrum. This function is advantageous in the comparison between theory and experiment and is plotted in Fig. 2.

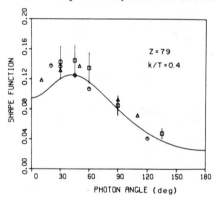

Fig. 2. Shape function S vs photon angle for 50-keV electrons on gold for k/T = 0.4. Squares, circles and triangles are data from Ref.[11,12,13], respectively. The curve is an interpolation of theoretical data from Tseng et al[10] (from Quarles and Heroy[11]).

An example for the photon energy spectrum of gold from the work of Quarles and Heroy is shown in Fig. 3. The presentation chosen by the authors and described by them in detail has the advantage of allowing one to plot all the data for a given energy and target regardless of photon emission angle on a single curve. In effect the authors are using the theoretical shape function to transform the doubly differential cross section data into a measure of the singly differential photon energy spectrum.

Further experiments were undertaken to test the Pratt calculations at lower electron energies. Thereby, thin gas targets were used so that both any solid-state effects and multiple-scattering effects were minimized.

Semaan and Quarles[14] measured the bremsstrahlung photon spectrum at 90° originating from the bombardment of atoms of neon, argon, krypton and xenon by 6-10-keV electrons. The experimental data are found to agree well with the calculations of Tseng et al[10] over the photon energy range from 4 keV to the endpoint of the spectrum.

394

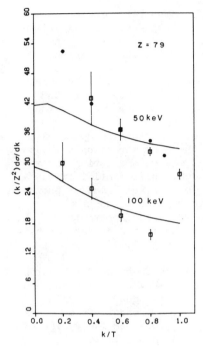

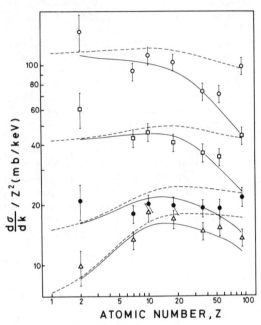

Fig. 3. Plot of $(k/Z^2)d\sigma/dk$ for 50- and 100 keV electrons on gold. The data points (squares) are averages of the data at different photon angles. The circles are data from Motz and Placious[13] (from Quarles and Heroy[11]).

Fig. 4. Scaled bremsstrahlung cross section $(d\sigma/dk)/Z^2$, vs atomic number Z at an incident electron energy T=10 keV and photon energies of 2 keV (open circles), 4 keV (squares), 7 keV (solid circles), and 9 keV (triangles). The solid lines are the theoretical calculations of Pratt et al[16] dashed lines are from the Sommerfeld-Maue formula[17] (from Hippler et al[15]).

Hippler et al[15] measured the relative bremsstrahlung cross section for gas atoms with atomic numbers in the range Z=2 to Z=92 at incident electron energies of 2.5 and 10 keV. The photons were observed by a Si(Li) detector. The results agree reasonably with the calculation of Pratt et al[16] except at the largest Z number, where differences of up to a factor of 2 at low photon energies occur (Fig. 4). Hippler et al[15] suggested that two-photon processes, which might be expected to become increasingly important at high Z, may be contributing to the enhancement. Such two-photon processes were not included in the theory. However, in a most recent investigation of the Z dependence, Semaan and Quarles[18] found the bremsstrahlung spectrum for Z=80 to be in very good agreement with theory and found no effects attributable to two-photon processes. Further experiments are desirable to resolve the question.

3.2   TRIPLY DIFFERENTIAL CROSS SECTION

In the atomic-field bremsstrahlung process the initial momentum
of the incident electron becomes shared between the momenta of three
particles: the decelerated outgoing electron, the atom, and the emitted photon. Therefore, the photon may be radiated in any direction,
regardless of the direction of the outgoing electron. In bremsstrahlung measurements made without regard to the outgoing electrons, the
results are necessarily averaged over all electron scattering angles.
When, on the other hand, the photons are detected in coincidence with
the outgoing electrons scattered in a fixed direction, the elementary
bremsstrahlung process can be investigated. The quantity measured in
the coincidence experiments is the triply differential cross section
$d^3\sigma/d\Omega_\gamma d\Omega_e dk$, differential in photon energy and angle and in the
scattering angle of the outgoing electron. If one does not observe
the polarization of electrons and photons, then this cross section
is the most detailed independently observable quantity.

Although such a coincidence experiment has been suggested[19] as
early as 1932, the first electron-photon correlation measurement[20]
was reported only in 1966. Subsequent measurements of relative and
absolute triply differential cross sections including angular and
spectral distributions performed by the groups of Aehlig and Scheer,
Nakel, and Quarles have been reviewed by Nakel[21] and will not be repeated here. In these experiments the photon angular distribution is
no longer rotationally symmetric about the direction of the incoming
electron as in the noncoincident case. The measurements show strong
electron-photon correlations, with the maximum emission probability
of the photon occurring on the same side of the initial beam into
which the outgoing electron is scattered.

For the triply differential cross section of atomic-field bremsstrahlung only approximate calculations are available. The
classical Bethe-Heitler formula[22] describes the process in the first
Born approximation. The region of validity is roughly $\alpha Z/\beta \ll 1$, where
$\alpha = 1/137$ is the fine-structure constant, β the velocity of the outgoing electron in units of the velocity of light, and Z is the atomic
number of the target atom. The region of validity of the calculations
of Elwert and Haug[23] using Sommerfeld-Maue wavefunctions can be expressed by $\alpha Z \ll 1$. This relation does not depend on the energy of the
incident electron or of the photon energy.

At present, there are no calculations which claim to be valid
for higher Z in the medium energy region. Numerical calculations of
the triply differential cross section using the partial-wave method
are not available.

Recently, further measurements of the absolute triply differential cross section have been reported by Komma and Nakel[24]. Targets
of carbon, copper, silver and gold were used (for the measurement on
carbon see Fig. 11 in Sec. 4.2). Good agreement with the Elwert-Haug
formula is obtained for the present parameters even for gold (Z=79).

3.3 PHOTON AND ELECTRON POLARIZATION CORRELATIONS

The role of the electron spin in the production of atomic-field bremsstrahlung and the polarization behaviour of the emitted photons has been the subject of several investigations. A survey of the types of polarization correlations which exist between incident electron and emitted photon assuming that the final electron is not observed is given by Tseng and Pratt[25].

3.3.1 LINEAR POLARIZATION OF BREMSSTRAHLUNG PHOTONS FROM UNPOLARIZED ELECTRONS

(a) Measurements without regard to the outgoing electrons

Atomic-field bremsstrahlung produced by a beam of unpolarized electrons is in general partially linearly polarized parallel or perpendicular to the emission plane. Many measurements were made to investigate the dependence of the polarization on the photon energy, the photon emission angle, the initial electron energy, and the atomic number of the target material. The behaviour of the polarization as a function of the parameters was found to be rather complex. Earlier experiments and theory were compared in the review article of Motz and Placious[26]. Most of the subsequent experiments were discussed in the review of Tseng and Pratt[25].

After that, an extensive experimental study was performed by Lichtenberg et al[27]. At incident electron energies of 0.5-1.5 MeV, bremsstrahlung linear polarization was measured for thin targets of Be, Al, Ag, and Au at emission angles of 10°-122° as a function of photon energy. Corrections were made for electron scattering in the target and for multiple scattering of photons in the Compton polarimeter employed. For low atomic numbers the measurements are in excellent agreement with the results of Born calculations including screening and with calculations using Sommerfeld-Maue eigenfunctions. The results for high atomic number agree only with partial-wave calculations after Tseng and Pratt[25]. An example is shown in Fig. 5.

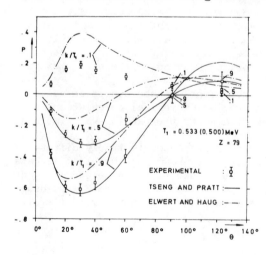

Fig. 5. Photon linear polarization for gold. Experimental results for incident electron energy T_1=0.533 MeV and data from Tseng and Pratt for T_1=0.500 MeV, and Elwert and Haug for T_1=0.533 MeV (from Lichtenberg et al[27]).

(b)   Measurements for fixed direction of outgoing electrons

In bremsstrahlung experiments without observation of the decelerated outgoing electrons in coincidence with the photons, the resulting polarization is produced by averaging over many elementary bremsstrahlung processes with different directions of outgoing electrons.
Thus, the polarization behaviour provides only a very limited amount
of information about the elementary radiation process.

An electron-photon coincidence experiment which involves measuring the angular dependence of the photon linear polarization for fixed direction of outgoing electrons was reported in 1978 by Behncke
and Nakel[28] and has been continued by Bleier and Nakel[29]. A schematic
view of the experimental arrangement is shown in Fig. 6.

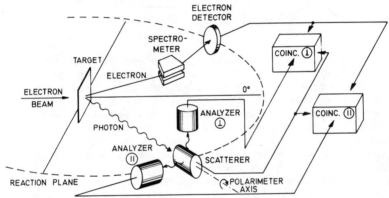

Fig. 6. Schematic view of the experimental arrangement
        (from Behncke and Nakel[28] and Bleier and Nakel[29])

Since, according to a simple classical picture, the orbital plane
of the radiating electron is determined by the coincidence measurement, one would expect the radiation to be always completely linearly
polarized in this experiment. However, at relativistic energies the
effect of electron spin becomes important and the emission of bremsstrahlung can take place not only as a result of a change of momentum, but as a result of a change of spin orientation. A detailed interpretation of the contributions of the so-called orbital and spin
currents to the intenstiy and polarization of the bremsstrahlung produced was made by Fano et al[30]. In measuring the angular distribution
of the polarization for a carbon target (Fig. 7) it was possible to
discern the appearance of the spin-flip radiation by a deviation from
the nearly complete linear polarization. The decrease of the polarization takes place at the minimum of the triply differential cross
section. A detailed discussion is given in the paper of Behncke and
Nakel[28]. The measurement is in good agreement with the calculation of
Elwert and Haug[23,31] for the case examined (Z=6).

Most recently Bleier and Nakel[29] have extended the measurements
to higher atomic numbers (Z=29 and 79). The result for Z=79 is shown
in Fig. 8. According to the calculations of Elwert and Haug[23,31], the
decrease of the polarization at the minimum of the cross section

should diminish strongly with increasing target atomic number Z. This is not the case for the experimental value. However, Zα<<1 is required for the calculation which is not fulfilled for higher Z. Numerical calculations according to the method of Pratt[3] would be highly desirable for comparison.

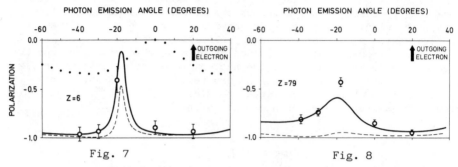

Fig. 7   Fig. 8

Fig. 7. Linear photon polarization in the elementary bremsstrahlung process as a function of the photon emission angle for outgoing electrons of +20° and 140 keV. Primary electrons of 300 keV are used incident on a carbon target. The theoretical curve (solid line) has been calculated for the present experimental situation with the Elwert-Haug theory. The dashed curve shows the theoretical polarization without corrections for the finite solid angles, the finite energy width of the magnetic electron spectrometer, and the plural scattering in the target. In addition, the correction for a contribution of electron-electron bremsstrahlung has been made. For comparison the dotted curve gives the calculated polarization integrated over all directions of the outgoing electron (from Behncke and Nakel[28]).

Fig. 8. Same as Fig. 7 except for gold (from Bleier and Nakel[29]).

3.3.2 BREMSSTRAHLUNG FROM POLARIZED ELECTRONS

There are only a few experiments in which bremsstrahlung production by polarized electron beams has been studied up to now. This is merely due to the lack of powerful sources of polarized electrons in the past.

(a) Azimuthal Asymmetry of Bremsstrahlung from Transversely Polarized Electrons

As a consequence of spin orbit interaction in analogy to Mott scattering one expects an azimuthal asymmetry in the spatial distribution of the bremsstrahlung around the direction of the incident electrons if these are transversely polarized with their polarization vector perpendicular to the production plane.

In earlier experiments, radioactive β-emitters[32,33] and Mott-scattering of fast electrons by heavy nuclei[34] were used to generate the polarized beams.

In a recent experiment of Schaefer et al[35] the polarized electron beam was produced by Fano effect on Rubidium, i.e., the photoionization of unpolarized alkali atoms with circularly polarized light. This arrangement allows one to change the orientation of the electron spin easily by reversing the sense of the rotation of the circularly polarized light without any influence on the electron beam. Therefore, the bremsstrahlung emission asymmetry was measured for fixed emission angles but with opposite orientation of the electron spin. By observing only the produced photons and disregarding the scattered electrons, the asymmetry can be described by the formalism introduced by Tseng and Pratt[25] who made detailed calculations to atomic-field bremsstrahlung based on a partial waves method. In this formalism the azimuthal asymmetry is described by a coefficient C_{20}, which is used by the authors[35] to compare their measurements with the theoretical predictions. An electron beam of 127.75 keV was used with a polarization degree of about 70%. The bremsstrahlung produced by the inelastic scattering from a gold target was detected with a Ge detector for special emission angles ($\theta = 60^\circ$, 100°, 145°) where the angular dependence of C_{20} was expected to be either maximum, crossing zero or minimum respectively. In Fig. 9a and 9b the results for the angles of 60° and 145° are shown together with previous experimental data and theoretical values of partial wave analysis[25,34]. The discrepancy for $\theta = 145^\circ$ has not yet been understood.

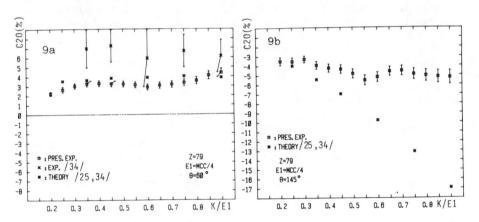

Fig. 9. Bremsstrahlung asymmetry factor C_{20} in dependence of the relative photon energy k/E_1 with a bin width of $\pm$ 7 keV for photon emission angles of (a) $\theta = 60^\circ$ and (b) $\theta = 145^\circ$ (from Schaefer et al[35]).

(b) Circular Polarization of Bremsstrahlung from Longitudinal Polarized Electrons

The circular polarization of atomic-field bremsstrahlung from longitudinal polarized electrons was measured by several authors (see for example Galster[36]) using β-rays.

4. ELECTRON-ELECTRON BREMSSTRAHLUNG

An electron beam passing through matter can produce bremsstrahlung not only in the Coulomb field of the nucleus but also in a collision with an atomic electron which then absorbs the recoil momentum and is ejected.

In contrast to the electron-nucleus system the electron-electron system has no electric dipole moment. Therefore, the (nonrelativistic) e-e bremsstrahlung consists predominantly of electric quadrupole radiation and the cross section for the process is very small. With increasing electron energy the cross section increases but even for energies of several 100 keV the e-e bremsstrahlung in general gives such a small contribution to the total bremsstrahlung emission that it is not taken into account in most measurements of atomic-field bremsstrahlung. Especially in the case of high-Z targets, the experiments give almost pure atomic-field bremsstrahlung, since it is roughly proportional to Z^2, whereas the e-e bremsstrahlung at best is proportional to the number of electrons.

An important guide to discern the e-e contribution in the total bremsstrahlung spectrum is the maximum photon energy. Whereas the energy the nucleus receives can be neglected, the energy of the recoil electron may be considerable. Therefore, the e-e bremsstrahlung spectrum is distributed up to a maximum value less than the endpoint of the atomic-field bremsstrahlung spectrum; moreover, it is angle dependent. The different upper bounds of the spectra were used to isolate the e-e contribution from the total spectrum in noncoincidence experiments (see f.e. Rester[37]). However, an analysis of Haug[38] shows that the experimental results are still inconclusive. It is interesting to note that Hackl[39] did not find any evidence for a contribution of e-e bremsstrahlung using 2-MeV electrons and lithium (Z=3) targets. These experimental investigations show that it is difficult to isolate the e-e bremsstrahlung from the total spectrum when only the noncoincident photon spectrum is observed.

4.1 TRIPLY DIFFERENTIAL CROSS SECTION AND ANGULAR DISTRIBUTION OF PHOTONS

In 1972 Nakel and Pankau[40] used the electron-photon coincidence technique to differentiate completely between e-e bremsstrahlung and atomic-field bremsstrahlung and measured the triply differential cross section[41,42] and the angular correlation[43]. 300-keV electrons were used incident on a carbon target. As the energy transfer to the electron is much greater than the binding energy, one may neglect the binding and treat the problem as a collision with a free electron initially at rest. The measuring method was based on a coincidence observation between the bremsstrahlung spectrum emitted in a definite direction and outgoing electrons of definite energy and direction. The e-e bremsstrahlung photons are well separated from the atomic-field bremsstrahlung photons since the former have an energy that is reduced by the recoil energy of the second unobserved electron.

In fig.10 a photon angular distribution of e-e bremsstrahlung for a fixed direction of outgoing electrons is shown. In contrast to the

atomic-field bremsstrahlung the photon energy of the e-e bremsstrahlung is dependent on the photon emission angle for kinematical
reasons.

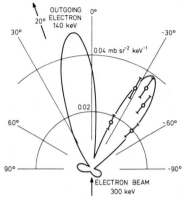

Fig. 10. Photon angular distribution of e-e bremsstrahlung for
fixed direction of outgoing electrons. The solid curve is the
theoretical prediction[44,45] .
Atomic number Z = 6. The cross
section is given per atom, i.e.,
for six target electrons (from
Nakel and Pankau[43]).

In all the theoretical work on e-e bremsstrahlung both the binding of the scatterer to the nucleus and its screening by the other
atomic electrons is neglected. The calculation of the cross section
in Born approximation is achieved by a straightforward application of
quantum electrodynamics. However, compared with the Bethe-Heitler
formula of the corresponding process in the Coulomb field of a nucleus, these expressions are extremely complicated owing to recoil and
exchange effects. Mack and Mitter[44] who solved the problem by using
computer programs for formula manipulations and Haug[45] who used the
traces calculated by Anders[46] obtained identical results.

4.2 TRIPLY DIFFERENTIAL CROSS SECTION AND SPECTRAL DISTRIBUTION

The bremsstrahlung processes have continuous spectra both in photon energy (k) and outgoing electron energy (T). In a recent experiment by Komma and Nakel[24] an electron beam of 300 keV was directed onto thin target foils of carbon, copper, silver, and gold to investigate the processes of atomic-field bremsstrahlung, e-e bremsstrahlung,
and K-shell ionization the latter not being considered in the present
paper. Using a two-parameter arrangement the outgoing electrons from
all these processes were detected in coincidence with the emitted photons as a function of outgoing electron and photon energies for
fixed detector angles. In Fig. 11 the result of the measurement on
carbon is shown. With increasing atomic number of the target the
ratio of e-e bremsstrahlung to atomic-field bremsstrahlung decreases.
Also for copper, the e-e bremsstrahlung cross section was found to be
still in good agreement with the theoretical prediction for free electrons. However, the measurements on silver and gold targets were not
evaluated because the number of e-e bremsstrahlung events on the kinematical curve was too low. Here, due to the strong binding and broad
momentum distribution of the inner-shell electrons a smear out of the
free electron kinematics will appear. The result of a rough estimate
of the smear out is given by Komma and Nakel[24] . A theoretical treatment of the bremsstrahlung process on bound electrons was performed
recently by Haug and Keppler [47].

402

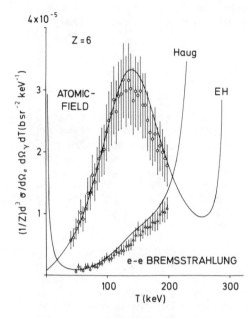

4×10^{-5}

Z = 6

Haug

ATOMIC-FIELD

EH

$(1/Z)d^3 \sigma/d\Omega_e \, d\Omega_\gamma \, dT(b \, sr^{-2} \, keV^{-1})$

e-e BREMSSTRAHLUNG

100   200   300

T (keV)

Fig. 11. Triply differential cross section for e-e bremsstrahlung and atomic-field bremsstrahlung on carbon. The solid lines are theoretical curves calculated after formulas of Haug [45] for free electrons and Elwert-Haug [23] (EH) (from Komma and Nakel [24]).

4.3 LINEAR PHOTON POLARIZATION FOR FIXED DIRECTION OF OUTGOING ELECTRONS

The linear polarization of e-e bremsstrahlung for fixed direction of outgoing electrons has been measured recently by Bleier and Nakel [29]. Agreement was found with the calculation of Mack [48] for the present parameters. 300-keV electrons were used incident on a carbon target. Coincidences between the two outgoing electrons and the emitted photon were observed. For this purpose, the experimental arrangement shown in Fig. 6 was modified by adding a detector for the second electron.

REFERENCES

1. M. Ya. Amusia, Comments At. Mol. Phys. 11, 123 (1982)
2. H. K. Tseng and R. H. Pratt, Phys. Rev. A3, 100 (1971)
3. R. H. Pratt, in Inner-Shell and X-ray Physics of Atoms and Solids, edited by D. J. Fabian, H. Kleinpoppen, and L. M. Watson (Plenum, New York, 1981) p. 367
4. S. Datz, in Inner-Shell and X-ray Physics of Atoms and Solids, edited by D. J. Fabian, H. Kleinpoppen, and L. M. Watson (Plenum, New York, 1981) p. 176
5. G. Diambrini Palazzi, Rev. Mod. Phys. 40, 611 (1968)
6. H. W. Koch and J. W. Motz, Rev. Mod. Phys. 31, 920 (1959)
7. S. T. Stephenson in Encyclopedia of Physics, edited by S. Flügge, Vol. XXX (Springer, Berlin 1957)
8. H. S. W. Massey, E. H. S. Burhop, and H. B. Gilbody, Electronic and Ionic Impact Phenomena, Vol. II
9. C. M. Lee, L. Kissel, R. H. Pratt, and H. K. Tseng, Phys. Rev. A13, 1714 (1976)

10. H. K. Tseng, R. H. Pratt, and C. M. Lee, Phys. Rev. A19, 187 (1979)
11. C. A. Quarles and D. B. Heroy, Phys. Rev. A24, 48 (1981)
12. D. H. Rester, N. Edmonson, and Q. Peasley, Phys. Rev. A2, 2190 (1970)
13. J. W. Motz and R. C. Placious, Phys. Rev. 109, 235 (1958)
14. M. Semaan and C. A. Quarles, Phys. Rev. A24, 2280 (1981)
15. R. Hippler, K. Saeed, I. McGregor, and H. Kleinpoppen, Phys. Rev. Lett. 46, 1622 (1981)
16. R. H. Pratt, H. K. Tseng, C. M. Lee, L. Kissel, C. McCallum, and M. Riley, At. Data Nucl. Data Tables 20, 175 (1977)
17. A. Sommerfeld and A. W. Maue, Ann. Phys. (Leipzig) 23, 489 (1935)
18. M. Semaan and C. A. Quarles, Phys. Rev., to be published
19. O. Scherzer, Ann. Phys. (Leipzig) 13, 137 (1932)
20. W. Nakel, Phys. Lett. 22, 614 (1966)
21. W. Nakel, in Coherence and Correlation in Atomic Collisions, edited by H. Kleinpoppen and J. F. Williams (Plenum, New York, 1980), p. 187
22. H. Bethe and W. Heitler, Proc. R. Soc. (London) Ser. A 146, 83 (1934)
23. G. Elwert and E. Haug, Phys. Rev. 183, 90 (1969)
24. M. Komma and W. Nakel, J. Phys. B: At. Mol. Phys. 15, 1433 (1982)
25. H. K. Tseng and R. H. Pratt, Phys. Rev. A7, 1502 (1973)
26. J. W. Motz and R. C. Placious, Nuovo Cimento 15, 571 (1960)
27. W. Lichtenberg, A. Przybylski, and M. Scheer, Phys. Rev. A11, 480 (1975)
28. H.-H. Behncke and W. Nakel, Phys. Rev. A17, 1679 (1978)
29. W. Fleier and W. Nakel, this conference, to be published
30. M. Fano, K. W. McVoy, and J. R. Albers, Phys. Rev. 116, 1159 (1959)
31. E. Haug, private communication; cf. Phys. Rev. 188, 63 (1969)
32. K. Güthner, Z. Phys. 182, 278 (1965)
33. R. E. Pencynski and H. L. Wehner, Z. Phys. 237, 75 (1970)
34. A. Aehlig, Z. Phys. A294, 291 (1980)
35. H. R. Schaefer, W. v. Drachenfels, and W. Paul, Z. Phys. A305, 213 (1982)
36. S. Galster, Nucl. Phys. 58, 72 (1964)
37. D. H. Rester, Nucl. Phys. A118, 129 (1968)
38. E. Haug, Phys. Lett. 54A, 339 (1975)
39. P. Hackl, Dissertation, University of Vienna, 1970 (unpublished); H. Aiginger and E. Unfried, Acta Phys. Austriaca 35, 331 (1972)
40. W. Nakel and E. Pankau, Phys. Lett. 38A, 307 (1972)
41. W. Nakel and E. Pankau, Phys. Lett. 44A, 65 (1973)
42. W. Nakel and E. Pankau, Z. Phys. 264, 139 (1973)
43. W. Nakel and E. Pankau, Z. Phys. A274, 319 (1975)
44. D. Mack and H. Mitter, Phys.Lett 44A, 71 (1973)
45. E. Haug (unpublished); cf Z. Naturforsch. 30a, 1099 (1975)
46. T. Anders, Dissertation, Universität Freiburg (1961); Nucl. Phys. 59, 127 (1964)
47. E. Haug and M. Keppler, to be published
48. D. Mack, private communication

BREMSSTRAHLUNG INDUCED BY HIGH-ENERGY, HEAVY CHARGED PARTICLE IMPACT

K. Ishii
Cyclotron and Radioisotope Center, Tohoku University, Sendai 980,
Japan

S. Morita
Research Center of Ion Beam Technology, Hosei University,
Kajinocho, Koganei, 184 Tokyo, Japan

ABSTRACT

Here, we introduce quasifree electron bremsstrahlung (QFEB) process which is interpreted in terms of radiation induced by a target electron scattered in the projectile frame. The mechanism of QFEB is discussed and the theory based on the impulse approximation, which has recently been advanced by ourselves, is described. The theory is compared with our recent experiment on a Be target. It is shown that the theory can reproduce well the experiment. Further, it is pointed out that the difference in the production cross sections of QFEB, normalized by the projectile charge, for incident particles of different charges with a same velocity reflects directly the velocity distribution of target electrons, and the contribution from REC process is also found in proton and $^{3}He^{2+}$ ion impact.

INTRODUCTION

When heavy charged particles or heavy ions bombard solid or gas targets, in addition to characteristic x rays, continuum x rays are emitted. Up to the present, the continuum x rays from the following processes have been observed: secondary-electron bremsstrahlung (SEB)[1], molecular-orbital rays (MO), radiative electron capture (REC)[2], nuclear bremsstrahlung and γ rays from nuclear reactions. SEB is a main component of the continuum x rays for light ion impact[1], while MO x-rays and REC are predominant components for heavy ion impact[2]. These x rays have been studied mainly in cases of low-energy ion bombardments. Recently[3,4], we have observed a new continuum x-ray component in bombardments with high-energy heavy-charged particles; here the high energy means the projectile velocity v_p larger than that of the target electron v_e. This component is characterized by the relative kinetic enerty $T_r = \frac{1}{2}m_e v_p^2$ (m_e is the electron mass) between the projectile and the target electron of velocity of $v_e = 0$. Figure 1 shows a typical continuum x-ray spectrum obtained by the bombardment of a Be target with 20-MeV protons. We have interpreted these x rays in terms of bremsstrahlung produced by target electrons scattered in the projectile Coulomb field. Namely, if velocity of the projectile is large enough in comparison with velocity of the target electron, the target electron can be considered as free and at rest. In the center-of-mass frame, the electron collides with the projectile with the relative kinetic energy T_r and bremsstrahlung is produced by the Coulomb interaction

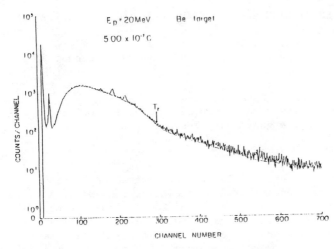

Fig. 1. A typical continuum x-ray spectrum for the Be target bombarded with 20-MeV protons, measured in the direction 135° to the proton (taken from Ref. 3).

between the projectile and the target electron. We called the radiation from this process quasifree-electron bremsstrahlung (QFEB).

Here, we will introduce recent developments on QFEB. First, the theory of QFEB based on impulse approximation is described and then will be discussed in comparison with recent experimental results[3-5].

THEORY

Theoretical treatments of QFEB were first developed by Jakubassa and Kleber[6] and they called this process radiative ionizations. On the basis of the plane-wave Born approximation (PWBA) they calculated the cross section of QFEB for a case where the projectile velocity is larger than the velocity of the orbital electron. For a case of $v_p < v_e$, Anholt and Saylor[7] have made calculations on the basis of the binary-encounter approximation (BEA). Here, we treat this process in the case where v_p is much larger than v_e. In such a case, the orbital electron can be regarded as free as at rest. So that the cross sections of QFEB can be estimated from the bremsstrahlung produced by a free electron in the projectile frame. Under these conditions, the retardation effect and the Doppler effect can easily be introduced into this process.

The production cross section for electron bremsstrahlung, including the first-order relativistic effect, is generally given by[8]

$$\hbar\omega \cdot \frac{d^2\sigma^{QFEB}}{d\Omega d(\hbar\omega)}(T_r,\hbar\omega,\theta) = \begin{cases} 0 \quad \text{for} \quad \hbar\omega \geq T_r \\ \frac{z_p^2}{\pi} z_T r_0^2 \alpha \frac{m_e c^2}{T_r} g(\xi_0,\xi) \{ \\ [\sin^2\theta + \frac{1}{4}(1+T)(3\cos^2\theta-1)] \ell n(\frac{1+\sqrt{T}}{1-\sqrt{T}}) - \frac{1}{2}\sqrt{T}(3\cos^2\theta-1) \end{cases}$$

$$+ \frac{\beta}{2}\cos\theta\,[\,(7-T)\sin^2\theta + \frac{1}{2}(\cos^2\theta - \frac{3}{2}\sin^2\theta)\,(10T+3-3T^2)\,] \times \ln(\frac{1+\sqrt{T}}{1-\sqrt{T}})$$

$$- 2\beta\sqrt{T}\,\cos\theta\,(\cos^2\theta - 2\sin^2\theta)\} \quad \text{for} \quad \hbar\omega \leq T_r. \tag{1}$$

Here, Z_p and Z_T are atomic number of the projectile, and the target nucleus, respectively, α is the fine-structure constant, $m_e c^2$ is the rest energy of electron, $\hbar\omega$ is the x-ray energy, θ is the emission angle of x rays with respect to the incident electron, and T_r is the kinetic energy of the orbital electron relative to the projectile. The notations T and β are defined by

$$\beta = (\frac{2T_r}{m_e c^2})^{1/2}, \quad T = \frac{T_r - \hbar\omega}{T_r} \tag{2}$$

Further, $g(\xi, \xi_0)$ is the correction term for the Coulomb deflection and has been given by Sommerfeld by[8]

$$g(\xi_0, \xi) = \frac{\xi}{\xi_0}\,\frac{1-e^{-2\pi\xi_0}}{1-e^{-2\pi\xi}},$$

with

$$\xi_0 = Z_p(\frac{R}{T_r})^{1/2}, \quad \xi = Z_p(\frac{R}{T_r - \hbar\omega})^{1/2}, \tag{3}$$

where R is the Rydberg constant.

The cross section $d^2\sigma^{QFEB}/d\Omega d(\hbar\omega)$ given by Eq. (1) has a finite value at the high-energy end $\hbar\omega = T_r$ because of the correction factor $g(\xi, \xi_0)$, whereas the PWBA calculation of $g(\xi_0, \xi) = 1$ gives $d^2\sigma^{QFEB}/d\Omega d(\hbar\omega) = 0$ at $\hbar\omega = T_r$. It will be shown below that this difference in the two calculations plays an important role in the behavior of QFEB spectrum near the end-point energy.

Equation (1) represents the QFEB formula in the center-of-mass system for the projectile and the orbital electron. In order to compare with the experiment, this equation must be transformed to that of the laboratory system by the Lorentz transformation[9] and is expressed by

$$\hbar\omega\frac{d^2\sigma^{QFEB}}{Ld\Omega_L d(\hbar\omega_L)}\,(T_r, \hbar\omega_L, \theta_L)$$

$$= \frac{1-\beta^2}{1-\beta\cos\theta_L}\,\hbar\omega\frac{d^2\sigma^{QFEB}}{d\Omega d(\hbar\omega)}\,(T_r, \hbar\omega, \theta),$$

$$\hbar\omega = \frac{1-\beta\cos\theta_L}{(1-\beta^2)^{1/2}}\,\hbar\omega_L, \tag{4}$$

408

and

$$\cos\theta = -\frac{\cos\theta_L - \beta}{1 - \beta\cos\theta_L} \ .$$

Equation (4) has been obtained by assuming that the orbital electron is free and at rest. Now, we will derive the QFEB formula for the orbital electron having the velocity components (v_x, v_y, v_z). The relative kinetic energy T_r' of the electron with respect to the projectile is

$$T_r' = T_r[\ (\frac{v_x}{v_p})^2 + (\frac{v_y}{v_p})^2 + (1 - \frac{v_z}{v_p})^2]\ , \tag{5}$$

where z axis is taken in the direction of the incident beam. Energy conservation of this radiation process in the projectile frame is expressed by[10]

$$T_r' + V_0 = \hbar\omega + T_f \tag{6}$$

where V_0 is the potential energy of the orbital electron in the target atom and T_f is the final kinetic energy of the electron in the projectile frame. Here, we note the relation between the velocity of orbital electron and the potential energy;

$$\frac{1}{2}mv_e^2 + V_0 = -I, \tag{7}$$

with

$$v_e^2 = v_x^2 + v_y^2 + v_z^2\ ,$$

where I denotes the binding energy of the orbital electron. Using Eq. (4), the QFEB formula taking account of the velocity distribution of the orbital electron is expressed by

$$\hbar\omega\sigma_{C.M.}^{QFEB}\ (Z_p, \hbar\omega, \theta) = \frac{1}{Z_T}\ \Sigma_i n_i \int_{-\infty}^{v_i^{max}} dv_z \int_{-\infty}^{+\infty} dv_x \int_{-\infty}^{+\infty} dv_y\ \rho_i(v_x, v_y, v_z)$$

$$\times\ \hbar\omega\frac{d^2\sigma^{QFEB}}{d\Omega d(\hbar\omega)}(T_r', \hbar\omega', \theta') \times (\frac{\hbar\omega}{\hbar\omega'})^2 \tag{8}$$

with

$$\hbar\omega' = \hbar\omega + I_i + \frac{1}{2}mv_e^2\ ,$$

$$\cos\theta' = \cos\theta\cos\theta_e + \sin\theta\sin\theta_e\cos\psi_e\ ,$$

and

$$\cos\theta_e = (v_p - v_z)/|\vec{v}_p - \vec{v}_e|\ .$$

Where, n_i, I_i and $\rho_i(v_x, v_y, v_z)$ are the number of electrons, the binding energy and the velocity distribution function for the i-shell, respectively. The function $\rho(v_x, v_y, v_z)$ is calculated from the momentum representation of the hydrogenic wave function. In Eq. (8), v_i^{max} is the upper limit of the integration on v_z and is determined from the condition $T_r' \geq \hbar\omega'$ and the energy conservation given by Eq. (6):

$$v_i^{max} = \frac{1}{2}v_p(1 - \frac{\hbar\omega + I_i}{T_r}) .$$ (9)

As seen from Eq. (5), the z component of the momentum relative to the projectile is effective for production of the radiation in the case of $v_p \gg v_e$, and Eq. (8) is approximated by

$$\hbar\omega\sigma_{C.M.}^{QFEB}(z_p, \hbar\omega, \theta) \approx \frac{1}{Z_T}\sum_i n_i \int_{-\infty}^{v_i^{max}} dv_z \rho_i(v_z) \times$$

$$\hbar\omega \cdot \frac{d^2\sigma^{QFEB}}{d\Omega d(\hbar\omega)}(T_r', \hbar\omega', \theta) \times (\frac{\hbar\omega}{\hbar\omega'})^2$$ (10)

with

$$\hbar\omega' = \hbar\omega + I_i + \frac{1}{2}mv_z^2 , \text{ and } T_r' \approx T_r(1 - \frac{v_z}{v_p})^2,$$

where $\rho_i(v_z)$ is defined by

$$\rho_i(v_z) = \int_{-\infty}^{+\infty} dv_x \int_{-\infty}^{+\infty} dv_y \, \rho_i(v_x, v_y, v_z)$$ (11)

and, for example, the expressions for the 1s and 2s electrons are given by[11]

$$\rho_{1s}(v_z) = \frac{32}{3\pi} \frac{v_{1s}^5}{(v_{1s}^2 + v_z^2)^3}$$

and

$$\rho_{2s}(v_z) = \frac{128}{3\pi} \frac{\frac{2}{5}v_{2s}^4 - v_z^2 \cdot v_{2s}^2 + v_z^4}{(v_{2s}^2 + v_z^2)^5} v_{2s}^5 ,$$ (12)

where v_{1s} and v_{2s} is the mean velocity of 1s and 2s electrons, respectively. When the potential energy V_0 is neglected, v_i^{max} and $\hbar\omega'$ in Eq. (10) are given by

$$v_i^{max} = v_p(1 - \sqrt{\frac{\hbar\omega}{T_r}}) ,$$ (13)

and

$$\hbar\omega' = \hbar\omega.$$

It was pointed out on Eq. (3) that the electron-bremsstrahlung cross section has a finite value at $\hbar\omega' = T_r'$. This behavior gives a significant result for the QFEB. Here, we consider the QFEB formula at the x-ray emission angle of 90° where the Doppler shift in the laboratory system and the retardation effect are not effective and can be neglected. The electron-bremsstrahlung cross section can be approximated for a small value of T ($=\delta T$) by

$$\hbar\omega \frac{d\sigma}{d\Omega d(\hbar\omega)}^{QFEB} (T_r',\hbar\omega',90°) \doteqdot \frac{z_p^2}{\pi} \cdot z_T \cdot r_0^2 \cdot \frac{m_e c^2}{T_r} T\{(\frac{3}{4} - \frac{1}{4}T) \ln\frac{1+\sqrt{T}}{1-\sqrt{T}} + \frac{1}{2}\sqrt{T}\} \quad (14)$$

for $T > \delta T$

$$\doteqdot \frac{z_p^2}{\pi} \cdot z_T \cdot r_0^2 \cdot \alpha \cdot \frac{m_e c^2}{T_r'} \cdot \frac{1}{\sqrt{T_r'}} \cdot 4\pi z_p \cdot \sqrt{R} \quad (15)$$

for $0 \le T \le \delta T$.

By replacing the electron bremsstrahlung cross section in Eq. (10) by those of Eqs. (14) and (15), the QFEB formula at $\theta = 90°$ is obtained by

$$\hbar\omega_L \cdot \sigma^{QFEB} (z_p,\hbar\omega_1,90°) = \frac{z_p^2}{\pi} \cdot r_0^2 \cdot \alpha \cdot m_e c^2 \cdot \Sigma_i n_i \{$$

$$\int_{-\infty}^{v_i^{max}} [(\frac{3}{4} - \frac{T}{4}) \ln\frac{1+\sqrt{T}}{1-\sqrt{T}} + \frac{1}{2}\sqrt{T}] \frac{\rho_i(v_z)}{T_r'} dv_z$$

$$+ \frac{2\pi \cdot \delta T \cdot z_p \cdot \alpha c}{\hbar\omega_L + I_i + \frac{1}{2}m(v_i^{max})^2} \times \rho_i(v_i^{max})\} . \quad (16)$$

The second term of Eq. (3) directly reflects the velocity distribution of orbital electrons as a function of v_i^{max} and is proportional to the third power of the projectile charge z_p in contrast to the z^2-dependence of the first term. From Eq. (16), the difference between the QFEB cross sections for the two kinds of projectile of equivelocity with the atomic charge z_1 and z_2 can be calculated and is expressed by

$$\hbar\omega_L \sigma^{QFEB} (z_2,\hbar\omega_L,90°)/z_2^2 - \hbar\omega_L \sigma^{QFEB} (z_1,\hbar\omega_L,90°)/z_1^2$$

$$\approx (z_2-z_1) \times 2 r_0^2 \alpha^2 m_e c^3 \cdot \delta T \Sigma_i n_i \frac{\rho_i(v_i^{max})}{\hbar\omega_L + I_i + \frac{1}{2}m_e(v_i^{max})^2} . \quad (17)$$

The distribution function $\rho_i(v_z)$ has a peak at $v_z = 0$. Thus the difference in QFEB has a peak at $\hbar\omega = T_r - I_i$ ($v_i^{max} = 0$). The result is just same as that of REC process[12].

COMPARISON WITH EXPERIMENT AND DISCUSSION

Here, the theory of QFEB developed in the last section is compared with the recent experimental results[3-5] on a Be target bombarded by protons and $^3He^{2+}$ ions with several tens of MeV/amu.

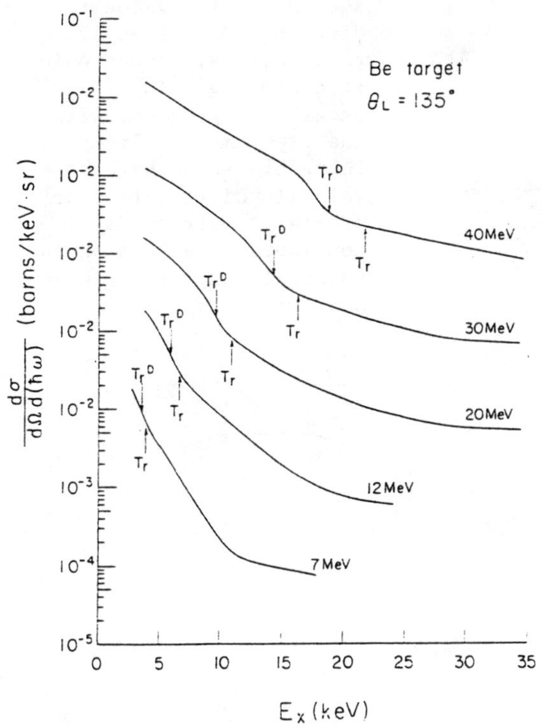

Fig 2. Projectile-energy dependence of QFEB spectrum (taken from Ref. 3).

The theory can be applied to the present case where the velocity of projectile is large enough in comparison with that of orbital electrons of Be target.

Figure 2 shows the continuum x-ray spectra from the Be target bombarded with 7-40 MeV protons[3] measured in the direction 135° to the proton beam. The end-point energy of QFEB varies with an increase in the proton energy and agrees with the value of T_r^D calculated by Eq. (4). The spectra[4] for the 20-MeV proton bombardment observed over the range of angle $\theta_L = 50°-148°$ are shown in Fig. 3. The end-point energy of QFEB varies with an increase in the angle: the Doppler shift. This fact proves that the QFEB process is a phenomenon in the projectile frame.

In Fig. 3, the smooth curves are obtained by fitting the continuum spectra in the region of $\hbar\omega = 15-30$ keV with a polynominal expression of $\hbar\omega$ using the least-squares method and the production cross sections of QFEB calculated from Eq. 4 are also presented with solid lines. It can be seen in this figure that the theory agrees well with the experiment except the region near the end point of QFEB. Figure 4 shows the angular distributions of QFEB obtained by subtracting the continuum background in Fig. 3 together with those calculated from Eq. (4). The discrepancy between the theory and the experiment seen in Figs. 3 and 4 is considered to come from the

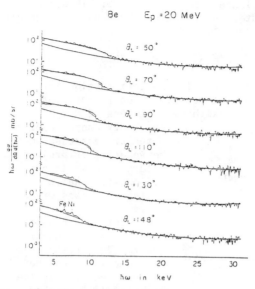

Fig. 3. Angular dependence of QFEB spectrum for 20–MeV proton bombardment (taken from Ref. 4).

Recently[5], we have measured the continuum x–ray spectra from the Be target bombarded with 20.14–MeV/amu–proton and $-^3He^{2+}$ ion in the direction 90° to the beam and have obtained the difference $\hbar\omega \cdot (\frac{1}{4}\sigma_h(\hbar\omega) - \sigma_p(\hbar\omega))$, where $\sigma_h(\hbar\omega)$ and $\sigma_p(\hbar\omega)$ are the production cross sections of continuum x rays for $^3He^{2+}$ ion and proton bombardments, respectively. The experimental result is shown together with theoretical calculations in Fig. 6. As predicted by Eq. (7), the continuum spectrum reflecting the velocity distribution of orbital electrons is obtained from difference in the cross sections. The theoretical prediction based on the formula Eq. (10), shown with a dashed

velocity distribution of target electrons. The spectrum[4] of QFEB obtained at θ_L = 90° is compared with theoretical calculations in Fig. 5, where the dot and dashed curve is obtained from Eq. (4) and the solid curve is obtained from Eq. (10) taking account of the velocity distributions of Be 1s and 2s electrons. The prediction from Eq. (10) is in excellent agreement with the experiment. It is therefore seen that the velocity distribution of orbital electrons has considerable effect on the spectral shape near the end point.

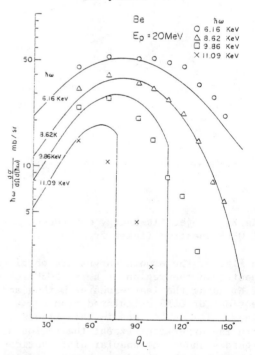

Fig. 4. Angular distribution of QFEB (taken from Ref. 4). The solid curves are theoretical predictions obtained from Eq. (4).

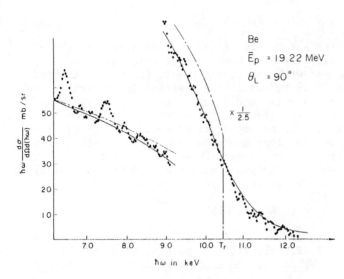

Fig. 5. The QFEB spectrum near the end-point
energy obtained at $\theta_L = 90°$ (taken from Ref. 4).

line in Fig. 6,
agrees well with
the experiment
while a small
difference is
seen at the
position of the
peak $\hbar\omega \approx T_r$.
This difference
at the peak
position may be
resolved by
taking account
of the REC
process which is
predominant in
the vicinity of
$\hbar\omega \approx T_r$. The
formula of the
cross section of
REC for the case
of $v_p \gg v_e$,
based on the

impulse
approximation,
have been given
by Kleber and
Jakubassa[12].
The solid line
in Fig. 6 shows
the difference
in the cross
sections taking
account of the
REC process for
protons and
$^3He^{2+}$ ions.
It can be seen
in Fig. 6 that
the agreement
between the
experiment and
the theory
corrected is
quite satis-
factory. The
REC process
have been well
studied in
heavy ion

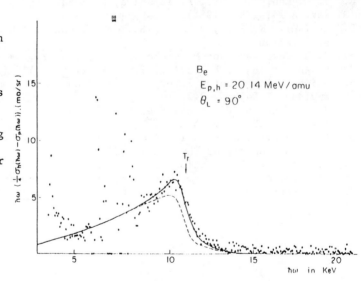

Fig. 6. The difference in the cross sections
$\hbar\omega(\frac{1}{4}\sigma_h(\hbar\omega) - \sigma_p(\hbar\omega))$ for proton and $^3He^{2+}$ ion
impacts (taken from Ref. 5). The peaks in the
region of $\hbar\omega = 6.4-8$ keV are due to K X rays from
impurities contained in the Be target.

impact but scarecely in light ion impact. Here, the contribution
of REC is clearly found for proton and $^3He^{2+}$ ion impacts.

414

SUMMARY

The theory of QFEB was developed and was compared with the experiments on Be target for the proton impact which have been previously reported, and with the recent result for the proton and $^3He^{2+}$ ion impacts at 20.14 MeV/amu. The spectrum in the region near the end point of QFEB was well reproduced by the theory taking account of the velocity distribution of the orbital electrons. It was pointed out that the difference in cross sections $\hbar\omega(\frac{1}{4}\sigma_h(\hbar\omega) - \sigma_p(\hbar\omega))$ for protons and ^3He ions bombardments of Be target, directly reflects the velocity distributions of Be 1s and 2s electrons. This result suggests that QFEB would provides a useful method for studying the state of valence electrons as REC process does.

REFERENCES

1. K. Ishii, S. Morita and H. Tawara, Phys. Rev. A 13, 131 (1976).
2. H. W. Schnopper, J. P. Delvaille, K. Kalata, A. R. Sohval, M. Abdulwahab, K. W. Jones, and H. E. Wegner, Phys. Lett. 47A, 61 (1974).
3. A. Yamadera, K. Ishii, K. Sera, M. Sebata and S. Morita, Phys. Rev. A 23, 24 (1981).
4. T. C. Chu, K. Ishii, A. Yamadera, M. Sebata and S. Morita, Phys. Rev. A 24, 1720 (1981).
5. K. Ishii, K. Sera, H. Arai, S. Morita, and K. Tokuda, (in preparation).
6. D. H. Jakubassa and M. Kleber, Z. Phys. A 273, 29 (1975).
7. R. Anholt and T. K. Saylor, Phys. Lett. 56A, 455 (1976).
8. Heither, The Quantum Theory of Radiation (Clarendon, Oxford, England, 1954), p. 242.
9. J. D. Jackson, Classical Electrodynamics, 2nd ed. (Wiley, New York, 1975), p. 522.
10. A. R. Sohval, J. P. Delvaille, K. Kalata, K. Kirby-Docken, and H. W. Schnopper, J. Phys. B 9, L25 (1976).
11. H. Bethe and E. E. Salpeter, Quantum Mechanics of One- and Two-Electron Atoms (Plenum, New York, 1977), p. 39.
12. M. Kleber and D. H. Jakubassa, Nucl. Phys. A252, 152 (1975).

Chapter 8   Soft X-Ray and Electron Energy-Loss Spectroscopy of
Atoms and Molecules

USX EMISSION SPECTROSCOPY OF ATOMS AND MOLECULES

Carl Nordling
Institute of Physics, University of Uppsala, Box 530
S-751 21 Uppsala, Sweden

ABSTRACT

Some developments of the experimental technique and several applications of USX (Ultra Soft X-ray) emission spectroscopy of atoms
and molecules are reviewed. Results are reported in the following
areas: Bond characteristics of molecular core hole states; binding
energies of core electrons; core hole lifetimes; valence electron
structure.

INTRODUCTION

In the November 1927 issue of the Physical Review [1] one finds an
article by T.H. Osgood with the title "X-ray Spectra of Long Wavelength". Here the concave grating grazing incidence technique for obtaining ultra-soft X-ray spectra (of solid phase samples) is introduced, Fig. 1 and Fig. 2.

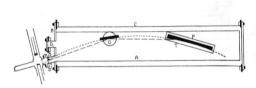

Fig. 1. Grazing incidence concave grating spectrograph used
by Osgood in 1927 for obtaining USX spectra [1]. The same
year Dauvillier recorded USX
spectra with a lead melissate
crystal as the dispersive
element [2].

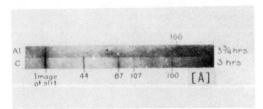

Fig. 2. Typical USX spectra
(of Al and C in the solid
state) recorded by Osgood [1].

Half a century later we are using the same technique to obtain
structural and dynamic information on free atoms and molecules.
Ultra-Soft X-rays (USX) have so long wavelengths (1 nm<λ<50 nm) that
in most of the region natural crystals cannot be used as dispersive
elements. Artificial soap film crystals can be made with atomic
spacings up to 2d≈16 nm, but for high resolution work in the USX
region the concave grating, used in grazing incidence, is the only
remaining expedient. There is considerable overlap in wavelength

416

between USX and XUV (Extreme Ultra Violet), the difference being the
type of physical processes involved. For the typical USX transition
the initial state involves a vacancy of one (or several) of the
outermost core electrons and the final state involves a valence
electron vacancy. High-resolution USX emission spectroscopy can
therefore be used as a probe of both core and valence electron
structure. There are several areas of interest at present time in
USX emission spectroscopy; the following examples derive from the
work done at our laboratory in Uppsala on gas phase samples. An
earlier report was given at the conference in Sendai in 1978 [3].

EXCITATION, DISPERSION, DETECTION

Considerable development of the experimental technique has
occurred over the last decade:

o Efficient means of exciting the sample gas by an electron beam
  have been used, Fig. 3.

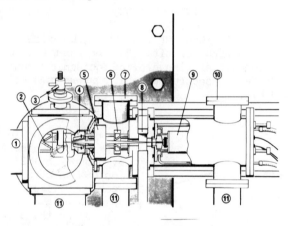

Fig. 3. Excitation of USX emission is made by electron beam in
a 10m-grazing incidence instrument at Uppsala.
1. To cryo pump
2. Water cooled Faraday cup used to dump the electron beam
3. Sample gas inlet
4. Water cooled slit
5. Magnetic quadrupole doublet for beam focusing
6. Water cooled aperture
7. Inspection window
8. Pneumatically operated vacuum valve
9. Two-stage electron gun (10 kV, 100 mA)
10. Flange for vacuum gauges
11. To oil diffusion pump

With synchrotron radiation (SR) one has made fluorescent excitation of solid samples [4] and the stronger SR sources now becoming available should be useful also for the study of USX
emission from free atoms and molecules.

o It has been possible to increase the efficiency of reflection gratings by almost an order of magnitude, from typically 1% to 10%; gratings can now be made by holographic techniques on toroidal backings etc. In our laboratory we are investigating the feasibility of using the radius of the grating as a parameter for the adjustment of the spectrograph, see preliminary data in Fig. 4.

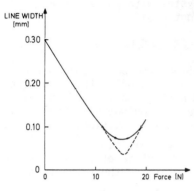

Fig. 4. By applying a force to the center of the concave grating one can change its radius of curvature slightly and use this change of radius as a parameter for adjusting the spectrograph. The figure shows the measured linewidth at λ=15 nm in the 10-m USX instrument vs applied force on the grating. The dashed curve corresponds to the linewidth predicted from ray tracing calculations. The difference is ascribed to Doppler broadening in the source and non-spherical shape of grating.

Fig. 5. Photon counting USX detector for the 10-m instrument in Uppsala.
1. USX-to-optical wavelength converter (e.g. Y_2O_3 ; Eu)
2. Fiber optics
3. Photo cathode layer
4. Grid electrode
5. Cylindrical acceleration electrodes
6. Electromagnetic focusing lens
7. Photodiode array with very thin (<1 μm) quartz window
8. Detector carriage with precision movement along Rowland circle
9. Cooling block to reduce dark current in diode array
10. Detail of diode array. Accelerated electrons (~15 keV) pass through thin window. Each electron gives sufficient charge on one of the diodes of the array to allow the recording of individual USX photons.

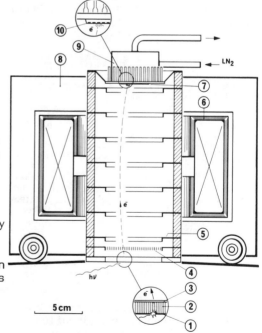

o It is hard to surpass the photographic plate as a multichannel
   detector for USX, but modern electronic devices, used as photon
   counters, may soon supersede photographic detection. Fig. 5
   shows a proposed photon counting USX detector for the 10-m instrument at Uppsala. Instead of the image intensifier with
   accelerating electrodes plus diode array shown in the figure it
   may be more advantageous to use a system with multichannel
   plates and position sensitive read-out.

BOND CHARACTERISTICS OF MOLECULAR CORE HOLE STATES

A high-resolution USX emission spectrum from a molecule often
shows vibrational excitation of the initial core hole states and the
final valence hole states. A Franck-Condon (FC) analysis of the vibrational bands of an X-ray transition has to include all three
electronic states which are involved in such a transition, namely

o the ground state from which the primary excitation is made
o the core hole state which is the initial state of the transition
o the valence state which is the final state of the transition

Each of these states can be represented by potential energy curves,
one for each vibrational mode. Since the ionization of a core electron induces reorganization of the valence electrons the core state
energy function may differ considerably from the ground state energy
curve, not only in the value of its energy minimum but also in the
corresponding bond length and vibrational frequencies (force constants).

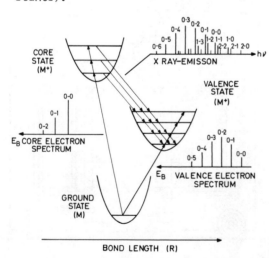

Fig. 6. Potential energy
curves of a diatomic molecule
and the formation of vibrational fine structure in UPS-,
XPS- and USX-spectra. Vibrational quantum numbers of the
two states involved in each
transition are indicated.

Likewise the valence state energy functions may have quite different bond lengths and vibrational energies. As a result one will
observe in the USX emission spectrum a vibrational structure which
contains several progressions, one for each vibrational excitation of

the core state. The forming of vibrational structure in USX spectra
and in the corresponding electron spectra (core- and valence electron spectra) is illustrated in Fig. 6. Spectra from five molecules
have till now been analysed in this fashion, see Table I.

The most recent analysis was made for the CO_2 molecule [5]. Here
C1s ionization leads to a bond length shortening of 2.0 pm while
O1s ionization mainly results in a displacement of the carbon atom
of 5 pm. Core ionization of one of the oxygen atoms therefore breaks
the $D_{\infty h}$ symmetry of the molecule, but as long as the FC analysis is
confined to the harmonic approximation one cannot tell whether the
carbon atom moves towards or away from the ionised oxygen atom. This
restriction of the present FC analysis could be relaxed when the
more accurate photon counting techniques are introduced for the recording of spectra.

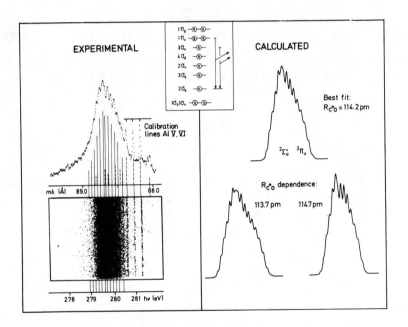

Figure 7. The C K emission spectrum of CO_2, recorded in
the second order of diffraction [5]. Lower left: the photographic plate; upper left: the densitometer spectrum, the
bars representing the most prominent peaks. To the right
are shown Franck-Condon fits of the vibrational fine
structure: At the top of the figure the best-fit spectrum
is shown and below this two calculated spectra demonstrating the sensitivity of the C-O bond length of the C1s
core state to the profile of the spectrum.

Fig. 7. shows the carbon K spectrum of CO_2 and three of its FC fits. Fig. 8 shows the oxygen K spectrum with FC fitted profiles. Three parameters were used to obtain the best fit of the carbon spectrum, which is a composition of the $(C1s)2\sigma_g^1 \rightarrow 3\sigma_u^1$ and $(C1s)2\sigma_g^1 \rightarrow 1\pi_u^3$ transitions:

o the bond lengths

o the $\tilde{A}\ ^2\pi_u/\tilde{B}\ ^2\Sigma_u^+$ intensity ratio

o the natural lifetime width of the $(C1s)\ 2\sigma_g^1$ state

Moreover it was assumed that the contribution from excitation of bending vibrations is negligible and that lifetime interference effects can be neglected [6]. To obtain a unique best fit a $^2\pi_u/^2\Sigma_u^+$ intensity ratio of 0.43 was used, and the lifetime width was determined to be 70 ± 20 meV, see below.

The oxygen K-emission spectrum of CO_2 was analysed in terms of symmetric ás well as antisymmetric stretching vibrations, while the influence of bending vibrations was assumed to be negligible. The parameters that were used to obtain the best fit were the changes in the symmetric and antisymmetric vibrational coordinates upon O1s ionisation.

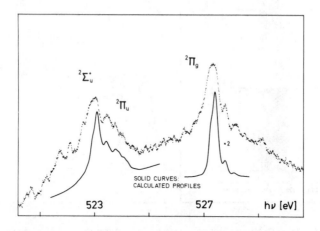

Figure 8. The oxygen K emission spectrum of CO_2, third order of diffraction [5]. The solid curves represent the Franck-Condon fit of the $^2\Sigma_u^+$, $^2\pi_u$, and $^2\pi_g$ bands for a change of 5 pm in the antisymmetric stretching coordinate, and an increase of 1 pm in the symmetric stretching coordinate upon O1s ionization. The intensity of the calculated $^2\pi_g$ band should be multiplied by 2.

BINDING ENERGIES OF CORE ELECTRONS

"Binding energies of the inner electrons of atoms have almost exclusively been studied by the method of X-ray absorption". This statement from one of the first doctoral theses on photoelectron spectroscopy is no longer valid when one considers the massive amount of data produced by electron spectroscopy (XPS and UPS) during the last two decades. This does not mean, however, that X-ray spectroscopy has become obsolete for the determination of electron binding energies. In fact, it has recently become feasible to use USX emission spectroscopy for accurate determinations of core electron binding energies in the 100-1000 eV range [10]. The procedure is to add, in a proper fashion, USX transition-energies and valence electron binding energies, the latter being obtained e.g. from UPS, see Fig. 9.

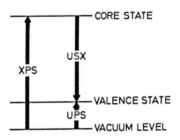

Fig. 9. Diagram of the relation between XPS, UPS and USX energies.

Calibration of spectra is in grazing incidence more elaborate than in normal incidence. The wavelenth dispersion, which is approximately constant in normal incidence, is strongly wavelength dependent at grazing angles and very accurate positioning is required of the optical elements of the instrument, particularly the image (detector) surface. In order to account for the deviations in the wavelength dispersion from the theoretical dispersion one can use well established XUV emission lines as calibration standards. The superposition of calibration lines from a discharge source can be carried out in a higher order recording of the X-ray spectrum which means that quite simple arrangements for producing the calibration line spectrum can be used. Thus, the design of a calibration source which allows convenient interchange with the ordinary X-ray source is facilitated.

Figure 10 shows the superposition of the second order carbon K emission spectrum of CO_2 and the corresponding range of first order XUV emission lines. The dispersion relation is established by a least square fit, and as can be seen in the inserted diagram the mean deviation is of the order of 0.1 pm. This corresponds in the present case to a mean deviation of about 3 meV in the X-ray transition energy. In fact, the error of 10 meV quoted for the adiabatic C K emission energy is mainly the error in the determination of the

position of the adiabatic transition in the spectrum. The resolved
vibrational peaks of the C K spectrum are not single-component peaks
but a superposition of several components. Therefore the position of
the adiabatic transition in the spectrum has to be established from
a fit of the vibrational fine structure. The quality of the fit is
in this case the limiting factor for the accuracy in the energy determination.

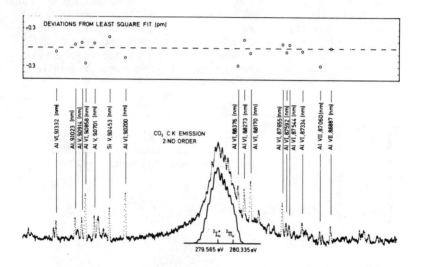

Figure 10. Calibration of the second-order C K emission
spectrum of CO_2 using XUV lines in the 9 nm region [10].
The mean deviation of the XUV wavelengths from the least-
squares fitted dispersion relation is of the order of
0.1 pm (corresponding to 3 meV in the C K spectrum).
The adiabatic X-ray transition energy is determined from
the Franck-Condon fit to the vibrational fine structure.

Table II contains core electron binding energies obtained in
the manner described for a number of atomic and molecular gases.
The quoted errors are typical at the present state-of-the-art for
the noble gases and for the C K and O K energies of CO_2. In the
other cases the errors are representative for similar measurements
at an earlier stage in the development of the technique.

CORE HOLE LIFETIMES

Lifetimes of USX core hole states are typically of the order of 10-100 fs and the corresponding natural lifetime widths are of the order of 10-100 meV. Only few measurements have been made of these lifetime widths because of the high resolution which is required. (Some data have also been provided by Auger- and EEL measurements.) In atoms the final state of the USX transition has in most cases negligible influence on the line width. In molecules where vibrational excitations are prominent, a detailed analysis of the vibrational fine structure has to be made. Eight core hole lifetimes of three atomic and molecular species have so far been determined from high-resolution USX spectra:

o   six L and LM vacancy states in argon [11]

o   K state in neon [12]

o   carbon K state in CO_2 [5]

Fig. 11 shows three fits of the CK emission spectrum of CO_2 with the width of the C1s state as the fitting parameter. The best fit (compare experimental spectrum in Fig. 7) is obtained for a Lorentzian lifetime width of

$$\Gamma_{1s} = 70 \pm 20 \text{ meV}$$

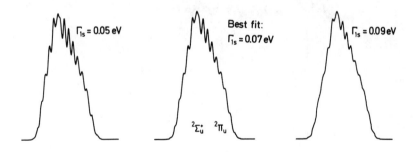

Fig. 11. The carbon K emission spectrum of CO_2 reproduced with different natural life time widths of the C1s state.

VALENCE ELECTRONIC STRUCTURE

When USX spectra are used for the study of the valence electron structure of molecules the dipole selection rules impose constraints with regard to symmetry and localization of the molecular orbitals. This necessity can often be made a virtue. For example, the long standing problem of the ordering of the valence shells of SF_6 seems to have been brought one step closer to its solution by the USX spectrum of the SF_6 molecule [13]. In Fig. 12 a comparison can be made between XPS and USX spectra of SF_6. Both show the valence orbital region; in particular both show a peak at 18.6 eV energy. Selection rules are not as restrictive in XPS, and both gerade and ungerade states are observed. One cannot therefore a priori say that the peak at 18.6 eV is of one symmetry or the other. In the USX spectrum, on the other hand, one can only reach final states of gerade symmetry. The peak at 18.6 eV (represented in the sulphur $L_{II,III}$ emission spectrum by two peaks at a distance equal to the sulphur 2p(1/2), 2p(3/2) spin-orbit splitting) therefore must have gerade symmetry. Concrete confirmation of the $3e_g$ assignment of this peak is therefore obtained from the USX spectrum. A discussion of the valence levels of SF_6 is given by Dehmer and coworkers in a forthcoming paper [14].

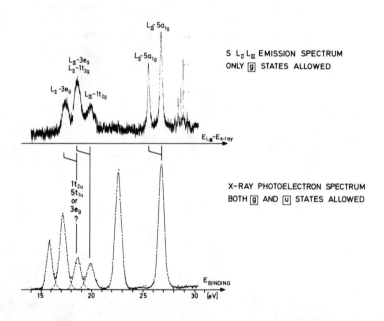

Fig. 12. The selection rules that govern the X-ray transitions can be used to facilitate assignment of molecular orbitals. For example, in the S L emission spectrum of SF_6 only final states of g symmetry are allowed.

The concept of molecular orbitals has been useful for the analysis of USX spectra. In the one-center LCAO model [15] one can relate the observed X-ray intensities to calculated local atomic contributions to the different molecular orbitals. Thus the ordinary dipole selection rules single out the 2p contributions in the K-emission of second row elements, and it has been shown that the use of the one-center intensity model as a guide for the assignment of these spectra is justified at both ab initio and semiempirical (CNDO) levels of approximation [16]. Calculations which include two-center contributions have been made for some molecules. In the investigation of the sulphur L-emission spectrum of SF_6 it was shown that the two-center contributions were significant [13], while this was not the case for the second row elements [16].

An example of the use of one-center calculations as an aid for the analysis of USX spectra is shown in Fig. 13. Here the carbon and nitrogen K-emission spectra of aminobenzene are compared with each other and with one-center ab initio calculations [17].

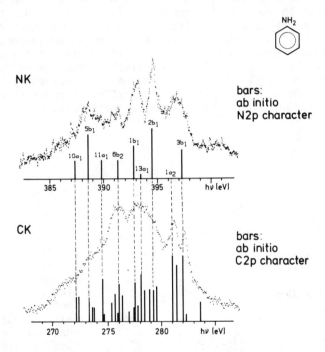

Fig. 13. The nitrogen K and carbon K emission spectra of aminobenzene. Calculated intensities are indicated by bars.

The dipole selection rules of the X-ray transitions assign considerable intensity to the N π-like orbitals, which facilitates the identification of these orbitals in the N K spectrum. Particularly, the nitrogen 2p "lone pair" orbital is identified ($2b_1$), as well as the $1b_1$ orbital. The latter has been assigned in UPS spectra by arguments of the "perfluoro effect" and the USX spectra support this assignment. An appreciable difference in width of the $3b_1$ band in the nitrogen K and carbon K spectra is interpreted in terms of geometric relaxation

of the N 1s and C 1s states. It is suggested that ionizing a C 1s
electron leads to a state of equilibrium geometry not very different
from the geometry of the $3b_1$ hole state, while the ionization of an
N 1s electron is associated with a considerable change, and the bond
geometry is further removed from the $3b_1$ geometry.

A comparison of the aminobenzene spectrum with the corresponding
spectrum of unsubstituted benzene confirms that the NH_2 radical introduces a splitting of the outermost $1e_{1g}$ orbital into its b_1 and a_2
components. Three recordings of the benzene USX spectrum are shown in
Fig. 14, two from the solid phase, excited with bremsstrahlung [18] and
synchrotron radiation [4], and one gas phase spectrum, excited by electron beam [19]. As can be seen in the figure the USX emission spectrum
of benzene remains essentially unchanged between the gas phase and the
solid phase. The main difference is a uniform shift of the solid
phase spectrum to lower photon energies. The shift amounts to 0.7 eV
and it is interpreted as the difference in extra-molecular relaxation
between the localized core state and the delocalized valence states.
The corresponding shift obtained as the difference between core and
valence solidification shifts in photoelectron spectra is 0.6 eV. The
solid and gas phase USX spectra show quite different satellite structure. This difference is related to the different modes of excitation
which are used, photons and electrons, respectively.

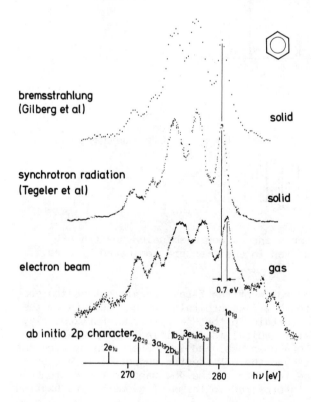

Fig. 14. USX emission
spectra of solid and
gas phase benzene ex-
cited with photons and
electrons, respect-
ively. Calculated in-
tensities are indi-
cated by bars.

Table I Core state bond characteristics derived from USX spectra

Molecule	USX-transition	change in bond length[a] (pm)	core state vibrational energy (eV)	
CO [7]	C1s-5σ	-5.5	0.34±0.01	
	O1s-1π/4σ	>0	-	
NO [7]	N1s-2π	+0.3[b]	-	
N$_2$ [7]	N1s-3σ$_g$/1π$_u$	-2.4	0.31±0.03	
NH$_3$ [8]	N1s-3a$_1$	-4.1[c]	-	
CO$_2$ [5]	C1s-3σ$_u$/1π$_u$	-2.0	0.17±0.05	
	O1s-3σ$_u$/1π$_u$/1π$_g$	+6.0(C-O*)[d] -4.0(O-C)[d]	0.32±0.05	0.14±0.05[e]

a) between neutral ground state and core hole state
b) see also discussion in ref. 7
c) core state (as well as valence state) is planar
d) sign based on calculations by Clark and Müller [9]
e) symmetric and antisymmetric stretching vibrations respectively

Table II Core electron binding energies obtained in USX emission [10]

Species	core state	valence state	core electron binding energy (eV)
SF$_6$	S2p$_{1/2}$	5a$_{1g}$	181.5±0.1
	S2p$_{3/2}$	5a$_{1g}$	180.2±0.1
Ar	2p$_{1/2}$	3s	250.777±0.010
	2p$_{3/2}$	3s	248.629±0.010
CO$_2$	C 1s	1π$_u$	297.651±0.010
	O 1s	1π$_g$	541.08±0.05
N$_2$	N 1s	3σ$_g$	409.52±0.10
Ne	1s	2p$_{1/2,3/2}$	870.21±0.05

REFERENCES

1. T.H. Osgood, Phys. Rev. 30, 567 (1927).
2. M.A. Dauvillier, J. de Physique et Rad. Ser VI T VIII No 1, 1 (1927).
3. C. Nordling, Japanese J. Appl. Phys. 17, Suppl. 17-2, 7 (1978).
4. See e.g. E. Tegeler, G. Wiech, A. Faessler, J. Phys. B 13, 4771 (1980).
5. J. Nordgren, L. Selander, L. Pettersson, C. Nordling, K. Siegbahn, J. Chem. Phys. 76 (8), 3928 (1982).
6. F. Kaspar, W. Domcke, L.S. Cederbaum, Chem. Phys. 44, 33 (1979).
7. H. Ågren, L. Selander, J. Nordgren, C. Nordling, K. Siegbahn, Chem. Phys. 37, 161 (1979).
8. H. Ågren, J. Müller, J. Nordgren, J. Chem. Phys. 72, 4078 (1980).
9. D.T. Clark, J. Müller, Chem. Phys. 23, 429 (1977).
10. L. Pettersson, J. Nordgren, L. Selander, C. Nordling, K. Siegbahn, J. El. Spectr. 27, 29 (1982).
11. J. Nordgren, H. Ågren, C. Nordling, K. Siegbahn, Physica Scripta 19, 5 (1979).
12. H. Ågren, J. Nordgren, L. Selander, C. Nordling, K. Siegbahn, J. Electron Spectr. 14, 27 (1978).
13. H. Ågren, J. Nordgren, L. Selander, C. Nordling, K. Siegbahn, Physica Scripta 18, 499 (1978).
14. J.L. Dehmer, A.C. Parr, S. Wallace, D. Dill, J. Chem. Phys., to be published.
15. R. Manne, J. Chem. Phys. 52, 5733 (1970).
16. H. Ågren, J. Nordgren, Theoret. Chim. Acta (Berl.) 58, 111 (1981).
17. J. Nordgren, L. Pettersson, R. Brammer, M. Bäckström, C. Nordling, H. Ågren, L. Selander, to be published.
18. E. Gilberg, M.J. Hanus, B. Foltz, Japanese J. Appl. Phys. 17 Suppl. 17-2, 101 (1978).
19. J. Nordgren, L. Pettersson, R. Brammer, M. Bäckström, C. Nordling, H. Ågren, L. Selander, to be published.

RECENT ADVANCES IN INNER-SHELL EXCITATION OF FREE MOLECULES BY ELECTRON ENERGY LOSS SPECTROSCOPY

C.E. Brion, S. Daviel, R. Sodhi and A.P. Hitchcock†
Department of Chemistry
The University of British Columbia
Vancouver, B.C. V6T 1Y6
Canada

ABSTRACT

A brief description is given of the underlying principles and important features of gas phase inner shell electron energy loss spectroscopy (ISEELS) Existing work in the field prior to 1980 is summarized. Recent ISEELS studies of Ne, Ar, HF, F_2, Cl_2, N_2 and NF_3 are discussed. The design and operation of a new, high performance, high resolution ISEELS spectrometer is presented.

INTRODUCTION

Electron energy loss spectroscopy (EELS) is a well established technique [1,2] for the study of gas phase atomic and molecular valence shell electronic spectra at both low and high resolution. In the past decade the technique has been extended [3,4] to the study of core electron excitation and inner shell electron energy loss spectroscopy (ISEELS) is now producing a large amount of new and interesting spectroscopic data. It is the purpose of the present article to briefly review the achievements of ISEELS to date and to report on recent progress and new developments in the field. The discussion will in general be restricted to studies of free (gas phase) atoms and molecules.

BACKGROUND

Electronic spectroscopy has been traditionally carried out mainly by studies of the absorption and emission of electromagnetic radiation. However, examination of the literature reveals that only limited information is available, particularly for molecules, in the far UV and X-ray regions of the spectrum, i.e. at energies above about 10 eV - that is at wavelengths shorter than about 1200 Å. These higher energy regions are important as they cover the range of many highly excited electronic states of valence electrons as well as all inner shell transitions and most ionization and fragmentation processes. Although a large body of information on ionization energies is available from photoelectron spectroscopy using UV and X-ray line sources there is very little absolute cross-section information on the photoabsorption or total and partial photoionization and fragmentation of molecules at continuous energies beyond 20 eV.

† Present address: Department of Chemistry, McMaster University, Hamilton, Ontario, Canada.

The main reason for this shortage of information has been the limited availability of continuum light sources at energies above 20 eV ($\lambda <$ 600 Å). Conventional light sources using hydrogen and noble gas continua have provided useful but relatively weak structured continua up to ~20 eV [5]. Until the relatively recent advent of tuneable synchrotron radiation there has been no effective source of far UV and X-ray continuum radiation with the exception of a few weak bremmstrahlung continua in restricted regions. Even where synchrotron radiation has been available, optical studies have not always been easy to carry out due to the problems involved in the dispersion of short wavelength radiation. For example the low reflectivity of mirrors and gratings at short wavelengths severely attenuates the useable photon intensity despite the high flux from the synchrotron itself. Also such sources are of enormous expense and remotely located from most users laboratories. Other difficulties, particularly at short wavelengths, include order overlapping, stray light corrections and absorption by optical components, particularly in the carbon K region due to surface contamination. Despite these difficulties great progress [6] is being made in the utilization of synchrotron radiation and many elegant experiments are now being done. However, the scope of many types of molecular measurements is still rather limited, particularly in the gas phase. This limitation seems to be mainly due to fears of violation of the ultra-high vacuum integrity of the storage ring.

In view of this situation what alternative methods exist for exciting and ionizing molecules? It has long been known that inelastic collisions of particles, in particular electrons, can also be used to achieve energy transfer to molecules, generally with larger cross-sections than with photons. Indeed the early experiments of Franck and Hertz [7] and others showed that in many ways an electron interacted with an atom or molecule in a similar fashion to a photon in that excitation, ionization and dissociation were observed. With electrons continuous energy transfer is possible over the whole electromagnetic spectrum. In 1930 Bethe [8], using the Born approximation showed that for fast electrons there was a quantitative relationship between the differential electron scattering cross-section and the generalized oscillator strength df/dE(K).

$$\frac{d\sigma}{dE} = \frac{2}{E} \cdot \frac{\overline{k}_n}{\overline{k}_0} \cdot \frac{1}{K^2} \frac{df(K)}{dE} \tag{1}$$

where $\overline{k}_0$, $\overline{k}_n$ and K are the incident, scattered and transferred momenta respectively and E is the energy loss.

Bethe also showed that the generalized oscillator strength could be expanded in a power series of K^2 with the first term equal to the optical or dipole oscillator strength.

$$\frac{df}{dE}(K) = \frac{df_0}{dE} + AK^2 + BK^4 + \dots \tag{2}$$

In the limit of zero momentum transfer it can be seen that the generalized oscillator strength becomes equal to the optical oscillator strength (OOS) df_o/dE. The OOS is simply related to the cross-section for photoabsorption or photoionization (σ_{ph}) by the relation

$$\frac{df_o}{dE} = \frac{mc}{\pi e^2 h} \sigma_{ph} = \frac{E}{2} \frac{\bar{k}_o}{k_n} K^2 \frac{d\sigma}{dE} \tag{3}$$

This important (Bethe-Born) relationship shows that not only are dipole transitions induced in atomic or molecular targets by fast electrons but that in principle their absolute intensity can be obtained by kinematic conversion of the differential electron scattering intensities. Thus it is possible to carry out quantitative optical spectroscopy without any photons! - i.e. "spectroscopy in the dark" - and at any energy transfer (or equivalent wavelength). In 1961 Bethe's theoretical work was discussed and clearly explained in Inokuti's noteable review article "The Bethe Theory Revisted" [9]. Such theoretical ideas were also discussed by Lassettre et al. [10] who then, in an elegant series of electron energy loss studies, obtained optical oscillator strengths for a number of discrete atomic and molecular valence shell transitions by extrapolating scattering intensity measurements at varying scattering angle to zero momentum transfer [1,2]. Meanwhile Van der Wiel had shown that it was possible to avoid the tedious procedure of extrapolation by making direct electron scattering measurements sufficiently close to the optical limit [11]. In practical terms this means using fast electrons (> 3 keV for valence electrons) at zero scattering angle (typically with a solid angle of 10^{-4} steradians centered about θ = 0°). Moving away from zero scattering angle increases the momentum transfer and thus the probability of non-dipole transitions.

Van der Wiel and his co-workers made a series of absolute measurements on the photoabsorption, photoionization and fragmentation of atoms and molecules using this so called "poor man's synchrotron" at modest resolution ($\sim$ 1eV) over an equivalent photon energy range (electron energy loss) up to 400 eV (i.e. λ down to 30 Å). This work was mainly concerned with ionization continuum processes involving both valence and core electrons although inner shell excitations were studied at modest resolution [12,13,14], including some absolute cross-section measurements [15]. These and later experiments [16] have shown that the use of fast electrons and coincidence techniques together with the Bethe-Born kinematic conversion produces quantitative results equivalent to those that would be obtained optically for a wide range of photoabsorption and photoionization phenomena.

The ability of fast electrons to induce dipole allowed (optical) transitions can be qualitatively understood in terms of the "virtual photon field". Figure 1 illustrates the principal effects occuring [16] when a fast electron interacts with a target molecule via a distant collision (large impact parameter and thus small scattering angle). As the electron passes by, the target experiences a sharply pulsed electric field. Ideally, in the limit the E field will approach a delta function which, if Fourier transformed into the frequency domain, would afford the perfect spectroscopic "light"

432

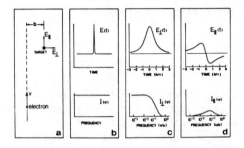

Figure 1: Virtual photon field

source consisting of a continuum composed of all frequencies at equal intensities. In practice, the pulse will have a narrow but finite width, and there will be a fall off in intensity of "pseudo-photons" at high frequencies. (It should be stressed that this method does not produce photons; rather, under the appropriate conditions, the electron-impact differential cross section is related to the optical cross-section by kinematic factors alone - see Equation 3). Nevertheless, a sufficiently wide spectral range can be achieved with impact energies readily obtainable in the laboratory [16].

It is of interest to compare the processes of resonant photo-absorption and electron impact excitation

$$h\nu + AB \rightarrow AB^* \qquad\qquad\qquad (4)$$
$$E$$

$$e + AB \rightarrow AB^* + e$$
$$E_0 \qquad\qquad (E_0 - E) \qquad\qquad\qquad (5)$$

It is obvious that the energy loss (E) of the incident electron of energy E_0 is analogous to the resonant photon energy (E) required to produce the excited state of quantum energy E. It is of importance to note that the non resonant property of electron excitation means that EELS studies are not subject to line saturation effects [9] which not infrequently result in spurious intensities in optical spectra. Thus it is possible to study the process of energy transfer [13] by electron energy loss spectroscopy (EELS). It is of particular advantage to use EELS in the UV and soft X-ray regions of the spectrum where continuum light-source availability is limited.

In actual practice only a <u>relative</u> photoabsorption cross-section is obtained from a Bethe-Born corrected electron scattering experiment since detector efficiencies and incident, scattered and target particle fluxes are not determined. However the flat virtual photon field (figure 1) of a fast electron (unlike photon continuua which always have an intensity which varies with energy, in a non straight-forward fashion) directly reproduces the true <u>shape</u> of the dipole oscillator strength (cross-section) after Bethe-Born kinematic correction. This relative curve, obtained out to very high energy loss, is readily put on an absolute scale by application of the TRK sum rule which states that the total sum of the oscillator strengths for all transitions is equal to the total number of electrons in the atom or molecule [16]. This procedure has been used effectively for a wide range of studies for valence and inner shell [4,16] spectra. In those few cases where absolute direct optical measurements have been made excellent agreement is found with the fast electron simulation [4,16-19]. Such absolute dipole oscillator strength data determined

by EELS for COS [19] and CS_2 [20] have even been used to put relative synchrotron data [21] on to an absolute intensity scale!

A detailed discussion of the background theory, technique,
applications and results of using electron impact "photon" simulation
experiments for optical oscillator strength measurements has been
given in recent review articles [4,16]. A comprehensive programme of
oscillator strength measurements for photoabsorption, photoionization
and fragmentation of molecules in both valence and core regions is
continuing in this laboratory over the energy loss (photon-energy)
range up to 1000 eV. As well as the basic spectral and oscillator
strength (cross-section) information, the data is being used to investigate the detailed dipole breakdown pattern of molecules. This work
finds application in areas including radiation induced decomposition,
radiation chemistry, biology and physics, dosimetry, aeronomy and
space physics, fusion, high temperature chemistry, electron microscopy and the development and evaluation of quantum mechanical procedures.

Using the fast electron EELS technique photoabsorption (excitation) studies can be carried out at a wide range of energy transfers.
Inner shell studies are particularly advantageous due to the superior
energy resolution advantages in electron impact spectroscopy at high
energy losses [16]. This advantage arises due to the inverse relationship between energy and wavelength. In EELS the energy resolution may be kept constant, independent of energy (energy loss), by
using electron optical lenses
and retardation whereas optically the energy resolution de-
grades with increase in energy.
This is illustrated in figure 2
which shows the equivalent value
for $\Delta\lambda$, in Å, of fixed EELS ΔE
resolutions from 0.01 eV (state
of the art using electron monochromators) to 0.5 eV (unmono-
chromated incident electron beam
using oxide cathodes). The
higher resolution achieved in
electron energy loss is well

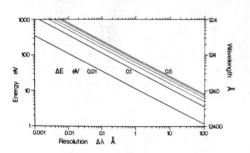

Figure 2: Resolution diagram

illustrated by recent studies of the K-shell excitation of N_2 [22,23]
(see fig. 16) compared with synchrotron photoabsorption work [24].

A further advantage of electron energy loss spectroscopy is that
a wide spectral range (IR → X-ray) can be covered in a single scan
with a single spectrometer. This is well illustrated in the cases of
CF_4 [25] and the methyl halides [26].

Electron energy loss measurements on molecules were made as
early as 1930 by Rudberg [27]. Using primitive apparatus Rudberg
observed valence shell transitions in N_2 but was unsuccessful in a
search for core excitations, probably due to the large background of
scattered electrons from other processes. However little work in
EELS was reported until the 1960's when a considerable resurgence of
interest resulted from the published works of Simpson and Kuyatt
[28], Schulz [29], and the group headed by E.N. Lassettre [1,2]. The

434

latter work was concentrated on valence shell molecular spectroscopy
at high resolution. Very high resolution (0.01 eV) EELS spectra were
also reported in 1966 by Boersch and Geiger [30]. However excitation
of inner shell electrons was not studied by electron impact until the
late 1960's when Van der Wiel and co-workers used electron-ion coincidence studies to make absolute dipole oscillator strength measurements for core ionization processes in a variety of targets including
noble gases [31], CO and N_2 [12]. In subsequent studies of CO and
N_2 [13] very intense bands were observed in the energy loss spectrum corresponding to 1s → π* excitation. Furthermore the ionic
fragmentation resulting from this K-shell excitation vacancy was
observed to be very different from that arising from the K-shell ionization continuum. More recently ionic fragmentation has been

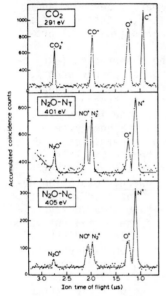

Figure 3: (e, e+ion)
coincidence [33]

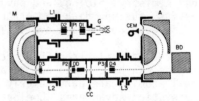

Figure 4 ISEELS spectrometer [26]

investigated from core-excited states of
SF_6 [32] and also as shown in figure 3
for CO_2 and N_2O [33]. Following this
work [13] in Amsterdam a systematic low
resolution (0.5 eV FWHM) ISEELS study of
molecular spectroscopy was initiated in
1970 at the University of British Columbia. In a series of studies Wight et al.
[3,25,34-38] reported ISEELS spectra
($ΔE ∼ 0.5$ eV FWHM) for some sixteen molecules at an impact energy of 2.5 keV and a
small scattering angle (∼1°). Under these
conditions the spectra are dipole-
dominated. Many new and interesting phenomena were observed including K-shell
resonances [3,23,39], potential barriers [25], core-analogy interpretations [3,40] and chemical shift phenomena [38].

In 1974 Hitchcock and Brion continued this programme and also
built a high resolution ISEELS spectrometer (figure 4) incorporating
an electron monochromator capable of energy resolution for valence
shells down to 0.02 eV [26]. A series of inner shell studies through
to 1981 at medium and high resolution were reported [23,26, 41-52].
Further investigation of resonance and potential barrier effects
[23,44,45,47,51] were made in a variety of compounds. Isotopic substitution [42], in some cases together with high resolution [48], was
used to explore isotope effects and to delineate between vibrational
and Rydberg structure. The carbon-K excitation region in a variety
of organic molecules [26,41-43, 48,49] was also studied. This energy
region is particularly challenging to study optically due to absorp-

tion of radiation by adsorbed carbon containing molecules on mirrors
and the diffraction grating (even in UHV systems). Nevertheless
studies have been made in this region using synchrotron radiation
[53,54]. Inner-shell electron energy loss spectra for the molecular
series CH_xCl_{4-x} indicated [45] extended energy loss fine
structure, EXELFS, (analogous to EXAFS and similarly due to interference between directly ejected electrons and electrons back scattered off surrounding atoms) in the chlorine L-shell continuum. Such
EXELFS patterns provide structural information in addition to the
spectral data from ISEELS.

In 1976 Tronc et al [55] at the University of Manchester published the first high resolution (0.070 eV FWHM) ISEELS spectra, for
the molecules CO and CH_4. This was followed in 1977 [22] by a
spectrum showing vibrational structure excited in the 1s → π* transition for N_2 observed earlier at low resolution by Wight et al [3].
The existence of this vibrational structure was later confirmed by
Hitchcock and Brion [23]. The resolution (0.075 eV FWHM) used by
King et al [22] at an energy loss of 401 eV corresponds to an optical
spectral resolution of 0.006 Å at 31 Å. This is approximately an order of magnitude higher than has currently been realized
optically [24]. The high resolution ISEELS spectrum permits an
accurate evaluation of the line width and thus the lifetime of the
state ($\Delta E.\Delta t = 7 \times 10^{16}$ eV secs). It was found that the lifetime
of the $(1s)^{-1}(\pi*)$ state of N_2 was only of the order of two
classical vibrational periods. The Manchester group has reported
further high resolution studies of Ar and Kr [56] which show both
dipole allowed and forbidden transitions, the carbon K-shell spectra
of a number of small molecules [57] and nitrogen K-shell spectra of
N_2, NO and N_2O [58]. King et al [61,62] have also reported negative ion resonances associated with inner shell excited states of
N_2, CO, NO, N_2O and CO_2.

Studies of the Bethe surface [9] of N_2 [85] and CO_2 [86],
including the inner-shell region, have been reported by Bonham,
Wellenstein and co-workers. This work gives an excellent overview of
the energy-momentum transfer surface and the angular dependences of
inner-shell excitation and ionization by electron impact.
Corresponding studies of noble gases had earlier been reported by
Afrosimov et al [87].

RECENT PROGRESS IN ISEELS

a) ISEELS SPECTRA

Shaw et al [59] have recently reported the $L_{2,3}$ electron
energy loss spectrum of Cl_2 obtained at an impact energy of 1.5 keV
and with an energy resolution of 0.065 eV. The spectrum (figure 5)
shows two series of lines leading to the L_3 and L_2 edges respectively. The equivalent core model suggests [59] that the relative
energies of core excited Cl_2 should be equivalent to those of
valence excited ArCl, a species of current interest in excimer lasers
[64].

436

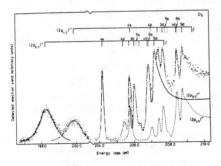

Figure 5: Cl_2 L-shell [59]

The fluorine K-shell excitation spectra of both F_2 and HF (figures 6 and 7) have been reported by Hitchcock and Brion [51]. These spectra are quite different in nature. The spectrum of F_2 displays a very prominent resonance due to a $1s \rightarrow \sigma*$ transition, with the higher Rydberg and continuum structure being of much lower intensity. In contrast HF shows a more normal, dominant Rydberg structure converging on a relatively more intense continuum. It is evident that HF is behaving in an essentially atomic-like fashion (not unlike the isoelectronic species Ne - see below) while in the case of F_2 the

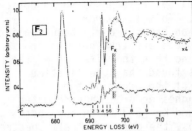

Figure 6: F_2 K-shell [51]

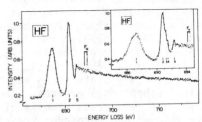

Figure 7: HF K-shell [51]

escaping core electron is drastically affected by the anisotropic molecular field giving rise to a resonance as predicted in the MSXα

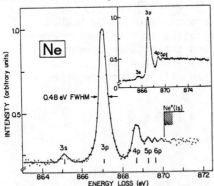

Figure 8: Ne K-shell [50]

calculations of Dehmer and Dill [65]. The equivalent core model applied to HF and F_2 results in predicted excitation energies for NeH and NeF respectively. The species NeF is of particular interest in that a very high energy (λ = 107 nm) excimer laser based on this system has been predicted [64]. The results for HF and F_2 have been combined with the earlier results for other second row diatomics and correlation plots were produced providing term value predictions for a variety of other diatomic species, both stable and unstable [51].

Hitchcock and Brion [50] have also studied the K-shell EELS spectrum of neon using 2.5 keV electrons in the energy loss range 863-872 eV with a resolution of 0.37 eV. The spectrum (figure 8), exhibits an atomic Rydberg series of dipole allowed $1s \rightarrow np$ transitions. The finite momentum transfer (2.6 au. at θ = 2×10^{-2}

radians) results in the observation of the (dipole forbidden) electric quadrupole allowed 1s → 3s transition at 865.1 eV. An upper limit of 0.31 eV for the 1s → 3p linewidth was obtained from considerations of the lineshape and the instrumental resolution.

Shaw et al [60] at the University of Manchester have very recently reported the first observation of an inner shell singlet to triplet transition, namely the transition from the ground state of N_2 to the configuration $(1s)^{-1}(\pi^*)$, $^3\Pi$. High resolution spectra obtained at variable impact energies down to 460 eV are shown in figure 9. The vibrational structure of the triplet state is

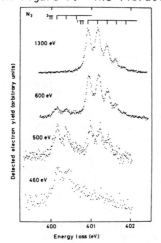

clearly visible below that due to the more familiar $^1\Pi$ excited state [22,23]. The parity forbidden transition to the $(2p_{3/2})^{-1}4p$ state of argon was also studied as a function of impact energy [60].

Recently in our own laboratory we have extended our ISEELS studies to the molecule NF_3 [66]. The nitrogen and fluorine K-shell spectra are shown in figures 10 and 11 respectively. Both spectra exhibit large σ^* resonances below the respective K-shell ionization edges while other resonance structure is apparent in the continuum. The spectra could also be interpreted on the potential barrier model [44,45]. In addition to the resonances the N_K spectrum shows sharp Rydberg type structure (peaks 2, 3 and 4). The insert

Fig 9: $N_2(1s \rightarrow \pi^*)$,$^3\Pi$ [60]

to figure 14 shows a detailed scan of the region below the nitrogen K-edge, with the lower trace being due to NF_3 alone. The upper

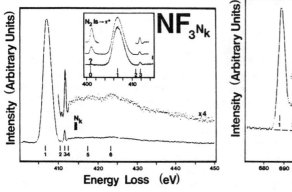

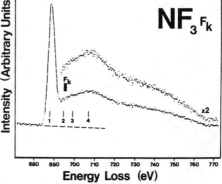

Fig 10: N(K) spectrum of NF_3 [66]   Fig 11: F(K) spectrum of NF_3 [66]

trace shows the ISEELS spectrum of a mixture of N_2 and NF_3. Earlier photoabsorption studies of NF_3 [67-69] show a peak in the N_K spectrum just below 401 eV which was ascribed to a spectral

438

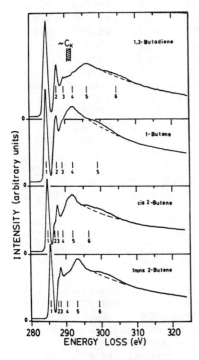

Figure 12: C(K) spectra of alkenes [63]

feature of NF_3. However our work [66] clearly shows (figure 10, upper trace) that the feature below 401 eV is in fact due to the $1s \rightarrow \pi^*$ transition in a molecular nitrogen impurity (N_2).

Most recently Hitchcock et al [63] have studied a number of unsaturated and conjugated hydrocarbons (figure 12) using a low resolution spectrometer located at McMaster University, Hamilton, Canada.

b) LITERATURE

Very recently Hitchcock [70] has published a detailed and comprehensive bibliography of atomic and molecular inner-shell studies for free molecules. This compilation includes photoabsorption and theoretical studies of core excitation as well as electron energy loss measurements. Core excitations are listed for atoms and molecules according to the atomic core level excited, the reference, energy range, method of study, and brief comments as to the contents. A notable feature is that the bibliography is available from the author on applicant supplied 8 in. floppy disks.

A number of useful short review articles have been published dealing with electron impact induced inner-shell electron excitation of free molecules by EELS [4,71-74].

c) ABSOLUTE ENERGY CALIBRATION FOR ISEELS

An existing problem in ISEELS (as well as in other electron spectroscopy experiments such as X-ray photoelectron and Auger spectroscopy) is the accurate measurement of voltages corresponding to the energy range 100-1000 eV, and beyond. Even a cursory examination of XPS [75] or ISEELS [70] compilations of absolute transition energies frequently reveals significant discrepancies between measurements in different laboratories for the same transition. Quite apart from systematic errors of measurement, the main source of error usually lies in the resolution and, more importantly, in the limited accuracy of commonly used digital voltmeters (DVM). For instance to attempt to measure a voltage of one thousand volts with a 4-1/2 digit DVM automatically restricts the readability to ± 0.1 eV while a manufacturer's quoted accuracy of ± 0.1% would limit accuracy to ± 1 volt! For accurate measurements at 1000 eV it would be desirable to have at least 6-1/2 digits (readability to ± 0.001 eV) and an accuracy of ± 0.001% (accuracy to ± 0.010 eV). In addition such a DVM should be capable of frequent and accurate recalibration and preferably be self calibrating via a built in microprocessor in order to compensate for short-term aging.

Conversely ISEELS in principle provides the possibility of accurate and direct high voltage (100-1000 volts) calibration standards once the energy losses of appropriate atomic and molecular transitions have been accurately determined. Thus in any ISEELS equipped laboratory the possibility should exist for accurate, in-house, self testing and recalibration of high voltage DVM's since energy loss is a direct observable.

As a first step towards this goal we have recently acquired a DATRON model 1071 6-1/2 digit DVM with a stated accuracy of 0.001% (90 day) or 0.002% (1 year) in normal operation. If the "averaging" mode is selected 7-1/2 digits are available and the accuracy is claimed to be increased by a factor of two. The manufacturers [76] claim that it is the most accurate DVM presently produced. It includes auto-recalibration via a microprocessor and also affords easy periodic recalibration at a standards laboratory.

Using this instrument together with our existing high resolution ISEELS spectrometer [26] we have recently redetermined the energy losses of a number of transitions below 900 eV with a view to their use as accurate calibrants in future ISEELS measurements and as a future check on DVM accuracy. The details are to be published elsewhere [77] and a summary of the results is shown in table 1.

INNER-SHELL TRANSITION	ENERGY - ISEELS This work	literature	[ref.]	ENERGY OPTICAL	[ref]
SF_6; $S2p \to t_{2g}$	184.54(5)	184.27	[44]	184.55	[83]
Ar; $2p_{3/2} \to 4s$	244.37(2)	244.39(1)	[56]	-	
CO; $C1s \to \pi^*(v=0)$	287.40(2)	287.31(5) 287.40(2)	[23] [55]	-	
N_2; $1s \to \pi^*(v=0)$	400.88(2)	400.70(5) 400.86(3)	[23] [22]	-	
CO; $O1s \to \pi^*$	534.21(9)	534.11(8)	[23]	534.2(3)	[88]
SF_6; $F1s \to a_{1g}$	688.27(15)	688.0 (2)	[44]	687.5 687.8	[89] [90]
Ne; $1s \to 3p$	867.23(7)	867.05(8)	[50]	867.13	[84]

In many cases, particularly at the higher energy losses, the peaks are broadened by lifetime considerations and so, for example in the O 1s → π* transition for CO, the vibrational structure is smeared-out resulting in a broad peak, even at high resolving power [23]. Likewise the natural line width of Ne 1s → 3p has been estimated to be 0.31 eV [50], resulting in a broad line which intrinsically limits the accuracy.

In the future we hope to prepare a gas mixture providing several calibration points in the range up to 900 eV. The ISEELS spectrum should then provide a ready test of DVM accuracy and linearity.

d) NEW INSTRUMENTATION and RESULTS

Existing ISEELS spectrometers suffer from a number of significant limitations. In particular all components, gun, monochromator, collision region, analyser and detector have been housed in a common vacuum chamber. In the presence of reactive gases oxide cathodes (and sometimes even metal filaments) in the electron gun are quickly poisoned while the decomposition products of many samples on the hot cathode and electron gun region frequently result in contamination of lens and analyser surfaces with a resulting loss of sensitivity and resolution. These factors frustrate and severely limit routine use of the spectrometer. Furthermore, the use of sensitive electron multipliers and delicate position sensitive detectors based on microchannel plates is compromised in the hostile environment of reactive gases.

A further serious limitation of existing ISEELS spectrometers has been the rather limited flexibility of the electron optics. Apart from reducing sensitivity and resolution, poor optics results in a significant portion of the relatively very intense main electron beam and valence scattered electrons reaching the detector and causing appreciable spurious backgrounds in ISEELS spectra at high energy losses. As a result existing work has been carried out not at zero degree scattering angle but at a small angle (typically 1-2°) in order to prevent the main (incident) beam from entering the analyser. Nevertheless spurious non-spectral scattered backgrounds are often troublesome, even with use of a beam dump behind a large slot in the outer analyser plate. This results in a serious loss of signal to noise, especially at higher energy losses. In addition the small but finite scattering angle increases the momentum transfer and thus the probability of nondipole transitions. While a non-zero scattering angle favours the observation of dipole forbidden transitions [50,56, 60] it precludes the direct measurement [15] of optical oscillator strengths (see introduction). Furthermore a higher impact energy (1-2.5 keV maximum at present) is desirable to probe deeper lying core levels and also to ensure that the momentum transfer is sufficiently small for optical oscillator strength measurements. Higher impact energies require better optics with multilens arrangements to provide the necessary high retardation ratios in order to obtain high resolving power in the electron analysers.

With the above considerations in mind we have designed, built and tested a new high resolution ISEELS spectrometer [78]. The main features are shown in figure 13. The gun, monochromator, collision chamber and analyser chambers are each differentially pumped with only small electron beam apertures interconnecting them. The differential pumping is so effective that no change in operating potentials is needed on gas introduction or on changing gases. The valves V_1, V_2 allow isolation of the source and monochromator. The large diameter (15 inches) of the hemispheres was chosen in order to allow

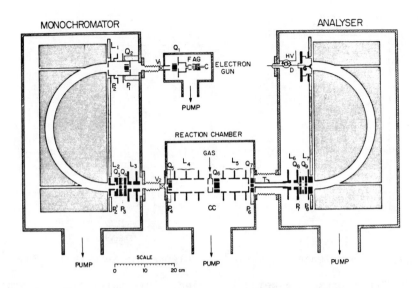

Figure 13: New ISEELS spectrometer [78]

high resolution at high pass energies, thus increasing transmission
and decreasing the retardation lens ratios needed with high impact
energy (1-10 keV).

The electron source is a black/white TV tube gun (Philips 6-
AW59) which provides a parallel electron beam (diameter 0.3 mm) at
the final beam energy. This beam is retarded by a double element
tube lens L_1 on to the real monochromator entrance slit P_2 (1
mm). Sets of quad deflectors Q_{1-9} are used throughout the spectrometer, in conjunction with apertures P_{1-8}, to align and focus
the primary electron beam. A virtual slit is formed at the monochromator exit by the accelerating lens L_2. A further acceleration
is provided by L_3 which focusses the beam onto P_4 at the entrance
of the stainless steel reaction chamber. The beam is transported to
the collision chamber, CC, by the einzel lens L_4. The beam then
passes out via a zoom energy-add lens, L_5, and through a tube T
floated at the energy loss. A virtual slit is formed at the analyser
entrance by L_6 and L_7 which are similar to L_3 and L_2 respectively. From the centre element of L_5, onwards, the whole analyser
system is driven at a potential equal to the energy loss. The lens
combination L_5, L_6, L_7 strongly discriminates against electrons
which do not have the selected energy loss with the result that inner
shell energy loss spectra can now be measured at zero degrees with
negligible background, and without using any beam dump. Furthermore
the very efficient optics have resulted in a marked improvement in
beam intensity.

At present a channeltron electron multiplier (Mullard B419AL) is
located behind the real (1 mm) exit slit. To date a programme (see
below) of testing has been completed using this single channel detector. A new multichannel detector similar to that described by Hicks

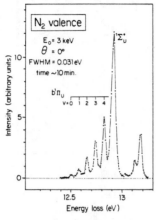

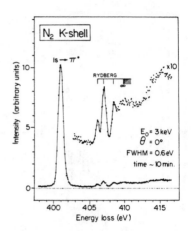

Figure 14: N_2 valence shell [78]   Figure 15: N_2 K-shell [78]

et al [80] is presently being installed. This multi-detector employs
a double (chevron) microchannel plate multiplier with an integral
phosphor screen and fibre optic block. A self scanning photodiode
array with fibre optic window (Reticon RL512SF) joins on to this
assembly. Full details of the new multi-detector will be published
later [81]. Magnetic shielding of the various regions of the spectrometer is provided by hydrogen annealed mumetal shields.

   The new spectrometer, with only the single channel multiplier
detector has already demonstrated significantly improved performance
for both valence and inner-shell spectra at low and high resolution.
Figures 14 and 16 show some results obtained at an interim stage of
development. Figure 14 shows part of the valence shell electronic
spectrum of N_2 obtained in a few minutes. The low resolution (0.6
eV FWHM) K-shell spectrum of N_2 (compare refs [3,23]) shown in
figure 15 was likewise obtained in a few minutes and is of note
because the scattering angle is zero degrees and there is almost zero
background (a very small intensity is expected [15] from the tail of
the valence shell continuum). In earlier instruments measurements
were not made at zero degree scattering angle due to very large
nonspectral backgrounds.

   A high resolution ISEELS spectrum of the N_2 1s → π^* transition
with the vibrational levels clearly resolved is shown in figure 16.
This spectrum, recorded at zero degrees scattering angle, illustrates
the spectacular improvement of the new instrument even without the
multichannel detector. There is negligible scattered background
which results in a large improvement in signal to noise even over
earlier small angle measurements [22,23]. The resolution is
comparable to that of the best published work [22] but the spectrum
was obtained at zero degrees in approximately one hour compared to 40
hours [22] or five days in our own earlier inferior result [23].
With such significant improvements it should be possible to make
accurate dipole oscillator strength measurements even for inner-shell
transitions at high resolution. The multichannel detector should

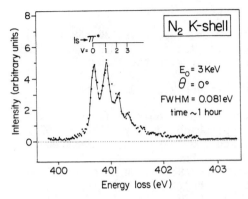

Figure 16: N_2 K-shell, high resolution [78]

provide significant further improvement.

With this spectrometer we plan to continue the systematic study of the spectroscopy of valence and inner-shell and in addition to attempt to measure the respective absolute oscillator strengths. Such data are of importance for instance in element sensitive imaging in electron microscopy. The stainless steel, differentially pumped collision region also affords the possibility of easily extending normal gas phase work to the study of very reactive gases as well as laser-excited atoms and molecules, transient species, radicals and ions. The microchannel plate detector will greatly facilitate the latter studies where target densities are expected to be quite low. A further possible development is to record wide range integrated ISEELS spectra. The wide spacing of the core levels, the energies of which are strongly element dependent, should permit quantitative analysis of low Z elements in molecular substances and mixtures. This might be of use for example in coal research and analysis [82] in conjunction with laser pyrolysis. ISEELS techniques may also benefit from the employment of field emission electron guns which provide narrow energy spread, high intensity and excellent focussing. This application would be particularly helpful in coincidence experiments [4,16, 32,33] involving inner-shell processes.

Inner-shell electron energy loss spectroscopy has opened up a new area of high resolution molecular spectroscopy particularly for low Z elements in the excitation energy range up to 1000 eV. It has already resulted in a considerable body of new spectroscopic data which in turn has led to new understanding of the processes occuring in and resulting from core excitation and ionization.

ACKNOWLEDGEMENT

We wish to thank Professor F.H. Read and Dr. G. King for helpful discussions of electron optics. We are grateful for the skilful technical assistance of the staff of the mechanical and electronics workshops of the Department of Chemistry, University of British Columbia. Financial support for this work has been provided by The Natural Sciences and Engineering Research Council of Canada.

REFERENCES

1. E.N. Lassettre, and A. Skerbele, Meth. Exp. Phys. $\underline{3B}$ 868 (1974).
2. E.N. Lassettre, in C. Sandorfy, P.J. Ausloos and M.B. Robin, Eds. Chemical Spectroscopy and Photochemistry in the Vacuum Ultraviolet, Reidel (Boston, 1974).
3. G.R. Wight, C.E. Brion and M.J. Van der Wiel, J. Electron Spectrosc. $\underline{1}$ 457 (1972/73).
4. C.E. Brion in Physics of Electronic and Atomic Collisions, S. Datz (Editor), (North Holland Publishing Company, 1982) p. 579.
5. J.A.R. Samson, "Techniques of Vacuum Ultraviolet Spectroscopy" John Wiley (New York, 1967).
6. H. Winick and S. Doniach, "Synchrotron Radiation Research" Plenum Publishing Corporation (New York, 1980).
7. J. Franck and G. Hertz, Verh. Deutsch Phys. Ges. $\underline{16}$ 10 (1914).
8. H. Bethe, Ann. Phy. (Leipzig) $\underline{5}$ (5) 325 (1930).
9. M. Inokuti, Rev. Mod. Phys. $\underline{43}$ 297 (1971).
10. E.N. Lassettre, Rad. Research (Supp) $\underline{1}$ 530 (1959).
11. M.J. Van der Wiel, Physica $\underline{49}$ 411 (1970).
12. M.J. Van der Wiel and Th.M. El-Sherbini, Physica $\underline{59}$ 453 (1972).
13. M.J. Van der Wiel, Th. M. El-Sherbini and C.E. Brion, Chem. Phys. Lett. $\underline{7}$ 161 (1970).
14. R.B. Kay, Ph. E. Van der Leeuw and M.J. Van der Wiel, J. Phys. B. $\underline{10}$ 2521 (1977).
15. R.B. Kay, Ph. E. Van der Leeuw and M.J. Van der Wiel, J. Phys B $\underline{10}$ 2513 (1977).
16. C.E. Brion and A. Hamnett, Continuum Optical Oscillator Strength Measurements by Electron Spectroscopy in the Gas Phase, in "The Excited State in Chemical Physics, Part 2", Adv. Chem. Physics, volume 45 (Ed. J.W. McGowan) John Wiley New York (1981).
17. C.E. Brion, K.H. Tan, M.J. Van der Wiel and Ph. E. Van der Leeuw, J. Electron Spectrosc. $\underline{17}$ 101 (1979). 18. K.H. Tan, C.E. Brion, Ph. E. Van der Leeuw and M.J. Van der Wiel, Chem. Phys $\underline{29}$ 299 (1978).
19. M.G. White, K.T. Leung and C.E. Brion, J. Electron Spectrosc. $\underline{23}$ 127 (1981).
20. F. Carnovale, M.G. White, and C.E. Brion, J. Electron Spectrosc. $\underline{24}$ 63 (1981).
21. T.A. Carlson, M.O. Krause and F.A. Grimm, J. Chem. Phys. in press.
22. G.C. King, F.H. Read and M. Tronc, Chem. Phys. Letters $\underline{52}$ 50 (1977).
23. A.P. Hitchcock and C.E. Brion, J. Electron Spectrosc. $\underline{18}$ 1 (1980).
24. A. Bianconi, H. Petersen, F.C. Brown and R.Z. Bachrach, Phys. Rev. A $\underline{17}$ 1907 (1978).
25. G.R. Wight and C.E. Brion, J. Electron Spectrosc. $\underline{4}$ 327 (1974).
26. A.P. Hitchcock and C.E. Brion, J. Electron Spectrosc. $\underline{13}$ 193 (1978).
27. E. Rudberg, Proc. Roy. Soc. (London) A$\underline{27}$ 628 (1930). 28. J.A.
28. Simpson and C.E. Kuyatt, J. Appl. Phys. $\underline{37}$ 3805 (1966); Rev. Sci.

Instr. 35 1698 (1964), 38 103 (1967).

29. G.J. Schulz, Rev. Mod. Phys. 45 378 (1973).
30. J. Geiger and M. Topchowsky, Z. Naturforsch, 21a 626 (1966); Z. Phys. 181 413 (1964).
31. M.J. Van der Wiel and G. Wiebes, Physica 53 225 (1971).
32. A.P. Hitchcock, C.E. Brion and M.J. Van der Wiel, J. Phys. B. 11 3245 (1978).
33. A.P. Hitchcock, C.E. Brion and M.J. Van der Wiel, Chem. Phys. Lett. 66 213 (1979).
34. G.R. Wight and C.E. Brion, J. Electron Spectrosc. 3 191 (1974).
35. G.R. Wight and C.E. Brion, J. Electron Spectrosc. 4 25 (1974).
36. G.R. Wight and C.E. Brion, J. Electron Spectrosc. 4 313 (1974).
37. G.R. Wight and C.E. Brion, J. Electron Spectrosc. 4 335 (1974).
38. G.R. Wight and C.E. Brion, J. Electron Spectrosc. 4 347 (1974).
39. J.L. Dehmer and D. Dill, Phys. Rev. Lett. 35 213 (1975); J. Chem. Phys. 65 5327 (1976).
40. G.R. Wight and C.E. Brion, Chem. Phys. Lett. 26 607 (1974).
41. A.P. Hitchcock and C.E. Brion, J. Electron Spectrosc. 10 317 (1977).
42. A.P. Hitchcock, M. Pocock and C.E. Brion, Chem. Phys. Lett. 49 125 (1977).
43. A.P. Hitchcock et al, J. Electron Spectrosc. 13 345 (1978).
44. A.P. Hitchcock and C.E. Brion, Chem. Phys. 33 55 (1978).
45. A.P. Hitchcock and C.E. Brion, J. Electron Spectrosc. 14 417 (1978).
46. A.P. Hitchcock and C.E. Brion, J. Electron Spectrosc. 15 401 (1979).
47. A.P. Hitchcock and C.E. Brion, Chem. Phys. 37 319 (1979).
48. A.P. Hitchcock and C.E. Brion, J. Electron Spectrosc. 17 139 (1979).
49. A.P. Hitchcock and C.E. Brion, J. Electron Spectrosc. 19 231 (1980.
50. A.P. Hitchcock and C.E. Brion, J. Phys. B. 13 3269 (1980).
51. A.P. Hitchcock and C.E. Brion, J. Phys. B. 14 4399 (1981).
52. A.P. Hitchcock and C.E. Brion, J. Electron Spectrosc. 22 283 (1981).
53. W. Eberhardt, G. Kalhoffen and C. Kunz, Chem. Phys. Lett. 40 180 (1976).
54. F.C. Brown, R.Z. Bachrach and A. Bianconi, Chem. Phys. Lett. 54 425 (1978).
55. M. Tronc, G.C. King, R.C. Bradford and F.H. Read, J. Phys. B9 L555 (1976).
56. G.C. King, M. Tronc, F.H. Read and R.C. Bradford, J. Phys. B10 2479 (1977).
57. M. Tronc, G.C. King and F.H. Read, J. Phys. B12 137 (1979).
58. M. Tronc, G.C. King and F.H. Read, J. Phys. B13, 999 (1980).
59. D.A. Shaw, G.C. King and F.H. Read, J. Phys. B13, L723 (1980).
60. D.A. Shaw, G.C. King, F.H. Read and D. Cvejanovic, J. Phys B. 15 1785 (1982).
61. G.C. King, J.W. McConkey and F.H. Read, J. Phys. B10 L541 (1977).
62. G.C. King, J.W. McConkey, F.H. Read and B. Dobson, J. Phys. B13

4315 (1980).

63. A.P. Hitchcock and S. Beaulieu, to be published.

64. J.J. Ewing, Physics Today, p. 32 May (1978).

65. D. Dill, private communication.

66. R. Sodhi and C.E. Brion, to be published.

67. T.M. Zimkina and A.S. Vinogradov. Bull. Acad. Sci. USSR Phys. Ser. 36 229 (1972) (Izv. Akad. Nauk. SSSR Fiz. Ser. 36 2481 (1972)).

68. R.L. Barinskii and I.M. Kulikova, J. Struct. Chem. 14 335 (1973) (Zh. Struk. Khim 14 3721 (1973)); (Izv. Akad. Nauk. SSSR. Ser. Fiz. 38 444 (1974)).

69. A.S. Vinogradov, T.M. Zimkina, V.N. Akimov and B. Shlarbaum, Bull. Acad. Sci. USSR Phys. Ser. 38 No. 3 69 (1974) (Izv. Akad. Nauk. SSSR Fiz. Ser. 38 508 (1974)).

70. A.P. Hitchcock, J. Electron Spectrosc. 25 245 (1982).

71. F.H. Read and G.C. King in Symposium on Electron- Molecule Collisions (University of Tokyo, Sept. 1979), Invited Papers, eds. I. Shimamura and M. Matsuzawa, p. 155.

72. M.J. Van der Wiel, in Electronic and Atomic Collisions, Invited Papers and progress reports, XI ICPEAC, Kyoto eds. N. Oda and K. Takayanagi (North Holland, 1980) p. 209.

73. F.H. Read, J. de Physique, Colloque C1, supp. 5 39 82 (1978).

74. M.J. Van der Wiel, in Physics of Electronic and Atomic Collisions, VII ICPEAC, p. 140 (1971) (North Holland).

75. A.A. Bakke, H.W. Chen and W.L. Jolly, J. Electron Spectrosc. 20 333 (1980).

76. Datron Instruments Inc.

77. R.S. Sodhi and C.E. Brion, to be published.

78. S. Daviel, C.E. Brion and A.P. Hitchcock, to be published.

79. G.C. King and F.H. Read, Private communication.

80. P.J. Hicks, S. Daviel, B. Wallbank and J. Comer, J. Phys. E. Sci. Instrum. 13 713 (1980).

81. S. Daviel and C.E. Brion, work in progress.

82. Report to the American Physical Society on Research Planning for Coal Utilization and Synthetic Fuel Production, Rev. Mod. Phys. 53 S29, October, (1981).

83. V.I. Baranovskii, T.M. Zimkina, V.A. Fomichev and B.E. Dzevitskii, Soviet Phys. JETP 3 260 (1967).

84. F. Wuilleumier, C.R. Acad. Sci. Paris B270 825 (1970).

85. H.F. Wellenstein, H. Schmoranzer, R.A. Bonham, T.C. Wong and J.S. Lee, Rev. Sci. Instrum. 49 92 (1975).

86. A.L. Bennani, A. Duguet and H.F. Wellenstein, Chem. Phys. Lett. 60 405 (1979).

87. V.C. Afrosimov, Yu.S. Gordeev, Y.M. Lavrov and S.G. Schelinin, Sov. Phys. JETP 28 821 (1968).

88. D.M. Barrus, R.L. Blake, A.J. Burek, K.C. Chambers and A.L. Pregenzer, Phys. Rev. A 20 1045 (1979).

89. R.E. LaVilla, J. Chem. Phys. 57 899 (1972).

90. A.S. Vinogradov, T.M. Zimkina and V.A. Fomichev, J. Struct. Chem. 12 823 (1971).

X-RAYS FROM LABORATORY AND ASTROPHYSICAL PLASMAS

U. Feldman
E.O. Hulburt Center for Space Research
Naval Research Laboratory, Washington, D.C. 20375

ABSTRACT

X-ray spectra of highly ionized elements are widely used in diagnosing high temperature plasmas from astrophysical as well as from laboratory sources. The emission lines in the X-ray region are produced primarily from two different types of transitions. One type involves the transitions 1s-2p in hydrogen-like ions, helium-like ions and their associated satellites. The second type involves the transitions $2\ell - n\ell'$ for $n > 2$. These transitions are sensitive to electron temperature, electron density and the ionization balance. High quality spectrometers in the X-ray region have instrumental line widths that in many cases are significantly less than the true line widths. As a result, it is quite convenient to use the physical information in the line profiles to determine additional properties of the plasma under investigation.

In the course of my talk I will describe results from recent experiments concerning solar plasmas as well as laboratory sources.

INTRODUCTION

In the next 45 minutes or so I will be discussing some aspects of spectroscopy of highly ionized atoms in the X-ray range. The X-ray range may mean different things to different people. Therefore, as a first step I would like to define the bounderies of the range as they may appear to a person concerned with emission spectra from highly ionized elements.

The simplest and most obvious type of spectra from highly ionized atoms to be found in the X-ray region will arise from He- and H-like ions. Because of the important role of such spectra in plasma diagnostics, these can be used to define the applicable spectral range. On the short wavelength side we will define the limit as the wavelength of the He-like spectra from the heaviest element so far ionized to the He-like stage. It can be safely assumed that the available technology will limit our capabilities to elements in the vicinity of, say, He-like krypton[1] (i.e., Kr XXVII). Therefore one would expect that for the time being 1 Å is a fairly good estimate for the short wavelength limit. The definition for the long wavelength limit is more subjective and depends on one's background or means of observations. For the sake of this talk I will define the upper limit at $\lambda = 22$ Å, which is the longest wavelength for the $1s^2-1s2\ell$ transitions of He-like oxygen. By doing so we have limited ourselves to He- and H-like spectra of some 28 elements from oxygen (Z=8) to krypton (Z=36).

0094-243X/82/940447-16$3.00 American Institute of Physics

We have used the H I and He I isoelectronic sequences which are fairly simple to define the limits of our wavelength range. However, it's obvious that they are not the only ones available in the above mentioned region. The next set of ions which have spectra in the defined wavelength region and may be of interest are those for which the ground configurations are of the type $1s^2 2s^k 2p^m$ where k = 2 and m = 6, i.e. the Li I, Be I, B I, C I, N I, O I, F I, and Ne I isoelectronic sequences. Under certain conditions transitions of the type $1s^2 2s^k 2p^m - 1s^2 2s^k 2p^{m-1}$ nℓ where n>2 may dominate the spectra in the region. To the best of my knowledge the shortest wavelength lines of such spectra produced so far are from Xe XLV (Z=54) at about 3 $\mathring{A}$[2]. On the long wavelength side we will be limited by spectra from elements heavier than Ar (Z=18)[1]. We could continue and look into spectra from isoelectronic sequences with ground configurations that involve electrons with n=3 where n is the prinicple quantum number. Recent reports of observed spectra of this kind become more and more available. However, because of lack of time I will not discuss them in this talk.

Instrumentation

High quality instruments for the X-ray region are well understood and are fairly simple to construct. Such instruments can provide not only reliable line intensities but also good line profile information[3,4,5]. As a matter of fact, over most of the defined wavelength range and for most of the elements in question the minimum instrumental line width that can be achieved is significantly less than the natural line width. Therefore plasma diagnostic techniques in the X-ray region involve studies of line shapes as well as line intensities.

Measurements from Line Widths and Shifts

Doppler line widths and line shifts provide us with information on random or net mass motions as well as on ion temperatures. Lately line profile measurements in the X-ray region were utilized extensively in laboratory and in astrophysical plasmas.

Figure 1a shows representative iron spectra from 3 large flares and Figure 1b shows the variation of the profiles of the Fe XXV resonance line in a solar flare as a function of time[6]. Figure 2a shows the Fe XXV resonance line in tokamak plasmas during and after a neutral beam injection[7], and Figure 2b shows the ion temperature determined from the line profiles. A typical electron temperature for the generation of the Fe XXV line is 2×10^7 K (or $\approx$ 2 kev)[7]. The effect of such a temperature on the Fe XXV line width is equivalent to the effect of a random mass motion of about 80 km/sec. Therefore random plasma motions of several tens of km/sec are easily detected by such measurements. In order to measure net mass motions it is necessary to establish an absolute wavelength reference system. However once such a system is available net motions of only a few tens of km/sec may be detected (see Figure 3)[7].

At very high electron densities the stark effect is a significant contributor to line widths in H-like spectra. Figure 4 shows the Ly β line of Ar XVIII and the $1s^2-1s4p$ line from Ar XVII

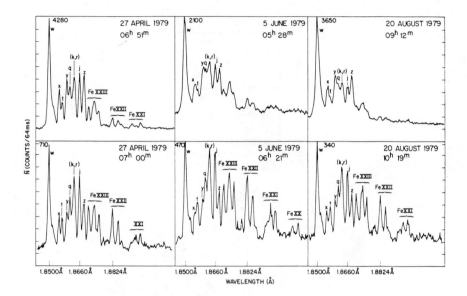

Fig. 1a – Representative iron spectra of 3 large solar flares. The top spectra are typical of rise phase and peak emission spectra, the bottom spectra are typical of decay phase spectra. Peak scale values of w are given in the upper left corner of each spectrum. (Ref. 6)

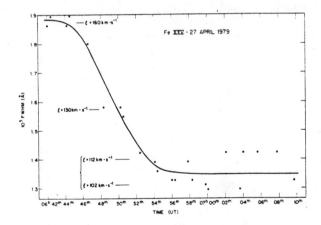

Fig. 1b – The full width at half maximum (FWHM) for line w of Fe XXV for the April 27 flare. (Ref. 6)

450

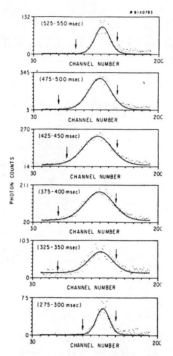

Fig. 2a - A time sequence of six Ti XXI K$_\alpha$ - lines profiles observed from PDX discharges with additional ion heating by injection of intense deuterium beams during the period from 300-400 m sec. The solid lines represent least squares fit of Voigt functions. The arrows indicate the limits used for the fit. (Ref.7)

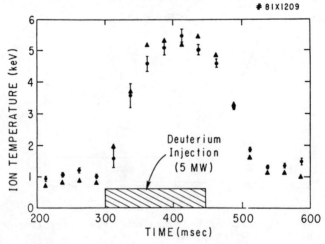

Fig. 2b - Ion temperature results obtained from the observed Ti XXI k$_\alpha$ line profiles (circles) and from measurements of charge-exchange neutrales (triangle) as a function of time. The bar represent the statistical error of experimental results (Ref.7)

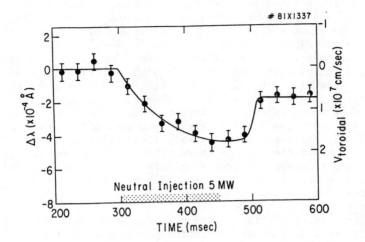

Fig. 3 – Observed Doppler shift of the Ti XXI k$_\ell$ line as a function of time. The Doppler-shift is due to a toroidal rotation of the plasma which results from momentum transfer by injected deuterium beams. (Ref. 7)

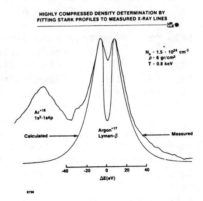

Fig. 4 – Argon lines from quartz target (Ref 8).

as measured from a laser-produced plasma during a compression experiment[8]. The experiment was done as follows: Argon filled plastic coated glass shells were imploded with short ($\simeq 50$ psec) pulse of energy $\approx 100 J$. X-ray spectra were measured with a flat crystal spectrometer. A spatial resolution of 10 μm was obtained by placing a narrow slit perpendicular to the spectral dispersion. The Lyman β line was fitted with a profile assuming an electron density of 1.5×10^{24} cm^{-3}, without any opacity correction. A small correction may improve the agreement with the experiment.

Figure 5 shows the O and F spectra produced with a blue laser. Notice the width of the stark broadened lines relative to some other lines not affected by the stark effect. The plasma density which produced this spectrum had an electron density of 5×10^{21} cm^{-3}. Again, electron densities from 10^{19} cm^{-3} and up can be studied from line shapes.

Measurements from Line Ratios

The line ratios within a particular degree of ionization or between different degrees of ionization provides significant information on electron density, electron temperature, and the state of ionization equilibrium. These techniques are widely used in plasma diagnostics.

Perhaps the most interesting and widely used spectra for diagnostics in the x-ray region originate from the He-like ions. Under some special conditions, one can expect to observe the following four lines originating from transitions between the configurations $1s^2 - 1s2\ell$. (see Figure 6). The lines are:

$$1s^2 \ ^1S_o - 1s2p \ ^1P_1 \quad \text{The resonance line}$$
$$1s^2 \ ^1S_o - 1s2p \ ^3P_1 \quad \text{The intercombination line}$$
$$1s^2 \ ^1S_o - 1s2p \ ^3P_2 \quad \text{The magnetic qradrupole line}$$
$$1s^2 \ ^1S_o - 1s2s \ ^3S_1 \quad \text{The forbidden line}$$

For light elements the magnetic quadrupole line is too faint to be seen. Its intensity relative to the intensity of the inter-combination line is rather small. The reason for this is primarily because of its small branching ratio. Most of the population of the $^3P_2$ level decays into the $1s^2 \ ^3S_1$ level. However, for the heavier elements the branching ratio starts to change in favor of the magnetic quadrupole line and eventually becomes the main decay channel for the $^3P_2$ level. If one plots the relative intensity of the four lines as a function of density for a particular ion; say O VII, the following picture emerges (see Figure 6). As was mentioned before, the magnetic quadrupole line is too faint and only three lines exist. The strongest is the resonance line, and at low densities the second strongest is the forbidden line. The least intense line is the intercombination line. Since the excitation is proportional to the electron density, there is a density at which the decay from the $1s2s \ ^3S_1$ level, which is the level responsible to the forbidden line, cannot keep up any more with the excitation by collisions and some of the population will be transferred to the $^3P_1$ level. As a result, the intercombination line

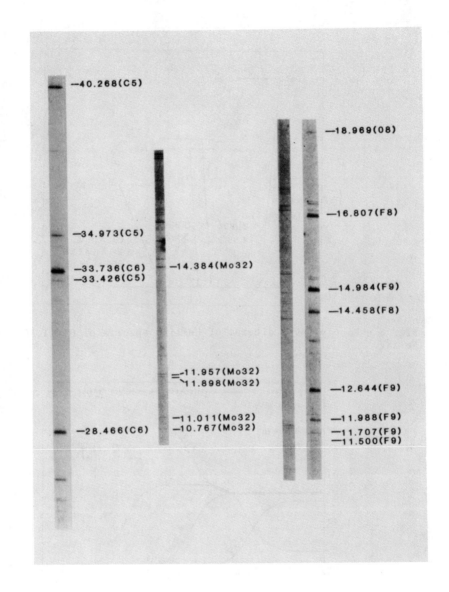

Fig. 5 – H and He-like lines of carbon, oxygen and fluorine. The narrow lines belong to highly ionized molybdenum.

454

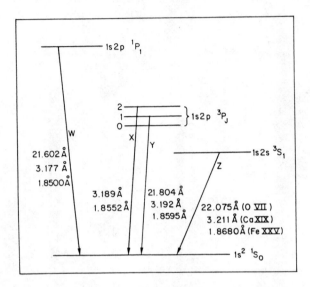

Fig. 6 - Energy level diagram of He-like spectra of O VII, Ca XIX, and Fe XXV.

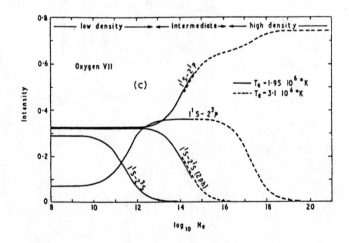

Fig. 7 - Predicted line intensities for helium-like ions over a wide range of densities are plotted on an intensity scale. (Ref. 9)

will begin to increase in intensity and become the second strongest line. As the density continues to increase, the same situation
occurs for the intercombination line and the only strong line
remaining will be the resonance line. As can be seen from the
qualitative description just given, the He-like ions are very
useful as density diagnostic indicators; in particular for astrophysical plasmas[9] (see Figure 7). Figure 8 shows an example of the
use of the O VII lines in solar plasmas[10]. O VII lines are emitted
from plasmas with temperatures near T_e = 2 x 10^6 K. The density
of the so called quiet solar atmosphere from which the O VII lines
are emitted is on the order of 10^9 cm^{-3}. At such densities, the
forbidden line is the second most intense in the spectra. As can
be seen from Figure 8, a flare is a phenomenon in which the density
increases significantly. As the flare fades away the disturbed
solar atmosphere returns to its "normal" state.

This same technique can also be used in low density laboratory
plasmas, although in the laboratory, one does have other ways to
measure the density, for example, scattering of a laser beam from
the plasma. However, at very high electron densities (N_e >10^{22}),
in order to penetrate the plasma, one needs to use a rather short
wavelength laser which is not readily available. The line ratio
technique is an easy to use substitute. For the very high density
plasmas, the intercombination to resonance line ratios of high z
elements will be most useful. A number of calculations of He-like
line ratios have been done by various groups.

Electron Temperatures and State of Ionization Equilibrium

It was shown by Gabriel and Jordan[11] that many of the Li-like
satellite lines in the vicinity of the He-like lines are produced
by dielectronic recombination of He I like ions, e.g.,

$$X (1s^2\ ^1S_0) + e \rightleftharpoons X (1s2p2\ell\ ) \rightarrow X (1s^2 2\ell\ ) + h\nu$$

On the other hand a few satellite lines are produced by collisional
innershell excitation of Li I like ions, e.g.,

$$X (1s^2 2s\ ^2S_{1/2}) + e \rightarrow X (1s2s2p) + e \rightarrow X (1s^2 2s\ ^2S_{1/2}) + h\nu + e$$

The ratios of the dielectronic Li-like satellite lines to the He-
like resonance lines could be used to determine the electron
temperature independent of whether or not the plasma is in ionization equilibrium. Once the electron temperature is determined
using dielectronic excited satellites, the departure from
ionization equilibrium can be determined using collisionally
excited satellite lines to resonance line ratios.

Figure 1a shows representative spectra of highly ionized iron
from a solar flare. The lower spectra are from the decay phase and
show most clearly the lines due to ions less ionized than Fe XXIV
(Li-like iron). In He-like spectra one of the lines formed by
electron impact excitation is the resonance line ($1s^2\ ^1S_0$-$1s2p\ ^1P_1$)
called line w (see Figure 1a). One of the unblended lines
formed by dielectronic recombination is called line j. This line
belongs to the Li-like spectra for Fe XXV. An analogous unblended
line for Ca XIX is called line k. The ratios j/w and k/w are
temperature sensitive because dielectronic recombination and
electron impact excitation have different temperature dependences.

456

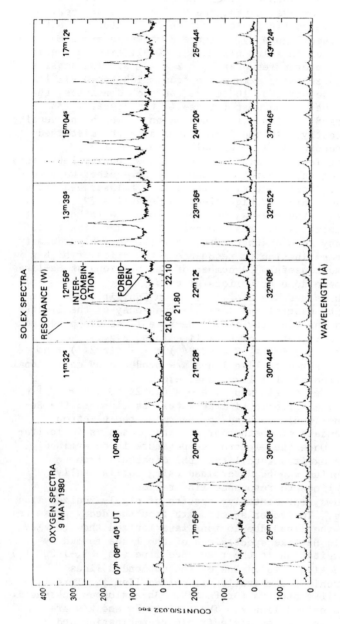

Fig. 8 - The O VII SOLEX spectra for the 1980 May 9 flare. The time at which the resonance line (w) was scanned are given in the upper right hand corner of each panel. (Ref. 10)

For the iron line ratio j/w the temperature dependence is approximately exp $(-4660/kT_e)/T_e$, where k is the Boltzmann constant. Figure 9 shows the time histories of the resonance line in Ca XIX and Fe XXV as well as the lines j/w and k/w.

For iron and calcium a strong unblended line formed by innershell excitation is the line q ($1s^2 2s\ ^2S_{1/2} - 1s(2s2p^3P)$ $^2P_{3/2}$). The ratio q/w can be used to determine the ratio of the number density of Fe XXIV to Fe XXV and Ca XVIII to Ca XIX and thus the state of ionization in the plasma can be determined (see Figure 9). In case the plasma is in ionization equilibrium the temperature determined from the state of ionization should equal the electron temperature determined from the dielectronic to resonance line ratios. In the solar flare spectra these two quantities are very near to each other, while in the tokamak plasma they deviate from each other[7]. The deviations are observed to depend in particular on the electron density. They are ascribed to changes of the charge state distribution due to impurity transport, which is disregarded in the model of coronal equilibrium.

Electron densities from Li I-F I like ions

As a general rule resonance lines emitted by ions formed in high temperature plasmas appear at shorter wavelengths than resonance lines emitted by ions formed in lower temperature plasmas. However, a close look into an energy level diagram of an ion with a ground configuration of the type $1s^2 2s^2 2p^k$ reveals a somewhat more complex picture. Figure 10 shows an energy level diagram for a typical ion with such an electronic configuration. In the X-ray region innershell emission between the n=1 and n=2 principal quantum numbers ($\Delta n=1$) occurs near 1.9 Å. Also in the X-ray region lines due to $1s^2 2s^2 2p - 1s^2 2s^2 3$ transitions (also $\Delta n=1$) are formed near 11.6 Å. However the $\Delta n=0$ transitions of the type $1s^2 2s^2 2p - 1s^2 2s 2p^2$ fall near 100 Å and the forbidden line $1s^2 2s^2 2p\ ^2P_{1/2} - 1s^2 2s^2 2p\ ^2P_{3/2}$ falls at 845 A. A similar situation holds for most of the ions from the Li I to the F I isoelectronic sequences. Although the wavelengths of the emission lines may span some three orders of magnitude, the relative intensity of some of the lines may show a density dependence. Moreover in some cases and in particular in low density plasmas such as in the solar atmosphere or in tokamak machines the reasons for the density dependence may be identical. Before I show a few examples of density effects in low temperature plasmas I would like to review some of the reasons for the effect.

In low density plasmas the excited levels are populated primarily by electron impact exitation. Density sensitivity of certain spectral line ratios occurs when the rate of depopulation of the excited levels due to collisions ($N_e C_{ij}^d$, where N_e is the electron density and C_{ij}^d is the deexcitation rate coefficient (cm^3/s) between levels i and j) becomes comparable to the spontaneous radiative decay rate A_{ij}. Two cases commonly occur with hypothetical three and four level ions. In these two cases application of detailed balance (the number of excitations into a level set equal to the number of deexcitations) leads to the

458

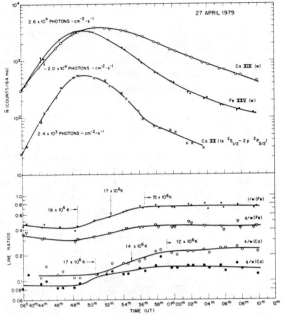

Fig. 9 - Intensity time histories for the resonance line of Fe XXV,
Ca XIX and Ca XX for the flare of 1979 April 27. The behavior with
time of j/w, k/w and q/w line ratios for Fe and Ca is also given.
Electron temperatures obtained from j/w, and k/w ratios are shown.

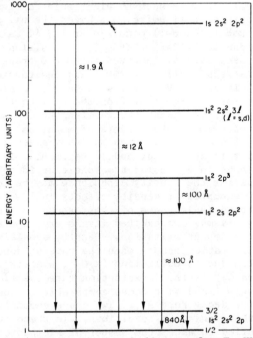

Fig. 10 - Approximate energy level diagram for Fe XXII showing
where the main groupings of iron lines appear in the X-ray and
EUV spectrum.

equations for the intensity ratios of lines. In one case, case A illustrated in the left in Figure 11, the equation for the ratio of lines I_{31} and I_{21} (where $I_{ij}=N_iA_{ij}$) is,

$$\frac{I_{31}}{I_{21}} = \frac{A_{31}(C_{13}/C_{12})}{N_e[(C_{32}C_{13}/C_{12}) + C_{32} + C_{31}] + A_{31}} , \tag{1a}$$

In the other case, case B, illustrated at the right in Figure 11, the equations are

$$\frac{I_{42}}{I_{31}} \rightarrow \frac{C_{14}}{C_{13}} \frac{A_{32} + A_{31}}{A_{31}} , N_e \rightarrow 0,$$

$$\frac{I_{42}}{I_{31}} \rightarrow \frac{C_{14} + (\omega_2 C_{24}/\omega_1)}{C_{13}} \frac{A_{32} + A_{31}}{(\omega_2 C_{23}/\omega_1)} \frac{A_{32} + A_{31}}{A_{31}} , N_e \rightarrow N_{eM} . \tag{1b}$$

These equations contain a number of simplifications that hold for low density and high temperature ions. For case A we are assuming that $A_{21} \gg A_{31} \gg A_{32}$; that is, level 2 is an allowed level. Also A_{21}/N_e $C_{12} \gg (C_{ij}/C_{kl})$ where ij are any permutation of levels 1, 2 and 3. For case B, $N_e C_{34} \ll A_{42}$, A_{32}, A_{31}. Also, $A_{41} \ll A_{42}$, A_{32}, A_{32}, A_{31} and $A_{41} \ll A_{42}$, A_{32}, A_{31}. That is, level 4 is not easily excited from level 1, and $C_{24} \gg C_{14}$ is true. w_2 and w_1 are the statistical weights of levels 2 and 1 and N_{eM} is defined as the maximum density for which $N_e C_{34}$ may be neglected.

The qualitative behavior of the intensity ratios as a function of density given by the preceding equations is shown in Figure 11. For case A the ratio is constant as long as the density dependent term in the equation is less than the radiative term (A_{32} + A_{31}). Above a certain critical density the collisional and radiative terms become comparable and when the collisional term dominates $I_{31}/I_{21} \propto 1/N_e$. This is the density region in which ions of case A are useful as diagnostics. For case B the ratio is constant for both low and high density limits. The case B ratios are useful in the intermediate regime where they vary with density.

The important point is that some line ratios have density sensitivity because for a certain range of densities collisional depopulation processes are comparable in magnitude to radiative decay rates. The reason for density sensitivity is fundamentally the same for both case A and case B ions and for other situations that cannot be classified as either purely case A or case B.

Now that we have outlined the principle of density sensitive line ratios we note as an example the Fe XX.

The density dependence of levels $1s^2 2s^2 2p^3$ are given in Figure 12. As a result of this behavior many intensities will depend on density in the range of 10^{11}–10^{15}. An example of such a case is given in Figure 13.

In conclusion I hope that during my short talk I was successful in my attempt to convey the idea that emission lines in the X-ray region may be used as tools in identifying conditions and parameters of high temperature plasmas.

460

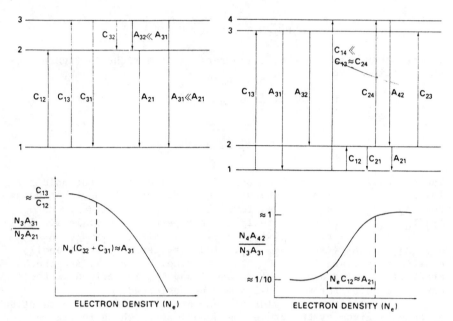

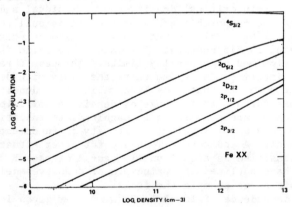

Fig. 11 – Behavior of density-sensitive line ratios as a function of electron density.

Fig. 12 – Relative populations of $2s^2 2p^3$ levels in Fe XX.

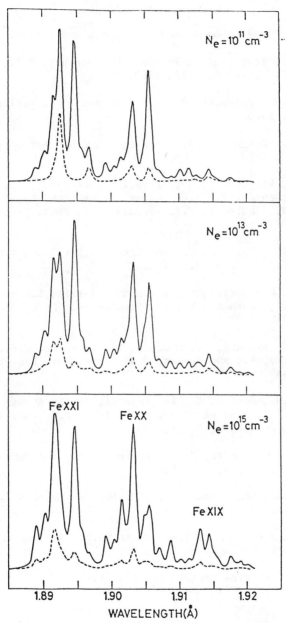

Fig. 13 – Calculated spectra of Fe XIX – XXI lines between 1.885 and 1.921 Å for electron densities of 10^{11}, 10^{13}, and 10^{15} cm^{-3}. The solid lines are spectral intensities including both dielectronic recombination and collisional excitation; the dashed lines collisional excitation alone (Ref 12).

REFERENCES

1. Kelly, R.L., and Palumbo, L.J. 1973, NRL Rept. 7599.

2. Conturi, Y., Yaakobi, B., Feldman, U., Doschek, G.A., and Cowan, R.D., J.O.S.A. 71, 1309, 1981.

3. Behring, W.E., Ugiansky, R.J., and Feldman, U., Applied Optics 12, 528, 1973.

4. Feldman, U., Doschek, G.A. and Kreplin, R.W., Astrophys. J. 238, 365, 1980.

5. Hill, K.W., von Goeler, S., Bitter, M., Campbell, L., Cowan, R.D., Fraenkel, B., Greenberger, A., Horton, R., Hovey, J., Roney, W., Sauthoff, N.R., Stodiek, W., Phys. Rev. A. 19, 1770, 1979.

6. Doschek, G.A., Feldman, U., and Kreplin, R.W., and Cohen, L., Astrophys. J. 239, 725, 1980.

7. Bitter, M., von Goeler, S., Goldman, M., Hill, K.W., Horton, R., Roney, W., Sauthoff, N., and Stodiek, W., PPPL-1891 UC20-F April 1982.

8. Yaakobi, B., Skupsky, S., McCrory, R.L., Hooper, C.F., Deckman, H., Bourke, P., and Souras, M.J., Phys. Rev. Letters 44, 1072, 1980.

9. Gabriel, A.H., Jordan, C., in "Case Studies in Atomic Collision Physics." (McDaniel, E.W., McDowell, M.R.C., eds.) North Holland, Amsterdam, 2, 211, 1972.

10. Doschek, G.A., Feldman, U., Landecker, P.B., and McKenzie, D.L. Astrophys. J. 249, 372, 1981.

11. Gabriel, A.H., and Jordan, C., Mon. Not. R. Astr. Soc. 145, 241, 1969.

12. Phillips, K.J.H., Leman, J.R., Cowan, R.D., and Doschek, G.A. to be published Astrophys. J. 1982.

13. Feldman, U., Physica Scripta 24, 681, 1981.

X-Ray Diagnostics of Tokamak Plasmas

Elisabeth Källne and Jan Källne

Harvard-Smithsonian Astrophysical Observatory

Cambridge, MA 02138 USA

1. INTRODUCTION

In this review, we will venture into a new arena for work in atomic X-ray spectroscopy which we can dub tokamak X-ray spectroscopy diagnostics (TOXRASD). Not only is the experimental development exciting, but the measurements explore areas of atomic and plasma physics which have been inaccessible until just recently. This, however, does not mean that spectroscopy of highly ionized atoms is lacking tradition; on the contrary, the pioneering work in this area goes back to the 40's and the soft X-ray studies of spark generated plasmas /1/. Even though much of the present experimental effort is oriented towards obtaining a TOXRASD for fusion conditions, the new results touch upon some very basic atomic physics questions as well. Much effort has gone into the study of few electron systems, i.e., H-, He-, and Li-like ions with the purpose of utilizing the characteristic X-ray emission for diagnosing laboratory and astronomical plasmas or with the aim of developing the diagnostics, i.e., extending our understanding of the relationship between X-ray line emission and the plasma conditions under which the ions are formed and their ground states excited. However, the emission from highly charged heavy ions, such as neon-like molybdenum, has also attracted interest. Several reviews have appeared describing the experimental programs and we will merely refer the reader to these /2-5/. Here we shall focus the discussion around results of the latest vintage from the TOXRASD project at the Alcator C tokamak at MIT.

2. PLASMA SOURCE

The plasma source can be characterized by the radial and temporal dependencies of the main parameters such as temperature (T_e) and densities (N_e, N_z) of electrons and ions. The plasma source is

0094-243X/82/940463-17$3.00 American Institute of Physics

464

confined by a toroidal magnetic field. However, through plasma-wall interactions, impurity ions are introduced to the plasma, which consists mainly of hydrogen, deuterium or helium. The impurity ions are successively ionized and distributed in the plasma volume determined by e.g. equilibrium conditions (e.g. coronal) and transport phenomena. The spatial distributions of these impurity ions can be characterized by shells of the individual ionization stages separated or overlapping in the plasma. Radial profiles of the emission from different ionization stages (see Fig. 1) will be dependent on plasma parameters /6-8/. In Fig. 1 the profiles of Cl^{15+} and Cl^{14+} at T_e = 1.4 keV are shown as predicted from coronal equilibrium and from a transport code /8/. The radial profiles of T_e and N_e are also included.

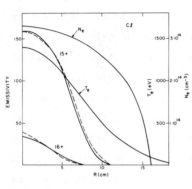

Figure 1.

Attempts to predict the distribution of ions in the plasma are made on the assumption that the ions are in coronal equilibrium, i.e. ionisation and recombination processes balance each other. With knowledge of the radial and temporal distributions of electron density and temperature, it is possible to predict the impurity ion distributions from rate equations. It is therefore crucial to know the atomic cross sections involved. The impurities are transported from the walls and a steady state distribution is achieved after a time comparable to the energy confinement time (20-30 msec). The ions are assumed to be excited and ionized from their ground states. Atomic processes of importance are electron impact excitation and ionization from the ground state followed by radiative transitions, dielectronic and radiative recombination. The relative importance of these processes for certain plasma conditions is a point of investigation for TOXRASD. Furthermore, the influence of other possible processes such as proton excitation, charge exchange and ion-ion collisions need to be further investigated.

3. EXPERIMENTAL TECHNIQUES

Probing the plasma conditions by means of the emitted photons in the X-ray region is performed with mainly four different techniques: X-ray imaging /9,10/, X-ray continuum measurements /11,12/, survey X-ray spectroscopy /2,13/ and high resolution, broad band line spectroscopy /14-17/. A typical time evolution of a plasma shot in

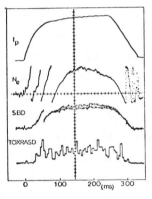

Figure 2.

Alcator C is shown in Fig. 2; each division along the abcissa is 10 ms. The fast rise of electron density (N_e) and plasma current (I_p) is followed by a stable period in which the plasma parameters are probed. The characteristic X-ray emission discussed in this paper is observed during this stable time period. However, during this apparently quiet period, there are internal instabilities developing with minor disruptions in the core of the plasma which can be imaged with the filtered X-ray diode arrays (SBD). With this technique, major m=1 (cos (mθ) symmetry in poloidal angle) magneto-hydrodynamical instabilities (so-called sawteeth) have been observed /18/. Furthermore, X-ray imaging can provide precursory information for major disruptions of the plasma (terminating the discharge) and can give microscopic information on the origin of these disruptions. Emission from a pure plasma is dominated by the bremsstrahlung-recombination continuum. It can be measured with solid state pulse height analyzing detectors and the electron temperature is thus determined. Departures from the pure hydrogen bremsstrahlung continuum emission occur because of line emission superimposed on the continuum and because of effects in the high energy tail of the electrons causing an electron distribution of non-Maxwellian shape /19/. These two X-ray diagnostic methods thus give information on the position, homogeneity and stability of the impurity content as well as presence of high energy run-away electrons, etc. To detect a wide wavelength range in the soft X-ray region during single plasma discharges, a fast scanning crystal

466

spectrometer has been developed at PLT /13/. With this diagnostic the radial distribution and the temporal evolution of emission from highly ionized low Z elements, such as oxygen, and from lower ionization stages of medium Z elements, such as iron, have been studied /20/. The first comparisons with predicted distributions, including transport effects, have shown general good agreement. The fourth technique of X-ray diagnostics is high resolution spectroscopy, which is in use at several major tokamak machines

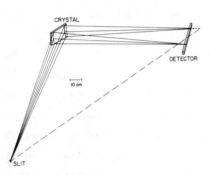

Figure 3.

/14-16/. Mainly two different geometries have been used, focussing Rowland circle and cylindrical van Hamos geometry. A schematic of the latter approach is shown in Fig. 3 /15/. We shall discuss some recent results obtained from the latter instrument at the Alcator C tokamak at MIT and their implications for future work.

4. ALCATOR RESULTS

The TOXRASD experiment has been taking data for about a year and the efforts have been directed towards good wavelength resolution $\Delta\lambda/\Delta \sim 1/3000$, large bandwidth, high sensitivity and high count rate. The data presented here was taken with a 10 cm long single wire proportional counter which limited the bandwidth to about 5% and the count rate to about 20 kHz. Much augmented spectroscopical and diagnostical performance can be achieved by detector improvements, which we shall discuss later. Our measurements of S, Cl and Mo impurities in the Alcator C tokamak /21/ (where these elements occur at trace element concentrations of about $10^{-5} N_e$) cover the wavelength region 4.3-5.3 A. This region can be covered in three detector settings from which a composite spectrum can be formed as show in Fig. 4. This spectrum shows characteristic X-ray emission from S^{13+}, S^{14+}, Cl^{14+}, and Cl^{15+} as indicated and a host of lines from molybdenum between 28+ and 32+; the plasma conditions of these discharges were such as to enhance the Mo concentration in the plasma. From the extensive data material of spectra from several thousand discharges, we shall select a few examples to illustrate some poignant features of the results.

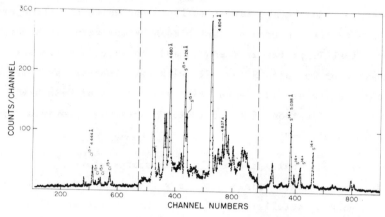

Figure 4.

4.1 LINE IDENTIFICATION OF FEW ELECTRON SYSTEMS

<u>H-Like Spectrum</u>. The 1s-2p transition gives rise to the resonance lines $2S_{1/2} - 2P_{1/2, 3/2}$ (W_H and W_H') and dielectronic satellite lines $1s2p^1P_1 - 2p^2 \, ^1D_2$. We observe these for S^{14+} (see Fig. 5) at positions consistent with the predicted wavelengths of

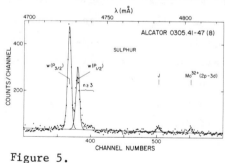

Figure 5.

4726.1, 4731.4 and 4783.6 mA /22,23/. Other satellites with a spectator electron in a higher orbit, i.e., transitions of the type 1sn -2pn, n≥3 are also indicated as a small foothill on the low energy side of the $2S_{1/2} - 2P_{1/2}$ resonance line.

<u>He-like Spectra</u>. The He-like spectrum was measured for both S and Cl (see examples in Figs. 6-9). These spectra are dominated by the $1s^2-1s2p$ resonance line (w) $^1S_0-^1P_1$, the spin forbidden $^1S_0-^3P_1$ and the intercombination line $^1S_0-^3P_2$, (y and x, will here both be referred to as intercombination lines). Transitions are referred to with letter symbols in the conventional way /23,24/. The 1s-2s

transition gives rise to the forbidden line (z) $^1S_o-^3S_1$. Several satellite lines to the resonance line with an extra (third) spectator electron in the 2s (lines q and r) or the 2p orbits (lines k and a) are observed. Some other satellite lines including the strongest, j, contribute to the spectrum but are hidden by the main lines (Fig. 8). The principal lines (w, x, y and z) and the satellite lines (q, r, a and k) are seen for both S and Cl with wavelengths as predicted. Excellent agreement is found with the results of Vainshtein /23/. Positive identification of these latter lines is provided by the spectra in Figs. 6-9.

We also note that while the Li-like charge states give rise to significant satellite lines to the w line in our S and Cl spectra, the corresponding satellites of Be-like charge states have escaped detection so far. They would be spread over a region of some 20 mA on the long wavelength side of the z line /27/. One such satellite has been observed in the spectrum of Fe from the plasma of the Princeton PLT tokamak /14/ which demonstrates the trend that the relative satellite intensity varies as Z^4, i.e., a factor of five increase between S and Fe when compared at temperatures of maximum emission for each atom.

The 1s-3p transition in sulphur is expected to produce a line at about 4.30 A which is just outside our range of measurements, but satellite lines to this resonance line (β) fall at somewhat longer wavelengths /25/. For certain plasma conditions as in Fig. 6 (low toroidal field, 40kG, low plasma current, 300 kA, N_e = 1.5 x 10^{14}cm^{-3}, and T_e = 0.75 keV), we observe a clear peak at 4.388$\pm$0.001 A of a width consistent with that of a single line. This

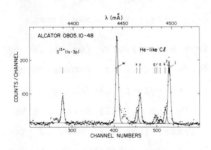

Figure 6

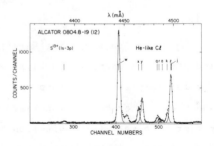

Figure 7

peak can be identified with the strongest satellite line of the 1s-3p transition, probably the transition $1s^22s-1s2s3p$ calculated to be at 4.3895 A. We observe other distinct satellite lines in this region and we find a remarkable similarity with the spectrum /25/ from the laser produced plasma at conditions of N_e-10^{19} cm^{-3} and $T_e = 350$ ev.

4.2 LINE IDENTIFICATION OF MANY ELECTRON SYSTEMS

In the wavelength range 4.3-5.3 A we observe strong emission from molybdenum (the limiter material) under certain plasma conditions (see Fig. 4). A point of practical consequence is that the Mo impurity concentration depends strongly on plasma conditions and becomes rapidly important for low N_e-values while the S or Cl emission does not vary very much. Therefore we can deliberately vary the relative intensities of spectral lines of Mo and S/Cl, which is crucial for the identification of the weakest S and Cl lines, for instance the J satellite (see Fig. 5). Based on measurements and calculated wavelengths, we have identified 2p-3d, 2s-3p and 2p-3s transitions in charge states between 28+ and 32+ /26/. The most interesting components of the Mo emission are the Ne-like spectrum of Mo^{32+}, which is dominated by $2p_{1/2,3/2}-3d$ transitions, and the strongest transitions identifying each charge state. We note that besides dominant principal transitions in the Mo spectrum, some contributions from dielectronic recombination satellites have been identified as well; for instance, the line at 4.837 A of Mo^{31+} /26/.

4.3 LINE INTENSITIES

T_e and N_e Dependences. Satellite line intensities (relative to the resonance line) depend on the electron temperature. For dielectronic satellites, this dependence goes approximately as $I'_s/I_w = 1/T_e$ and for inner shell excitation the $I'_s{}'/I_w$ ratio follows the abundance ratio N_{Li}/N_{He}. The N_{Li}/N_{He} ratio has a temperature dependence which can be calculated assuming a certain equilibrium for the plasma (coronal equilibrium) and certain ionization and recombination rates /27,28/. If the temperature T_e is known from the I'_s/I_w ratio or elsewhere, one can from the measured

I_s'/I_w ratio determine the actual N_{Li}/N_{He} ratio which is the calculated N_{Li}/N_{He} ratio for coronal equilibrium at the (ionization) temperature T_z. The difference T_z-T_e expresses whether the plasma is in a state of ionization ($T_z \langle T_e$) or recombination ($T_z \rangle T_e$), i.e., whether the considered ion of charge state Z is superheated or supercooled in the plasma environment of given electron temperature.

An example with particularly strong satellites in the He-like spectrum of Cl is shown in Fig. 6. From the intensity $I_k = 0.177 \pm 0.015$, we determine a temperature of $T_e = 760$ eV with a statistical error of ± 50 eV. This determination of temperature is based on the most accurate calculations by Safronova /29/, other calculations /30/ give slightly lower values, $T_e = 630$ eV. The thus obtained temperature is consistent with $T_e = 650 \pm 150$ eV measured from the X-ray continuum shape /31/. Furthermore, we note that a satellite ratio I_a/I_k of 0.27 ± 0.05 agrees with the calculated one of 0.23. Other dielectronic satellites (see Fig. 8) are too small to be measured except for the j line whose influence on the z line can be estimated from the measured I_k and the calculated ratio $I_j/I_k = 1.4$ /29/. In Fig. 8 we have presented the calculated intensities of the strongest dielectronic recombination satellites relative to the k line with normalization to the observed I_k. The dotted curves for q and r include the inner shell excitation contribution assuming $T_e = T_z = 760$ eV. The experimental spectrum is the same as in Fig. 6 apart from three-point smoothing of the raw data points. The calculated contribution of dielectronic recombination to the q line I_q' is 0.010 and that of inner-shell excitation is $I_q' = 0.029$ /29/. The total calculated intensity of q is 0.039 which amounts to only 50% of the measured ratio of $I_q = 0.073 \pm 0.015$. The situation is similar for the r satellite where $I_r' + I_r' = 0.02 + 0.012 = 0.032$ /29/ compared to a measured value of 0.053 ± 0.017. We have thus found that the calculated I'' values are too low indicating that the plasma is ionizing with a nominal T_z of about 420 eV with reservation, however, for possible difference in radial profiles between Cl^{14+} and Cl^{15+} ions. This is a general feature for most of our spectra at lower temperatures that $I_q(\exp) \rangle I_q(\text{theor})$ at given T_e, indicating that this is a characteristic feature of the plasma we observe or perhaps one should check for theoretical problems with the predicted rates of inner shell excitations.

Fig. 6 shows another example with significant satellite intensities but at a systematically lower relative level than for Fig. 7. From I_k/I_w, we determine $T_e = 940\pm80$ eV (compared to 1050 ± 150 eV from X-ray continuum measurements /31/) and the temperature difference is reflected in the overall satellite intensity difference. Another reason for comparing these particular spectra is, however, the change in density from 1.5 to 3.0×10^{14} cm^{-3} in going from Fig. 6 to 7, which is of interest with regard to the

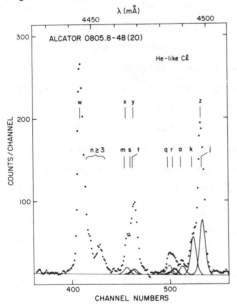

Figure 8

line intensity ratio $R = z/x+y$. We find this ratio to be 1.20 ± 0.14 and 0.97 ± 0.07 after corrections for the satellite interferences as discussed above and shown in Fig. 8. The calculated values are 1.37 and 0.92 where the difference comes from the collisional population transfer between the long lived triplet n=2 states, i.e. $^3S-^3P$, which is density dependent; the collisional and radiative rates are comparable at densities $N_e = 3 \times 10^{14}$ cm^{-3} /32,33/.

With the above two examples, we have demonstrated that the He-like spectra of Cl (and the same is true for S) contain information on T_e and N_e. The statistical accuracy with which we can measure the relative intensities of k/w and z/x+y is a matter of data accumulation rate per single discharge, or over how many constant discharges one can accumulate data. In the latter case, the crucial question is that of reproducibility of plasma discharges which we shall address below. With regard to the theory, the atomic rates computed and models used appear to give a good overall description of the plasma dependence of the present (as well as of previous) observations, except for the satellites due to inner shell excitation.

Comparison of 1s-2p and 1s-3p Transitions.

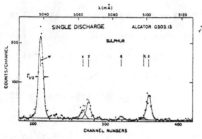

Figure 9

The He-like spectrum for sulphur of n=2 upper states appear in the λ range 5.03 to 5.11 A and is similar to that of Cl already discussed (see Fig. 9). The 1s-3p resonance line of the He-line S (β) is located at λ = 4.30 A /25/, which is just outside the spectrometer

bandwidth ending at 4.31 A. Since this line is of diagnostics interest we can estimate its intensity relative to the observed satellite peak as well as to the 1s-2p resonance. We use the results in Fig. 7 obtained at T_e = 960 eV. From the observed satellite to resonance intensity of 0.91 in the laser plasma at T_e = 350 eV and the calculated T_e dependence /25/ we determine the relative satellite intensity to be 0.09 at T_e = 960 eV; i.e., the 1s-3p resonance line in Fig. 7 should be 11.4 times the satellite line. The 1s-2p resonance line for S was measured during other but similar discharges of the same run 0805. Assuming constant S emission, we determine the 1s-3p to 1s-2p intensity ratio to be 0.5±0.2.

The Temperature of Electrons and Ions.

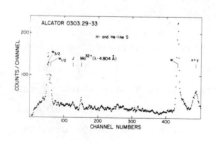

Figure 10

For the run 0303 we have determined the electron temperature in four ways. The spectrum in Fig. 10 shows the resonance lines of H-like and He-like S; these lines are broader than normal because of larger detector contribution at the extreme ends. We determine the ratio of

I_W/I_W = 0.75±0.07, which implies a temperature of T_e = 2.1±0.2 keV

assuming that the N_H/N_{He} abundance ratio is in coronal equilibrium
/35/. From the satellite ratio in the same figure, one determines as
I_J/I_W = 0.040$\pm$0.012 and a temperature of 1.7$\pm$0.3 keV. From another
but similar discharge we recorded the He-like S spectrum shown in
Fig. 9. From this single discharge spectrum we determine
I_k/I_W = 0.014$\pm$0.01 and the statistical error can be improved to
0.012$\pm$0.007 by adding three consecutive, similar discharges. The
corresponding temperature is 2.3$\pm$0.5 keV. Finally, the I_q/I_W ratio
is found to be 0.016$\pm$.006 which corresponds to a temperature of
1.8$\pm$0.6 keV with a N_{13+}/N_{14+} abundance ratio of S given by coronal
equilibrium. We thus find consistency between the four different
spectroscopic determinations of electron temperature. The results
can be used for testing the underlying atomic physics calculations
and to judge the ion abundance ratio relative to that of coronal
equilibrium, i.e., the ionization equilibrium or departure from the
same. At high temperatures, the results on q suggest a plasma closer
to ionization equilibrium between Li- and He-like sulphur than would
be the case for lower T_e. The intensity ratios and the deduced
temperatures are a first attempt to check consistency in these data
and explore the possibilities of TOXRASD. For a complete analysis
the radial distributions of the different ionization stages need to
be considered.

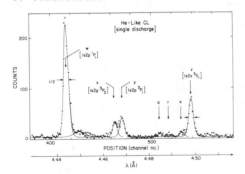

Figure 11

A parameter of crucial
importance for plasma fusion is
the ion temperature (T_i) or even
more so is information on the
relationship between T_i and T_e.
A well separated single line in
the X-ray spectrum can be used
to determine Doppler broadening
due to the thermal motion of the
ions. The resonance line w is a good candidate for this purpose,
especially for higher temperatures where the $n \geq 3$ foot hill satellites
are small (see Figs. 9 and 11), and the y and z lines can provide
complementary information. The line width in Fig. 9 (from run 0303)
is determined to be 7.3$\pm$0.3 channels which can be improved to

7.2±0.15 (corresponding to 3.1 mA) using an average of three discharges. From this we determine $T_i = 1.4 \pm .1$ keV having subtracted the estimated instrumental resolution corresponding to 630 eV. With this result, we have demonstrated that one can achieve simultaneous spectroscopial measurement of T_e and T_i of sufficient statistical accuracy to see differences between the two temperatures. With reservation for not yet unravelled systematic errors, our results suggest that the thermal coupling between electrons and ions is not 100% effective even for ohmically heated plasmas at relatively high electron densities.

<u>Ionization Equilibrium.</u> The line ratio $g = I_q / I_w$ is a predictable function of T_e for an ion abundance ratio N_{14+}/N_{15+} (to choose the Cl case) in coronal equilibrium. Knowing the temperature, the difference between experiment and calculation tells us about the departure from coronal equilibrium $(T_z - T_e)$. The ratio $G = x+y+z/w$ is temperature dependent because of a difference in temperature dependence of the triplet and singlet impact excitation rates /32/. G has also been predicted to depend on $T_z - T_e$ because of the larger recombination contribution to the triplet state, so G would reflect $T_z - T_e$ too. We thus have two line ratios relating to $T_z - T_e$, which we consider for the low and high temperature, $T_e = 0.7$ and 2 keV, cases of the He-like chlorine spectrum shown in Figs. 6 and 11. At the low temperature, we find G $= 0.94 \pm .05$ compared to the predicted /32,33/ value of G = 0.96 with no recombination. As mentioned already, we have determined $q_{exp} > q_{th}$ so both these results suggest a plasma under ionization or that the N_{14+}/N_{15+} ratio is larger than that of coronal equilibrium. At the high temperature, we determine G$=0.59 \pm 0.04$ compared to the predicted value of 0.45 which increases to 0.55 if recombination is included /33/. In this case, the experimental value of g (as stated above) is consistent with the prediction, so that both the g and G parameters indicate less departure from ionization equilibrium at the higher temperature. The parameters reflecting $T_z - T_e$ and their experimental determination are probably closely related to ion diffusion effects, i.e., this information has a bearing on the important question of transport in the plasma and hence confinement times.

Intensity Variations of Mo Lines. On the basis of wavelength, we have preliminarily identified lines from molybdenum in charge states from 28+ to 32+ /26/. It is clear that the temperature dependent intensity variation of lines will show the characteristics of the particular charge state involved which can help in the line identification. Assuming that we know the temperature of the plasma in the region of the emission studied, the line ratios can aid in the identification. Assuming instead that the line identification is known, plasma diagnostics information can be deduced. Examples of data that can be used in this context are shown in Fig. 12. The J satellite line has an intensity relative to the W resonance which determines $T_e = 1.6\pm.4$ and 1.9 ± 0.2 keV for the cases 12a and 12b. The intensity ratio of dominant 2p-3d lines in Mo^{30+} to Mo^{32+} changes between 0.5 and 2. It is clear that the observed variation of the Mo^{30+}/Mo^{32+} intensity ratio is not consistent with the identification made. However, the temperature given by J refers to the plasma region of

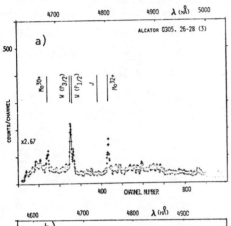

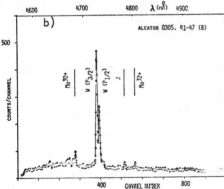

Figure 12a and b

maximum emission from the S^{13+} ions (cf. Fig. 1) which can differ from the radial distribution for $Mo^{30+,32+}$ and the two Mo species can be spatially separated too. Therefore, the line identification for many electron systems such as Mo is a complicated business and the temperature dependence must be used judicially. Once the identification is clear and a systematic of temperature dependences emerge, we can hope to use these complicated line spectra for detailed radial plasma mapping.

Intensity Variations of Fine Structure Components. We have furthermore measured the intensity ratio of the fine structure components of the H-like resonance lines for S ($^2S_{1/2}-^2P_{1/2,3/2}$) and of the intercombination transition of He-like S and Cl ($^1S_0 - ^3P_{1,2}$). The intensities are predicted to split according to statistical weights, i.e., without preference for angular momentum of the upper state of the transition; the predicted /32,33/ intensity ratios are $^2P_{1/2}/^2P_{3/2} = 0.5$, and $^3P_2/^3P_1 = 0.52$ and 0.65 for S and Cl. Our average values for many discharges are generally not far from these predictions but drastic variations are observed in shot-to-shot comparisons. Examples are shown in Figs. 13 and 14 The explanations

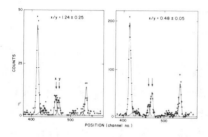

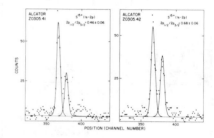

Figure 13   Figure 14

for these variations are still at large. It is neither clear which atomic population processes could be at play here nor do we know enough about the plasma conditions affecting these ratios. For the $^3P_2/^3P_1$ ratio the typical variation range is from 0.3 to 1.3 (similar for both S and Cl). Theoretically we have only been able to come up with a 10% difference in this ratio assuming proton collisonal coupling between the n=2 triplet states /33/. Proton collisons have previously been predicted to couple the 2s and 2p states of H-like ions and hence increase the $2P_{1/2}/2P_{3/2}$ intensity ratio with increasing proton (i.e., also electron) density /27/. However, the proton effects appear too small to explain the observed variations let alone that the change is only in one direction. Here, we need to make simultaneous measurements of these two ratios to see if the variations are correlated and hence answer the question whether there is a common cause affecting these line ratios. These results are both experimentally clear and interpretationally puzzling and should entice some stretching of the atomic physics imagination.

5. NEXT GENERATION TOXRASD MEASUREMENTS

The results from the new TOXRASD experiment at the Alcator C tokamak and those from the well established experiment at the Princeton tokamak have already yielded valuable diagnostics information and touched upon interesting atomic physics questions. However, they also indicate the potential for further development to probe the plasma and the atomic processes in greater detail. For our project the limit has been on the detector side which we now hopefully have solved with the development of a new fast delay line x–y detector /36/; similar, the Princeton Bragg spectrometer has been augmented with a new detector. Together with a fast CAMAC data acquisition we hope to achieve count rates of 0.2 to 1 MHz. The new detector is 25 cm long and can detect about 0.8 mA of our spectrum shown in Fig. 4 at an expected resolution of better than 0.2 mm or (<0.4 mA at 4.4 A). This will allow measurements of high resolution, broad band spectra with an expected statistical accuracy of better than 3% for dominant lines for single Alcator discharges. The objective here is to measure accurately the line width of several lines simultaneously (and hence the ion temperature) and its correlation with the electron temperature deduced from various line ratios. With the many line ratios one can then study simultaneously, the plasma presents a unique atomic physics laboratory where the excited states involved in the transitions observed are subjected to the same and (partially) controllable medium of hot electrons and ions which is sustained for a long period of time on the scale of atomic relaxation times. Once the detector system can handle high count rates it is natural to probe the time evolution of the line emission for single plasma discharges. Thanks to the high light collection efficiency of the van Hamos geometry of our spectrometer (and to H. Schnopper /37/ who originated this spectrometer idea for TOXRASD), we expect to achieve 5 to 10% statistical accuracy for strong lines for a time resolution of 10 ms, which can be augmented by increasing the crystal height and the viewing solid angle into the plasma, etc. With the extended band width and time information, one would be able to measure the time correlation between the intensity ratios x/y and $W_{1/2}/W_{3/2}$ and determine the cause of the variations we observe in these ratios for seemingly constant shots and hopefully reveal the atomic process affecting these states and the hidden parameter that varies for seemingly constant plasma shots. Results of the type presented (including measurements and atomic physics interpretations) and the realization of time resolved spectroscopy are important for plasmas heated by other measures than ohmic heating. The Princeton group has already measured He-like Fe spectra for plasmas heated by neutral ion injection /38/. Under these conditions one surely disturbs the plasma equilibrium and simultaneous measurements of T_i, T_e, and T_z as a function of time provides a real challenge for time resolved TOXRASD measurements and should be of great atomic physics interest, too. The highly ionized atoms are effective sensores distributed over the whole plasma volume. The information is encoded in the form of characteristic line-spectra and we are making rapid progress in increasing the accuracy of the measurements and finding the formula for decoding the TOXRASD spectra.

ACKNOWLEDGEMENTS

We have greatly benefitted from the stimulating support by Ron Parker and the Alcator group. We also wish to thank M. Bitter for sharing recent data with us, R. Petrasso for results on X-ray imaging, and John Rice for providing the results on radial profiles. The work was supported by the U.S. Department of Energy.

REFERENCES

1. B. Edlen, Symposium on Production and Physics of Highly Charged Ions, to appear in Physica Scripta, 1982.
2. K.W. Hill et al., AIP Proc. No. 75, 8 (1981) and PPPL-1887, April 1982.
3. N.J. Peacock, AIP Proc. No. 75, 101 (1981).
4. C. deMichelis and M. Mattioli, Nuclear Fusion 21, 677 (1981).
5. Equipe TFR, Nuclear Fusion 18, 647 (1978).
6. S. Sudkewer, Physica Scripta 23, 72 (1981).
7. T. Kato et al, Institute of Plasma Physics, Nagoya, IPPJ-544, (1981).
8. E.S. Marmar et al., MIT/PFC Report JA-82-12 submitted for publication (1982).
9. R. Petrasso et al., Nuclear Fusion 21, 881 (1981), and Rev.Sci.Instr. 51, 585 (1980).
10. S. van Goeler, W. Stodiak, and N. Sauthoff, Phys.Rev.Lett 33, 1201 (1974), and N.R. Sauthoff, SPIE, Vol. 106, 40 (1977).
11. M. Bitter et al., in Inner Shell and X-ray Physics of Atoms and Solids (Plenum Press) 1981, p.861 and PPPL-78 6492.
12. J.E. Rice, K. Molvig, and H.I. Helava, Phys.Rev. A25, 1645 (1981).
13. S. van Goeler et al, Bull.Am.Phys.Soc. 26, 812 (1981).
14. K.W. Hill et al., Phys.Rev. A19, 1770 (1979), M.Bitter et al., Phys.Rev.Letter 43, 129 (1979), and M. Bitter et al., PPPL-1891, April 1982.
15. E. Kallne and J. Kallne, submitted to Physica Scripta (1982).
16. P. Platz et al., J.Phys.E 14, 448 (1981) and J.Phys. B15, 1007 (1981).
17. M.G. Hobby, N.J. Peacock, and J.E. Bateman, Culham Laboratory, CLM-R103 (1980).
18. G.L. Jahns et al., Nuclear Fusion 18, 609 (1978) and M.A. Dubois, D.A. Marty and A. Pochelon, Nucl.Fusion 20, 1355 (1980).
19. H. Knoepfel and D.A. Spong, Nucl.Fusion 9, 785 (1979).
20. E. Meservey et al., Bull.Am.Phys.Soc. 26, 981 (1981).
21. B. Blackwell et al., in Plasma Physics and Controlled Nuclear Fusion Research Proc. 9th Int.Conf., Baltimore 1982.
22. U.I. Safronova, Physica Scripta 23, 241 (1981).
23. L.A. Vainshtein and U.I. Safronova, Atomic Data Nuclear Data Tables 21, 49 (1978).
24. A.H. Gabriel, Mon.Not.R.Astr.Soc. 160, 99 (1972).
25. V.A. Boiko, et al., Mon.Not.R.Astr.Soc. 185, 789 (1978).
26. R.D. Cowan, Symposium on Production and Physics of Highly Charged Ions, Physica Scripta (1982), and E. Kallne, J. Kallne, and R.D. Cowan, submitted for publication (1982).
27. V.A. Boiko et al., J.Quant.Spectr.Rad.Transf. 19, 11 (1978).
28. J. Dubau and S. Volonte, Rep.Prog.Phys. 43,199 (1980).
29. U.I. Safronova, A.M.Urnov, and L.A. Vainshtein, Dielectronic Satellites for High Charged Resonance Lines, Preprint.
30. C.P. Bhalla, A.H. Gabriel, and L.P. Presnyakov, Mon.Not.R.Astr.Soc. 172, 359 (1975).
31. D.S. Pappas et al., Bull.Am.Phys.Soc. 26, 885 (1981) and J.E. Rice, private communication.
32. A.K. Pradhan, Astroph.J.(in press), and Phys.Rev. A23, 619 (1981).
33. E. Kallne, J. Kallne, and A.K. Pradhan, submitted for publication (1982).
34. E. Kallne, J. Kallne, and J.E. Rice, Phys.Rev.Lett. 49, 330 (1982).
35. V.L. Jacobs et al., Astroph.J. 230, 627 (1979).
36. J. Kallne, E. Kallne, L. Atencio, C. Morris, and A. Thompson, Nucl.Instr.Methods, Fall 1982 (to appear) and ibid Forth APS Topical Conf. on High Temp. Plasma Diagnostics, Boston, Aug. 25-27, 1982.
37. H.W. Schnopper and P.O. Taylor, U.S. Department of Energy Report No. E4-76-502-4021 (1977).
38. R.J. Hawryluk et al., Phys.Rev.Lett. 49, 326 (1982).

MEASUREMENTS AND ANALYSES FOR SOFT X-RAY GAIN IN PLASMAS

R. C. Elton, R. H. Dixon and J. F. Seely
NAVAL Research Laboratory, Washington, DC 20375

ABSTRACT

The eventual extension of lasers to the x-ray spectral region utilizing non-equilibrium population densities for optical levels in plasma ions will require careful optimization of population inversions at high densities to achieve high gain at short wavelengths. Progress is being made in both direct observation of gain and in the spectroscopic measurement of absolute bound state densities for the inverted levels involved. The latter provides a value for the net gain from a plasma as a gauge for further parameter and geometry optimization to increase gain for lasing. Such gain values deduced for carbon ions are in good agreement with both basic physical analyses and with more detailed numerical modeling, particularly for electron-capture pumping on which most experimental results to date are based. Encouraging results towards scaling to higher population density and gain at increased pump power are described.

I. INTRODUCTION

For x-ray lasers, one thinks first of population density inversions under non-equilibrium conditions on K-transitions. However, unless a proper sequence of particular auxiliary higher-shell transitions[1,2] (e.g., Auger) follow rapidly to generate specific line shifts[1,2] lasing conditions are expected to self-terminate in approximately a decay period, which becomes extremely short at high photon energies. Efforts[2] have been more towards L- and M-transitions with rapid depletion of the lower laser state by electron decay to a vacancy in the K-shell, thereby maintaining the population inversion in a quasi-cw operating mode. The high pump powers required simply to overcome the spontaneous radiation losses for such transitions virtually assures that a high temperature plasma will be formed, and that the lasing will take place on multiply-ionized atoms. Hydrogenic as well as helium-like and lithium-like (1-, 2-, and 3- electron) ions are usually studied as possible lasants, with the first the simplest to analyze.

In Section II following, a general gain analysis is extended to specifically hydrogenic ions of varying nuclear charge Z, with a discussion of parameters limitations. Following this in Section III, gain measurements for hydrogenic C^{5+} ions are described, and then compared with an electron capture pumping model in Section IV. Initial experimental efforts towards increasing the measured C^{5+} gain in a systematic manner are described in Section V, followed by some concluding remarks in Section VI.

0094-243X/82/940480-11$3.00 American Institute of Physics

II. GAIN ANALYSIS

The achievement of significantly-enhanced beamed output from an ion laser at wavelengths shorter than 700 Å in the "x-ray" region[3] requires an exceptionally high gain medium to compensate both for the decrease in cavity efficiency and for the explicit (λ) and implicit wavelength dependences of the gain coefficient

$$G = N_u \sigma_{ind} - N_\ell \sigma_{abs} = \frac{\pi^2 r_o c \lambda f(g_\ell/g_u)}{2} \left(\frac{M}{2\pi kT}\right)^{1/2} N_u I, \qquad (1a)$$

where I is the inversion factor

$$I = \left[1 - \frac{N_\ell}{N_u} \cdot \frac{g_u}{g_\ell}\right]. \qquad (1b)$$

This relates to the single-pass amplification exp (GL) over a length L for upper and lower state densities and statistical weights N_u, N_ℓ, and g_u, g_ℓ, respectively, and for a Doppler-broadened spectral line in a plasma of temperature T and ion mass M. The oscillator strength is denoted by f, r_o is the classical electron radius, and σ_{ind}, σ_{abs} represent the respective cross sections for induced emission and absorption. There is an additional usually-strong implicit wavelength dependence for the particular pump mechanism that is selected to provide the desired upper state density N_u and the required population inversion I > 0. High gain is therefore obtained by a high density of ions in the initial pre-pumped state and by sufficient pumping rates to overcome the high depopulation rates.

The population inversion factor I > 0 here is crucial in at least two respects. If at high densities it becomes very small because of collisional mixing and radiative trapping (on the lower level K-depletion transition), then to keep the gain constant one must compensate by increasing the density according to Eq. (1), which becomes self-defeating. Secondly, verification at I $\approx$ 0 conditions necessitates direct amplification tests, which can be less than conclusive for marginal gain levels during the current early development phase for x-ray lasers. In contrast, a finite value such as I = 0.3, achievable at lower densities, results in population densities varying significantly from equilibrium and measurable directly from fluorescence on resonance K-transitions[4]. The absolute gain coefficient can then be deduced (even at values too low for lasing) using Eq. (1) for a known temperature (or line width) and a measured upper state density (see Section III below). The gain coefficient then becomes a valuable measurement gauge for advancing towards higher values in a systematic parameter study. This is an important point for present state-of-the-art experiments.

Plasma density and temperature limiting conditions for the two cases I $\approx$ 0 and I = 0.3, are illustrated in Fig. 1 and in Fig. 2 (Δ = 0 curves), respectively, for 2-3 LM and 3-4 MN transitions.

482

The decrease in permitted electron density at a particular temperature when I increases from 0 to 0.3 is readily apparent. These two figures actually show three effects on the inversion limits, namely, electron collisional mixing between level u and ℓ (Fig. 1, with dashed extension), the high temperature cutoff associated with electron collisional excitation from the K-shell, and the effect of radiative trapping of resonance K-radiation in a reduced depth Δ (Fig. 2) defined below. These results were derived from a simple 3-level steady-state analytical model[5] involving the upper (u), lower (ℓ), and ground state (1), so that the quantity (1-I) in Eq. (1b) becomes

$$\frac{N_\ell}{N_u} \cdot \frac{g_u}{g_\ell} = \frac{g_u}{g_\ell} \cdot \frac{N_e X_{u\ell}(T) + A_{u\ell} + \rho N_e X_{1\ell}(T)}{N_e X_{\ell u}(T) + g(\tau) A_{\ell 1}}, \tag{2}$$

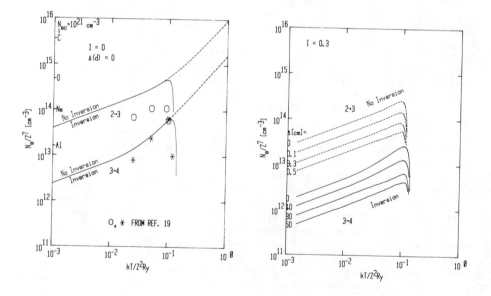

Fig. 1. Gain parameter space with & without (--) excitation cutoff.

Fig. 2. Gain parameter space for various (reduced) depths Δ.

where $\rho \equiv N_l/N_u$, N_e is the electron density, and X_{mn} represents the electron collisional excitation and deexcitation rate coefficients for transitions from level m to level n, for which the Bethe-Born effective Gaunt factor ($\bar{g}$=0.2) formulism[6] is used (the collisional) deexcitation rate $N_e X_{l1}$ is relatively negligible). The transition probability[7] is represented by A_{mn}. For an optical depth τ, the quantity $g(\tau)$ is the escape factor[8], which decreases from unity for increasing opacity on the resonance lines. For hydrogenic ions of nuclear charge Z, Eq. (2) becomes independent of Z when the reduced parameters N_e/Z^7, kT/Z^2Ry and $\Delta = dZ^{4.5}$ (N_l/N_e) are adopted for the electron density, temperature and scale length, respectively. The latter enters through the K-line optical depth, where d represents the characteristic depth, e.g., the diameter of a cylindrical plasma.

In order to include the high temperature cutoff effect of electron collisional excitation from the ground state into the upper laser state, a value for ρ is needed. A non-equilibrium population distribution between the laser levels must be present for population inversion to exist. Under such conditions, equilibrium relations (e.g., coronal) cannot be expected to apply between N_u and N_l. However, the upper level can approach Saha-Boltzmann equilibrium with the higher bound levels and the continuum (at least within a factor[9] of approximately 1/3 included below), which gives $N_s/N_u \propto Z^3/N_e$, where N_s represents the stripped-ion density. Writing $N_l/N_u = (N_s/N_u)(N_l/N_s)$, it is the second factor that remains to be evaluated. From a similarity analysis[10] for a plasma expanding with a characteristic time τ_{ex}, the ratio N_l/N_s can be expressed as $(\tau_{ex}/\tau_r)/(\tau_{ex}/\tau_i)$ for ionization and recombination times τ_i and τ_r, respectively, and this ratio is proportional to Z^4. Using this scaling and an average of frozen-ion state densities measured in expanding carbon (Z=6) laser-produced plasmas[11-16], we find that $(N_l/N_s) \approx 2.3 \times 10^{-3} Z^4$ is a reasonable value for the second factor above and that $\rho \propto (N_e/Z^7)^{-1}$ preserves the reduced density scaling in Eq. (2).

Opacity on the K-transition, leading to radiative trapping population of the lower laser level and inversion depletion, was incorporated into the formulism through the escape factor $g(\tau)$ modification[8] to the transition probability, i.e., A_{l1} becomes $g(\tau)A_{l1}$. The reduced-parameter scaling of density and temperature are preserved for a reduced depth Δ defined above. It is clear from Fig. 2 that the inversion limit is severely degraded at all temperatures up to collisional cutoff by opacity population of level l [compared to that for purely-collisional mixing (Δ= 0)] when Δ reaches $\sim$1 for the 2-3 LM transition and $\sim$100 for the 3-4 MN transition. These translate into maximum diameters using d = $(\Delta/Z^{4.5})(N_l/N_e)^{-1}$ cm, providing that the ratio of densities is specified. Obviously if N_l/N_e = 0 there is no opacity problem and d can be infinite in extent. On the other hand, if all ions were in this ground state the ratio would be simply 1/(Z-1). Again, based upon averaged measurements for carbon, N_l/N_e = 1/Z is more typical, so that d $\approx \Delta/Z^{3.5}$ cm can be used to estimate the maximum dimension.

Serious degradation of inversion conditions then occur for
d $\approx$ 1/Z$^{3.5}$ cm for the 3-2 LM transitions and d $\approx$ 100/Z$^{3.5}$ cm for the
4-3 MN transition. For carbon plasmas these correspond approximately
to 20 μm and 2 mm respectively[17], and for aluminum plasmas to 1.5 μm
and 150 μm. This is a major reason that laser-produced plasmas of
minute focal dimensions are used to produce the lasant for such very
short wavelengths as 656/Z^2 and 1875/Z^2 nm for 3-2 and 4-3 transitions,
respectively (an additional reason is the very high pump power density
required to maintain a population inversion against rapid spontaneous
radiation cooling).

Credibility for the simple physical model can be furthered by
comparing in Fig. 1 the present results for the I=0 threshold and
for d=0 (optically thin--pure collisional mixing) with the more
complete multi-level collisional-radiative numerical rate equation
computations (discrete points) of McWhirter and Hearn[19]. The agreement, including the high temperature cutoff, is very satisfactory
considering the use[19] of $\bar{g}$=0.4, the simplicity of the present model,
and inherent uncertainties of approximately X2 in collisional rates.

The lower density plasmas indicated in Figs. 1 and 2 at the lower
temperatures (desirable for recombination pumping[15], for example) are
consistent with the nature of expanding laser-produced plasmas
typically used for such experiments. The limiting electron densities
have been verified experimentally[9,18]. For laser-produced plasmas,
values of the critical electron density for maximum pump laser
absorption given by $N_{ec} = 10^{27}/\lambda_p^2$ (λ_p in nm) which equals 10^{21} cm^{-3}
for a 1 μm pump laser wavelength (λ_p) are also indicated in Fig. 1
on the left ordinate for various target materials.

An important additional point bearing on the above discussion of
density limitations is that it is usually very desirable to maintain
ionic conditions as pure as possible, i.e., preferably all ions
initially in the particular ionic state from which pumping proceeds,
with electrons contributed mainly from the lasant ions (minimal
impurities) and with pumping proceeding preferentially into the
upper laser state at as high a rate as possible.

III. GAIN MEASUREMENTS

In what follows we describe some recent gain coefficient values
determined using the methods outlined above, i.e., by applying
direct parameter measurements to Eq. (1). This has been done[14]
for the 3-4 MN transition in a hydrogenic C^{5+} laser-produced plasma,
where overpopulation of the lower n=3 level by self absorption
is minimized. Also, the application of auxiliary measurements (such
as N_e, N_s) to a reasonable guiding model for further scaling is
discussed.

In most of the experiments described here (see Fig. 3) the carbon plasma was produced by a focused Nd:glass laser of intensity 4.1×10^{11} W/cm^2 (8 J, 10 ns, 500 μm spot diameter). Some scaling measurements performed at higher intensities are described in Section V below. The plasma was viewed spectroscopically with both spatial and temporal resolution in the visible and soft x-ray spectral regions.

Experiments have shown[4,18] that at expansion distances of about 3 mm or more, inversion factors $I \sim 0.3$ are possible between n=4 and n=3 levels, in C^{5+}. At a temperature kT = 10 eV, Eq. (1) then indicates[20] that an upper state density $N_u = N_4 \sim 10^{14}$ cm^{-3} is required for a "threshold" gain coefficient of unity, i.e., a net gain of exp (1) over a 1·cm path. We have obtained[14] a time-resolved experimental value of $N_4 = 1.2 \times 10^{12}$ cm^{-3} from a measured value of N_7, where the relative distribution over bound states can be determined from the intensities in the soft x-ray Lyman series. This leads to a gain coefficient of $\sim$1%/cm from Eq. (1).

The population density N_7 of the n = 7 level (shown in Fig. 4 with $N_s = N^{6+}$ and N_e) was obtained from a measurement of the absolute

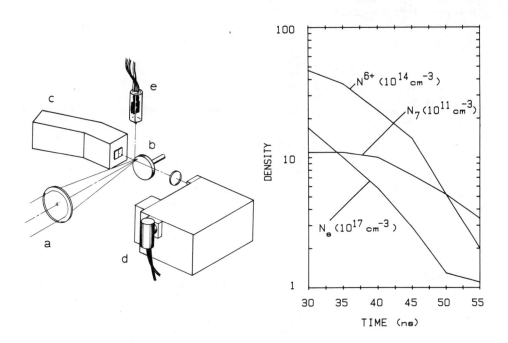

Fig. 3. Experiment: (a) laser, (b) target, (c) x-ray spectrograph, (d) uv/visible spectrograph, (e) deuterium calibrated lamp.

Fig. 4. Densities at 3 mm from the target.

emission from the 6-7 line at 343 nm wavelength in the near-uv spectral region. The spectrograph used was calibrated using a standard deuterium lamp.

IV. MODELING: ELECTRON CAPTURE PUMPING

So far we have derived gain values from measured parameters specified in Eq.   . We can also model this gain on an electron-capture (recombination) steady-state model using

$$N_4 = N_s \left(\frac{P}{D} \right) , \tag{3}$$

where N_s is the C^{6+} initial stripped ion density, D is the total depopulation rate and P is the mean pumping rate which can be written as[15]

$$P \sim \frac{10^{-31} N_e^2 (n')^6}{Z^6} \left(\frac{Z^2 Ry}{kT_e} \right)^2 \exp \left[\frac{Z^2 Ry}{(n'+1)^2 kT_e} \right] \sec^{-1} . \tag{4}$$

Here n' represent the collision limit[6,21]. This shows the strong inverse dependence on electron temperature that favors increased pumping with rapid cooling.

Further modeling requires values for N_e as well as one for N_s (which is sometimes optimistically taken to be $N_e/6$, i.e., a completely-stripped carbon plasma). At a distance of 3 mm from the target this is clearly not the case, because recombination occurs rapidly in the first 100 μm or so beyond which frozen-ion-state distributions follow in expansion. We have obtained[14] an experimental value of $N_s = 1.4$ X 10^{15} cm^{-3} from the absolute density N_7 described above, assuming local thermodynamic equilibrium. The localized electron density $N_e = 3$ X 10^{17} cm^{-3} required was obtained from the Stark width of the C^{5+} 7→6 line at 343 nm wavelength, based on a profile fit to recent detailed calculations of Kepple and Griem[22]. Density versus temperature plots analagous to Figs. 1 and 2 can be generated for Z=6, to which fixed-gain curves can now be added according to this modeling. Such constant-G curves are plotted in Fig. 5 for a fixed inversion factor I = 0.3 and for the N_s measured value described above. A fairly typical ion trajectory with distances measured from the target is included dotted in Fig. 5, beginning at the critical density; this indicates the present status in free expansion. Also included

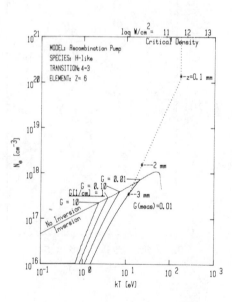

Fig. 5. Parameter space with gain curves and typical ion trajectory.

in Fig. 5 is our measured gain at 3 mm, and the agreement with the
present modeling is considered excellent. From this graph we can
reason that further progression towards matching plasma conditions
with a gain coefficient of unity include: (a) shifting the gain
curves in Fig. 5   towards lower N_e by increasing N_s, (b) rapid
cooling at a distance of 3 mm by some type of added "barrier" and/or
enhanced radiative cooling from higher-Z materials which would increase the pumping rate as seen in Eq. (4), (c) relaxing the requirement that the inversion factor I be as large as 0.3 for measurement, which would permit closer-in operation as shown in Fig. 1 [but
with a net upward shift in the fixed gain curves at smaller I, unless N_s were also increased at very close (< 100 μm) distances] ,
and (d) higher-Z materials with closer-in operating conditions
(scaling indicated in Figs. 1 and 2).

V.   INCREASING THE GAIN

The next question is how to best increase $N_u = N_4$ and thereby G,
preferably without having the inversion factor I become immeasurably
small so that our measurements of G described above can continue to
serve as a gauge of progress. Maintaining generality, this is accomplished both by increasing the population density N_s in the
initial source reservoir state (C^{6+}) and/or by increasing the
pumping rate into the upper n=4 state.

A threshold gain coefficient of 1 cm^{-1} is achieved in electron-
capture pumping by rapidly reducing the temperature in the lasing
region for increased recombination into excited states and for pre-
ferential population of the n=4 upper laser level. First, however,
it is important to maximize the stripped-ion density N_s at a level
hopefully $\sim$ 100 X higher than N_4. To achieve this we increased the
target irradiance to raise the temperature in the near-target, thereby producing more stripped ions near the target and increasing the
density N_s in the outer (mm's) region.

To accomplish this, the measurements described in the previous
sections have been repeated at higher irradiation. The energy of
the Nd:glass laser was increased to 18 J and the focal spot diameter
was reduced to 100 μm. Observations were made at a maximum intensity
of 5.7 x 10^{12} W/cm^2 and at four lower intensity levels by using neutral
density filters. The photomultiplier signal of the 343 nm C^{5+} line
center 3 mm from the target is shown in Fig. 6. The 343 nm emission
increases with irradiation intensity up to 1.5 X 10^{12} W/cm^2 and decreases at higher intensities.

The electron and ion densities, measured at the time (45 ns)
when the 343 nm emission is strongest, is shown in Fig. 7. At an
irradiance of 1.5 X 10^{12} W/cm^2 we obtain N_s = 1.2 X 10^{16} cm^{-3}, which
is 100 X the density of N_4 needed for a "threshold" gain coefficient
= 1 cm^{-1}, and a fractional stripping of 6N^{6+}/N_e $\approx$ 0.13. These values
are $\sim$9X and $\sim$4X those obtained at an irradiance of 4.1 X 10^{11} W/cm^2.
Yet to be measured from soft x-ray lines at the higher irradiance is
the ratio N_4/N_7, which is not expected to increase with higher irradiance and temperature; however its value is more important after

488

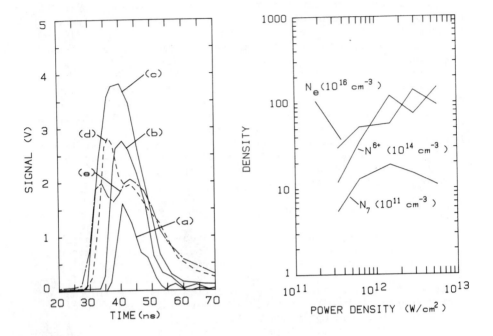

Fig. 6. 343 nm emission for inten- Fig. 7. Densities at 3mm from
sities: (a) .34, (b) .63, (c) 1.5, target and 45 ns into pulse.
(d) 2.9, (e) 5.7 X 10^{12} W/cm^2

the pumping is enhanced with rapid cooling, now that the stripped-
ion density is optimized. How much additional cooling is needed?
Starting with these measurements of N^{6+} and assuming LTE with highly-
excited states [e.g., $N_7/N^{6+} \alpha T_e^{-3/2} \exp (10/T_e)$], an exponential
scaling of the lower (e.g., n=4) bound state distribution, a T_e^{-2}
dependent pumping rate, and thermal Doppler broadening (all of which
favor increasing N_4 with decreasing temperature), threshold gain
could be reached by lowering T_e from approximately 20 eV to 10 eV,
i.e., a factor-of-2, for the 1.5 X 10^{12} W/cm^2 irradiance. It is
interesting to note that the same factor applies at the lower irradiance (4.1 X 10^{11} W/cm^2), i.e., 9 eV to 5 eV; however, again, the
absolute stripped ion density is marginal at this level.

VI. CONCLUSIONS

In this paper we use a straightforward physical model to first
define plasma boundary conditions for achieving population inversion
on short wavelength transitions in hydrogenic ions. Earlier analyses
are expanded-upon here to include electron collisional and opacity-
pumped population from the ground state, and maximum plasma dimensions are derived for 2-3 LM and 3-4 MN lasing transitions. The
opacity limit on dimensions along with the excitation limit at high

temperature remain major limitations on suitable minute-plasma pumping sources, at least until some mechanism can be developed for further depletion of the ground state population compared to that which has been measured to far. To operate at the maximum density for high gain, a plasma expanding from a high density region, such as at a laser-heated target, can be used and the pump laser wavelength λ_p can be adjusted to suit the region of critical electron density (see Fig. 1) where the dimensions are small, particularly for moderate-Z elements.

We have also applied some recent experimental observations of essential plasma parameters to an assessment of presently achieved gain on a 4→3 transition in a hydrogenic carbon plasma radiating at 52 nm wavelength. We show that these measurements can be used as a gauge of progress, as certain variables such as pump power density are adjusted in an attempt to increase the gain towards threshold for lasing. Extended analytical modeling based on the measured plasma parameters assuming recombination pumping is in excellent agreement with the measured gain coefficients, which is very encouraging as far as applying basic physical ideas to extending the gains measured so far.

It is clear that the gain already being measured in laser-produced plasmas would achieve lasing threshold if cavities were available in the vuv region which were as efficient as those in the visible. Continued research resulting in parallel progress towards higher gains with both an elongated (perhaps channeled[23]) plasma and with more efficient cavities could lead to vuv laser operation in the forseeable future, followed by continuation isoelectronically towards shorter wavelengths with higher Z-materials.

Assuming that such L- and M- series lasers become a reality, a challenge to x-ray physicists remains to translate the knowledge gained to the vastly more complicated innershell K-transitions at much shorter x-ray wavelengths, where higher efficiencies may be possible compared to stripped plasma ions and where rapid lower laser state depletion by, for example, Auger transitions might eliminate both the rapid self-termination of the population inversion and the opacity pumping discussed above[1].

ACKNOWLEDGEMENTS

It is a pleasure to acknowledge the expert technical assistance of J. L. Ford in all phases of the experiment described, including laser operation, and calculational assistance from Ms. A. K. Surr. Appreciation is expressed to H. R. Griem for illuminating technical discussions.

REFERENCES

1. R. C. Elton, Appl. Optics 14, 2243 (1975).
2. R. W. Waynant and R. C. Elton, Proc. IEEE 64, 1059 (1976); R. C. Elton, "X-Ray Lasers" in Handbook of Laser Science and Technology, M. J. Weber, ed. (CRC Press, Inc., 1982).

3. J. A. Bearden, "X-Ray Wavelengths", AEC Report No. NYO-10586, John Hopkins University, 1964. The longest wavelength in the finding list is the 692 Å MN (3-4) line of potassium. LM (2-3) lines extend to 407 Å (Na) and KL (1-2) lines to 228 Å.

4. R. C. Elton and R. H. Dixon, Phys. Rev. Lett., $\underline{38}$, 1072 (1977); also R. H. Dixon, J. F. Seely, and R. C. Elton, Phys. Rev. Lett., $\underline{40}$ 122 (1978).

5. R. C. Elton, Comm. At. and Mol. Phys. (to be published).

6. R. C. "Atomic Processes" in Methods of Experimental Physics, Plasma Physics, H. R. Griem and R. H. Lovberg, eds., Vol. 9A, Chapter 4 (Academic Press, New York, 1970).

7. W. L. Wiese, M. W. Smith, and B. M. Glennon, "Atomic Transition Probabilities", NSRDS-NBS-4, Vol I (U. S. Government Printing, Washington, DC 1966).

8. R. W. P. McWhirter, "Spectral Intensities", in Plasma Diagnostic Techniques, R. H. Huddlestone and S. L. Leonard, eds., Chapters 3 and 8, (Academic Press, N. Y., 1965).

9. F. E. Irons and N. J. Peacock, J. Phys. B $\underline{7}$, 1109 (1974).

10. G. Pert, J. Phys. B, $\underline{9}$, 3301 (1976).

11. F. E. Irons and N. J. Peacock, J. Phys. B $\underline{7}$, 2084 (1974).

12. A. M. Malvezzi, L. Garifo, E. Jannitti, P. Nicolosi, and G. Tondello, J. Phys. B, $\underline{12}$, 1437 (1979).

13. D. Jacoby, G. Pert, S. Ramsden, L. Shorrock, and G. Tallents, Optics Comm., $\underline{37}$, 193 (1981).

14. R. C. Elton, J. F. Seely, and R. H. Dixon, "AIP Proc. Topical Meeting on Laser Techniques for Extreme Ultraviolet Spectroscopy", (Am. Inst. of Physics, 1982), invited and postdeadline papers.

15. R. C. Elton, Opt. Engineering $\underline{21}$, 307 (1982); also SPIE $\underline{279}$, 90 (1981). The inversion analysis here is favorably improved upon in the present article and in Ref. (14).

16. S. Suckewer and H. Fishman, J. Appl. Phys. $\underline{51}$, 1922 (1980).

17. The 2 mm maximum depth for the 4-3 transition in C^{5+} is in agreement[9,14,15,18] with experiments and typical trajectories (see Fig. 5).

18. R. C. Elton, T. N. Lee, R. H. Dixon, J. D. Hedden and J. F. Seely, in "Laser Interaction and Related Plasma Phenomena, v. 5, p. 135, H. Schwarz, H. Hora, M. Lubin and B. Yaakobi, eds. (Plenum Publ. Co., New York, 1981).

19. R. W. P. McWhirter and A. G. Hearn, Proc. Phys. Soc. $\underline{82}$, 641 (1963). See also Proc. Roy. Soc. $\underline{207}$, 303 (1962), $\bar{g}$=0.4.

20. For thermal Doppler broadening, i.e., streaming is not included for elongated medium[23].

21. H. R. Griem, Plasma Spectroscopy, McGraw-Hill, p. 160, 1964.

22. P. C. Kepple and H. R. Griem, Phys. Rev. A $\underline{26}$, 484 (1982).

23. J. F. Reintjes, R. H. Dixon and R. C. Elton, Optics Letters $\underline{3}$, 40 (1978); also T. N. Lee (to be published).

Laser Generation of Light in the Extreme Ultraviolet and Soft X-ray Regime

by

R. R. Freeman and R. M. Jopson
Bell Laboratories, Murray Hill, NJ 07974

J. Bokor and P. Bucksbaum
Bell Laboratories, Holmdel, NJ 07733

Laser based techniques for extreme ultraviolet and x-ray spectroscopy fall into two categories: production of coherent light by direct lasing or nonlinear optical mixing, and generation of incoherent light using laser-induced plasmas.

INTRODUCTION

The uses for a pulsed, laboratory scale source of light for wavelengths shorter than 100nm fall naturally into two categories: narrow band, high spectral brightness sources for spectroscopy of gas phase species, and widely tunable sources of somewhat larger bandwidth for studies of condensed matter.[1] High powered, pulsed lasers in the visible and the near UV can be used to produce either type of light in the XUV (defined here as $\lambda \leq 100$nm). For narrow band, high spectral brightness sources, nonlinear mixing and multiphoton absorption processes are used, while for wider bands, continuum radiation from a laser induced plasma is employed. Recent advances in laser techniques have produced coherent light as short as 38nm, and incoherent light as short as 0.3nm.

COHERENT SOURCES

Narrow band, high spectral brightness light is normally associated with lasing. The search for an XUV laser has been going on since the demonstration of the first visible laser. However to date, no verifiable laser at wavelengths shorter than 100nm has been demonstrated. The production of coherent narrow band light in the XUV has been limited to up conversion of visible lasers by nonlinear mixing. In these processes, radiation from lasers in the visible and near UV is summed together to produce light in the XUV. In the mixing process, the output contains the coherence and the spectral and temporal characteristics of the primary laser. A well established technology exists for producing 10-100μ watts average power (up to 10^3 watts peak power) essentially over the entire VUV (100nm < λ < 200nm). Recently attention attention has turned to producing coherent output in the XUV.

Below 100nm, window materials become opaque so that the technology for confining the nonlinear optical material becomes much more complex.[2] This problem is made all the more severe in the region shorter than 100nm because for many materials the nonlinear material must be confined to a small volume in which the input lasers are focused.

One way of accomplishing this confinement is shown in Figure 1 where the apparatus for generating coherent light between 85 and 100nm is schematically represented. This Figure illustrates several generic requirements for an apparatus to generate coherent light shorter than 100nm. The nonlinear material (mercury vapor) is confined by a few Torr of He buffer gas.

492

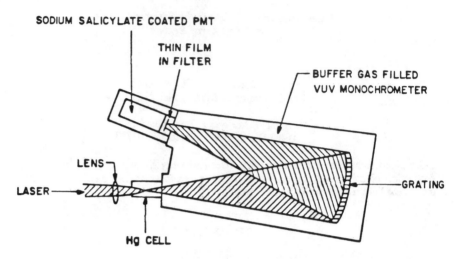

SODIUM SALICYLATE COATED PMT

THIN FILM
IN FILTER

BUFFER GAS FILLED
VUV MONOCHROMETER

LENS

LASER

GRATING

Hg CELL

Fig. 1 Apparatus for tripling and four-wave mixing in Hg vapor.

There is no output window on the mercury cell; rather the He is allowed to fill the 1 meter normal incidence monochromator. The monochromator is used to disperse the high powered fundamental laser beams (which are necessarily copropagating with the generated light) from the weak XUV radiation. The thin-film film In filter blocks any scattered laser light from the grating.

Figure 2 shows the variation in output power of the four-wave mixing output as a function of the input wavelength. For this work, two lasers were employed: one laser was tuned to two-photon resonance with the $6s^2$-6s6d transition in Hg at a laser wavelength of $\lambda_1 = 280.3$nm; a second laser was spatially and temporally overlapped with the first laser. This laser was tuned between $\lambda_2 = 286.0$nm and 294.0nm. The output, at a frequency of $2/\lambda_1 + 1/\lambda_2$, which corresponds to a wavelength tunable to 94.6nm, is shown in Figure 2. The output increased when the sum of the input lasers was coincident with an autoionizing resonance, in this case a $J=1$, $5d^9 6s^2 7p$. We demonstrated that this increase is due to the autoionizing resonance by directing the laser beams to intersect an atomic beam of Hg atoms between the plates of an ion detector. Figure 2 shows the variation of ion yield as a function of tuning of the second laser. The variation in the ion signal matches the output power variation with the exception of the extra peak on the low frequency side of the ion signal which is probably a $J \geq 2$ state: these states are allowed in three-photon ionization but not in four-wave mixing.

We measured an output of approximately $0.01\mu J$/pulse at the maximum of the autoionizing peak. At high densities of Hg ($>10^{16}$cm^3) the four-wave sum frequency was substantial even off the autoionizing peak, yielding approximately $0.005\mu J$/pulse. (The absolute output intensity is subject to some uncertainity due to unknown grating efficiency and phototube response.) The tunability was limited by the dye range of the tunable (second) laser. Assuming an effective fundamental laser range of 1000nm to 200nm (making use of standard IR and UV nonlinear optics to extend the dye laser range), the four-wave mixing output, using the 6s6d two-photon resonance, covers 123nm to 82.4nm. Tomkins and Mahon[3] have recently reported high-conversion efficiencies in Hg vapor over an extended range in the VUV using lower lying two-photon resonances; thus, Hg vapor has been shown to be a highly efficient, convenient material for the generation of coherent light.

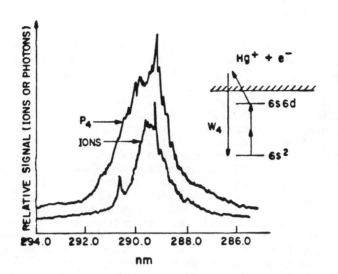

Fig. 2 Relative output of four-wave mixing (P_4) near 94.5nm and relative output of ions measured in an atomic beam for the same laser frequencies.

The technique described above is limited to the generation of light wavelengths longer than approximately 50.5nm due to the onset of continuum absorption in He which fills the spectrometer. Recently, we have been studying the applicability of a new device for the confinement of the nonlinear medium with no windows or background buffer gas. This device is a pulsed supersonic jet expansion of rare gas, triggered synchronously with the input laser, which produces a short (100μsec) burst of high-density gas. The pumping system of the chamber and monochromator easily produces a background pressure lower than 10^{-4} Torr even when the pulsed jet runs at 10pps. We have been studying the generation of the harmonics of a short pulse 248nm KrF* laser. Both the third and fifth harmonics have been generated at 83nm and 50nm, respectively, using Xenon gas as the nonlinear medium in the jet expansion.

INCOHERENT SOURCES

When a high-intensity laser is focused onto the surface of a heavy metal, a well defined plasma is initiated which, depending upon the intensity of the laser light, can reradiate a substantial fraction of the incident light in the extreme ultraviolet and soft x-ray portion of the spectrum. For low z target materials and high intensities ($>10^{14}$ W/cm^2) the radiation is mostly in the characteristic K and L resonance lines. For the heavier materials, especially the rare earth metals, the output has been found to be a continuum with a well- defined black-body radiation profile. The black-body temperature depends on the intensity of the laser; the peak temperatures observed with I$>10^{15}$W/cm^2 are in the vicinity of 1keV.

Until recently, these measurements have been made using very high energy, essentially single-shot lasers. With the development of 10pps lasers with lower energy, but very much shorter pulse duration new interest has arisen in laser induced plasmas as a source continuum soft x-

494

rays.[4] We have studied the light produced from a laser-induced plasmas on the surface of Hg in the spectral region of 30 to 80nm. We find that a liquid metal target is ideal for a high- repetition rate system because, unlike solid metal targets used in the past, the Hg surface "self-heals" after each laser shot. Thus, we do not need to reposition the target material after each shot, making a repetition rate of 10pps possible. In these initial experiments, we produced the plasma with a doubled Nd:YAG laser at 532nm. We estimate the intensity is 10^{11} W/cm^2. Figure 3 shows a recording of the spectrum from 20 to 60nm. The spectrum was recorded using a 1 meter normal incidence monochromator. The short wavelength cutoff of the recorded spectrum was due to loss reflectivity of the grating. We found a substantial number of HgII and HgIII lines in the recorded spectrum for wavelengths longer than 80nm, yet smooth continuum below 80nm. Estimates of the conversion indicate that approximately .1% of the incident laser light is converted into light between 80 and 30nm. Since we expect that in the neighborhood of 1% of the laser light is converted to XUV, most of the output is at wavelengths shorter than 30nm. We plan to utilize a grazing incidence monochromator to investigate the spectra in the soft x-ray domain. We also will replace the Nd:YAG laser with a short pulse KrF* laser which should yield an intensity of nearly 10^{15} W/cm^2 at 10pps. Under these circumstances, we expect to observe black-body radiation with an electron temperature of 1keV with nearly 1/2 watt of average power in the soft x-ray region of the spectrum.

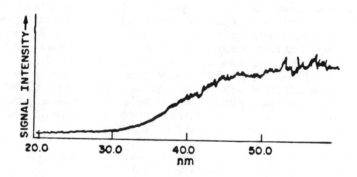

Fig. 3 The recorded signal from a laser induced plasma in Hg from 20 to 60nm when no He is used in the monochromator. The fall off below 40nm is due to loss of reflectivity of the normal incidence monochromator.

REFERENCES

1. T. J. McIlrath and R. R. Freeman, "Laser Techniques for Extreme Ultraviolet Spectroscopy", AIP series on Optical Science and Engineering, Vol. II (to be published).

2. R. R. Freeman, R. M. Jopson and J. Bokor, "Generation of Light Below 100nm in Hg Vapor", AIP series on Optical Science and Engineering, Vol. II (to be published).

3. F. S. Tomkins and R. Mahon, Optics Lett. *6*, 179 (1981).

4. P. K. Carroll, T. J. McIlrath and M. L. Ginter, Appl. Optics *20*, 3243 (1981).

DYNAMIC EFFECTS IN CORE SPECTRA

G. Wendin
LURE, Bât 209c, Université Paris-Sud, F 91405 Orsay, France
and
Institute of Theoretical Physics, Chalmers University of
Technology, S-412 96 Göteborg, Sweden

ABSTRACT

This paper gives an overview of various excitation and decay
processes in atoms, molecules and solids, discussing the physics
underlying electron and photon spectra excited with e.g. synchrotron
light or electron beams. We shall describe the conditions under
which a system will show strong many-body effects in core level
spectra and discuss some explicit examples of collective effects,
breakdown of one-electron pictures for core holes, resonance
excitation-emission and threshold effects.

INTRODUCTION

Dynamic effects are natural concepts in descriptions of the
time development of physical systems. However, the physical
processes are often observed experimentally by way of various
spectroscopies, giving information about the abilities with which
the system will be found in different final levels. The measured
spectral distributions clearly must reflect the history of the
system but the information about dynamic effects is then implicit:
The time development of the system from excitation to emission and
detection is given by the intensity distribution over the spectrum
of stationary levels in the final state. In order to find the
explicit time dependence, one must perform a transformation back to
time space. However, what one usually does is to draw conclusions
about the dynamics by looking at the spectral distributions themselves, e.g. photoionization and photoelectron spectra.

From a theoretical point of view, one normally works in energy
space using a basis of stationary states. One can then calculate
transition matrix elements, excitation energies, binding energies,
etc. and obtain e.g. photoabsorption spectra as a function of photon
energy ω and photoelectron spectra as a function of electron kinetic
energy ϵ and photon energy ω. There is no explicit reference to
time, and any conclusions about time-related dynamic effects have to
be inferred from the shapes of the calculated spectra. This applies
to all methods working in energy space, including various forms of
many-body theories. It is therefore perfectly possible to calculate
various excitation and emission spectra without ever interpreting
the results in terms of dynamic behavior, and this has led to
greatly different views of the problem and to different opinions
about the role and nature of dynamics and many-electron effects in
core spectra.

It is the purpose of this talk to try to bridge this gap, in a
small way. We shall consider a limited number of cases where
excitation spectra are characterized by very strong many-electron

effects, present the basic elements of a many-body treatment in energy space, and then interpret the results in a real space and time picture. In this way the physics becomes independent of the method of calculation, and it is relatively easy to understand the similarities and differences between excitations in different systems.

EXAMPLES OF DYNAMIC EFFECTS

Dynamic effects are associated with the response of a system to time dependent external or internal perturbations involving energy transfer between the field and the system and between different parts of the system itself. Examples of dynamic effects are

(i) Photoabsorption:[1-12] A time dependent external field induces charge displacements, which results in a frequency-dependent effective field (dynamic screening).

(ii) All kinds of relaxation and excitation effects following ejection of an electron.[1-7,9,12-18]

(iii) Decay of an electron-hole pair (autoionization) or of a core hole (Auger and x-ray emission).

(iv) Interference between excitation/ionization and decay processes, e.g. post-collision interaction (PCI),[3,6,18] incomplete relaxation[2,14-15,19] and plasmon gain.[14,15].

(v) Resonance excitation of photoelectron satellites and Auger lines.[3,18,20-26]

We emphasize again that <u>dynamic effects</u> are connected with the <u>distribution</u> of intensity over the spectrum of energy levels of the system. The <u>position</u> in energy of each of these final levels is a <u>stationary</u> property and therefore does not move in energy if the time scale of the excitation is varied.

In order to illustrate these points, let us briefly discuss an important example, namely photoelectron emission in the case of high and low kinetic energies ε of the photoelectron (Fig. 1).

For $\varepsilon \gg 0$ (Fig. 1a) the electron is emitted in a <u>sudden</u> manner. This perturbation has a very wide energy spectrum causing shake-up and shake-off to all symmetry-allowed final state levels. When the photon energy ω, and therefore also the photoelectron kinetic energy ε, is lowered the high-energy part of the shake-off spectrum will be successively cut off and the intensity transferred to the core hole structure at lower binding energies. Finally, as $\varepsilon \to 0$ the time scale of ejection will be long in comparison with the longest response time $\tau = \hbar/\Delta E$ of the system (Fig. 1a). There is then not enough energy for excitation of the lowest satellite, and the system will relax <u>adiabatically</u>. We conclude that the shape of the satellite intensity distribution reflects the time scale of the perturbation (velocity of the photoelectron). On the other hand, the position of the photoelectron peaks will not depend on photo-electron energy, provided they really correspond to stationary states, i.e. do not decay and therefore have no width. An example of such a sharp level could be a 5p-hole level in atomic Xe.

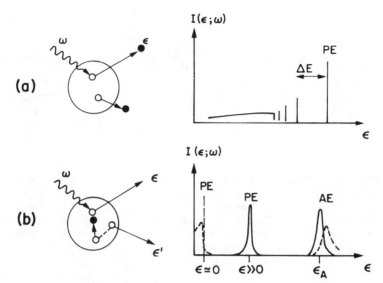

Fig. 1. (a) Schematic photoelectron (PE) spectrum for
relatively high PE kinetic energy ε
(b) Interference between PE emission and Auger
electron (AE) emission: Post-collision interaction (PCI)[18] (dashed curves).

Let us now assume that the core level can decay via an Auger
process and therefore has a width Γ (Fig. 1b). The discussion above
is valid as long as $\varepsilon \gg \Gamma$, i.e. there is no level shift if the
photoemission process is fast in comparison with the time for decay.
However, when $\varepsilon \rightarrow 0$ the decay takes place while the photoelectron is
still in the neighbourhood of the ion core and feels the ion
potential. The line profile then becomes distorted from a
Lorentzian (Fig. 1b), and the same is true for the Auger peak. This
is the well known PCI phenomenon[3,6,18] and is clearly an example of
a <u>dynamic</u> effect since it depends on the <u>time scales</u> in the problem.
The dynamics is reflected in the redistribution of line strength
from a Lorentzian to an asymmetric shifted profile. Had the core
level been sharp, there could not be any redistribution and therefore not any shift. In the present case, however, the stationary
states are the levels of the two-hole-one-electron Auger continuum
(assuming these holes to be sharp), and it is over these levels that
the line intensity becomes redistributed, leading to an asymmetric,
shifted line profile.

A DYNAMIC PICTURE OF PHOTOIONIZATION

Let us consider a time-space picture and imagine that an
excitation is created in the system at time $t = 0$. In Figure 2 this
is illustrated in the case of photoionization: At short times (Fig.
2a) the hole and the photoelectron are very close together. The
excitation then looks like a dipole and the electrons flow to screen
this dipole. At intermediate times (Fig. 2b) the electron and the
hole have separated enough that the system can respond also to the

 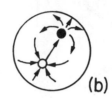 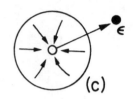

Fig. 2.   Schematic pictures of the photoionization process
in real space and time: (a) Strongly overlapping hole and
electron, dipolar response.   (b) Hole and electron begin
to separate; relaxation and excitation in response to the
individual hole and electron.   (c) Photoelectron outside
ionic charge density but still in the potential of the
ion; ionic relaxation mainly.

individual charges of the electron and the hole. Clearly, this is a
very complicated stage of the response because the screening
electrons are shared between the hole and the electron: There will
be very important interference effects between the action of the
hole and that of the photoelectron. Finally, at large times (Fig.
2c) the photoelectron has totally separated from the hole, and the
system has relaxed around the hole. It is at this stage that differences between various systems show up in a decisive manner: In
an atomic system the photoelectron will rapidly escape into vacuum,
losing its screening cloud, while in a solid the photoelectron will
propagate together with its screening cloud as a quasi-electron
until it escapes through the surface into the vacuum. In an atomic
system, the photoelectron feels an ionic potential because the core-
hole cannot be screened out at large distances, while in a solid
there is indeed long-range screening. Finally, when the photoelectron escapes into vacuum it always has to leave its proper
screening cloud behind. In an atom this leads to a polarization
potential which is usually a small correction to the ionic
potential. In a solid (metal) this polarization potential ·
corresponds to the image potential and is extremely important.
     The different stages of time-development shown in Figure 1 can
be viewed as representing different physical situations and requiring different levels of theoretical descriptions. Of course the
system has to develop through all of the stages, but the relative
importance of these stages may vary between different systems and
regions of different excitation energy, and may also depend on the
type of measurement, e.g. photoabsorption vs photoelectron spectroscopy.
     Viewed in a time-dependent picture, all rearrangement and
relaxation processes are dynamic: The external field is switched on
at a certain time and energy starts to become transferred to the
system via a number of different processes. In the simplest possible case, a photoelectron is directly emitted with high velocity,

leaving behind a core hole that can be approximated by a spherically symmetric, static charge distribution, acting like a classical test charge at the nucleus. As the photoelectron wave packet propagates out through the ion core, the shells become successively exposed to the potential of the core hole and respond by relaxation, including excitation and ionization (shake-up and shake-off). The relaxation of the shells behind the photoelectron will set up a fairly long-range polarization field that acts back on the photoelectron and ensures that it "knows" the dynamic state of the core. In principle there are also polarization effects induced by the photoelectron, but in the high-velocity limit these vanish (cf. the semi-classical discussion by Noguera[13]).

In the present case, as t → ∞ the system develops into a final ionic stationary state with a _statically_ screened core hole, or to one of the various excited ionic states. The dynamic evolution is finished and the distribution over final ionic states established. There can no longer be any energy transfer between the core hole and the rest of the system. Note however, again, that formation of the satellite spectrum is a dynamic problem: To calculate satellite intensities, one must use frequency-dependent response functions describing _dynamic_ screening.

In the general case, the above picture represents always a slight simplification and sometimes a gross oversimplification. Even for times long in comparison with the time for radial relaxa-tion of the atomic shells the system is not stationary due to radiative and non-radiative (Auger) _decay_ processes. This is a _dynamic_ process because it involves non-stationary states and energy transfer between the hole and the medium, and it introduces another time scale into the problem.

The dynamic aspects introduced by decay processes can be generalized in the following way:[4,27] In an interacting system, a core-hole is not strictly confined to a particular $n\ell$ subshell and does not strictly have the symmetry of that subshell. Instead it can be viewed as a _dynamically screened wave packet_ with a possi-bility to _move_ throughout the core region and acting as a time-dependent perturbation, with a frequency spectrum consisting of all transition energies $\omega_{ij} = E_j - E_i$ between different subshells. The charge distribution of the "moving" hole can be expressed in terms of a set of static and dynamic multipole moments: The _static mono-pole_ moment describes the hole as a spherical shell, and the associ-ated monopole relaxation can be obtained from a ΔSCF calculation. The higher static moments (all even) describe deviations from a spherical core-hole distribution, and the associated relaxation energies are small. Nevertheless, in some cases the effect on the core-hole spectrum can be dramatic (outermost p shells in Ca, Sr and Ba).[4,28,29]

Among the dynamic multipole moments, the _dynamic dipole_ moment is of particular importance: Now the hole wave packet can be built up _within a main shell_ by mixing $n\ell$ and $n\ell+1$ wave functions which may lead to strong deviation from spherical symmetry of the core-hole charge distribution and of the associated distribution of screening charge. Note however that the quasi-hole, i.e. core hole plus screening charge, must fulfill the requirements of the overall spherical symmetry. It is only the subshell symmetry that can break down.

The dynamic dipole moment of the core hole (actually all dynamic multipole moments) may decay radiatively by emitting photons or decay non-radiatively by ejecting electrons from outer shells (Auger emission). However, it is important to realize that the dynamical dipole moment is <u>screened</u>: Not only can this give a very important contribution to the relaxation energy but it will also influence radiative and non-radiative transition rates. In particular, this means that radiative transition rates can be strongly influenced by electronic dynamic screening.

In order to compare the time-dependent picture of the photo-ionization process with a stationary-state formulation, we show in Figure 3 the fundamental diagrammatic process. This describes in a direct way the excitation and development of an electron-hole pair while at the same time providing a description in terms of stationary-state quantities, namely a transition amplitude $t_\epsilon(\omega)$ and a core-hole spectral function $A_{\underline{i}}(\epsilon-\omega)$ (see refs. 4, 27).

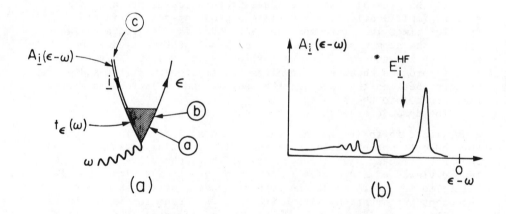

$$(a) \qquad\qquad (b)$$

Fig. 3. (a) Basic diagram for the photoionization process (see text). The encircled letters refer to Fig. 2 a, b and c resp.
(b) Schematic picture of the spectral function for a core hole. $E_{\underline{i}}^{HF}$ is the frozen HF (Koopmans') energy.

The photoelectron intensity is given by

$$I(\varepsilon_i \omega) \sim | t_\varepsilon(\omega) |^2 A_{\underline{i}}(\varepsilon-\omega) \tag{1}$$

where

$$t_\varepsilon(\omega) = <i|z(\omega)\varepsilon>_{eff} \tag{2}$$

$z(\omega)$ is the effective field, i.e., the dipole operator $z = r \cos\theta$ modified by dynamic screening, and $|\varepsilon\rangle_{eff}$ contains effects of screening of the potential seen by the photoelectron. Note however that this division is not unique: One can in principle obtain the amplitude and the cross section by referring the dynamic effects entirely to the effective operator $z(\omega)$ or to the effective wave function $|\varepsilon\rangle_{eff}$. From an effective wave function one can work backwards and derive an energy-dependent effective one-electron potential.[3,30] This is closely related to calculating final-state wave functions using LS-dependent Hartree-Fock potentials.[31]

WHEN DO CORE SPECTRA SHOW STRONG DYNAMICAL EFFECTS?

A general answer to this question is: When there are filled and empty orbitals with similar spatial dimensions (mean radii), resulting in large radial overlaps and large polarizabilities. External and internal perturbing fields will then induce large dipole moments and strong screening fields, and dynamic many-electron effects will become important. However, this makes it necessary for us to introduce the concept of an open main shell and to distinguish between two types of core levels, to be referred to as shallow and deep levels. As long as one or several subshells are not filled, one can have transitions between subshells within the same main shell. Core levels belonging to such an open main shell will then have large overlap with empty levels and will be called shallow core levels. On the other hand, core levels belonging to a deeper closed main shell will have much weaker overlap with empty levels and shall be referred to as deep core levels. Due to this difference in overlap, many-electron effects are an order-of-magnitude more important for transitions within an open main shell, than between different main shells.

As an example, let us consider a schematic one-electron level structure for a 5f element, as shown in Figure 4.

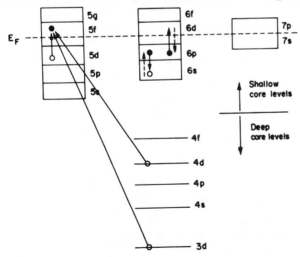

Fig. 4. Schematic picture of the one-electron level structure in the region of 5f elements; illustrates the difference between shallow and deep core levels due to difference in overlap between empty and filled levels.

There are three open main shells, n = 5,6,7. The 7s/7p, 6d and 5f levels represent <u>valence</u> levels, and are involved in strong many-electron effects in molecules and solids,[4] e.g. in optical absorption, screening of core holes, screening of valence holes with band narrowing and localization, etc. The 6s/6p and 5s/5p/5d levels represent <u>shallow</u> core levels; nevertheless they can be as deep as about 300 eV. It is therefore clear that the binding energy is not a good measure of the depth of a core level. The <u>deep</u> core levels begin with the n = 4 shell (Fig. 4), because from then on there are no empty levels that overlap strongly with the core levels.

We can now conclude that 5d→f (5f) and 6p→d (6d) transitions are going to be strongly influenced by many-electron effects and simple one-electron models will break down in descriptions of e.g. the cor- responding photoionization cross sections. These are the cases of the so-called <u>giant dipole resonances</u>, showing collective behaviour.[32] Furthermore, we may conclude that 6s and 5s/5p holes will be strongly influenced by dynamic many-electron effects via intra-main-shell Auger type processes (as shown in Fig. 4 for a 6s hole). These have been called <u>giant Coster-Kronig</u> processes,[4] and lead to large relaxation shifts and sometimes to rapid decay. On the other hand, decay of a deep core level involves at least one transition between different main shells (Auger, super Coster-Kronig processes), or interaction between transitions within different main shells (Coster-Kronig processes). The transition rates are then smaller or much smaller than in the case of intra-main-shell processes.

Finally it should be noted that localization (collapse) of the empty levels plays an important role. The empty 5f levels start to become important as very broad resonances in the continuum already in the elements around mercury (Z = 80), influencing the dynamics of the 5d→εf transitions[33] and the structure of 5s and 5p holes.[4] The effects increase dramatically with increasing atomic number Z due to the sharpening up of the f resonance in the continuum (shape resonance), until the 5f levels become occupied in the potential of two 5d holes,[4,34] and finally also in the ground state in the 5f elements. Increasing Z further will cause a gradual reduction of the many-electron effects due to filling of the 5f levels. However, in this limit there are many complications due to term level structure caused by the open 5f subshell.

PHOTOABSORPTION/PHOTOIONIZATION

The strongest many-electron effects in photoabsorption spectra are associated with the so-called giant dipole resonances or collective resonances, which characterize e.g the 4d absorption spectra of Ba, La, Ce and surrounding elements and the 5d absorption spectra of Th and surrounding elements (Fig. 5). This problem is of central importance for understanding the dynamics of core excitations in atoms, molecules and solids and requires a brief

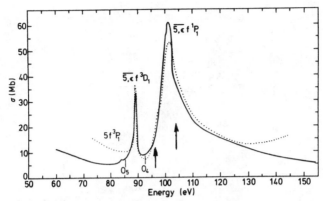

Fig. 5. Photoabsorption of the metal[35] (—) and ThF$_4$[36]
(· · ·). The vertical arrows denote estimated atomic 5d
thresholds).

discussion. Here we shall only discuss a number of particular
aspects and we refer to literature for more detailed discussions.[1,7,8,35-38]

A spatial picture of the photoionization process has already
been given in Fig. 2, describing different stages of the photoionization process: Screening of the external field and absorption
of a photon (Fig. 2a), dynamic screening of the propagating
electron-hole pair (Fig. 2b), and escape of the photoelectron in the
potential of the dynamically screened ionic system (Fig. 2c). The
basic physics of the giant dipole resonance is connected with Fig.
2a, and can be discussed in a number of different ways, in terms of
effective dipole transition matrix elements, effective electromagnetic driving field or effective photoelectron wave function
(effective potential).

In terms of dielectric screening of the external driving field,
we have the picture shown in Figure 6a. The external field induces
a displacement of the orbitals of the 5d subshell. This sets up an
r-dependent induced electric field and gives rise to a space and
frequency dependent effective electric field that affects both the
core electrons and the valence electrons. To some extent, there are
always some effects of dynamic screening in any case of photoexcitation. What makes the giant dipole resonance so special is
that the charge displacement is enormous and roughly corresponds to
displacing an entire subshell at a certain frequency. This means
that the external electric field can be completely screened out in
the region of the n=5 main shell in the frequency range below the
giant dipole resonance (in the region where the single-particle
oscillator strength is maximal). Passing through the resonance, the
field suddenly penetrates and the absorption goes through a giant
maximum. This situation is analogous with that of a metal, where
the field is kept out and the metal is totally reflecting below the
plasma frequency and where the field penetrates above the plasma
frequency. In fact, the original approaches to collective behaviour

504

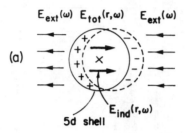

Fig. 6. (a) Schematic picture of dynamic screening of the
         external field.
         (b) Diagrammatic picture of the screening process
         in (a): The random phase approximation (RPA).

of atomic shells[39] were focused on electronic plasma behaviour,
atomic plasmons, and these concepts have continued to attract considerable interest.[39]   The term giant dipole resonance was proposed
by Wendin[32] for Ba to emphasize the very close relationship between
collective resonances in atoms and in nuclei. The Th 5d resonance
has been discussed in similar terms by Connerade et al.[36,37]
      There is evidently a direct connection between large oscillator
strengths in a narrow frequency region in the one-electron model and
strong dynamical many-electron effects: The large oscillator
strength is necessarily associated with large charge displacements
and strong induced fields and this must necessarily modify the   .
dynamics, i.e. the distribution of oscillator strength. It should
be noted that the dynamic interactions in terms of induced fields
discussed above are closely related to those taken into account by
going from a configuration-average picture to considering $^1S \to {}^1P$
transitions in a many-electron term level picture.
      Let us now discuss the dielectric approach in some detail in
order to be able to make some comments on relaxation effects, local
density functional approach, atomic vs solid state effects, and
spin-orbit interaction and statistical weights. The induced field
in Fig. 6a can be associated with a non-local, frequency-dependent
dielectric function $\epsilon^{-1}(r,r';\ )$ but since in practice the external
field is homogeneous, we are effectively left with a local
dielectric function

$$E_{tot}(r,\omega) = \varepsilon^{-1}(r,\omega) \, E_{ext}(\omega); \quad \varepsilon^{-1}(r,\omega) = \int dr' \varepsilon^{-1}(r,r';\omega) \qquad (3)$$

from which we obtain the transition amplitude

$$t_\varepsilon(\omega) = \langle i | r\varepsilon^{-1}(r,\omega) \, \cos\theta \, | \, \varepsilon \rangle. \qquad (4)$$

The mean-field formulation above is the essence of the random-phase approximation (RPA) and is equivalent to first order time-dependent perturbation theory. Mean-field theory accounts for the gross distribution of oscillator strength in the neighbourhood of a giant dipole resonance, regardless of whether the distribution is discrete or continuous or mixed. The collective resonance occurs in a certain range of <u>excitation</u> energies and does not really care about whether the <u>ionization</u> threshold lies below, in the middle of, or above the resonance maximum. Exactly this situation occurs in 4d absorption going from Ba to Ba^{2+}, a problem that has recently attracted considerable attention:[9,37,38,40-42] With increasing ionization stage, the 4d-ionization energy moves some distance up through the resonance, revealing very prominent discrete structure in particular in Ba^{2+}. However, put on a continuum normalization, one finds that the oscillator-strength distribution is closely the same in all three cases. A similar situation also occurs in going from <u>an atom to a metal</u>: The extra-atomic screening of the core hole by the conduction electrons will lower the binding energy (relative vacuum) by up to 10-15 eV, and it will modify the potential seen by the photoelectrons. However, it will not change much the position or shape of the oscillator-strength distribution. In the case of 4d absorption in Cs, Ba, La the atomic 4d threshold is already below the resonance in the atom, and the metal spectrum is therefore nearly identical to the atomic one. In Th (Fig. 5) only the atomic $5d_{5/2}$ threshold is below the main resonance around 100 eV. The $5d_{3/2}$ threshold lies above and this is probably the reason for the narrowness of the giant resonance, which looks nearly the same for the metal[35] and ThF_4[36] and most probably also for the vapour.

However, in the case of the 5p (6p) absorption spectrum in Ba (Th), the situation is entirely different. The spectra are still dominated by a giant-dipole type of resonance [5p→5d,(6p→6d)] but now the atomic 5p (6p) binding energy lies essentially above the resonance while the 5p (6p) binding energy in the solid lies below. The solid threshold will show a large, broad continuum resonance in the same region where the atom has extremely prominent discrete structure. There are not yet any systematic experimental results to support this, only in a few cases like Ca, Ba, Yb. Nevertheless it seems likely that the 3p, 4p, 5p and 6p absorption spectra in the 3d, 4d and 5d transition metals and the 4f and 5f elements should be characterized by mainly discrete giant-dipole resonance structure in the atoms in contrast to the continuum distribution in the solids.

The dipole transition amplitude can be calculated in typically three ways:

506

$$t_{\varepsilon}(\omega) \begin{cases} \rightarrow (<5d|r\cos\theta|\varepsilon f>)_{eff} & \text{(a)} \\ \leftrightarrow <5d|r\varepsilon^{-1}(r,\omega)\cos\theta|\varepsilon f> & \text{(b)} \\ \rightarrow 5d|r\cos\theta(|\varepsilon f>)_{eff} & \text{(c)} \end{cases} \quad (5)$$

Eq. (5a) refers to a direct calculation of the matrix element via an integral-equation approach (e.g. the RPAE)[7,43,44] or a perturbation treatment,[9,44] indicated in Fig. 6b. Eq. (5b) denotes the effective-mean-field approach. The dielectric function $\varepsilon^{-1}(r,\omega)$ can be obtained with an integral-equation technique[38,45] or from a differential equation 8. Finally, Eq. (5c) describes the effective-wave-function approach. The effective wave function can be found from the dielectric function acting upon the one-electron continuum wave function, from perturbation theory[30] or from direct calculations using LS-dependent Hartree-Fock[1,30,31] plus ground state correlations.[3,42]

Within the effective-field approach [Eq. (5b)] one can now immediately take one step further by considering photoionization of any outer shell, $i \rightarrow \varepsilon$, driven by the effective dynamically screened field, as schematically shown in Figure 7.

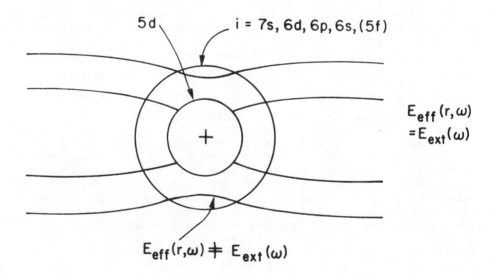

Fig. 7.   Schematic picture of effective field $E_{eff}(r,\omega)$ due to strongly polarizable inner shell, as seen by outer-shell electrons.

The transition amplitude simply becomes

$$t_\varepsilon(\omega) = <i \,|r\varepsilon^{-1}(r,\omega) \cos\,\theta|\varepsilon>. \tag{6}$$

In a self-consistent description, all the shells to be ionized are also involved in the calculation of $\varepsilon^{-1}(r,\omega)$. In the neighbourhood of a giant resonance the contributions to $\varepsilon^{-1}(r,\omega)$ mainly come from a single shell. However, in general inter-shell coupling is important, in particular for weak channels (e.g. a 5s shell coupled to 5p and 4d).

The mean-field approach is standard procedure in solid state physics, where it is obvious that the electric field inside e.g. a piece of metal is going to be very different from the external applied field. This is expressed implicitly or explicitly in some of the previous many-body work on atoms but has not been used in actual atomic calculations until the recent work of Zangwill and Soven,[8] using the so-called time-dependent local-density approxima-tion (TDLDA). This is equivalent to the time-dependent Hartree-Fock method except for replacing the Fock exchange potential by a Kohn-Sham local-density exchange/correlation potential, and the method has had some remarkable success in describing atomic photoionization spectra in the regions of giant-dipole resonances. Note that the calculation is done in energy space, even though the TDLDA is derived from a time-dependent Schrödinger equation.

It is quite clear that an effective-field approach is the essential ingredient for obtaining a correct average distribution of oscillator strength, continuous or discrete. However, the LDA (local density approximation) one-electron potential is somewhat different from the HF (configuration average) potential, and as a consequence the TDLDA gives results that can only be obtained if relaxation effects are included in the HF-based schemes. In fact, the LDA potential shows similarities with the relaxed HF potential in the inner-well region where the average shape of the oscillator-strength distribution is determined.[42] In the outer region it looks more like a screened Coulomb potential, not supporting any Rydberg levels, and it could very well be that the atomic TDLDA approach directly gives a reasonable description of inner-shell photoionization spectra of solids (metals).

Since the TDHF method is equivalent to the RPAE (random phase approximation with exchange), one can imagine an RPAE-based method equivalent to TDLDA. In the LDRPA (local density RPA) approach by Wendin,[38,45] the RPA bubble series of diagrams is evaluated in the LDA basis, and the effective dynamically screened field is obtained from an integral equation. In addition, spin-orbit split subshells are treated as individual subshells, statistically weighted and separated in energy by the spin-orbit splitting. The result for the partial photoionization cross sections in the 4d region of atomic Ba agrees very well with that of Zangwill and Soven.[8] A preliminary result for 4d photoionization of La is shown in Figure 8.

508

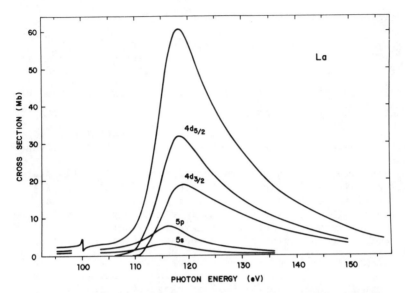

Fig. 8. Partial photoionization cross sections for
atomic La(Z = 57) in the LDRPA (local density random
phase approximation; see text); Wendin 1982, preliminary
result; the resonance at 100 eV appears in all the
partial cross sections.

In contrast to the case of a single 4d-shell in Fig. 6a, we now
have <u>two</u> 4d shells, $4d_{5/2}$ and $4d_{3/2}$, and therefore <u>two modes</u> of
collective motion: In one of the modes, the shells move together <u>in
phase</u>, acting essentially like a single shell and giving rise to the
giant-dipole resonance at around 120 eV photoenergy (Fig. 8). In
the other mode the spin-orbit split shells move in <u>opposite phase</u>
and the associated oscillator strength is very weak. The intensity
of this peak is a function of the ratio of the spin-orbit splitting
Δ_{so} to the exchange splitting $[\sim\Delta_{so}/2R^1(4d4f;4f4d)]$. In La 4d
ionization this ratio is about 1/6 and in the 5d-ionization (Fig. 5)
the ratio is about 1/2, and clearly there is more intensity in the
low-energy absorption peak in Th than in La. However, the ratio
must become very large before one approaches the j-j limit: A ratio
of ~ 3 gives about equal intensity for the two peaks. This situation
may apply to the $3d_{5/2}\rightarrow 4f$ and $3d_{3/2}\rightarrow 4f$ absorption peaks in La.

It should be noted that in an LS-term level picture, the low-
energy peaks in the experimental cross sections (Figs. 5, 8) correspond to ^3P and ^3D levels mixed with some ^1P character due to
spin-orbit interaction. It appears that in the treatment above we
obtain approximately the total ^1P character mixed into triplet
levels without considering the actual term level structure in this
region.

DYNAMIC EFFECTS IN CORE-HOLE SPECTRA

In this section we shall discuss some particularly interesting effects of core-hole relaxation and their manifestations in photoelectron spectra, Auger spectra and x-ray emission spectra. This has been discussed in great detail in refs. 4, 27, 46, 47 and here we shall only try to develop a feeling for the physical aspects of the problem in real space and time.

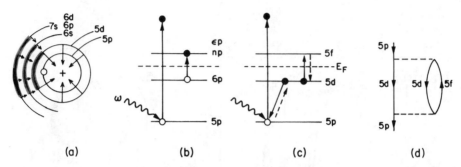

Fig. 9. (a) Schematic picture of the general relaxation process. → denotes radial (monopole) relaxation; ⬚⬚⬚ denotes angular (dipole) relaxation.
(b,c) One-electron level pictures of monopole shake-up (relaxation) (b) and dipole (giant Coster-Kronig) relaxation and "shake-up" (c).
(d) Lowest order 5p self-energy diagram for the giant Coster-Kronig process in (c).

In Fig. 9a we represent a 5p core hole in e.g. Th (Z = 90) by a wave-packet "localized" within the 5p subshell and relaxing by attracting a cloud of screening charge. In a metal, a hole in the valence band would be surrounded by a nearly spherical screening cloud and could move as a quasi-hole through the solid. In the core region of an atom, molecule or solid the quasi-hole is more complicated in a way but if we make an expansion in a basis of spherical harmonics around the nucleus (multipole expansion), we note the following two major effects:

(i) There is a radial contraction (monopole relaxation) of the atomic shells, as if the hole charge had been distributed over a spherical shell (Fig. 9a,b).

(ii) There is an angular flow of screening charge along the "surfaces" of the atomic shells, breaking the spherical symmetry of the individual subshells. The main effect is associated with the dipole term of the multipole expansion (Fig. 9a,c). In the case of a 5p hole in Th (4p hole in Ba) or in the surrounding elements, relaxation due to angular redistribution is more important than relaxation due to radial contraction, and we have in fact a situation of breakdown of the one-electron orbital (subshell) picture of a core hole: We have a quasi-hole involving several strongly coupled subshells. This is the same effect that causes

510

breakdown of the one-electron orbital picture in inner valence-shell spectra in molecules.[4,12,48]

A one-electron level scheme for the dipole relaxation process is shown in Fig. 9c. In the present case of Th, it takes the form of a virtual Auger process, a giant Coster-Kronig process,[4,34] entirely within the n = 5 main shell, and it corresponds to the lowest order self-energy diagram for the 5p hole shown in Fig. 9d.

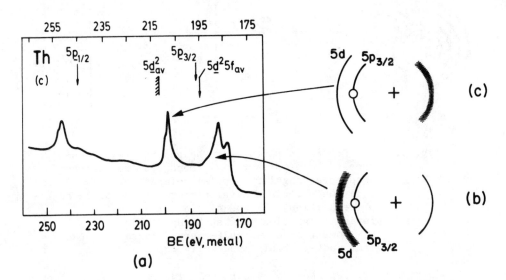

Fig. 10. (a) 5p XPS spectrum of metallic Th (Z = 90).[34] (b,c) Schematic picture of possible behaviour of a "5p" $^2P_{3/2}$ quasi-hole in its ground (b) and excited (c) state.

Fig. 10 shows the experimental result (XPS) for Th metal[34] together with a qualitative interpretation of the peak structure of the $^2P_{3/2}$ hole quasi-hole in real space. The coupling (5$\underline{p}$ ↔ 5$\underline{d}^2$5f configuration mixing[4,34,49] is so strong that a one-electron orbital picture becomes meaningless, and we may regard the quasi-hole as having two well-defined levels: A ground-state level corresponding to screening and an excited level corresponding to antiscreening. This is analogous to the case of charge-transfer screening and satellites in molecules[4,48] and to the concept of well screened and poorly screened core holes and satellites in transition-metal alloys and compounds.[14,15] Note that in the $^2P_{3/2}$ the giant Coster-Kronig process is virtual, which can be thought of as the quasi-hole not moving fast enough to be able to ionize the 4d shell. However, the angular motion of the quasi-hole is fast enough to ionize the outer 6s, 6p, 6d and 7s subshells, leading to Coster-Kronig decay.

The $5p\ ^2P_{1/2}$ hole is affected in a similar although less dramatic manner. The clearly visible peak in the 250-eV region (Fig. 10) should be analogous to the excited level of the $^2P_{3/2}$ quasi-hole, while the "ground state" level should be spread over discrete and continuum levels at lower binding energies. Thus, some of the structure in the 200-eV region could be due to the $^2P_{1/2}$ quasi-hole. Judging from the positions of the relevant configurations ($5\underline{s} \leftrightarrow 5\underline{p}_{3/2}5\underline{d}5f$, $5\underline{p}_{1/2}5\underline{d}5f$),[4,49] a $5\underline{s}_{1/2}$ quasi-hole will be in a situation intermediate to that of a $^2P_{3/2}$ and a $^2P_{1/2}$ hole.

For lower atomic number Z, around Rn, the 5f level (in the presence of two 5d holes) becomes a shape resonance in the continuum. The screening cloud is then no longer necessarily stable and there can be rapid giant Coster-Kronig decay. There is no experimental data in this region for elements. However, in the analogous and well-known case of the $4\underline{p}$ levels in Xe (Z = 54),[4,27,50] the lowest ("well screened") level of the $4\underline{p}\ ^2P_{3/2}$ hole (cf. Fig. 10) appears as a fairly sharp (Coster-Kronig broadened) peak structure, while most of the remaining part of both the $^2P_{3/2}$ and $^2P_{1/2}$ spectra has turned into an essentially structureless continuum. The only existing strong peak in the spectrum corresponds to a $4\underline{p}\ ^2P_{3/2}$ hole which has lowered its energy so much through angular relaxation that its angular velocity no longer is sufficient for ionizing the 4d shell. This then is the "ground state" of a $4\underline{p}\ ^2P_{3/2}$ hole in Xe in analogy with the corresponding case of a 5p hole in Th.

The effects of dynamic screening described above also affect multiple vacancies containing one or several quasi-holes. There are many examples of complete breakdown of the one-electron picture in Auger spectra having 4s or 4p holes in the final state, e.g. $3\underline{p}\ 4\underline{p}4\underline{d}$ ($M-N_{2,3}N_{45}$) Auger spectra in Pd-Xe,[47] and there must be many analogous cases in the 5f and preceding elements. One should also remember that the strong angular charge redistribution in Figs. 9, 10 heavily screens the $5\underline{p} \rightarrow 5\underline{d}$ dipole transition, and therefore must strongly influence all related Auger and x-ray emission rates, as well as the term level structure of e.g. a $5\underline{p}5\underline{d}$ double vacancy (or a $4\underline{p}4\underline{d}$ vacancy in e.g. Xe[47]).

Finally, we shall briefly consider a completely different core-hole interaction process, involving 4d ionization of Ce, as illustrated in Fig. 11a.

512

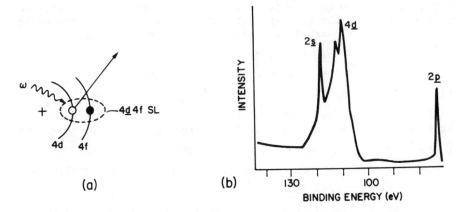

Fig. 11. 4d-ionization of Ce metal. (a) Schematic orbital picture showing 4d-4f LS-dependent Coulomb interaction. (b) 4d-XPS from $CeAl_2$ (plasmon losses deconvoluted).[51]

   In the relaxed final state, the 4f electron will not only see the spherical part of the 4d core-hole potential: There will also be very strong correlation of the orbits of the 4d hole and the 4f electron due to the large 4d-4f exchange interaction, leading to a wide spectrum of energy levels (LS-term level structure). The maximum triplet-singlet splitting is about 20 eV (^3P-^1P splitting in La photoabsorption, Fig. 8). However, since the 4f electron is not oriented and is free to couple in all possible ways, there are a large number of levels in between. Of particular interest are the high-lying singlet levels, for which the attractive potential of the 4d hole is partly or completely compensated for by a repulsive exchange interaction. Since the 4f levels are lying only a few electron volts below the Fermi level, it is quite possible that a 4f electron could find itself in the continuum above the Fermi level after 4d ionization. This represents a kind of shake-off process due to angular correlation (the usual term level structure would correspond to shake-up), and might be the explanation for the prominent tail on the high binding-energy side of the 4d core-hole spectrum in Fig. 11b.[51] Similar aspects might also be important for describing the structure of a 4f-like screening cloud in La, Ce, etc., in particular the differences between screening of a 3d hole and screening of a 4d hole.
   The above discussion should apply also to 5d5f interaction and 5f screening in the 5f elements. In the case of metals, however, there is also the effect of itinerancy of the 5f electrons, as well as effects of localization due to presence of core holes.[34]

PERSPECTIVES

Although many aspects of dynamic processes in core spectra are reasonably well understood, it is fair to say that in general our knowledge is fragmentary, especially in what concerns molecules and solids. Even in a free atom the dynamics of double excitation, ionization plus excitation, and double ionization processes is poorly understood in many respects. In view of the continuous developments in light sources, electron and particle beams, detection techniques and target preparation methods, it should be possible and desirable to perform systematic experimental investigations connected with a corresponding theoretical effort. An example of such a study would be an investigation of core-level spectra of a given metal atom, e.g. Ba, as a function of

    (i)   Core level depth
    (ii) Kinetic energy of the photoelectron
    (iii) Final-state products (detection of electrons, photons, atomic and molecular fragments, if possible in coincidence)
    (iv) Chemical composition (free atom, metal, small clusters, free molecular complexes, solids and absorbate systems).

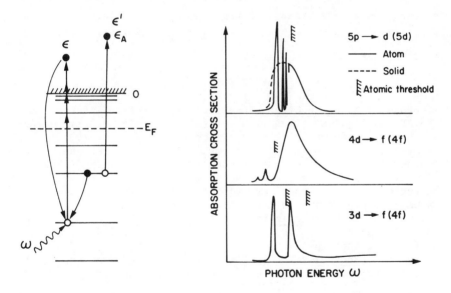

Fig. 12. (a) Resonance and threshold photoionization processes.
          (b) Schematic pictures of La photoabsorption spectra.

By studying experimentally the photoionization process in a systematic way from below to above various ionization thresholds (Fig. 12), one would be able to map the dynamics in a systematic way, and determine experimentally the relative importance of various excitation and emission processes. Although the general phenomena can be quite well understood by theoretical models, the range of possibilities in real systems is enormous, and careful experiments

514

are necessary for establishing the hard facts. It is also very important to determine to what extent resonance photoemission experiments can be used as a diagnostic tool, e.g. to get information about the character of initial and final levels, without elaborate theoretical interpretation. At present, the field is developing very rapidly, with numerous experimental applications to atoms,[22,23,25,28,29,52] molecules,[53-55], solids[23,24,56-58] and absorbates.[59-61]

Finally I would like to draw attention to a subject that has been omitted due to lack of time and space, namely excitation, ionization and decay processes following electron impact. The dynamics of these processes is similar to the case of photo-ionization but is richer and more complicated, in particular in threshold regions where the incident electron really takes part in the dynamics.[19] For a discussion of bremsstrahlung and ionization processes in atoms and ions we refer to papers presented at this conference.[62,63] In the case of solids the resonance phenomena can be much more pronounced than in atoms due to the final-state band structure, with narrow 4f bands in La, Ce, etc., resulting in strong resonances in e.g. the bremsstrahlung tip,[64,65] in fluorescence and Auger emission,[19] and in the x-ray yield (Appearance Potential Spectroscopy, APS[65,66]).

REFERENCES

1. *Photoionization and Other Probes of Many-Electron Interactions*, Ed. F. Wuilleumier (Plenum, NY, 1976). Contains a large number of highly relevant articles and references.
2. *Many-Body Theory of Atomic Systems*, Eds. I. Lindgren and S. Lundqvist, Physica Scripta (Sweden), vol. 21, no. 3/4 (1980) (Proceedings of Nobel Symposium 46). An important volume which gives a cross section of descriptions of dynamical phenomena in atoms, molecules and solids.
3. G. Wendin, in *Photoionization of Atoms and Molecules*, Ed. B. D. Buckley (Daresbury Laboratory report DL/SCI/R11 (1978), and references therein.
4. G. Wendin, *Photoelectron Spectra*, Structure and Bonding, 45, 1 (1981) (Springer Verlag), and references therein.
5. A. F. Starace, Appl. Optics 19, 4051 (1980).
6. M. Ya. Amusia, Appl. Optics 19, 4042 (1980).
7. M. Ya. Amusia, in Advances in Atomic and Molecular Physics, vol. 17, Eds. D. Bates and B. Bederson (Academic Press, NY, 1981).
8. A. Zangwill and P. Soven, Phys. Rev. Lett. 45, 204 (1980) A. Zangwill, Proceedings of the 8th ICAP, 2-6 August 1982, Göteborg, Sweden.
9. H. P. Kelly, Proceedings of the 8th ICAP (invited talks) 2-6 August 1982, Göteborg, Sweden.

10. G. R. J. Williams and P. W. Langhoff, Chem. Phys. Lett. <u>78</u>, 21 (1981).

11. R. R. Lucchese, G. Raseev and V. McKoy, Phys. Rev. A <u>25</u>, 2572 (1982).

12. P. W. Langhoff, S. R. Langhoff, T. N. Rescigno, J. Schirmer, L. S. Cederbaum, W. Domcke and W. von Niessen, Chem. Phys. <u>58</u>, 71 (1981).

13. C. Noguera, these Proceedings.

14. T. C. Fuggle, these Proceedings.

15. O. Gunnarsson, these Proceedings.

16. J. W. Wilkins, these Proceedings.

17. C. Nordling, these Proceedings.

18. V. Schmidt, these Proceedings.

19. J. Kanski and G. Wendin, Phys. Rev. B <u>24</u>, 4977 (1981), and references therein.

20. L. C. Davis and L. A. Feldkamp, Phys. Rev. B <u>23</u>, 6239 (1981), and references therein.

21. R. E. Dietz, Y. Yafet, G. P. Williams, G. J. Lapeyre and J. Anderson, Phys. Rev. B <u>24</u>, 6820 (1981).

22. D. Chandesris, J. Lecante and Y. Pétroff, these Proceedings.

23. G. E. Ice, G. S. Brown, G. B. Armen, M. H. Chen, B. Crasemann, J. Levin and D. Mitchell, these Proceedings.

24. F. Gerken, J. Barth and C. Kunz, these Proceedings.

25. R. Brulin, E. Schmidt, H. Schröder and B. Sonntag, Abstracts of contributed papers, 8th ICAP, 2-6 August 1982, Göteborg, Sweden, and to be published (Cr, Mn and Cu vapour PES).

26. F. Combet Farnoux, Phys. Rev. A <u>25</u>, 287 (1982) and references therein.

27. G. Wendin and M. Ohno, Phys. Scripta <u>14</u>, 148 (1976).

28. J. M. Bizau, F. Wuilleumier, P. Dhez and G. Wendin, Abstracts of contributed papers, 8th ICAP, 2-6 August 1982, Göteborg, Sweden.

29. J. M. Bizau, F. Wuilleumier, G. Wendin and P. Dhez, Abstracts of contributed papers, X-82, 23-27 August 1982, Eugene, Oregon.

30. G. Wendin and A. F. Starace, J. Phys. B <u>11</u>, 4119 (1978), and references therein.

31. J. E. Hansen, in ref. 2, p. 510.

32. G. Wendin, Phys. Lett. <u>46</u> A, 119 (1973).

33. F. Keller and F. Combet Farnoux, J. Phys. B, in press.

34. T. K. Sham and G. Wendin, Phys. Rev. Lett. <u>44</u>, 817 (1980).

35. M. Cukier, P. Dhez, B. Gauthé, P. Jaeglé, C. Wehenkel and F. Combet Farnoux, J. Physique <u>39</u>, L315 (1978).

36. J. P. Connerade, M. Pantelouris, M. A. Baig, M. A. P. Martin and M. Cukier, J. Phys. B <u>13</u>, L357 (1980).

37. J. P. Connerade, <u>The Physics of Non-Rydberg States</u>, in New Methods in Atomic Physics, Les Houches Summer School, 28 June-29 July 1982 (North Holland), and references therein.

38. G. Wendin, <u>Many-Electron Effects in Atomic Physics</u>, in New Methods in Atomic Physics, Les Houches Summer School, 28 June-29 July 1982 (North Holland), and references therein.

39. S. Lundqvist and G. Mutchopadhyay, in ref. 2, p. 503, and references therein.

516

40. T. Lucatorto, T. J. McIlrath, W. T. Hill III, and C. W. Clark, these Proceedings and references therein; T. Lucatorto et al., Phys. Rev. Lett. <u>47</u>, 1124 (1981).
41. K. Nuroh, E. Zaremba and M. Stott, to be published.
42. Ž. Crljen and G. Wendin, to be published.
43. G. Wendin, in ref. 1, p. 61.
44. H. P. Kelly, in ref. 1, p. 82.
45. G. Wendin, Abstracts of contributed papers, 8th ICAP, 2-6 August 1982, Goteborg, Sweden.
46. M. Ohno, in ref. 2, p. 589.
47. M. Ohno and G. Wendin, Proceedings of X-80, Stirling 180; Solid State Commun. <u>39</u>, 875 (1981).
48. L. Cederbaum, W. Domcke, J. Schirmer and W. von Niessen, in ref. 2, p. 481.
49. M. Boring, R. D. Cowan and R. L. Martin, Phys. Rev. B <u>23</u>, 445 (1981).
50. S. Svensson, N. Mårtensson, E. Basilier, P. Å. Malmquist, U. Gelius and K. Siegbahn, Phys. Scripta <u>14</u>, 141 (1976).
51. J. C. Fuggle, F. U. Hillebrecht, Z. Zolnierek, R. Lässer and Ch. Freiburg, to be published.
52. U. Becker, S. S. Southworth, P. H. Kobrin, C. M. Truesdale, D. W. Lindle, H. G. Kerkhoff and D. A. Shirley, Abstracts of contributed papers, 8th ICAP, 2-6 August 1982, Göteborg, Sweden.
53. D. A. Shirley, these Proceedings.
54. W. Eberhardt and H. J. Freund, J. Chem. Phys, in press.
55. P. Morin, M. Y. Adam, I. Nenner, J. Delwiche, M. J. Hubin-Franskin and P. Lablanquie, Proc. of the Int. Conf. on Instrumentation for Synchrotron Radiation, Hamburg, August 9-13, 1982, to be published in Nuclear Instr. Methods.
56. I. Lindau, these Proceedings.
57. J. Barth, private communication (Ca metal; 3d metals).
58. T. M. Zimkina, Trieste Int. Symposium on Core Level Excitations in Atoms, Molecules and Solids, 22-26 June 1981, and to be published.
59. M. L. Knotek, these Proceedings.
60. G. Loubriel, T. Gustafsson, L. I. Johansson, and S. J. Oh, Phys. Rev. Lett <u>49</u>, 571 (1982).
61. W. Eberhardt and A. Zangwill, to be published (4d-resonance photoemission from Xe absorbed on a metal).
62. See e.g. the papers by R. H. Pratt and coworkers, these Proceedings, and references therein.
63. R. A. Falk, G. Dunn, D. C. Griffin, C. Bottcher, D. G. Gregory, D. H. Crandall, and M. S. Pindzola, Phys. Rev. Lett. <u>47</u>, 494 (1981), and X-82 Abstracts volume.
64. G. Wendin and K. Nuroh, Phys. Rev. Lett. <u>39</u>, 48 (1977), and references therein.
65. K. Nuroh and G. Wendin, Phys. Rev. B <u>24</u>, 5533 (1981), and references therein.
66. See e.g. X-82 Abstracts volume, session on Appearance Potential Spectroscopy.

ONE-STEP MODEL OF X-RAY PHOTOEMISSION AND AUGER PROCESSES

O. Gunnarsson

Max-Planck Institut für Festkörperforschung,
7000 Stuttgart 80, Federal Republic of Germany

and

K. Schönhammer

I. Institut für Theoretische Physik, Universität
Hamburg, 2000 Hamburg 36, Federal Republic of
Germany

ABSTRACT

We present a formalism which describes the X-ray photoemission (XPS) process and the following Auger decay as one quantum mechanical event. It is shown that this can lead to a non-Lorentzian broadening of the XPS spectrum. We discuss the validity of a two-step model where the XPS and Auger processes are treated as independent. Sum rules for the Auger spectra are derived and it is shown that these predict that the relative weights of the Auger peaks depend strongly on the life-time of the XPS hole. We treat the plasmon gain satellites in free-electron like metals and trace the unusual shape of these peaks to interference effects. Finally, Auger-photoelectron coincidence spectroscopy (APECS) is considered. We show that for a finite system, APECS can give information about the spatial properties of the excited states, while it is harder to obtain such information for a system with a continuous spectrum.

I. INTRODUCTION

In core level X-ray photoemission spectroscopy (XPS) an electron is emitted due to the interaction with a photon. If the core level is not too deep, it is filled through an Auger process, and a second electron is emitted. It is usually assumed that the Auger decay has no influence on the XPS spectrum apart from a Lorentzian life-time broadening. Similarly, in most calculations it is assumed that the XPS process provides a core hole for the Auger process and causes a broadening of the spectrum, but that otherwise there is no coupling between the two processes.

In a simplified picture, one can argue that the excitations created in the XPS process propagate away and decay. If the core hole life-time is long, these excitations can therefore not influence the Auger process. In the calculation of the Auger spectrum, we can then assume that at the time of the Auger transition the valence electron system has locally relaxed to the lowest state in the presence of a core hole. This leads to a decoupling of the XPS and Auger processes and we have arrived at a two-step model. Even if these arguments are slightly oversimplified, we show that the two-step model is valid for long core hole life-times, provided that certain additional conditions are fulfilled.

In many situations the core hole life-time is not very long compared with the relaxation times of the excitations created. Transition metals, for instance, can have core hole life-times which are of the same order as the inverse width of the d-band, and we then expect large deviations from a two-step model. We have studied the free-electron like metals Na, Mg and Al, which are particularly simple from a theoretical point of view. Although the 1s core hole life-time is fairly long, so-called plasmon gain satellites have nevertheless been observed. In the XPS process plasmons are excited, and in the following Auger process they are deexcited so that the Auger electron can gain the plasmon energy. These processes can not be described in a two-step model.

In this paper, we discuss the conditions under which the two-step model is valid. We treat some cases where a one-step model is necessary to describe the XPS or Auger spectra. We also consider Auger-photoelectron coincidence spectroscopy (APECS), where the XPS and Auger experiments are performed in coincidence. We discuss what information can be obtained from APECS in addition to what separate XPS and Auger experiments give.

In Sec. II we briefly present our formalism, and in Sec. III we consider the XPS spectrum. The validity of the two-step model for the Auger spectrum is discussed in Sec. IV and sum rules are derived in Sec. V. The plasmon gain satellites are treated in Sec. VI, and in Sec.VII we discuss the information contained in APECS.

II. FORMALISM

Most calculations of Auger spectra have been based on a two-step model[1], but there has also been some work using a one-step description[2-4]. We have developed such a formalism[4] using the quadratic response formalism. We treat the coupling of the system to the external electromagnetic field to lowest order. This field is given by

$$V(t) = V e^{\eta t} (e^{i\omega t} + e^{-i\omega t}) \tag{1}$$

where ω is the frequency and $\eta \to 0$ describes how the field is switched on slowly at large negative times. The total Hamiltonian is

$$H_T = H + V_A + T + V(t) \tag{2}$$

where V_A describes the Auger decay, T the emitted electrons and the remaining part H models the valence and core electrons. We can now calculate the lowest order change $|\phi_1(t)\rangle$ of the ground-state wave-function $|E_o\rangle$ due to the external field $V(t)$

$$|\phi_1(t)\rangle = f e^{\eta t} \frac{1}{E_o + \omega - H - V_A - T + i\eta} V |E_o\rangle \tag{3}$$

where we have neglected the term with $-\omega$. We calculate the current j_k of electrons in a scattering state $|k\rangle$,

$$j_k = \frac{d}{dt} \langle \phi_1(t) | n_k | \phi_1(t) \rangle \tag{4}$$

This expression contains both the XPS and Auger electrons, and in general they can not be separated. We can, however, use the fact that when the photon energy goes to infinity, the same is true for the energy of the XPS electron, while the energy of the Auger electron stays constant. Thus, we assume that the XPS and Auger electrons are in different energy ranges, which simplifies the calculation of (4).

We use an approximation similar to the sudden approximation for XPS. Thus we assume that the kinetic energies of the emitted electrons is so high that we can neglect the interaction of the emitted XPS electron and the rest of the system altogether, and that the Auger electron interacts with the rest of the system via V_A, only. It is important to treat V_A to infinite order to obtain the right number of Auger electrons, i.e. one Auger electron for each XPS electron, if cascade Auger processes are neglected. If V_A is treated just to low-est order the Auger current diverges.

With the assumptions specified above, we can derive[4] an expression for the Auger current. If the levels involved in the Auger decay have no dynamics we obtain

$$j_A(\varepsilon) \sim \frac{\Gamma}{\pi} \sum_n \left| \sum_m \frac{<E_n^{(2)}|\psi_i\psi_j\psi_c^+|E_m^{(1)}><E_m^{(1)}|\psi_c|E_o>}{\varepsilon - E_m^{(1)} + E_n^{(2)} + i\Gamma} \right|^2 \qquad (5)$$

where ψ_c, ψ_i and ψ_j are annihilation operators for the levels emptied in the XPS and Auger processes, respectively. The eigenstates in the presence of the XPS hole are given by $|E_m^{(1)}>$, and those in the presence of the Auger holes by $|E_n^{(2)}>$. The life-time broadening of the levels participating in the XPS process is 2Γ.

The assumption that the states involved in the Auger process have no dynamics should be good if these states are core levels. However, if a valence level makes an important contribution both to the screening of the core hole created in the XPS process and to the Auger decay, Eq. (5) is not valid. In this case the c-number Γ has to be replaced by an operator, and a more complicated formula must be used[4]. In a similar way, formulas for the XPS current can be derived[4].

III. THE XPS SPECTRUM

It is normally assumed that the Auger decay leads to a broadening of the XPS spectrum but does not influence the spectrum in any other way. While this should be a good assumption in many cases, it is not always justified.

To discuss this, we consider a system where the screening can be described by the charge transfer model introduced by Lang Williams[5] and by Shirley[6]. In this model, the screening of a core hole takes place through the transfer of an electron to an orbital which, in the initial state, was at least partly empty.

The model is, for instance, applicable to adsorbates such as CO and N_2, where the screening orbital is the 2π orbital, and to transition metals, where the d-orbital is the screening orbital. To represent this model we use a slightly modified Anderson model

$$H = \sum_k \varepsilon_k n_k + \varepsilon_a n_a + \sum_k V_{ak} (\psi_a^+ \psi_k + h.c.)$$

$$+ (1-n_c) U_{ac} n_a + \varepsilon_c n_c$$

(6)

where ε_a is the screening orbital and ε_k are the conduction states of the system. The hopping between these states is given by V_{ak}, and U_{ac} describes how the screening orbital is "pulled down" when the core hole ε_c is created. For, e.g., CO on a metal $|a\rangle$ is the 2π molecular orbital and $|k\rangle$ is a one-particle state of the free semi-infinite metal. In this case a realistic value for U_{ac} may be 13 eV[7].

We can proceed as in Sec. II[4] and derive an expression for the XPS current

$$j_{XPS}(\varepsilon) \sim \frac{1}{\pi} Im \langle E_o | \frac{1}{\varepsilon - \omega - E_o + H - i\hat{\Gamma}} | E_o \rangle$$

(7)

As before $\hat{\Gamma}$ is in general an operator. We assume that there are two dominating channels for Auger decay; one KLL process involving just core levels, described by Γ_{KLL}, and one KLa process involving the screening (valence) orbital a, described by Γ_{KLV}. This leads to[4]

$$\hat{\Gamma} = \Gamma_{KLL} + \Gamma_{KLV} n_a$$

(8)

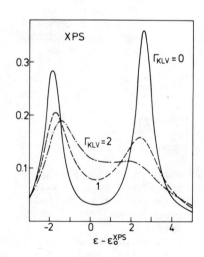

In the following we shall for simplicity often drop the subscript KLL.

Fig. 1 The XPS spectrum (Eq.(7)) for the zero band width limit of the model (6). We have used the parameters $\varepsilon_a=6$, $\varepsilon_b=0$, $V=2$, $U_{ac}=8$ and $\Gamma_{KLL}=0.5$ eV. The energy ε_o^{XPS} gives the position of the peak if there were no relaxation effects ($U_{ac}=0$).

We use a simplified version of the model (6) where the band width is zero and the conduction band is replaced by one level ε_b. The result[4] is shown in Fig. 1. For $\Gamma_{KLV}=0$ the operator Γ becomes a c-number and the spectrum in Fig. 1 consists of two delta-functions broadened by a Lorentzian. As Γ_{KLV} is increased, however, the broadening of the high energy peak increases faster than for the low energy peak. To understand this we consider the final states. In the presence of the core hole the a-level is located at $\varepsilon_a-U_{ac}=-2$ eV, i.e. below the b-level. The lowest final state therefore has mainly a-character while the excited final state is mostly b-like. The leading peak in Fig. 1 corresponds to the lower a-like state. If Γ_{KLV} is large, this state decays rapidly via the KLa channel, leading to a large broadening of the high energy peak. The low energy peak, on the other hand, corresponds to the b-like final state, which has a slower decay, and this peak has a smaller broadening.

IV. VALIDITY OF THE TWO-STEP MODEL

As we discussed in the introduction, we may expect the two-step model to be valid in the limit when the core hole created in the XPS process has a long life-time. In this section we want to study this issue more closely.

To be able to perform most of the calculations analytically, we consider a model where the core holes couple to bosons. These bosons could be plasmons or, following Tomonaga[3], we could assume that they are electron-hole pair excitations. Thus we assume that the valence electron Hamiltonian in the presence of the XPS hole is given by

$$H_1 = \int_0^\infty \rho(\omega) \ \{\omega b^+(\omega)b(\omega) + \lambda_1(\omega) \ [b^+(\omega) + b(\omega)]\} \ d\omega \tag{9}$$

where $\rho(\omega)$ is the density of boson excitations and $\lambda_1(\omega)$ is the coupling between the bosons and the core hole created in the XPS process. After the Auger process there are two holes, and we assume that in (9) $\lambda_1(\omega)$ is replaced by $2\lambda_2(\omega)$, where $\lambda_2(\omega)$ describes the coupling to an Auger hole.

Straightforward but lengthy calculations show[4] that the two-step model is valid in the limit where the life-time of the XPS core hole goes to infinity, provided that

$$(2\lambda_2(\omega) - \lambda_1(\omega)) \ \lambda_1(\omega)\rho(\omega) \tag{10}$$

contains no delta functions and is well-behaved for small and large ω (see Ref. 4 for the details). For instance, the forms of $\lambda(\omega)\rho(\omega)$, often used to describe electron-hole pair excitations or to describe plasmons with dispersion, satisfy these conditions. On the other hand, dispersion less plasmons would lead to a delta-function in $\rho(\omega)$ and the two-step model is invalid even in the limit of a long core hole life-time.

Analytical and numerical calculations[4] strongly suggest that the two-step model is valid also for other Hamiltonians in the limit

522

of a long core hole life-time, provided that the spectrum of XPS states has no bound states.

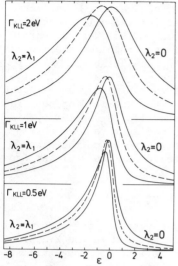

Fig. 2 The KLL Auger spectrum in the Tomanage model with the parameters $\alpha = 0.2$ eV^{-1} and $\lambda = 0.7$ eV. The full curves show the results for $\lambda_2 = \lambda_1$ and for $\lambda_2 = 0$, and the dashed curve shows the results in the two-step model. The life-time broadening of the XPS level is given by 2 Γ_{KLL}.

In Fig. 2 we show results using

$$\rho(\omega) \, \lambda_i(\omega)^2 = \lambda_i^2 \, \omega e^{-\alpha\omega} \tag{11}$$

which is appropriate for electron-hole pair excitations. For an Auger process involving just core levels, $\lambda_2 = \lambda_1$ is probably most appropriate, but we show $\lambda_2 = 0$ as well. The dashed curve shows the result in a two-step model.

The figure illustrates how the result of the two-step model is approached when Γ is reduced. However, it also shows that there can be a substantial difference for non-zero Γ between the one and two-step model and, in particular, there may be a fairly substantial shift in the peak position.

V. SUM RULES

For core level XPS there is a sum rule[9] which states that, under certain assumptions, the center of gravity is given by the core level eigenvalue in a Hartree-Fock calculation

$$\int_{-\infty}^{\infty} \varepsilon \, j_{XPS}(\varepsilon) d\varepsilon = \hbar\omega + \varepsilon_c + V(0) \tag{12}$$

where ε_c is the Hartree-Fock energy of the core level in a free atom and $V(0)$ is the (chemical) shift of the level due to its environment.

This sum rule plays an important role for the qualitative understanding of XPS, in that it gives a relation between screening and relaxation, on one hand, and the satellite structure, on the other hand. We show here that the corresponding sum rule for the Auger spectrum is more complicated, and interesting, since the center of gravity depends on the dynamics of the system.

We have derived expressions for the center of gravity for a general Auger process[4]. Since these formulas are rather complicated, it is instructive to consider a case where only core levels are involved in the Auger process. We neglect the dynamics of the core electrons and focus on how the dynamics of the screening of the XPS hole influences the center of gravity. This quantity contains a term[4] which depends on the life-time broadening 2Γ of the XPS hole

$$\int_0^\infty V(t) \; 2\Gamma \; e^{-2\Gamma t} \; dt \qquad\qquad (13)$$

When the XPS core hole is created, there is a flow of screening charge towards the hole. This charge changes the potential at the core levels and causes a time dependent shift $V(t)$. For simplicity we assume that all core levels are shifted the same amount. There is a formal similarity between (15) and (16). The Auger process takes place at a time t after the XPS process, with a probability $p(t) = 2\Gamma \exp(-2\Gamma t)$. The analyis[4] shows that to obtain the proper sum rule we take the potential $V(t)$ at this time, multiply by $p(t)$ and integrate. Similarly, (15) contains the potential $V(0)$ at the time of the XPS process.

The integral (16) involves two time-scales, the decay time Γ^{-1} and the screening time T, where T describes the time scale of $V(t)$. We illustrate this schematically in Fig. 3, where we show $V(t)$ and $\exp(-2\Gamma t)$ for two values of Γ. If Γ is large Eq.(13) approaches $V(0)$

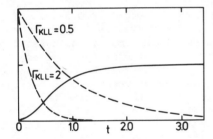

Fig. 3 The time-dependent screening potential $V(t)$ (full curve) for the parameters in Fig. 4 and the function $\exp(-2\Gamma_{KLL}t)$ which enter Eq. (13).

and if Γ is very small it approaches $V(\infty)$. In general we expect the screening charge to contain one electron. Then we obtain

$$V(\infty) - V(0) = \frac{e^2}{<r>} \qquad\qquad (14)$$

where $<r>$ is the spatial extent of the screening charge measured from the core hole. If $<r> \simeq 3a_0$, $e^2/<r> \simeq 9$ eV, which is the difference in the center of gravity between the limits $\Gamma \to 0$ and $\Gamma \to \infty$.

If the Auger spectrum shows pronounced peaks, the peak positions often have a weak dependence on Γ and mainly depend on the final state excitation energies. Since, however, the center of gravity depends strongly on Γ the weights of the peaks have to change dramatically with Γ. This is illustrated in Fig. 4, where we show

524

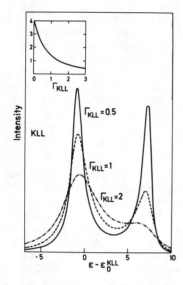

Fig. 4. The KLL Auger spectrum for the model (6) of the valence electrons. The life-time broadening of the XPS level is $2\Gamma_{KLL}$. We have used the parameters $\varepsilon_a=8$, $U_{ac}=8$ and V=2 eV. The insert shows the center of gravity. For the details of the model see Ref. 4.

the Auger spectrum for different valus of Γ. When Γ is increased the center of gravity is decreased and weight is shifted to the lower peak.

VI. PLASMON GAIN SATELLITES

The 1s core hole life time of Na, Mg and Al is long ($2\Gamma \sim 0.3$-0.4 eV) compared to the screening time scale ($\omega_p \sim 5$-15 eV). We therefore expect the deviations from the two-step model to be small, which is also essentially what is observed. Because of their simplicity, however, these systems are well suited for a theoretical study. We can construct reasonably realistic models and the parameters of these models can be obtained from other independent experiments. In the experimental spectra, small but distinct deviations from the two-step model can be observed in the form of so-called plasmon gain peaks. Such a peak corresponds to a process where a plasmon , created in the XPS event, is annihilated in the Auger process, and the Auger electron gains the plasmon energy. This effect has been discussed theoretically in several papers[2,3] and Almbladh has calculated the weight of the plasmon gain peak[3].

To study the plasmon gain process we can use the simple Hamiltonian

$$H = \varepsilon_c n_c + \varepsilon_1 n_1 + \varepsilon_2 n_2 + \sum_q \omega_q b_q^+ b_q +$$

$$(3-n_c-n_1-n_2) \sum_q g_q (b_q^+ + b_q) \tag{15}$$

where ε_c describes the XPS core hole and ε_1 and ε_2 the Auger core holes. The plasmon dispersion is given by ω_q, where q is the wave vector of the plasmon, and the coupling to a core hole is g_q. In the numerical calculations we also include the decay of the plasmons into electron-hole pairs and the coupling between the core holes and low energy electron-hole pair excitations.

Using Eq. (5) we can calculate the Auger spectrum. The results[10] for Na are compared with experiment in Fig. 5. The figure shows that there is a quite satisfactory agreement between theory and experiment. We now want to analyze the shape of the peak. In this discussion we neglect electron-hole pair excitations (Eq. (18)) and only consider

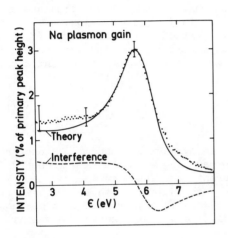

Fig. 5 The Auger spectrum of Na in the energy range of the plasmon gain peak, with the contribution from the primary peak subtracted. The full curve shows the theoretical calculation, the dots the experimental results and the dashed curve the contribution due to interference between gain and no loss events. To facilitate comparison between theory and experiment we have reduced the peak height of the theoretical curve by about 13 per cent and shifted the peak position 0.1 eV so that the peak height and position agree with experiment. The error bars show estimated experimental uncertainty.

states in Eq. (5) with zero and one plasmons. Apart from life-time broadening effects, we may then expect zero weight between the main peak (at the energy zero in Fig. 5) and ω_p. Since the density of states for the plasmons goes as $(\omega-\omega_p)^{-1/2} \Theta(\omega-\omega_p)$ we would, however, expect gain events above ω_p. It is clearly crucial to keep Γ finite, as the weight of the plasmon peak otherwise goes to zero. The life-time broadening can therefore not be neglected. However, the arguments above may suggest that the plasmon peak should be asymmetric, with most of the weight above ω_p. Both theory and experiment show that this is not the case and that the assymetry is the opposite. To demonstrate that this asymmetry is not due to the Lorentzian tail of the main peak, we have subtracted this tail in Fig. 5.

To understand this effect we use Eq. (5). The gain peak results from terms where the final state $|E_n^{(2)}>$ is the ground-state and the intermediate (XPS) state $|E_m^{(1)}>$ contains a plasmon. However, the final ground state can also be reached over the intermediate ground-state $|E_o^{(1)}>$. In Eq. (5) we sum all such terms and the sum is squared. This leads to interference terms, which are products of two terms with different intermediate states. The existence of such terms was first pointed out by Watts[2]. In Fig. 5 the dashed curve shows the contribution $j(\varepsilon)$ from interference terms, where one factor corresponds to a gain event (intermediate state contains a plasmon) and the other corresponds to a no loss event (intermediate state contains no plasmon).

$$j_I(\varepsilon) \sim \frac{\Gamma}{\pi} \sum_q \frac{<o^2|q^1><q^1|o><o|o^1><o^1|o^2>}{(\varepsilon-\omega_q + i\Gamma)(\varepsilon - i\Gamma)} + c.c. \qquad (16)$$

where we have used the position of the main peak as energy zero. The initial, intermediate and final ground-states are denoted $|o>$, $|o^1>$ and $|o^2>$, and $|q^1>$ is the intermediate state with a plasmon with wave vector q. From Eq. (16) it is obvious that the dashed curve should change sign above ω_p, since that is where the individual terms in (16) change sign.

These interference terms are only part of the total story. If we neglect them we may expect to find the asymmetric curve discussed below Fig. 5. The calculations, however, show that if (16) is neglected we find a gain peak

$$j_G(\varepsilon) \sim \frac{\Gamma}{\sqrt{(\varepsilon-\omega_p)^2 + \Gamma^2}} \qquad (17)$$

for $|\varepsilon-\omega_p|$ small, i.e. a result which is symmetric around ω_p. Since the weight below ω_p is due to life-time broadening we might have expected (17) to be some function broadened by a Lorentzian. The slow decay $\Gamma/|\varepsilon-\omega_p|$ for $\varepsilon<\omega_p$ and $|\varepsilon-\omega_p|>>\Gamma$, however, excludes such an interpretation. These effects are due to interference between gain events involving different plasmons with wave vectors q and q^1. This leads to a non-exponential behaviour

$$P_G(t) \sim \frac{1}{t} e^{-2\Gamma t} \qquad\qquad |t|>>\omega_p^{-1} \qquad (18)$$

for the probability of observing a gain event a time t after the core hole was created. This behaviour can be related to the energy dependence (17) in an intuitive way using a time-energy uncertainty relation. For a detailed discussion see Ref. 10.

VII. AUGER-PHOTOELECTRON COINCIDENCE SPECTROSCOPY

It has been demonstrated[11] that the XPS and Auger measurments can be performed in coincidence. In this new spectroscopy, Auger photoelectron coincidence spectroscopy[11] (APECS), the Auger spectrum $j(\varepsilon_p,\varepsilon_k)$ is measured for a given XPS energy ε_p. It is obvious that if we disregard possible resolution or intensity problems, APECS gives at least as much information as separate XPS and Auger experiments. Here, we want to discuss what additional information it can give. Haak et al[11] demonstrated that APECS can be used to separate the Auger spectra from closly spaced core levels excited in the XPS process. We shall here focus on the situation when there is just one core level in the energy range of interest, and see if APECS still can provide additional information. We first consider a process cij where only core levels are involved. Here and in the following, we assume that the main contribution to the broadening Γ of the level c comes from such processes, so that Γ can be treated as a c-number. We calculate the center of gravity of the Auger spectrum for a given XPS energy ε_p. Thus we study $\int \varepsilon_k j_{cij} (\varepsilon_p, \varepsilon_k) d\varepsilon_k / \int j_{cij} (\varepsilon_p, \varepsilon_k) d\varepsilon_k$ for different values of ε_p. The part which depends on ε_p can be written

$$- \frac{<\psi_{ij}(\varepsilon_p)|\delta H|\psi_{ij}(\varepsilon_p)>}{<\psi_{ij}(\varepsilon_p)|\psi_{ij}(\varepsilon_p)>} \tag{19}$$

where

$$|\psi_{ij}(\varepsilon_p)> = \sum_n \frac{1}{E_o + \omega - \varepsilon_p - E_n^{(1)} + i\Gamma} |E_n^{(1)}> <E_n^{(1)}|\psi_i \psi_j \psi_c^+ |E_o> \tag{20}$$

is a wave packet of XPS states and δH is the change of the valence Hamiltonian due to the Auger process. We assume that the core holes have negligible size. Then δH simply describes that one extra core hole is created in the Auger process and it samples the electrostatic potential in the core region formed from the wave packet (19).

$$\delta H = -\sum_\nu \frac{e^2}{|r_\nu - R|} \tag{21}$$

where r_ν is the position of the νth electron and R is the location of the core holes.

We next consider a process cia, where two core levels c and i and the screening orbital a participate in the Auger process. If we can identify these processes in the spectrum we add the weights of all such events for a given energy ε_p and divide the sum $P_{cia}(\varepsilon_p)$ by the XPS current $j_{XPS}(\varepsilon_p)$. We find

$$P_{cia}(\varepsilon_p)/j_{xps}(\varepsilon_p) \sim \frac{<\psi_i(\varepsilon_p)|n_a|\psi_i(\varepsilon_p)>}{<\psi_i(\varepsilon_p)|\psi_i(\varepsilon_p)>} \tag{22}$$

where the definition of $|\psi_i(\varepsilon_p)>$ is analogous to Eq. (20). This measurement gives information about the occupancy of the a-orbital.

Finally we can focus on caa processes, where the spin up and spin down orbitals are involved. Again we sum all such processes and devide by the XPS current. The ratio is proportional to

$$\frac{<\psi(\varepsilon_p)|n_{a\uparrow} n_{a\downarrow}|\psi(\varepsilon_p)>}{<\psi(\varepsilon_p)|\psi(\varepsilon_p)>} , \tag{23}$$

where the definition of $|\psi(\varepsilon_p)>$ is similar to Eq. (20). This measurment gives information about the probability for double occupancy of the a-level.

We now discuss how to extract useful information from (19,22, 23). First we consider a system with discrete levels where the spacing of the relevant levels is substantially larger than Γ. If we now tune ε_p at one peak in the XPS spectrum, only one term in (20) gives an important contribution to the wave packet. We can from (19) and (21) see that the center of gravity gives information about the charge distribution around the hole. By varying ε_p we can compare the charge distribution for different excited states $P\;|E_n^{(1)}>$. In Fig. 1, for instance, we showed the XPS spectrum for a simple model, which has one valence orbital a on the same atom as the core holes and another orbital b on the surrounding atoms. We assigned the leading peak to an XPS state in which a is mainly occupied, and the satellite to a state which is mainly b-like. If this assignment is correct, the APECS quantity (19) should give a higher center of gravity for the Auger spectrum if we focus on the leading XPS peak than if we look at the XPS satellite. This procedure provides a means of checking the interpretation of XPS spectra and makes it possible to study the spatial character of the final XPS states. Alternatively, we can use Eq. (22) and study the cia spectra. In this case we do not know the prefactor in Eq. (22), since it involves matrix elements which may be hard to estimate. However, these matrix elements should have a weak dependence on ε_p. Thus we can obtain the relative occupancy of a in different XPS states. In our example above (Fig. 1) this would be sufficient to decide whether the leading peak or the satellite in the XPS spectrum is mainly a-like. In a similar way we can use Eq. (23).

On the other hand, if the spectrum $E_n^{(1)}$ is continuous and Γ is small, the quantities (19, 22, 23) become independent of ε_p. Then the applications discribed above give no additional information[12]. We have studied[13] the different limits in a model which is a direct generalization of the two-level model in Fig. (1) to a N-level model. As the initial state valence Hamiltonian, we use

$$H = \varepsilon_b \sum_{i=1}^{N-1} n_i + t \sum_{i=1}^{N-2} (\psi_{i+1}^+ \psi_i + h.c.) + \varepsilon_a n_a + V(\psi_a^+ \psi_{N-1} + h.c.)$$

$$(24)$$

In Fig. 6 we show the quantity (22). We have an ε_p corresponding to the leading peak and the largest satellite. For N=2 the result is fairly independent of Γ and for Γ not too large we obtain a good estimate of the occupancy of the states. However, for N=96 and Γ small the quantity (22) is almost independent of ε_p. To obtain any new information for large N, we have to use $\Gamma \geq 0.2$, which is roughly of the same size as the inherent width W of the peaks. This is consistent with the intuitive idea that the excitations created in the XPS process stay close to the core for a time W^{-1}. To study them, we must consider an Auger process which takes place at a shorter or similar time, Γ^{-1}.

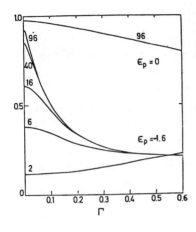

<u>Fig. 6</u> The quantity $P_{cia}(\varepsilon_p)/j_{XPS}(\varepsilon_p)$ Eq. (22) as a function of Γ for the leading peak ($\varepsilon_p=0$) and the largest satellite ($\varepsilon_p=-1.6$). The numbers at the curves show the number of levels N. The results for $\varepsilon_p=0$ have a weak dependence on N. We have used the parameters $\varepsilon_a=0.1$, $\varepsilon_b=0$, $U_{ac}=1.5$, $|V_a|=0.45$ and $|t|=1$.

REFERENCES:

1. For a review see, e.g., D. Chattarji,
   The Theory of Auger Transitions (Academic, New York, 1976)

2. C.M. Watts, J. Phys. F 2, 574 (1972); T. McMullen and B.Bergersen,
   Can. J, Phys. 52, 624 (1974); S. Abraham-Ibrahim, B. Caroli,
   C. Caroli and B. Roulet, Phys.Rev. B18, 6702 (1978); J. Physique
   40, 681 (1979); M. Ohno and G. Wendin, J. Phys. B12, 1305 (1979)

3. C.O. Almbladh, Nuovo Cimento B23, 75 (1974)

4. K Schönhammer and O. Gunnarsson, Surf. Sci. 89, 575 (1979);
   O. Gunnarsson and K Schönhammer, Phys. Rev. B22, 3710 (1980).

5. N.D. Lang and A.R. Williams, Phys. Rev. B16, 2408 (1977)

6. D.A. Shirley, Chem. Phys. Lett. 16, 220 (1972)

7. O. Gunnarsson and K Schönhammer, Phys. Rev. Lett. 41, 1608 (1978)

8. S. Tomonaga, Prog. Theor. Phys. 5, 544 (1950)

9. B. I Lundqvist, Phys. Kondens. Mater. 9, 236 (1969); R. Manne and
   T. Åberg, Chem. Phys. Lett. 7, 282 (1970)

10. O. Gunnarsson, K Schönhammer, J.C. Fuggle and R. Lässer, Phys.
    Rev. B23, 4350 (1981)

11. H.W. Haak, G.A. Sawatzky and T.D. Thomas, Phys. Rev. Lett. 41,
    1825 (1978)

12. O. Gunnarsson and K Schönhammer, Phys. Rev. Lett 46, 859 (1981)

13. O. Gunnarsson and K Schönhammer, Phys. Rev. (in press) (1982)

ELECTRON CORRELATION MANIFESTED BY SATELLITES IN ATOMIC SPECTRA

F.P. Larkins
Department of Chemistry, Monash University, Clayton, Australia, 3168.

ABSTRACT

Satellite structure has been observed for a wide range of atoms in all branches of high resolution atomic spectroscopy. The observed structure results mainly from electron correlation including relaxation effects. Some recent theoretical work is discussed with particular emphasis on photoionisation phenomena in the helium, lithium and argon systems.

INTRODUCTION

The satellite structure observed in high resolution photo-electron (PES), Auger (AES), (e, 2e) and x-ray emission (XES) spectroscopies provide valuable information on the role of electron correlation including relaxation effects in atomic systems. In recent years significant progress has been made towards a quantitative understanding of satellite line energies and intensities.[1-3]

This paper summarizes some of the recent theoretical work by my group with particular emphasis on photoionisation processes. Some correlation satellites originate from electronic rearrangement following core ionisation. They are often adequately accounted for by relaxed initial and final state Hartree Fock (HF) calculations. Such is the case for photoionisation in helium and lithium atoms. Relaxation effects manifest one special aspect of electron correlation. The satellites would not occur if a frozen orbital model, which preserves the orthogonality of one electron orbitals from initial to final states, was adopted. Other correlation satellites result from configuration interaction (CI) type effects, such as the $[ns] \leftrightarrow [np^2]n'd$ correlations in the rare gases. CI effects can also perturb satellites predominantly due to relaxation effects.

THEORETICAL CONSIDERATIONS

Qualitative Approach: The origin of correlation satellites may be schematically explained within a single manifold framework with the aid of the diagram presented in Figure 1. There is a series of available atomic states of appropriate symmetry associated with the initial and final manifolds which may in principle be populated during the transition process. For clarity, in Figure 1 a single excited state is shown for each manifold. The intensities of the relevant spectral lines depends upon the relative transition probabilities for the various competing pathways and the population distribution of the initial states. For the core ionisation process, consistent with energy conservation requirements, the lines in the photoelectron spectrum (PES) associated with the population of final excited core ion states will be observed at lower kinetic energies than the diagram line. Such lines are designated <u>final state correlation satellite</u> (FSCS) lines. In the limiting case double ionisation will result.

ORIGIN OF CORRELATION SATELLITES

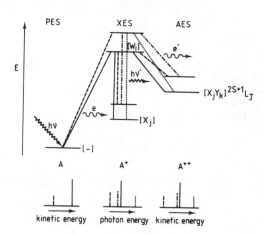

Fig. 1. A schematic representation of correlation satellites within a single manifold framework, —— diagram line; --- FSCS line; -·-·- MECS line; ..... ISCS line.

In XES and AES in addition to FSCS lines, <u>initial state correlation satellite</u> (ISCS) and <u>multiply-excited correlation satellite</u> (MECS) lines exist as discussed elsewhere.[1c]

This terminology should not be confused with the usage of the terms <u>initial state configuration interaction</u> (ISCI) and <u>final state configuration interaction</u> (FSCI) by Martin et al.[2,4] All correlation satellite lines (ISCS, MECS and FSCS) may in principle derive their intensity from both ISCI and FSCI effects. Similar principles can be applied to the classification of correlation satellites accompanying multiple ionisation of the atomic system.

<u>Quantitative Approach:</u> A framework for the determination of energies and intensities of correlation satellite lines for PES, XES and AES including relaxation and CI has been developed elsewhere.[1] Here we concentrate on the photoionisation process.

In general spectroscopic terms an N electron atomic system may be considered to make a transition from the mth member of the initial state manifold, described by the exact many-body wave function $\psi_m^i(N)$, to the nth member of the final state manifold described by the exact wavefunction $\psi_n^f(N)$. In practice, the initial state for the photoionisation process may be considered as the ground state of the system ψ_o^i (N). This correlated wavefunction may be approximated by ϕ_o^i (N) expressed as a linear combination of configuration state functions (CSF) χ_{ok}^i (N) involving the N electrons in bound orbitals:

$$\psi_o^i \, (N) \simeq \phi_o^i \, (N) = \sum_k b_{ok} \, \chi_{ok}^i \, (N) \tag{1}$$

The energy for the state is E_o^i determined from the equation:

$$H^i \, (N) \, \phi_o^i \, (N) = E_o^i \, (N) \, \phi_o^i \, (N) \tag{2}$$

The final state involves the N − 1 electrons of the ion core in bound orbitals plus the photoelectron in the continuum, and this state ψ_n^f (N) is represented by the approximate function ϕ^f (N). The (N − 1) ion-core wavefunction ϕ_n^f (N − 1) may be expressed, using FSCI, as

$$\phi_n^f(N-1) = \sum_k c_{nk} \, \chi_{nk}^f \, (N-1) \tag{3}$$

The energy for this state is E_n^f (N − 1) determined from the equation

$$H^f \, (N-1) \, \phi_n^f \, (N-1) = E_n^f \, (N-1) \, \phi_n^f(N-1) \tag{4}$$

The kinetic energy of the electron corresponding to the nth satellite line in the PE spectrum is given by

$$\varepsilon_n = h\nu - (E_n^f \, (N-1) - E_o^i(N)) \tag{5}$$

where $h\nu$ is the photon energy.

The CSFs χ_{ok}^i (N) and χ_{nk}^f (N−1) are antisymmetrized products of one electron functions of appropriate symmetry. b_{ok} and c_{nk} are the coefficients of the CSFs in the initial and final ASFs respectively.

To account for the interaction with the continuum electron represented by a wavefunction ϕ_ε (1), continuum state configuration interaction (CSCI) must be used. The final state corresponding to the ion in state, ϕ_n^f (N − 1), and the continuum electron in orbital $\phi_{\varepsilon_n}^f$ (1) may be represented as

$$\psi_n^f \, (N) \simeq \phi_n^f \, (N) = \sum_j d_{nj} \, (\varepsilon) \mathscr{A} \, \phi_j^f \, (N-1) \, \phi_{\varepsilon_j}^f(1) \tag{6}$$

d_{nj} (ε) are the expansion coefficients for the N-electron CSF. These functions are antisymmetrized and must have the correct symmetry of the N electron final states. This procedure of continuum CI is frequently referred to as interchannel coupling or close coupling[5,6] since interaction between the exit channels is included.

Calculations of PES satellite lines for helium,[7,8] for example, have included interchannel coupling.

If this interaction is ignored then

$$\Phi_n^f (N) = \mathscr{A} \Phi_n^f (N-1) \; \phi_{\varepsilon_n}^f (1) \qquad (7)$$

The continuum functions, $\phi_{\varepsilon_n}^f$, required may be generated in the $V(N-1)$ potential of the final state ion core via the numerical Hartree-Fock method with or without inclusion of the exchange interaction.[9,10,11] In this approach the continuum orbital is not determined self-consistently with the remaining $(N-1)$ electrons, but within the framework of a frozen ion core i.e, it is partly decoupled.

If only relaxation effects are considered then from (1) and (3) respectively we have

$$\Phi_o^i (N) = \chi_o^i (N) \text{ and } \Phi_n^f (N-1) = \chi_n^f (N-1) \qquad (8)$$

The cross-section for photoionisation of a system by an unpolarized photon beam of energy $h\nu$ (in atomic units) is given[12,13] by

$$\sigma(h\nu) = \frac{4\Pi^2 \alpha a_o^2}{3g_i} \; h\nu \left| M_{if} \right|^2 \qquad (9)$$

where α is the fine structure constant, a_o is the Bohr radius and g_i is the statistical weight of the initial discrete state. σ is in units of Mb ($10^{-18} cm^2$). Within the dipole approximation [13], the transition moment M_{if} is given by

$$M_{if} = \langle \Psi_o^i(N) | \sum_j \underset{\sim}{d}_j | \Psi_n^f(N) \rangle \qquad (10)$$

where $\underset{\sim}{d}_j$ is the one-electron dipole operator.

When the wave-functions $\Psi_o^i(N)$ and $\Psi_n^f(N)$ are eigenfunctions of a single Hamiltonian H, application of the off-diagonal hypervirial theorem

$$(E_n^f - E_o^i) \; \langle \Psi_o^i | A | \Psi_n^f \rangle = \langle \Psi_o^i | [A,H] | \Psi_n^f \rangle \qquad (11)$$

leads to the formal equivalence of the length, velocity and acceleration forms of the dipole operator.[13] That is,

$$\underset{\sim}{d}_j = \underset{\sim}{r}_j = \frac{1}{\Delta E_{if}} \; \underset{\sim}{\nabla}_j = \frac{Z}{(\Delta E_{if})^2} \; \frac{\underset{\sim}{r}_j}{|\underset{\sim}{r}_j|^3} \qquad (12)$$

where the energy term ΔE_{if} in a.u. is given by $E_n^f(N) - E_o^i(N) = h\nu$ for photoionisation processes. The relative merits of various forms when approximate wavefunctions are used has been widely discussed,[14-16] but the problem remains unresolved.

When the approximate wavefunctions $\Phi_o^i(N)$ and $\Phi_n^f(N)$ are used, with (1) and (7) then the transition moment M_{if} is given by

$$M_{if} = <\Phi_o^i(N) | \sum_j \underset{\sim}{d}_j | \Phi_n^f(N)>$$

(13)

$$= \sum_{k\ell} \sum b_{ok} c_{n\ell} <\chi_{ok}^i(N) | \sum_j \underset{\sim}{d}_j | \mathscr{A} \chi_{n\ell}^f(N-1)\phi_{\varepsilon_n}^f(1)>$$

When the initial and final ion states are represented by single HF type configurations, then from (7) and (8)

$$M_{if} = <\chi_o^i(N) | \sum_j \underset{\sim}{d}_j | \mathscr{A} \chi_{n\ell}^f(N-1)\phi_{\varepsilon_n}^f(1)>$$

(14)

This is the situation when either a frozen orbital model or a relaxed orbital model is used.

Cohen and McEachran (17) have shown that for approximate wavefunctions such as $\Phi_o^i(N)$ and $\Phi_n^f(N)$, corresponding to the approximate effective Hamiltonians H^i and H^f respectively, (11) must be replaced by

$$(E_n^f - E_o^i)<\Phi_o^i|A|\Phi_n^f> = <\Phi_o^i|[A,H]|\Phi_n^f>$$

(15)

$$+ <\Phi_o^i|A(H-H^f) - (H-H^i)A|\Phi_n^f>$$

It follows that for our purposes the relationship between the three dipole operator forms must be established in the context of (15) rather than (11). The second term on the RHS of (15) should either be evaluated or the problem formulated in a manner which minimizes the effect of this term. The latter will be the case with a frozen orbital model, but not with a relaxed orbital model required for satellite structure. Richards and Larkins[18,19] have discussed the problem in the context of helium PES satellite transitions at the relaxed HF level. They conclude that for diagram and monopole 'shake up' satellite transitions it is reasonable to apply (14) directly with the relationship between the various forms of d_j given by (12). The bound state orbital and the continuum orbital for the electron directly involved in the transition are determined in the potential field of the other N-1 electrons of the system.

The potentials have the same symmetry and therefore the second term on the RHS of (15) is considered to be small and is therefore neglected. However, for dipole 'conjugate shake-up' satellite transitions, where two bound state orbitals are involved in the one electron dipole matrix element, this approach is not satisfactory. The initial and final state bound orbitals are generated in the potential field of the remaining N-1 and N-2 bound electrons respectively. This is because from (7) the continuum electron is considered to be decoupled from the (N-1) electron ion core. The second term on the RHS of (15) is non-negligible. If neglected without modification the ratio of the length to velocity matrix elements for conjugate shake-up processes is not independent of $h\nu$ as is required.

The problem may be reformulated at the relaxed HF level to minimize the second term in (15) by redefining the energy term ΔE_{if} as

$$\Delta E_{if} = E_n^f(N-1) - E_o^{i'}(N-1) \qquad (16)$$

instead of $\Delta E_{if} = E_n^f(N) - E_o^i(N) = h\nu$. The $E_n^f(N-1)$ term is that obtained from (4), while $E_o^{i'}(N-1)$ is determined from the equation

$$H^{i'}(N-1) \; \chi_o^i(N-1) = E_o^{i'}(N-1)\chi_o^i(N-1) \qquad (17)$$

where $\chi_o^i(N-1)$ is an initial state N-1 electron wavefunction obtained by projecting out the electron and orbital formally associated with the photoionisation. $H^{i'}(N-1)$ and $H^f(N-1)$ are of the same form and ΔE_{if} is now independent of $h\nu$. This reformulation of (12) influences the velocity and acceleration forms, however, the length form of the matrix element is independent of the energy term.

Extension of this approach to CI wavefunctions and to other many electron atomic systems is not however straightforward. For multiconfiguration treatments the single particle energies lose their validity. Therefore, ideally one should use total N electron atomic state functions including coupling to the continuum with $\Delta E_{if} = h\nu$. This problem warrants further investigation especially for conjugate shake up satellite lines.

RESULTS AND DISCUSSION

The helium system: The importance of hole relaxation effects, initial state correlation effects and the relative merits of the three forms of the dipole operator have been investigated in detail for $n = 2$ and $n = 3$ cross-sections[18,19] The He system is convenient for study because of the availability of i) experimental data of high quality and ii) excellent neutral atom correlated wavefunctions. The He$^+$ wavefunctions are also exact one-electron functions.

The σ_{1s} photoionisation cross-section curve ($1s^2 \rightarrow 1s\epsilon p$) is well accounted for using relaxed Hartree-Fock wavefunctions for the initial and final ion states without continuum exchange with $\Delta E = h\nu$. The length and velocity curves agree with experiment[20] to within 2% near threshold and show a maximum deviation of 10% around 200 eV. The acceleration curve lies 10-40% higher below 120 eV. As a consequence of the inclusion of continuum exchange for which the helium system is the optimum test, the agreement between any two dipole forms or between any one form and experiment is ±20% in the photon energy range from threshold to 200 eV. When correlated He CI wavefunctions by Nesbet and Watson[21] (20 terms, 97.7% of correlation energy) and Taylor and Parr[22] (4 terms, 85.0% of correlation energy) are used for the He atom the calculated σ_{1s} cross-sections are similar to the relaxed HF ones, but agreement between the length and velocity curves is typically only ±5% and between either form and experiment only ±10%.[19] In this work interchannel coupling is ignored. Similiar calculations have been undertaken for σ_{2s}, σ_{2p}, σ_{3s} and σ_{3p}. Wide variations between the results obtained with the various approaches have been obtained.[18,19] Here we consider mainly the σ_{2s}, ($1s^2 \rightarrow 2s\epsilon p$) and σ_{2p}, ($1s^2 \rightarrow 2p\epsilon s$) at the relaxed HF level with the continuum orbital being generated without exchange. The 2s cross section is shown in fig. 2. Above 130 eV the three dipole forms agree to within 1-2% while below this energy they diverge. The agreement with the data of Woodruff and Samson,[23] for the velocity and acceleration dipole forms, is at least as good as that reported by Jacobs and Burke,[7] except very near threshold, and superior to the theoretical predictions of Chang.[8] The previous theoretical studies have used sophisticated correlated wavefunctions with interchannel coupling.

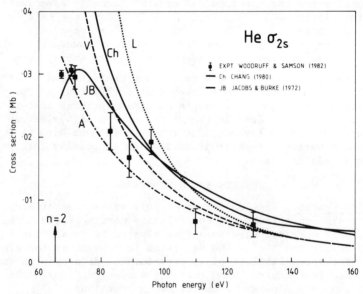

Fig. 2. Helium $1s^2 \rightarrow 2s\epsilon p$ photoionisation cross-section ΔE_{if} = hν: length form L, velocity form V, acceleration form A. Experimental data, ref 23; Theory, ref. 7,8.

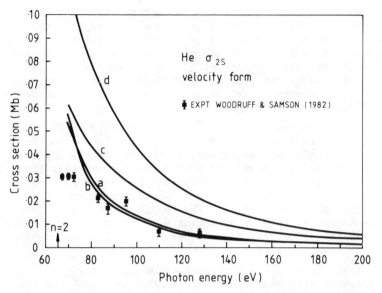

Fig. 3. Helium σ_{2s} velocity form. a) relaxed HF no continuum exchange. b) relaxed HF with continuum exchange. c) Nesbet-Watson[21] CI wavefunction d) Taylor-Parr[22] CI wavefunction. Experimental data ref. 23.

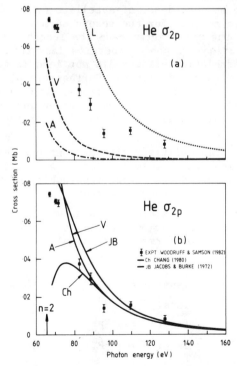

Fig. 4.

Helium $1s^2 \rightarrow 2p\epsilon s$ photo-ionisation cross-section

a) $\Delta E_{if} = h\nu$

b) $\Delta E_{if} = E_{2p}^f - E_{1s}^{i'}$.

L,V,A, see Caption Fig.2.

The sensitivity of σ_{2s} to the inclusion of continuum exchange and to the use of initial state CI wavefunctions is illustrated in figure 3 using the velocity form of the dipole operator, while σ_{2s} is influenced little by the inclusion of exchange, there is poor agreement with experiment using correlated wavefunctions. This calculation underlines the point that the quality of a wavefunction as measured by an energy criterion may not be a reliable guide to its suitability for intensity calculations.

The σ_{2p} results for the cases $\Delta E_{if} = h\nu$ and $\Delta E_{if} = E^f_{2p}(N-1) - E^{i'}_{1s}(N-1)$ are shown in figures 4a and 4b respectively. In figure 4a the length curve is considerably too large, whereas the velocity and acceleration forms become vanishingly small for photon energies greater than 120 eV. However, when ΔE is redefined as discussed previously then the velocity and acceleration curves agree within 0.5% and are in excellent agreement with experiment above 75 eV. The length curve remains unchanged from figure 4a and is in poor agreement with experiment. This is by definition in the reformulation and does not resolve the problem of which dipole form is superior.

The relaxed HF total $n = 2$ cross-section and the σ_{2p}/σ_{2s} ratio are in good agreement with experiment and provide support for the direct fluorescent decay measurements of Woodruff and Samson[23] rather than the indirect photoelectron measurements of Bizau et al.[24]

In summary, a relaxed HF procedure can be satisfactorily used with the velocity and acceleration forms of the dipole operator to explain the major features of the photon dependence of satellite cross-sections in helium provided the need to correct for the second term in (15) is recognized. In contrast with the n=1 diagram case the length form is not suitable for satellite calculations at photon energies less than 150 eV. Inclusion of initial state correlation effects did not lead in general to improved cross-sections for satellite lines. Furthermore all calculations reported to date deviate significantly from experiment within approximately 15eV of threshold. There is a need to reconsider how the N electron wavefunction is formulated along with validity of the dipole approximation.

The Lithium system: Relaxed HF calculations without continuum exchange have been performed to determine the photoionisation cross-sections for the main processes in lithium, extending our previous published work[11] and using the experience gained from the helium study.

Photoionisation cross-sections have been calculated for the following processes with $n = 2$ or 3[18,19] from threshold to 200 eV using the length, velocity and acceleration forms of the dipole operator

$$Li \ 1s^2 \ 2s \ ^2S \xrightarrow{h\nu} Li^+ \ 1s^2 \ (^1S)\epsilon p \ ^2P^o$$
$$Li \ 1s^2 \ 2s \ ^2S \xrightarrow{h\nu} Li^+ \ 1sns(^1S)\epsilon p \ ^2P^o$$
$$1sns(^3S)\epsilon p \ ^2P^o$$
$$1sns(^1P^o)\epsilon s \ ^2P^o$$
$$1snp(^3P^o)\epsilon s \ ^2P^o$$

Experimental data is limited to a low precision photoabsorption measurement of the 2s photoionisation cross-section[25] at photon energies between 5 and 25 eV and recent 1s photoionisation data at several discrete energies up to 150 eV.[26] The results for n = 2 cross-sections are shown in figure 5 using the velocity form of the dipole operator, and in figure 6 for the n = 3 case. Conjugate shake up cross-sections were calculated with the modified ΔE_{if} value as discussed previously.

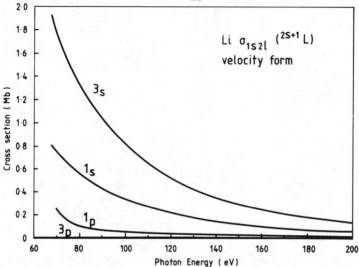

Fig. 5. Lithium $1s^2 2s \rightarrow 1sn\ell\varepsilon\ell'$, n = 2 photoionisation cross-sections velocity form, see text for details.

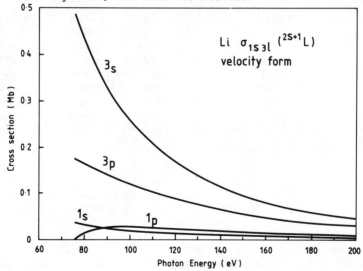

Fig. 6. Lithium $1s^2 2s \rightarrow 1sn\ell\varepsilon\ell'$, n = 3 photoionisation cross-sections velocity form, see text for details.

Principal features of this work are that $\sigma_{1s2s^3S}/\sigma_{1s2s^1S}$ cross-section ratio is predicted to be constant at 2.4 to within ±0.1% over the energy range from ~70 to 200 eV, while the ratio $\sigma_{1s3s^3S}/\sigma_{1s3s^1S}$ is also constant at 13.8 ± .2 from 85 to 200 eV. For n = 2 the triplet : singlet ratio is less than the statistical ratio (3:1), however, for the n = 3 satellites it is very much greater. The photoelectron spectrum is predicted to be strongly energy dependent below 100 eV, because for n = 2 the ^1P satellite increases in intensity relative to the n = 2 ^1S and ^3S lines at lower energies whereas for n = 3 the ^3P increase and the ^1P satellites decrease relative to the ^1S satellite. The n = 2, ^3P intensity is predicted to be negligible at all photon energies relative to the other multiplets, however, for n = 3 the ^3P line has the second largest cross-section. The results are generally in good agreement with the limited experimental data available although direct comparison is difficult because of the presence of overlapping molecular features in the lithium spectrum that has been reported.[26]

Rare Gas Atomic Systems: In a series of papers we have investigated satellite structure in the argon system from a theoretical viewpoint with numerical relaxed HF and CI wavefunctions. Photoelectron spectra for the n = 3 valence shell,[1a,b] and for the n = 2 core shell[27] have been analysed. Furthermore, the L x-ray emission spectrum [1c] and the L-MM Auger spectrum[1d] have been calculated and compared with experiment. The valence shell PES of neon, krypton and xenon have also been calculated.[1b]

These calculations highlight the important role of relaxation and other electron correlation effects in determining satellite structure. They also provide an insight into the complexity of the theoretical approaches required.

In the valence shell PES the FSCS of ^2S, $^2P^o$ and ^2D symmetry are the major contributions within an LS coupling framework. The ^2S satellites dominate the spectrum of Ar, Kr and Xe at high photon energy (1487 eV). They arise principally from strong FSCI between the CSF $[ns]^2S$ and $[np^2]n'd^2S$. The distribution of the $[ns]^2S$ CSF over various ASF is the major ^2S populating mechanism for excited states. CI has the central role in any theoretical model to account for the intensity distribution. The $^2P^o$ FSCS are well accounted for by electronic relaxation effects following p-shell ionisation. ASF dominated by CSF of the kind $[np^2]n'p$ $^2P^o$ are populated by a 'shake-up' mechanism. Much of the $^2P^o$ satellite intensity may be accounted for using relaxed Hartree Fock calculations. The relative importance of $^2P^o$ to ^2S FSCS increases with decreasing photon energy. ^2D satellites populated by a 'conjugate shakeup' mechanism of the kind $ns^2np^6(^1S) \rightarrow ns^2np^4n'd(^2D)\epsilon p(^1P)$ also become important at low photon energies. A combination of electronic relaxation effects and initial state correlation effects (ISCI) involving interaction between the ns^2np^6 ground state CSF and the $ns^2np^4n'd^2$ CSF with direct ionisation from the n'd orbital are the populating mechanisms. Other mechanisms resulting in the population of satellites are possible, but the above are considered to be the most important.

The results of calculations, including relaxation effects and FSCI and ISCI effects, for the intensities of the valence PES of argon using the length form are summarized in Table. 1.   Photon energies of 151 and 1487 eV are considered.   Full details are reported elsewhere.[28]

Table I   Argon Valence Shell Photoelectron Spectrum

	Intensity[a]	
	$h\nu$ = 151 eV	$h\nu$ = 1487 eV
$^2P^o$ main line	585	168
^2S main line	100	100
Total ^2S satellite	22.4	26.6
Total $^2P^o$ satellite	17.5	4.9
Total ^2D satellite	6.2	0.0

[a]Only satellite lines within 14 eV of the ^2S diagram line are considered.

The increase in total $^2P^o$ and ^2D FSCS intensity relative to the ^2S intensity with decreasing photon energy is clearly illustrated.

The FSCS structure of the argon $2p^2$[7a] and $2s^2$[7b] core photoelectron spectra at high photon energy is dominated by population of $[2\ell 3p]np$ CSF by relaxation processes of the 'shake-up' type, although in the 2s spectrum $[2s3s]ns$ FSCS also contribute some intensity.   The total calculated 2s and 2p FSCS intensities at high photon energies where the shake model is valid are 10.3 and 9.2 per cent respectively relative to the corresponding main photoelectron line.   The origin of argon n = 2 PE satellites is consistent with the 'shake-up' type origin of the neon 1s PE satellite spectrum.   The dominance of relaxation effects over other possible correlation effects occurs because with core ionisation, valence shell electrons experience a greater change in effective nuclear charge than for valence shell ionisation.   The strong ^2S correlation effects observed in valence shell PES of rare gases result predominantly from $[ns] \to [np^2]n'd$ interactions.   Such effects are expected to be largely absent in inner-shell spectra, because of the greater energy separation of core-hole levels.   Hence, satellite structure arising from FSCI effects, as distinct from relaxation effects, is expected to be weak in core-hole PE spectra.

The model developed for rare gas systems in general reproduces experimental valence shell PES energies to within 1 eV and intensities at x-ray photon energies to within a factor of 1.5. The major deficiences are the lack of relativistic effects for high atomic numbers and the incomplete relaxation of the core for excited states.

CONCLUSION

Satellite structure in atomic spectra provides a sensitive test of the quality of theoretical wavefunctions and the formalism used to evaluate intensity distributions as a function of photon energy. While significant progress has been made, several problems remain to be overcome before quantitative predictions for a range of satellite processes can be made with a high degree of confidence.

ACKNOWLEDGEMENT

Most of the work summarized in this report is based upon collaborative research with Dr. K.G. Dyall and Mr. J.A. Richards. Their enthusiasm and stimulating discussions, along with the continued financial support of the Australian Research Grants Scheme are gratefully acknowledged.

REFERENCES

1a K.G. Dyall and F.P. Larkins, J. Phys. B. $\underline{15}$, 203 (1982); 1b, $\underline{15}$, 219 (1982); 1c, $\underline{15}$, 1811 (1982); 1d, $\underline{15}$, Sept. (1982); and references therein.

2. R.L. Martin and D.A. Shirley, J. Chem. Phys. $\underline{64}$, 3685 (1976); Phys. Rev. A13, 1475 (1976).

3. Wendin G and Ohno M, Phys. Scr. $\underline{14}$, 148 (1976).

4. R.L. Martin, P. Kowalczyk and D.A. Shirley, J. Chem. Phys. $\underline{68}$, 3829 (1978).

5. G. Howat, T. Åberg, O. Goscinski, J. Phys. B: Atom. Molec. Phys., $\underline{11}$, 1575 (1978).

6. S.T. Manson, J. Elec. Spec. and Relat. Phenomena, $\underline{9}$, 21 (1976).

7. V.L. Jacobs and P.G. Burke, J. Phys. B: 5, L67 (1972).

8. T.N. Chang, J. Phys. B. $\underline{13}$, L551 (1980).

9. J.W. Cooper, Phys. Rev. $\underline{128}$, 681 (1962).

10. M.H. Chen, B. Crasemann, M. Aoyagi and H. Mark, Phys. Rev. A18, 802 (1978).

11. F.P. Larkins, P.D. Adeney and K.G. Dyall, J. Electron Spectrosc. Relat. Phenom. $\underline{22}$, 141 (1981).

12. D.R. Bates, Mon. Not. Roy. Astr. Soc. $\underline{106}$, 432 (1946).

13. S.T. Manson, Adv. Electronics Electron Phys. $\underline{41}$, 73 (1976).

14. A.F. Starace, Phys. Rev. A $\underline{3}$, 1242 (1971); $\underline{8}$, 1141 (1973).

15. I.P. Grant, J. Phys. B. Atom. Molec. Phys. $\underline{7}$, 1458 (1974).

16. I.P. Grant and A.F. Starace, J. Phys. B. Atom Molec. Phys. $\underline{8}$, 1999 (1975).

17. M. Cohen and R.P. McEachran, Chem. Phys. Letters $\underline{14}$, 201 (1972).

18. J.A. Richards and F.P. Larkins, J. Phys. B. to be published.

19. J.A. Richards, Honours Thesis, Monash University, 1981.
20. J.A.R. Samson, Phys. Rep. 28, 303 (1976).
21. R.K. Nesbet and R.E. Watson, Phys. Rev. 110, 1073 (1958).
22. G.R. Taylor and R.G. Parr, Proc. Nat. Acad. Sci. 38, 154 (1952).
23. P.R. Woodruff and J.A.R. Samson, Phys. Rev. A25, 846 (1982).
24. J.B. Bizau, F.J. Wuilleumier, P. Dhez, D.L. Ederer, T.N. Chang,
     S. Krummacher and V. Schmidt, Phys. Rev. Lett. 48, 588 (1982).
25. R.D. Hudson and V.L. Carter, J. Opt. Soc. Am. 57, 651 (1967).
26. J.M. Bizau, F.J. Wuilleumier, D. Ederer, P. Dhez,
     S. Krummacher and V. Schmidt, J. Phys. B: Atom. Molec. Phys.
     to be published.
27a. K.G. Dyall, F.P. Larkins, K.D. Bomben and T.D. Thomas,
     J. Phys. B. 14, 2551 (1981); b, K.G. Dyall and F.P. Larkins,
     J. Phys. B. 15, 1021 (1982).
28. K.G. Dyall and F.P. Larkins. J. Phys. B. to be published.

POST-COLLISION INTERACTION IN INNER-SHELL IONIZATION

Volker Schmidt
Fakultät für Physik, Universität Freiburg, D-7800 Freiburg, FRG

ABSTRACT

Within a classical description, the post-collision interaction in inner-shell ionization processes by photon and electron impact is explained. Based on this model, the main effects of post-collision interaction as observed so far experimentally are described in detail. Some theoretical descriptions of the phenomena beyond the classical model are also discussed.

INTRODUCTION

When the energy transfered to an atom or molecule in a collision process (i.e. photon or electron impact) is close to the threshold for the ejection of an inner-shell electron, the subsequent Auger decay may be influenced by the presence of the slow ejected electron (and possibly also by the slow scattered particle). This is a special class of long-range Coulomb interactions between charged particles which have been produced in the primary ionization process. It is called post-collision interaction (PCI). PCI has been studied first in the decay of outer-shell autoionizing states following excitation by slow ion impact[1,2] (for review papers see Ref. 3-5). In inner-shell ionization processes PCI-phenomena have generated growing interest since the first experimental[6-8] and theoretical[9] investigations. The situation of PCI is characterized by the emission of one fast (Auger) electron while at least one slow electron is still present around the target ion. This situation is quite different from the highly correlated motion of two slow electrons (compare Ref.10). The most direct case for the investigation of PCI in inner-shell ionization is that of photoionization followed by Auger decay. Here only one slow electron (photoelectron close to threshold) and one fast (Auger) electron are subject to PCI. It is more complicated in the case of electron-impact ionization when there are two slow electrons which can bring in the highly correlated motion of the two slow electrons into the PCI phenomena. In the following, we will discuss PCI phenomena occuring in photon- and electron-impact processes. The basis for the discussion will be provided by the simple classical model for PCI and by means of selected experimental investigations. The following subjects will be considered in more detail: PCI as the natural link between ionization processes above and excitation processes below the ionization threshold; PCI as the origin for a considerable change in the energy distribution of the ejected electrons; PCI as the source of peculiarities in the cross section near the ionization threshold. Guided by theoretical considerations that are beyond the simple classical model, further phenomena of PCI will be described finally.

THE CLASSICAL MODEL OF PCI

Several theoretical formulations for PCI in inner-shell ionization have been developed.[9], [11-17] However, for our forthcoming discussion of PCI phenomena observed so far experimentally, the simple classical description will provide even for a quantitative understanding of the PCI phenomena. It should be noted that the classical model yields the same result as that of Niehaus[9] provided interference effects can be neglected.

All PCI models are based on the sudden approximation. The idea of this approximation is that the time of interaction between the fast and slow electron is so short that the slow electron makes a sudden transition from its initial to its final state[18] i.e. PCI can be described as a relaxation process. This is illustrated in Fig. 1 for the simple case of photoionization and subsequent Auger decay with one initial and one final state only. We consider a photoelectron with its nominal excess energy $E_{exc} > 0$ which is the kinetic energy it would have at infinity without PCI i.e. $E_{exc} = h\nu - E_B$; $h\nu$ = photon energy; E_B = inner-shell ionization energy of the atom. This photoelectron moves in the field of the remaining other electrons of the ion which adapt adiabatically to the varying distance $R(t)$ of the slow receding photoelectron. This field is described before the Auger decay by the potential energy curve $V(E^+, t)$, where E^+ is the relaxed initial state energy of the ion and $E^+ = E_B$. At each time t the photoelectron has an instantaneous kinetic energy $\varepsilon^{inst}(t)$ which is connected with its velocity by $v(t) = (2\varepsilon^{inst}(t))^{1/2}$ where $\varepsilon^{inst}(t) = E_{exc} + 1/R(t)$. When τ is the lifetime of the inner-shell vacancy state, at some time t^* the Auger decay takes place. In the sudden approximation the Auger electron leaves the system suddenly and the remaining electrons of the doubly-charged ion relax immediately and produce the potential curve $V(E^{++}, t \geq t^*)$. This implies the Auger transition is a vertical one with an energy

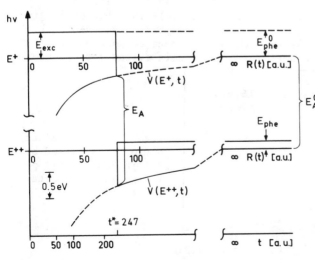

Fig. 1. Potential curve model describing PCI for photoionization (for details see text).

$E_A = V(E^+, t^*) - V(E^{++}, t^*)$, and the slow photoelectron takes its kinetic energy $\varepsilon^{inst}(t^*)$ to the new potential curve and continues there receding from the ion. As a consequence, the Auger electron gains the energy $\varepsilon(t^*)$ balanced by an energy loss for the photoelectron where

$$\varepsilon(t^*) = 1/R(t^*)$$

according to Fig. 1. In the following we will distinguish the energies without PCI from that with PCI by a zero as upper subscript i.e.

$$E_A = E_A^0 + \varepsilon \quad \text{and} \quad E_{phe} = E_{phe}^0 - \varepsilon.$$

The energy transmission between Auger and photoelectron links the ionization and the decay process into a united one.[14]

From the data given above the spectral intensity (energy distribution) $P(\varepsilon)$ of the energy gain (loss) ε for the Auger electron (photoelectron) can be obtained as follows: For t between 0 (time of production of the inner-shell hole) and t^* (time of Auger decay) the velocity v of the photoelectron is known for each value of R. Therefore, one can evaluate

$$t^* = \int_o^{t^*} dt = \int_{R(0)}^{R(t^*)} (1/v(R))\, dR$$

the solution of which gives t^*. In the resulting expression of t^* one then replaces $R(t^*)$ by $1/\varepsilon(t^*)$ according to the given situation of the two potential curves. The $P(\varepsilon)$-distribution then follows from simple manipulations of the decay law $N = N(0) \exp(-t/\tau)$:

$$dN/dt = (dN/d\varepsilon)(d\varepsilon/dR)(dR/dt) \quad \text{at} \quad t = t^*.$$

From this relation one gets

$$P(\varepsilon) = (dN/d\varepsilon) = N(0) \exp(-t^*/\tau) / (\tau \varepsilon^2 v(t^*)).$$

The most probable energy shift ε_p of the $P(\varepsilon)$-distribution is given by the condition $dP(\varepsilon)/d\varepsilon = 0$. Fig. 2 shows an example of the $P(\varepsilon)$-distribution. Clearly one sees three phenomena because of PCI. Firstly, all Auger electrons suffer an energy gain ε, the most probable

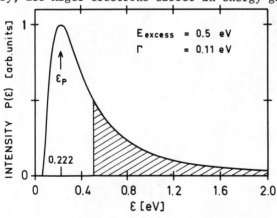

Fig. 2. Energy distribution $P(\varepsilon)$ caused by PCI in inner-shell photoionization (for details see text).

value ε_p is 222 meV in this example (for comparison, the value $\varepsilon(t^* = \tau)$ is 343 meV). Secondly, the $P(\varepsilon)$-distribution has a very asymmetric shape and is quite broad (fwhm = 400 meV which has to be compared with the width Γ of the state with the inner-shell hole, Γ = 110 meV). Thirdly, there exist energy gains ε larger than the excess energy E_{exc}; this region is marked by the shaded area. For these cases, the photoelectron will not appear in the continuum since $E_{exc} - \varepsilon < 0$, rather it is "shaken down"[19] to a bound Rydberg orbital. This process gives rise to singly-charged ions and not doubly-charged ions as expected after the Auger transition. The shake-down probability can be calculated by taking the ratio of the shaded area with respect to the total $P(\varepsilon)$-distribution. For the example of Fig. 2 it gives $P_{shake\ down}$ = 43 %.

So far we have discussed PCI only for the case of inner-shell photoionization and subsequent Auger decay. However, the model can be applied to the case of electron-impact ionization with subsequent Auger decay where one has to deal with two slow electrons. These electrons will share the excess energy E_{exc} individually having then the energies E' and E'' with E' + E'' = E_{exc}. The existence of two slow electrons requires two extensions: Firstly, the process now moves on a multi-dimensional potential curve which depends on the radial distances $R_1(t)$, $R_2(t)$ and on the radial correlation $\theta(R_{12}, t)$ between both slow electrons and, secondly, the probability distribution $d\sigma/dE'$ (and $d\sigma/dE''$) of the energy sharing between both slow electrons has to be taken into account. Since these additional requirements are not known in general, only limiting cases can be considered until now. The one case is the limit of uncorrelated motion of both slow electrons (E_{exc} large), the other case is the threshold value (E_{exc} = 0). The threshold value might be extended into the region $E_{exc} \approx 0$ by the application of the threshold laws of the Wannier theory.[10] From experimental[21] and theoretical[22] investigations it has been learned that screening effects lead to an unexpected large energy range over which the threshold theory holds.

For large values of E_{exc} a two-dimensional potential diagram can be plotted (Fig. 3), and the process follows a specific line (open circles) on the upper potential diagram until the Auger decay takes place. The transition to the lower potential curve yields the energy shift $\varepsilon(t^*) = 1/R_1(t^*) + 1/R_2(t^*)$. The radial distances $R_i(t^*)$ have to be calculated accordingly to the paths on the potential curve which is determined by the energy sharing to E' and E''. This sharing of the excess energy between both slow electrons is rather unsymmetrical (see Fig. 4, and compare Ref. 23). As example, Fig. 3 shows the path for E_{exc} = 250 eV with E' = 240 eV and E'' = 10 eV, and for t^* = 100 a.u. From the expression for $\varepsilon(t^*)$ one already sees that larger shifts ε can be expected as compared to photon-impact because here one has the interaction of the fast Auger electron with each one of the two slow electrons. For the example of Fig. 3 one has $1/R_1(t^*)$ = 65 meV and $1/R_2(t^*)$ = 316 meV, i.e., $\varepsilon(t^*)$ = 381 meV.

The other case for PCI effects with two slow electrons and one fast Auger electron occurs at threshold. Here both slow electrons move extremely correlated accordingly to Wannier's threshold laws.[10]

548

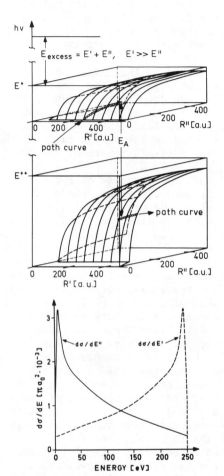

Fig. 3. Potential curve model describing PCI for electron-impact ionization. $\cos(\theta_{12}, t)$ has been set equal to 0.5 (this quantity appears in both potential curves; therefore it has no influence on the energy shift ε). For details see text.

Fig. 4. Differential cross sections $d\sigma/dE'$ and $d\sigma/dE''$ (from Ref. 23) for the occurence of an electron with energy E' or E'', respectively, whereby $E' + E'' = E_{exc} = 250$ eV; the energy of the primary electron beam is 500 eV in this example (Ar-2p-ionization).

Just at threshold ($E_{exc} = 0$) the calculation for the PCI effect can be done straightforwardly. Especially, for the most probable energy shift $\varepsilon_p^{e^-}$ one gets $\varepsilon_p^{e^-}(E_{exc}=0) = (8\sqrt{3}/(15\tau))^{2/3}$, Ref. 20, 25, 26. Again, it can be seen that this value is larger than for photon-impact. One has $\varepsilon_p^{e^-}(E_{exc} = 0) / \varepsilon_p^{hv}(E_{exc} = 0) = 2.20$.

PCI AS NATURAL LINK BETWEEN IONIZATION AND EXCITATION

The discussion above has shown how PCI produces an energy gain for the fast Auger electron compensated by an energy loss for the slow electron(s) when the threshold for the inner-shell ionization is reached. For simplicity, we will consider further the case of photon-impact only. Fig. 5 shows how the spectrum of ejected electrons in the region of the Auger peak changes when the photon energy is lowered towards, across and below the threshold for inner-shell ionization (compare Ref. 13, 27, 28). At high photon energy ($E_{exc} \gg 0$) one finds the normal Auger peak (Lorentzian distribution with fwhm = Γ = 110 meV). It gets unsymmetrical and broader towards threshold ($E_{exc} > 0$). Close at threshold the energy loss for the slow photoelectron can exceed the value of E_{exc}. In this case the slow photoelectron is captured in a bound orbital (shake down; shaded area in Fig. 5). This effect then causes the energy distribution $P(\varepsilon)$ of the Auger peak to get a discrete structure at higher kinetic energy. These discrete lines can correspond to final states produced by a different mechanism, namely outer-shell ionization and simultaneous excitation. Similarly, the smooth contribution of the Auger peak then overlaps with outer-shell double-ionization processes. If both processes, inner-shell ionization with PCI and outer-shell double-photoprocesses, yield the same final states, they are undistinguishable, and interference effects are possible.

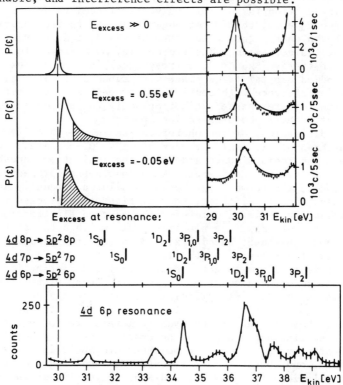

Fig. 5. Compilation for the spectrum of ejected electrons in the region of the xenon $N_5-O_{23}O_{23}$ $^1S_0$ Auger peak following photon impact. The corresponding experimental spectra are from Ref. 27, 28. The symbol $\underline{4d}6p \rightarrow \underline{5p}^2 6p$ stands for: the resonance $4d^9(^2D_{5/2})5s^25p^66p$ $^1P_1$ decays into the final ionic state $4d^{10}5s^25p^4(^{2S+1}L_J)6p$ $^1P_1$; and the corresponding positions for the kinetic energies are marked in the figure (compare Ref. 27, 28). It should be noted that in this energy region, there exist also other decay channels, for example to $4d^{10}5s^25p^4(^{2S+1}L_J)7p$ $^1P_1$ which are seen clearly in the experimental spectrum, but have been omitted for simplicity in the theoretical bar diagram.

The PCI phenomena discussed so far within the classical picture can be extended even to the situation just below threshold i.e. $E_{exc} \leq 0$.[9] Here the excitation process with PCI taken into account can be described by the movement of the slow "ejected" electron that moves towards its boundary at the ion-potential curve $V(E^+,t)$ or $V(E^{++},t)$ after the Auger decay, respectively. The energy gain for the Auger electron is subjected to two restrictions in this case. The "ejected" electron can move out only as far as to the classical turning point; from $\varepsilon^{inst} = 1/R(t) - |E_{exc}| \geq 0$ together with $\varepsilon = 1/R(t^*)$ follows the condition $\varepsilon \geq |E_{exc}|$. The boundary at the potential curve $V(E^{++},t)$ requires standing wave solutions (which correspond to truly bound Rydberg states); this gives discrete values ε (i.e. the interaction is possible only at selected values of time t^*). As a consequence, the former Auger peak starts disappearing when the photon energy goes below the threshold, and other discrete lines appear. This is also indicated in Fig. 5. At specific values of the photon energy ($E_{exc} < 0$, at resonance) inner-shell excitation to a truly bound orbital takes place with subsequent autoionizing decay (and interference with double-photoprocesses in the outer-shell). Below this resonance excitation, only the photoelectron spectrum from outer-shell photoionization appears.

In the discussion for the photoprocesses close to an inner-shell ionization threshold it could be seen how the PCI-phenomena are the link between the regions above and below the ionization threshold: Despite effects due to interference with double-photoprocesses in the outer-shell, the former Auger peak is diluted gradually and replaced by electrons at different energy. First this process results in a smooth spectral distribution for the ejected electrons; however, it ends up with very distinct electron peaks which occur at resonance excitation and subsequent autoionizing decay of truly bound inner-shell photoexcited states, and the latter electron peaks vanish very abruptly. It is possible to get a criterion for a distinction whether one is in the smooth region or not. The Auger decay with lifetime τ imposes (via the uncertainty principle) a limit on the possible energy resolution of $\Delta E \geq 1/\tau = \Gamma$ for the Rydberg states produced by the incident photon. Thus if the Rydberg levels are to be observed, the Rydberg spacing $\Delta E_{Rydberg}$ must be grater than ΔE. Otherwise, when the photon energy is just below threshold and this criterion is not satisfied, one observes just a continuum resulting from the unresolved population of highly excited Rydberg levels. In the case of 4d - excitation in xenon one has $\tau = 247$ a.u. corresponding to $\Gamma = 0.110$ eV. This value is small in comparison to $\Delta E_{Rydberg}$ (6p, 7p) = 1.26 eV, $\Delta E_{Rydberg}$ (7p,8p) = 0.47 eV and $\Delta E_{Rydberg}$ (8p,9p) = 0.24 eV. The above criterion explains that excitations to these orbitals are excitations to truly bound orbitals, and one can see them very clearly in the photoabsorption spectrum.[30]

PCI AS ORIGIN OF ENERGY CHANGES

As has been illustrated in Fig. 2, PCI results in an asymmetric energy gain in the observed Auger spectra (energy loss in the observed photoelectron spectra, compare Ref.31). Fig. 6 shows two

examples, photoionization in the 4d-shell of xenon and subsequent
Auger decay and electron-impact ionization in the 2p-shell of argon
with subsequent Auger decay. With respect to other experimental investigations of this kind (Ref. 32 - 36 for photon impact, Ref. 7,
20, 36-41 for electron impact) these selected examples give in addition to the experimental points a fit curve which is based on the
PCI energy distribution $P(\varepsilon)$. This allows more detailed information
than the usual determination of the most probable energy shift ε_p
alone. For details about the fitting procedure the reader is referred to the original literature because of individual contributions
from other processes than the one under consideration (in the case
of photon impact[27]) and because of the adoption of a $P(\varepsilon)$-distribu-
tion according to the model of Niehaus[9] with ε_p set equal to the
experimental value (in the case of electron impact[20]). Nevertheless,
the fit curve based on the PCI-model reproduces well the main
features of the experimental spectra. From such spectra the most
probable energy gain ε_p for the Auger electrons can be extracted.
For example, the gain ε_p for each of the lowest spectra in Fig. 6 is
marked below this figure.

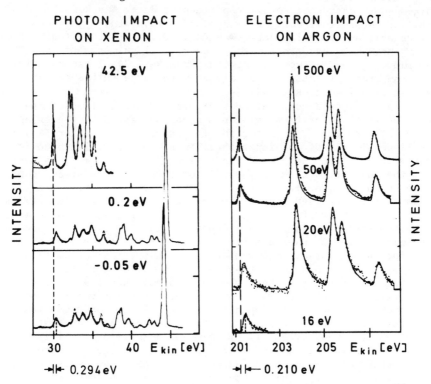

Fig. 6.   PCI - effect in the $N_{45}-O_{23}O_{23}$ Auger spectra of xenon[27]
and $L_{23}-M_{23}M_{23}$ of argon[20] following inner-shell ionization.
The energy parameter of the individual spectra is the
excess energy E_{exc}. The solid line is a fit to the experimental data based on the PCI model.

552

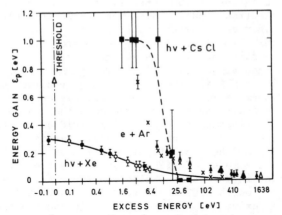

Fig. 7. Energy gain ε_p as function of the excess energy for three selected examples. The energy axis is quadratic in E_{exc} in order to cover the region from around the threshold up to high values. Xenon (N_{45}-$O_{23}O_{23}$ Auger electrons): experimental data: ● = Ref. 27, ○ = Ref. 35; theoretical curve — = Ref. 27. Argon (L_{23}-$M_{23}M_{23}$ Auger electrons): experimental data: ▲ = Ref. 20, X = Ref. 40 (only data for L_2-$M_{23}M_{23}{}^1D_2$ have been used); theoretical values: △ = Ref. 20. Caesium (N_{45}-$O_{23}O_{23}$ Auger electrons) from Ref. 34. These examples have comparable lifetimes. For a case with large shifts because of smaller lifetime see Ref. 49.

Fig. 7 gives a compilation of ε_p-values as function of the excess energy E_{exc} for these elements xenon and argon (for caesium see later). The relevant parameter for the energy distribution $P(\varepsilon)$ is the level width Γ of the inner-shell hole. Because this value is nearly the same for xenon and argon (110 meV and 130 meV, respectively), the most probable energy shift for both elements produced by different ionization mechanisms can be compared directly. Fig. 7 shows that for a given excess energy the energy shift ε_p is always larger by a factor of the order of 2 for electron impact as compared to photon impact. This is consequence of two slow electrons seeing the doubling in the nuclear attraction after the Auger decay. The data for photon impact follow nicely the theoretical prediction (full line) and cross the threshold smoothly. For the electron-impact data it is not yet clear whether and how they reach the threshold value predicted so far only by theory (open triangle). At high excess energy, the theory (open triangles) yields good agreement with the experimental values. From Fig. 7 it can be seen that the energy shift caused by PCI is by no means a small or negligible quantity: For the example of electron impact on argon the predicted energy shift at threshold is 732 meV, and even at an excess energy of 1750 eV the energy shift amounts 25 meV. For photoionization, generally, the effect is smaller. However, as is illustrated for the case of photoionization in caesium, there can exist special conditions that cause very strong energy shifts due to PCI. In caesium the reason is the wave-function collapse of the continuum wave function (f-channel) of the escaping photoelectron (compare Ref. 34 and further references therein). The collapse of the wave function means that the continuum wave function gets a quasibound or quasiresonant behaviour. This quasiresonant behaviour of the slow photoelectron shows up also in a delay time $\tau_{quasibound}$ of the slow photoelectron in the region of the ion. This changes the situation for the potential curve diagram considerably as can be seen in Fig. 8. The Auger decay takes place when the photoelectron is much closer to the core as compared to the uncollapsed case of the photoelectron wave function (dotted line in Fig. 8). As a consequence, a huge energy shift ϵ occurs. In the example shown in Fig. 8 the energy gain ϵ is 1.09 eV as compared to 0.34 eV in the uncollapsed case (E_{exc} = 0.5 eV, t^* = τ = 250 a.u.). While for this model description it has been assumed that the wave function of the photoelectron close at threshold is a collapsed one, there exists also the possibility that the wave function collapse occurs as a result of the Auger decay.[43] As a consequence an enlarged probability exists for shake down to a truly bound nf-orbital which produces also huge energy shifts ϵ.

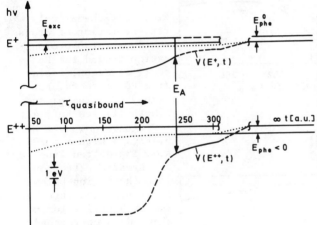

Fig. 8. Schematical potential curve model (solid line) describing PCI for photoionization with a collapsed (quasibound) wave function for the slow photoelectron. Hypothetical uncollapsed state potentials are indicated by dotted lines (compare Fig. 1).

PCI AS SOURCE FOR PECULIARITIES IN THE CROSS SECTIONS

The discussion of the PCI-phenomena as a link between ionization processes above and below the threshold for inner-shell ionization has shown that just around threshold shake-down processes play a dominant role and represent a different mechanism for processes of ionization and simultaneous excitation in the outer-shell. It is obvious that this additional mechanism can produce quite strong peculiarities in the outer-shell cross section.[14] When interference effects are neglected this can be described most easily: The cross section σ^{+*} (outer shell) for ionization and excitation in the outer-shell receives via the Auger effect with PCI the contribution $P_{shake\ down}$ from the inner-shell ionization cross section σ(inner-shell) i.e.

$$\sigma^{+*}_{total}(\text{outer-shell}) = \sigma^{+*}(\text{outer-shell}) + \sigma(\text{inner-shell}) \cdot P_{shake\ down}.$$

The upper part of Fig. 9 shows in a schematic way the relevant cross sections for photoionization in the region of the $2p_j$-shell in argon together with the outer-shell cross section σ(M-shell) which is due mainly to single-ionization processes only. In the middle part of Fig. 9 the quantity $\sigma(2p_j) \cdot P_{shake\ down}$ is plotted; the dotted line gives a schematic extension into the region below the ionization threshold. The energy positions for outer-shell excited states are given in the figure, too. For energies even smaller than discussed so far, the process of inner-shell excitation to truly excited states with subsequent autoionizing decay becomes particularly important for also feeding the individual final states $2p^5 3s^2 3p^6 n\ell$.

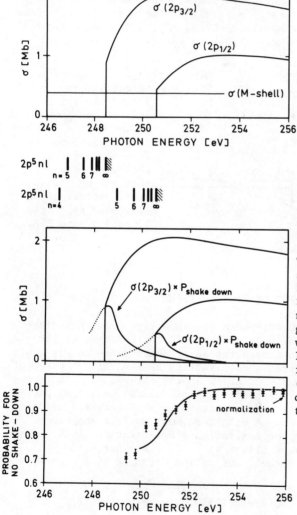

Fig. 9. Schematical representation of cross sections for photoionization around the $L_{2,3}$-shell of argon, including PCI effects. For $\sigma(2p_{1/2})$, $\sigma(2p_{3/2})$ and σ(M-shell) see Ref. 42; for $P_{shake\ down}$ see Ref. 9, Fig. 4 with $\tau = 200$ a.u. The dotted line is a schematic extension into the region below the ionization threshold. The probability for no shake-down corresponds i) to $Ar^{2+}/(Ar^{1+} + Ar^{2+})$ in the simple model neglecting higher charge state contributions which gives the solid line or, respectively, ii) to $(Ar^{2+} + Ar^{3+})\ /$ $(Ar^{1+} + Ar^{2+} + Ar^{3+})$ as determined experimentally.[6]

This mechanism brings the resonance phenomena into the outer-shell cross section, expecially into σ^{+*}(outer-shell) when the inner-shell truly excited states are reached by the photon energy.

From Fig. 9 it can be seen clearly that the outer-shell ionization cross section σ(M-shell) of argon is influenced strongly due to PCI because both contributions, $P_{shake\ down} \cdot \sigma(2p_j)$ and σ(M-shell), are of equal order of magnitude. This is even more dramatic when one considers that σ(M-shell) contains mostly single-ionization but fewer ionization and excitation processes while the shake-down processes contribute directly to σ^{+*}(M-shell). Although these individual cross sections show special behaviour in the region of inner-shell photoprocesses, the detailed analysis given above shows that this behaviour is not really peculiar any longer when the logical decomposition of the total cross section into partial cross sections of individual processes is considered.

The influence of PCI on the outer-shell photoionization cross section σ(M-shell) in argon has been investigated experimentally by Van der Wiel et al. 1976.[6] The idea of this investigation is that the above considerations must reflect themselves in the observed intensities for different charge states. Without PCI, 2p-ionizations in argon yield Ar^{2+}-ions because of the subsequent Auger decay, and M-shell-ionizations yield Ar^{1+}-ions (just for simplicity we will neglect here higher charge-state contributions). With PCI the relative ion-intensities will change accordingly to

$$\frac{Ar^{2+}}{Ar^{1+}+Ar^{2+}} = \frac{(1-P_{shake\ down}) \cdot \sigma(2p_{1/2}) + (1-P'_{shake\ down}) \cdot \sigma(2p_{3/2})}{\sigma(M\text{-shell}) + \sigma(2p_{1/2}) + \sigma(2p_{3/2})}$$

The lower part of Fig. 9 shows this quantity together with the experimental data. The reduction of Ar^{2+}-ions which is due to the shake-down process can be seen clearly. A quantitative comparison with theoretical data is complicated because of the Ar^{3+}-contributions, because the effects from $2p_{3/2}$-ionization processes extend into the region of $2p_{1/2}$-ionization, and because the charge-ratio depends also on the true shape of the individual cross sections involved. Therefore, the agreement between theoretical and experimental data is only qualitatively (see Ref. 11; in Ref. 9 the good agreement is somewhat accidental because of the neglect of $2p_{3/2}$-contributions and of the adaption of the lifetime τ).

CONCLUSION

In the foregoing discussion PCI-phenomena in inner-shell ionization processes have been considered that are based on the energy exchange between one fast and at least one slow electron. It has been shown that this energy exchange is the consequence of a relaxation process imposed onto the slow electron(s). Based on this relaxation mechanism, the simple classical model for PCI allowed a nearly quantitative interpretation of PCI-phenomena observed so far in inner-shell ionization processes. This simple classical model for PCI facilitated the discussion of PCI-phenomena considerably, and it provided for an analytical expression for the energy distribution $P(\varepsilon)$. However, there are limitations for such a simple description of PCI which shall be discussed now.

Firstly, for $E_{exc} \to \infty$ the energy distribution $P(\varepsilon)$ as calculated for the most clear case of photon-impact does not converge to the required Lorentzian line shape. This defect can be removed (for example Ref. 15), but no analytical expression for the distribution $P(\varepsilon)$ has been given. The latter would be rather helpful for experimentalists.

Secondly, as was pointed out several times in the foregoing discussion, outer-shell ionization and excitation processes can be coupled directly to the PCI-phenomena in inner-shell processes. Therefore, these interference effects should be taken into account i.e. the energy distribution $P(\varepsilon)$ should be calculated accordingly to Ref. 9, 15

$$P(\varepsilon) = \left| A + C(\varepsilon)e^{-i\alpha(\varepsilon)} \right|^2$$

with A = amplitude for direct outer-shell ionization and excitation process that competes with the results of inner-shell processes including PCI, $C(\varepsilon)$ and $\alpha(\varepsilon)$ are the amplitude and phase, respectively, of these inner-shell processes. Especially, in the absence of interference, one has $P(\varepsilon) = |C(\varepsilon)|^2$ which is equal to the energy distribution $P(\varepsilon)$ of the discussion above. The interference effects can produce a remarkable oscillatory structure in the $P(\varepsilon)$-distribution that should be visible in the spectra when the energy spread in the incoming photon- or electron beam is much smaller than the width of the states with the inner-shell hole.[15] In addition to this kind of interference, Niehaus[9] has discussed also the case of interference between distinct states with the inner-shell hole that become undistinguishable because of PCI (see Fig. 7 of Ref. 9).

Finally, in addition to the energy exchanges between the fast and the slow electron(s), a change of angular momentum might be possible, too. For such an exchange two cases should be considered. One is based on the exchange between the slow electron and the final states of the ion, the exchange being caused by the in general non-spherical symmetry of the final ionic state.[15] The other case is based on the angular momentum exchange between the escaping electrons themselves. It is possible to give estimates when such exchanges of angular momenta will be important. In the first case, the slow electron has to realize during the relaxation process the difference between a spherically symmetric or unsymmetric ionic field i.e. it must be localised near to the atom (quasi-resonance excitation or extremely short lifetime of the inner-shell hole). In the second case the interaction time between the escaping electrons and the torque exerted on the electrons give the relevant criterion. As upper bound for the angular momentum exchange Δl between two interacting electrons one can estimate[44, 45] $\Delta l \leq 1/v_f$ where v_f is the velocity of the faster electron. This criterion yields different results for photon- and electron-impact, respectively, when appropriate numbers are inserted. When one fast and one slow electron interact (photon-impact) then the most probable value is $\Delta l = 0$. However,

when two slow electrons interact (electron-impact) they bring into
the PCI-phenomena the other class of long-range Coulomb-interactions,
namely the highly correlated motion of two slow electrons[10, 44, 46].
Especially with respect to the angular momentum exchange $\Delta\ell$, here
both slow electrons may acquire significant orbital momenta. Such an
effect has been observed in electron-impact excitation[47] (probably
also Ref. 48) when at threshold the motions of the scattered and
excited electron become highly correlated.

It is a pleasure to thank A. Starace and W. Mehlhorn for
valuable comments on the manuscript.

REFERENCES

1. H. W. Berry, Phys. Rev. 121, 1714 (1961).
2. R. B. Barker and H. W. Berry, Phys. Rev. 151, 14 (1966).
3. F. H. Read, in: Proceedings IX. ICPEAC, Seattle 1975, p. 176.
4. A. Niehaus, in: Proceedings X. ICPEAC, Paris, 1977, p. 185.
5. H. G. M. Heideman, in: Coherence and Correlation in Atomic Collisions, ed. by M. Kleinpoppen and J. F. Williams (New York,
   Plenum, 1980), p. 493.
6. M. J. Van der Wiel, G. R. Wight and R. R. Tol, J. Phys. B 9,
   L 5 (1976).
7. S. Ohtani, H. Nishimura, H. Suzuki and K. Wakiya, Phys. Rev.
   Lett. 36, 863 (1976).
8. V. Schmidt, N. Sandner, W. Mehlhorn, M. Y. Adam and F. Wuilleumier, Phys. Rev. Lett. 38, 63 (1977).
9. A. Niehaus, J. Phys. B 10, 1845 (1977).
10. G. H. Wannier, Phys. Rev. 90, 817 (1953).
11. M. Ya. Amusia, M. Yu. Kuchiev, S. A. Sheinerman and S. I. Shef-
    tel, J. Phys. B 10, L535 (1977).
12. G. Wendin, in: Photoionization of Atoms and Molecules, Pro-
    ceedings of the Daresbury Meeting, ed. by B. D. Buckley, Report
    No. DL/SCI/R 11 (1978), p. 1.
13. G. Wendin, in: VI. VUV Radiat. Phys., Charlottesville, 1980,
    Extended Abstracts II-87.
14. M. Ya. Amusia, Appl. Optics 19, 4042 (1980).
15. M. Ya. Amusia, M. Yu. Kuchiev and S. A. Sheinerman, in: Cohe-
    rence and Correlation in Atomic Collisions, ed. by H. Klein-
    poppen and J. F. Williams (Plenum, New York, 1980), p. 297.
16. T. Åberg, Phys. Scr. 21, 495 (1980).
17. T. Åberg, in: Proceedings of X. Internat. Conf. on X-Ray Pro-
    cesses and Inner Shell Ionization, Stirling 1980, p. 251.
18. W. van de Water, H. G. M. Heideman and G. Nienhuis, J. Phys.
    B 14, 2935 (1981).
19. G. C. King, H. F. Read and R. C. Bradford, J. Phys. B 8, 2210
    (1975).
20. R. Huster and W. Mehlhorn, Z. f. Physik, A 307, 67 (1982).
21. H. P. Schmitt, Diplom - Thesis, Univ. Würzburg, Germany (1978).
22. H. Klar, J. Phys. B 14, 3255 (1981).
23. F. Pichou, A. Huetz, G. Joyez and M. Landau, J. Phys. B 11,
    3683 (1978).

24. S. Manson, private communication (1980).

25. J. Mizuno, T. Ishihara and T. Watanabe, in: Abstracts of Papers, XII. ICPEAC, Gatlinburg 1981, p. 253 (misprint in the enhancement factor, see Ref. 26).

26. T. Watanabe, T. Ishihara and J. Mizuno, private communication (1982).

27. V. Schmidt, S. Krummacher, F. Wuilleumier and P. Dhez, Phys. Rev. 24, 1803 (1981).

28. V. Schmidt, Appl. Opt. 19, 4080 (1980).

29. J. E. Hansen and W. Persson, Phys. Rev. A 20, 364 (1979).

30. D. L. Ederer and M. Manalis, J. Opt. Soc. Am. 65, 634 (1975).

31. G. R. Wight and M. J. Van der Wiel, J. Phys. B 10, 601 (1977).

32. M. K. Bahl, R. L. Watson and K. J. Irgollic, Phys. Rev. Lett. 42, 165 (1979).

33. H. Hanashiro, Y. Suzuki, T. Sasaki, A. Mikuni, T. Takayanagi, K. Wakiya, H. Suzuki, A. Danjo, T. Hino and S. Ohtani, J. Phys. B 12, L 775 (1979).

34. T. C. Chiang, D. E. Eastman, F. J. Himpsel, G. Kaindl and M. Aono, Phys. Rev. Lett. 45, 1846 (1980).

35. S. H. Southworth, Ph.D. thesis, Lawrence Berkeley Lab. (1982).

36. W. Hink, H. P. Schmitt and T. Ebding, J. Phys. B 12, L 257 (1979).

37. W. Hink, L. Kees, H. P. Schmitt and A. Wolf, in: Proceedings of X. Internat. Conf. on X-Ray Processes and Inner-Shell Ionization, Stirling 1980, p. 327.

38. H. Suzuki, M. Muto, T. Takayanagi and K. Wakiya, in: Abstracts of XI. ICPEAC, Kyoto 1979, p. 258.

39. G. N. Ogurtsov, V. M. Mikoushkin and I. P. Flaks, in: Abstracts of I. Europ. Conf. Atom. Phys., Heidelberg 1981, p. 759.

40. K. Wakiya, M. Suzuki, T. Takayanagi, M. Muto, S. Ito, Y. Iketaki and S. Ohtani, in: Abstracts of XII. ICPEAC, Gatlinburg 1981, p. 247.

41. A. Yagishita, H. Hanashiro, S. Ohtani and M. Suzuki, J. Phys. B 14, L 777 (1981).

42. M. Ya. Amusia, in: Atomic Physics, vol. 5, ed. R. Marrus, M. Prior and H. Shugart, Plenum (New York, 1977), p. 537.

43. A. F. Starace, private communication (1982).

44. U. Fano, J. Phys. B 7, L 401 (1974).

45. H. G. M. Heideman, T. van Ittersum and G. Nienhuis, J. Phys. B 8, L 26 (1975).

46. W. van de Water, F. B. Kets, L. G. J. Boesten and H. G. M. Heideman, J. Phys. B 11, L 465 (1978).

47. H. G. M. Heideman, W. van de Water and L. J. M. van Moergestel, J. Phys. B 13, 2801 (1980).

48. R. D. DuBois, L. Mortensen and M. Rødbro, J. Phys. B 14, 1613 (1981).

49. G. S. Brown, M. H. Chen, B. Crasemann and G. E. Ice, Phys. Rev. Lett. 45, 1937 (1980).

MANY-ELECTRON EFFECTS IN PHOTOELECTRON CORE SPECTRA

I. Lindau
Stanford Electronics Laboratories, Stanford University
Stanford, California 94305, U.S.A.

ABSTRACT

Many electron effects in photoelectron core spectra are now well-established phenomena. This paper will be centered around experiments where the tunability of synchrotron radiation has been utilized to study: 1. Photoelectron emission enhancement of the 5s and 5p partial photoionization crossections at and above the photoionization threshold for the 4d subshell of Ba. These many-electron effects can be studied as the photon energy is continuously tuned through the ionization thresholds of inner core levels; 2. The plasmon loss intensity as a function of the excitation energy of the photons for a few free-electron like elements (Al, Si).

A strong enhancement is observed in the 5s and 5p partial crossections at and above the 4d ionization threshold in Ba. These resonances cannot be described in the framework of a one-electron model but are strong many-body effects. However, the more advanced many-body calculations can reproduce the 4d excitation as well as the enhancement in the outer 5s and 5p shells quite accurately.

The intensity of the bulk plasmon loss structure was studied over a broad photon energy range from core levels of Si and Al. A clear trend is observed when the excitation energy is tuned from the ionization threshold and up. The plasmon intensity increases rapidly with increasing kinetic energy up to about 100 eV where it levels off to a slow but steady increase. A model based on extrinsic plasmon losses cannot describe the intensity variations close to threshold satisfactorily. A discrepancy is also observed for the onset energy of the plasmon creation - the experimental data fall 10-20 eV higher than the theoretical ones. The importance of a model which takes into account the transition from the adiabatic to the sudden approximation will be discussed.

I. INTRODUCTION

Many-electron effects in photoelectron core spectra in the broadest sense have been the subject of extensive theoretical and experimental studies over the years. The proceedings from this conference and the previous one[1] give a good survey of the most relevant topics in the field. In this paper we will concentrate on two aspects: 1) many-electron effects in the 4d, 5s, and 5p partial photoionization crossections of solid Ba; 2) plasmon losses from core lines in free-electron like metals.

The discussion of the many-electron phenomena in partial crosssections will center around the effects observed in more shallow core levels when a deeper lying core level is photoionized. With the advent of synchrotron radiation as an excitation source, a large number of phenomena in core levels, satellite structures, etc., have now been studied for various systems (gas[2], metal vapors[3], and solids[4]).

We will focus our attention on solids and the particular experimental and theoretical aspects with these systems. The plasmon work is also very heavily dependent on the utilization of synchrotron radiation, since we are particularly interested in studying how the plasmon loss intensity depends on the photon excitation energy[5,6]. These studies will address the transition from the adiabatic to the sudden approximation and the intrinsic versus extrinsic nature of collective excitations in solids.

II. MANY-ELECTRON EFFECTS IN THE 4d, 5s, AND 5p SUBSHELLS OF Ba

Absorbtion spectra of solids and gases are generally similar except for a smearing out of fine structure details in the solid. This similarity is not surprising for deep core levels or narrow bands when when electrons are excited far above the Fermi level. Ba provides an interesting test of the limits of applicability of this atomiclike excitation model, since details of the 4d excitation are critically dependent on unoccupied levels just above the Fermi level. Photoemission is a particularly well-suited technique for these kind of studies of the excitation process, since the contributions from different decay channels can be separated.

4d excitations in Xe-like atoms are of particular theoretical interest because under certain circumstances there can be a large overlap of the 4d electron with the final-state f-type wave function as a result of many-electron effects. The importance of the Ba system as a test of many-body theory is underscored by the fact that the 4d cross section alone has been the subject of at least eighteen different calculations as reported by five sets of authors in recent papers[8-13] (Fig. 1). Only the most sophisticated of these calculations have been able to adequately describe previously measured absorption spectra.[13,14]

We present here measurements of the angle-integrated 5p, 5s, and 4d subshell photoionization cross sections for polycrystalline Ba around and above the 4d threshold.[5] The energy of soft x-rays absorbed in a solid or gas is converted principally into photoelectron energy. The kinetic energy of outgoing photoelectrons provides information about the initial and final states of the atomic excitation responsible for

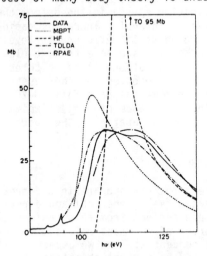

Fig. 1. Total Ba 4d cross section compared to several calculations by various authors: generalized random phase approximation with exchange (RPAE) (Ref. 8), many-body pertubation theory (Ref. 10), time-dependent local density approximation (Ref. 11), and Hartree-Fock (Ref. 12).

the absorption. The experiment described here employs an electron energy analyzer in conjunction with a tunable soft x-ray source, so that we are thus able to determine the partition of the absorption into several types of excitations. This is an important advantage over experiments where only the total cross section is measured (absorption).

The measurements were all performed on the 4-deg beam line of the Stanford Synchrotron Radiation Laboratory (SSRL).[15] The samples were prepared by in situ evaporation in an ultrahigh vacuum environment. The monochromator transmission was determined by using a calibrated photodiode, by measuring the photoyield for Au,[16] and by using a numerical method to remove contributions from nonmonochromatic light. The cylindrical mirror analyzer was run in a constant retarding ratio ($\Delta E/E$=const) mode to take advantage of the resulting constant transmission function.[17] The sample normal was oriented at less than 30 deg from the light beam to avoid reflection and refraction from the sample surface. The area under each photoemission peak of interest was then measured as a function of photon energy.[18] No correction was made for direct two-electron continuum emission under the assumption that high ionization states are more easily reached by photoemission and subsequent Auger decay. Satellite structure, which is a small fraction of the total emission, was included in the photoemission peak when possible. It should be noted that neither of these effects would reduce the rate of Auger decay of the 4d hole.

The Auger electron yield, which should be proportional to the rate of core hole creation (and thus, in most cases, the photoelectron yield), was similarly measured. The total absorption cross section was measured using the partial-yield technique, i.e., by measuring the yield of low-energy electrons as a function of photon energy. Both Auger and partial-yield measurements were corrected for contributions from harmonics of the fundamental photon energy.

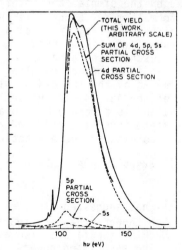

Fig. 2. Total yield, partial cross sections, and sum of partial cross sections of Ba (this work).

The results are extremely sensitive to corrections for electron-kinetic-energy-dependent effects, i.e., escape depth,[19] surface refraction (which narrows the escape cone), and surface reflection of electrons.[20] Justification of the corrections applied will be presented in a future publication.

Evidence for the localized nature of the 4d excitation is provided by Rabe's absorption measurements on Ba vapor and films, which are similar in all but the details of the fine structure, which are smeared for the solid.[13] Our photoelectron-yield data (Fig. 2) were indistinguishable from Rabe's thin-film absorption data except for a sloping background which we did not observe. The photoemission results are displayed in Fig. 3. The absolute cross sections were assigned by fitting the tail of the 4d cross section with the theoretical curve. The error bars refer to statistical erors only. The 4d partial cross section (Fig. 3) has been determined both by a measurement of the direct photoemission line and the associated Auger decay lines.

The broad, delayed onset of the 4d emission can be described as the result of a centrifugal barrier which can confine the f-type wave functions and thus produce a large overlap with the 4d hole.[21].

Somewhere between Xe and La, depending upon the details of the atomic configuration, the final-state wave function collapses into the inner potential well. Connerade notes that this collapse is extremely sensitive to excited atomic configurations,[21] a fact that should be considered in interpreting the solid-state results.

The final state of the 4d excitation is a continuum wave function $\overline{4, \varepsilon f}$. (There has been some confusion about the identification of this level, which has been resolved by the finding that all the nf wave functions lose a node upon collapse.[21,22]

The enhancement of the outer shells at the 4d threshold is usually described[9] as an autoionization phenomenon resulting from a process such as

$$4d^{10}5s^25p^6 \rightarrow 4d^95s^25p^6\overline{4,\ \varepsilon f} \rightarrow \begin{cases} 4d^{10}5s^25p^5 + e^- \\ 4d^{10}5s^15p^6 + e^- \end{cases}$$

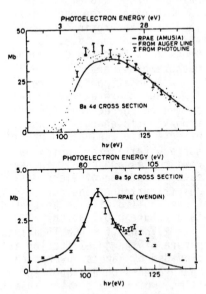

interfering with the direct outer-shell photoemission. This is in analogy to the decay of a discrete Rydberg state, in which case the resulting profile would be described by the simple Fano interference formalism.[23] Since this intermediate state is, in fact, a continuum level, it is more rigorous to describe the enhancement as purely an interchannel coupling effect caused by the induced dipole field at the $4d \rightarrow 4, \varepsilon f$ threshold models, and is probably due to a final state involving a 4d and a 5p hole, according to the excitation

$$4d^{10}5s^25p^6 \rightarrow 4d^95s^25p^5nln'l'.$$

Wendin[23] has suggested $nln'l'$ = (5d5d), although other configurations such as (6p4f) are also allowed. The associated outer-shell enhancements are further evidence of the discrete nature of this intermediate state.

Fig. 3. 4d and 5p partial cross sections of Ba compared to RPAE and GRPAE calculaitons (Refs. 8 and 22). The 4d cross section is normalized to the calculation at 125 eV.

These results are different from similar measurements for vapor
-phase xenon by West et al.[2] and Adam et al.[3] in two noteworthy ways
The Xe measurements showed no such structure from two-electron excitation. In addition, the peaks of the 4d, 5p, and 5s cross section were
quite a bit closer in energy than in the present measurements. Thus
it appears that the coupling in Ba is significantly more complex than
in Xe, perhaps as a result of the collapse of the 4f level in Ba.

It can be seen in Fig. 2 that the sum of the 5p and 4d partial
cross sections closely resembles the total cross section as measured
with constant final-state spectroscopy. In addition, except for the
secondary peak at 120 eV, the shapes of the subshell cross sections
are in good agreement with the theoretical curves of Wendin.[9,24]

It is clear that photoexcitation in Ba is extremely complex, resulting in significant rearrangement of the initial-state atom and a
variety of interacting excitation and decay channels. This measurement demonstrates that advanced many-body techniques such as the
RPAE can reproduce general trends of the principal 4d excitation as
well as the outer-shell enhanced excitation. The details of multi-
electron excitation are, however, important and have not as yet been
adequately calculated.

It is, in addition, clear that the details of the excitation in
the solid are both qualitatively and quantitatively consistent with a
totally atomic description. Thus it would appear that, at least in
elemental metals, we are justified in using atomic models for excitation of levels which are corelike or localized in the initial state.
Evidence is still scarce about the extent to which formation of compounds affects the excitation, though some recent results for Ba, Ba+
and Ba++ have been reported by Lucatorto et al.[25] In summary, we
note that the study of photoemission in solids appears to hold promise for determining precisely the value of subshell cross sections
and related quantities.

III. THE EXCITATION ENERGY DEPENDENCE OF THE PLASMON IN SOLIDS

In this section we will discuss the variation in intensity of
the plasmon loss structure in solids as a function of electron kinetic
energy, in particular we want to examine the variation in loss intensity close to threshold. It has been a long standing problem to obtain more knowledge about the plasmon loss processes[26-47] that photoelectrons can suffer in the photoexcitation event itself (intrinsic
processes) or on their path out of the material (extrinsic effects).
These two mechanisms give rise to loss structure that coincide in a
photoelectron spectrum so they are not readily separated experimentally.

Using synchrotron radiation it is possible to tune the photon
energy and to follow in detail how the intensity of the plasmon loss
is changing as a function of the excitation energy. There are a number of theoretical predictions that the dynamical screening of the
photoexcited hole will cause a suppression in the plasmon intensity
close to the threshold.[35,36] This can also be expressed as a negative interference between intrinsic and extrinsic processes occuring
on the same atomic site. Plasmon loss intensity studies[6,7] may thus
offer an opportunity to follow the transition from the adiabatic to

564

to the sudden approximation, a phenomenon that has received much theretical attention but for which there is very scarce experimental data.[48,49]

The samples selected for the studies reported here were Si and Al which have experimentally determined bulk plasmon energies of, respectively, 16.7 eV and 15.7 eV.[6] These values are close to their calculated free electron plasmon energies. They have thus free electron densities which are nearly equal and should consequently exhibit very similar trends in the loss intensity with kinetic energy. We have studied the loss intensity very carefully from the ionization of the 2p core levels to several hundred eV above threshold.

The experiments were performed on the 4° line[15] of Beam Line I at the Stanford Synchrotron Radiation Laboratory. Monochromatized radiation in the photon energy range 100 to 600 eV was used for excitation. The emitted electrons were energy analyzed with a double pass cylindrical mirror analyzer. The relative energy resolution of the analyzer is 1.6%. A pass energy of 100 eV was normally selected for these experiments. The optical axis of the analyzer was at an angle 75° relative to the direction of the incident light, while the samples were aligned so that the surface normal was at an angle of about 65° ($\neq \beta$).

The silicon samples were in situ cleaved single crystals, of (111) orientation. The Al sample was a film, with a thickness varying from 10 to 1000 Å, prepared in situ by evaporation onto a GaAs substrate. Binding energies of 99 eV (Si 2p) and 73 eV (Al 2p) were used when assessing kinetic energies above the Fermi level for the analyzed electrons. A simple graphical deconvolution technique was used in extracting the intensity of the bulk plasmon loss structure from the background of inelastically scattered electrons.

Recordings of the Si 2p doublet and the first bulk plasmon loss structure (labelled P^1) are shown in Fig. 4 for four different photon energies. The parameter to be extracted from the data is primarily the ratio in intensity between the bulk loss structure and the elastic 2p photoelectron peak at the various photon energies used. In order to do that, the area of, respectively, the loss structure and the elastic peak need to be determined which means that an assumption about the proper background to be subtracted must be made. A smoothly varying background is assumed which is illustrated by the dashed lines in Fig. 4. The increased width of the peaks with photon energy is

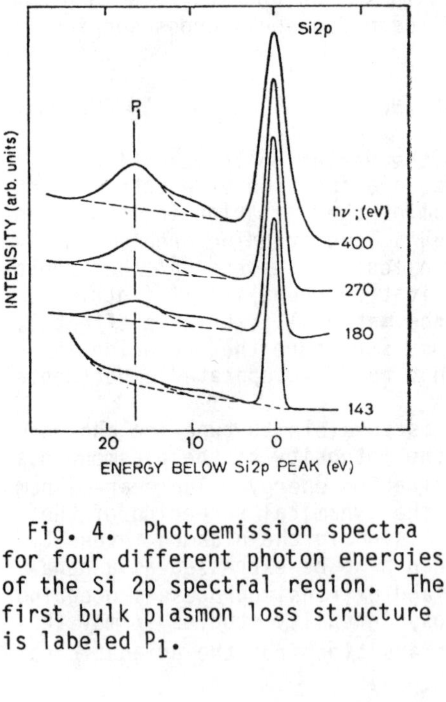

Fig. 4. Photoemission spectra for four different photon energies of the Si 2p spectral region. The first bulk plasmon loss structure is labeled P_1.

ue to the increased width of the excit tion radiation with photon energy. When taking peak area ratios, this effect is cancelled, however, however, so it does not affect our results. The weak structure between the elastic peak and the bulk plasmon loss peak is ascribed to to surface plasmon losses. This structure is more pronounced in the Al pectra than in the Si spectra shown. he bothersome strong influence of secondary electrons in extracting the plasmon intensity close to threshold is seen in the 143 eV spectrum in Fig. 4.

The experimental results for the normalized bulk plasmon loss intensity (the peak area ratio) are shown in Fig. 5 as curves (c) and (d) for Si and Al, respectively. The trends of the curves are, as seen, very similar. The normalized bulk loss intensity increases rapidly with electron kinetic energy from the onset and up to 80 - 100 eV kinetic energy. It levels off thereafter and shows a much

slower increase. The data has been corrected for the variation in analyzer efficiency with kinetic energy. It is essential that this correction factor is included because it varies significantly, especially close to threshold. The bars on the experimental data illustrates an estimated uncertaintly in the area ratio determination.

Curves (a) and (b) in Fig. 5 represent calculated values of the normalized bulk loss intensity for extrinsic plasmon creation. The model by Mahan,[28] of random spatial emission for bulk extrinsic plasmons,[43,44,46] predicts for a thick sample a normalized intensity of

$$\frac{P_1}{P_0} = \left(1 + \frac{\ell}{L}\right)^{-1}$$

where ℓ is the mean free path for plasmon emission and L is the mean free path for other processes (single particle excitations). Using the mean free values calculated for Al by Tung and Ritchie,[29] in the case of no damping, curve (b) in Fig. 5 is obtained. Curve (a) gives the values when plasmon damping effects are taken into account.

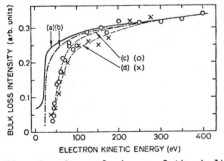

Fig. 5. General shape of the bulk loss intensity as a function of electron kinetic energy above the Fermi level. Curve (c) represents data for for Si and (d) for Al. The calculation curves, (a), and (b) are based on tung and Ritchie's[29] calculated electron mean free path for Al.

In a comparison between the calculated and experimental results, one notices there is a discrepancy in the onset energy, extracted from the experimental results by extrapolation, falls at 10 to 20 eV higher energy than the theory predicts.[29] For energies below 150 eV, the experimental curves are seen to fall off faster with decreasing energy than the calculated curves. Fig. 5 clearly displays this effect as well as the discrepancy in the onset energy (which falls at about twice the plasmon energy). Our data agree well with the trend observed in earlier work on Al.[44,45] Our data also support the conclusion made by Flodstrom et al.[44] that the bulk loss intensity for Al should indeed be very small at UPS energies.

For the moment, we have no unambiguous interpretation of the observed faster decrease of the loss intensity with decreasing kinetic energy and of the higher onset energy than predicted by the extrinsic model. It should be noted that inclusion of the plasmon dispersion does not significantly improve the comparison between theory and experiment, though the correction goes in the right direction.[50] The discrepancy may be caused by the proposed interference effects[27,31,35,36] between intrinsic and extrinsic loss mechanisms. However, it should be noticed that our results indicate that these effects seem to be of relatively little importance at kinetic energies above 150 to 200 eV, where the experimental data agree very well with a theoretical model based solely on extrinsic losses, see Fig. 5. Our work may be one of the first experimental manifestations (see also refs. 48,49) of the transition from the adiabatic to the sudden approximation and indicates in what electron excitation energies this occurs.

A systematic study of the bulk loss intensity close to threshold for materials with different free-electron densities would be of great importance to clarify the existing discrepancy between experiment and theory. It would also be useful to study the intensity distribution of multiple plasmon losses (both bulk and surface plasmons) as a function of excitation energy. The 1s core levels (of Si, Al and other free-electron like metals) would be more suitable than the 2p core level where interference with the 2s level occurs for multiple loss peaks. On the theoretical side the next logical step would be to refine and apply, for instance, the model by Noguera et al.[35] which includes the interference between intrinsic and extrinsic loss mechanisms. Ultimately one and the same theory should include the line-shape of the photoline, satellite structure electron-electron scattering losses, plasmon losses and other loss mechanisms for a complete representation of the experimental data.

The work reported in this paper was done in collaboration with M.H. Hecht and L.I. Johansson, whose contributions are gratefully acknowledged.

The work was supported by the National Science Foundation under contracts No. DMR 77-02519 and No. DMR 79-13102. The experiments were performed at the Stanford Synchrotron Radiation Laboratory which is supported by the National Science Foundation under Contract No. DMR 77-27489 in cooperation with the Stanford Linear Accelerator Center and the U.S. Department of Energy.

REFERENCES

1. D.J. Fabian, H. Kleinpoppen and L.M. Watson (eds): "Inner-Shell and X-Ray Physics of Atoms and Solids"; Plenum Press, New York, 1981.
2. J.B. West, P.R. Woodruff, K. Codling and R. Houlgate, J. Phys. B 9, 407 (1976); and references therein.
3. M.Y. Adam, F. Wuilleumier, N. Sandner, V. Schmidt, and G. Wendin, J. Phys. (Paris) 39, 129 (1978); and references given therein.
4. R. Bruhn, E. Schmidt, H. Schroder and B. Sonntag, Phys. Lett. 90A, 41 (1982); P.H. Kobrin, U. Becker, S. Southworth, C.M. Truesdale, D.W. Lindle, and D.A. Shirley, Phys. Rev. A 26, 842 (1982); and references therein.
5. M.H. Hecht and I. Lindau, Phys. Rev. Lett. 47, 821 (1981); and references therein.
6. L.I. Johansson and I. Lindau, Solid State Commun. 29, 379 (1979); and references therein.
7. R.Z. Bachrach and A. Bianconi, Solid State Commun. 42, 529 (1982); and references therein.
8. M. Ya. Amusia, V.K. Ivanov, and L.V. Chernysheva, Phys. Lett. 59A, 191 (1979).
9. G. Wendin, in Photoionization and Other Probes of Many-Electron Interactions, edited by F.J. Wuilleumier (Plenum, New York, 1976), p. 61.
10. A.W. Fliflet, R.L. Chase, and H.P Kelly, J. Phys. B 7, 1443 (1974).
11. A. Zangwill and P. Soven, Phys. Rev. Lett. 45, 204 (1980).
12. F. Combet Farnoux, in Proceedings of the International Conference on Inner Shell Ionization and Future Applications, Atlanta, Georgia, 1972, edited by R.W. Fink et al. (U.S. Atomic Energy Commission, Oak Ridge, Tenn., 1973), Vol. 2, p. 1130.
13. R. Rabe, K. Radler, and H.W. Wolff, in VUV Radiation Physics, edited by E.E. Koch et al. (Vieweg-Pergamon, Berlin, 1974), p. 247.
14. J.P. Connerade and M.W.D. Mansfield, Proc. Roy. Soc. London Ser. A 341, 267 (1974).
15. F.C. Brown, R.Z. Bachrach, and N. Lien, Nucl. Instrum. Methods 152, 72 (1978).
16. H.J. Hagemann, W. Gudat, and C. Kunz, J. Opt. Soc. Am. 65, 742 (1975).
17. P.W. Palmberg, J. Vac. Sci. Technol. 12, 379 (1975).
18. P.R. Woodruff, L. Torop, and J.B. West, J. Electron. Spectrosc. Relat. Phenom. 12, 133 (1977).
19. M.P. Seah and W.A. Dench, Surf. Interface Anal. 1, 2 (1979).
20. C.S. Fadley, Prog. Solid State Chem. 11, 265 (1976).
21. J.P. Connerade, Contemp. Phys. 19, 415 (1976).
22. G. Wendin and A.R. Starace, J. Phys. B 11, 4119 (1978).
23. U. Fano, Phys. Rev. 124, 1866 (1961).
24. G. Wendin, in VUV Radiation Physics, edited by E.E. Koch et al. (Vieweg-Pergamon, Berlin, 1974), p. 225.
25. T.B. Lucatorto, T.J. McIlrath, J. Sugar, and S.M. Younger, Phys. Rev. Lett. 47, 1124 (1981) and T.B. Lucatorto et al. (these

568

proceedings).

26. B.I. Lundqvist, Phys. Kondens, Mater. 9, 236 (1969).
27. J.J. Chang and D.C. Langreth, Phys. Rev. B8, 4638 (1973); B5, 3512 (1971).
28. G.D. Mahan, Phys. Status Solidi B55, 703 (1973).
29. C.J. Tung and R.H. Ritchie, Phys. Rv. B16, 4302 (1977).
30. D.R. Penn, Phys. Rev. Lett. 38, 1429 (1977); 40, 568 (1978).
31. M. Sunjic and D. Sokcevic, J. Electron Spectrosc. 5, 963 (1974).
32. M. Sunjic and D. Sokcevic, Solid State Commun. 18, 373 (1976).
33. J.W. Gadzuk and M. Sunjic, Phys. Rev. B12, 524 (1975).
34. J.W. Gadzuk, J. Electron Spectrosc. 11, 355 (1977).
35. C. Noguera, D. Spanjaard, and J. Friedel, J. Phys. F 9, 1189 (1979).
36. S.M. Bose, P. Kiehm, and P. Longe, Phys. Rev. B23, 712 (1981).
37. P.L. Longe and S.M. Bose, Solid State Commun. 38, 527 (1981).
38. S.M. Bose, S. Prutzer, and P. Longe, Phys. Rev. B26, 729 (1982).
39. D. Chastenet and P. Longe, Phys. Rev. Lett. 44, 91 (1980).
40. J.F. Inglesfield, Solid State Commun. 40, 467 (1981).
41. Y. Baer and G. Busch, Phys. Rev. Lett. 30, 280 (1973).
42. R.A. Pollak, L. Ley, F.R. McFeely, S.P. Kowalczyk, and D.A. Shirley, J. Electron Spectrosc. 3, 381 (1974).
43. W.J. Pardee, G.D. Mahan, D.E. Eastman, R.A. Pollak, L. Ley, F.R. McFeely, S.P. Kowalczyk, and D.A. Shirley, Phys. Rev. B11, 3614 (1975).
44. S.A. Flodstrom, R.Z. Bachrach, R.S. Bauer, J.C. McMenamin, and S.B.M. Hagstrom, J. Vac. Sci. and Technol. 14, 303 (1977).
45. R.S. Williams, P.S. Wehner, G. Apai, J. Stohr, D.A. Shirley, and S.P. Kowalczyk, J. Electron Spectrosc. 12, 477 (1977).
46. D. Norman and D.P. Woodruff, Surface Sci. 79, 76 (1979).
47. J.C. Fuggle, D.J. Fabian, and L.M. Watson, J. Electron Spectrosc. 9, 99 (1976).
48. J.C. Fuggle, R. Lasser, O. Gunnarsson, and K. Schonhammer, Phys. Rev. Lett. 44, 1090 (1980).
49. F.J. Himpsel, D.E. Eastman and E.E. Koch, Phys. Rev. Lett. 44, 214 (1980).
50. E. Tosatti, Private communication.

RESONANCE AND THRESHOLD EFFECTS IN PHOTOEMISSION UP TO 3500 eV*

D.A. Shirley, P.H. Kobrin, D.W. Lindle, C.M. Truesdale,
and S.H. Southworth [†]
Materials and Molecular Research Division
Lawrence Berkeley Laboratory
and
Department of Chemistry
University of California
Berkeley, California 94720

U. Becker and H.G. Kerkhoff
Technische Universität Berlin,
Fachbereich Physik,
Berlin, West Germany

ABSTRACT

Beam Lines at the Stanford Synchrotron Radiation Laboratory (SSRL) now provide photon beams throughout the entire energy range 5-5000 eV, with a pulse structure very well-suited to time-of-flight (TOF) photoelectron spectroscopy. We have used this facility, together with a TOF spectrometer, to measure photoemission cross sections $\sigma(\varepsilon)$ and asymmetry parameters $\beta(\varepsilon)$ for several interesting systems. A summary of early results is given.
Metal vapors (Ba, Cd, Mn, Hg) were studied using a high-temperature oven. Resonant photoemission was observed in several cases. Both $\sigma(\varepsilon)$ and $\beta(\varepsilon)$ showed resonant behavior at 21.1 eV for several lines in Cd. The 4d, 5s, and 5p $\sigma(\varepsilon)$ line profiles differed dramatically, illustrating the detailed information about continuum states that is available from photoemission.
Correlation satellites in photoemission from rare gases have been observed over a very wide energy range, including those seen in the K-shells of He, Ne and Ar and in the L-shell of Ne. The structure and preliminary intensity variations of these satellites will be discussed.
Molecular shape resonances in C(1s), N(1s), and O(1s) photoemission were observed for the first time, in the molecules CO, CO_2, OCS, CF_4, N_2 and NO. Both the π and σ resonances were observed in KVV Auger emission, and the σ resonances were studied by photoemission. The asymmetry parameters were measured in all cases. The results are in fair agreement with theory, but show systematic deviations and trends.

INTRODUCTION

Over the last four and one-half years, our group has been involved in gas-phase photoelectron spectroscopy using synchrotron radiation at the Stanford Synchrotron Radiation Laboratory

(SSRL). We have used three beam lines at SSRL which together
span the photon energy range 5 to 4000 eV: The 8° line, which
uses a Seya-Namioka monochromator, up to ~32 eV; the new 4° line,
a grasshopper monochromator, from 25 to 600 eV; and the 2° line,
which has a double-crystal monochromator with interchangable
crystals, between 800 and 4000 eV.

The pulsed light through each of the monochromators is used
to measure the kinetic energy spectrum of photoelectrons by the
time-of-flight (TOF) method. This method of energy analysis is
feasible at SSRL because of the large circumference of the SPEAR
storage ring, which provides SSRL with photon pulses separated by
as much as 780 nsec. The advantage of measuring the time spectrum is that all photoelectron energies are sampled simultaneously, making the analyzer more efficient than a scanning deflection analyzer.

The differential cross section for emission of photoelectrons from a randomly oriented sample by linearly polarized
light into a given solid angle, Ω, is given by

$$\frac{d\sigma}{d\Omega} = \frac{\sigma(\epsilon)}{4\pi} \left(1 + \beta(\epsilon)P_2(\cos \theta) \right) , \qquad (1)$$

where θ is the angle between the polarization of the light and
the momentum of the photoelectron, $\sigma(\epsilon)$ is the energy dependent
total cross section, $\beta(\epsilon)$ is the energy dependent angular distribution asymmetry parameter, and $P_2(\cos \theta)$ is the second
Legendre polynomial. To measure $\sigma(\epsilon)$, we use a TOF analyzer
positioned at $\theta=54.7°$, the "magic angle", where $P_2(\cos\theta)=0$. In
order to determine $\beta(\epsilon)$, we need only measure $d\sigma/d\Omega$ at one other
angle. This scheme is known as the double-angle time-of-flight
(DATOF) method and is schematically represented in Fig. 1 and
described in Refs. 1 and 2.

There is an additional advantage to the DATOF method besides
the collection efficiency. When measuring a branching ratio,
which is the ratio of two
peak areas in the magic-
angle detector, or a $\beta(\epsilon)$,
which is the ratio of two
peak areas, one from each
detector, the sample
pressure, photon flux and
even the collection time
are unimportant. Thus,
the pressure and photon
flux do not need to be
monitored except to com-
pare cross sections at
different photon energies.
We have used our

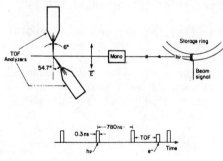

Fig. 1. Schematic of the
Double-Angle Time-of-Flight
(DATOF) method.

apparatus to probe several interesting systems, some of which are described in this paper. A resistively heated oven has been built to produce metal vapor beams of Ba, Cd, Mn, and Hg. These metal vapor experiments are described in Sec. II. The efficiency of the TOF method was utilized in several experiments on low intensity satellite lines in He, Ne and Ar which are described in Sec. III. The ability to measure angular distributions at the C, N, and O K-edges was used to observe shape resonances from small molecules. These experiments are discussed in Sec. IV. Sec. V describes recent work on CH_3I as a molecular analog to our previous study of the Xe 4d subshell. Conclusions are presented in Sec. VI.

METAL VAPOR EXPERIMENTS

Our research group has had an historic interest in the photoemission of metal vapors. These early electron correlation studies were based upon the Perkin-Elmer PS-18 Photoelectron Spectrometer, which was used to investigate mostly group IIA, group IIB and lanthanide elements at the HeI and NeI line source energies. This work led us to the use of synchrotron radiation and the study of resonant photoemission.

Fig. 2 shows the total electron yield from atomic Ba. We see that there are several strong absorption features which are due to discrete states of neutral Ba embedded in the ion plus electron continuum. To examine these resonances further, the photon energy was tuned to the two photon energies that have been

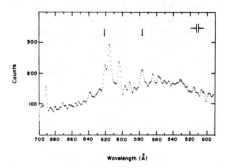

Fig. 2. Total electron yield spectrum of atomic Ba.

Fig. 3. Photoelectron spectra of atomic Ba taken at two autoionizing resonances.

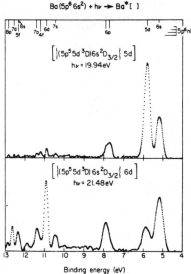

denoted by arrows in Fig. 2. These two photoelectron spectra are shown in Fig. 3. A direct photoemission spectrum would show the 6s photoelectron peak to be the largest, yet at 19.94 eV the 5d peak is the largest, and at 21.48 eV the 6d peak is greatly enhanced. These enhancements are a fingerprint of the auto-ionizing states that are excited and illustrate the utility of the resonance photoemission method. The nature of the decay can be interpreted in terms of an "Auger decay" model in which the excited nd electron acts as a spectator to the Coster-Kronig transition $5p^5 5d6s \rightarrow 5p^6 + e^-$.[3]

It is well known that absorption features above the first ionization threshold are not always lorentzian as are those of Ba in Fig. 2. In fact, it has been shown by Fano that the inter-ference between direct photoemission and photoexcitation followed by autoionizing decay can lead to a family of absorption shapes now known as Fano profiles.[4] These profiles have the form

$$\sigma(\varepsilon) = \sigma_t \left[\rho^2 \frac{(q+\varepsilon)^2}{1+\varepsilon^2} + 1-\rho^2 \right] \tag{2}$$

where ε is a reduced energy equal to $2(E-E_0)/\Gamma$, E_0 is the resonance position, Γ its width, σ_t is the cross section away from the resonance, and q and ρ^2 are constants. While Eq. (2) gives the shape of the total absorption or total electron emis-sion, we may ask what the shapes of the partial cross section profiles for each photoelectron peak are.

To answer this question we looked at the broad autoionizing resonance in Cd I at 21.1 eV. Photoelectron spectra were taken over this asymmetric resonance, and the partial cross section for producing each of the final ionic states was determined. These partial cross sections are shown in Figs. 4 and 5. We find that the partial cross section profiles differ dramatically at this resonance. These different shapes can be understood by con-sidering the details of the autoionization. A theoretical formalism developed by Starace[5] and by Davis and Feldkamp[6] has been applied to this system and parameters have been extracted.[7].

The absorption spectrum of Mn I is dominated by a "giant resonance" near 50 eV which is based upon a $3p \rightarrow 3d$ transition into the half-filled 3d shell. Resonant photoemission from Mn shows that several ionic states are populated from the auto-ionizing state. In addition to the main $3d^{-1}$ line, there are several two-electron satellites with a 3d hole. These satellites show enhancements at the giant resonance and also at the 3p threshold 5 eV higher. The nature of the enhancement at the 3p threshold is interesting in that the main $3d^{-1}$ line does not undergo a corresponding enhancement.

In addition to the above mentioned resonant work, we have

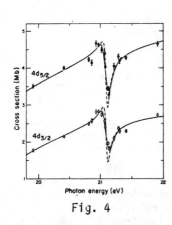

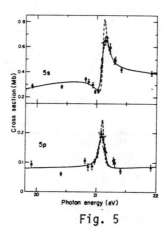

Fig. 4   Fig. 5

Fig. 4. Partial cross section measurements of the $3d^94s^2 \; ^2D_{5/2}$ (filled circles), and $3d^94s^2 \; ^2D_{3/2}$ (open circles), photoelectrons. The solid curves are fits to a theoretical lineshape convoluted with the monochromator bandpass. Dashed curves show fits with monochromator broadening removed.

Fig. 5. Same as Fig. 4, except for $3d^{10}4s \; ^2S_{1/2}$ and $3d^{10}4p \; ^2P_{3/2,1/2}$ photoelectrons.

studied the photoionization of Hg. Because of its high atomic number, we expect both relativistic and many body effects to be important in Hg. We measured the relative cross sections, sub-shell branching ratios and angular distributions of the 5d, 5p and 4f subshells between 50 and 275 eV.

Fig. 6 shows the angular distribution asymmetry parameter for the 5d photoelectron and a calculation using the RRPA theory. The RRPA calculation includes intershell correlations with the 4f and 5p channels as well as relativistic effects. The drop in $\beta(\epsilon)$ near 190 eV is caused by a Cooper minimum in the $5d \rightarrow \epsilon f$ channel.

Fig. 7 shows the asymmetry parameter of the 4f subshell. The cross section and branching ratio as well as the $\beta(\epsilon)$ parameter show the effects of a large centrifugal barrier near threshold in the $4f \rightarrow \epsilon g$ channel.

The $5p_{3/2}$ and $5p_{1/2}$ photoionization channels show large variations in all of the measurable parameters due to a large spin-orbit splitting (18.6 eV) and a Cooper minimum 100 eV above threshold.

These experiments with metal vapors highlight the advantages of the DATOF method: the insensitivity of the branching ratio and angular distribution measurements to the sample pressure and photon flux.

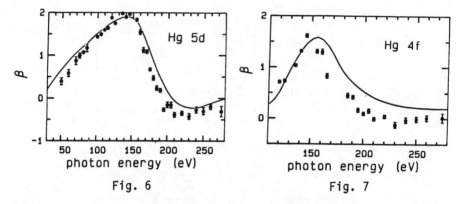

Fig. 6

Fig. 7

Fig. 6. Angular distribution asymmetry parameter of the Hg 5d^{-1} photoelectrons. The solid curve is a relativistic random-phase approximation (RRPA) calculation from Refs. 8 and 9.

Fig. 7. Angular distribution asymmetry parameter of the Hg 4f^{-1} photoelectrons. The solid curve is an RRPA calculation from Ref. 9.

CORRELATION SATELLITES

The photoionization of He provides the simplest example of electron correlation in atomic physics. For photon energies above the second ionization limit of He(65.4 eV), it is possible to produce a He^{+} ion in the ground state (1s) or in one of two excited (satellite) states (2s, 2p). We have studied the production of the n=2 satellites by the measurement of the partial cross section $\sigma(\varepsilon)$, the branching ratio to the ground ionic state, and the angular distribution asymmetry parameter $\beta(\varepsilon)$ as functions of photon energy, for the combination of the 2s and 2p final ionic states, which are effectively degenerate in a photoemission experiment. These studies were made with photon energies from near threshold to 90 eV, and are comprised of two distinct sets of data. The first set, off-resonance, was taken with photon energies from just above threshold to 69.5 eV and from 75 to 90 eV. The second set of data, on-resonance, was taken with photon energies (69.5-73 eV) capable of exciting members of a Rydberg series leading to the He^{+}(n=3) ionization threshold at 73 eV. For both sets of measurements, the DATOF method allowed us to simultaneously determine the partial cross sections and asymmetry parameters for both the one-electron ionization to He^{+}(n=1) and the simultaneous, two-electron ionization and excitation to He^{+}(n=2).

In the off-resonance regions, the partial cross section and branching ratio (to the ground ionic state) measured show excellent agreement with previous experimental results.[10,11] The

off-resonance $He^+(n=2)$ asymmetry parameter, $\beta_{n=2}(\epsilon)$, measured
here, also shows good agreement with previous results above 75 eV
photon energy (Fig. 8),[12] and has determined, for the first
time, the behaviour of $\beta_{n=2}(\epsilon)$ near threshold.

The asymmetry parameter measurable in this experiment,
$\beta_{n=2}(\epsilon)$, is an average of the asymmetry parameters for ionization to $He^+(2s)$ and $He^+(2p)$, weighted by the respective
partial cross sections, as shown in Eq. (3).

$$\beta_{n=2}(\epsilon) = \frac{\sigma_{2s}(\epsilon)\beta_{2s} + \sigma_{2p}(\epsilon)\beta_{2p}(\epsilon)}{\sigma_{2s}(\epsilon) + \sigma_{2p}(\epsilon)} \tag{3}$$

The value of β_{2s} is always two. Rearranging Eq. (3), we find
an expression for the relative populations of $He^+(2p)$ and
$He^+(2s)$;

$$R = \frac{\sigma_{2p}(\epsilon)}{\sigma_{2s}(\epsilon)} = \frac{2 - \beta_{n=2}(\epsilon)}{\beta_{n=2}(\epsilon) - \beta_{2p}(\epsilon)} . \tag{4}$$

Thus, by measuring $\beta_{n=2}(\epsilon)$ and using theoretical values for
$\beta_{2p}(\epsilon)$, the intensity ratio R of the excited ionic state can be
determined. Fig. 8 shows two theoretical calculations of
$\beta_{2p}(\epsilon)$[13,14] as well as experimental[12] and theoretical[12] values
for $\beta_{n=2}(\epsilon)$. The solid circles represent the present data,
which indicate that the value of $\beta_{n=2}(\epsilon)$ approaches zero near
threshold, in disagreement with the calculation of Bizau et
al.[12] From Eq. (4), it is possible to calculate R, the ratio
of the cross sections for production of $He^+(2p)$ to $He^+(2s)$,
from the values of the satellite asymmetry parameter. This has
been done, using $\beta_{2p}(\epsilon)$ values calculated by Jacobs and
Burke,[13] for both the present data and those of Bizau et
al.[12] The results are shown in Fig. 9, where we see that the
close-coupling theory of Jacobs and Burke[13] has the correct
qualitative shape, in contrast to the indication from the
previous data[12] that Chang's theory[14] is more correct.

In the resonance region, dramatic changes in the $He^+(n=2)$
partial cross section and asymmetry parameter were seen as functions of photon energy (see Figs. 10 and 11). This structure is
attributed to the $He^*(3s3p)$ and $He^*(sp, 3n+, n=4,5)$[15] autoionizing levels leading to the $He^+(n=3)$ ionization threshold.
As discussed in Sec. II, it is possible to derive, from the cross
section data, parameters describing these Fano profiles[4] in
terms of the matrix elements governing the autoionization
process. This has been done by Woodruff and Samson,[16] and the
fit to their data is shown with the present results in Fig. 10.
Interpretation of the on-resonance results for the asymmetry
parameter, however, awaits further theoretical development of the
behaviour of the angular distribution of photoelectrons from

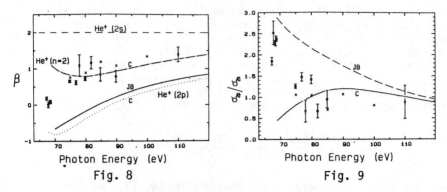

Fig. 8   Fig. 9

Fig. 8. Angular distribution asymmetry parameter for He+(n=2) photoelectrons: ● - present results; x - Ref. 12. Theoretical curves: (JB) - Jacobs and Burke, Ref. 13 and dotted curve (C) - Chang, Ref. 14 are calculations for He+(2p) only. The dot-dash curve (C) is a calculation of He+(n=2) by Bizau et al.[12]

Fig. 9. Ratio, R, of partial cross sections for photoionization to He+(2p) and He+(2s): ● - present results; x - Ref. 12. Note that these values are calculated using values for $\beta_{2p}(\epsilon)$ calculated in Ref. 13. Theoretical curves: (JB) - Ref. 13; (C) - Ref. 14.

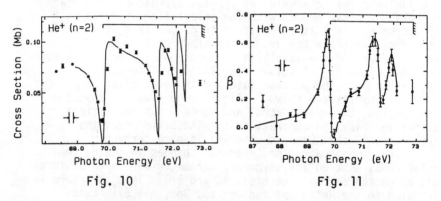

Fig. 10   Fig. 11

Fig. 10. Partial cross section for photoionization to He+(n=2). The oscillations are due to the autoionizing states He+(3s3p) at 69.8 eV and He*(sp, 3n+, n=4,5) at 71.55 and 72.13 eV. The solid curve is a fit to the data of Ref. 16.

Fig. 11. Angular distribution asymmetry parameter of the He+(n=2) photoelectrons. See caption of Fig. 10 for explanation. The solid curve is drawn only as a visual aid.

individual subshells in the vicinity of autoionization resonances.

After He the simplest and perhaps best understood satellite spectrum is that of Ne. The high resolution Al Kα spectrum of the Ne K-shell satellites[17] has been analyzed in detail using the sudden approximation and configuration interaction theory by Martin and Shirley.[18] The applicability of the sudden approximation greatly simplifies the analysis.

With the availability of the double-crystal monochromator at SSRL, our group has begun exploratory work on the threshold behavior of the Ne and Ar K-shell satellites. In the sudden approximation, the satellite to main line ratio should be constant while threshold effects may change this ratio. Preliminary experiments on Ne show a satellite to main line ratio below the sudden limit. The Ar K-shell spectrum, which has not been previously recorded, shows several satellite lines, the first of which increases in intensity by a factor of two relative to the main line between 25 and 85 eV kinetic energy.

We have also measured the asymmetry parameters and branching ratios of Ne(2s,2p) correlation satellites referenced to the 2s main line to further investigate electron correlation effects. The Ne 2s and 2p photo-peaks were used to calibrate the measurement of asymmetry parameters and correct the cross section data for the transmission of the 54.7° detector. The counting statistics were sufficient that satellite structure was observable. Figure 12 shows a TOF spectrum converted to photoelectron energy taken at 66.36 eV photon energy showing satellites derived from the configurations 2p^{4}ns, 2p^{4}nd, 2p^{4}np (S1,S2,S3,S4,S5), and the 2s line. Table I lists the configurations[19,20] and the comparable assignments given by Wuilleumier and Krause.[21] Fig. 13 shows the β(ε)'s for satellites S1, S3, S4, and S5. We observe that the β(ε)'s for S3, S4, and S5 follow the same general trend as β(ε) for the Ne 2p main line. In contrast, S1

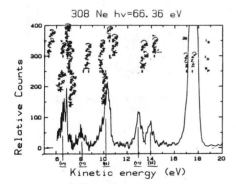

Fig. 12. Time-to-energy converted spectrum of Ne at a photon energy of 66.36 eV.

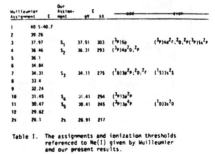

Table I. The assignments and ionization thresholds referenced to Ne(I) given by Wuilleumier and our present results.

appears somewhat different. The asymptotic value of S1 does not agree well with the other three satellite peaks. This suggests that S1 contains $2p^4nd$ configurations mixed with $2p^4np$ and $2p^4ns$ configurations, while S3, S4, and S5 contain large $2p^4np$ admixtures with a smaller $2p^4ns$ contribution.

The branching ratios of satellites 1-5 are shown in Fig. 14. S3-S5 show a decrease in intensity at 61.5 ± 0.3 eV. Satellite 1 shows a weak minimum at ~ 74 eV. A preliminary explanation for the decrease observed in the branching ratio for S3 is that it is due to autoionization of a doubly-excited Rydberg level.

Equation (5) represents excitation to a resonant bound state. Equations (6a) and (6b) represent this state's subsequent autoionization.

$$(He)2s^2 2p^6 + h\nu \rightarrow Ne^* \qquad (5)$$

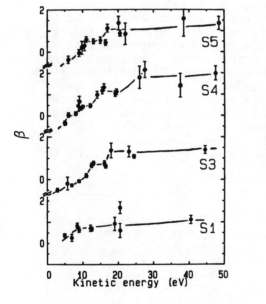

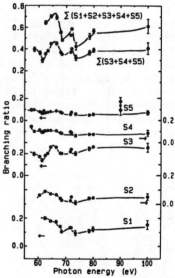

Fig. 13. Asymmetry parameters for satellites 1, 3, 4, and 5.

Fig. 14. Branching ratios of Ne satellites which are referenced to the $(2s)^{-1}$ core.

$$Ne* \rightarrow (He)2s^2 2p^4 4s + e^- \qquad (6a)$$

$$Ne* \rightarrow (He)2s^2 2p^4 4p + e^- \qquad (6b)$$

The occurrence of this resonant excitation could influence the intensity of S2 (Eq. (6b)). Further measurements must be performed to clarify the behavior of S3. Theoretical calculations including initial and final state configuration correlations and transitions to doubly-excited Rydberg levels of Ne(I) that predict the intensities and asymmetry parameters of Ne satellites would complement these experimental measurements.

K-SHELL EXCITATIONS OF SMALL MOLECULES

There is great interest in studying discrete and continuum transitions in the K-shells of small molecules for exploring absorption, potential barriers, and photoionization processes theoretically[22-25] and experimentally.[26-29] Shape resonances have been observed previously in photoabsorption and electron impact experiments.[26-29] In this paper we report the first observations of shape resonances in free molecules in the photoemission and Auger channels following core-level excitation. For the past two years, our group has measured K-shell asymmetry parameters $\beta(h\nu)$ and partial cross sections $\sigma(h\nu)$ for Auger and photoelectrons from the C(1s) shell in CO, CO_2, OCS, and CF_4, the O(1s) shell in CO and CO_2, and the N(1s) shell in N_2 and NO.

The excitation of the C K-shell in CO, CO_2, OCS, and CF_4 clearly demonstrates that the asymmetry parameters of KLL Auger and photoelectrons can be used to explore C(1s) discrete transitions, molecular shape resonances, and the influence of shape resonances on the KLL Auger decay process. Fig. 15 shows $\sigma(h\nu)$ and $\beta(h\nu)$ for the carbon KLL Auger peaks of CO. A discrete transition is responsible for the Auger resonance, clearly seen in the cross section at 287.3 eV. Dill et al.[30] have used the multiple-scattering method (MSM) to calculate the orientation parameter $\beta_m(h\nu)$ for $\sigma \rightarrow \pi$ discrete transitions. The excited $CO*(^1\pi)$ molecule should orient perpendicular to the photon $\vec{E}$ vector following excitation. Note that β (287.3) is ~ 0. This may be a consequence of the experimental averaging over the KVV Auger transitions. At resonance, β is essentially zero for the KVV Auger peaks in CO_2, OCS, and CF_4. $\beta_m(h\nu)$ is related to $\beta(h\nu)$ by a constant factor ($\mathcal{A}$) that is independent of the excitation energy and is only dependent on the dynamics of the Auger decay process as shown in the following equation

$$\beta(h\nu) = \beta_m(h\nu)*\mathcal{A} \qquad (7)$$

where $-1 \leqslant \beta_m(h\nu) \leqslant 2$ and $-1/2 \leqslant \mathcal{A} \leqslant 1$.[30]

Fig. 16 shows the C(1s) $\sigma(h\nu)$ and $\beta(h\nu)$ for CO. The solid curve is the MSM calculation performed by Dill et al.[22] and the dashed curve pictured with our results (•) for the C(1s) $\beta(h\nu)$ is the MSM calculation of Grimm.[23] We show excellent agreement with Dill et al.[30] for the shape of $\beta(h\nu)$ as a function of photon energy, but the minimum occurs between the two calculations. The manifestation of the continuum shape resonance ($\sigma \to \sigma^*$) results in a minimum for the $\beta(h\nu)$ of the C(1s) photoelectron. The $\sigma(h\nu)$ clearly shows the shape resonance as a broad peak centered at 305.0 eV. We have excellent agreement with the Stieltjes-Tchebycheff calculation of Padial et al.[24] Our relative cross sections for the C(1s) photoelectron were scaled to Kay et al.[26] electron-ion coincidence experimental cross sections to yield C(1s) absolute partial cross sections. An expansion of the C(KVV) Auger cross section over the shape resonance region demonstrates that the Auger cross section is influenced by the continuum resonance.

The solid line pictured in Fig. 15 with the C(KLL) Auger $\beta(h\nu)$ is the calculation of $\beta_m(h\nu)$ by Dill et al.[30] Our results show less variation than their calculation.

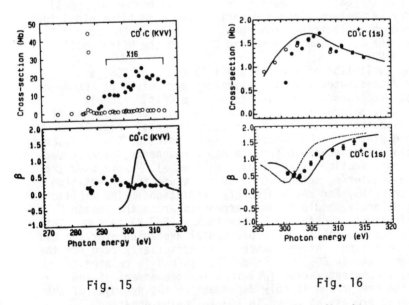

Fig. 15   Fig. 16

Fig. 15. Partial cross section and angular distribution asymmetry parameter for C(KVV) Auger electrons from CO.

Fig. 16. Partial cross section and angular distribution asymmetry parameter for C(1s) photoelectrons from CO.

As a comparison, Fig. 17 shows the $\beta(h\nu)$ for the C(1s) photoelectron in CO_2. Here the shape resonance is ~ 20 eV broad and located more than 1 Rydberg from threshold. The dashed curve in Fig. 17 is a calculation by Grimm[23] and the solid curve is a Hartree-Fock calculation by Lucchese and McKoy.[25]

The O K-shell is also predicted to have shape resonances and discrete transition resonances. Most of our O K-shell results were obtained from the second order light present during the measurement of the C K-shell systems. Unfortunately, the shape resonance region was not done in detail. One can infer from our data that the O(1s) shape resonance is present. Fig. 18 shows the $\sigma(h\nu)$ and $\beta(h\nu)$ of the O(1s) photoelectron of CO. The O(1s) data were scaled such that the maximum in the O(1s) cross section was 2/3 that of the C(1s) cross section. The solid curve is a calculation by Padial et al.[26] There are data missing for the true maximum in the cross section at the O(1s) shape resonance, but because the cross section does not have a maximum at threshold, we can infer that a centrifugal barrier is present as the O(1s) photoelectron scatters through the molecular potential in CO.

Our $\beta(h\nu)$ measurements, also shown in Fig. 18, cannot be used to confirm if the theory of Grimm[23] (solid curve) or that of Dill et al.[22] (dashed curve) is more capable of predicting the asymmetry parameter. More work is necessary for any strong conclusions.

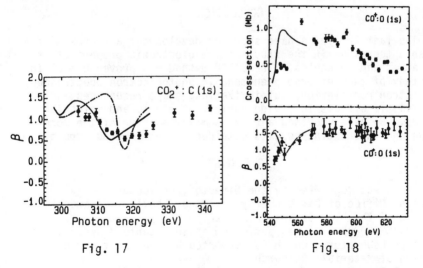

Fig. 17   Fig. 18

Fig. 17. Angular distribution asymmetry parameter for C(1s) photoelectrons from CO_2.

Fig. 18. Partial cross section and angular distribution asymmetry parameter for O(1s) photoelectrons from CO.

582

In addition to the C and O K-shell measurements, we have measured partial cross sections and asymmetry parameters for photoemission and Auger electron emission from the N K-shell in N_2 and NO. Similar effects to those seen in the C K-edge experiments are apparent. One significant conclusion is that the σ shape resonance above threshold produces a much larger effect in N_2 than NO, despite the relative similarity of the two molecules.

INNER-VALENCE STUDIES

Recent measurements[31] by this group have confirmed the predicted oscillatory nature of the asymmetry parameter $\beta(\varepsilon)$ for the Xe 4d subshell. The behaviour seen is primarily due to non-relativistic, one-electron effects such as the variation of Coulomb phase shifts in outgoing photoelectron channels and the change in sign of radial matrix elements which occurs at a Cooper minimum. In the past few months, we undertook a similar experiment on methyl iodide (CH_3I) as a molecular analog of atomic Xe. We find that $\beta(\varepsilon)$ for the I 4d subshell in CH_3I exhibits almost identical behaviour to $\beta(\varepsilon)$ for the Xe 4d subshell over a wide photon energy range. We conclude from this that the 4d subshell on the iodine atom in CH_3I is primarily atomic and that the methyl group has little interaction with the outgoing photoelectron waves.

CONCLUSION

The last few years have seen the development and growth of a valuable technique in the study of the electronic properties of free atoms and molecules. The DATOF method has proven useful in the study of such diverse phenomena in photoelectron spectroscopy as electron correlation, autoionization, shape resonances and relativistic effects in a wide variety of atomic and molecular systems. We have provided a brief summary of some of the more unique and interesting results obtained recently by our group.

REFERENCES

*This work was supported by the Director, Office of Energy Research, Office of Basic Energy Sciences, Chemical Sciences Division of the U.S. Department of Energy under Contract No. DE-AC03-76SF00098. It was performed at the Stanford Synchrotron Radiation Laboratory, which is supported by the NSF through the Division of Materials Research.
†National Bureau of Standards, Washington, DC 20234.
 1. M.G. White, R.A. Rosenberg, G. Gabor, E.D. Poliakoff, G. Thornton, S.H. Southworth, and D.A. Shirley, Rev. Sci. Instrum., 50, 1268 (1979).

2. S.H. Southworth, C.M. Truesdale, P.H. Kobrin, D.W. Lindle, W.D. Brewer, and D.A. Shirley, J. Chem. Phys. 76, 143 (1982).
3. R.A. Rosenberg, M.G. White, G. Thornton, and D.A. Shirley, Phys. Rev. Lett., 43, 1384 (1979).
4. U. Fano, Phys. Rev., 124, 1866 (1961).
5. A.F. Starace, Phys. Rev. A 16, 231 (1977).
6. L.C. Davis and L.A. Feldkamp, Phys. Rev. B 23, 6239 (1981).
7. P.H. Kobrin, U. Becker, S.H. Southworth, C.M. Truesdale, D.W. Lindle, and D.A. Shirley, Phys. Rev. A 26, 842 (1982).
8. W.R. Johnson, V. Radojević, P. Deshmukh, and K.T. Cheng, Phys. Rev. A 25, 337 (1982).
9. V. Radojević and W.R. Johnson, private communication.
10. M.O. Krause and F. Wuilleumier, J. Phys. B 5, L143 (1972). F. Wuilleumier, M.Y. Adam, N. Sandner, and V. Schmidt, J. Physique-Lett., 41, L373 (1980).
11. P.R. Woodruff and J.A.R. Samson, Phys. Rev. Lett., 45, 110 (1980).
12. J.M. Bizau, F. Wuilleumier, P. Dhez, D.L. Ederer, T.N. Chang, S. Krummacher, and V. Schmidt, Phys. Rev. Lett. 48, 588 (1982).
13. V.L. Jacobs and P.G. Burke, J. Phys. B 5, L67 (1972).
14. T.N. Chang, J. Phys. B 13, L551 (1980).
15. R.P. Madden and K. Codling, Astrophys. J., 141, 364 (1965).
16. P.R. Woodruff and J.A.R. Samson, Phys. Rev. A 25, 848 (1982).
17. U. Gelius, J. Elect. Spect., 5, 985 (1974).
18. R.L. Martin and D.A. Shirley, Phys. Rev. A 13, 1475 (1976).
19. C.E. Moore, Atomic Energy Levels, National Bureau of Standards 457, 1949.
20. W. Persson, Physica Scripta, 3, 133 (1971).
21. F. Wuilleumier and M.O. Krause, Phys. Rev. A 10, 242 (1974).
22. J.L. Dehmer and D. Dill, Phys. Rev. Lett. 35, 213 (1975); J.L. Dehmer and D. Dill, J. Chem. Phys. 65, 5327 (1976); D. Dill, S. Wallace, J. Siegel, and J.L. Dehmer, Phys. Rev. Lett. 42, 411 (1979).
23. F.A. Grimm, Chem. Phys. 53, 71 (1980).
24. N. Padial, G. Csanak, B.V. McKoy, and P.W. Langhoff, J. Chem. Phys. 69, 2962 (1978).
25. R.R. Lucchese and B.V. McKoy, to be published.
26. R.B. Kay, Ph.E. Van der Leeuw, and U.J. Van der Wiel, J. Phys. B 10, 2513 (1973).
27. M. Tronc, G.C. King, R.C. Bradford, and F.H. Read, J. Phys. B 9, L555 (1976).
28. A. Hamnett, W. Stall, and C.E. Brion, J. Elect. Spect., 8, 367 (1976).
29. G.R. Wight, C.E. Brion, and M.J. Van der Wiel, J. Electr. Spectr., 1, 457 (1972).
30. D. Dill, J.R. Swanson, S. Wallace, and J.L. Dehmer, Phys. Rev. Lett. 45, 1393 (1980).
31. S.H. Southworth, P.H. Kobrin, C.M. Truesdale, D.W. Lindle, S. Owaki, and D.A. Shirley, Phys. Rev. A 24, 2257 (1981).

THE 4d-PHOTOABSORPTION OF Ba, Ba$^+$, and Ba^{++}:
A VIEW OF SHELL COLLAPSE vs. CONTRACTION

T.B. Lucatorto, T.J. McIlrath,[*] W.T. Hill III[*], and C.W. Clark

National Bureau of Standards, Washington, DC 20234

ABSTRACT

Ba with Z=56 is at the edge of 4f collapse; neutral ground state atoms with Z<56 have 4f orbitals which are hydrogenic with $r_{av} \sim 17a_0$ while elements with Z>56 have a "collapsed" 4f orbital with $r_{av} \sim 1a_0$. Since the 4d orbital is collapsed ($r_{av} \sim 1a_0$) the nature of the 4d-absorption is expected to depend critically on whether the 4f orbital can be considered "collapsed" or not. Using a laser technique to prepare dense homogenous columns of Ba$^+$ and Ba^{++} we have obtained the 4d photoabsorption spectra of Ba, Ba$^+$ and Ba^{++}. This technique thus allows us to observe the effects on the 4f orbitals of the increased nuclear attraction experienced in the absence of screening by the 6s electrons. It is found that exchange effects are critically important in configurations of the type 4d^{9}4f, and that progressive "contraction" rather than "collapse" is a more appropriate description under such conditions.

INTRODUCTION

The 4d-photoabsorption of Ba^{++}, shown in Fig. 1 beneath the spectra of Ba and Ba$^+$, manifests several surprising and interesting features not seen in any previously observed 4d-photoabsorption spectrum of atoms, molecules or solids. The previously observed spectra for 54≤Z≤68 (Figs. 2 and 3) are similar to spectra of Ba and Ba$^+$ and are all characterized by the bulk of the 4d-absorption being localized in a single broad feature having some relatively weak structure and ranging in width from about 70eV for the

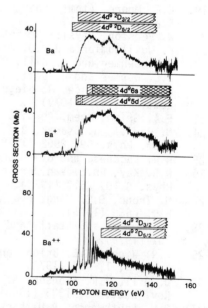

Fig. 1 Spectra of 4d photoabsorption of Ba (Z=56), Ba$^+$ and Ba^{++}, from Ref. 1.

0094-243X/82/940584-18$3.00 AMerican Institute of Physics

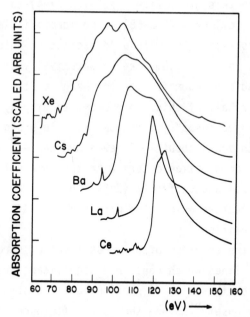

Fig. 2   Spectra of 4d photoabsorption for the (solid) elements
Z=54 - Z=58, from Ref. 2.

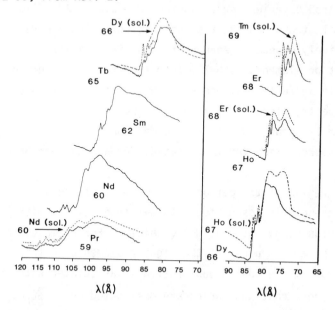

Fig. 3 Spectra of 4d photoabsorption for the elements Z=59 - Z=68,
from Ref.3.   Spectra obtained in both the vapor phase (solid line)
and solid phase (dashed line) are shown. The wavelengths correspond
to the vapor phase spectra. (See Ref. 3 for details.)

lightest element in this group (i.e., Xe with Z=54) to about 15 eV for the heaviest (i.e., Er with Z=68). The identifiable discrete autoionizing states associated with a 4d excitation which lie below the $4d^9$ ion limits in these atoms are comparatively very weak. In addition, the broad continuum feature for each of the atoms in roughly the lighter half of this group seems to display what has been termed a "delayed onset": a gradual rise in photoionization cross section which peaks about 10-20 eV above the $4d^9$ ionization limit.[4] An abrupt demarcation occurs with the completion of the 4f shell in the final state; for Tm (Z=69) there is a single relatively narrow feature associated with the $4d^{10} \, 4f^{13} \, ^2F_{7/2} \rightarrow 4d^9 \, 4f^{14} \, ^2D_{5/2}$ transition.[5]

In contrast, the spectrum of Ba^{++} has a set of strong, well-defined absorption lines which contain a significant fraction of the oscillator strength and a $4d^9$ photoionization continuum characterized by a sharp onset.[6] Since the main difference in the 4d-photoabsorption in going from Ba to Ba^{++} can be attributed to the excited electron experiencing an increased nuclear attraction when the 6s electrons and their associated screening are removed, one would naively expect to encounter a spectrum similar to Ba^{++} in an element or elements with Z larger than that of Ba or in an ionic molecule where the valence electron(s) of the cation (with Z>55) would be transfered to the anion. The fact that so far the Ba^{++} spectrum is unique is not fully understood. However, we do have a good appreciation for the great difference between Ba and Ba^{++}.

In this region of the periodic table, most of the 4d-absorption oscillator strength is associated with d→f transitions, with only a small fraction going into the d→p channel.[4] For example, in Ba, which has a spectrum typical of the lighter elements in this group, over 90% of the 4d-photoabsorption is associated with transitions into the f continuum which are designated $4d^{10}5s^25p^66s^2 \, ^1S_0 \rightarrow 4d^95s^25p^66s^2 \, \epsilon f \, ^1P_1$.[6] We thus expect the dramatic difference between the absorption spectra of Ba and Ba^{++} to be correlated with a significant change in the f orbital of the final state. This

expectation is bolstered by the existence of a very similar phenomena in the same neighborhood of the periodic table: the well-known
"collapse" of the 4f shell for Z>56, which is responsible for the
chemical similarity of the rare earth elements. Indeed, the concepts
derived from the study of f shell collapse provide a useful basis for
discussion of the barium absorption spectra; and, in turn, interpretation of these spectra yields insight relevant to the general problem of shell collapse.

COLLAPSE OF THE F SHELL

We proceed with a brief review of the phenonenon of shell
collapse, which was first explained theoretically by Fermi.[7] Goep-
pert-Mayer[8] elaborated on Fermi's work, showing that the summed
Coulomb interactions (approximated by the Thomas-Fermi potential[7])
and the centrifugal repulsion combine to yield a potential function
with a double well structure for the f electrons of rare earth atoms.
One may gain a qualitative understanding of the origin of double
well structure by defining an effective charge $Z_{eff}(r)$ which incorporates all of the electrostatic interactions. The effective potential (i.e., electrostatic plus centrifugal terms) for an f electron
is then (in Rydbergs)

$$V_{eff}(r) = \frac{-2Z_{eff}}{r} + \frac{\ell(\ell+1)}{r^2} \tag{1}$$

where $\ell=3$. In Fig. 4, V_{eff} for H and Ce(Z=58) are compared. We see
that in Ce the effective potential is essentially hydrogenic for
$r>5a_o$, reflecting the fact that the 57 residual electrons then
perfectly screen the nucleus from the f electron, i.e., $Z_{eff}\sim1$.
As r decreases, the electronic screening is reduced, making $Z_{eff}>1$.
Starting at about $r=3a_o$ this reduction in screening is sufficient
for the Coulomb interactions, $-2Z_{eff}/r$, to overcome the centrifugal
repulsion $12/r^2$. (Recall that $Z_{eff}\to58$ as $r\to0$). This gives rise to
the deep potential well in the core of the atom, bounded at small r
by the centrifugal barrier which must predominate as $r\to0$.

The "collapse" of the f shell refers to an abrupt change in
the character of 4f wavefunctions which occurs between Ba(Z=56)

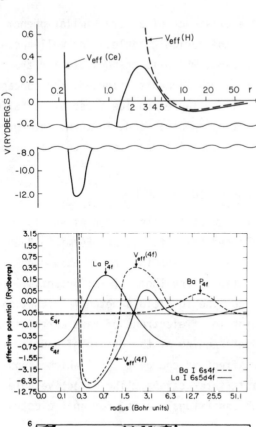

Fig. 4 Potentials for the 4f electron of H and Ce (Z=58), courtesy of W.C. Martin. The Ce potential is obtained from the Herman-Skillman model. Ce displays a double well structure. Note the enourmous disparity in the size and strength of the potential wells; if this curve were plotted on a linear scale of the same size, the outer well would be imperceptible.

Fig. 5 Wavefunctions and effective potentials for excited states of Ba (Z=56) and La(Z=57), as computed in the HFX approximation, from Ref. 9.

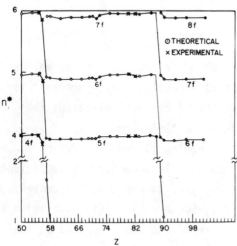

Fig. 6 The effective quantum number n^* associated with the nf orbital, as a function of nuclear charge Z. From Ref. 9. The effective quantum number of the 4f state, which is related to the binding energy of the 4f orbital, makes a sudden departure from its hydrogenic value of 4 at La (Z=57) at the beginning of the rare earth period. A similar departure in the 5f states is seen at the beginning of the actinides (Z~90).

and La(Z=57). For Ba, the 4f orbital is practically hydrogenic: it lies in the outer potential well and has a mean radius $<r_{4f}> \sim 17a_0$. For La the 4f orbital is localized in the inner well and has $<r_{4f}> \sim 1.2a_0$. (See Fig. 5.) The small relative change in the nuclear Coulomb attraction in going from Z=56 to Z=57 increases the depth and width of the inner well just beyond the "limit" before which a compact, tightly bound solution is not possible. Beyond this limit the inner well can support just a single solution, the 4f, until approximately Z=90 where 5f collapse occurs. Collapse is also strikingly evident in a plot of effective quantum number vs. Z as in Fig. 6. The collapse of the 4f shell is associated with the existence of the series of chemically similar rare earth elements, and the collapse of the 5f shell with the actinides.

THE 4d-PHOTOABSORPTION

In addition to its important chemical consequences, f shell collapse has significant implications for photoabsorption studies. In particular, we note that the 4d orbitals in this region have mean radii on the order of $1a_0$ (see Fig. 7). Since the cross section for an optical transition between any two states i and j is dependent on their spatial overlap through the dipole matrix element $|<i|r|j>|$, the 4d→f absorption spectrum will be markedly influenced by the collapse of the f orbital. For instance, in Xe (Z=54) the 4f orbital in the excited configuration $4d^9 4f 5s^2 5p^6$ is hydrogenic (i.e., uncollapsed), as are all successive f orbitals in the Rydberg series $4d^9 nf$, n>4. As the overlaps of bound f orbitals with the 4d orbital are thus very small, most of the 4d→f oscillator strength is associated with transitions to continuum states. The centrifugal barrier prevents very low energy continuum f waves from penetrating the inner well so that maximum 4d→εf overlap is attained when the continuum energy ε exceeds the barrier height (see Fig. 7). For ε much greater than the barrier height, however, the overlap decreases because of the oscillatory behavior of the continuum wave in the inner well. Thus, for atoms without

collapsed f orbitals, the
double well structure is
responsible for a delayed
onset of the photoioni-
zation continuum and
for a peak in the photo-
absorption cross section
at energies of the order
of the barrier height
($\sim$10 eV) above threshold.

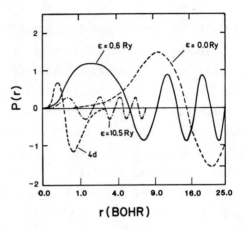

Fig. 7 Radial wavefunctions for the 4d and εf orbitals of Xe (Z=54) from Ref. 10.

If, on the other hand, the 4f orbital is collapsed, the 4d-photoabsorption spectrum should be dominated by a strong $4d^{10} \rightarrow 4d^9 4f$ resonance
line. The traditional Hartree-Fock average of configuration
model (HF_{av}), in which the electron-electron interaction is
approximated by its average over all LS terms of a given configuration, gives a collapsed f orbital for the configuration
$4d^9 4f 5s^2 5p^6 6s^2$ of Ba(Z=56) (See Fig. 8). The collapse occurs at
Ba in this case rather than at La(Z=57), the point of collapse
for a $4d^{10} 4f$ configuration, because of the removal of a screening

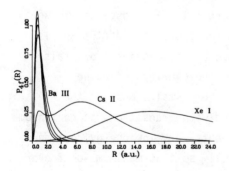

Fig. 8 The 4f orbitals calculated in the HF_{av} approximation
for the configuration $4d^9 4f 4s^2 5p^6$
XeI (Z=54) - PrVI (Z=59). The
collapse occurs at Ba III. The
HF_{av} 4f orbital of Ba I
$4d^9 5s^2 5p^6 6s^2$ is essentially the
same as that of Ba III.

4d electron from the core. It is apparent from Fig. 1, however, that the absorption spectrum of Ba is more characteristic of an atom with uncollapsed f orbitals. Two models have been used to account for this discrepancy.

The first explanation was based on a model developed by Dehmer et al.[11] to explain the general characteristics of $4d^{10}4f^N \rightarrow 4d^94f^{N+1}$ absorption spectra in the rare earths. In that model the energies of the LS terms of $4d^94f^{N+1}$ configurations are determined by diagonalizing the inter-electron and spin-orbit interactions in a basis of orbitals computed in the HF_{av} approximation. The collapse of the 4f shell in the rare earths has two significant consequences which are due to the resulting large overlap between 4f and 4d orbitals: first there is a very large exchange integral $[G^1(d,f)$; see following section] between the 4d and 4f shells, which is responsible for "raising" some of the optically allowed terms of $4d^94f^{N+1}$ above some of the $4d^94f^N$ ionization limits into the "far continuum"; and secondly, the 4f state carries the bulk of $4d \rightarrow f$ oscillator strength. Thus, in this model, the 4d absorption spectra are characterized by a few strong features which are enormously broadened by autoionization into the $4d^94f^N\epsilon f$ continuum. In applying this model to Ba, Ederer et al.[12] showed that it gave the energy of the optically allowed $4d^95s^25p^66s^24f$ $^1P_1$ term to be 10.5 eV above threshold, in good qualitative agreement with the peak of the broad photoabsorption feature.

Expressing dissatisfaction for the Ba model[12] with its $4d^94f$ $^1P_1$ state autoionizing into its own $4d^9$ continuum, Hansen et al.[13] showed that the Ba spectrum could be understood in a model employing a Hartree-Fock term-dependent (HF_{TD}) basis set. This basis set is constructed by solving the coupled set of the variational equations for the energy specific to the term of interest and not for the average of configuration value as in the HF_{av} procedure. The term-dependence in the Ba $4d^94f$ configuration is striking: the radial functions associated with the ^3P and ^3D terms are collapsed (with $\langle r_{4f} \rangle \sim 1.2a_0$ as in the HF_{av} 4f orbital), but the function assocated with the optically allowed ^1P term is uncollapsed ($<r_{4f}> \underset{\sim}{\sim} 17$ a_0, the hydrogenic value). Thus for the optically allowed ^1P channel in Ba the situation in the HF_{TD} basis is similar to that in Xe and Cs: the optically allowed bound 4f state has little oscillator strength because of the small 4f-4d overlap, and the bulk of the absorption occurs in the $4d^9\epsilon f$ ^1P continuum.

These two views were later reconciled in the cases of Ba and La by Wendin and Starace,[14] who demonstrated an equivalence between the first order approximation wavefunctions derived by applying perturbation theory to the HF_{av} basis and the HF_{TD} wavefunctions. The extreme term dependence exhibited in HF_{TD} calculations manifests itself as a strong mixing of states in the HF_{av} nf and ϵf basis. The strong mixing leads to a relabeling of states in the ^1P channel with the lowest bound state having essentially the character of a $5f_{av}$ orbital (minus a small amount of $4f_{av}$ orbital which removes the node) and with the $4f_{av}$ orbital becoming the major component in the first order perturbed ϵf wavefunction in the vicinity of the "raised" eigenvalue obtained through diagonalization of the $4d^94f$ terms in the HF_{av} basis. Since, within the single particle picture, the HF_{TD} approximation accounts for intrachannel coupling to all orders, it is probably the more suitable model for this problem.

THE TERM-DEPENDENT MODEL FOR Ba, Ba$^+$ and Ba^{++}

The Ba, Ba$^+$ and Ba^{++} spectra provide a unique opportunity to observe some of the important aspects of 4f shell collapse and its effect on 4d-photoabsorption. In Ba and Ba^{++} there are three J=1 states that are accessible from the $^1S_0$ ground state:

$$\text{Ba} - (4d^94f\ 5s^25p^6\ 6s^2)\ ^1P,\ ^3P \text{ and } ^3D$$

$$\text{Ba}^{++} - (4d^94f\ 5s^6\ 5p^6)\ ^1P,\ ^3P \text{ and } ^3D.$$

Since LS coupling is a fairly good approximation for these $4d^94f$ configurations, the ^3P and ^3D have very little oscillator strength. The Ba$^+$ absorption spectrum in Fig. 1 was obtained from a plasma with approximately equal amounts of $5p^66s$ and $5p^65d$ population;

for this absorption spectrum, which is a superposition of spectra originating from both the 6s and 5d states, we must consider two sets of optically favored 4f states:

$[(4d^9 4f) \ ^1P \ 5s^2 5p^6 5d]^2F, \ ^2D, \ ^2P$ and $[(4d^9 4f) \ ^1P \ 5s^2 5p^6 6s]^2P$

In the non-relativistic approximation, the best single particle wavefunction (in the variational sense) is obtained from a term-dependent Hartree-Fock procedure in which a separate radial function is calculated for each term. For most cases there is little difference between the various radial functions of a given configuration, and a single radial function obtained from the minimization of the configuration average energy can be used for all subsequent calculations involving the term structure without any loss of accuracy. (This is the HF_{av} model.) However such is not the case in regions of the periodic table which are near points of shell collapse. As was indicated by our introductory discussion of the $4d^9 4f$ configuration in Ba, the conditions for collapse are so delicate that the collapse itself is term-dependent, and the basic assumption underlying the use of a single radial function in the traditional HF_{av} model is not fulfilled.

The sum of interactions between a single electron and all the electrons in another shell is term-dependent only for open shells. The present cases have a 4d open shell in addition to the singly-occupied 4f shell. (Ba^+ also has an open 6s shell, but the 4f-6s interactions do not lead to a fundamental change in interpretation.) The interaction energy between 4d and 4f shells, averaged over all LS terms of the $4d^9 4f$ configuration is:

$$E(d^9 f, \text{ av}) = 9F^0(d,f) - \frac{27}{70} G^1(d,f) - \frac{6}{35} G^3(d,f) - \frac{15}{77} G^5(d,f),$$

and the energies of the terms with J=1 components are:

$$E(d^9 f, \ ^1P) = E(d^9 f, \text{ av}) - \frac{8}{35} F^2(d,f) - \frac{22}{231} F^4(d,f) + \frac{137}{70} G^1(d,f)$$
$$- \frac{2}{105} G^3(d,f) - \frac{5}{231} G^5(d,f)$$

$$E(d^9 f, \ ^3P) = E(d^9 f, \ ^1P) - 2G^1 (d,f)$$

$$E(d^9 f, \ ^3D) = E(d^9 f, \ ^3P) + \frac{6}{35} F^2(4d,4f) + \frac{5}{21} F^4(4d,4f)$$

where $F^k(d,f)$ are the Slater direct integrals and $G^k(d,f)$ the exchange integrals.[15]

These integrals depend very strongly on the amount of overlap as can be seen when comparing the values associated with collapsed vs. uncollapsed 4f functions. For example in $(4d^9 4f)^{3,1}P \; 5s^2 5p^6$ of Cs^+:

	^3P (collapsed)	^1P (uncollapsed)
$F^0(4d,4f)$	0.852	0.134
$F^2(4d,4f)$	0.389	0.003
$F^4(4d,4f)$	0.245	0.0005
$G^1(4d,4f)$	0.460	0.0004
$G^3(4d,4f)$	0.283	0.0002
$G^5(4d,4f)$	0.199	0.0002

(In atomic units; evaluated using the non-relativistic MCHF-77 code of C. Froese-Fischer[16])

The Hartree-Fock equation for a given orbital is obtained by requiring that the total energy be stationary with respect to variations of that orbital. As we have seen, the sole difference in the energies of ^3P and ^1P terms of a $d^9 f$ configuration arises from the exchange integral $G^1(df)$ which appears in the total energy with a coefficient -3/7 for the ^3P term, and +11/7 for the ^1P term. As the exchange integral itself is a necessarily positive quantity which increases with increasing overlap between the d and f orbitals, it corresponds in the ^1P term to a repulsive interaction between d and f shells. In the cases considered here, this additional repulsion is sufficient to delay the collapse of the singlet term.

A more detailed dynamical description of the exhange interaction is complicated by its non-local character. In the equation for the 4f orbital, $P_{4f}(r)$ of the $4d^9 4f \; ^1P$ term for instance, the exhange potential appears in the form of a non-local potential acting on the 4f orbital $P_{4f}(r')$:

$$\frac{11}{7} \left[\int_0^\infty dr' \; \frac{r_<}{r_>^2} \; P_{4f}(r') \; P_{4d}(r') \right] P_{4d}(r). \qquad (3)$$

Because of this non-local property and because the Hartree-Fock equations must accomodate the requirments of self-consistency, it is rather difficult to predict just how the single term of Eq. 3 will affect the final solution. However, once the solution to the

Hartree-Fock equations has been obtained, one can construct an effective potential for each orbital by inverting the single particle Schrödinger equation. Results of this procedure applied to the 4f orbitals of the $4d^9 4f$ $^{3,1}P$ terms of Ba, are shown in Fig. 9. Since the equations of motion for the two 4f orbitals differ only in the coeffi-cient of the exchange potential, the strength

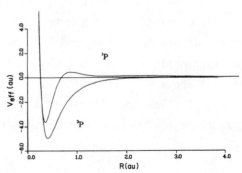

Fig. 9 Potentials for the 4f orbital of $^{3,1}P$ terms of the $4d^9 4f 5s^2 5p^6 6s^2$ configuration of Ba derived from HF_{TD} wavefunctions.

of the exchange potential for the singlet state can be regarded as the distance between the potential curves of Fig. 9. At its maximum it is about 50 eV, which is large compared to the binding energy of either 4f electron. Moreover, the exchange interaction results in the formation of a potential barrier for the singlet state at $r \sim 1a_o$. The height of this barrier, about 10 eV, is the same size as the delayed onset of the photoabsorption cross section as seen in Fig. 1.

In Fig. 10, we show the 4d orbital of Ba and the 4f orbitals of the configurations, $(4d^9 4f)$ ^1P $5s^2 5p^6 6s^x$ of Ba, Ba$^+$, Ba^{++} as cal-culated in the HF_{TD} approximation. The 4f orbitals of the corre-sponding triplet terms are all collapsed and are essentially iden-tical to the Ba orbital shown in Fig. 8. As the 6s electrons of Ba are removed, the screening of the nucleus is reduced, and the 4f orbital contracts gradually. A marked increase in 4d→4f overlap between Ba and Ba^{++} is apparent; this is consonant with the transfer of oscillator strength from the continuous to the discrete spectrum which is seen in Fig. 1. Detailed interpretation of the discrete spectrum of Ba^{++} remains a challenge, although some preliminary work has been done.[17,18] The existance of several strong features can be associated with similarly contracted term-dependent 5f and 6f orbitals.[18]

SYSTEMATICS OF SHELL CONTRACTION

If we consider, in the term dependent framework, the behavior of the 4f orbitals along and isoelectronic sequence, we find a contraction similar to that observed in the Ba isonuclear sequence. Fig. 11 shows HF_{TD} 4f orbitals for the same isoelectronic sequence as in Fig. 8. The transitions from hydrogenic to collapsed behavior is seen to proceed more gradually then is indicated in the HF_{av} picture.

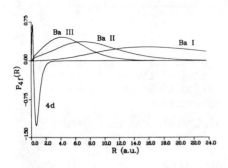

Fig. 10 The 4f orbitals of the states $(4d^9 4f)$ ^1P $5s^2 5p^6 6s^x$ of Ba (x=2), Ba$^+$ (x=1), and Ba^{++} (x=0); and the 4d orbital which is essentially independent of the ionization stage. From HF_{TD} calculations.

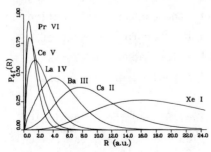

Fig. 11 The 4f orbitals for the state $4d^9 4f 5s^2 5p^6$ ^1P calculated in the HF_{TD} approximation, along the same isoelectronic sequence shown in Fig. 8. Collapse is replaced by contraction.

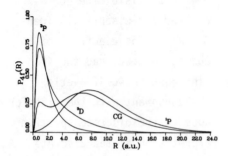

Fig. 12 The 4d orbitals for the J=1 terms of Cs II (Z=55) $4d^9 4f 5s^2 5p^6$, as calculated in HF_{av} (C.G.) and HF_{TD} (^3D, ^3P, ^1P) approximations.

Fig. 13 Measured (———)
vs. theoretical prediction
(-----), using method of
Starace[20] for metallic
La. Arrow indicates
position of calculated
$4d^9 4f \, ^1P$ eigenvalue in the
HF_{av} model.

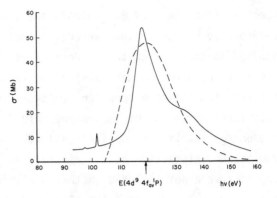

In atomic inner shell processes, therefore, the concept of
"contraction" better describes the trends associated with increasing
nuclear attraction than does "collapse". It should be noted that the
anomalous, bimodal feature in the HF_{av} orbital of Cs^+ at $\sim 2a_0$ (Fig. 8)
disappears in the HF_{TD} solution. From Fig. 12, it is apparent that the
HF_{av} approximation is attempting to accomodate two incompatible forms
of behavior, which are clearly distinguished in the HF_{TD} results.

It is clear from these figures that Ba is in the region of
transition between hydrogenic and fully collapsed behavior. We
also see that the effects of exchange, which are responsible
for the barium contraction, have pronounced influence in
this region which extends over a range of $\Delta Z \sim 3$. Beyond this
region (for instance, by Pr VI which can be seen in Figs. 8 and 11),
the exchange interaction does not have sufficient strength to
cause significant term dependence in the orbitals. For these cases,
then, a picture built on the HF_{av} approximation should provide an
adequate description of the single particle aspects of photoabsorption. Such a model has not been constructed in full detail, but
the semi-quantitative description of Dehmer et al.[11,19] appears to
indicate the course that should be taken.

AN INTERPRETATION OF Ba BASED ON HF_{av} WAVEFUNCTIONS

In view of the potential usefulness of the HF_{av} model for the
heavier elements and in an effort to clarify some persistent
misunderstandings associated with the early explanation of the Ba
4d photoabsorption,[12] it is instructive to look at the continuum

spectra of Ba, Ba^+ and Ba^{++} within the framework of this model as interpreted by Wendin and Starace.[14] As the HF_{av} model was originally applied to the rare earths by Dehmer, et al.[11] and by Sugar[19] the procedure used was to make a simple direct identification of the features in a given absorption spectrum with the energies and oscillator strengths of the appropriate $4d^9 4f^{N+1}$ terms derived from the energy matrix diagonalization in the HF_{av} representation. Effects due to the interaction of the discrete $4d^9 4f^{N+1}$ states and the $4d^9 4f^N \varepsilon\ell$ continua were not included in this preliminary set of calculations.

Starace[20] outlined a formal method, based on an extension of Fano's theory of autoionization,[21] to incorporate the model of Dehmer et al.[11] into a rigorous description of the 4d absorption process. By treating the interaction of the discrete and continuum states in the $\{nf, \varepsilon f\}$ manifold, Starace described how to calculate the positions and autoionizing widths associated with the "raised" $4d^9 4f^{N+1}$ states. The method was then applied to the case of metallic La (assumed to be trivalent with excited state configuration $4d^9 4f$).[22] The results of the calculation showed that the bulk of the La photoionization cross section can be attributed to the "raised" $4d^9 4f$ ^1P wavefunction which becomes mixed with continuum εf_{av} wavefunction over a band $\Delta\varepsilon \sim 40eV$. The calculated and observed photoionization for solid La are compared in Fig. 13.

While the formalism presented by Starace[20] is mathematically correct, the subsequent work of Wendin and Starace[14] showed that for Ba and La an interpretation consistent with the HF_{TD} model required the aforementioned relabeling of states amongst the first order perturbed HF_{av} wavefunctions. In this reinterpretation, the "raised" $4d^9 4f_{av}$ ^1P eigenvalue is no longer associated with an autoionizing resonance, but becomes associated with a resonant enhancement of the energy denominator of the $4f_{av}$ coefficient in the perturbation expansion for the continuum wavefunction. However, we expect that in the heavier rare earths, in which the 4f orbital is only slightly term-dependent, many terms associated with the $4d^9 4f^{N+1}$ configuration will lie above one or more of the complex set of limits associated with the $4d^9 4f^N$ configuration in the ion.

In such cases these $4d^9 4f^{N+1}$ terms can be considered to undergo proper autoionization into the neighboring $4d^9 4f^N \epsilon f$ continuum.

As can be seen from Fig. 13, the observed La photoionization is in qualitative agreement with the calculated curve, with the position of the "raised" $4d^9 4f_{av}$ ^1P eigenvalue near the energy of the central absorption peak. We may use the HF_{av} model to gain a qualitative understanding of the difference between the Ba^{++} and the Ba, Ba^+ spectra by noting the position of the appropriate HF_{av} eigenvalue relative to the corresponding continuum threshold. Table I contains the positions of the appropriate optically favored terms of Ba, Ba^+ and Ba^{++} relative to the calculated HF_{av} $4d^9$ limits. As can be seen, the $4d^9 4f_{av}$ ^1P position in Ba^{++} is below the 4d limit whereas all the optically favored terms in Ba and Ba^+ are raised above their respective limits. Because the $4d^9 4f_{av}$ ^1P of Ba^{++} does not lie in the continuum, the component of $4f_{av}$ in the expansion for the perturbed continuum wavefunction is not resonantly enhanced. Thus, we can expect that a significant fraction of the 4d-absorption oscillator strength will appear below the $4d^9$ ionization limit in Ba^{++}.

TABLE I. Calculations for Ba, Ba^+, and Ba^{++} based on the HF_{av} model.

	Initial state	Strongest allowed terms (zero order, LS coupling)	$E_{av}(4d^9 4f)$ [a] (eV)	$G^1(4d, 4f)$ [b] (eV)	Energy (eV) [a] terms column 3
Ba	$4d^{10} 6s^2\ ^1S_0$	$4d^9 4f 6s^2\ ^1P$	-6.5	9.4	$+9.1$
Ba^+	$4d^{10} 6s\ ^2S_{1/2}$	$4d^9 4f(^1P)6s\ ^2P$	-12.5	9.4	$+3.1$
	$4d^{10} 5d\ ^2D_{5/2,3/2}$	$4d^9 4f(^1P)5d\ ^2P, ^2D, ^2F$	-9.8	9.3	$+5.6^c$
Ba^{++}	$4d^{10} 5p^6\ ^1S_0$	$4d^9 4f\ ^1P$	-19.4	9.5	-3.7

[a] **Energies** relative to calculated limit; Ba: $E_{av}(4d^9 6s^2)$ of Ba^+; Ba^+: $E_{av}(4d^9 6s)$ or $E_{av}(4d^9 5d)$ of Ba^{++}; Ba^{++}: $E_{av}(4d^9)$ of Ba^{+++}.
[b] Scaled by a factor of 0.67, to account for additional configuration interaction. Cf. J. Sugar (Ref. 19).
^{c} $^2P, ^2D$, and ^2F terms located in a region which is centered near this value.

CONCLUSIONS

We have seen that term dependent effects dominate the behavior of atomic configurations with inner shell vacancies in the region near the collapse of the f shell. These term dependent effects stem from the strong exchange interaction and replace "collapse" by a more gradual contraction. We expect additional

600

implications for the interpretation of solid molecular spectra.

In Fig. 3, we see that in the lanthanides there is a close correspondence between atomic and solid spectra. For divalent Ba compounds, on the other hand, this correspondence at the least needs further elucidation. The correlation of spectra of the type shown in Fig. 14 remains a problem for the future.

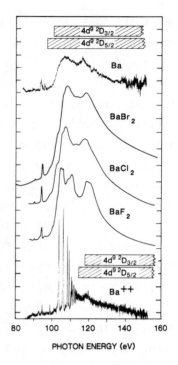

PHOTON ENERGY (eV)

ACKNOWLEDGEMENTS

We are deeply grateful to A.W. Weiss, S.M. Younger, W.C. Martin and J.W. Cooper for many spirited discussions and helpful advice. J.E. Hansen, B. Sonntag, G. Wendin, R. Karazija, and A. Karosiene have, through correspondence, provided much helpful information. Thanks also to the authors who generously allowed figures from their work to be reproduced here and to R.P. Madden for his constant support and encouragement. The excellent assistance of J. Bowles, C. Anthony and D. Dulik in preparing this manuscript is greatly appreciated.

Fig. 14   Spectra of BaBr$_2$, BaCl$_2$ and BaF$_2$ from work of Rabe[22] compared with spectra of Ba and Ba^{++} from Ref. 1

REFERENCES

1. T.B. Lucatorto, T.J. McIlrath, J. Sugar and S.M. Younger, Phys. Rev. Lett. 47, 1124 (1981).
2. E.E. Koch, C. Kunz and B. Sonntag, Phys. Repts. (Sect. C of Phys. Letts.) 29, 153 (1977).
3. E.-R. Radtke, J. Phys. B 12, L77 (1979); and Dissertation, Universität Bonn (unpublished, 1980).
4. U. Fano and J.W. Cooper, Rev. Mod. Phys. 40, 441 (1968).
5. E.-R. Radtke, J. Phys. B 12, L71 (1979).
6. H.P. Kelly, S.L. Carter, and B.E. Norum, Phys. Rev. A 25, 2052 (1982).
7. E. Fermi, Leipziger Vortrage, 95 (1928).
8. M. Goeppert Mayer, Phys. Rev. 60, 184 (1941).

9. P.C. Griffin, K.L. Andrew and R.D. Cowan, Phys. Rev. 177, 62 (1969).
10. B.F. Sonntag, DESY Report SR-81/05 (1981) (unpublished).
11. J.L. Dehmer, A.F. Starace, U. Fano, J. Sugar, J.W. Cooper, Phys. Rev. Lett. 26, 152 (1971).
12. D.L. Ederer, T.B. Lucatorto, E.B. Saloman, R.P. Madden and J. Sugar, J. Phys. B 8, L2 (1975).
13. J.E. Hansen, A.W. Fliflet, and H.P. Kelly, J. Phys. B 8, L127 (1975); see also: S.A. Kuchas, A.V. Karosens and R.I. Karaziya, Izv. Acad. Mauk. SSSR Ser. Phys. 40, 270 (1976).
14. G. Wendin and A.F. Starace, J. Phys. B 11, 4119 (1978).
15. R.D. Cowan, "The Theory of Atomic Structure and Spectra", (University of California Press, Berkeley and Los Angeles, California, 1981).
16. C. Froese-Fischer, Conp. Phys. Comm. 14, 145 (1978).
17. J.-P. Connerade and M.W.D. Mansfield, Phys. Rev. Lett. 48, 131 (1982).
18. S. Kuchas, A. Karosiene and R. Daraziyu, Lietuvos Fizikow Rinkinys 23, (in press) (Soviet Physics Collection, in press).
19. J. Sugar, Phys. Rev. B 5, 1785 (1972).
20. A.F. Starace, Phys. Rev. B 5, 1773 (1972).
21. U. Fano, Phys. Rev. 124, 1866 (1961).
22. J.L. Dehmer and A.F. Starace, Phys. Rev. B 5, 1792 (1972).
23. P. Rabe, thesis, DESY Intern Beicht No. F41-74/2, 1974 (unpublished).

*Permanent address: Institute of Physical Science and Technology, University of Maryland, College Park, MD 20742

PARTIAL PHOTOIONIZATION CROSS-SECTIONS OF RARE EARTHS METALS IN THE REGION OF THE 4d RESONANCE

F. Gerken, J. Barth and C. Kunz

II. Inst. f. Exp. Phys., Universität Hamburg, Luruper Chaussee 149

2000 Hamburg 50, FRG

ABSTRACT

The partial photoionization cross-sections of the 4d-, 5p-, 4f-shells and the valence band of Ce, Pr, Nd, Eu and Gd are measured in the region of the 4d→4f excitation. The sum is compared with the corresponding absorption spectrum in order to estimate the importance of different decay channels of the excited $4d^9 4f^{N+1}$ configurations. For the heavy rare earth metals, which show a large 4f multiplet splitting, we demonstrate that the coupling of the 4f ionization to the 4d→4f excitations strongly depends on the particular $(4f^{N-1})\ ^{2S+1}L_J$ - multiplet lines. This effect is also discussed for the 5p multiplet lines in Eu which arise from the coupling of the 5p hole with the 4f electrons.

INTRODUCTION

The partially filled and localized 4f shell in the rare earths gives rise to many interesting problems in different physical fields. The absorption spectra in the region of the 4d excitation show a number of sharp lines followed by a large maximum over an energy range of 10 to 20 eV [1]. These spectra, similar for rare earth atoms and solids [2], could be explained by excitations into bound 4f states [3] with a large multiplet splitting of the excited $4d^9 4f^{N+1}$ configuration [2,4]. Strong intershell interactions in the region of the 4d→4f excitation have been well established in photoemission experiments by use of synchrotron radiation ("resonant photoemission"). The partial photoionization cross-sections (PPCS's) show intensity variations which can be described by the Fano-theory [5,6,7]. For barium, which has no 4f electron in the ground state, the PPCS's of the 4d, 5p and 5s shell have been measured by Hecht and Lindau [8] at the 4d threshold and the results are compared with several calculations. For the 4d- and 5p-cross-sections the best agreement was found for RPAE calculations by Amusia [9] and by Wendin [10]. The theoretical understanding of the subshell cross-sections at the 4d threshold is quite satisfactory for the elements without 4f electrons (Ba and La) for which a consistent interpretation of alternative calculation procedures could be given by Wendin and Starace [11].

For the rare earths with a partially filled 4f shell the theoretical description is much more complicated. For Ce, which has one electron in the 4f shell in the ground state, Zangwill and Soven [12] presented a calculation of the different PPCS in a time dependent local density approximation. We have measured the 4d-, 5p-, 4f-shells and valence band cross-sections for Ce, Pr, Nd, Eu and Gd which show a relatively simple 4f-multiplet structure in photoemission spectra. Since the sum of all PPCS's gives the total absorption coefficient, the individual importance of the different contributions can be estimated. For the rare earths with a large 4f multiplet splitting our high resolution photoemission spectra demonstrate that the coupling of the 4f ionization to the 4d→4f excitation strongly depends on the particular $(4f^{N-1})$ $^{2S+1}L_J$-multiplet line. This is shown as an example for Er and Tm. This effect is also observed for the 5p multiplet lines which arise from the coupling of the 5p hole with the partially filled 4f shell and will be discussed for Eu.

EXPERIMENTAL PROCEDURE

Our measurements were performed at the Hamburger Synchrotron-strahlungslabor HASYLAB with the monochromator FLIPPER. Details of the experimental setup are given elsewhere [13]. The samples were evaporated in situ from tungsten baskets under UHV conditions onto stainless steel substrates. The spectra presented here are normalized to the incoming photon flux which was monitored with a Au-photo-diode [14]. The cylindrical mirror analyser was used in the retarding mode (ΔE=const.) to allow the measurement of high resolution CIS-spectra from narrowlyspaced structures. The spectra are corrected for the analyser transmission by a new method described in ref. 14. The influence of the mean free path to the PPCS has been omitted since the variation of the mean free path is small in the energy range of interest [15]. The different PPCS's are measured continuously by use of Constant-Initial-State (CIS) spectroscopy.

Although we used ultrapure materials (rare earth products 99.99 %) which we evaporated in ultrahigh vacuum ($\sim 1.10^{-10}$ Torr) a small contamination signal at about 6 eV binding energy was detectable on all samples. Auger measurements of the samples show that this contaminations are mainly oxygen and chlorine which are probably residuals from the technical production process [16]. We can estimate the coverage to be less than 0.1 monolayer if we take into account that our measurements are extremely surface sensitive. Therefore we do not expect an influence of the contaminations to the PPCS's of interest.

RESULTS AND DISCUSSION

The data analysis for the PPCS-measurements is the same for all materials and shall be described in detail only for Ce. Fig. 1 shows a series of Electron Distribution Curves (EDC) for photon energies in the region of the 4d excitation. The Ce spectra show a double structure with maxima at 0.5 eV and 2 eV binding energy. It is still a matter of discussion whether only the 2 eV peak or both structures have 4f character [6,17,18,19,20]. Both peaks show drastic intensity variations when the photon energy is tuned to the region of the 4d→4f excitation. This phenomenon is usually described as a direct

604

recombination of the excited state resulting from a process such as

$$4d^{10}5s^25p^64f^N V \rightarrow 4d^95s^25p^64f^{N+1}V \rightarrow 4d^{10}5s^25p^64f^{N-1}V\epsilon\ell$$

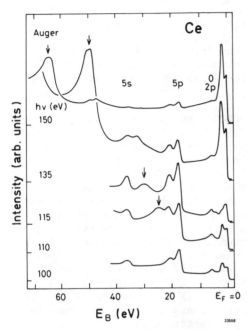

Fig. 1: EDC's of Ce metal in the region of the 4d → 4f excitation. The spectra are normalized to the photon flux. The arrows indicate the main Auger peak.

V stands for the valence band 5d- and 6s-electrons. The process given above interferes with the direct photoemission from the 4f shell. The resulting intensity profile can be described by the Fano interference formalism[6,17].

Apart from this process the energy from the direct recombination may be transferred to electrons from the other outer shells resulting in processes such as

$$4d^95s^25p^64f^{N+1}V \rightarrow$$

$$\rightarrow \begin{cases} 4d^{10}5s\ 5p^64f^N V\epsilon\ell \\ 4d^{10}5s^25p^54f^N V\epsilon\ell \\ 4d^{10}5s^25p^64f^N V^{(-1)}\epsilon\ell \\ 4d^95s^25p^64f^N V\epsilon\ell \end{cases}$$

The last decay is equivalent to the direct photoemission of the 4d electron. The 4d hole decays via an Auger process leading to an Auger structure in the spectra at fixed kinetic energy. In Fig. 1 the main Auger peak is indicated by an arrow. The in-

tensity of the Auger peak should be directly proportional to the number of 4d holes. From Fig. 1 it is obvious that this autoionisation process is an important competition compared to the direct recombination processes.

In order to determine the relative intensities of all PPCS's we have measured CIS-spectra from the structures with $E_B = 0,5$ eV and $E_B = 2$ eV and the 5p structure with the lowest binding energy and normalized these spectra to the corresponding peak areas as shown in Fig. 2. For each structure the background of inelastically scattered electrons which is created by the other structures at lower binding energies is considered by subtraction of a CIS-sprectrum of the preceding background at lower initial energy.

Since the direct 4d emission near threshold is superimposed on the very steep background of scattered electrons at low kinetic energies, this channel could not be obtained with high accuracy by CIS-spectroscopy. In order to solve this problem we have measured a series of Constant-Final-State (CFS) spectra, where the electron

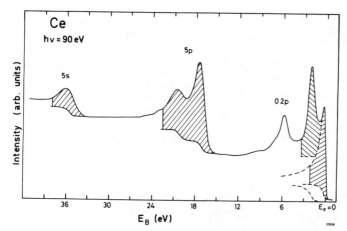

33550

Fig. 2: EDC of Ce metal measured with a photon energy of 9o eV indicating the peak areas which are used for the correction of the CIS-spectra

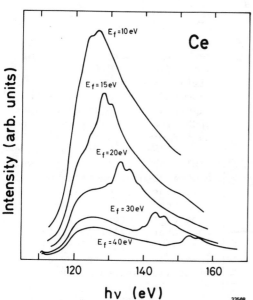

33568

analyser is fixed to a constant final energy E_f while the photon energy is varied. The result for five different final energies is shown in Fig. 3. In such spectra the shape of the background is proportional to the absorption structure independently of the analyser's final energy. The contribution from the direct 4d emission appears at different photon energies superimposed on this background and can be much easier extracted. At threshold the Auger intensity serves to estimate the 4d-PPCS. In Fig. 4 the different PPCS's are compared with the absorption structure which was measured using the partial yield technique, i.e., by measuring the yield of low-energy electrons as a function of photon energy. The 4f-, 5p, E_B = o.5 eV- and E_B = 2 eV PPCS's which are continuously mea-

Fig. 3: Series of CFS-spectra of Ce metal with different final energies E_f. The 4e emission is superimposed on the absorption structure and shifts on the photon energy scale when changing the final energy

sured by CIS- spectroscopy are subtracted from the total yield ("a-b" in Fig. 4). The remaining part agrees well with the data points obtained for the 4d-PPCS. The 5s-PPCS could not be determined because of overlapping Auger structures for excitation energies within the giant resonance. This fact may explain the remaining small

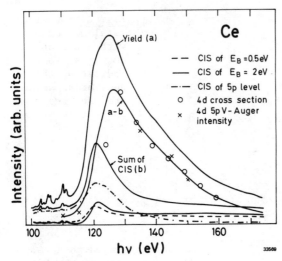

Fig. 4: Different partial photoionization cross-sections (PPCS's) for Ce metal. The sum of the $E_B = 0.5$ eV-, $E_B = 2$ eV- and $5p$- PPCS's (b) is substracted from the yield spectrum (a) and the remaining part (a-b) is in good agreement with the obtained points for the 4d-PPCS.

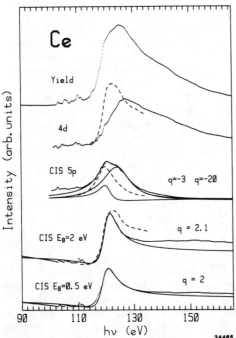

Fig. 5: Line shapes of the different PPCS's for Ce metal on an arbitrary scale ....:experimental data, ———: fitted Fano-profiles with given q-values, ----:theory of Zangwill and Soven[12].

deviations between the sum of all measured PPCS's and the yield since the 5s PPCS must be expected to be of comparable magnitude to the valence band-PPCS. The 4d-PPCS is by far the most dominant part of all PPCS's for Ce and is obviously responsible for the high energy part of the absorption structure while the preceding shoulder is reproduced by the enhancement of the cross-sections of the outer levels. Apart from the relative intensities it is interesting to note the dissimilar line shapes of the different PPCS's from Fig. 5. While the two structures with $E_B = 0.5$ eV and $E_B = 2$ eV can be fitted quite well by a single Fano-profile with a positive q-value, the 5p-PPCS consists of at least two curves with negative q-values and peak-positions corresponding to the double-structure of the absorption profile. Our results can be compared with the calculation of Zangwill and Soven[12] in the TDLDA. The relative inten-

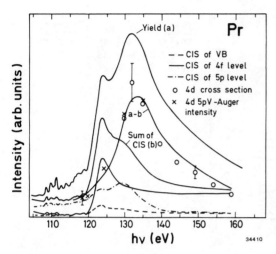

Fig. 6: Different PPCS's for Pr metal analysed in the same way as for Ce metal (see Fig. 4)

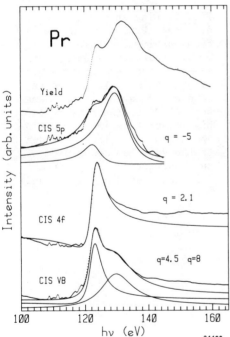

Fig. 7: Like shapes of the different PPCS's for Pr metal on an arbitrary scale ....: experimental data, ———: fitted Fano profile with given q-values

sities of their calculation are in agreement with our results with respect to the experimental error which is mainly caused by the difficult determination of the exact peak areas for the normalization of the CIS spectra. This is due to the rich multiplet splitting of the 5p- and 4d core levels in the photoemission spectra. On the other hand the calculated line shapes and peak positions are in distinct contradiction to our results (see Fig. 5). The 4d- and 4f PPCS's differ in line shapes as well as in the position of the maxima (~ 4 eV) while Zangwill and Soven emphasize the similarity of the energy dependence of both PPCS. Furthermore, the theory is unable to explain the double-structure of the 5p-PPCS which is obviously coupled to different multiplet lines of the absorption curve. We can compare our results for Ce with an analysis of our Pr data. Pr is the following element of the rare earths with two 4f electrons in the ground state. The

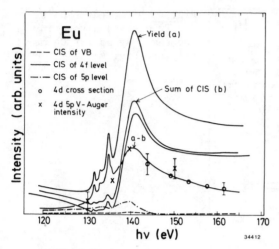

Fig. 8: Different PPCS's for Eu metal analysed in the same way as for Ce metal (see Fig. 4)

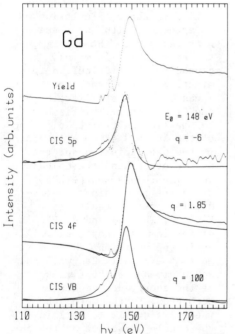

Fig. 9: Line shapes of the different PPCS's for Gd metal on an arbitrary scale ....: experimental data, ———: fitted Fano profile with given q-values. The given resonance energy E_0 is the same for all PPCS. The relatively high statistics in the 5p CIS sprectrum above 15o eV is due to the very low count rate of the subtracted CIS sprectra (5p and preceding background).

4f emission appears at about 3.5 eV binding energy in the photo-emission spectra. The different PPCS's in the region of the 4d → 4f excitation are shown in Fig. 6.

Similar to Ce the absorption-curve of Pr consists of a double structure in the giant maximum but the separation of the peaks is significantly larger in Pr. While the relative intensities of the different PPCS's changed from Ce to Pr, the line shapes are comparable (see Fig. 6 and Fig. 7). Again the 4d-PPCS coincides with

	valence band				4f				5p			
	relative intensity	E_o (eV)	Γ (eV)	q	relative intensity	E_o (eV)	Γ (eV)	q	relative intensity	E_o (eV)	Γ (eV)	q
Ce	80	119.25	2.9	2	100	120.05	2.8	2.1	200	120.6	2.5	- 3
										124.3	6	-20
Pr	20	122.6	2.2	4.5	100	122.6	2.4	2.0	60	123	3.3	- 5
		129	6	8						130.5	5.0	- 5
Eu	2.5	138.2	3.7	5	100	138.9	3.0	1.45	13	140.5	2.75	- 3
Gd	3	148	3	100	100	148	3	1.85	10	148	3	- 6

Table 1: Fit parameter for the Fano-profiles for Ce, Pr, Eu and Gd. The relative intensities of the different PPCS refer to the intensity in the maximum of the giant resonance relative to the intensity of the 4f level (E_B = 2 eV in Ce) which is arbitrarely chosen to be 100 in each case.

the high energy maximum of the yield. The 4f-PPCS only contributes to the onset of the giant resonance and can be well fitted by a single Fano-profile. The valence band- and 5p-PPCS's are coupled to the whole absorption structure and can be fitted by a sum of two Fano-profiles each with negative q-values for the 5p- and positive q-values for the valence band PPCS. The significantly lower q-value for the first Fano-profile of the valence band-PPCS in Pr compared to the second one may be caused by an overlap of the relatively broad 4f multiplet lines (HWHM $\cong$ 0.4 eV) with the preceding valence band resulting in a mixture of 4f- and VB-PPCS in the measured CIS-spectrum.

The line shapes for the different PPCS's discussed above appear to be independent of the material. This is confirmed by the analysis for Nd, Eu and Gd: negative q-values for the 5p-PPCS's, low positive q-values for the 4f-PPCS's and high positive (nearly Lorentzian-like) q-values for the valence-band PPCS's (see Table 1). Especially the elements in the middle of the rare earths series, Eu and Gd which both have a half filled 4f shell in the ground state, are a good example for the different PPCS-line shapes. The width of the absorption structure decreases from 2o eV to about 1o eV for the heavy rare earths (from Eu to Tm) resulting in PPCS's with a single peak in the giant absorption maximum. Here, the peak positions are nearly the same for all different PPCS's. Their relative intensities change continuously from Ce to Gd. The data analysis for Eu in Fig. 8 shows the increasing importance of the 4f-PPCS compared to the lighter rare earths. In Fig. 9 the different line shapes of the PPCS's in Gd are presented on an arbitrary intensity scale. In Gd the valence band and the 4f level are well separated by about 8 eV in the photoemission spectrum. This provides an easy determination of the PPCS's without mixing of different levels. In this case the valence band PPCS can be fitted with an almost Lorentzian curve.

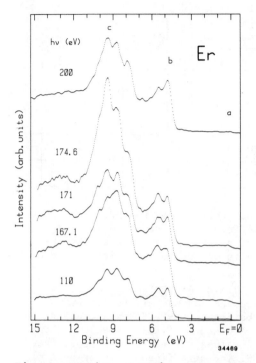

Fig. 1o: Series of EDC's of Er metal showing the valence band region with the rich multiplet structure of the $4f^{11} \rightarrow 4f^{10}$ transition between 4 eV and 15 eV binding energy. A drastic change in the relative intensities of the different multiplet structures appears when the photon energy is scanned to the region of the $4d \rightarrow 4f$ excitation

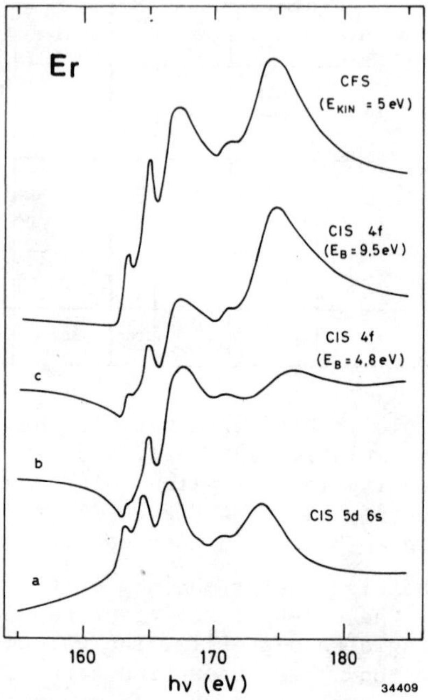

Fig. 11: CIS spectra of two different 4f multiplet lines (b and c from Fig. 1o) and from the 5d6s-valence band (a) compared with a CFS which is proportional to the absorption structure

Comparing the results of Ce, Pr, Nd, Eu and Gd it is striking that only for Ce the line shapes of the two structures with $E_B = 0.5$ eV and $E_B = 2$ eV are very similar and both show a 4f-like resonance, e.g. there is no structure in the Ce spectra which shows the same behaviour as the 5d6s-valence electrons from the other materials. Since we have no reasons to expect the 5d6s-electrons in Ce to be totally different from the other elements we come to the conclusion that there is also 4f character in both the $E_B = 0.5$ eV and $E_B = 2$ eV peaks. Indeed, it is also possible to fit the Ce CIS-spectrum of the peak with $E_B = 0.5$ eV with a sum of a Lorentzian and a Fano profile which corroborates with the expectation that the $E_B = 0.5$ eV peak consists of a 5d6s- and a 4f-part.

The data analysis presented here comprises only the most important decay channels in the $4d \rightarrow 4f$ excitation range. We neglect

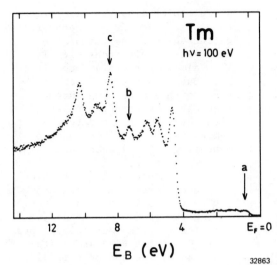

Fig. 12: EDC of Tm metal at 100 eV photon energy showing the multiplet structure of the $4f^{12} \to 4f^{11}$ transition (experimental resolution: 0.2 eV). The letters refer to the CIS spectra in Fig. 13

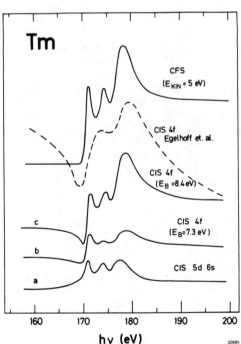

Fig. 13: CIS spectra of two different 4f multiplet lines (b and c) and from the 5d6s-valence band (a) from Tm metal compared with a CFS and a 4f CIS spectrum from Ref. 22 (dashed curve).

smaller effects such as the appearance of additional multiplet lines [21] and different PPCS's for the different $(4f^{N-1})^{2S+1}L_J$ multiplets which influence the sum of all PPCS's mainly in the region of the absorption fine structure preceding the giant maximum. The latter effect can easily be studied on heavy rare earth metals if monochromator and spectrometer are operated with high resolution. Fig. 10 shows a series of EDC's of Erbium (ground state $4f^{11}$) around the $4d \to 4f$ resonance. A dramatic change in the relative intensities of the 4f multiplets lines occurs when the photon energy is tuned to the region of the $4d \to 4f$ excitation. Fig. 11 presents CIS-spectra of the valence band and two different 4f multiplet lines as indicated in Fig. 10 compared with an absorption curve (CFS in Fig. 11). Equivalent measurements for Tm are presented in Fig. 13. The different 4f-CIS spectra refer to two neighbouring

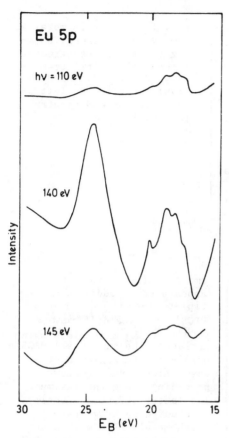

Eu 5p

hν = 110 eV

140 eV

145 eV

Intensity

E_B (eV)

30   25   20   15

Fig. 14: EDS's of the 5p multiplet lines in Eu for photon
energies below and within the
4d resonance. The spectra are
measured with zero-suppresion.
The amplitudes are normalized
to the incoming photon flux.
A dramatic change in the relative multiplet intensities
appears in the region of the
4d resonance

4f multiplet lines in Fig. 12. We note that also for the heavy rare earths the CIS spectra of the 5d6s valence electrons clearly show almost Lorentzian profiles.

Our results for Tm are compared with a recent measurement from Egelhoff et al.[22] obtained with obviously reduced resolution demonstrating the importance of measuring these spectra with high resolving power.

The theoretical description of these effects is complicated but Davis[23] succeeded in calculating the PPCS's of the different $(4f^{12})^{2S+1}L_J$ multiplet-lines in Yb_2O_3 in the 4d → 4f excitation range. The sum of this different PPCS's is in good agreement with an experimental result from Johansson et al.[24]

Finally Fig. 14 shows a series of EDC's of the Eu-5p-core level for excitation energies below and within the 4d-resonance. The photoemission of a 5p-electron in the RE results in a coupling of the 5p-hole to the partially filled 4f-shell which leads to a complicated multiplet structure. The simplest case is given for the half filled 4f-shell in Eu with 6 possible final states: $4f^7(^8S_{7/2})5p^5$ $^{7,9}P_J$. It is obvious from Fig. 14 that the coupling of the individual 5p multilet lines in the EDC's to the individual 4d → 4f resonant states differs. In this respect the result is similar to that obtained for the 4f-multiplets on the heavy rare earths. For Gd which

has the same number of 4f electrons the result is completely equivalent to Eu. For the other rare earths we find the same behaviour.

CONCLUSION

We have shown that the main absorption structures of the rare earths in the region of the 4d → 4f excitation can be composed by the sum of the different outer shell PPCS, e.g. there are no important decay channels which lead to satellite structures of considerable intensity as observed in the elements of the 3d-series.[15,25,26] The relative intensities of the different $2S+1_{L_J}$ final state multiplet lines are strongly influenced by the intershell interaction. This has been demonstrated for the $4f^{10}$ multiplets in Er and the $4f^{11}$ multiplets in Tm as well as for the 5p multiplet lines in Eu which arise from the coupling of a 5p-hole with the partially filled 4f shell. These effects as well as the appearance of additional multiplet lines[19] influence the sum of the PPCS mainly in the fine structure of the absorption.For Tm it was possible to correct previous errors in the 4f resonance spectra of other authors[22] by our high resolution measurements. A further improvement of our experimental set-up which is now in preparation will in future allow an even more detailled analysis including all effects described above in the measurements of the PPCS and therefore will provide even deeper insight into the intershell interactions in the rare earth resonances. The theoretical understanding of the resonant photoemission in the rare earth is in the beginning stage and we hope that our results may stimulate a futher progress in this field.

REFERENCES

1. T.M. Zimkina, V.A. Fomichev, S.A. Gribovskii and I.I. Zhukova, Fiz. Tverd. Tela 9, 1447 (1967) Sov. Phys. Solid State 9, 1128 (1967)

2. H.W. Wolff, R. Bruhn, K. Radler and B. Sonntag, Phys. Lett. 59A, 67 (1976)

3. J.L. Dehmer, A.F. Starace, U. Fano, J. Sugar and J.W. Cooper, Phys. Rev. Lett. 26, 1521 (1971)

4. J. Sugar, Phys. Rev. B5, 1785 (1972)

5. W. Lenth, F. Lutz, J. Barth, G. Kalkoffen and C. Kunz, Phys. Rev. Lett. 41, 1185 (1978)

6. L.I. Johansson, J.W. Allen, T. Gustafsson, I. Lindau and S.B. Hagström, Solid State Commun. 28, 53 (1978)

7. F. Gerken, J. Barth, K.L.I. Kobayashi and C. Kunz, Solid State Commun. 35, 179 (1980)

8. M. Hecht and I. Lindau, Phys. Rev. Lett. 47, 821 (1981)

9. M.Y. Amusia, V.K. Ivanov, L.V. Chernysheva, Phys. Lett. 59A, 191 (1979)

614

1o. G. Wendin in VUV Radiation Physics, edited by E.E. Koch et al. (Vieweg-Pergamon, Berlin, 1974) p. 225

11. G. Wendin and A.F. Starace, J. Phys. B11, 4119 (1978)

12. A. Zangwill and P. Soven, Phys. Rev. Lett. 45, 2o4 (198o)

13. J. Barth, F. Gerken, J. Schmidt-May and C. Kunz, Contribution to the Synchrotron Radiation Instrumentation (SRI) Conference, Hamburg, 1982, to be published in Nucl. Instr. and Methods

14. J. Barth, F. Gerken and C. Kunz, Contribution to the SRI Conference, Hamburg, 1982, to be published in Nucl. Instr. and Methods

15. J. Barth, Thesis, University of Hamburg (1982)

16. A. Platau, Thesis, University of Linköping, Sweden (1982)

17. G. Kalkoffen, Thesis, University of Hamburg, Germany (1978)

18. D. Wieliczka, J.H. Weaver, D.W. Lynch and C.G. Olson, to be published

19. S.H. Lin and K.-M. Ho, to be published

2o. S. Hüfner and P. Steiner, Zeitschrift f. Phys. B46, 37 (1982)

21. F. Gerken, J. Barth and C. Kunz, Phys. Rev. Lett. 47, 993 (198.)

22. W.F. Egelhoff, E.E. Tibbetts, M.H. Hecht and I. Lindau, Phys. Rev. Lett. 46, 1o71 (1981)

23. C. Davis, to be published

24. L.I.Johansson, J.W. Allen, I. Lindau, M.H. Hecht and S.B.M. Hagström, Phys. Rev. B21, 14o8 (198o)

25. J. Barth, F. Gerken, K.L.I. Kobayashi, J.H. Weaver and B. Sonntag, J. Phys. C13, 1369 (198o)

26. J. Barth, F. Gerken and C. Kunz, to be published

PHOTOIONIZATION OF ATOMS IN EXCITED STATES

François J. Wuilleumier

Laboratoire de Spectroscopie Atomique et Ionique and LURE, Université
Paris-Sud, 91405- Orsay, France

ABSTRACT

This paper gives a review of photoionization studies in atoms in
which an outer electron has been promoted onto an excited orbital.
The various experimental methods presently available for such studies
are briefly described. Examples are given to illustrate the type of
information which can be extracted from the few experiments carried
out so far. Special emphasis is put on the first successful experiment combining the use of a cw dye laser tuned to the first resonance
line and of synchrotron radiation to observe electrons produced by
inner shell and outer shell photoionization of excited Na and Ba a-
toms.

INTRODUCTION

This paper presents a short review of photoionization studies
in free atoms in which an outer electron has been promoted onto an
empty excited orbital. Only <u>single</u> photon absorption by atoms brought
into such excited states will be considered here.

Photoionization experiments on atoms in the ground state contributed, to a large extent, to a better knowledge of the geometrical and
dynamical properties of the electronic motion. In particular, the expanding availability of high flux of monochromatized synchrotron radiation over a broad photon energy range (at present from 10 to 200
eV, typically) played an important role in determining atomic parameters such as the binding energies of atomic electrons, the individual
subshell photoionization cross sections and oscillator strengths of
discrete transitions, the angular distribution assymetry parameter
of Auger and photoelectrons, and quite recently, the spin polarization of the photoelectrons. This long list serves to illustrate the
importance of photon impact experiments as a key to our understanding of atomic structure. Parallel to this experimental activity,
continuous progress has been made in the development of theoretical
methods. This interplay between theory and experiment has provided
unique information about the characteristics of the wavefunction
which map out the photoionization process. The behaviour of these
wavefunctions, their amplitude and phase are sensitive indicators of
the potential that describes the systems. In this way, the role played by electron correlations and relativistic effects is better and
better understood. For a review of the present state of the experimental and theoretical situation in this field, the reader is referred to recent articles[1-7] and the references quoted therein.

Up to a recent past, most of the photoionization measurements
had been made when the initial state of the atom under investigation
was the ground state. However, the information obtained from these
experiments, while having a great value, is somewhat restrictive.

Because of the dipole selection rules, only a limited class of states can be probed. Furthermore, the initial state is often an ensemble of nearly degenerate levels, particularly in open shell atoms. With the recent advent of high power-frequency tunable dye lasers, it has become possible to prepare the initial state in a specific way. Interaction of photons with such excited states is of fundamental interest in atomic physics. In particular, the so-called "complete" or perfect experiment[8] in which one wants to determine not only the moduli of the transition amplitudes that describe the photoionization process, but also their relative phases might be easier to interprate when starting from a well defined excited state. In fact, photoionization of such excited states are probably among the most easily interpretable experiments that may be performed. In particular, new effects have been predicted[9] for the energy dependence of the corresponding photoionization cross sections which are expected to present, in some cases, up to three minima in the $\ell \rightarrow \ell+1$ channel and one in the $\ell \rightarrow \ell-1$ channel, at variance with the ground state. In addition to the capability of preparing a system in a specific state, the excitation of an outer electron can modify dramatically the effective potential experienced by inner electrons with high orbital quantum numbers; in this way, the so-called collapse[10] or alternately, the contraction[11,12] of atomic wavefunctions could be induced and controlled. Furthermore, when one starts from an optically excited state, it is also possible to produce by step-wise excitation autoionizing states having the same parity as the ground state and to obtain experimental information on the geometrical and dynamical properties of atoms from states hitherto inaccessible.

The new experimental possibilities, initially demonstrated by the measurements of a few excited state photoionization cross sections[13-16], motivated theoreticians to calculate such cross sections, although it should be pointed out that such calculations are not yet numerous. Hartree-Slater potential was used to examine first the photoionization of alkali metals[17] and, later on, to extend these calculations to the 3d-4d and 5d excited states across the periodic table.[18] A single-particle parametric potential was applied to the calculations of cross sections for s, p and d rydberg states of Li, Na and K.[19] Photoionization of inner-shell electron under changing outer-shell electronic environment in alkali atoms was recently investigated.[20] Photoionization cross sections of excited states in rare gas atoms were also calculated in a few cases.[21,22]

In the following, we will give some indications on the actual possibility to study experimentally photoionization of excited states. Then, we will present a few examples of previous experiments in which laser radiation was used to photoionize excited electrons. Finally, we will describe in more details recent experiments in which the use of a continuum source of photons tunable over an extended energy range in the VUV region (20-150 eV) made possible to get the first information on inner-shell absorption-ionization and autoionization processes in laser excited atoms.

EXPERIMENTAL METHODS

The experimental studies of photoionization of excited states require the combination of two sets of techniques: the usual techniques developed for photoionization of atoms in their ground state and special methods to prepare the atoms in a specific excited state.

Various techniques of increasing sophistication have been used to explore the photoionization of ground state species. In a simple photoabsorption experiment, carried out with a continuous source of photons, the excited states of the atom are populated at discrete photon energies: they show up as resonance excitations in the photoabsorption spectrum. In addition, the total photoabsorption cross section can be measured: it is the sum of partial cross sections corresponding to the various excitation and ionization processes that are possible, depending upon the photon energy. When one process is dominant, the variation of the total photoabsorption cross section gives a good indication of the energy dependence for this particular process. Below the second ionization threshold of the atoms under investigation, the photoabsorption cross section is stricly equivalent to the photoionization cross section for the subshell whom binding energy is lower than the photon energy (outer electron in the ground state, excited electron in the excited state). But, as soon as the photon energy exceeds this second ionization threshold in a many-electron atom, a photoabsorption experiment is unable to distinguish betwwen the various photoionization processes, since electrons belonging to different subshells contribute to the total photoabsorption cross section. Electron spectroscopy has then to be used to analyse individually the exit channels open in the continuum. Since a well defined kinetic energy of the photoelectrons corresponds to each final state of the residual positive singly charged ion, the various photoionization processes can be clearly identified in the spectrum and analyzed subshell by subshell. In addition, when a hole is created in inner subshell or when two outer electrons are simultaneously excited to produce an autoionizing state, the subsequent Auger decay or autoionization give rise to additional electron peaks appearing usually at constant kinetic energies in the electron spectrum. Above the double ionization threshold, photoion spectroscopy enables one to measure the relative abundance of singly and multiply charged ions. Combining the relative data obtained in electron and ion spectroscopies, it is possible to derive, on an absolute basis, the partial cross section for each subshell by normalization to the measured total photoabsorption cross section.

The characteristics required from a monochromatic photon beam continuously tunable over an extended energy range are quite different for photoabsorption and for the more selective spectroscopies. In photoabsorption, it is possible to confine the sample in a cell with windows, since one has only to count, on a relative scale, how many photons are transmitted through the sample. High pressures can be used, typically 1 to 10 torr in the VUV region. Low intensity levels of radiation are sufficient for such a measurement. For electron spectroscopy, a much more intense and well collimated photon beam is required, because, in these experiments. the pressure in the interaction zone has to be lower by typically four orders of magnitude: no window can be used between this zone and the particle detector; furthermore, the transmission of an electron spectrometer is low, e.g. 1% for a Cylindrical Mirror Analyzer, and even less when angular distribution studies are being made. For this type of experiments, synchrotron radiation is the most suitable photon source in the VUV range. The reader is referred to recent review papers for more detailed information about the use of synchrotron radiation for photoionization studies.[2,3,5]

Because of the very narrow energy range (typically 1 to 4 eV) in which dye lasers can be tuned to produce atoms in excited states, most of these experiments have to deal with metallic vapors in which the binding energy of the outer electrons is lower than 5 to 6 eV. For absorption studies, the heat pipe system with two windows and a buffer gas to confine the vapor has been widely used. In case of photoemission experiments, the requirements to combine a high temperature vapor source with an electron spectrometer add to the experimental difficulties. The vapor source has to provide a collimated atomic beam of sufficient density (10^{12} to 10^{13} /cm³), operate stable for hours, be easily removable, cleanable and rechargeable, cause low electric and magnetic stray fields.

We turn now to the experimental techniques available to produce excited states. Only the advent of high power—narrow band width—frequency tunable dye lasers has made practically possible the preparation of atoms in specific non-metastable excited states. In the first experiments, the lasers were pulsed lasers; later on, the development of cw dye lasers allowed to maintain a steady population of excited states over a long period of time. When the laser preparing the excited state is a pulsed laser, with pulse duration ranging for the nsec to the μsec region, the ionizing agent has also to be pulsed. Thus the first series of experiments were limited to laser ionization of laser excited states. With this mode of excitation, the energy range explored was restricted to absorption measurements a few electron volts above threshold. It was also the case for photoionization of metastable excited states produced by electron bombardment of a beam of rare gas atoms: the density of atoms prepared in such metastable excited states was so low that only the very intense photon beam from a laser could produce detectable effects. In both type of experiments, of course, the photoionization of only the excited electron could be probed.

The energy range of photoabsorption experiments was considerably expanded by using a pulsed source of continuum radiation, produced by a discharge between two uranium electrodes, to probe the absorption of atoms brought into some excited states with a pulsed laser. In this way, absorption measurements have been made over a broad photon energy range (30 eV to 100 eV), providing the first information on inner-shell photoabsorption in laser excited atoms.

Finally the wide spectrum of photon energies available with synchrotron radiation was an obvious extension of such experiments. Combining the use of a cw dye laser and of a monochromatic beam of synchrotron radiation, energy resolved photoemission spectra from inner and outer subshells in optically excited atoms were recently obtained; in addition, a quantitative study of autoionization processes from states having the same parity as the ground state was also achieved.

In the three following parts, we will review the various experimental works corresponding to these three categories: laser ionization of excited atoms, inner-shell photoabsorption of laser excited atoms, electron spectrometry of laser excited atoms using synchrotron radiation.

LASER IONIZATION OF EXCITED ATOMS

The first experiments combining the use of high power-frequency tunable dye lasers to selectively excite, and subsequently ionize, the outer electron in alkali metals[14,23,24] and alkaline earths[13,16,25-28] took place more than ten years ago. Nd glass lasers, delivering typical pulses of 30J in 30 nsec, were usually used to produce the primary laser beam at 1.06 μ. Frequency doubling in ADP crystal provided the second harmonic beam to pump a dye cell and to produce the excited state. The fourth harmonic generated in another frequency doubling ADP crystal serves to produce the background continuum necessary for photoabsorption transitions from the excited states. Fig. 1 shows, as an example, the microdensitometer trace of the photographic plate corresponding to the $3s3p^1P_1 \rightarrow 3snd^1D_2$ absorption series obtained from selectively excited MgI.[26] Later on, the same group made the first absolute measurement of the cross section for photoionization from a selectively excited short lived atomic state: using two simultaneous tunable dye laser pulses, they recorded the number of photoelectrons produced by a measured laser flux![16] Their result, corresponding to the $3s3p^1P_1 \rightarrow 3p^2\ ^1S_0$ autoionizing transition in MgI

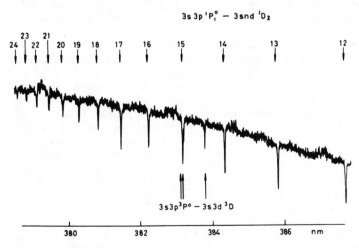

Fig. 1.- Microdensitometer trace showing higher terms of the $3s3p^1P_1 \rightarrow 3snd^1D_2$ absorption series from selectively excited MgI (from Ref 26)

620

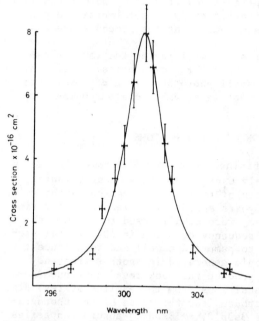

Fig.2- Photoionization cross section from the $3s3p^1P_1$ state of magnesium. The second laser is tuned to the $3s3p^1P_1 \rightarrow 3p^{2} {}^1S_0$ autoionizing transition (from Ref. 16).

is shown in Fig.2. Variations of the state of polarization of the laser beams allowed to select a particular angular momentum state in the continuum, enabling the relative contributions of different states to the total photoionization cross section to be measured.

This combination of two dye lasers was, for a long time, the only way to measure directly photoionization cross sections for non metastable excited states. The 4D and 5S states of Na (with two cw dye lasers pumped by a Nd-glass laser)[29], the 5P[30] and 6P[31] states of Rb, the 2P state of Li[32], the 6S[33] and 7P[34] states of Cs were investigated in this way, at one photon energy. Because of the high photon flux available from such laser beams, angular distribution of the photoelectrons could also be determined for ionization of the 3P states of Na(with a N_2 laser)[35] and 7P states of Cs[34]; even the spin polarization of the photoelectrons was measured in the case of Cs.

Another type of experiment deserves special interest. Using electron bombardment to produce the metastable excited states and a dye laser pumped by a N_2 laser, the absolute cross sections for the photoionization of He(2^1S) and He(2^3S) metastable atoms was measured from threshold to 2400 Å.[15] The results for He(2^1S) atoms are presented in Fig.3, showing excellent agreement with some theoretical calculations[36-38]. Only recently, a similar experiment was carried out again, for polarized Ne $3p^3D_3$ atoms.[39]

In ten years, the collection of pulsed and cw dye lasers available has grown up in an impressive way. However, in spite of their great interest, measurements of the type described in this part are severely limited by the narrowness of the energy range in which dye lasers are presently tunable or can be expected to become tunable. Such lasers are, indeed, irreplaceable to create a high density of

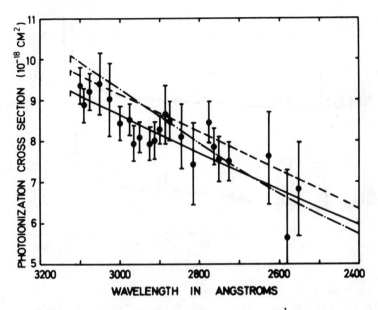

Fig.3.- Photoionization cross section for He(2^1S) atoms as a function of the wavelength. Experimental results are compared with theoretical calculations of Jacobs, solid line (Ref.36), Norcross, dash-dotted line (Ref.37) and Burgess and Seaton, dashed line (Ref. 38). (from Ref. 15).

excited atoms, but, even though the flux available from a continuum source of radiation, such as synchrotron radiation, is lower by many orders of magnitude, the very broad energy range in which such monochromatic photon beams can be provided makes their use obligatory to study inner shell and outer shell photoionization from excited atoms.

INNER-SHELL PHOTOABSORPTION OF LASER EXCITED ATOMS

The first observation of the photoabsorption spectrum of core excited autoionizing states having the same parity as the ground state were carried out on Li (from the 1s core)[40] and sodium (from the 2p core)[41]. In these experiments, a flash-lamp pumped dye laser tuned to the first resonance line of Li ($2s \rightarrow 2p$) or Na ($3s \rightarrow 3p$), with an output power of typically 1MW in 1Å band width, delivered pulses of 800 nsec into a long heat pipe oven (density of about 10^{16} /cm^3)placed in front of the entrance slit of a grazing incidence spectrograph. The photoabsorption of the atoms brought into the first excited state was probed with a triggered vacuum spark which provided a 100 nsec continuum pulse in the region 100-300 Å. The delay time between the initiation of the laser pulse and the peak of the UV probe pulse was controlled within 50 nsec and the UV probe pulse could be adjusted to sample the laser excited vapor at various times after the initiation of the laser pulse. This possible control of the time intervall between the laser pulse and the UV pulse was of fun-

622

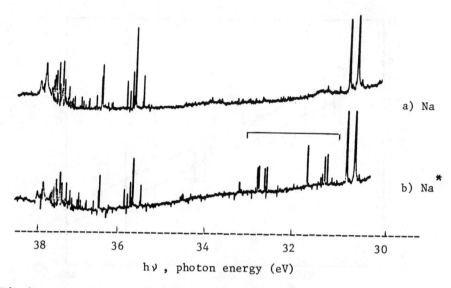

a) Na

b) Na*

hν , photon energy (eV)

Fig.4.- Absorption spectra of Na: a) before initiation of laser pulse and b) 150 nsec after initiation of the pulse (from Ref. 41).

damental importance; in the initial experiment on Na,[42] the laser was fired about 500 nsec before the BRV source and full ionization of Na atoms happened, as demonstrated by the observation of the absorption spectrum of Na$^+$ on the photographic plate. This full ionization was satisfactorily explained, later on, in taking into consideration several mechanisms producing seed electrons in the medium (mainly multiphoton ionization, associative ionization, energy pooling collisions); these primary electrons are then heated in super-elastic collisions   and become able   to fully ionize the medium about 700 nsec after the laser pulse. Here, we concentrate on atoms left in the excited state, whose maximum density was obtained (about 20%) between 100 and 150 nsec after the laser pulse. In Fig.4, we show a microdensitometer trace of the absorption spectrum of Na without (upper trace) and with (lower trace) the laser pulse. The additional features observed in the photoabsorption spectrum of Na* are due to resonance excitations of a 2p electron in the $2p^63p$ atoms to $2p^53s3p$ autoionizing states (from 31.19 eV and 31.24 eV for the $^4P_{5/2,3/2}$ to the $^2P_{1/2}$ at 32.90 eV).

Such experiments were made possible by the fact that the number of photons emitted in one pulse of the BRV source is high enough (10^{10} to 10^{11} photons) to produce absorption during the time intervall the atoms are still in the excited state. Such an experiment would not yet be possible with synchrotron radiation, because the number of photons in one pulse of synchrotron radiation is extremely small (10^4 to 10^5 at present time); the high flux obtainable from a synchrotron beam is due to the high repetition rate of the emission (1 to 10 MHz).Experiments combining pulsed laser and synchrotron radiation beams have already been made, but on solid samples.[53]

SYNCHROTRON RADIATION ELECTRON SPECTROMETRY OF LASER EXCITED ATOMS

The first photoemission experiments combining the use of a cw dye laser to create a steady population in an excited state and of synchrotron radiation to photoionize the excited atoms were successfully carried out in Orsay with the synchrotron radiation emitted by the ACO storage ring.[43-46] Due to the complexity of such measurements several requirements have to be fulfilled to provide optimum conditions. Four different experimental systems must work simultaneously: the high temperature oven, the electron spectrometer, the laser beam and the production of monochromatic synchrotron radiation. Fig.5 shows the experimental set up.[47] A laser beam irradiates a weakly collimated effusive beam of atoms in the vapor phase at right angle to its mean direction. The oven is mounted on the axis of a cylindrical mirror analyzer (CMA). The synchrotron radiation beam comes from the exit slit of the A61 toroïdal grating monochromator[48] and is colinear with the CMA axis (see Fig.6). It delivers high flux of monochromatic photons (10^{12}/sec in 1% band pass) in the energy region between 20 and 150 eV, with band pass varying from 0.1 to 1 eV.

The laser is a cw ring dye laser operating in the monomode regime. It is pumped by an Argon ion laser (maximum power of 18 watts) and is linearly polarized in the horizontal plane. The laser beam is adjusted to fit the shape of the source volume seen by the CMA, located closed to the circular exit of the oven.

In a first series of experiments, photoionization out of excited Na atoms was studied.[43-45] The laser was locked to the $3s^2S$ F=2→ $3p^2P_{3/2}$ F=3 hyperfine component of the D_2 resonance line at 5890 Å and delivered up to 400 mW (about 3 watts/cm^2). The atomic density of sodium in the ground state was about 1.5x 10^{13}/cm^3 in the active volume.

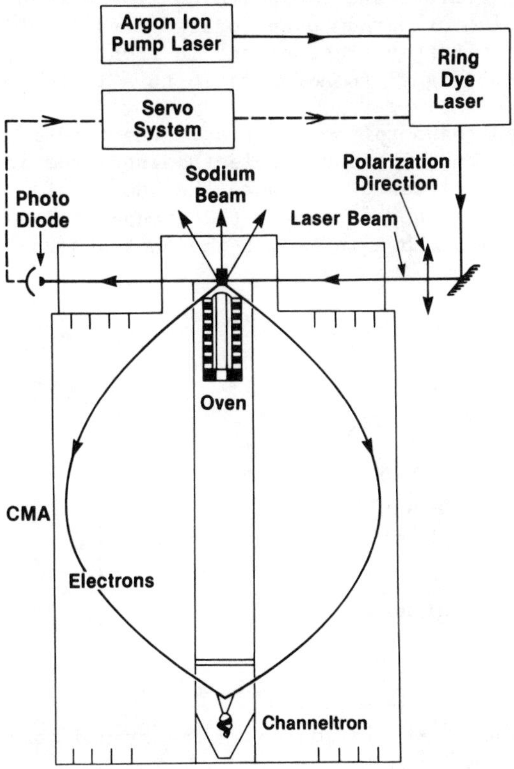

Fig.5.- Experimental set up. The monochromatic synchrotron radiation beam (not shown in the figure) comes from the top of the figure and is colinear with the CMA axis (Ref.47).

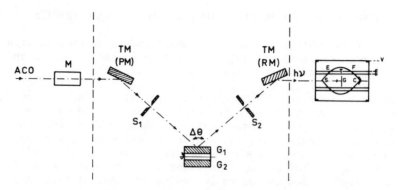

Fig.6.- Schematic view of the A61 toroïdal grating monochromator and of the CMA electron analyzer. M= plane mirror, TM = toroïdal mirrors, G_1 G_2 holographically ruled toroïdal gratings, S_1= entrance slit, S_2 = exit slit, S = interaction zone, C = channeltron (from Ref.48).

In a second series of experiments[46], photoelectron spectra from laser excited sodium and barium atoms were obtained. The laser power could be increased up to 1.2 watt (about 10 W/cm^2). The density of atoms in the ground state was varied between 10^{11} and 10^{13}atoms/cm^3. At that time, the laser was stabilized and locked to D_2 line within 20 MHz, using an auxiliary sodium or barium oven located outside of the CMA. In addition, a liquid nitrogen trap, placed in front of the oven, contributed to improve the signal to noise ratio to a large extent.

A good knowledge of the photoabsorption spectra in the ground state and in the excited state, and of the photoelectron spectrum in the ground state is of primary importance to insure the succes of such experiments. The photoabsorption spectrum of the ground state was measured some years ago and is shown in Fig.7; the photoabsorp-

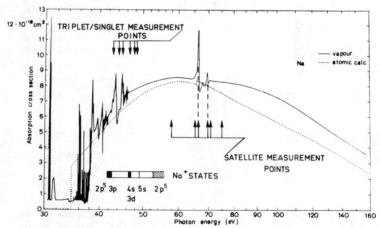

Fig.7.- Photoabsorption spectrum of atomic sodium in the ground state (from Ref.49).

tion spectrum of the excited state has also been measured (see Fig. 4); extensive photoionization studies of Na in the ground state had also been carried out before the excited state experiment was done 50.

For the experiments on Ba, the laser was locked to the $6s^2 {}^1S_o \rightarrow 6s6p^1P_1$ resonance line at 5535 Å.

Two main features have been observed during the first set of experiments on inner shell photoionization and autoionization processes in laser excited sodium.[43-45] Fig.8 shows a photoelectron spectrum of Na atoms taken at 75 eV photon energy, in a flat region of the continuum absorption (see Fig.7). In the absence of laser radiation (upper part of the figure), one observes photolines corresponding to ionization of sodium atoms in the 2p subshell, with the atoms initially in the ground state: the main peak at 38 eV binding energy corresponds to a positive ion being left in its lowest $2p^53s^{1,3}P$ states, the other peaks of lower intensity are satellite peaks corresponding to ions left in the $2p^53p^{1,3}S, {}^3P, {}^{1,3}D$ excited states via final state electron correlations. In the lower part of the figure (laser on), an additional photoelectron line appears, whose energy corresponds to the binding energy of a 2p electron in an excited $2p^63p^2P_{3/2}$ atom. This peak, at about 40 eV binding energy, is a clear manifestation of the presence of excited atoms in the vapor. Using

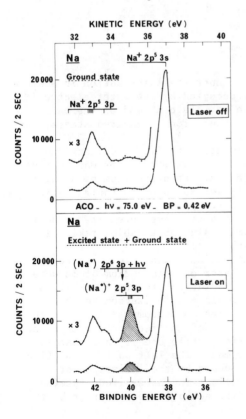

Fig.8.- Photoelectron spectron spectrum of atomic sodium obtained with 75 eV synchrotron radiation photons. Upper frame: the laser is off, the large peak is due to photoionization of a 2p electron with the ion being left in ist ground state; the small peaks are due to ionization of a 2p electron with simultaneous excitation of the 3s electron to a 3p orbital Lower frame: the laser is on; in addition to the normal photoelectron spectrum from the ground state, a new peak appears (hatched area) due to ionization of a 2p electron in laser excited Na atoms(43)

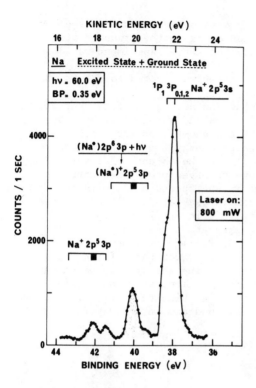

Fig.9.- Photoelectron spectrum of sodium with the laser on, taken at 60 eV photon energy. Compared to the spectrum of Fig.8, the background due to Na atoms hitting the channeltron has been significantly reduced, the density of atoms in the ground state is lower by one order of magnitude, and the intensity of the peak due to ionization of a 2p electron in the excited atom is higher by a factor 2.5. (from Ref. 46).

the fact that the cross section for inner shell ionization in sodium depends very little on the orbital occupied by the outer electron[20], a comparison of the integrated area of this peak and of the main peak allows to obtain the relative population of excited states, here close to 10%.

In the second series of experiments with improved laser and sodium beam conditions, a higher density of excited atoms was routinely obtained. Fig. 9 presents the photoelectron spectrum of Na atoms taken at 60 eV photon energy under laser and synchrotron radiation impact. Now, the photoelectron line corresponding to ionization of a 2p electron in the excited atom has a relative intensity close to 25%. This is not far from the maximum value obtainable, taking into consideration the polarization of the laser.[51]

Fig.10 shows an example of an autoionization process observed in a doubly excited atom, in which laser excited sodium atoms were additionally excited by synchrotron radiation to the $2p^5(^2P)3s3p$ $^3P)^4P_{5/2}$ state;[41]. This state decays via autoionization to the $2p^6$ $^1S_0$ ground state of Na^+ion[43,46]. The monochromator is set here at 31.19 eV photon energy, the position of the resonant transition. The presence of second order radiation, while usually a nuisance, was important in this type of experiment. The left part of the figure displays the photoelectron lines produced by the ionization of sodium atoms in the ground state as well as in the excited state(hatched peak) with 62.38 eV photons diffracted in second order by the

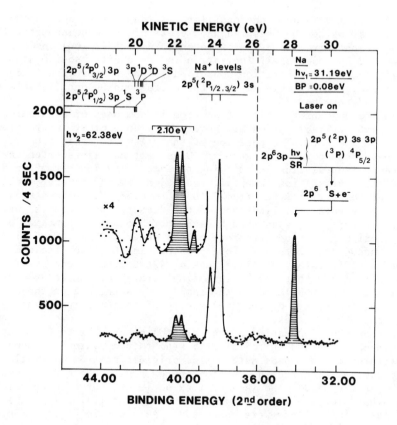

Fig. 10.-Photoelectron spectrum of Na produced by 31.19 eV photons diffracted in first order and by 62.38 eV photons diffracted in second order by the monochromator. The hatched structures appear only when the laser is on and therefore manifest the presence of excited $2p^6 3p$ atoms in the vapor. In this spectrum, one can observe simultaneously photoionization in the 2p subshell, in the ground state and in the excited state of Na (left part of the figure) and the autoionization line due to the decay of doubly excited sodium atoms produced by absorption of 31.19 eV photons by the laser excited $2p^6 3p$ Na atoms. (Spectrum obtained during the first series of experiments, Ref. 43-45).

monochromator (the energy of first order photons is too low to produce ionization of 2p electrons with 38 eV binding energy). The same features as in Figs 8 and 9 are observable, with a better resolution. In the right part of the figure, an intense line (hatched) appears, corresponding to the decay of the autoionizing state excited by 31. 19 eV photons. Located at the binding energy of the 3p excited electron, it may be considered also as the result of the resonant photoemission of this 3p excited electron. From the integrated area of this line and of the photoelectron lines produced by second order photons, it is possible to deduce the oscillator strength for such resonance transitions, using the measured parameters of the monochromator (band pass, relative intensity of second order radiation) and the cross section for 2p inner shell ionization in the excited sodium atom.[20] The main resonance transitions observed in photoabsorption[41] have been investigated in this way and the corresponding oscillator strenghts were determined.[46]

The $2p^53p$ configuration of Na$^+$ ion is composed of a number of fine structure levels primarily governed by spin-orbit interaction within the $2p^5$ core. The position of these levels, taken from optical data, have been noted on Figs.8 to 10 as vertical lines extending below the horizontal line identifying the configuration. From Fig.8 to Fig.10, the overall instrumental resolution (monochromator + electron spectrometer)improved from 0.50 eV to 0.25 eV. In comparing the shape of the satellite lines and of the "laser " peak, it is interesting to notice that, even though each corresponds to the same ionic configuration, the fine structure of the laser peak at 40 eV binding energy, associated with a 2p hole produced in the laser excited atom is different from the satellite peak at 42 eV binding energy. This seems to indicate that the various final states of the $2p^53p$ configuration are not populated in the same way via the one photon and the two-photon route, or that the nature of the coupling might be different in both cases.

Direct photoionization of the excited electron into the continuum was not observed in sodium with synchrotron radiation. However, in searching for a photoelectron line originating from ionization of the 3p excited electron in a photon energy region where the cross section is supposed to be larger, i.e. at lower photon energy,closer to threshold,low energy electrons produced by collisional processes under laser impact only were observed[47], as shown in Fig.11. The me-

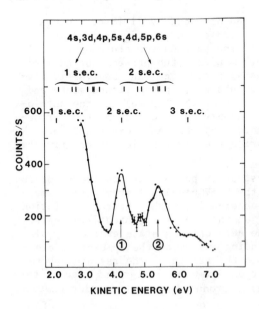

Fig.11.- Energy spectrum of the electrons ejected from the laser irradiated sodium vapor, in the absence of synchrotron radiation.The upper group of bars indicates the energy positions of the electrons created by collisional ionization from the various excited states and heated by either one or two superelastic collisions(sec); the lower group of bars shows the energy positions of the electrons created by associative ionization and having undergone one, two or three superelastic collisions (from Ref.47).

chanisms responsible for producing these electrons were recently described.[47] Here, we would like only to note that the existence of these low kinetic energy electrons could be a troublesome source of background that would mask the expected spectrum arising from photoionization of the excited electron.

The photoelectron spectrum of Ba in the ground state is more complex, because many subshells can be ionized in the energy region covered by these experiments. Fig.12 shows such an electron spectrum obtained with photons of 130 eV, i.e. 30 eV above the 4d ionization thresholds. Photoelectron lines arising from ionization in the 6s-5p-5s-and 4d subshells are observed as well a Auger lines corresponding to the decay of 4d and 5p vacancies.[52] Fig.13 presents part of the photoelectron spectrum obtained when the laser was tuned to the $6s^2 \rightarrow 6s6p$ resonance line of Ba.[46] In our experimental conditions, this excited state decays rapidly to the 6s5d metastable state of Ba, 1.40 eV above the ground state. The spectrum recorded without the laser (upper part) displays the 4d photoelectron spectrum emitted from ground state atoms by the the radiation diffracted in 4th order by the monochromator ($h\nu_4$= 119.8 eV) and the photoline due to ionization of the outer 6s subshell by photons diffracted in the 1st order by the monochromator ($h\nu_1$ = 29.95 eV).In this way, the photoelectron spectrum from the 4d subshell is obtained with an improved resolution compared to the spectrum in Fig.12. When the laser is turned on (lower part), the photoelectron spectrum changes dramatically: all 4d-components become doublets , since the binding energies on these inner electrons are shifted by about 1.5 eV in the excited atom.   In addition, an intense photoline appears at the binding energy of the 5d excited electron ionized by 1st order photons, illustrating the enhancement of the photoionization cross section for the excited electron.

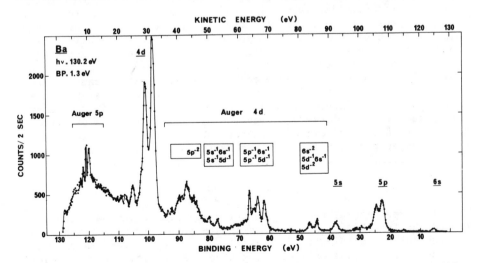

Fig.12.- Electron spectrum following ionization of atomic Ba by 130 eV photons. $\underline{n\ell}$ =photoelectron line; $n\ell^{-1} n'\ell'^{-1}$ = Auger line(from 52)

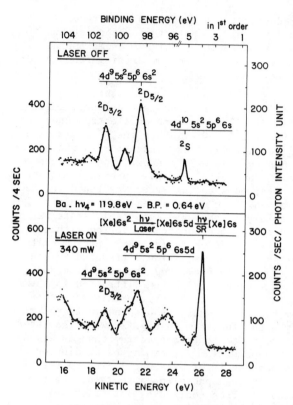

Fig.13.– Photoelectron spectrum of atomic baryum ionized in the 4d subshell by 119.8 eV photons diffracted in 4th order by the monochromator and in the 6s outer subshell by 29.95 eV photons diffracted in 1st order. In the upper part the usual photoelectron spectrum of Ba in the ground state is observable (laser off); in the lower part (laser on), the various photoelectron lines originate from ionization of atoms in the ground state and in the 6s5d excited state, as indicated (from Ref. 46).

A similar splitting of the photoelectron lines has also been observed for the 5s and 5p electrons. For these subshells, the cross sections for ionization in the ground state and in the excited state are not expected to be quite different.[20] On the contrary the 5d photoionization cross section seems to be quite different from the 6s cross section; in addition, the cross section for 4d ionization in excited Ba atoms may present some interesting features, since the potential experienced by the 4d electrons may vary significantly from the ground state to the excited state (see paper by Lucatorto et al. in this volume), although the major effects are expected for the excited states of Ba+ ions. Photoelectron spectra of excited Ba have been measured over an extended photon energy range. The values of the 4d and 5d photoionization cross sections can be extracted from these measurements and will be presented elsewhere.

CONCLUSION

The experimental results presented in this review have demonstrated the feasibility of combined laser-synchrotron radiation experiments for photoionization studies of atoms in excited states. The success of this new technique opens wide possibilities for the future. Large progress should be made in optimizing the various parameters involved in these experiments, in particular the choice of the geometry, of the electron spectrometer, of the electron detection, the possibility to study angular distribution in relation with the polarization of the laser and the alignment or the orientation of the excited states.Many new schemes combining the use of various cw or pulsed lasers, in connection with synchrotron radiation may be imagined.(see paper by P. Koch in this volume). In addition to the intrinsic interest of such experiments for atomic physics, other fields such as astrophysics or laboratory plasma physics require data on excited states of atoms and ions. In the future, when photon beams emitted by undulator devices mounted on new storage rings will be available, their intensity might be high enough to allow photoionization studies of excited states in which the excited electron would not come from laser excitation of an outer electron of low binding energy, but possibly from undulator excitation of an outer electron in a rare gas atom or even from the promotion of an inner electron onto some empty excited orbital.

ACKNOWLEDGEMENTS

For the experiments combining the use of laser and synchrotron radiation, my coworkers were J.M. Bizau, B. Carré, P. Dhez, D. Ederer, J.C. Keller, P. Koch, J.L. LeGouët, J.L. Picqué. The success of this venture would not have been possible without this closed cooperation between synchrotron radiation people and laser men.

REFERENCES

1." Photoionization and Other Probes of Many-Electron Interactions", ed. F.Wuilleumier (Plenum Press, New York, 1976).
2. M.O. Krause, in "Synchrotron Radiation Research", ed. H. Winick and S. Doniach (Plenum Press, New York, 1980), p.101.
3. F.J. Wuilleumier, Atomic Physics, vol.7, 482 (1981).
4. F.J. Wuilleumier, "Inner-Shell and X-Ray Physics of Atoms and Solids" ed. D.J.Fabian et al. (Plenum Press, New York, 1981),p.395.
5. B. Sonntag and F. Wuilleumier, Proceedings of the Intern. Conf.on Instrumentation for Synchrotron Radiation, Hamburg, 1982, to be published in Nucl. Instr. Meth., 1983.
6. V. Schmidt, Appl. Opt. 19, 4080 (1980).
7. A.F. Starace, Appl. Opt. 19, 4051 (1980).
8. U. Heinzmann, Appl. Opt. 19, 4087 (1980).
9. J. Lahiri and S.T. Manson, Phys. Rev. Lett. 48, 614 (1982).
10. J.P. Connerade, Contemp. Phys. 19, 414 (1978).
11. T.B. Lucatorto, T.J. McIlrath, J. Sugar and S.M. Younger, Phys. Rev. Lett. 47, 1124 (1981).
12. T.B. Lucatorto, T.J. McIlrath, W.T. Hill III, C.W.Clark,this vol.
13. J.L.Carlsten,T.J.McIlrath and W.H.Parkinson,J.Phys.B7,1020(1974).
14. K.J. Nygaard, IEEE J. Quantum Electron. 9, 1020 (1973).
15. R.F.Stebbings, F.B.Dunning,F.K.Tittel and R.D.Rundel, Phys. Rev. Lett. 30,815(1973); F.B.Dunning,R.F.Stebbings,Phys.Rev.A9,2378(74)
16. D.J. Bradley, C.H. Dugan, P. Ewart and A.F. Purdie, Phys. Rev.A 13, 1416 (1976).
17. A. Msezane and S.T. Manson, Phys. Rev. Lett. 35, 364 (1975).

632

18. A.Z. Msezane and S.T. Manson, Phys. Rev. Lett. 48, 473 (1982).
19. M.Aymar,E.Luc-Koenig and F.Combet Farnoux, J.Phys.B 8,1279(1976).
20. T.N. Chang and Young Soon Kim, contribution in this volume.
21. M.S. Pindzola, Phys. Rev. A 23, 201 (1981).
22. T.N.Chang and Young Soon Kim, Phys. Rev. A 26,(1982).
23. D.J. Bradley, Physics Bulletin, 21, 116 (1970).
24. D.J. Bradley, P. Ewart, J.V. Nicholas and J.R.D. Shaw, Int.Quan-
    tum Electronics Conference (1972), digest of Technic.Papers, 58-9.
25. R.V.Ambartzumian and V.S. Lethokov, Appl. Opt. 11, 354 (1972).
26. D.J.Bradley,P.Ewart,J.Nicholas,J.Shaw,D.Thompson,Phys.Rev.Lett.
    31, 263 (1973).
27. T.J. McIlrath, Appl. Phys. Lett. 15, 41 (1969).
28. D.Bradley,P.Ewart,J.Nicholas,J.Shaw, J.Phys.B 6, 1594 (1973).
29. A.V.Smith,J.E.M.Goldsmith,D.E.Nitz,S.Smith,Phys.Rev.A22,577(1980).
30. A.N. Klyucharev and N.S. Ryazanov, Opt. Specktrosk. 32,1253(1972).
31. R.V. Ambartzumian et al. Appl. Phys. 11, 335 (1976).
32. N.V.Karlov,B Krynetskii,O.M.Stel'makh,Kvantovaya Electron. Moscow,
    4, 2275(1977) (Sov. J. Quantum Electron. 7, 1305 (1977).
33. A.N. Klyucharev and V. Yu. Sepman, Opt.Spectrosk. 38, 712 (1975).
34. H. Kaminskii, J. Kesler and K.J. Kollath,Phys.Rev.Lett.45,1161(80)
35. J.C. Hansen, J.A. Duncanson, Jr., R.L. Chien and R. Stephen Ber-
    ry, Phys. Rev. A 21, 222 (1980) and references therein.
36. V. Jacobs, Phys. Rev. A 4, 939 (1971).
37. D.W. Norcross, J. Phys. B4, 652 (1971).
38. A. Burgess and M.J. Seaton, Mon. Notic. Astron. Soc.120,121(1960).
39. A. Siegel, J.Ganz,W.Bassert,H.Hotop,B.Lewandowski,M.W.Ruf,H.War-
    bel, 8th Intern.Conf. on Atom. Phys., Göteborg,1982, Abstr. B14.
40. T.J. McIlrath and T.B. Lucatorto, Phys. Rev. Lett. 38, 1390(1977).
41. J.Sugar,T.B.Lucatorto,T.J.McIlrath,J.Sugar, Opt. Lett.4,109(1979).
42. T.B. Lucatorto and T.J. McIlrath, Phys. Rev. Lett. 37, 428 (1976).
43. J.M.Bizau,J.L.LeGouët,D.Ederer,P.Koch,F.Wuilleumier,J.L.Picqué,
    P.Dhez, 12th Intern. Conf. on the Physics of Electronics and Ato-
    mic Collisions, Gatlinburg,1981,Abstr. post-post dead line pap.1.
44. J.M. Bizau, F. Wuilleumier, P. Dhez, D. Ederer, J.L. LeGouët,
    J.L. Picqué, P. Koch, Bull. Amer. Phys. Soc. 26, 1300 (1381).
45. J.M.Bizau, F.Wuilleumier, P. Dhez, D.Ederer, J.L.Picqué, J.L.Le-
    Gouët, P.Koch, Proceedings of the Topical Meeting " Laser Techni-
    ques for ExtremeUltraviolet Spectroscopy",Boulder,1982,to be pub.
46. J.M. Bizau, F. Wuilleumier, P. Dhez, D. Ederer, J.L. Picqué, J.L.
    LeGouët, J.C. Keller, P. Koch, B. Carré, to be published.
47. J.L. LeGouët, J.L. Picqué, F. Wuilleumier, J.M. Bizau, P. Dhez,
    P. Koch, D.L. Ederer, Phys. Rev. Lett. 48, 600 (1982).
48. P.K. Larssen, W.A.M. van Beers, J.M. Bizau, F. Wuilleumier,S.Kru-
    mmacher,V.Schmidt,D.Ederer,Nucl.Instr.Meth.195,245(1982).
49. H.Wolff,K.Radler,B.Sonntag and R.Haensel, Z.Phys.257,353(1972).
50. S.Krummacher,V.Schmidt, J.M. Bizau, D. Ederer, P. Dhez and F.
    Wuilleumier, J. Phys. B, to be published.
51. A. Fischer and I.V. Hertel, Z. Phys. A 304, 103 (1982).
52. F. Wuilleumier, J.M. Bizau, G. Wendin, P. Dhez, Intern. Conf. on
    X-Ray and Atomic Inner-Shell Phys. Eugene, 1982, Abstr. p.153.
53. V. Saile, Appl. Opt. 19, 4115 (1980).

PHOTOIONIZATION OF THE INNER-SHELL ELECTRON UNDER CHANGING OUTER-SHELL ELECTRONIC ENVIRONMENT IN ALKALI ATOMS[*]

T. N. Chang[**] and Young Soon Kim
Physics Department, University of Southern California,
Los Angeles, California 90089-1341 U.S.A.

ABSTRACT

Recent experimental study at LURE on the photoelectron spectrum of the 2p-subshell of sodium atom with its valence electron excited by laser has made it possible to examine in detail the photoionization of the inner-shell electron under changing outer-shell electronic environment. In this paper, we report the result of a comprehensive theoretical study of this atomic process for alkali atoms with the valence electron at various stages of excitation. The contribution from important many-body interactions to the photoionization cross section is examined. The calculated cross section and the angular distribution of the ejected photoelectron are presented.

INTRODUCTION

A small change in the height of the potential barrier experienced by an atomic electron in a high-ℓ orbit could lead to the collapse of the atomic wavefunctions.[1] Physically, this height change in potential barrier could be controlled by exciting the valence electron into various bound excited orbits. The experimental investigation along this direction is made possible recently by Bizau et al[2] at LURE in their study on the photoelectron spectrum of the 2p-subshell of sodium atom in the presence of laser radiation with its wavelength corresponding to the excitation of an electron from the 3s orbit to a 3p orbit. Although the study of the atomic transition of the 2p electron is of less interest than that of an electron in a high-ℓ orbit, such an investigation, nonetheless, provides an important initial step towards a more comprehensive investigation on the atomic transition involving electron from high-ℓ orbit. In this paper, we report the quantitative result of a corresponding theoretical study of the photoionization of a np electron from a laser-excited alkali atom A, i.e.,

$$A^*(1s^2...ns^2np^6n_e\ell_e) + h\nu \rightarrow A^+(1s^2...ns^2np^5n_e\ell_e) + e^- \qquad (1)$$

in the energy region between np and ns thresholds.

*Work supported by NSF under Grant No. PHY80-09146.
**On leave at the National Central University in Taiwan until December 31, 1982.

CALCULATIONAL PROCEDURE

The calculational procedure employed in the present work is similar to those used in our previous studies for the rare gas atoms.[3,4] The dominant many-body effects included in the present calculation are i) the ground state configuration interaction among electrons in the np subshell and ii) the final state interchannel interaction between the np → kd and np → ks transitions.

The photoionization cross sections for the np → kℓ transition in the dipole-velocity and dipole-length approximations are

$$\sigma^V = \frac{16\pi\alpha}{kE} (5.29 \times 10^{-9})^2 |D^V|^2 \text{ cm}^2 \tag{2}$$

and

$$\sigma^L = \frac{4\pi\alpha E}{k} (5.29 \times 10^{-9})^2 |D^L|^2 \text{ cm}^2 \tag{3}$$

where

$$E = k^2 - \varepsilon_{np} \tag{4}$$

is the photon energy in Rydberg unit, ε_{np} is the energy eigenvalues of the single-particle Hartree-Fock np orbit, α is the fine structure constant, and k is the photoelectron momentum. The transition amplitude D is given by Eqs. (3) and (4) of Ref. 3. The single particle radial functions ψ_ℓ and ϕ_ℓ required in the evaluation of the transition amplitude D are obtained by solving a set of coupled integrodifferential equations similar to Eqs. (5a)-(5f) given in Ref. 3. The present calculational procedure is the same as that of Ref. 4 with two modifications: i) in the present calculation the solution ψ_p (see, Eq. (5) of Ref. 3) is set to zero for photon energies between np and ns thresholds, and ii) the Hartree-Fock hamiltonian H_ℓ^N for an electron with orbital quantum number ℓ is modified by adding a static-with-exchange term due to the excited valence electron in the $n_e\ell_e$ orbit, i.e.,

$$H_\ell^N f_\ell(r) = [-\tfrac{1}{2}\frac{d^2}{dr^2} - \frac{Z}{r} + \tfrac{1}{2}\frac{\ell(\ell+1)}{r^2}$$

$$+ \sum_{n'\ell'}^{np} \frac{[\ell']^{\frac{1}{2}}}{[\ell]^{\frac{1}{2}}} 2(\ell\|v^0(\chi_{n'\ell'},\chi_{n'\ell'};r)\|\ell)] f_\ell(r)$$

$$- \frac{1}{[\ell]} \sum_{n'\ell'}^{np} \sum_k (-1)^k (\ell\|v^k(\chi_{n'\ell'},f_\ell;r)\|\ell') \chi_{n'\ell'}(r)$$

$$+ V_{ex}(r)f_\ell(r), \tag{5}$$

where

$$V_{ex}(r)\, f_\ell(r) = [\ell_e,\ell]^{-\frac{1}{2}}\, (\ell\| V^0\, (\chi_{n_e\ell_e}, \chi_{n_e\ell_e}; r)\, \|\ell)\, f_\ell(r)$$

$$- \frac{1}{[\ell]}\, \sum_k\, (-1)^k\, (\ell\| V^k(\chi_{n_e\ell_e}, f_\ell; r)\, \|\ell_e)\, \chi_{n_e\ell_e}(r) \qquad (6)$$

and $[\ell] = 2\ell+1$. The interaction term V^k is defined by

$$(\ell\| V^k(a,b,;r)\, \|\ell') = (\ell\| C^{[k]}\|\ell')\, (\ell_a\| C^{[k]}\|\ell_b)$$

$$x\, \int_0^\infty dr'a(r')b(r')\, \frac{r_<^k}{r_>^{k+1}} \qquad (7)$$

The matrix elements of $C^{[k]}$ are standard.[5] The single-particle numerical Hartree-Fock wavefunction for the initial state is obtained with the program MCHF72.[6] With $H_\ell^N(r)$ thus defined, the corresponding effective single particle hamiltonian $H_\ell^{N-1}(r)$, defined by Eq. (6) of Ref. 3, is then used to solve the set of the coupled differential equations similar to Eqs. (5) of Ref. 3.

The angular distribution of the ejected photoelectron due to a polarized incident photon beam is expressed in terms of the differential cross section by[7]

$$\frac{d\sigma_{np}}{d\Omega} = \frac{\sigma_{np}}{4\pi}\, (1 + \beta\, P_2\, (\cos\theta)) \qquad (8)$$

where σ_{np} is the total photoionization cross section for the np subshell, P_2 is the Legendre polynomial, and θ is the angle between the photoelectron momentum $\vec{k}$ and the polarization vector of the incident light. The asymmetry parameter β is given by[8]

$$\beta = \frac{|S_2|^2 - 2\sqrt{2}\, Re\, (S_2\, S_0^*)}{|S_2|^2 + |S_0|^2} \qquad (9)$$

$$S_\ell = (-i)^\ell\, e^{i\delta_\ell}\, D_{1\ell} \qquad (10)$$

with δ_ℓ and D the same as those defined in Ref. 3.

RESULT AND DISCUSSION

In the present study, we have included the photoionization cross section calculation for sodium and potassium from several different initial states. They are (i) the ground state of the neutral sodium and potassium, ii) the excited state of the neutral sodium and potassium with valence electron in 3p and 4p orbit res-

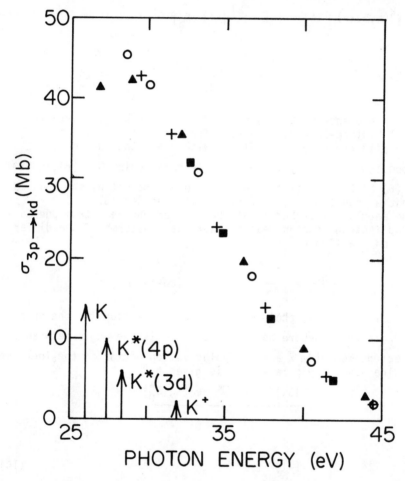

Figure 1. The calculated photoionization cross section for the 3p→kd transition for potassium initially in i) the ground state (▲), ii) the excited state with valence electron in 4p orbit (O), iii) the excited state with valence electron in 3d orbit (+) and iv) the ground ionic state (■). The arrows indicate the calculated ionization thresholds.

Table I.   The calculated Hartree-Fock energy for the 2p
orbit of sodium and the 3p orbit of potassium.

Initial State	ε_{2p}^{HF}(in Ryd.)	Initial State	ε_{3p}^{HF}(in Ryd.)
Na	− 3.036	K	− 1.909
Na*(3p)	− 3.196	K*(4p)	− 2.016
Na*(3d)	− 3.371	K*(3d)	− 2.085
Na+	− 3.594	K+	− 2.343

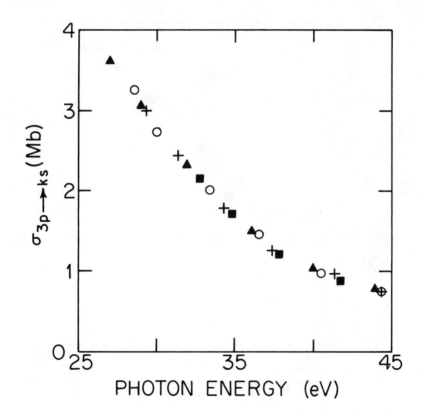

Figure 2.   The calculated photoionization cross section for the
3p→ks transition for potassium initially in i) the ground state
(▲), ii) the excited state with valence electron in 4p orbit (O),
iii) the excited state with valence electron in 3d orbit (+), and
iv) the ground ionic state (■).

638

pectively, iii) the excited state of the neutral sodium and potassium with valence electron in 3d orbit, and iv) the ground state of
the ionic sodium and potassium. Table I gives the calculated
Hartree-Fock orbital energies for the 2p orbit of sodium and 3p
orbit of potassium. The difference between the calculated energy
and experimental ionization energy ranges from 8.7% for Na to 3.4%
for Na$^+$ and from 4.0% for K to 0.8% for K$^+$.

   The calculated partial photoionization cross sections (in dipole
velocity approximation) as function of photon energy for the 3p→kd
and 3p→ks transitions for potassium from different initial states
are given in Figs. (1) and (2) respectively. The difference between

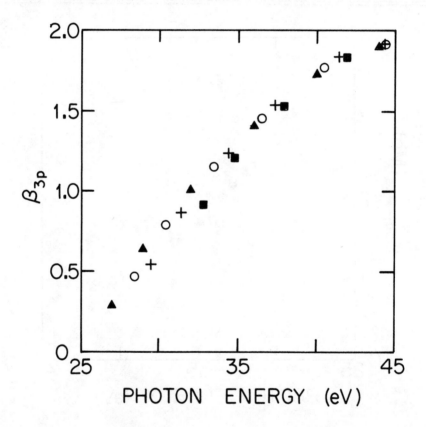

Figure 3.   The asymmetry parameter β for the photoionization of
the 3p electron of potassium initially in i) the ground state (▲),
ii) the excited state with valence electron in 4p orbit (o), iii)
the excited state with valence electron in 3d orbit (+), and iv)
the ground ionic state (■).

the velocity and length calculation is less than 3% for all energies.
The result of the angular distribution for potassium in velocity
form is shown in Fig. 3 in terms of the asymmetry parameter β.
Again, the difference between the velocity and the length calculation
is amall for all energies.
    Figs. (4) and (5) represent, respectively, the calculated

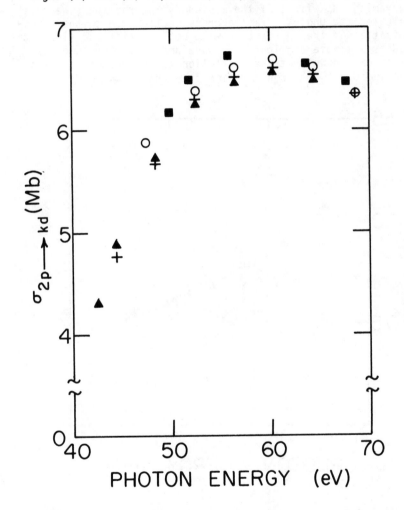

Figure 4.   The calculated photoionization cross section for the
2p→kd transition for sodium initially in i) the ground state (▲),
ii) the excited state with valence electron in 3p orbit (+), iii)
the excited state with valence electron in 3d orbit (O), and iv)
the ground ionic state (■).

photoionization cross sections for the 2p→kd and 2p→ks transition for sodium from different initial states. Again, only the velocity result is given as the difference between the length and velocity calculation is less than 2% for all energies. Fig. 6 gives the calculated asymmetry parameter β (in velocity form) for the photoionization of the 2p electron for sodium from different initial states.

To examine the effect of the many-body interactions, we have also evaluated the photoionization cross section due to the direction ionization of the np electron with both initial and final state described by a single configuration wavefunction. Our numerical calculation has shown that the correction to the photoionization cross section due to the contribution from those many-body inter-

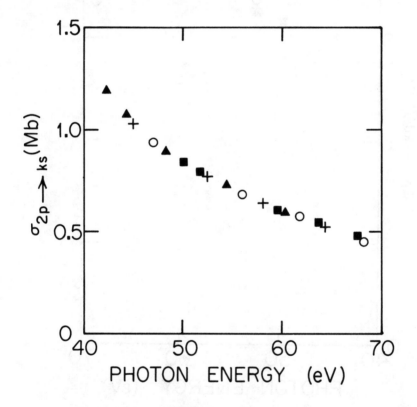

Figure 5. The calculated photoionization cross section for the 2p→ks transition for sodium initially in i) the ground state (▲), ii) the excited state with valence electron in 3p orbit (○), iii) the excited state with valence electron in 3d orbit (+), and iv) the ground ionic state (■).

actions included in the present study, i.e., the initial state configuration interaction among electrons in the np subshell and the final state interchannel interaction between the np→kd and np→ks transitions, is significant. For example, in the velocity calculation, the combined contribution from these many-body interactions

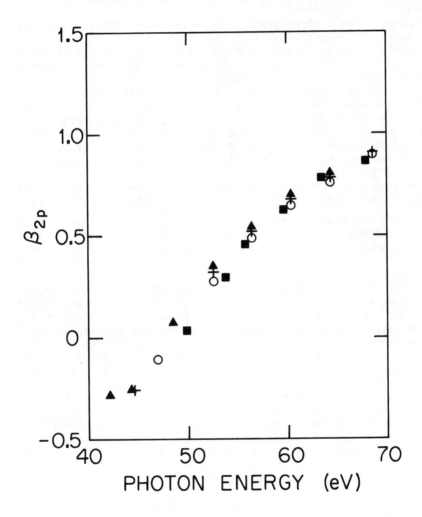

Figure 6. The asymmetry parameter β for the photoionization of the 2p electron of sodium initially in i) the ground state (▲), ii) the excited state with valence electron in 3p orbit (O), iii) the excited state with valence electron in 3d orbit (+), and iv) the ground ionic state (■).

642

increases the 3p→kd photoionization cross section by as high as 75% for K to about 50% for K^+ at photon energy about 1 eV above the 3p ionization threshold. For 3p→ks transition, this correction becomes smaller (40-50%) but remains significant. For sodium, as we reported in an earlier paper,[3] this correction ranges from about 20% for the 2p→kd transition to about 30% for the 2p→ks transition in the velocity calculation.

A more detailed breakdown of the individual contribution from these many-body interactions indicates that for the np→kd transition, the cross section correction comes primarily from the contribution due to the initial state configuration interaction among electrons in the np subshell. Also consistent with what was found in our earlier study on argon[3] and neon,[4] the contribution from the final state interchannel interaction is small for the np→kd transition. On the other hand, for the np→ks transition, the cross section correction comes from the combined contribution due to the initial state configuration interaction and the final state interchannel interaction. Again, this conclusion is consistent with our earlier study on argon[3] and neon.[4]

We have also estimated the contribution from the interactions between the np electron and the excited $n_e \ell_e$ valence electron by calculating the lowest order many-body perturbation terms. A more detailed discussion was given elsewhere.[9] Our numerical calculation has shown that the total contribution due to those interactions is small.

From Figs. (1)-(6), it becomes clear that unlike what was expected for the high-ℓ state, the photoionization of the inner-shell np electron at a fixed photon energy is practically independent of the atomic orbit that the valence electron occupies. Consequently, one can conclude that experimentally the relative population of the excited atoms in the atomic beam such as the one prepared in the experiment by Bizau et al[2] can be determined by a detailed measurement of the relative intensity of the photoelectron from its respective initial atomic state. With the relative population thus determined, and with its cross section normalized against the calculated value, it would become possible to carry out the absolute cross section measurement for other more interesting processes such as the autoionization of the doubly excited state generated with the resonant absorption of the UV radiation by the laser excited atoms.

REFERENCES

1. J. P. Connerade, J. Phys. B11, L381 (1978).
2. J. M. Bizau, J. L. Le Goüet, D. Ederer, P. Koch, F. Wuilleumier, J. L. Picqué, and P. Dhez, XIIth ICPEAC, Abstracts of post-deadline papers, S. Datz, editor, P. 1 (1981).
3. T. N. Chang, Phys. Rev. A18, 1448 (1978).
4. T. N. Chang and T. Olsen, Phys. Rev. A23, 2394 (1981).

5. A. R. Edmonds, Angular Momentum in Quantum Mechanics (Princeton University Press, Princeton, N.J., 1957), P. 76.
6. C. F. Fischer, Comp. Phys. Commu. 4, 107 (1972).
7. C. N. Yang, Phys. Rev. 74, 764 (1948).
8. D. Dill, Photoionization and Other Probes of Many-Electron Interactions, ed. F. Wuilleumier (Plenum: New York, 1976) P. 387.
9. T. N. Chang and Y. S. Kim, J. Phys. B15, LXXX (1982).

ATOMIC PHYSICS EXPERIMENTS COMBINING SYNCHROTRON RADIATION AND LASERS: PRESENT CAPABILITIES AND FUTURE POSSIBILITIES

Peter M. Koch

J.W. Gibbs Laboratory, Yale University, New Haven, CT 06511 USA
[Present address: Physics Department, State University of New York at
Stony Brook, Stony Brook, NY 11794 USA]

ABSTRACT

This (occasionally speculative) paper considers the range of "hybrid experiments" in atomic physics that are now or may soon be possible with combined use of two rather different photon sources, monochromatized synchrotron radiation (SR) in the vacuum ultraviolet (VUV) below about 300 eV and lasers below about 5-10 eV. A wide range of experiments will employ one or more continuous lasers to create an appreciable stationary fraction of valence-excited target atoms in a beam and will draw heavily on the techniques previously developed to create aligned or oriented targets for electron scattering experiments. The very low duty factor of most pulsed lasers mitigates against their use with SR, but it should be possible to develop specialized, mode-locked, pulsed dye lasers whose pulse train is synchronized to that of the SR source. The development of "free-electron" lasers at SR facilities will provide an ideal tunable laser source in the same laboratories, certainly in the infrared and perhaps into the UV. SR and photoelectron spectroscopy can be used to probe photo-excitation and -ionization of electrons in inner or outer shells, either directly or via doubly-excited resonances. Triply-excited states may also be open to study. Analogous experiments with electronically or vibronically excited molecules will be challenging. In laser-excited alkali and alkaline-earth vapors, collisional interactions among the excited atoms can produce a weakly ionized plasma emitting "hot electrons" that can confuse SR photoelectron spectra, especially near ionization thresholds.

INTRODUCTION

Tunable lasers are having a profound effect on experimental atomic physics. For those ranges of wavelengths available, one is able to match the laser(s) to the atomic transition(s) of interest. First used for spectroscopic studies, they have more recently been used to produce excited atomic gas or vapor targets in a cell or atomic beam for particle or photon impact experiments. Both pulsed and continuous tunable lasers have been used, but certain experiments favor one type over the other. This paper considers situations where those tunable lasers that are now or may soon be commercially available will be mated with photons from a SR source. We arbitrarily concentrate on VUV SR sources below about 300 eV photon energy. Let us use the term "hybrid experiments" for those which combine two very different photon sources such as lasers and SR.

Such atomic physics experiments have already begun at the storage ring ACO at the SR facility LURE, Orsay, France. This author is a member of the collaboration of physicists[1] that is studying photoionization and autoionization of laser-excited Na and Ba atoms with use of SR radiation between about 20 and 150 eV. Since these first experiments are reviewed in another talk of this Conference,[2] they will not be repeated here. Rather, the goal of this paper is to consider laser–SR hybrid experiments in a more general context.

LASERS AND SYNCHROTRON RADIATION AS PHOTON SOURCES

Table I compares the general properties of laser and SR sources. Taken from manufacturers' catalogs, the laser parameters are typical of those for commercially available systems that can be moved without undue effort from one laboratory to another. Of course, much larger, high power, pulsed (and continuous) lasers have been built for specialized research needs such as experiments in "laser fusion," but these are hardly portable. The SR parameters are typical of those at operational or planned SR facilities around the world.[3] SR provides in short pulses, but with a reasonable duty factor, a broadband, polarized continuum up to rather high photon energy. A monochromator must be used to narrow the bandwidth. The monochromatized SR photon beam is generally much less intense and spectrally much broader than those available from continuous tunable lasers, but the latter are restricted to rather low photon energies.

Table II amplifies this point by summarizing the properties of commercially available, dye-solution, tunable lasers pumped by various pump lasers. Continuous dye lasers with output powers up to ~1 W are available over the rather restricted photon energy range between ~1-3 eV. As is covered in more detail in another talk of this Conference,[4] pulsed pump lasers allow one to push tunable, narrowband, coherent and incoherent sources to somewhat 10 eV and higher, but their very low duty factor makes them a poor match for the much less intense SR. If such a ~10 ns wide, ~10 Hz pulse-laser-pumped source were used simultaneously in an experiment with a ~1 ns wide, ~50 MHz pulsed SR source, it would be difficult to synchronize the two sources. Even if they were, each laser pulse would shine on only one out of five million SR pulses. These lasers are poorly suited for hybrid experiments with SR. Rather, they are better suited to pulsed VUV (or x-ray) light sources whose repetition rate matches and can be synchronized with that of the laser. A flashlamp-pumped, pulsed dye laser and pulsed BRV continuum source have been mated in a pioneering series of hybrid photoabsorption experiments on alkali and alkaline-earth ions and excited atoms at the US National Bureau of Standards.[5] Some of these are reviewed in another talk of this Conference.[6] Another possibility would be to use one pulsed laser as a pump source both for a tunable dye laser and for a shorter wavelength source, either coherent or incoherent. This would easily solve the problem of synchronization.

An attractive pulsed laser for hybrid laser-SR experiments that has not yet been tried, as far as this author is aware, is the synchronously pumped, mode-locked ring dye laser.[7] These devices emit short (~ps) pulses of tunable radiation (see Table II) at a repetition frequency given by the inverse of the round-trip transit time of a pulse inside the dye laser cavity. This must be matched to the pump laser pulse train, which is related to the length of the pump laser optical cavity. Commercially available systems (Spectra-Physics and Coherent) operate near 80 MHz. It would be possible to make a specialized system whose optical cavities and optical modulator allowed it to operate at a somewhat different, but fixed frequency. One attractive value would be the 58.75 MHz repetition rate for the (nine) bunches of electrons in the VUV storage ring of the Brookhaven

TABLE I

A limited comparison of tunable laser and synchrotron radiation photon sources

ITEM	TUNABLE LASER	SYNCHROTRON RADIATION
Energy/photon	meV to ~10 eV	up to tens of keV
Spectral width	can be very narrow (<<1 MHz)	must be monochromatized (0.1-1% BW typical)
Intensity	high-enormous (10^{17}-10^{24} (peak)/s)	much lower (<<10^{13}/s monochromatized in 1% BW)
Time structure	continuous or pulsed (ps to µs pulses)	~1 ns pulses
Rep. rate	low power- up to tens of MHz high power- Hz to kHz	~10-50 MHz, but value is fixed
Duty factor	continuous- 1 pulsed- typically 10^{-6}	~0.1-5%
Polarization	easily controlled-linear or circular	elliptic: linear in orbital plane some circular off orbital plane not easily controlled
Portable	~yes	no
Expensive	can be (≤1K to >$100K/system)	yes, national facility typical

TABLE II

A compilation of approximate operating parameters for commercially available, dye solution, tunable lasers

LASER	LAMBDA(nm)	BANDWIDTH	REP. RATE	PUL.WIDTH	DUTY FACTOR	PK. PWR.	AVG. PWR.
Ring dye (doubled)	400-800 (~300)	10 MHz	continuous	continuous	1	-----	0.05-few W (<<0.05 W)
Mode-locked dye (doubled)	570-620;700-780 (285-310;350-390)	few cm^{-1}	~80 MHz	~few ps	~10^{-4}	<few kW	<1 W (<<10 mW)
N$_2$-dye (doubled)	355-970 (217-360)	0.1-1 cm^{-1}	0-2 kHz	~1-10 ns	<8x10^{-6}	<200 kW (<1 kW)	<1 W
Flashlamp-dye (doubled)	420-760 (217-380)	~1 cm^{-1}	0-20 Hz	~1 μs	~10^{-6}	<100 kW (<1 kW)	<1 W
Excimer-dye (doubled)	320-980 (217-355)	0.1-1 cm^{-1}	0-100 Hz	~10 ns	<10^{-4}	0.1-1 MW (5-100 kW)	0.1-1 W

Other lasers: diode (near and intermediate IR), color-center (near IR), optical parametric oscillator (near and intermediate IR); CO$_2$-laser-pumped molecular lasers (far IR). No attempt has been made to be complete; rather, the laser parameters listed are representative of those generally available in 1982. For an up-to-date compilation, consult trade journals and compilations such as the LASER FOCUS BUYING GUIDE published annually or data sheets available from manufacturers (e.g. Spectra-Physics, Coherent, Lambda Physik, etc.).

National Synchrotron Light Source (NSLS). As is shown in Table II, the average power of the mode-locked ring dye laser is comparable to that of the continuous one, but the energy is compressed into the short ps pulses. Therefore, the peak energy per pulse is in the kW range. Depending on the phase jitter of the SR pulses, one might have to phase-lock the mode-locked laser source to the rf of the SR ring, but such a feedback loop presents practical, not fundamental, problems. By varying the optical path between the output of the mode-locked dye laser and the point where the laser and SR beams interact with the (atomic, molecular, surface, or bulk) sample being studied, the time delay between the photon pulses from the different light sources could be carefully controlled.

We tentatively conclude, therefore, that a specialized, synchronously pumped, mode-locked ring dye laser system matched to the repetition frequency of an existing SR source would give experimenters using that source a very powerful system for time-resolved studies. One could imagine using either the SR or the laser to prepare an optically excited system and the other to probe it as a function of time-delay between the pulses. This would allow measurements of excited state lifetimes, chemical reaction rates, etc. on at least the ns time scale. At present, the time resolution is limited by the ~1 ns width of the SR pulses, but it may be possible to reduce this with future sources. It would be worthwhile for interested experimenters working at a given SR ring, such as the NSLS VUV ring, to explore further the building up of such a specialized, common facility available to all users of the ring.

ATOMIC TRANSITIONS BEST MATCHED TO AVAILABLE LASERS

Table III lists main resonance transitions for the alkali and alkaline-earth elements.[8] All have large oscillator strengths and, with the exception of Mg, can be pumped by available continuous ring dye lasers (Table II). (Mg is barely within the tuning range of some frequency-doubled, mode-locked, ring dye lasers.) Other elements in the periodic table have transitions connected to the ground state that are within the range of dye lasers, but their oscillator strengths are generally weaker. This explains why most previous experiments using continuous dye lasers have concentrated on the alkali and alkaline-earth metals. This necessitates the use of ovens and metal vapors in vacuum, and in a SR experiment, one must be careful to avoid contamination of the VUV monochromator optics if a window[9] is not used.

A rich literature exists on the use of lasers to pump the transitions listed in Table III. In the doublet spectra of the alkalis, each fine structure line is split by the hyperfine interaction caused by non-zero nuclear spin. This can produce unwanted optical pumping effects which tend to limit the stationary fraction of excited state population that one can produce in the atomic beam, even with a very intense continuous laser. These depend on the effective optical resolution, which in turn depends on the spectral width of the laser, the natural width of the excited level, the Doppler width associated with the atomic motion and the laser-atomic beam crossing geometry, power broadening caused by the laser, and the finite interaction time with the laser beam. They are further influenced by the degree of radiation trapping in the atomic vapor. This depends on the vapor density and on the geometry of the interaction volume.

Hertel and co-workers have made extensive experimental and theoretical

TABLE III

Main resonance transitions between ground states (G.S.) and excited states (E.S.) in alkali and alkaline-earth atoms. Metastable states (M.S.) that can be filled by spontaneous radiative decay of the E.S. are also listed.

ATOM	G.S.	eV	E.S.(J)	eV	λ(vac): G.S.-E.S.	M.S.(J)	eV	M.S. FILLED FROM E.S.?
Li	$2s\,^2S(1/2)$	0	$2p\,^2P(1/2;3/2)$	1.85;1.85	671.0;671.0	No	---	----
Na	$3s\,^2S(1/2)$	0	$3p\,^2P(1/2;3/2)$	2.10;2.10	589.8;589.2	No	---	----
K	$4s\,^2S(1/2)$	0	$4p\,^2P(1/2;3/2)$	1.61;1.62	770.1;766.7	No	---	----
Rb	$5s\,^2S(1/2)$	0	$5p\,^2P(1/2;3/2)$	1.56;1.59	795.0;780.2	No	---	----
Cs	$6s\,^2S(1/2)$	0	$6p\,^2P(1/2;3/2)$	1.39;1.45	894.6;852.3	No	---	----
Mg	$3s\,^1S(0)$	0	$3s3p\,^1P(1)$	4.34	285.3	$3s3p\,^3P(0,1,2)$	2.71	No (LS coupling)
Ca	$4s\,^1S(0)$	0	$4s4p\,^1P(1)$	2.93	422.8	$4s4p\,^3P(0,1,2)$ $4s3d\,^3D(1,2,3)$ $4s3d\,^1D(2)$	1.89 2.52 2.71	No (LS coupling) No (ditto) Yes
Sr	$5s\,^1S(0)$	0	$5s5p\,^1P(1)$	2.69	460.9	$5s5p\,^3P(0,1,2)$ $5s4d\,^3D(1,2,3)$ $5s4d\,^1D(2)$	1.83 2.27 2.50	No (LS coupling) No (ditto) Yes
Ba	$6s\,^1S(0)$	0	$6s6p\,^1P(1)$	2.24	553.7	$6s5d\,^3D(1,2,3)$ $6s5d\,^1D(2)$ $6s6p\,^3P(0,1,2)$	1.16 1.41 1.62	Weakly (intercomb.) Yes No (parity)

studies of the production of a beam of laser-excited Na atoms for use in electron scattering experiments. For details, the interested reader should consult various recent articles and reviews by members of that group.[19,11] Here we mention only one representative result of their studies: With use of plane-polarized laser radiation pumping the

$$Na(3\,^2S_{1/2}\,(F=2) \rightarrow 3\,^2P_{3/2}\,(F=3)$$

transition, it is possible to produce significant alignment of the excited atoms, and with circularly-polarized radiation, significant alignment and orientation. It was found, however, that the degree of alignment and orientation was influenced by the factors mentioned in the previous paragraph that influence the resolution, and most particularly, that they fall rapidly to zero when the atomic beam density is greater than about 10^{11} cm^{-3}.[11] This is caused primarily by radiation trapping; the unpolarized photons from spontaneous decay are imprisoned in the vapor and can dominate the effect of the polarized laser photons.

Since for many laser-SR hybrid experiments one is usually interested in maximizing the density of excited atoms, the effect just mentioned places severe constraints on the attainable density, and, therefore, on the signal rate that one would be able to obtain in experiments probing alignment and orientation effects. For example, the first experiments studying SR VUV photoionization of laser-excited Na atoms[1,2] used Na beam densities as high as 10^{13} cm^{-3}. Thus, despite the use of polarized laser radiation, it is likely that there was no atomic alignment in the laser-excited state.

For alkaline-earth atoms there are additional problems associated with metastable triplet P- and D-states and singlet D-states that lie between the singlet ground S- and laser-excited P-states. These are listed in Table III. If one pumps continuously the main singlet S-P transition, radiative cascade (or, perhaps, collision processes[12] in the vapor as well) can fill these metastable states and trap most of the excited population there. For the case of Ba, intense continuous pumping of the 6s-6p transition is calculated[13] to transfer about 85% of the Ba atoms to the singlet D-level lying 1.41 eV above the ground state. Less than 5% of the atoms are in the singlet P-level; the rest are in the ground state. In the alkaline-earths, singlet-triplet mixing (breakdown of LS coupling) is largest in Ba. A high-power ($\sim$1 MW/cm^2) pulsed laser has been used to drive strongly the lowest singlet-S to triplet-P transition at 791 nm.[14] Radiative decay then dumped as much as 78% of the population into the metastable triplet-D term lying 1.16 eV above the ground state. Intercombination processes would be much weaker in Ca and Sr.

For use in gas-phase, chemical reaction experiments, a plane-polarized, blue dye laser with maximum intensity $\sim$100 mW/cm^2/GHz pumping the main Ca transition (Table III) has been used[15] to prepare an atomic beam of aligned Ca(5s5p)^{1}P$_1$ atoms. (The excitation fraction was not given.) In order to eliminate the effects of optical trapping, the atomic beam density was kept below 10^{11} cm^{-3}. For the conditions of the experiment, the population of the metastable singlet-D level (Table III) was estimated to be only $\sim$1%, although changes in the pumping geometry could have been made to increase it.

Thus, techniques have already been demonstrated for production of appreciable fractions of laser-excited (and aligned or oriented if the atomic beam density is kept low enough to avoid radiation trapping) or radiative-cascade-filled excited states of most of the elements listed in in Table III. These are suitable targets for hybrid experiments using SR to study inner- or valence-shell photoionization of excited atoms.

ATOMIC PROCESSES AFFECTING ELECTRON SPECTRA

When, as has been done to date, electron spectroscopy is used to distinguish electrons produced by photoionization, autoionization, or Auger decay of the (various) excited and ground state atoms excited by SR in the interaction volume, several important processes which depend on the presence of the laser beam and the excited atoms have been observed to affect the measured electron spectra.

(1) Collisional production of "hot electrons." The laser-SR collaboration at Orsay [1,2] observed that laser-excited Na vapor in the absence of SR emitted "hot electrons" in a number of lines with energies up to nearly 7 eV. [12] One such spectrum is shown in Fig. 1. The various lines

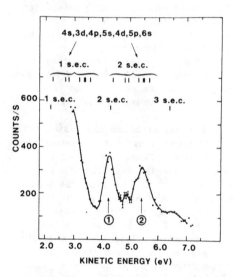

Fig. 1. Typical energy spectrum of electrons ejected from laser-excited Na vapor. The upper group of bars show the energy positions of electrons created by collisional ionization of the various excited states and heated by either one or two super-elastic (s.e.c.) collisions with Na(3p) atoms. The lower group of bars refers to electrons created by associate ionization and heated by one, two, or three super-elastic collisions, respectively. [from Ref. 12]

were satisfactorily explained quantitatively in terms of a model involving collisions between laser-excited atoms which produce low-energy "seed" electrons. These are heated in quantized, 2.1 eV steps by "super-elastic" collisions which transfer the Na(3p) internal energy to kinetic energy of the scattered electron. The higher the energy of the "hot" electron, the more collisions involving laser-excited Na(3p) atoms are needed. The small bump near 6.35 eV in Fig. 1 has been confirmed in more recent, unpublished experiments at Orsay; according to the model, [12] the intensity of this hot-electron line, as well as those between 5 and 6 eV, increase as the fifth power of the Na(3p) density [Na(3p)]. As mentioned earlier, this depends simultaneously on the Na(3s) density [Na(3s)], the laser power, the effective optical resolution, and optical pumping effects. These collision processes are a subset of those used to explain previous observations [16] of efficient ionization of dense columns of alkali and alkaline-earth metal vapors illuminated by intense, pulsed laser radiation tuned to main resonance lines (Table III). At Orsay, the much weaker, continuous laser illumination of Na produced orders of magnitude smaller fractional ionization. Photoionization by SR coupled with electron spectroscopy were combined to measure [Na(3p)]/[Na(3s)], which varied between ~1-25% as the various parameters mentioned on the bottom of the previous page were varied. In some Na experiments, the intensities of the strongest hot-electron lines were more intense than the strongest SR photolines. Near threshold, where some photolines could overlap the hot-electron lines, this contamination of the photoline spectra could be serious.

In general, it was much more accurate to measure [Na(3p)]/[Na(3s)] using SR-induced photoelectron spectra than to calculate it. Since the sodium vapor that could be optically thick to the resonant laser radiation was optically thin to the VUV SR, one had a linear diagnostic tool for measuring relative densities. Thus, we see that laser-SR hybrid experiments can investigate VUV photoionization processes as well as provide a very useful diagnostic tool for low-energy collision processes involving excited atoms. Avoiding details, it is useful to mention that recent measurements [17] of Na(3p)-Na(3p) energy-pooling collision cross sections do not agree with several earlier, less accurate experiments but do confirm the estimated values used by the authors of Ref. 12.

In recent hybrid experiments on laser- and cascade-excited Ba atoms, the laser-SR collaboration at Orsay observed hot-electron lines that were even more intense than in Na. Much less is known at present about the atomic and molecular states of barium that are important for ionizing collisions in laser-excited barium vapor. This is obviously a fertile topic for future experimental and theoretical study.

A possible solution to the "hot-electron" problem in laser-SR hybrid experiments could be to use a "cold" atomic beam produced by a supersonic expansion through a nozzle.[18] The lower the temperature, the lower would be the collision rate for particles with unequal velocities in the beam.[19] This might cause other problems, particularly in metal vapor beams: a dimer concentration as high as 20% has been reported for a supersonic sodium beam having a molecular rotational temperature of 40-70 K.[18] Of course, if one were interested in studying the dimers (trimers?), this could be used to advantage.

(2) Electron energy shifts induced by the metal vapor and the laser. In electron spectroscopy, contact potentials and other effects necessitate calibration of the energy scale of the electron spectrometer. When metal vapors are used, the coating of slits can change continuously the contact potentials and retard photoelectrons by 1 eV or more until an equilibrium is reached after many minutes. In hybrid laser-SR Na experiments at Orsay,[1,2] which used an oven and a cylindrical mirror electron analyzer (CMA) which have been described,[20] the presence of the laser tuned to a Na 3s-3p transition was observed to retard electrons emitted from the vapor by as much as an additional 1 eV; how much depended on the laser power. (Similar effects were seen with different magnitudes in preliminary hybrid laser-SR experiments with Ba.[2]) One contributing cause could be a positive plasma potential produced by the escape of collisionally-produced "hot" electrons. A second one could be laser-light-induced changes in the contact potential of sodium-covered surfaces in the electron spectrometer. The shifts with and without the laser beam could be measured[1,2] by admitting a reference gas such as Ne into the interaction volume and observing a known photoline produced by SR. Of course, one must assume that the presence of the

reference gas does not affect the shift, which is probably reasonable for a rare gas. In any laser-SR hybrid experiment using electron spectroscopy for detection, it will be necessary to account for these effects, particularly with metal and other reactive vapors.

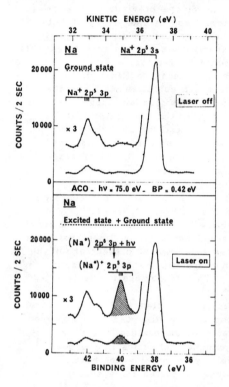

Fig. 2. Photoelectron spectra for 2p-subshell ionization of sodium obtained at a photon energy of 75 eV. Upper frame: "laser off" spectrum for ionization only of atoms in the 3s-ground state. The main peak near 38 eV binding energy and the first satellite peaks between 41-42 eV correspond to the residual ion being left in the two configurations shown. Lower frame: "laser on" spectrum which shows the above peaks plus a new (hatched) peak near 40 eV binding energy which is produced by 2p-ioinization of laser-excited Na(3p) atoms. Even though it leaves the residual ion in the same configuration as the first satellite for Na(3s), the different shapes of the peaks show that different distribution of fine-structure levels are produced in each case. [From Ref. 1]

EXAMPLES OF POSSIBLE LASER-SR HYBRID EXPERIMENTS

Having outlined above mostly the range of available technology for laser-SR hybrid experiments and some types of problems encountered in the first such experiments,[1,2] let us consider a few types of novel experiments that could be performed now or in the near future.

(1) Photoionization of excited atoms. This is the first type of laser-SR hybrid atomic physics experiment to be performed and is discussed in more detail in Refs. 1,2. One of the first photoelectron spectra obtained by the laser-SR collaboration at Orsay for photoioionization of the 2p-subshell of Na(3s) and laser-excited Na(3p) atoms by 75 eV photons is shown in Fig. 2. The peak near 38 eV binding energy is the main photoline corresponding to 2p-ionization of Na(3s). The blended peaks near 41-42 eV binding energy belong to the first satellite of the main line. A peak (shown hatched) near 40 eV binding energy appears only in the spectrum in the lower frame, which was taken with a laser pumping a 3s-3p(J=3/2) atomic transition. Notice that this main photoline for 2p-photoionzation of Na(3p) and the first satellite of the main Na(3s) line leave the residual ion with different distributions of fine-structure levels of the same configuration Na ($2p^5 3p$). This reflects the different

production mechanisms and is being investigated further.[2]

Now we consider a more ambitious experiment in sodium to investigate a recent prediction that multiple minima will occur for photoionizing transitions in excited atomic d states over a broad range of atomic number Z.[21] For Na(3d), Fig. 3 of Ref. 21 indicates that a zero in the 3d-kp dipole matrix element for Z=11 (sodium) will occur at a kp-photoelectron energy near 0.5 Rydberg, or about 6.8 eV. Since the 3d-electron in sodium is bound by 1.52 eV, this corresponds to a photon energy near 8.3 eV. This is within the reach of SR as well as some pulsed laser-based systems.[5] A beam of Na(3d) atoms could be produced by step-wise excitation of Na(3s) with two tunable dye lasers (Table II): the first laser at 589.2 nm would pump the 3s-3p(J=3/2) transition while the second at 819.7 nm would pump the 3p(J=3/2)-3d(J=3/2 or 5/2) transition. Since the laser-SR collaboration at Orsay[1,2] has been able to produce as much as 25% Na(3p) excitation in a weakly-collimated beam pumped by the first laser, one could hope that addition of the second laser could raise 25% or more of the 3p-atoms to the 3d-level, or perhaps as much as 10% of the Na(3s) atomic beam. For photoionization by 8.3 eV photons, photoelectrons would appear at kinetic energies of 3.16(3s), 5.26(3p), and 6.78(3d) eV, for the respective initial states of the valence electron shown in parentheses. These lines could be resolved easily. Theory[21] predicts that the 3d-kp channel, but not the 3d-kf channel, would have a minimum. The influence of the former would produce an overall minimum in the total 3d-subshell cross section. One could extract the contributions of each channel by measuring the angular distribution of the emitted photoelectrons.

For photoionization of excited 5d-electrons, theory[21] predicts that the 5d-kp channel will have one minimum whereas the 5d-kf channel will have two. To test this theoretical prediction, one could produce Na(5d) atoms by changing the second laser step above to 3p(J=3/2)-5d at 498.4 nm.

In these experiments, one would face many of the problems mentioned earlier in this paper. Most noteworthy, collisional processes among the excited atoms would produce "hot electrons" with kinetic energies as high as 6 eV.[12] This could confuse the observation of the predicted minima in this photoelectron energy range.

(2) A recent paper[4] by the NBS group (see also Ref. 6) discusses the very different features observed in the 4d-photoabsorption spectra of Ba^{++} between 80-150 eV, as compared to those in Ba^{+} and neutral Ba, in terms of increasing contraction (vs. collapse) of the 4f-wavefunction as the ionic charge state is raised. Since at Z=56 Ba is just on the edge of 4f-wavefunction "collapse" from an outer to an inner potential well, the presence or absence of outer electron(s) can have a much larger effect than one might ordinarily expect on the structure of this inner shell. As has already been predicted via calculations for Cs I,[22] one could exert a kind of amplified control of the collapse of the inner-shell wavefunction by changing the excitation of the valence electron. For the case of the charge states of Ba, it would be especially interesting to study 4d-photoabsorption in the singly-charged ion $Ba^{+}(nlm)$ as a function of its nlm quantum numbers. The NBS pulsed laser technique[4,6,16] could be used to produce efficiently $Ba^{+}(6s)$ (and $Ba^{+}(5s)$) ions in a Ba heat pipe. This would be followed by 6s-6p excitation with a slightly time-delayed 493.5 nm laser, and further step-wise excitation with various combinations of other, suitably time-delayed lasers operating at 452.6 nm (6p-7s), 389.3 nm (6p-6d), 536.2 nm (6d-6f), etc. Control of polarization and time-delays might

even make it possible to have some degree of alignment and/or orientation in the excited ionic states. One would expect that as the nl-values of the excited electron were made larger, thereby decreasing its overlap with the Ba^{++} core, the 4d-photoabsorption spectrum measured with a suitably time-delayed VUV photon pulse from a BRV lamp or other source would change continuously from that of $Ba^+(6s)$ to that of Ba^{++}.

With use of a fast (keV energy range) $Ba^+(6s)$ ion beam crossed by a photon beam from a monochromatized SR source, one could perform a similar experiment with SR that would, in principle, admit the possibility of photoelectron energy analysis.

(3) Triply-excited states. Neutral Ba has a "5d7d" $^1D_2$ level[23] lying 41841.5 cm^{-1} above the ground state, or just below the first ionization limit. A recent analysis employing multichannel quantum-defect theory showed that its makeup is determined predominantly by two configurations, 31% doubly-excited 5d7d and 69% singly-excited 6snd Rydberg character with an effective quantum number $n^*=23.8$.[23] It has been prepared by step-wise excitation with two tunable dye lasers: $(6s^2)^1S_0-(6s6p)^1P_1$ at 553.7 nm followed by $(6s6p)^1P_1-("5d7d")^1D_2$ at 420.3 nm.[24] It would be interesting to use SR to study 4d-, 5s-, and 5p-subshell photoionization of the atom in its "5d7d" state, particularly via odd-parity (three photons added to the even parity ground state), triply-excited, autoionizing resonances. As far as this author knows, the position of these resonances has not been calculated, but they may be observable in a photoabsorption experiment out of this state. Loosely speaking, for 31% of the time the "5d7d" level has doubly-excited character, so that excitation of an inner-shell electron to an excited state of the complex would depend on the triply-excited nature of the final state.

It is worth emphasizing that total VUV photoabsorption spectra of laser-excited atoms are a useful if not essential prerequisite for laser-SR hybrid experiments that use monochromatized SR to investigate multiply-excited, autoionizing resonances. The latter experiments are wavelength selective, and if one does not know where to look for a particular resonance of interest, the search time could be make such experiments tedious, if not impractical, to perform. For example, availability of the positions of the even-parity, autoionizing resonances in laser-excited $Na(3p)$[25] greatly aided the laser-SR collaboration at Orsay in their experiments.[1,2]

(4) Shape-resonance effects in vibrationally-excited molecules. Thus far, we have avoided mention of possible laser-SR hybrid experiments with molecules. The types of atomic experiments mentioned above could easily be extended to molecules when one important problem is dealt with. At room or elevated temperatures associated with thermal beams from effusive oven sources, many rotational states and, possibly, several vibrational states of some molecules will be appreciably populated. For driving electronic transitions or vibrational-rotational transitions, standard tunable lasers are sufficiently narrow-band that they will excite molecules out of only a single or a small number of rotational levels in a vibrational manifold. This will greatly reduce the fraction of molecules in the thermal Boltzmann distribution that can be laser-excited into a specific level unless one reduces the number of initially populated levels. This can be done by expanding the molecular beam through a nozzle, thereby cooling it.[10,26] For example, NO has been cooled to a rotational temperature of 30-40 K for multiphoton ionization experiments with a pulsed, tunable dye laser.[26]

Attention was first called theoretically[27] to the effect that coupling of electronic and nuclear motions via shape resonances could have in molecular photoionization spectra and in electron-molecule scattering experiments. This was first observed in 5σ-photoionization of CO[28] as non Franck-Condon vibrational intensity distributions in the CO^+ final state.

As far as this author is aware, these shape-resonance effects have not been studied as a function of the vibrational (and/or rotational) excitation of the molecule above the ground vibrational state. Such studies could be carried out in laser-SR hybrid experiments with molecules such as NO. For example, a pulsed IR F-center laser has been used[29] to drive v=0 to v=2 transitions to produce state-selected NO molecules in one of the spin-orbit levels of the ground electronic state. It would be interesting to study shape-resonance-enhanced photoionization of these vibronically excited molecules.

Vibrational and rotational transitions in molecules are in the IR and far-IR, respectively, where few tunable laser sources exist. This may change in the future, however, in such a way as to enhance the prospects for IR laser-SR hybrid experiments on molecules. Research efforts on "free-electron" lasers are underway at a number of SR research laboratories around the world. Since these naturally tunable devices will be much more efficient in the IR than in the visible or the UV,[30] they may prove to be a very important and intense laser source for molecular studies. If they can be to operate simultaneously with "bending magnet" SR sources, or the more intense wigglers and undulators[31] that will become common in the future, the experimenter will have naturally under one roof the two sources needed for the hybrid experiments. Future planning for free-electron lasers at SR users' facilities should take this exciting possibility into account.

(5) "Inverse" SR-laser hybrid experiments. The examples of possible experiments mentioned so far in this paper have all had the atom or molecule absorbing a laser photon before the VUV SR photon. Of course, this is not necessary; there are other types of experiments in which one could profitably reverse the order. We mention here only one example. Consider the UV photo-dissociation/ionization of H_2[32] by SR

$$h\nu + H_2 \rightarrow H(nl) + H(n'l')$$

which dissociates into a final state with at least one excited hydrogen atom. One or both could also be in the continuum. Low-nl excited states can be detected by decay fluorescence, but higher-nl levels have much smaller decay rates and branching-ratio problems that makes difficult their detection by this method. Instead, one could use a tunable laser to induce bound-bound or bound-free transitions that would be used to detect the individual final states.

(6) Spin effects. The partial circular polarization of SR photons emitted out of the electron orbital plane[5] has been used[33] to study spin-effects in photoionization of unpolarized heavy atoms. Since, as mentioned earlier, lasers can be used to produce aligned and/or oriented, excited atoms,[10,11,15] it would be interesting to extend these spin-dependent studies[33] to such excited atoms (and molecules).

CONCLUSIONS

Mindful of the caveat that experimentalists should talk about experiments they have done, and leave the ones they might do to proposals or conversations over wine or beer with colleagues, this author has risked ignoring this advice in the present paper. The ideas presented herein would occur to most physicists familiar with the rudiments of laser and SR spectroscopies if they took the time to think about them. Since it afforded him the opportunity to engage seriously in this thinking process, this author appreciates the invitation from the organizers of this conference to speak on this subject.

It is clear that whenever experimentalists combine different techniques, new experiments are made possible. The combination of laser and SR photon sources will be no different. At the present time, the typical tunable laser is a much brighter source than monochromatized SR, but the latter provides much higher photon energies. Each year tunable lasers or laser-pumped photon sources push farther into the UV (and beyond), at first with small, but later, ever-increasing intensities. Electron storage ring photon sources, conversely, are becoming ever brighter as the development of wigglers and undulators continues. Free-electron lasers at storage rings will offer additional possibilities, first in the IR and visible portions of the spectrum, perhaps later in the UV. Paraphrasing a colloquial American expression, "Photons is photons." If one is interested in certain physics best studied with photons, it doesn't make any difference where the photons come from as long as they are useful and available. At the present time, there is a moving, fuzzy dividing line between the laser and the SR photon sources. Unfortunately, that frequently extends also to the physicists; the laser and SR communities have been somewhat separated. It is clear from this conference that one is making efforts to knock down the wall and that such efforts should continue. It is a goal of this paper to aid that process.

ACKNOWLEDGEMENTS

The author appreciates support of this work by the US National Science Foundation and, during 1981, by the French Conseil National de la Recherche Scientifique, as well as receipt of an Alfred P. Sloan Fellowship and a Yale University Junior Faculty Fellowship. He was priveleged to be able to spend part of a sabbatical leave during 1980-81 at the Laboratoire pour l'Utilisation du Rayonnement Electromagnetique (LURE), Orsay. F. Wuilleumier, D. Ederer, and P. Dhez, and a capable graduate student, J.M. Bizau, were enthusiastic and able guides for this author into the world of physics with synchrotron radiation and shared the excitement as we and our collaborators from Laboratoire Aime Cotton, J.L. Le Gouet, J.C. Keller and J.L. Picque, and B. Carre from Saclay began to bridge the gap between the SR and laser communities. Some of the ideas in this paper were stimulated by conversations with these people. The author also appreciates the invitation to speak at the Conference and the patience of its co-Chairman B. Crasemann during his wait for a late manuscript.

REFERENCES

1.   J.M. Bizau, F. Wuilleumier, P. Dhez, D.L. Ederer, J.L. Picque, J.L.

LeGouet, and P. Koch, in Topical Meeting on Laser Techniques for Extreme Ultraviolet Spectroscopy, Boulder, 8-10 March 1982 (AIP Conference Proceedings, in press).

2. F. Wuilleumier, invited talk B14-2 of this Conference, elsewhere in this volume.

3. See, for example, relevant chapters of SYNCHROTRON RADIATION RESEARCH, H. Winick and S. Doniach, eds.(Plenum Press, New York, 1980).

4. R.R. Freeman, invited talk AG-2 of this Conference, elsewhere in this volume.

5. T.B. Lucatorto, T.J. McIlrath, J. Sugar, and S.M. Younger, Phys. Rev. Lett. 47, 1124(1981), and references therein.

6. T.B. Lucatorto, T.J. McIlrath, and W.T. Hill III, invited talk B9-2 of this Conference, elsewhere in this volume.

7. Literature of manufacturers (Spectra-Physics and Coherent) are good sources of current information on commercially available systems.

8. C.E. Moore, ATOMIC ENERGY LEVELS, NBS Circular 467 (U.S. Government Printing Office, Washington, D.C., 1949).

9. R.A. Rosenberg, M.G. White, C. Thornton, and D.A. Shirley, Phys. Rev. Lett. 43, 1384(1979), and references therein.

10. I.V. Hertel and W. Stoll, Adv. At. Mol. Phys. 13, 113(1978); I.V. Hertel, in PHYSICS OF ELECTRONIC AND ATOMIC COLLISIONS, S. Datz, ed.(North-Holland Publ. Co., Amsterdam, 1982).

11. A. Fischer and I.V. Hertel, Z. Phys. A 304, 103(1982).

12. J.L. LeGouet, J.L. Picque, F. Wuilleumier, J.M. Bizau, P. Dhez, P. Koch, and D. Ederer, Phys. Rev. Lett. 48, 600(1982).

13. The author is indebted to J.L. LeGouet for this calculation.

14. J.L. Carlsten, J. Phys. B 7, 1620(1974).

15. C.T. Rettner and R.N. Zare, J. Chem Phys. 77, 2416(1982).

16. T.B. Lucatorto and T.J. McIlrath, Appl. Opt. 19, 3948(1980).

17. J.P. Huennekens, PhD Thesis (Univ. of Colorado, 1982, unpublished).

18. K. Bergmann, U. Heffer, and J. Witt, in ELECTRONIC AND ATOMIC COLLISIONS, N. Oda and K. Takayanagi, eds.(North-Holland Publ. Co., Amsterdam, 1980).

19. W.E. Baylis, Can. J. Phys. 55, 1924(1977).

20. CMA: M.Y. Adam, PhD Thesis (Univ. Paris-Sud, 1978, unpublished); CMA and ovens: J.M. Bizau, Thesis, 3e Cycle (Univ. Paris VII, 1981, unpublished) and S. Krummacher, PhD Thesis (Albert-Ludwigs-Univ., Freiburg, 1981, unpublished).

21. A.Z. Msezane and S.T. Manson, Phys. Rev. Lett. 48, 473(1982); see also J. Lahiri and S.T. Manson, Phys. Rev. Lett. 48, 614(1982).

22. J.P. Connerade, J. Phys. B. 11, L381(1978).

23. M. Aymar and O. Robaux, J. Phys. B 12, 53(1979).

24. S.A. Bhatti, C.L. Cromer, and W.E. Cooke, Phys. Rev. A 24, 161(1981).

25. J. Sugar, T.B. Lucatorto, T.J. McIlrath, and A.W. Weiss, Opt. Lett. 4, 109(1979).

26. P.M. Johnson, Appl. Opt. 19, 3920(1980).

27. J.L. Dehmer and D. Dill, in ELECTRONIC AND ATOMIC COLLISIONS, N. Oda and K. Takayanagi, eds.(North-Holland Publ. Co., Amsterdam, 1980)

28. R. Stockbauer, B.E. Cole, D.L. Ederer, J.B. West, A.C. Parr, and J.L. Dehmer, Phys. Rev. Lett. 43, 757(1979).

29. Aa.S. Sudbo and M.M.T. Loy, J. Chem. Phys. 76, 3646(1982).

30. C. Pellegrini, in SYNCHROTRON RADIATION RESEARCH, H. Winick and S. Doniach, eds.(Plenum Press, New York, 1980); C. Pellegrini, invited

660

      talk P4-1 of this Conference, elsewhere in this volume.

31. G.S. Brown, H. Winick, and T. Pate, invited talk A12-1 of this
      Conference, elsewhere in this volume.

32. M. Glass-Maujean, J. Breton, and P.M. Guyon, Phys. Rev. Lett. 40,
      181(1978), and references therein; S. Strathdee and R. Browning, J.
      Phys. B 12, 1789(1979); I.V. Komarov and V.N. Ostrovskii, Pis'ma Zh.
      Eksp. Teor. Fiz. 28, 446(1978)[Sov.Phys.-JETP Lett. 28, 413(1978)].

33. U. Heinzmann, Appl. Opt. 19, 4987(1980).

CONFIGURATION INTERACTION AND SCREENING IN CORE-LEVEL
SPECTROSCOPY OF SOLIDS

J.C. Fuggle
Institut für Festkörperforschung, der KFA Jülich,
D-5170 Jülich, W. Germany.

ABSTRACT

This paper deals with systems which show complicated
line-shapes in core-level spectroscopies. The emphasis is
on photoelectron spectroscopy. The aim is to give a simple
picture of the processes behind such complex line-shapes
and to describe their relevance as a possible diagnostic
tool for study of the electronic structure of solid state
materials. It will be shown that the complex line-shapes,
with two or more peaks, arise when core-hole screening
involves nearly localized valence electrons: i.e. electrons for which the ratio of the effective Coulomb interaction, U_{eff}, to the band-width, W, is nearer to infinity
than to one. This condition is commonly met in the early
lanthanides, the actinides, and some adsorbate systems.
The phenomena will be explained within the formalism
of configuration interaction, and a screening model which
is physically, if not formally closely related.

CONFIGURATION INTERACTION AND SHAKE UP

The many-body wave function of an atom, molecule, or
solid can be written as a certain distribution, or configuration, of electrons in separable wave functions, or
single-electron orbitals. However such independant
single-electron orbitals do not really exist because of
interaction between the electrons. As a consequence of
these many-body interactions between the electrons, the
most realistic description of the system, with the best
total energies obtainable in such a formalism, includes
small weightings of configurations with electrons excited
into the unoccupied orbitals. For instance, even the He
$2s^2$ wave function is improved by addition of small amounts of 2p character to allow for correlation effects
which keep the two electrons apart. This mixing in of
higher configurations is known as configuration interaction. Because of the large computing effort involved
its use is mostly restricted to atoms and small molecules, but the concepts involved are often relevant to
larger systems. In core level spectroscopies one can,
and should, consider configuration interaction in both
the initial and final states. The consequences of configuration interaction (CI) for the energies of transitions involving non-convseration of core-holes may be
small, but its inclusion can sometimes explain drastic

effects on the line-shapes and on satellites and their
intensities. The purpose of this paper is to give a conceptual description of examples of these effects in
solids.

In atoms configuration interaction and its effects
are, to a first approximation, limited to configurations
with the same term designation. For instance in XPS of
Kr, ejection of a 4s electron would lead to the configuration .....$4s^1 4p^6$ (^2S). This configuration can interact with the configuration ....$4s^2 4p^4 4d^1$ (^2S), but not
$4s^1 4p^5 4d^1$ ( ) which has very different terms. In solids
and molecules two new effects may be important. The
first is simply that the surroundings may affect the
radial distribution of the unoccupied valence orbitals and
the overlap integrals between initial and final states.
The second effect is that the atomic term symbols may
not be strictly valid any more, even for strongly localized orbitals, because of hybridization. For instance we
will discuss later the case of 3d core hole creation in
XPS of lanthanum compounds and talk of ....$3d^9$...$4f^0$...
and ....$3d^9$...$4f^1$... final states being created from the
ground state ...$3d^{10}$...$4f^0$.. Here we imply that

1) there can be either intra-atomic or inter-atomic
transfer of an electron to the 4f level

2) configuration interaction between the $d^9 f^0$ and
$d^9 f^0$ and $d^9 f^1$ states is allowed because of mixing and
hybridization of the f and the other orbitals. i.e.
correlation effects need not be the major driving force
behind CI, as in atoms.

The concept of configuration interaction and its
application to core level spectroscopies includes one
refinement of the idea of shake-up that I wish to discuss.
Once CI has been applied the occupation coefficients of
the atomic one-electron valence orbitals are generally
not equal to 1 and the differences between different
final states cannot be accurately expressed in terms of
a unit change in the occupancy of a single atomic orbital.
My discussions with non-specialists have led me to believe that this complication is not always sufficiently
appreciated so I wish to illustrate it here in this
general overview. Consider figure 1 which shows a typical diagram of a shake-up process in photoelectron
spectroscopy. The idea behind this is that in sudden
ionization of a core hole there is a finite probability
of an electron being shaken up from an occupied to an
unoccupied orbital. In the most simple formulation of
shake up the valence electron configurations are written
as in equation 1

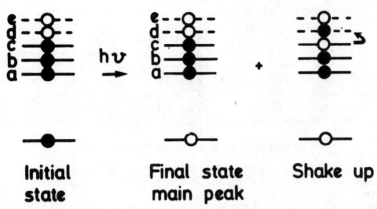

Fig. 1. A standard illustration of shake-up.

$$\text{ground state} = a^1 b^1 c^1 d^0 e^0$$
$$\text{lowest final state} = a^1 b^1 c^1 d^0 e^0 \qquad (1)$$
$$\text{first shake-up} = a^1 b^1 c^0 d^1 e^0$$

etc.

When CI is taken into account the occupation coefficients are not equal to one and we must generalize equation to

$$\text{ground state} = a^v b^w c^x d^y e^z$$
$$\text{lowest final state} = a^{v'} b^{w'} c^{x'} d^{y'} e^{z'} \qquad (2)$$
$$\text{first shake-up} = a^{v''} b^{w''} c^{x''} d^{y''} e^{z''}$$

etc

In most applications of equation 2

$$v, \quad w \quad \text{and } x \qquad y \quad \text{and } z$$
$$\text{and} \qquad v', \quad w' \text{ and } x' \qquad y' \text{ and } z' \qquad (3)$$

although this is not always the case, as we shall see later. Also we should remember that because of relaxation the occupation coefficients of the orbitals are not the same in the initial and final states, i.e. $v \neq v'$, $x \neq x'$, etc. In other words the molecular orbitals of the system change in response to the new core potential. In many cases we may find that $x' \sim 1$ and $y' \sim 0$ whilst $x'' \sim 0$ and $y'' \sim 1$ so that the shake-up picture given by equation 1 becomes a reasonable approximation. However this is by no means universal /1/ and some cases are known in which the few-particle description of shake-up breaks down completely.

Some authors have tried to distinguish between shake-up and configuration interaction effects, saying that shake-up involves only one valence electron and the label "CI

effects should be reserved for cases where more than one
valence electron is involved. I find this division artificial in the solid state areas that I have worked on
because the changes of occupation number are seldom, if
ever, exactly zero or one in shake-up. Also it is usually
clear that more than two orbitals change their occupation number, as would be the case if only one electron
were being "shaken up". Thus I have come to regard all
the satellites as "shake up" and CI as a suitable scheme
in which to discuss the physics of the processes involved.

RELAXATION AND SCREENING

When a core hole is created all the wave functions
of the other N-1 electrons of the system change, or
relax, to adjust to the new potential. In a metallic
solid the creation or destruction of the impurity charge
must be screened from the rest of the metal, otherwise
the potential would be changed throughout the metal.
The valence electrons thus build up their density around
an atom where a core hole has been created, as is the
case in XPS for example. The distribution of this extra
charge may look rather like that in an atomic orbital,
so that one can speak of "screening orbitals" and screening processes /2/. The concept of a screening orbital
is, however, not restricted to metals and yields good
results for some insulators, such as La_2O_3 /3/ and
for some adsorbates /4/.

Within the simple one-electron orbital scheme describing shake up in figure 1 one would equate the screening orbital with a change in the order of the one-
electron orbitals, as described by figure 2.

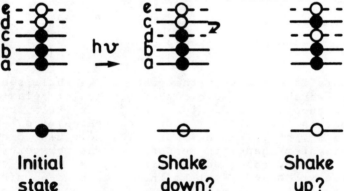

Fig. 2. The role of a screening orbital in shake-up

In this scheme one could describe the transfer of an
electron from the c to the d orbital as a "shake-down"
process. However at this stage I argue that the "shake"

terminology becomes too limited by its simplifications and I will not use the term "shake down".

The major problem with schemes like figure 2 is that they disregard the mixing of the orbitals, or the CI, and insist that an orbital be completely full, or completely empty. Another scheme has been put forward by several authors /3-9/ for systems with quite well localized screening orbitals, and this is more suitable for much solid state work because it incorporates partial occupation of the screening orbitals and hybridization effects as well as the dynamics of photoemission. This is done by giving the screening orbital a width and energy in both the initial and final states, as in figure 3 which illustrates the results of numerical calculations by Schönhammer and Gunnarsson (S.G. model) /7/.

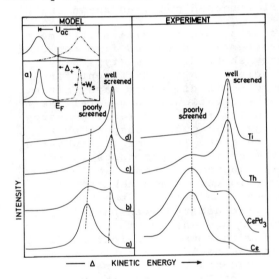

In the initial state of photoemision the screening orbital is at an energy Δ_+ above the Fermi level and has a width, W, which represents the coupling to the other levels of the system. In the final state, in the presence of a core hole, the screening level is pulled down U_{ac} with respect to the Fermi level by the core hole potential. The level is now Δ_- below E_F. In this model the probability that the screening orbital will be occupied on core level photoemmission is related to the position of the orbital and its coupling to the other orbitals. Thus if the screening orbital is narrow and far above E_F in the initial state, as shown in inset a, then its occupancy is low.

Fig. 3. Left: Illustration of the Schönhammer-Gunnarsson model of screening and its implications for the core level line-shapes. The values of W/$_+$ were d=0.94, c=0.75, b=0.56, a=0.38 and U_{ac} was 1.5 $_+$ (Ref. 7).

Right Ce $3d_{5/2}$, Th $4f_{7/2}$, and Ti $2p_{3/2}$ XPS peaks from Ce, $CePd_3$ Th and Ti. The peak binding energies are 883, 333 and 454 eV for the Ce, Th and Ti levels respectively.

There is then very little overlap between the initial state and the final state with the screening level occupied. This means that the probability of a transition to the "well-screened" final state is small. Most of the XPS intensity will be found in the "poorly screened" peak lying approximately Δ_- to higher BE and corresponding to the transition to a final state with the screening level almost unoccupied as shown in curve a. If, on the other hand, the width W of the screening level is of the same order as its position Δ_+, the level is strongly coupled to the system and its extensive tail below E_F can be interpreted as considerable occupancy, even in the ground state. The probability that it becomes occupied in the final state, as a result of core-level photoemmission, is then high and most of the XPS intensity is found in a so-called well-screened peak, as shown in curve d).

All the final states of a core transition in a metal have the change in core potential screened from the rest of the metal within a very short distance (typically about the Wigner-Seitz radius). In the different final states this screening is done in different ways, often with accompanying plasmon excitations. In general the state with the smallest radius for the screening charge will have the lowest energy, although I suspect that this need not always be the case. In order to emphasize the common features of the screening processes in XPS from different systems we adopted the name "well screened" to describe the peak due to transitions to the final state of lowest energy and "poorly screened" to describe the peaks at higher binding energy /3/. We debated with ourselves for a long time on what terms to use and eventually adopted this nomenclature to stress that the core-ionized final state of lowest energy in lanthanides or actinides was one where the screening cloud had a smaller radius and was energetically more favourable. Unfortunately some workers understood the term poorly screened to imply a state where the core hole was incompletely screened from the metal and this is incorrect. It has since been suggested to us that "locally screened" and "diffusely screened" might have been better labels /12/. This is a good suggestion but to change the names now would only add to any confusion. I am also sure that some people would have chosen to be confused whatever names were used.

There is an important difference between the SG model and the simple "shake" model of figure 2. Notice that in models of the SG type the occupation coefficient of the screening level is more than zero in the initial state and less than one in the final state. This could also be expressed by saying that there is some configur-

ation interaction in both the initial and the final
states. Numerical calculations of this type show that
the observed line shape is very sensitive to such CI /5-11/.
The relative intensities in the well and poorly screened
peaks is definitely not just linearly dependant on the
screening orbital occupation in the ground state.

In the discussion that follows we will often use the
approximate occupation of the screening orbital as a
shorthand label for the initial and final states. It
should be noted that these labels are approximate because of CI effects as described above.

EXPERIMENTAL EXAMPLES

There are several examples now known in which the
trend in core-level XPS line-shapes follows that predicted within impurity models of the SG type. Figure 3
illustrated the case of the Ti $2p_{3/2}$, Th $4f_{7/2}$ and Ce
$3d_{5/2}$ XPS peaks from reference 3. The screening levels
in these cases are the Ti 3d, Th 5f and Ce 4f levels
respectively. These levels are increasingly localized
and decoupled from the other valence levels in the order
Ti 3d >Th 5f >Ce 4f so that the weight in the well-
screened peak also decreases in that order. It should
be recognized that the SG model uses an impurity-like
screening level. There have been few attempts to study
what happens when the screening level is part of a band
and one is safest extrapolating from the SG results when
the direct interaction between screening orbitals on different sites is small. This is certainly the case for
the Ce 4f and Th 5f levels but must be less true for the
Ti 3d levels.

The core level XPS line-shapes of about sixty La, Ce,
Pr, Th and U intermetallic compounds have now been studied. A principle aim of many of these studies was to use
the data to obtain trends in the degree of localization
of the 4f and 5f levels /3,13-17/. In those compounds
whose magnetic properties were known, a good correlation
was found between the occurence of "localized" magnetic
behaviour and large XPS intensity in the poorly screened
peaks; i.e. electrons which are poorly coupled to the
system for XPS purposes are also poorly coupled for magnetic properties. Also Bremsstrahlung isochromat (BIS)
studies of U and Ce compounds usually show 5f or 4f levels
near E_F /16,18/ when the core level XPS line-shapes
indicate good coupling of the screening levels to the
system. In contrast, in substances like UO_2 or Ce where
the poorly screened peaks are relatively strong, the
screening levels are found several eV above E_F by BIS.
These observations thus support the use of the SG model
in such cases.

CONFIGURATION FLUCTUATION AND SCREENING PROBLEMS

In solids it is easy to imagine non-integral occupation for the valence orbitals when the different configurations are strongly coupled- that is when the electrons are weakly correlated. There is, however, a class of compounds including some of the rare earths that have very narrow f-bands (10 eV is typical) but which still show many of the phenomena of non-integral occupation of the f levels. These compounds have unusual properties in magnetism, specific heats, etc /21/. It has been suggested that their configurations effectively <u>fluctuate</u> on a rather long time scale (10^{-13} seconds or longer are often mentioned /21/). Hence the rather long name "Inter-Configurational Fluctuation (or ICF) compounds". The compounds are also often called "Mixed Valence" compounds, despite the almost audible protests of the nineteenth century chemists who must be turning in their graves at the misuse of the word "Valence". The interest of core level spectroscopists in this field arises because there is a need to determine the "f-count", or weights of the different configurations in ICF compounds and it was hoped that core level spectroscopy might yield the answers. The most desperate need was thought to be for Ce and its compounds so I shall consider this case.

The configuration suggested for Ce takes the form

$$\text{Ce...4f}^0(5d6s)^4 \longleftrightarrow \text{Ce...4f}^1(5d6s)^3$$

if such fluctuations occur they must be much slower than the processes of core level spectroscopy. The energy of the two configurations must be nearly equal if they are to be allowed to fluctuate in the initial state, but in the presence of a core hole the $\bar{c}4f^0(5d6s)^4$ and $\bar{c}4f^1(5d6s)^3$ configurations are separated by about 10 eV. (Here $\bar{c}$ represents a core hole). One might then think that in any spectroscopy creating a core hole, such as L_3 x-ray absorption spectroscopy or XPS of the 3d core levels, one should see two peaks arising from transitions to the f^0 and f^1 final states. It would then be proposed that the intensity of these two peaks would be equal to the weights of the two configurations in the ground state, as indicated in figure 4a. In practise the situation is rather different because of screening effects and the dynamics of photoemission. As described above, we also see a peak arising from the well screened f^2 final state in the 3d XPS spectrum. The new situation is described in figure 4b, where transitions to the f^2 final state are indicated by a diagonal arrow. In the ICF picture of Ce one would state that if transitions from $3d^{10}4f^1$ to $3d^9 4f^2$ must be considered, then one must also consider

the possibility of transitions from $f^0 \to f^1$, as indicated
by the dotted diagonal line in figure 4b. In the presence of such transitions the $3d^9 4f^0$ peak intensity is
not a reliable indicator of the weight of f^0 character
in the ground state. A more accurate model might stress
the integration of the two configurations in the ground
state more than the ICF picture, but the point remains
that one has no justification for assuming that the f^0
peak intensity reproduces the amplitude of f^0 character
in the ground state.

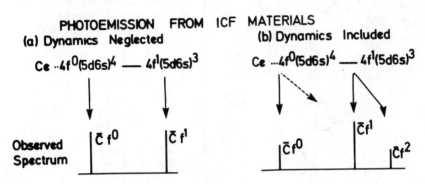

Fig.4a. If screening effects and their dynamics can be
neglected, core level photoemission from a Ce compound
with equal amplitudes of f^0 and f^1 in the initial state
gives a spectrum with two equal peaks.
Fig.4b. In the presence of a screening level with dynamics three peaks are found whose intensities do not
directly reflect the ground state f^0, f^1 and f^2 amplitudes.

It was at one time thought that some Ce intermetallic compounds contained Ce in the pure f^0 configuration.
A typical example is CeNi$_5$, whose 3d XPS spectrum is shown
in figure 5. The three peak structure shown in figure 4b
is doubled in figure 5 because of the 3d spin-orbit
splitting, but the peaks due to transitions to f^0, f^1-
and f^2-like final states are clearly discernable. The f^0
peaks are rather small and although we have insisted
that its intensity is not a quantitative guide to the
f^0 amplitude in the ground state, it is inconceivable
that Ce could be nearly pure f^0 in the ground state.
Calculations have recently been attempted using models
to simulate spectra like that in figure 5. One can
adopt the procedure of using the f^1:f^2 peak ratio to
determine the hybridization width of the 4f levels and
then using this the f-count can be varied to find the
value nessecary to reproduce the f^0 peak in the spectrum.
preliminary results suggest that within the model used,
the error involved in taking the f^0 peak intensity as

proportional to the f^0 ground state amplitude is only about 20% in this case /19,20/.

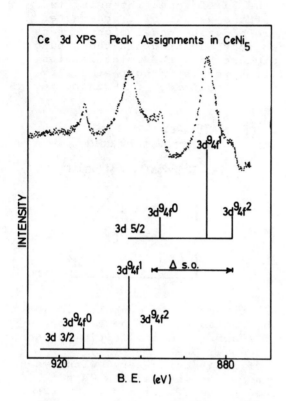

Fig.5. The XPS spectrum of CeNi$_5$ with the peak assignments for the 3d region. The features correspond to transitions to final states with the approximate f-counts given. It can be assumed that the other (sd) valence electrons adjust to give approximate charge neutrality. The f^0 peaks in CeNi$_5$ are amongst the largest found in intermetallic compounds /15/.

The reason that the Ce f^0 peak intensity gives a comparatively good measure of the f^0 amplitude in the case of Ce intermetallics can be explained with the help of figure 6. This figure shows the energy of the Ce system as a function of f count for the initial state and in the 3d core-ionized final state. The actual ground state is a mixture of mainly f^1 and f^0 configurations with a few % of f^2 admixed. If the 3d-ionized states had no configuration interaction and were pure f^0, f^1, or f^2, then the distribution of XPS intensity would exactly mirror the ground state amplitudes. There is some configuration mixing but it is reduced as the energy separation of the configurations is increased. As shown in figure 6, the f^0 configuration is 10 eV above the f^1 configuration in the presence of a 3d hole and is even further above the f^2 state. As the Ce 4f hopping integrals are rather small anyway this large energy separation makes the $3d^9 4f^0$ state quite pure with respect to the f^1 and f^2 states. Thus the overlap matrix elements between the ground state and the f^O final state is quite closely related to the f^O amplitude in the initial state.

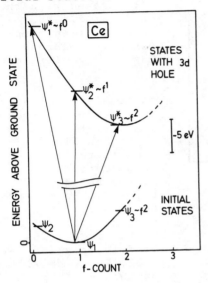

Fig.6. Schematic diagram of the 3d XPS transitions in Ce and its compounds. Note that the true states are mixtures of the f^O, f^1 and f^2 configurations and the f-counts are not perfectly integral in any states shown. The "f^O" final state is probably quite pure, as discussed in the text.

INTERACTION BETWEEN THE SCREENING LEVEL AND THE OTHER VALENCE LEVELS IN THE FINAL STATE

Until now we have ignored the effects of structure in the occupied valence levels in the final state. This is not always jusified! Gunnarsson and Schönhammer showed in 1978 that the strong structure in the Cu valence band was one possible source of a <u>three</u> peak structure in the XPS spectrum of CO chemisorbed on Cu. Recently we have found related effects in the spectra of La and Ce compounds which can be explained with the help of figure 7. On the left hand side we show again the lorentzian form of the screening level when it is pulled below E_F by the core hole potential. It is, in fact only lorentzian in the absence of interaction with the other valence levels. As shown schematically on the right, if the screening level in the final state falls in a region with a high density of states it can be distorted. The strength of the distortion depends on the strength of the interaction and the density of states. We regard the energy of the $3d^9 4f^{n+1}$ states as fixed and consider the $3d^9 4f^n$ states to be derived from them by ionization from the screening level, which can then be regarded as a virtual bound state. This means that distortion of the virtual bound state will be reflected in the shape of the $3d^9 4f^n$ core level XPS peaks /23/.

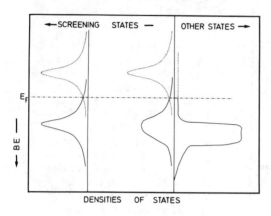

Fig.7. Schematic illustration of the influence of the DOS of the matrix on the final state screening level.

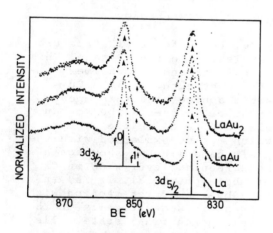

Fig.8. La 3d XPS spectra from La and its compounds with Au. Note the large broadening in the intermetalllic compounds.

The effects described above are illustrated with the example of LaAu$_x$ intermetallic compounds in figure 8. The energy difference between the $3d^9 4f^0$ and $3d^9 4f^1$ final states is about 5 eV. We should thus look for the effect described in figure 7 when there is a large density of states about 5 eV below E_F. In La and Ce this is not the case at all, as the width of the 5d6s band is only 1.5-2ev /18/. Thus the widths of the $3d^9 4f^0$ peaks in La are equal to the sum of the natural linewidth and the instrumental broadening.(≈1.8 eV /14/) However the same peaks in LaAu$_x$ compounds are up to 3 eV broader and the lines do not have simple shapes, but appear to involve partially resolved structure. We attribute this to distortion of the virtual bound state /screening level in the final state by the large density of gold states at ∼5 eV in the matrix. Our assignment of the broadening is backed by studies of about 30 other La and Ce

intermetallic compounds where the trends in core level
linewidths fit our hypothesized mechanism very well
/14,23/.

CONCLUDING REMARKS

This paper has attempted to explain configuration
interaction and screening effects on core level line-
shapes in a simplified way. The effects have been
illustrated with examples from core level XPS, but the
way of thinking of the problems is also applicable to
x-ray absorption and emission. The most dramatic effects
arise when the "screening" orbitals are nearly, but not
completely localized, i.e. in systems where one is
closer to the limit of infinitely large Coulomb interactions than to the band limit with infinitely small
Coulomb interactions between the valence electrons. The
multiple peak structures can be used to derive parameters
for the screening (valence) level widths and occupation
coefficients, so long as only a few configurations
interact strongly in the initial and final states. We
regard the parameters derived by such metods to be only
a semiquantitative guide for further studies, rather
than reliable and quantitative.

The valence band structure of the matrix plays a
profound role in determining core level line-shapes when
the screening levels are only weakly coupled to the
system. Here again we forsee possibilities for determining the strengths of solid state interactions and hopping
integrals between orbitals of different symmetries and
sites in a way that is only semi-quantitative but very
direct.

ACKNOWLEDGEMENTS

This paper is a summary of ongoing research that has
been in progress for several years. The people involved
in cooperations on these general themes included C.R.
Brundle, M.Campagna, O.Gunnarsson, F.U.Hillebrecht, R.
Lässer, D.Menzel G.A.Sawatzky. K.Schönhammer, E.Umbach,
A.Platau and K.Wandelt. I thank them for their various
roles as mentors, discussion parners, coworkers, and
fellow-slaves and for their very real contributions to
this work.

FURTHER READING

The references below mainly concern screening effects. Configuration Interaction is handled in standard texts on Quantum Mechanics and in particular in those on Quantum Chemistry. A more formal treatment for photoelectron spectroscopy is given by R. L. Martin and D. A. Shirley in "Electron Spectroscopy" Vol. I. Ed. C. R. Brundle and A. D. Baker, Academic Press (1977). Other early work on shake up worth reading includes

1. M. O. Krause, M. L. Vestal, W. H. Johnson, and T. A. Carlson, Phys. Rev. A 133, 385 (1964).
2. M. O. Krause, T. A. Carlson, and R. O. Dismukes, Phys. Rev. 170, 37 (1968).
3. K. Siegbahn et al., "ESCA applied to free molecules," N. Holland (1969).
4. T. A. Carlson, "Photoelectron and Auger Spectroscopy," Plenum Press, (1975).
5. R. Manne and T. Åberg, Chem. Phys. Lett. 7, 282 (1980).

Other aspects of relaxation and screening are handled in

6. P. M. Citrin and T. D. Thomas, J. Chem. Phys. 57, 4446 (1972).
7. S. P. Kowalczyk, F. R. McFeely, L. Ley, R. A. Pollak, and D. A. Shirley, Phys. Rev. B9, 381, 3573 (1974); and references therein.
8. N. D. Lang and A. R. Williams, Phys. Rev. B20, 1369 (1979) and references therein.
9. R. Lässer and J. C. Fuggle, Phys. Rev. B22, 2637 (1980) and references therein.

REFERENCES

1. See e.g. L. S. Cederbaum and W. Domcke, Adv. Chem. Phys. 36, 205 (1977).
2. In the description that follows we find the "screening" terminology more appropriate than the more general term "relaxation." This does not mean we regard screening and relaxation as different things. In fact I am at times embarrassed by questions from people who seem to think I regard screening and relaxation as different things (I do not) and that I was one of the inventors of the term "screening" (it was in use in the XPS literature for at least 3 years before we first used it in print).
3. J. C. Fuggle, M. Campagna, Z. Zolnierek, R. Lässer, and A. Platau, Phys. Rev. Lett. 45, 1597 (1980).
4. J. C. Fuggle, E. Umbach, D. Menzel, K. Wandelt, and C. R. Brundle, Solid State Commun. 27, 65 (1978).
5. A. Kotani and Y. Toyozawa, Jpn. J. Phys. 35, 1073, 1082 (1973) and 37, 912 (1974).
6. S. Hüfner and G. K. Wertheim, Phys. Lett 51A, 299 (1975).
7. K. Schönhammer and O. Gunnarsson, Solid State Commun. 23, 691 (1977) and 26, 147, 399 (1978), and Z. Phys. B 30, 297 (1978).

8. A. Kotani, Jpn. J. Phys. 46, 488 (1979).
9. J. C. Parlebas, A. Kotani, and J. Kunamori, Solid State Commun. 41, 439 (1981).
10. S. Larsson, Phys. Scr. 16, 378 (1978).
11. G. Wendin, Structure and Bonding, 45, 1 (1981).
12. G. Wendin, private communication.
13. J. C. Fuggle and F. U. Hillebrecht, to be published.
14. R. Lasser, J. C. Fuggle, M. Beyss, M. Campagna, F. Steglich, and F. Hulliger, Physica 102B, 360 (1980); J. C. Fuggle and F. U. Hillebrecht, Phys. Rev. B25, 3550 (1982) and references therein.
15. J. C. Fuggle, F. U. Hillebrecht, Z. Zolnierek, R. Lässer, and Ch. Freiburg, submitted to Phys. Rev. B.
16. Y. Baer, Physica 102B, 104 (1980; Y. Baer, H. R. Ott, J. C. Fuggle, and L. E. DeLong, Phys. Rev. B24, 5384 (1981) and references therein.
17. W.-D. Schneider and C. Laubschat, Phys. Rev. Lett. 46, 1023 (1981).
18. J. K. Lang, Y. Baer, and P. A. Cox, J. Phys. F 11, 121 (1981).
19. S.-J. Oh and S. Doniach, in press.
20. O. Gunnarsson and K. Schönhammer, private communication, to be published.
21. "Valence Fluctuations in Solids," Ed. L. M. Falicov, M. Hanke, and M. B. Maple, N. Holland, Amsterdam (1980).
22. O. Gunnarsson and K. Schönhammer, Phys. Rev. Lett. 41, 1608 (1978).
23. G. A. Sawatzky, J. C. Fuggle, and F. U. Hillebrecht, to be published.

DYNAMICAL SCREENING OF CORE HOLES BY CONDUCTION ELECTRONS

C. Noguera
Laboratoire de Physique des Solides - Bât. 510
Université Paris-Sud - Centre d'Orsay
91405 Orsay (France)

ABSTRACT

When an x-ray photon ejects an electron from a deep level of an atom situated in a metal, there is a relaxation of the conduction electrons of the metal around the deep hole and around the photo-electron. Depending upon the velocity of the photoelectron the electron gas may remain in its ground state (adiabatic screening) or not (sudden limit). Limiting ourselves to the possibility of excitation of plasmons, we describe a simple model which accounts both for the intensity of plasmon satellites in x-ray photoemission and for the velocity dependent potential acting on the photoelectron which is needed in the calculation of the phase shifts entering the EXAFS (extended x-ray absorption fine structure) theory.

INTRODUCTION

After a core hole excitation the conduction electron gas relaxes around the two charges that have been created: the hole on the deep atomic level and the photoelectron. This relaxation gives rise to several features that can be observed in photoemission or absorption experiments, for example: (i) the shift of the threshold energy of absorption relative to its value in the atom, (ii) the occurrence of plasmon satellites in photoemission, (iii) the divergence at the threshold of absorption and the asymmetry of the elastic peak of photoemission due to electron hole pair excitation.

This problem of relaxation of conduction electrons in a metal is an intricate many body problem which has been so far either avoided or partially solved by means of elaborate diagrammatic methods. Nevertheless these methods are of a difficult use when one wants to combine them with the complexity of the diffraction of the photoelectron by the atoms of the solid; and one is generally obliged to neglect the relaxation processes when trying to determine atomic arrangements by means of photoemission or EXAFS (extended x-ray absorption fine structure).

We have thus built up a model of the relaxation of the conduction electrons in a metal which reduces the full many body problem to the determination of a one electron complex and time dependent potential. This model allows to account in a coherent way and at the same time for (i) all the expected characteristics of the screening of an impurity, (ii) the shift of the threshold of absorption, (iii) the choice of the potential of the central atom necessary for the calculation of the phase shifts of EXAFS, and (iv) the occurrence of plasmon satellites in the photoemission spectrum and the dependence of their intensity as a function of photon frequency.

The paper is built along these lines and in each section the stress will be put on the hypotheses, their physical significance and their consequences. For more details of the calculation the reader is referred to the references 1-4.

I. MODEL OF RELAXATION OF CONDUCTION ELECTRONS

When an x-ray photon is absorbed by an atom it may excite an electron on a deep level. After the absorption, the resulting atom is positively charged and a photoelectron escapes with a kinetic energy equal to the difference between the photon energy and the threshold energy of creation of the deep hole. We assume that this process is very rapid compared to all other characteristic times of the problem, and we describe the hole on the inner shell of the atom as a point positive charge: this may be reasonable if the hole is deep enough since the dimensions of the nucleus are negligible and the spatial extension of e.g. a 1s orbital is very small compared to the characteristic lengths that will appear later. If the electron is ejected from a bound level, the hole has little recoil and we will assume also that its lifetime τ is sufficiently long $(1/\tau \stackrel{<}{\sim} 1 \text{ eV})$. The photoelectron is considered as a point particle moving with a constant velocity, v, related to its kinetic energy. This description neglects the quantum aspect of the particle which should rather be represented as a wave packet extending with time, and it will not apply for photon energies close to threshold since there, diffraction effects of the photoelectron by the atoms are important. Finally one should notice that the energy losses due to the interaction of the electron with the conduction gas that will be introduced in the following are not treated self consistently but rather up to first order in perturbation. To sum it up, when the photon is absorbed, it appears in the metal a density of charge due to the dipole: hole plus photoelectron:

$$n_o(\vec{r},t) = e[\delta(\vec{r}) - \delta(\vec{r}-\vec{v}t)]\theta(t). \qquad (1)$$

The photoexcited atom determines the point r = 0, δ is the delta function and $\theta(t)$ the Heavyside step function such that

$$\theta(t) = 0 \text{ if } t < 0$$
$$\theta(t) = 1 \text{ if } t > 0.$$

When this dipole appears all the other electrons of the system are attracted by the positive charge and repelled by the negative one. We shall not try to describe here the relaxation of the bound electrons of the photoexcited atom. This is an interesting problem of atomic physics which has not been solved till now within the present context. We will mainly focus on the relaxation of the conduction electrons of the metal. Within the framework of the linear response theory we have used the dielectric constant $\varepsilon(q\omega)$ of the Fermi gas which allows to calculate the total density of charge

n($\vec{r}$t) relevant to the x-ray excitation problem; n(rt) is equal to

the bare charge of the dipole $n_o(\vec{r},t)$ plus the displaced charge of conduction electron. After Fourier transform in space and time $n(rt)$ is related to $n_o(rt)$ by the expression

$$n(q\omega) = \frac{n_o(q\omega)}{\varepsilon(q\omega)} .$$

(2)

We have chosen a simplified form of the Lindhard[5] dielectric function, namely the one pole approximation of Lundqvist[6] which takes into account only the plasmon excitations, that should be predominant for large kinetic energies of the photoelectron.

Once $n_o(rt)$ and $\varepsilon(q\omega)$ are chosen, it is possible to determine several quantities. We will first describe the potential created by the hole and the Fermi gas, acting on the photoelectron which will give the characteristics of the screening process. Then we will relate it to the potential of the central atom in EXAFS and discuss the resulting phase shifts in the context of distance and coordination number determination. Then we will derive the potential of interaction of the hole with its cloud of screening, and the excitation spectrum of the hole. Finally combining both effects we will obtain the intensity of plasmon satellites in XPS (x-ray photoemission spectroscopy).

II. DESCRIPTION OF THE POTENTIAL ACTING ON THE PHOTOELECTRON AT $\vec{r} = \vec{v}t$

The photoelectron interacts with the hole and the displaced charge of conduction electrons. It feels a potential $V(rt)$ obtained as a function of the density of charge by Poisson's law. After double Fourier transform in time and space $V(rt)$ is found under an analytic expression. We will first study it along the classical trajectory of the photoelectron i.e. at $t = r/v$. It may then be written in the following way:

$$V(r, t=r/v) = V_{scr}(r) + V_{XC} + \alpha(r)$$

$$\times [V_{unscr}(r) - V_{scr}(r)] + iV_{im}(r).$$

(3)

The first two terms are real and represent the permanent part of the interaction, i.e. the one that would subsist after the relaxation process is achieved at $t = \infty$. Here, $V_{scr}(r)$ is the potential of the hole completely screened by conduction electrons (in a Thomas-Fermi approximation of the static dielectric function it would be equal to $e^{-\lambda r}/r$, λ being the Thomas-Fermi wave vector). V_{XC} is the potential of interaction of the photoelectron with its own cloud of screening, equal to what is generally called the exchange correlation potential (although in our model there is no exchange).

The following real term is the transient part of the potential, due to the sudden perturbation of the system at the time $t = 0$ of absorption of the photon. Before studying it at $t = r/v$ one can show that it vanishes at time t = ∞ in agreement with the usual meaning of the word transient. In fact the transient potential has a complicated analytical expression but we have shown that it is possible to write it as in Eq. (3) where $V_{unscr}(r)$ denotes the potential due to bare hole (unscreened potential) and $\alpha(r)$ is a function of the position and the velocity of the photoelectron which has interesting properties. The first one is the existence, for each velocity, of a distance r_o beyond which $\alpha(r)$ vanishes; this means that for $r > r_o$:

$$W(r) = V_{scr}(r) + \alpha(r)[V_{unscr}(r) - V_{scr}(r)] \approx V_{scr}(r). \qquad (4)$$

One can show that, except at very low velocities, for which anyhow the model is not valid, r_o is proportional to the velocity v. This allows to define a characteristic time $t_o = r_o/v$ and because $W(r) \approx V_{scr}(r)$ for $r > r_o$, i.e. $t > t_o$ this time t_o represents the time necessary for the full relaxation of conduction electrons. With the approximations of the model it is equal to a fraction of the inverse of the plasmon frequency ω_p. The second interesting feature of $\alpha(r)$ is that, in the range $0 < r < r_o$, $\alpha(r)$ is always less than or equal to unity. For this reason we have called it percentage of ionicity of the hole, since for $\alpha = 0$ the hole is neutralized $[W(r) = V_{scr}(r)]$ while for $\alpha = 1$ the hole has a positive charge +e $[W(r) = V_{unscr}(r)]$.

As a function of velocity and distance, $\alpha(r)$ has the behavior shown in Fig. 1. For large velocities $\alpha(r = 0)$ is nearly equal to 1

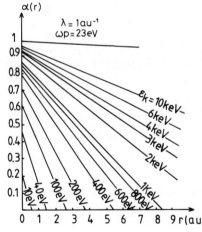

Fig. 1. Percentage of ionicity $\alpha(r)$ as a function of r (in atomic units) for the written values of the photo-electron kinetic energy – $\lambda = 1$ au^{-1} and $\omega_p = 23$ eV.

i.e. there is a large transient potential. This is due to the sudden disturbance of the system when the hole and the photoelectron are created and we will show in a following section that this corresponds in the photoemission spectrum to large plasmon satellites. It may be called the sudden limit. On the contrary for low velocities $\alpha(r = 0)$ is very small, the transient potential is nearly zero for all values of r, which means that the conduction gas is able to remain in its ground state at each moment and to react adiabatically to the slow separation of the dipole hole plus photoelectron. In photoemission no plasmon satellites are present.

$V_{im}(r)$ is the imaginary part of the potential due to the true excitation of plasmons in the system. At time t = 0 (i.e. r = 0) no loss has had time to occur and consequently $V_{im}(r) = 0$. At time t = ∞, $V_{im}(r)$ takes its permanent value which describes the losses on the trajectory of the photoelectron. With our model we find $V_{im}(r \rightarrow \infty)$ independent of r; this means that far from the photoexcited atom, one can attribute a mean free path to the photoelectron; we find it equal to the one calculated by Pines and Bohm.[7]

If one admits that in the final state of the absorption process the photoelectron feels the potential $V(r, t = r/v)$ that we have just described, this is precisely the potential of the central atom needed in EXAFS.

III. IMPORTANCE OF THE RELAXATION OF THE CONDUCTION ELECTRONS FOR EXAFS

It is well known that the oscillations of the absorption coefficient as a function of photon frequency above the threshold of absorption are due to the diffraction of the photoelectron by the atoms located in the neighbourhood of the photoexcited atom, this latter being called central atom. More specifically they may be interpreted as due to constructive or destructive interferences between the electronic wave which goes outside the central atom and the one which comes back after scattering on another atom. The mathematical formulation is the well known expression[8] and we send back the reader to reference 8 for the explanation of the notations:

$$\chi(k) = - \sum_{j} \frac{N_j}{kR_j^2} \operatorname{Jm} f(\pi) \, e^{2i(kR_j + \delta_1)} \, e^{-2\sigma_j^2 k^2}. \tag{5}$$

We are mainly interested here in the phase shift δ_1 (of angular momentum $\ell = 1$ if the hole has an s character) of the central atom, for one could wonder: with which potential has δ_1 to be calculated? Till this study it was generally admitted[9] that, as concerns the conduction gas relaxation, the potential of the central atom should be taken unscreened, i.e. δ_1 was calculated using $V_{unscr}(r)$. Nevertheless one used to add to $V_{unsc}(r)$ an imaginary part independent of r representing the losses of energy of the electron on its trajectory, which yielded an imaginary part of δ_1 equal to:

$$\mathcal{I}m\delta_1 = -\frac{R_j}{L_t(k)}, \qquad (6)$$

where $L_t(k)$ was the total mean free path.

We emphasize here that the use of the potential of Eq. (3) introduces corrections to both the real and the imaginary part of δ_1. In consequence both the phase and the amplitude of the oscillations are affected by the relaxation of the conduction electrons and we will show that this may be important for the determination of the distance R_j and the coordination numbers N_j. As an example we have shown on Fig. 2 the total phase shift of EXAFS $\phi(k) = 2Re\delta_1 +$

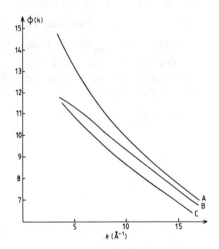

Fig. 2 EXAFS phase shifts of aluminum $\phi(k) = 2\delta_1 + \arg f(\pi)$ calculated with the three hypotheses, A: no screening, B: dynamic screening (present model) and C: static screening of the central atom, at the distance $R_j = 2.86$ Å of the first neighbours.

$\arg f(\pi)$ as a function of the wave vector k for the first nearest neighbours in aluminum (curve B). At low velocity the phase shift is more screened like since (i) there are no transient effects and (ii) during the time of travelling the distance $2R_j$ the relaxation has had time to be achieved. At high energy the phase shift is more unscreened-like due to a percentage of ionicity of the hole closer to one. We have then tried to analyze an hypothetical experimental spectrum built with this $\phi(k)$ and the true distance R_j^o, with the help of unscreened phase shift $\phi_{unscr}(k)$ similar to the one existing in the literature. Letting the reference energy E_o vary we have determined the first neighbour distance R_j and the error in the determination $\Delta R_j = R_j - R_j^o$. For aluminum the results are shown on Fig. 3 for different distances R_j^o. ΔR_j depends crucially on the number of oscillations in the experimental spectrum (k_{max} is the maximum wave vector of the spectrum for which the ratio signal/noise is non zero). If k_{max} is large, in a large range of wave vectors $\phi(k)$ is not too different from the tabulated phase shifts. The precision on the determination of distance is thus good, less than 0.01 Å. Nevertheless if k_{max} is small, due to thermal smearing or backscattering by light atoms, then the error increases and becomes more and more important for further and further shells of neighbours.

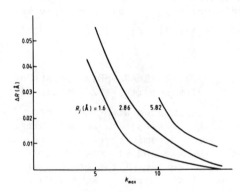

Fig. 3. Error ΔR_j (in Å) in the determination of the distance R_j in aluminum as a function of k_{max} ($Å^{-1}$) (extent of the EXAFS spectrum) when one neglects the relaxation of the Fermi gas.

As concerns the imaginary part of the phase shift we have stressed in the preceding section that $V_{im}(r) \approx 0$ for small distances since the losses have not time to occur. On the other hand, when one uses the expression $\Im m \delta_1 \approx -R_j/L_t(k)$ one assumes that the losses happen at a constant rate along the trajectory, i.e. this last expression overestimates the losses by a factor equal to $e^{-2\Im m \delta_1}/e^{-2R_j/L(k)}$. This factor is plotted on Fig. 4 as a function of wave vector k; it is always greater than one indicating that the theoretical amplitude $e^{-2R_j/L(k)}$ of the usual formula of EXAFS [Eq. (5)] is always too small compared to its real value (at least when plasmon losses are concerned). This was noticed by number of

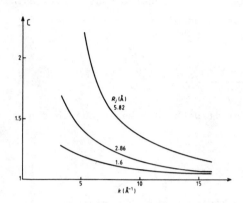

Fig. 4. Ratio of the dynamic versus static amplitude factors computed under the same conditions as Fig. 3.

experimentalists and of course it affects the determination of the coordination number N_j.

IV. EXCITATION FUNCTION OF THE HOLE

We will show briefly in this section that within the framework of our model, we are able to display the major features of the excitation function of the hole: $A_h(\omega)$ equal to the Fourier transform of the time dependent Green's function of the hole. The potential $V_h(r,t)$ acting on the hole is created by the charge of conduction electron displaced by the density $n_0(r,t)$; it can be calculated by means of Poisson's law. Its time dependence is very simple since we have retained only the possibility of plasmon excitation:

$$V_h(r,t) = f_1(r) + f_2(r)\, e^{-i\omega_p t}. \tag{7}$$

Assuming that the hole does not recoil under the interaction, its Green's function is easily obtained, together with its Fourier transform in time and one obtains the well known expression:

$$A_h(\omega) = \sum_n \frac{e^{-a}\, a^n}{n!}\, \delta[\omega - (E_0 + \Delta E_0) - n\omega_p]. \tag{8}$$

The excitation spectrum consists in delta functions separated in energy by ω_p and their intensities are given by a Poisson's law of coefficient a. The no loss peak is displaced by:

$$\Delta E_0 = \langle \psi_0 | f_1(r) | \psi_0 \rangle. \tag{9}$$

E_0 is called the extraatomic shift; it depends only on the wave

684

function ψ_o of the hole and on the static dielectric constant $\varepsilon(qo)$.
One is thus able within this context to obtain realistic values of
ΔE_o. The coefficient of the Poisson's law is equal to:

$$a = \frac{\langle \psi_o | f_2(r) | \psi_o \rangle}{\hbar \omega_p} .$$

(10)

To obtain an analytic expression for it we have taken $\varepsilon(q0)$ in the
Thomas-Fermi approximation. Then

$$a = \frac{e^2 \lambda}{2 \hbar \omega_p} \left(1 - \frac{\omega_p}{\lambda v} \text{ arctg } \frac{\lambda v}{\omega_p} \right).$$

(11)

The quantity a varies with the velocity v of the photoelectron. At
high velocity $a = e^2 \lambda / 2\hbar\omega_p$ which is close to the value found by
Langreth[10] while at low velocity a ≠ 0: this is the so called
interference effect, due to disappearance of the transient effects
in the relaxation process.

   An interesting quantity is the mean energy value of the excitation function:

$$E = \int_{-\infty}^{+\infty} \omega d \omega A_h (\omega) = E_o + \Delta E_o + a \cdot \hbar \omega_p .$$

(12)

By studying it one finds that at high velocity $E = E_o$ while at low
velocity $E = E_o + \Delta E_o$; at high velocity the mean energy has the
unrelaxed value: everything happens as if the electron gas had no
time to relax and remained frozen (Koopman's theorem for the sudden
limit). On the contrary at low velocity, due to the slow separation
of the dipole hole + photoelectron, the Fermi gas screens
adiabatically the two charges and we find E at the relaxed position.
The variation of E with the velocity v displays the continuous
transition from the adiabatic limit to the sudden limie and Eq. (12)
combined with Eq. (11) represents a sum rule valid in the whole
kinetic energy range.

V. PLASMON SATELLITES IN X-RAY PHOTOEMISSION (XPS)

   In order to get the intensity of plasmon satellites in XPS, it
is necessary to calculate the time-dependent Green's function of the
photoelectron[11] in presence of the interaction with the electron
gas. This is a more complicated task than for the hole since the
photoelectron cannot be assumed without recoil. We have thus made a
perturbative calculation up to first order which is believed to give
at least correctly the first excitation peak. There are now two
contributions: one is due to the sudden appearance of the charges
in the Fermi gas b_{trans} (due to the transient part of the potential)
and the other to the excitation of plasmons along the trajectory

(imaginary part of the permanent potential) b_{perm}. Here, b_{trans} displays the same features as the quantity a described previously, while b_{perm} is proportional to the distance z travelled by the photoelectron, the coefficient of proportionality being the inverse of the plasmon mean free path.

To get the total first plasmon peak intensity in normal photo-emission, one has to integrate over all the possible planes of photoexcited atoms, taking into account the total mean free path L_t of the photoelectron. One thus gets the relative intensity of the first plasmon peak relative to the elastic peak:

$$I_1/I_o = a + b_{trans} + \frac{L_t}{L} .$$

There has been a controversy over the years to know whether the observed plasmon peaks were intrinsic (part $a + b_{trans}$ due to the transient process) or extrinsic (part L_t/L). Of course it should be very interesting to be able to subtract the quantity L_t/L which plays a role also in electron energy loss spectroscopy, in order to find the quantity $a + b_{trans}$ which is specific of the x-ray process. Unfortunately till now one is limited by the imprecision on the total mean free path L_t: if one plots L_t/L as a function of v with the values found in the literature one already gets a value greater than the experimental I_1/I_o. So for the moment a comparison between calculated $a + b_{trans}$ and values extracted from experiment is impossible.

VI. CONCLUSION

We have briefly described here how the relaxation of conduction electrons around a photoexcited atom and a photoelectron plays a role in x-ray absorption spectroscopy and x-ray photoemission spectroscopy. We have also described a very simple model which yields qualitatively and sometimes quantitatively the observed specific quantities within a coherent framework.

This is only a first step in the complete knowledge of relaxation processes: improvements could be made to include the possibility of electron hole pairs excitation of the Fermi gas which would make the model valid at lower velocities. In order to get better agreement with experiment, for example for the phase shifts of EXAFS, one would also need a similar description of the relaxation of the bound electrons of the photoexcited atoms, including shake up and shake off excitation.

From an experimental point of view we have tried to show that in EXAFS one needs to be careful in using the tabulated non relaxed phase shifts especially if the spectrum does not extend far above threshold: this could yield imprecisions on distances larger than 0.01 Å; and there is still a lot to be done to understand the amplitude of the oscillations, necessary to determine coordination numbers, absorption sites . . . . Similar comments would apply to photoelectron diffraction experiments which also try to derive these same quantities.

REFERENCES

1. C. Noguera, D. Spanjaard, J. Friedel, J. Phys. F $\underline{9}$, 1189 (1979).
2. C. Noguera, D. Spanjaard, J. Phys. F $\underline{11}$, 1133 (1981).
3. C. Noguera, Ph.D. Thesis, Orsay, France (1981).
4. C. Noguera, J. Friedel, accepted in J. Phys. F.
5. J. Lindhard, K. Dansk. Vidensk Selsk Mat. Fys. Meddr. $\underline{28}$ (1954).
6. B. I. Lunqvist, Phys. Kondens. Mater. $\underline{6}$, 206 (1969).
7. D. Pines and D. Bohm, Phys. Rev. B $\underline{5}$, 388 (1952).
8. D. E. Sayers, E. A. Stern and F. W. Lytle, Phys. Rev. Lett. $\underline{27}$, 1204 (1971).
9. P. A. Lee and G. Beni, Phys. Rev. B $\underline{15}$, 2862 (1977).
10. D. C. Langreth, Phys. Rev. B $\underline{1}$, 471 (1970).
11. D. Chastenet and P. Longe, Phys. Rev. Lett. $\underline{44}$, 91 (1980).

X-RAY EDGE IN METALS

John W. Wilkins
Laboratory of Atomic and Solid State Physics
Cornell University, Ithaca, NY 14853

ABSTRACT

The x-ray absorption and emission spectra near the threshold energy ω_T have the asymptotic form $(D/|\omega - \omega_T|)^{\alpha_\ell}$ where D is an energy of order the Fermi energy. The threshold exponents α_ℓ have been deduced by using experimental data in the range $0 < |\omega - \omega_T| < 0.1$ D. Recent theoretical work on simple models suggests that deviations from the asymptotic theory are of order of ten percent in the upper part of that range; a similar result applies for the singularity in the XPS line shape. Other sources of deviations are reviewed, especially that associated with the dynamic screening of the core hole.

The asymptotic form in the case of absorption, for example, can be thought of simply as resulting from a Golden Rule calculation involving a dipole matrix element between an initial many-body wavefunction for the ground state and the possible many-body final states appropriate to a core hole in the photoexcited atom. Consequently the understanding of the singularity has implications for the entire x-ray spectra, both the near and the extended x-ray absorption fine structure.

INTRODUCTION

As long ago as 1959 there was concern[1] about the effect of the core hole produced in x-ray absorption upon the valence electrons and the absorption spectrum. A decade passed before it became clear[2] that the potential of the core hole could affect not only the "effective wave function" of excited electrons[2] but also all the valence wave functions.[3] In particular, near the threshold energy ω_T for absorption the absorption spectrum $\mu_\ell(\omega)$ associated with exciting an electron to a state with angular momentum ℓ (defined about the excited atom) has the <u>asymptotic</u> form

$$\mu_\ell(\omega) = \mu_\ell \left(\frac{D}{\omega - \omega_T}\right)^{\alpha_\ell} , \quad \omega - \omega_T \ll D. \qquad (1)$$

While the strength μ_ℓ and the characteristic energy D are subjects of this review, the threshold exponent, derived in 1969, is given by

$$\alpha_\ell = \frac{2\delta_\ell}{\pi} - \alpha \qquad (2)$$

where
$$\alpha = 2 \sum_\ell (2\ell + 1) \left(\frac{\delta_\ell}{\pi}\right)^2 \qquad (3)$$

where the δ_ℓ are the phase shifts of the core hole potential
evaluated at the Fermi energy of the conduction electrons.

In these three equations lie questions which have just started
receiving answers in the last few years. Basically we want to know
how $\mu_\ell(\omega)$ behaves away from the asymptotic region. There are three
general sorts of questions:
(1) For how large an $\omega - \omega_T$ is the asymptotic theory valid?
(2) Outside this asymptotic region how important is the core hole on
     the absorption spectrum?
(3) How important is the dynamic response of the other core and
valence electrons to the x-ray ionization of the core electron?
(This last question will receive little attention here since it is
the subject of the preceeding talk.[5]) While these questions are
posed for the absorption spectrum, they have their counterparts for
the x-ray emission spectrum, and both shall be discussed here.

Before these questions are addressed some fundamentals of the
asymptotic theory and of the experimental situation need to be
reviewed. Incidentally an excellent review[6] of the theoretical
aspects of inner-level spectroscopy by Kotani and Toyozawa should be
consulted for topics covered and omitted in this review.

PHYSICS OF THE ASYMPTOTIC THEORY

If the finite lifetime of the core hole can be neglected,[7] then
the x-ray absorption spectrum can be calculated via the Fermi golden
rule

$$\mu(\omega) = \frac{2\pi}{\hbar} \sum_F |\langle F|H_x|I\rangle|^2 \delta(\omega + E_I - E_F) \tag{4}$$

where $|I\rangle$ and $|F\rangle$ denote the initial and final many-body configura-
tions, respectively, and H_x the perturbation due to the x-ray
having energy ω. In essentially all treatments[8] the interactions
between the conduction electrons are neglected; the threshold exponents would not be affected[9] by including such interactions but of
course the non-asymptotic spectrum would be.
The standard theorists' model. There are several other approximations that are routinely made in order to simplify the calculation
even at the risk of decreased relevance to actual experiments.

(1) The dipole nature of the perturbation due to the x-ray is
neglected in the calculation although it often belatedly recognized
at the end where the appropriate angular momenta are subscripted on
the phase shifts. The only function of H_x is to transfer a core
electron (denoted by the operator d) to the conduction band (denoted
by the operator C_k) with some amplitude T:

$$H_x = T \sum_k C_k^+ d + \text{h.c.} \tag{5}$$

(2) Since most calculations are done at zero temperature, the
initial state is a ground state for non-interacting electrons--i.e. a
Slater determinant of the core orbital and the filled orbitals,

$\phi_k^I(r)$, of the conduction band. As such it is a many body state. Typically the conduction band has a constant density of states,

$$\rho(\varepsilon) = \rho, \quad -D < \varepsilon < D, \tag{6}$$

band width 2D, and the Fermi energy at zero (i.e., half-filled band). Usually the spin degrees of freedom are ignored. When they matter they will be specifically pointed out. Generally the Hamiltonian for the initial state for absorption (or the final state in the case of emission) is written

$$H_I = \sum_k \varepsilon_k n_k + \varepsilon_c d^\dagger d \tag{7}$$

where ε_c is core energy and the single particle energy ε_k are indexed by some parameter k which is often a wavevector.
(3) The effect of the core hole on the final states in the absorption process is to introduce a potential $\overline{V_{kk'}}$ which scatters the conduction electrons[10]

$$H_F = \sum_k \varepsilon_k n_k + \sum_{kk'} V_{kk'} c_k^\dagger c_{k'}. \tag{8}$$

In most cases the k-dependence of $V_{kk'}$ is ignored while its attractive nature is recognized (i.e., $V_{kk'} = -U$)

$$H_F = \sum_k \varepsilon_k n_k - U(\sum_k c_k^\dagger)(\sum_{k'} c_{k'}). \tag{9}$$

The terms in parentheses in (9), such as $(\sum_k c_k^\dagger)$, correspond to states localized about the core hole. Thus the core hole, when k-dependence of $V_{kk'}$ is neglected, looks like a delta-function potential affecting every conduction state the same way.[11]
Depending on the density of states the potential term in (9) may create a bound state below the band. In any case its effect can be summarized by a t-matrix $t(\varepsilon) = |t(\varepsilon)| \exp(i\delta(\varepsilon))$ with phase shift $\delta(\varepsilon)$. For the asymptotic theory only the phase shift at the Fermi surface

$$\delta = \delta(\varepsilon = 0) \tag{10}$$

will be needed. In a more realistic model there will be phase shifts $\delta_\ell(\varepsilon)$ for all angular momenta ℓ. That these phase shifts are non zero means that single particle orbitals, $\phi_k^F(r)$ (say), in the final state will be different from those in the initial state.[12] In particular the overlap $(\phi_k^F(r), \phi_k^I(r))$ is less than unity even though the individual orbitals are normalized.
Contributions to the x-ray singularity. We can use all these ingredients for H_x, $|I\rangle$ and $|F\rangle$, in an argument due to Friedel[13], to characterize the contributions to the x-ray edge singularities. In the expression (4) for $\mu(\omega)$ it important to recognize (i) that

690

$\langle F|H_x|I\rangle$ is a determinant of elements of the form $(\phi_c^I, H_x\phi_{k'}^F)$ in the first row and $(\phi_k^I, \phi_{k'}^F)$ in all other rows and (ii) that there are many possible final states $|F\rangle$ consistent with the conservation of energy in (4). Both of these points are illustrated in Figure 1.

Consider the <u>direct</u> process, Figure 1a, where the core electron goes into the final orbital state ϕ_q^F(say). This corresponds to keeping one term in the determinant $\langle F|H_x|I\rangle$, namely

$$(\phi_c, H_x\phi_q^F) \; \mathrm{Det}\left|(\phi_k^I, \phi_{k'}^F)\right| \tag{11}$$

where only those final single-particle states below the Fermi level are in the determinant in (11). If this determinant were unity, this would correspond to a single-particle picture of x-ray absorption. In fact it might be called the "final-state" rule

$$\mu_{fsr}(\omega) = \frac{2\pi}{\hbar} \sum_q \left|(\phi_c|H_x|\phi_q^F)\right|^2 \delta(\omega - \omega_T - \varepsilon_q^F) \tag{12}$$

where ε_q^F is the single-particle energy calculation in the presence of the core hole--i.e., the final state single-particle energy--and the matrix element is a single-particle dipole matrix element involving the final state orbital $\phi_q^F(\vec{r})$. The possible range of validity of this final state rule will be discussed later, especially in comparison with the much more easily calculated (at least for absorption) "initial-state" rule

$$\mu_{isr}(\omega) = \frac{2\pi}{\hbar} \sum_q \left|(\phi_c|H_x|\phi_q^I)\right|^2 \delta(\omega - \omega_T - \varepsilon_q^I) \tag{13}$$

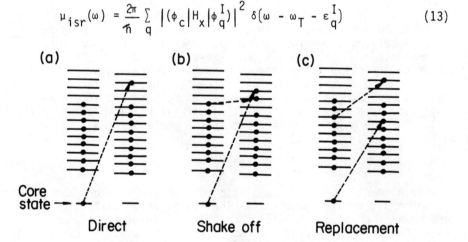

Fig. 1. Characteristic[13] terms in the x-ray absorption singularity. In the text the initial states are indexed with k while the final states below the Fermi level are indexed with k'; in (a) and (c) the highest lying final state is indexed q. The dashed lines indicate the physical processes associated with each diagram.

where the superscript I refers to single-particle energies and orbitals of the initial Hamiltonian (7).

Leaving this extended subjunctive ("if this determinant were unity"), we note that the determinant in (11) is far less than unity since it is the product of many overlap integrals $(\phi_k^I, \phi_{k'}^F)$ each less than unity. For a simple model of a finite system[12] of band width D for which the average spacing between levels is Δ --i.e, the number of electrons is $N \simeq D/\Delta$ --the overlap can be shown[3,13,14] to be

$$\mu_{\text{direct}} \quad \alpha \quad \text{Det}\left|(\phi_k^I, \phi_{k'}^F)\right| \sim \left(\frac{D}{\Delta}\right)^{-(\delta/\pi)^2} \sim N^{(-\delta/\pi)^2} \tag{14}$$

which goes to zero in the continuum limit. This zero just reflects the fact the initial and final ground states are orthogonal. Nonetheless this effect has been given the grand name of the "orthogonality catastrophy".[3]

While the direct part of the absorption spectrum is always zero independent of the excitation energy $\omega - \omega_T$, the other parts depend strongly on $\omega - \omega_T$. Consider for example the <u>replacement</u> terms shown in Figure 1c. Note that the final state is the same as Figure 1a. Accordingly the replacement term results from expanding the same total determinant used for the direct term about elements of the first row other than $(\phi_c, H_x\phi_q^F)$. The resulting determinant contains in each row one overlap matrix element that is special: (ϕ_k^I, ϕ_q^F) which is proportional to $\delta/(\epsilon_q^F - \epsilon_k^I)$. This special term in each of N - 1 determinants leads to a divergence of the form[2]

$$\mu_{\text{replacement}} \sim \left(\frac{D}{\omega - \omega_T}\right)^{2\delta/\pi} . \tag{15}$$

If δ is positive, as it will be for an attractive core hole potential, this divergence will dominate the asymptotic spectrum. It is easy to insert the correct δ_ℓ phase shifts into (15). By dipole selection roles the final state(s) for the K-edge is $\ell = 1$ and for the $L_{2,3}$ edge are $\ell = 0$ and $\ell = 2$.

The <u>shake off</u> term, see Figure 1b, involves different final states in which particle-hole pairs have been excited. As more and more particle-hole pairs are excited, the final state, which with no such pairs was orthogonal to the initial state, has greater and greater overlap with initial state. As a result this overlap will grow with $(\omega - \omega_T)$ or, more specifically,

$$\mu_{\text{shake off}} \sim \left((\omega - \omega_T)/\Delta\right)^{(\delta/\pi)^2} . \tag{16}$$

If we include all angular momenta, then the exponent of the direct and shake off term becomes

$$\alpha = 2 \sum_{\ell} (2\ell + 1)\left(\frac{\delta_\ell}{\pi}\right)^2 \tag{3}$$

where $(2\ell + 1)$ is the degeneracy of each ℓ channel and factor of two is for spin. Accordingly we see that

$$\mu_\ell(\omega) = \mu_\ell \left(\frac{D}{\omega - \omega_T}\right)^{\alpha_\ell} \tag{1}$$

where $\alpha_\ell = 2\delta_\ell/\pi - \alpha$. Note the energy scale D and the strength μ_ℓ are not predicted by the asymptotic theory.[2-4]

XPS singularity. There is one limiting case where it is natural to consider only the direct and shakeoff term. In x-ray photoemission spectroscopy (XPS) a high energy emitted electron is detected at some energy E outside the metal. Accordingly the overlap (ϕ_k^I, ϕ_q^F) is zero and the replacement term can be neglected. Clearly the differential cross section $d\sigma(E;\omega)/dE$ will be zero for any electron energy greater than $E_0 = \omega - \omega_T$, but it will be finite at $E < E_0$ since the rest of the energy $(E_0 - E)$ can be taken up by particle-hole pairs in the final state. The integral of $d\sigma/dE$ over all possible energies is clearly proportional to the x-ray absorption (omitting replacement), that is

$$\left(\frac{D}{\omega - \omega_T}\right)^{-\alpha} \sim \int_0^{\omega - \omega_T} dE \left(\frac{d\sigma}{dE}\right) . \tag{17}$$

The derivative of (17) gives the result that

$$\frac{d\sigma}{dE} \sim \left(\frac{D}{E_0 - E}\right)^{1-\alpha} , \quad E < E_0, \tag{18}$$

where α is called the XPS singularity index. This singularity[15] is of course smeared out by the core hole lifetime but nevertheless there is a discernible asymmetric line.

Relevance of this approach.[16,17] One reason for doing these arguments in such detail is that they provide a natural basis for discussion of non-asymptotic calculations. Another reason is that the arguments can be easily applied to other cases.

One famous example[18] is how to deduce the x-ray edge singularity from the XPS singularity. The argument starts from the observation that the interaction H_x adds an electron localized about the core hole (see $\sum C_k^+$ in (5)). Accordingly the x-ray problem is the same as XPS problem except the conduction electrons see the core hole potential "screened" by this extra electron. By the Friedel sum rule[19] this shifts the sum of phase shifts down by π. One then claims that the excited electron feels this entire shift. Accordingly the XPS exponent must change to mimic the x-ray process:

$$1 - \alpha \rightarrow 1 - [(\frac{\delta_\ell}{\pi} - 1)^2 + \sum_{\ell' m_\ell m_s}{}' (\frac{\delta_{\ell'}}{\pi})^2] = \frac{2\delta_\ell}{\pi} - \alpha$$

where the prime on the sum excludes that one particular channel whose phase shift is reduced by π.

Another example,[20] shown in Figure 2, is the x-ray spectrum that results when the core hole potential is strong enough to create a bound state split-off from the bottom of the band (which can happen in the model calculations). In a model for spinless electrons with only a single phase shift, the first threshold (when this state is filled) has the standard exponent $2\delta/\pi - (\delta/\pi)^2$. However there will be a second threshold at an energy $\omega_1 - \omega_0$ above the first sufficient to empty this state. Since this state is empty the filled final states look like they have been shifted up compared to the filled initial states--i.e., as if they felt a phase shift $\delta - \pi$. With this effective phase shift the exponent α_1 in $(\omega - \omega_1)^{\alpha_1}$ becomes

$$\alpha_1 = 2(\frac{\delta - \pi}{\pi}) - (\frac{\delta - \pi}{\pi})^2 = -3 + 4(\frac{\delta}{\pi}) - (\frac{\delta}{\pi})^2$$

which tends to be negative. By the way, for all examples up to now the emission spectra were symmetric about $\omega - \omega_T$ with the absorption spectra. In this case while the absorption spectrum has two threshold, the emission has only one.

(a)   (b)

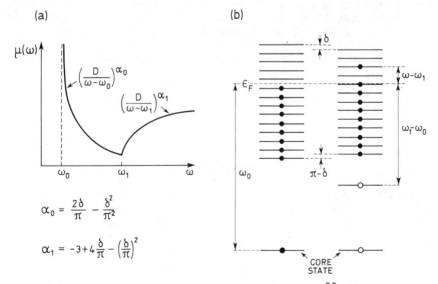

$$\alpha_0 = \frac{2\delta}{\pi} - \frac{\delta^2}{\pi^2}$$

$$\alpha_1 = -3 + 4\frac{\delta}{\pi} - (\frac{\delta}{\pi})^2$$

Fig. 2.(a) Schematic representation of the results[20] for the absorption spectrum in the case of core hole potential strong enough to create an exciton state. (b) Energy level structure for (a). The exponent for the second threshold, corresponding to exciton ionization, i.e. the new bound state being empty, is negative.

Table I Summary of x-ray edge and XPS exponents

See reference 22 for a detailed discussion of each of these abbreviated entries.

Element	Spectrum	K-edge	$L_{2,3}$-edge	XPS
Li	absorption emission	temp. dep. rounding incomplete phonon relaxation	no 2p core no 2p core	$\alpha(.23\pm.02)$
Na	absorption emission	no data no data	$\alpha_0(.37\pm.03)$ incomplete phonon relaxation	$\alpha(.20\pm.015)$
Mg	absorption emission	no data no useable data	$\alpha_0(.23\pm.02)$ $\alpha_0(.18\pm.02)$	$\alpha(.13\pm.015)$
Al	absorption emission	$\alpha_1(.095\pm.015)$ $\alpha_1(.12\pm.015)$	$\alpha_0(.18\pm.02)$ $\alpha_0(.17\pm.02)$	$\alpha(.115\pm.015)$

EXPERIMENTAL VALUES OF EXPONENTS

Only cores with widths less than a few tenths of an eV are appropriate to an analysis of the threshold exponents. While the higher lying core levels of the heavier elements might be studied,[21] to date only Li, Na, Mg, and Al have been extensively studied. In Table I the results of extensive analysis[22] are briefly summarized. A few points are worthy of mention. (1) The same value of the XPS exponent α is deduced for all core levels of a given metal.[22a] (2) The emission data is particularly difficult to analyse because of the fact that the sample can reabsorb the emitted x-rays. (3) The analysis is restricted to $(\omega - \omega_T)$ less than ~ 0.1 D in Na and ~ 0.04 D in Al (where D is of order the Fermi energy).

An important question is whether the exponents are self-consistent. Also one might wonder whether the exponents can be predicted. Historically the answer to the second question is largely no (see reference 22b for details) and to first yes. The quest for self-consistency between $\alpha_\ell = 2\delta_\ell/\pi - \alpha$ and $\alpha = \sum_\ell (2\ell + 1)(\delta_\ell/\pi)^2$ is aided by the Friedel sum rule[19] which for the processes of x-ray absorption, emission, and photoemission requires that

$$2 \sum_\ell (2\ell + 1) \frac{\delta_\ell}{\pi} = 1 . \qquad (19)$$

If one supposes $\delta_\ell = 0$ for $\ell > 1$, then the XPS exponent predicts α_0 and α_1. If one includes δ_2 then there is a curve given by the loci of (α_0, α_1) compatible with α.

Such curves are shown for the case of Al in Figure 3. The two values of α correspond to the extreme ranges deduced from XPS. The shaded rectangles show the range of α_0 and α_1 resulting from the edge analyses. Note that there is a mutual intersection of the rectangles and XPS compatiblity curves, indicating that all experiments are consistent. This is the best case; in Mg the emission and absorption rectangles don't quite overlap, although each overlaps the XPS compatibility curves. Since reference 22b a new analysis[23] of the XPS line of Al has been undertaken using the results of a numerical calculation[24] to be discussed later. This yields an $\alpha = 0.14$ which, while outside the limits of the previous determination, does not require a new consistency analysis as can be seen from Figure 3.

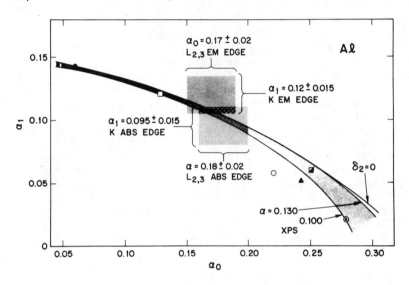

Fig. 3. Plot showing self consistency between x-ray threshold exponents α_0 and α_1 and the XPS singularity index α for Al. No compatibility curve can lie above $\delta_2 = 0$ curve as then δ_2 would be imaginary. The circles, squares, and triangle refer to various calculations for the phase shifts; see ref. 22b for details.
© 1979 APS. Reprinted, with permission, from "One-electron and many-body effects in x-ray absorption and emission edges of Li, Na, Mg, and Al metals," by P.H. Citrin, and G.K. Wertheim, and M. Schlüter, Phys. Rev. B20, 3067 (1979).

In summary the asymptotic theory has been applied to the experimental data with consistent results for $\omega - \omega_T$ up to 0.1 D. The question is: should the asymptotic theory work that far?

NUMERICAL CALCULATIONS BEYOND ASYMPTOTIC THEORY

That numerical calculations were a natural way to study x-ray absorption and emission spectra was recognized quite early[25] and then later perceived[26,27,24a] for the XPS differential cross section $d\sigma/dE$. Many of the calculations utilize a band with a constant density of states in which the single-particle states are evenly spaced or in one case[17,28] logarithmically spaced about the Fermi level. The core hole potential scatters all states with the same strength U as in Equation (9). The other alternative is to construct the wave functions in a spherical box of radius R with core hole potential, centered at the origin, being a square well[26,29,30] or a delta shell.[30] Usually only $\ell = 0$ wave functions are retained, although in one case $\ell = 1$ wave functions were used.[31] In all calculations to date only spinless fermions are employed. As a result the asymptotic forms against which these calculations should be compared have the exponents $2(\delta/\pi) - (\delta/\pi)^2$ for x-ray absorption (c.f. Equation (1)) and $1 - (\delta/\pi)^2$ for XPS line shape (c.f. Equation (18)).

X-ray edge singularity. The difficulty with all the calculations using an even spacing of single-particle levels[25,24a] or particles in a spherical box[30] is that the energy resolution depends on the number of states retained. Thus for 50 states in a band of width 2D an energy resolution of $\Delta\varepsilon \sim .04$ D is really too large to make very precise statements about deviations from the asymptotic theory. The calculation[28] using a logarithmic mesh is more helpful since for only 25 states the resolution $\Delta\varepsilon$ can be less than 10^{-5} D. The results of a typical calculation are shown in Figure 4. The calculated spectrum is divided by the asymptotic form (including the strength μ_0 which is calculated at $(\omega - \omega_T)/D < 10^{-6}$) in order to expose deviations. (It would be useful if all calculations were reported this way.) One interesting feature is the sensitivity of the deviations to the form of the core hole potential which is modeled by $\int d\varepsilon d\varepsilon'\{G_0 + G_1(\varepsilon + \varepsilon')/D\}C_\varepsilon{}^+C_{\varepsilon'}$ where the energy ε is measured from the Fermi level. In all cases the coefficient G_0 and

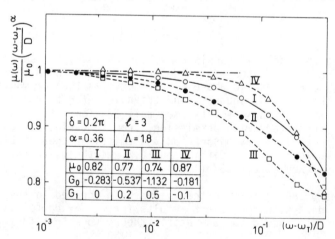

Fig. 4. Absorption spectrum normalized by the asymptotic form (1) for four values of the parameter G_1 in the energy-dependence core-hole potential. In each case G_0 was adjusted to give the same phase shift $(0.2\ \pi)$ of the Fermi level. Note that the strength μ_0 depends on G_1. For discussion of the parameters ℓ and Λ used in the calculation, see refs. 28 and 17.

G_1 are adjusted to keep the same phase shift at the Fermi level. These results suggest that the fractional deviation from the asymptotic form at $\omega - \omega_T = 0.1$ D is of order ten percent. To confirm this speculation it would be useful to have careful calculations with differing conduction bands and with such details as the dipole matrix element included.

One feature of these numerical calculations that has surprised all the practitioners is how few final states are required in order to have convergent results. Typically the contribution of a final state containing a particle-hole pair is two orders of magnitude smaller than a final state with no particle-hole pair.[32]

XPS singularity. Here the situation is not as clear as for the x-ray edges. The quantity usually calculated, called $I(\omega)$ here, is proportional to the XPS differential cross section:

$$I(\omega) = \sum_F \left| \langle F | I \rangle \right|^2 \delta(\omega + E_I - E_F). \tag{20}$$

Note that no electron is excited from the core hole into the conduction band but that the final and initial states are calculated using the initial and final Hamiltonians (c.f. Equations (7) and (8)), respectively. More useful for comparison with the x-ray absorption spectrum is the integral of $I(\omega)$, called $J(\omega)$ here,

$$J(\omega) = \int_0^\omega d\nu \; I(\nu). \tag{21}$$

Clearly $J(\omega)$ is the shake off term and has the asymptotic form

$$J_A(\omega) \sim \omega^\alpha$$

where $\alpha = (\delta/\pi)^2$ for a single phase shift, spinless model.

Unfortunately it is difficult to compare the various calculations[24a,27,31] since the nature of the model used can have strong effects on $I(\omega)$.[33] For purposes of comparison, it is convenient to discuss the model[24a] with a constant density of states and an attractive delta-function core potential. One consistency test for such calculations is that $J(\omega)$ must saturate to unity at high ω. [This follows since $\int d\omega \; I(\omega) = \langle I | I \rangle = 1$.] This doesn't happen until ω exceeds the energy necessary to ionize the bound state.[34] One extremely fascinating result[24a] is that for $\delta/\pi = .45$, $I(\omega) - I_A(\omega) = -.13 \, (\omega/D)$, $0 < \omega < .3$ D. This means that there is a negative step in the XPS differential cross section, a fact which as mentioned earlier has already been used to produce a better fit to the XPS line shape for the Al 2p excitation.[23] This result can also be used to estimate the deviation from the asymptotic result. At $\omega = 0.1$ D the deviation is of order three percent, see Table II. Clearly more studies are needed.

ANALYTIC RESULTS BEYOND ASYMPTOTIC THEORY

It might seem surprising, if there were any analytic results, to put them after the numerical calculations. Not only is that the correct historical order, but the analytic results are not as neat as one might like and lean on the computer to some degree. It turns[35] out that something very like the replacement term in Figure 1c can be summed exactly. One can calculate the response (i.e. the propagator) of the excited electron in the presence of the core hole potential. Furthermore it is possible to do this even when there are a fixed number of particle-hole pairs present. The imaginary part of this response, call it a transition rate R, can be written $\sum R_n(\omega - \omega_T)$ where the index n denotes the number of particle-hole pairs present. The zero pair rate can be calculated analytically:[36]

$$\frac{R_0(\omega)}{2\pi\rho} = \left| \frac{D}{\omega} - 1 \right|^{2\delta(\omega)/\pi} e^{f(\omega)} \tag{22a}$$

where
$$f(\omega) = \frac{2}{\pi} \int_0^D d\nu \frac{\delta(\nu) - \delta(\omega)}{\nu - \omega} . \tag{22b}$$

Note that R_0 goes to zero at the edge of the band as it should and, for a monotonically decreasing phase shift, f is negative. It is also possible with some effort to evaluate $R_1(\omega)$. A value of $R_0 + R_1$ at $\omega - \omega_T = 0.1$ D is used to estimate the deviation from the asymptotic result for the replacement term in Figure 1c (see Table II).

Some analytic efforts has been made on the shakeoff term (Figure 1b).[38,39] Unfortunately they have not been carried far enough

Table II Fractional Changes With Respect to Asymptotic Theory

The fractional changes of the x-ray absorption[17,28] $\mu(\omega + \omega_T)$, the shake off term[24a,*] $J(\omega)$, and the replacement term[35] $R_0(\omega) + R_1(\omega)$ with respect to their asymptotic forms are estimated at $\omega = 0.1$ D. One might hope that, for such small deviations, the second column might be the sum of the third and fourth.

Phase shift	X-ray	Shakeoff	Replacement
$\delta = .2\ \pi$	$- .066^{28}$	$- .002*$	$- .096^{35}$
$\delta = .4\ \pi$	$- .025^{17}$	$- .023*$	n.a.
$\delta = .45\ \pi$	n.a.	$- .032*$ $- .027^{24\,a}$	n.a.

* Calculated by D.L. Cox.

that one can use them to estimate when the shakeoff terms deviate from the asymptotic theory. However as one of the references[38] points out, some of the intermediate results of the analytic theory could be computed numerically as a check. In Table II is a summary of those fractional changes from the asymptotic theory that can be extracted from the literature. The blanks and obvious inconsistencies are comment enough.

DYNAMIC RESPONSE TO THE CORE HOLE

There is one specific model for the core hole potential that permits an examination of the time evolution of the system in response to core hole. Specifically, if the bare core potential $V_{kk'}$ is assumed to be only a function of the momentum difference $\vec{q} = \vec{k} - \vec{k}'$ (i.e. V_q) then it couples into density waves of the conduction electrons. It turns out for a suitably weak function[40]

$$\alpha(\omega) = - \frac{2e^2}{\pi^2 \omega} \int_0^\infty dq \left| \frac{V_q q^2}{4\pi e^2} \right|^2 \mathrm{Im} \frac{1}{\varepsilon(q,\omega)} , \tag{23}$$

where $\varepsilon(q,\omega)$ is the dielectric function of the electron gas, that it is possible to show[41] that $I(\omega)$[42] satisfies the following integral equation (for $\omega > 0$)

$$I(\omega) = \frac{1}{\omega} \int_0^\omega d\nu \, \alpha(\omega - \nu) \, I(\nu). \tag{24}$$

A few comments are in order. (i) That $\alpha(\omega)$ has been chosen as the symbol in (23) is no accident. Suppose $\alpha(\omega)$ were a constant α, then the integral equation (24) has the asymptotic form $I(\omega) \sim \omega^{\alpha-1}$. (ii) $I(\omega)$ is, of course, proportional to the XPS cross section $d\sigma(\omega)/d\omega$. (iii) In general we expect $\alpha(0) \sim (\delta/\pi)^2$ to be small so that the conditions for the derivation[40] of the integral equation are satisfied.

For any specific $\alpha(\omega)$ Equation (24) is easily solved. Accordingly there have been model calculations[43,44] for $\alpha(\omega)$. For a range of bare core hole potentials[44] $\alpha(\omega)$ is remarkly constant varying no more than 15 percent for energies as large as the plasma frequency. For $\omega \sim 0.1$ D the variation is at most a few percent. This model for the dynamic screening of the core hole potential thus yields $I(\omega)$ remarkably close to the asymptotic theory for $\omega < 0.1$ D. Unfortunately no one has ever extended this work to a study of replacement terms nor to a study of whether $\alpha \sim 0.2$ (a typical value) is small enough to justify the theory.

Alternately if $I(\omega)$ is known, then the integral equation can be solved to yield $\alpha(\omega)$. This has been done once[34] (see footnote 13 of reference 24a) for a case where $\delta/\pi = 0.45$ and yielded $\alpha(\omega)/\alpha(0) \sim 1 - (\omega/D)$. This is a surprisingly strong dependence in view of the small deviation in $J(\omega)$ at $\omega = .1$ D (see Table II). Incidentally such strong frequency dependences in α and α_ℓ have been suggested by some low order perturbation calculations of the exponents.[45]

Finally there are aspects of the dynamics which should be mentioned even though they cannot be covered in any detail. First there is a question of whether in absorption or XPS the decay processes of the hole can affect the spectrum.[46] Fortunately it turns[47] out that the conventional procedure of convoluting the theoretical spectrum with a Lorentzian core hole line shape is appropriate. Second, the emission spectrum can be influenced by the response of lattice to the core hole. When the core hole width and the phonon frequencies are comparable, incomplete phonon relaxation can produce broadening of the emission edge[48,49] (and also of the Auger line[50]).

FINAL STATE RULE

Given our imperfect knowledge on the range of validity of the asymptotic theory, there still remains the question: what effects does the core hole potential have on the rest of the spectrum?

Consider, for example, the x-ray absorption spectrum at high excitation energies in the EXAFS regime. There the replacement term (Figure 1c) can be ignored while the shake off terms (and shake up terms, involving transitions into an occupied bound state) can be included in the net EXAFS spectrum.[51,52] Then the absorption spectrum is proportional to

$$|<c|T|q> S|^2 \tag{25}$$

where the single-particle final state $|q>$ will, in the case of EXAFS, involve multiple scattering off of adjacent unexcited atoms. The important point is that the amplitude is reduced by

$$S \sim <\psi^I_{atom}|\psi^F_{atom}>, \tag{26}$$

the overlap of initial and final atomic wave functions of the so-called "passive" electrons of the photoexcited atom. Since the final state atomic orbitals will be affected by the core hole, S will be less than unity; a typical value is 0.8.[53] Clearly many-body corrections to the final state rule, present even in the EXAFS regime, will be important in the lower energy spectrum.

There have been a number of studies[54-58] touching on the final state rule. One early calculation[54b] illustrated several interesting aspects. The model is characterized by a core hole potential (9) of strength $U/2D = 0.2$ and 0.45 acting on an initial conduction band whose density of state, shown as a dashed line in Figure 5,

$$\rho(\varepsilon) = \frac{2}{\pi D} \left[1 - \left(\mu + \frac{\varepsilon}{D}\right)^2 \right]^{1/2} \tag{27}$$

allows the consideration of the effect of band filling (μ assumes three values 0.0, and $\pm$ 0.6). Both absorption and emission spectra

are shown. Three features in Figure 5 should be noted:
(i) Even for a symmetric half-filled band the emission and
absorption curves are not symmetric, reflecting no doubt that the
final state for emission is the unperturbed band while for absorption
it is the conduction electron in the presence of a core hole; the
density of final states is shown by a dashed-dot curve in Figure 5.
(ii) The effect of the edge singularity is so large that it must be
included. Accordingly one could define[59] a final state rule which
included the asymptotic theory as factor.
(iii) Most surprisingly the validity of the final state rule, even
if modified by (ii), depends on band filling. In the case of
absorption spectra, for partial to half-filled bands the final state
rule works well[55] but for a nearly filled band the initial state rule
seems more appropriate as is indicated by the right hand panels of
figure 5. In the case of emission spectra the transition from final
state to initial state rule occurs as the band filling decreases.

   This last point, (iii), clearly deserves more attention. The
only relevant published result[57] has pointed out, as a calculational
feature, that the final state rule (12) mimicks the exact calculation if the final excited orbit $\phi_q^F(r)$ is replace by one orthogonalized to all the filled initial state orbitals--namely, by using

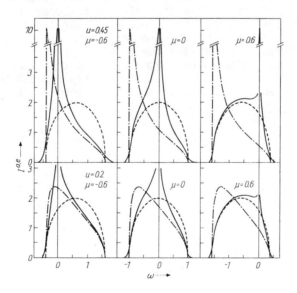

Fig. 5. Combined absorption and emission spectra (solid line) for
model described in text (u = U/D). Initial-state density of states
is indicated by dashed line and final (i.e., one with core hole) by
dashed-dot line.
Reprinted from "Extra-atomic relaxation and x-ray spectra of
narrow-band metals" by V.I. Grebennikov, Yu. A. Babanov, and O.B.
Sokolov, physica status solidi (b) <u>80</u>, 73 (1977).

$$\tilde{\phi}_q^F(r) = \phi_q^F(r) - \sum_{\text{occ. } k} \phi_k^I(r) \, (\phi_k^I, \phi_q^F) \tag{28}$$

in (12). For a nearly empty band, the $\tilde{\phi}$ orbitals are simply the original final state orbits themselves. However, as the band fills up, these $\tilde{\phi}$ orbitals must more and more resemble the high lying initial orbits, since they are orthogonal to all the rest. This interesting possibility for understanding point (iii) could be explored[60] by using orthogonalized $\tilde{\phi}$ orbits constructed from ϕ via a unitary transformation. But this is still in the future.

Given all these general remarks about the final state rule, what is known about the x-ray spectra of metals, besides the edge singularities discussed earlier? First, with the advent of synchrotron radiation, many have been measured over a wide energy range and more will be. Second, the final state rule for absorption is so hard to apply in practice that almost all calculations have used the initial state rule (13). This is not the place to review those calculations; there is only space to point out one persistent anomaly.

Figure 6 displays the measured and computed K-edge spectra of palladium to 150 eV above the threshold. The reader must take on trust that the measurement[61] and the initial state calculation[62,63] are competently performed. Note that all the features in one spectra are seen in the other but that the features in the experimental spectrum lie at increasingly higher energies with respect to the theoretical features. In fact the theoretical spectrum can be aligned with the experimental one by multiplying the excitation energy relative to threshold by a factor slightly greater than unity (~ 1.06). This need to scale up $\omega - \omega_T$ of the theoretical

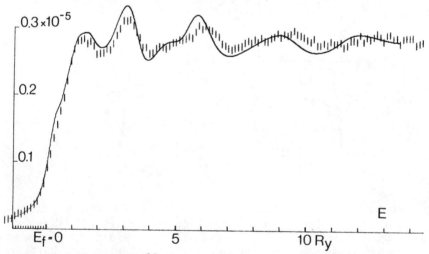

Fig. 6. The calculated[63] (solid line) and measured[61] (short vertical bars) K-edge absorption spectrum of palladium as a function of energy in Ry. The thresholds of the two spectra have been aligned at E_f.

spectrum has been observed both in the heavy rare earths[64] and in the transition elements Ti, Fe, and Ni.[65] It seems very likely that a general explanation of this effect must have something to do with core hole potential; exactly what remains to be seen.

CONCLUSIONS

After all these words, perhaps a terse summary is in order.
(1) The asymptotic theory is an essential ingredient to understanding the x-ray spectra of Na, Mg, and Al and the XPS line shapes of many metals. The fits to the asymptotic theory extend out to an excitation energy on the order of one-tenth of the Fermi energy.
(2) Numerical calculations for very simple models suggest that the deviations from the asymptotic theory at one-tenth of the Fermi energy should be on the order of ten percent. There is a need for calculations of more realistic models and for more thoughtful analysis of the results.
(3) There seems hope that analytic calculations may provide considerable insight into the deviations from the asymptotic theory and into the dynamic response of the conduction electrons to the core hole.
(4) The major challenge is to understand the effect of the core hole (or its filling, in the case of emission) on the near and extended fine structures of the absorption spectrum.

ACKNOWLEDGEMENTS

I am indebted to Daniel Cox and Charbal Tannous for their considerable assistance in compiling Table II. My long-term collaborators Jorge Müller and Luiz Oliveira have taught me all I know. Also conversations with L.C. Davis, J.D. Dow, and M.D. Stiles were useful. This work was supported in part by the National Science Foundation under grant no. DMR-80-20429.

REFERENCES

1. L.G. Parratt, Rev. Mod. Phys. 31, 616 (1959); see, for example, the discussion starting at the top of p. 623.
2. G.D. Mahan, Phys. Rev. 163, 612 (1967).
3. P.W. Anderson, Phys. Rev. Lett. 18, 1094 (1967).
4. P. Nozieres and C.T. De Dominicis, Phys. Rev. 178, 1097 (1969).
5. C. Noguera, "Dynamical screening of core holes by conduction electrons" in these proceedings.
6. A. Kotani and Y. Toyozawa in Synchrotron Radiation, ed. C. Kunz (Springer-Verlag, Berlin, 1979), p. 169.
7. If the core hole lifetime is very short, then the spectrum is broadened so much that the singularities are unobservable.
8. See, for example, two excellent reviews by G.D. Mahan, (a) Solid State Physics 29, 75 (1974) and (b) Many-Particle Physics (Plenum, New York, 1981), pp. 743-775.
9. The "one-line argument" for this result in renormalization group language is that the perturbation due to the core hole is a

marginal interaction while the interactions between the conduction electrons (due to the coulomb and electron-phonon interactions and to bond structure) are irrelevant ones.

10. The potential can be thought of as arising from a coulomb matrix element of the form
$\int d^3 r d^3 r' \; \phi_c(\vec{r}) \; \phi_c(\vec{r}') \; \psi_k(\vec{r}) \; \psi_{k'}(\vec{r}')/|\vec{r} - \vec{r}'|$. The second term of (8) or (9) is the marginal operator referred to in ref. 9.

11. This very simple model ignores several topics, some of which are discussed later in this review: (i) the dynamics of the hole creation and/or its subsequent screening by the conduction electrons may affect the spectrum, (ii) different models for the potential (e.g., repulsive barrier, square well, delta function shell, etc.) may introduce k-dependence into the potential which, in addition to the density of states, can affect whether or not a bound state is formed, (iii) impurity nature of potential destroys periodicity of solid [by the way, for a band containing N electrons, an explicit recognition of the normalization would require that -U/N be the coefficient of the second term of (9)].

12. For example for the case of a constant potential in a sphere of radius R an initial orbital function will have (for $k_F r \gg 1$) an asymptotic form $\sin(kr - \ell\pi/2)$ while a final orbital function due to the additional potential at origin with phase shifts δ_ℓ will have the asymptotic form $\sin(kr + \delta_\ell - \ell\pi/2)$.

13. J. Friedel, Comm. Solid State Phys. 2, 21 (1969).

14. The calculation proceeds by calculating the overlap matrix elements using the functions in reference 12 and then by applying some results of determinant theory (see refs. 34 and 38).

15. S. Doniach and M. Sunjic, J. Phys. C3, 285 (1970).

16. Much of the organization of the arguments in this entire section on the physics of the x-ray edge singularity are due to ref. 17.

17. Luiz Nunes de Oliveira, Ph.D. thesis, Cornell University, 1981.

18. K.D. Schotte and U. Schotte, Phys. Rev. 182, 479 and 183, 509 (1969). A summary of the argument is in J.J. Hopfield, Comm. Solid State Phys. 2, 40 (1969) and on p. 122 of ref. 8a.

19. J. Friedel, Adv. Phys. 3, 440 (1954).

20. M. Combescot and P. Nozieres, J. de Phys. 32, 913 (1971); see p. 925.

21. In fact it will be hard to study 2p absorption for the transition and rare earth metals since the large d density of states will tend to dominate the spectra.

22. (a) P.H. Citrin, G.K. Wertheim, Y. Baer, Phys. Rev. B16, 4256 (1977) [XPS]; (b) P.H. Citrin, G.K. Wertheim, and M. Schlüter, Phys. Rev. B20, 3067 (1979) [edges]. A particularly valuable part of ref. 22b is the analysis of how and why the deduced exponents have changed with time.

23. G.K. Wertheim, Phys. Rev. B25, 1987 (1982).

24. L.A. Feldkamp and L.C. Davis, Phys. Rev. B22, 4994 (1980) (a), and B23, 4269 (1981) (b).

25. A. Kotani and Y. Toyozawa, J. Phys. Soc. Japan 35, 1073 (a) and 1082 (b) (1973).

26. A. Kotani, J. Phys. Soc. Japan 46, 488 (1979). The calculations

reported here and in ref. 25b are for bands composed of s and d electrons. The core hole potential shifts only the center of the unhybridized d bond.

27. J.D. Dow and C.P. Flynn, J. Phys. C13, 1341 (1980).
28. L.N. Oliveira and J.W. Wilkins, Phys. Rev. B24, 4863 (1981).
29. G.D. Mahan, Phys. Rev. B21, 1421 (1980).
30. C.A. Swarts, J.D. Dow, C.P. Flynn, Phys. Rev. Lett. 43, 158 (1979).
31. M.A. Bowen and J.D. Dow, Phys. Rev. B22, 220 (1980). This calculation used both square well and delta function shell potentials.
32. See pp. 121-123 in ref. 17.
33. It is possible to give a qualitative explanation for the differing $I(\omega)$ and $J(\omega)$ in refs. 27 and 31, where the density of states is free-electron like. The core hole potential $V(r)$, if attractive, will generally not have a bound state [except for phase shift larger than ~ .8 π], while if repulsive there is a broadened high-lying "bound" state. This means that for an attractive potential $J(\omega)$ will saturate to unity at a much lower energy than in the case of a repulsive potential where saturation will be delayed until all excitations can sample the high-lying "bound" state. The difficulty is ref. 31 of comparing the results of a square well and delta function shell potentials with those of an impenetrable barrier (put in the initial state to mimic a positive final-state phase shift) are largely obviated by interchanging Figures 2 and 5 so that the comparison is between phase shifts of the same sign. Then it would seem that the deviations at $\omega - \omega_T = 0.1$ D from the asymptotic theory may be no more than ten percent (see footnote 16 in ref. 31 and footnote 11 in ref. 24a).
34. L.C. Davis (private communication).
35. D.R. Penn, S.M. Girvin, and G.D. Mahan, Phys. Rev. B24, 6971 (1981). Note the band width is unity, and so the zero pair result must go to zero in Figure 7 at the highest possible single particle energy $\omega = 0.5$ D. The number in Table II was supplied by S.M. Girvin.
36. U.v. Barth and G. Grossmann, Phys. Rev. B25, 5150 (1982). Ref. 34 points out that ref. 35 had obtained the same results as their equation (2.42) for the zero-pair propagator.
37. In addition the value of $\exp(2f(0))$ is an estimate for the strength μ_0. When this estimate is reduced by the contribution from multi-pair terms, it agrees well with the result of refs. 17 and 28 ($\mu_0 \simeq .8$).
38. G.D. Mahan, Phys. Rev. B25, 5021 (1982).
39. W. Hänsch and W. Ekardt, Phys. Rev. B24, 5497 (1981) and B25, 7815 (1982).
40. D.C. Langreth, Phys. Rev. B1, 471 (1970) and in Interactions of Radiation with Condensed Matter (IAEA, Vienna, 1977), Vol. 1, p. 295.
41. P. Minnhagen, Phys. Lett. A56, 327 (1976).
42. In refs. 40 and 41 $I(\omega)$ is called $A(\omega)$.
43. P. Minnhagen, J. Phys. F7, 2441 (1977).
44. K. Shung and D.C. Langreth, Phys. Rev. B23, 1480 (1981).

706

45. Y. Ohmura and K. Ishikawa, J. Phys. Soc. Japan 48, 1176 (1980) [α], 49, 1829 (1980) [α_ℓ].
46. M. Sunjic and A. Lucas, Chem. Phys. Lett. 42, 462 (1962).
47. C.O. Almbladh and P. Minnhagen, Phys. Rev. B17, 929 (1978).
48. G.D. Mahan, Phys. Rev. B15, 4587 (1977).
49. C.O. Almbladh, Phys. Rev. B16, 4343 (1977).
50. S. Abraham-Ibrahim, B. Caroli, C. Caroli, and R. Roulet, Phys. Rev. B18, 6702 (1978).
51. J.J. Rehr, E.A. Stern, R.L. Martin, and E.R. Davidson, Phys. Rev. B17, 560 (1975).
52. See a review of EXAFS by P.A. Lee, P.H. Citrin, P. Eisenberger, and B.M. Kincaid, Rev. Mod. Phys. 53, 769 (1981) for a discussion of this point, pp. 785-6.
53. E.A. Stern, B.A. Bunker, and S.M. Heald, Phys. Rev. B21, 5521 (1980).
54. V.I. Grebennikov, Yu. A. Babanov, and O.B. Sokolov, phys. stat. sol. (b) 79, 423 (1977) (a) and 80, 73 (1977) (b).
55. U.v. Barth and G. Grossmann, (a) Solid State Comm. 32, 645 (1979); (b) Physica Scripta 21, 580 (1980); (c) Phys. Rev. B25, 5150 (1982).
56. G.D. Mahan, Phys. Rev. B21, 1421 (1980).
57. L.C. Davis and L.A. Feldkamp, Phys. Rev. B23, 4269 (1981).
58. D.E. Ramaker, Phys. Rev. B25, 7341 (1982) discusses a final-state rule for Auger line shapes.
59. See, for example, equation (4.2) of reference 55c.
60. E.A. Stern and J.J. Rehr (private communication).
61. V.O. Kostroun, R.W. Fairchild, C.A. Kukkonen, and J.W. Wilkins, Phys. Rev. B13, 3268 (1976).
62. J.E. Müller, O. Jepsen, O.K. Andersen, and J.W. Wilkins, Phys. Rev. Lett. 40, 720 (1978) [4d K-edges]; J.E. Müller, O. Jepsen, and J.W. Wilkins, Solid State Comm. 42, 365 (1982) [3d K- and L-edges, 4d L-edges].
63. J.E. Müller, Ph.D. thesis, Cornell University, 1980.
64. G. Materlik and J.E. Müller (private communication).
65. L. Grunes, Ph.D. thesis, Cornell University, 1982.

3d-4f RESONANT RAMAN SCATTERING IN RARE-EARTH METAL

Takeshi Watanabe, Akimasa Sakuma and Hiroshi Miyazaki

Department of Applied Physics, Faculty of Engineering
Tohoku University, Sendai, Japan

ABSTRACT

We have studied theoretically the soft x-ray scattering by La
metal in the 3d excitation region. The Hamiltonian used includes
the terms characterizing a lowering of the 4f level by the 3d hole,
Anderson-type mixing of the 4f orbital with the conduction band,
and electron-photon coupling between the 3d and 4f levels. Using
the U-operator formalism we have obtained the transition probability
of Raman scattering as well as fluorescence. Because of the form of
interactions considered and approximations made, the mathematical
expression of the present theory is in resemblance with that of
Nozieres and Abrahams. However, since our main interest is in near
resonance excitation, the life time broadening of the 3d hole and
the mixing width of the 4f level are explicitly included in the
scattering probability, and analytical expressions are derived for
the regions near and far from the incident photon energy. As a result of calculations it is shown that the spectra for incident
photon energies above the threshold consist of Raman and fluorescence components in contrast to the spectra obtained by incident
energies below the threshold which show only a weak Raman component.
It is emphasized that the fluorescence component is not an ordinary
"absorption followed by luminescence" but the resonance fluorescence
induced by the 4f-conduction electron mixing.

INTRODUCTION

The electronic energy states of rare-earth metals have been a
subject of much interest in recent years because of their rather
diverse electrical and magnetic properties. It has been believed
that some of these properties are attributed to the unfilled 4f
level which is spacially localized around the rare-earth ion.

X-ray photoemission spectroscopy has been widely used to study
the electronic energy distribution of these metals, and useful and
interesting information has been obtained.[1-5] However, XPS measurements are usually sensitive to surface conditions and the final
state of photoemitted electrons, and it is often questioned whether
or not the observed results are intrinsic to the bulk nature of
samples.[6-9] On the contrary, x-ray emission spectra are less sensitive to surface conditions, and the observed spectra are more
intrinsic to the bulk property of materials.[10-14] However, in this
case neither the initial nor the final state is the ground state of
the system, and it is difficult to obtain information on the ground
state unless the effect of the core hole is explicitly clarified.

In the present work we aim at studying theoretically the core hole effect on resonant Raman scattering in the 3d–4f excitation region of rare-earth metals, particularly lanthanum metal. One of the features of inelastic scattering experiments is the possibility of choosing momentum and energy transfers in much a wider range in comparison to ordinary absorption and emission spectroscopy, and observed data provide, in principle, more information on the dynamical nature of electronic excitations of the system, However, owing to its experimental difficulties such as a low S/N ratio it was only recent that one can make useful and meaningful comparison between observed inelastic scattering spectra and theoretical expectation. The use of bright x-ray sources has overcome to a large extent these difficulties, and there have been a number of experimental and theoretical investigations of inelastic x-ray scattering these days.

Another feature of the present problem is that the 4f level is localized and the 3d–4f transition is dipole-allowed. Therefore, one can expect a relatively large matrix element for Raman transitions. Furthermore, the 4f level of lanthanum metal, which is empty and located about 5 eV above the Fermi edge in the ground state, is pulled down into the Fermi sea upon the production of the 3d core hole. It is therefore of great interest to clarify how the 3d and 4f holes affect the resonant Raman scattering spectrum in that the 4f orbital couples with the conduction band in the intermediate state.

In the course of the present study we introduced in the Hamiltonian the terms characterizing lowering of the 4f level due to the 3d hole, mixing of the 4f level with the conduction band, and electron–photon coupling through the p·A term. The formulation adopted follows the Keldysh-Schwinger time-dependent perturbation[15] and the obtained result is mathematically in resemblance with that of Nozieres and Abrahams(hereafter called NA).[16]

SPECTRA OF La METAL IN 3d EXCITATION REGION

Several types of spectroscopic studies of La metal in the 3d excitation region have been carried out. Quantitative and precise measurements of the M-series spectrum were performed by Liefeld et al.[17] In their experiment the electron excitation energy with a finite width was scanned in the vicinity of the M threshold region, and the observed spectrum exhibits several structures which are different from the ordinary emission lines obtained by off-resonance excitation. After careful analyses of their data they proposed the possibility of $4f^2$ configurations in the intermediate state which can decay radiatively into the 3d hole with the second 4f electron lift back to above the Fermi level. As is mentioned later this channel gives rise to the M emission line different from the ordinary 4f–3d radiative transition and can take place only in the case of electron excitation. Furthermore, it has been confirmed experimentally that the unfilled 4f level locates 5.5 eV above the Fermi edge in the ground state.

The "bount 4f^2 configuration" was further studied by means of different types of APS experiments by Kanski et al.[18] According to their results the electron excited x-ray appearance potential spectrum(EXAPS) reveals two peaks reflecting the ordinary 4f–3d transition and the $4f^2$–3d4f transition(835.6 eV), while the x-ray excited electron appearance potential spectrum(XEAPS) exhibits only one dominant peak(834.0 eV) which should be compared with the first peak of EXAPS. Since the 3d binding energy was determined as 835.8 eV by their own XPS measurement and the bremsstrahlung isochromat indicated the unfilled 4f level at 5.5 eV above the Fermi level, they concluded that the 3d vacancy should pull the 4f level as much as 7.3 eV down into the conduction band. These APS data of La metal were theoretically analyzed by Wendin and Nuroh in terms of the bremsstrahlung resonant with the 3d–4f excitation.[19] In their analysis the transition matrix element was taken to be the sum of those for the ordinary bremsstrahlung and for the 3d–4f polarization due to the incident electron captured by the 4f level, and use of the Hartree-Fock wave functions has lead to even quantitative agreement.

The electron excited Auger electron appearance potential spectrum(EAAPS) can supply information on the decay channels of the 3d hole state. The spectrum obtained by Kanski and Nilsson together with the characteristic electron energy loss spectrum(CEES) suggested that the second peak of EXAPS is not necessarily determined as the bremsstrahlung resonant with $3d^94f^2$ configuration, but is most likely due to characteristic x-ray emission.[20]

The photoabsorption spectrum in the vicinity of the 3d excitation region of La metal was measured by Bonnelle et al,[21] which is characterized by the two distinct peaks, M_V(834.9 eV) and M_{IV}(850.9 ev), separated by the 3d spin-orbit interaction, and a faint peak near 830 eV. In the case of triply ionized La ion the possible optical transitions are $^1S_0 \rightarrow {}^3P_1$, $^1P_1$, $^3D_1$, and they were able to make satisfactory assignment of the observed peaks. On the other hand, the M_{IV} and M_V emission spectra were measured by Mariot and Karnatak,[11] and it was found that the main emission lines appear at the same energy positions as the absorption peaks. Therefore, these emission lines are likely to be due to resonance emission associated with the 3d electron being excited to the highly localized 4f level by electron excitation. It is also important to notice that the resonance emission occurs at the photon energy of 834.9 eV·which is close to the 3d–4f excitation energy estimated from the APS experiment.

In summary the 3d excitation spectrum of La metal seems to indicate that i) the 3d resonance excitation to the unfilled 4f level is likely to occur in the vicinity of the 3d edge regardless of electron excitation or photon excitation, ii) the 3d vacancy pulls the 4f level down below the Fermi edge and the energy gained amounts to as much as 7 eV in reference to the unfilled 4f level of the ground state which locates 5.5 eV above the Fermi level, iii) the M emission and absorption spectra have their peaks(corresponding to J = 5/2 and 3/2) at the same photon energies indicating strong resonance emission due to a highly localized 4f level, and iv) in the case of electron excitation there is an appreciable probability of the $3d^94f^2$ excitation which appears a few eV higher energy side of the 3d–4f excitation.

In the soft x-ray scattering with a tunable monochromatic beam
the 3d electron can be resonantly excited to the empty 4f level without producing bremsstrahlung and the scattered photon can reflect
the possible decay channels of the intermediate state. Therefore,
this method provides the advantages possessed by XEAPS and EAAPS,
except for the experimental difficulties. It is also expected that
the mixing of the 4f level with the conduction band should appear in
the scattered photon intensity.

X-RAY RESONANT RAMAN SCATTERING

From a theoretical stand point x-ray scattering occurs through
the A^2 term of electron-photon interaction in first order and the
p·A term in second order. When both incident and scattered photons
have much higher energies than electronic excitation energies the
A^2 term gives rise to dominant contribution(Compton scattering).
Resonant Raman scattering takes place when either incident or scattered photon energy is close to one of the characteristic energies of
electronic excitations, and the contribution of the p·A term is resonantly enhanced and dominates that of the A^2 term.

In the present work we are interested in the resonant Raman
scattering in the 3d-4f excitation region of lanthanum metal. In
the course of formulation we consider only a resonant term and make
use of the time-dependent perturbation theory following NA.[16]

First write the Hamiltonian as a sum of the electronic part H^e
and the electron-photon interaction part H', i.e.

$$H = H^e + H',$$

where H' is written in second quantized form as

$$H' = P\, a_f^+ a_d \sum_q (\alpha_q^+ + \alpha_q) + h.c.. \tag{1}$$

Here, a_f^+, a_d are the creation and annihilation operators of the f
and d electrons respectively, α_q^+ the creation operator of photon
with wave vector q, and P is the matrix element of the 3d-4f dipole
excitation. According to the Keldysh-Schwinger perturbation formalism the probability of finding a photon q' ($= \omega'/c$) at time τ is
expressed by

$$P_{q,q'}(\tau) = |P|^4 \int_{-\infty}^{\tau} du' \int_{-\infty}^{u'} dt' \int_{-\infty}^{\tau} du \int_{-\infty}^{u} dt \times$$

$$\times\, S(t',u',u,t)\, \exp[i\omega'(u-u') - i\omega(t-t')], \tag{2}$$

where the kernel $S(t',u',u,t)$ is equal to

$$S(\ ,\ ,\ ,\ ) = \langle a_d^+(t')a_f(t')a_f^+(u')a_d(u')a_d^+(u)a_f(u)a_f^+(t)a_d(t)\rangle, \tag{3}$$

and depends only on the electronic part of the Hamiltonian. Here, the time-dependent operators are in the Heisenberg representation generated by H^e, and the expectation value is to be taken in the ground state of the electronic system. The transition probability $W(\omega,\omega')$ is then written as

$$
\begin{aligned}
W(\omega,\omega') &= \lim_{\tau \to \infty} \frac{d}{d\tau} P_{q,q'}(\tau) \\
&= |P|^4 \int_{-\infty}^{\infty} du' \int_{-\infty}^{u'} dt' \int_{-\infty}^{0} dt \times \\
&\quad \times S(t',u',0,t)\, e^{-i\omega'u'-i\omega(t-t')}.
\end{aligned} \tag{4}
$$

Now, the problem is reduced to calculate the four-body correlation function $S(t',u',u,t)$ with a given Hamiltonian H^e for the electron system.

EVALUATION OF $S(t',u',u,t)$

We write the Hamiltonian H^e for the electron system as

$$
H^e = \varepsilon_f a_f^+ a_f + \varepsilon_d a_d^+ a_d + \sum_k \varepsilon_k a_k^+ a_k + U_a a_d a_d^+ a_f^+ a_f +
$$

$$
+ V \sum_k (a_f^+ a_k + a_k^+ a_f). \tag{5}
$$

Here, we assume a flat dispersion for the 3d as well as 4f electron. The fourth term is introduced to account for the lowering of the 4f level due to the presence of the 3d hole and a parameter U_a is taken to be negative. Spinless Anderson-type mixing of the 4f level with the conduction band is included by the last term which is treated as perturbation. It should be mentioned that the Hamiltonian given by eq. (5) does not contain the core hole-conduction electron scattering term which is responsible for the edge anomaly. In this respect the present Hamiltonian is different from the one used by NA, although the obtained result is quite similar. It is also noticed that Kotani and Toyozawa[22] have used a similar Hamiltonian for the study of x-ray absorption in transition metals, but in their case mixing takes place between the d-level and the s-band both of which are dipole-allowed state from the p core level. Therefore, they observed a Fano-type interference in the final state. In the present case, on the contrary, only the f level is a dipole-allowed level from the 3d core level.

First, the fourth term is put into the first term by taking

$$
\tilde{\varepsilon}_f = \varepsilon_f + U_a < a_d a_d^+ > , \tag{6}
$$

712

where $\langle a_d a_d^+ \rangle$ is taken to be unity in the intermediate state. Then, we put

$$H_o = \tilde{\varepsilon}_f a_f^+ a_f + \varepsilon_d a_d^+ a_d + \sum_k \varepsilon_k a_k^+ a_k , \qquad (7)$$

and

$$H'' = V \sum_k (a_f^+ a_k + a_k^+ a_f) . \qquad (8)$$

Introducing the time-evolving operator $U(t,t')$ by

$$U(t,t') = T \exp[-i \int_{t'}^{t} \hat{H}''(s)\, ds] , \qquad (9)$$

we convert the time-dependent operators in eq. (3) to the interaction representation which is denoted by the symbol $\hat{}$. Then, $S(t',u',u,t)$ is expressed as

$$S(t',u',u,t) = \langle |P\{U(-\infty,t')\hat{a}_f(t')\hat{a}_d^+(t')U(t',u')\hat{a}_d(u')\hat{a}_f^+(u') \times$$

$$\times\ U(u',u)\hat{a}_f(u)\hat{a}_d^+(u)U(u,t)\hat{a}_d(t)\hat{a}_f^+(t)U(t,-\infty)\}| \rangle , \qquad (10)$$

where P denotes the Keldysh path-ordering operator. The state $|\rangle$ is taken to be the ground state of lanthanum metal with the 4f level lowered into the Fermi sea. By using Wick's theorem and Keldysh's procedure we write the four-body correlation function as the product of four one-body propagators and a linked cluster part as shown in Fig. 1.

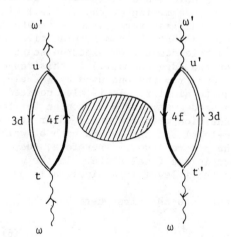

Fig. 1 : Diagrammatic representation for the transition probability.

With this approximation eq. (10) can be written as

$$S(t',u',u,t) = \phi_d(u-t)\overset{\backsim}{\phi}_d(t'-u')\psi_f(u-t)\overset{\backsim}{\psi}_f(t'-u') \times$$

$$\times \exp C(t',u',u,t) , \qquad (11)$$

where $\phi_d(\ )$ and $\psi_f(\ )$ are the 3d hole and 4f electron propagators, respectively, and the exponential factor is the contribution from the linked cluster part indicated by the shaded area in Fig. 1.

The 3d hole propagators, $\phi_d(u-t)$ and $\phi_d(t'-u')$, are defined by

$$\phi_d(u-t) = <T \hat{a}_d^+(u)\hat{a}_d(t)> ,$$

$$\overset{\backsim}{\phi}_d(u'-t') = <\overset{\backsim}{T} \hat{a}_d^+(t')\hat{a}_d(u')> .$$

Here, $\overset{\backsim}{T}$ denotes the product ordered with respect to the reversed time. By introducing Γ_d, the life time broadening of the 3d hole, these propagators are explicitly given by

$$\phi_d(u-t) = \exp[i\varepsilon_d(u-t) - \Gamma_d(u-t)], \qquad (12)$$

$$\overset{\backsim}{\phi}_d(t'-u') = \exp[-i\varepsilon_d(u'-t') - \Gamma_d(u'-t')] . \qquad (13)$$

In order to obtain the 4f electron propagator we take the diagrams shown in Fig. 2(a,b), which give the most dominant contribution of $\overset{..}{H}''(t)$, and further assume a rectangular density of states of the conduction band with a full width D and the Fermi edge ε_F in the middle. Then, we include the contribution of Fig. 2(a) by introducing the 4f level width Γ_f in the propagator, which is approximated by

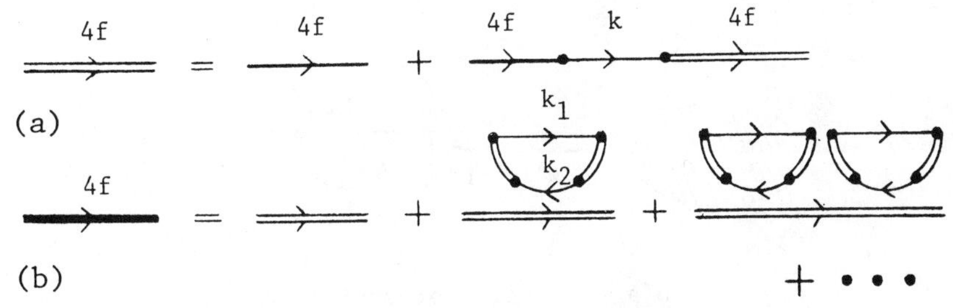

(a)

(b)

Fig. 2 : Diagrammatic representation for the 4f electron propagator, (a) only with the mixing width Γ_f, and (b) with the linked cluster part.

$$\Gamma_f = \pi \rho_F V^2 . \tag{14}$$

The density of states at the Fermi edge is denoted by ρ_F. The contribution of Fig. 2(b) can be included in $\psi_f(u-t)$ in a form of an exponent by the use of the linked cluster theorem, where the exponent $C(u-t)$ is shown to behave asymptotically as

$$C(u-t) \to -g^2 \ln [iD(u-t)] \qquad \text{for } |u-t| \to \infty . \tag{15}$$

The factor g^2 is given by

$$g^2 = \frac{1}{\pi^4} \frac{\Gamma_f^2}{(\tilde{\varepsilon}_f - \varepsilon_F)^2 + \Gamma_f^2} [\cot^{-1} \frac{\tilde{\varepsilon}_f - \varepsilon_F}{\Gamma_f}]^2 . \tag{16}$$

Summing up the contributions of Fig. 2(a,b) we obtain an asymptotic form for the 4f electron propagators:

$$\psi_f(u-t) = \exp[-i\varepsilon_f(u-t) - \Gamma_f(u-t)] \times [iD(u-t)]^{-g^2} , \tag{17}$$

$$\tilde{\psi}_f(u'-t') = \exp[i\varepsilon_f(u'-t') - \Gamma_f(u'-t')] \times [iD(u'-t')]^{-g^2} . \tag{18}$$

Finally we approximate the shaded area of Fig. 1 by Fig. 3, i.e. the linked cluster theorem is again used as in the evaluation of the 4f electron propagator. As a result of calculations it is shown that $C(t',u',u,t)$ approaches asymptotically to

$$C(t',u',u,t) \to g^2 \ln [\frac{(u-t'+i\varepsilon)(t-u'+i\varepsilon)}{(t-t'+i\varepsilon)(u-u'+i\varepsilon)}] . \tag{19}$$

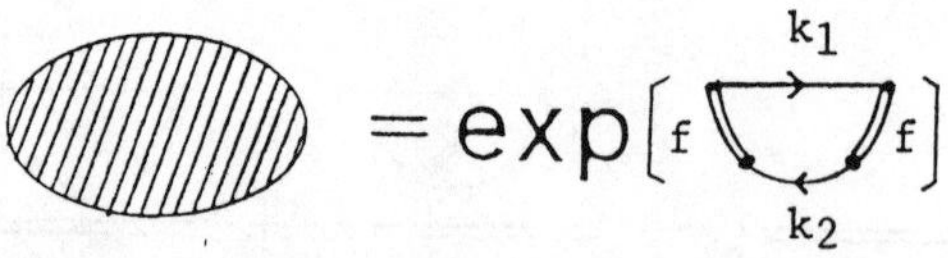

Fig. 3 : Diagrammatic representation for the linked cluster part.

Substitution of eqs. (12), (13), (17), (18) and (19) into eq. (11) leads to

$$S(t',u',u,t) = \frac{1}{D^2 g^2} \left[\frac{(u-t'+i\epsilon)(t-u'+i\epsilon)}{(t-t'+i\epsilon)(u-u'+i\epsilon)(u-t)(t'-u')}\right]^{g^2} \times$$

$$\times \exp[-i(\tilde{\epsilon}_f - \epsilon_d)(u-t+t'-u') - (\Gamma_f + \Gamma_d)(u-t+u'-t')] . \tag{20}$$

ANALYTICAL EXPRESSION FOR TRANSITION PROBABILITY

The transition rate per unit time is now given by eq. (4) with eq. (20). For convenience we measure the incident and scattered photon energies relative to the threshold energy $\tilde{\epsilon}_f - \epsilon_d$, denote them by Ω and Ω', and also put $t' = u' + v$. Then it follows that

$$W(\Omega,\Omega') = \frac{|P|^4}{D^2 g^2} \int_{-\infty}^{\infty} du' \int_{-\infty}^{0} dv \int_{-\infty}^{0} dt \left[\frac{(u'+v-i\epsilon)(t-u'+i\epsilon)}{(u'-i\epsilon)(t-v-u'+i\epsilon)tv}\right]^{g^2} \times$$

$$\times \exp[-i\Omega'u' - i\Omega(t-u'-v) + \gamma(t+v)] , \tag{21}$$

where the sum of Γ_f and Γ_d is denoted by γ. From the analyticity of the integrand in the lower half plane of u' the integral is shown to vanish for $\Omega - \Omega' < 0$. Furthermore it is noticed that for $g^2 = 0$, i.e. $V = 0$, the transition probability tends to

$$W(\Omega,\Omega') = \frac{2\pi|P|^4}{\Omega^2 + \gamma^2} \delta(\Omega - \Omega') , \tag{22}$$

which is the Rayleigh scattering intensity caused by a two-level system. We further change variables to obtain[23)]

$$W(\Omega,\Omega') = \int_{-\infty}^{\infty} du\, dv\, f(u,v)[(u-i\epsilon)(v+i\epsilon)]^{-g^2} \exp[-i(\Omega'u + \Omega v)] , \tag{23}$$

$$f(u,v) = \int_{|u+v|/2}^{\infty} dx \left[\frac{4x^2 - (u-v-i\epsilon)^2}{4x^2 - (u+v)^2}\right]^{g^2} \exp(-2\gamma x) . \tag{24}$$

We first study the case of a small value of γ, i.e. the mixing width of the 4f level as well as the intrinsic life time broadening of the 3d hole is small enough so that the time variation of the system is characterized by Ω and Ω'. For this purpose f(u,v) given by eq. (24) is divided into two parts as follows:

$$f(u,v) = \frac{1}{2\gamma} \exp[-|u+v|\gamma]$$

$$+ \frac{|u+v|}{2} \int_1^\infty dy \; [[\frac{y^2 - (z-i\varepsilon)^2}{y^2 - 1}]^{g^2} - 1] \; \exp(-\gamma|u+v|y), \quad (25)$$

where $z = (u-v)/(u+v)$. It is easily seen that the first term equals the value of $f(u,v)$ for $g^2=0$ and the second term gives the remainder. In the case of a small γ only the first term is considered. Then, by changing the variables u and v to $\xi = v$ and $\eta = u+v$, we obtain the transition probability $W_1(\Omega,\Omega')$ given by

$$W_1(\Omega,\Omega') = \int_{-\infty}^\infty d\eta \; \frac{1}{2\gamma} \; Z(\eta) \; \exp(-i\Omega'\eta - \gamma|\eta|). \quad (26)$$

The function $Z(\eta)$ is defined by

$$Z(\eta) = \int_{-\infty}^\infty d\xi \; [(\eta-\xi-i\varepsilon)(\xi+i\varepsilon)]^{-g^2} \exp[-i\xi(\Omega - \Omega')], \quad (27)$$

and vanishes for $\Omega - \Omega' < 0$. For $\Omega - \Omega' > 0$, noticing that $0 < g^2 < 1$, we carry out a contour integration along a path in the complex ξ-plane which starts from $+\infty$ below the real axis, encircles the poles at $-i\varepsilon$ and $\eta-i\varepsilon$ clockwise, and returns to $+\infty$ above the real axis. As a result of the integration $Z(\eta)$ is written by using the Gamma function $\Gamma(1-g^2)$, the Bessel function $J_\nu(\eta\Delta\Omega/2)$, where $\nu = g^2 - 1/2$ and $\Delta\Omega = \Omega - \Omega'$, and other simple functions. Finally the integration over η leads to

$$W_1(\Omega,\Omega') = \frac{\sqrt{\pi}}{\gamma} \frac{\Gamma(1-g^2)}{\Gamma(g^2+1/2)} \frac{4^{1-g^2} \sin(\pi g^2)}{(\Omega - \Omega')^{1-2g^2}} \theta(\Omega - \Omega') \times$$

$$\times \; \text{Re} \; [F(1/2, g^2-1/2, g^2+1/2; \frac{\beta^2}{\alpha^2 + \beta^2}) / \sqrt{(\alpha^2 + \beta^2)}] . \quad (28)$$

Here, $F(\; , \; , \; ; \;)$ denotes the hypergeometric function and $\alpha = \gamma + i(\Omega + \Omega')/2$, $\beta = (\Omega - \Omega')/2$. This expression is exact as long as $f(u.v)$ can be represented by the first term of eq. (25).

In the case of $\Omega-\Omega' \lesssim 2\gamma$, it is necessary to use the exact form for $f(u,v)$. Introducing a new variable ρ by $\rho = |u+v|/2$, we first rewrite eq. (23) and eq. (24) in terms of ρ and z, and then integrate over ρ. After performing path integrals we obtain the transition probability as follows:

$$W(\Omega,\Omega') = \frac{8\Gamma(3-2g^2)\sin(\pi g^2)}{(\Omega-\Omega')^{3-2g^2}} \theta(\Omega-\Omega') \times$$

$$\times \ \mathrm{Re}[\exp(-i\pi g^2) \int_1^\infty dy\{Y_-(\overset{\backsim}{z}) - Y_+(\overset{\backsim}{z})\}], \qquad (29)$$

where the function $Y_\pm(z)$ is defined by

$$Y_\pm(\overset{\backsim}{z}) = \int_1^y dz \ [\frac{(y^2-z^2)}{(y^2-1)(z^2-1)}]^{g^2}(\pm z - \overset{\backsim}{z})^{-3+2g^2}, \qquad (30)$$

with $\overset{\backsim}{z}(y) = (\Omega+\Omega'-2i\gamma y)/(\Omega-\Omega')$. We will use eqs. (28) and (29) for the numerical calculations of the transition probability for arbitrary values of Ω and Ω'. For a given value of g the constant factor omitted in these equations(see 23)) should make no effects on the value of $W(\Omega,\Omega')$.

RESULTS OF NUMERICAL CALCULATIONS

For the numerical calculations it is necessary to determine the exponent g^2 which depends on both U_a and V. However, it is noticed from its definition given by eq. (16) that its value is always positive, a smoothly varying function of $(\overset{\backsim}{\epsilon}_f - \epsilon_F)/\Gamma_f$ with one maximum, and tends to zero as its argument approaches to $\pm\infty$. The maximum value turns out to be 0.034 and appears at $(\overset{\backsim}{\epsilon}_f - \epsilon_F)/\Gamma_f = -0.5$. Here, we consider two cases of g^2, its maximum value, 0.034, and 0.01 for trial. The total width γ is taken to be 1 eV.

In Fig. 4 the transition probability is plotted against the scattered photon energy Ω' for different values of the incident photon energy Ω with $g^2 = 0.01$. It is seen that the spectrum with excitation energy below the threshold exhibits only a very weak and broad Raman component whereas those excited by energies above the

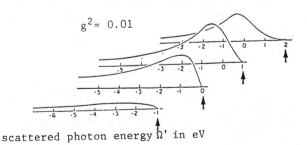

scattered photon energy Ω' in eV

Fir. 4 : Transition probability for $g^2 = 0.01$.
Positions of Ω are indicated by arrows.

threshold are found to be composed of Raman and fluorescence components. However, in the latter case the two components are hardly
separable because of an extremely small contribution of the Raman
component.

In Fig. 5 the transition probability is plotted for the case of
$g^2 = 0.034$. In comparison to Fig. 4 all the spectra are enhanced by
a factor of more than 3. This fact means that the fluorescence component is significantly influenced by the amount of mixing and is
expected to disappear as $g^2 \to 0$. It is also noticed from the spectra
with lower excitation energies ($\Omega < 0$) that the Raman component increases linearly with ($\Omega - \Omega'$) in the vicinity of Ω, reaches its
maximum, and decreases to zero as Ω' departs from Ω. In this case
the spectra are not accompanied by the fluorescence component.

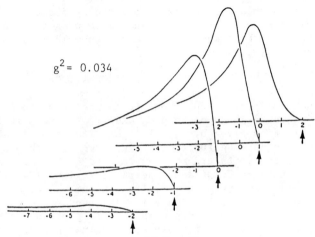

$g^2 = 0.034$

scattered photon energy Ω' in eV

Fig. 5 : Transition probability for $g^2 = 0.034$.
Arrows indicate the energy positions of incident
photon. The ordinate is in the same scale as in
Fig. 4.

DISCUSSION

From the numerical calculations it is shown that the transition
probability exhibits distinct features depending on the incident
photon energy Ω. When Ω is lower than the threshold ($\Omega < 0$), the
spectrum shows only the Raman component which is interpreted as being caused by electron-hole excitations through the fourth order
term of the Anderson-type mixing. Diagrammatically this is represented by the shaded area of Fig. 1. The contribution of zero=
energy pair excitations is much suppressed in the present case as

a result of the finite life time of the 3d and 4f holes, and the Raman component approaches to zero at $\Omega' = \Omega$.

On the other hand, when Ω is above the threshold ($\Omega > 0$), the spectrum indicates the presence of fluorescence component which has its maximum slightly below the threshold. It is found that the total intensity decreases as the exponent g^2 decreases and vanishes for $g = 0$. Therefore, this fluorescence component is thought to be essentially caused by the mixing of the 4f level with the conduction band, and the intensity increases noticeably as Ω approaches to the threshold. It is also shown from eq. (28) that the probability $W(\Omega,\Omega')$ for the case $\Omega \gg \gamma$ and $\Omega' \sim 0$ behaves asymptotically as

$$W_1(\Omega,\Omega') \sim \frac{2}{\gamma} \frac{[\Gamma(1-g^2)]^2 \sin(\pi g^2)}{[(\Omega^2+\gamma^2)(\Omega'^2+\gamma^2)]^{(1-g^2)/2}} \sin[(1-g^2)\Phi(\Omega,\Omega',\gamma)],$$

(31)

with $\Phi(\Omega,\Omega',\gamma) = \tan^{-1}[\gamma(\Omega+\Omega')/(\Omega\Omega'-\gamma^2)]$ and $0 \le \Phi < 2\pi$. Equation (31) indicates that the probability is represented mainly by the fluorescence component which is given by the product of absorption probability, emission probability and the life time of the core hole. In this case the scattering can be considered as two independent quantum processes, i.e. absorption followed by emission.

We also studied the case that the incident photon has a white spectrum below a given maximum energy. For this purpose $W(\Omega,\Omega')$ given by eq. (29) is numerically integrated over Ω. The probability of observing photons with energy Ω' is plotted in Fig. 6 for $g^2 = 0.034$ and $\gamma = 1$ eV with a maximum energy of 10 eV. It is seen

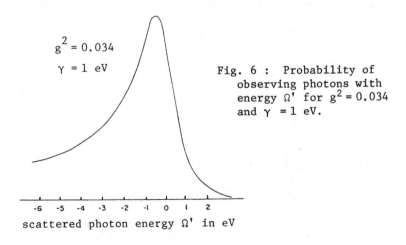

$g^2 = 0.034$

$\gamma = 1$ eV

Fig. 6 : Probability of observing photons with energy Ω' for $g^2 = 0.034$ and $\gamma = 1$ eV.

-6 -5 -4 -3 -2 -1 0 1 2

scattered photon energy Ω' in eV

that the fluorescence peak appears slightly below the threshold. Since the absorption peak is known to appear at the energy $\tilde{\epsilon}_f - \epsilon_d$, it is expected that the M emission spectrum excited by fluorescence should exhibit two peaks, one at the energy where the absorption peak is observed and the other in the lower energy side of the first peak which is induced by the mixing of the 4f level with the conduction band. However, because of small separation of these two peaks observed spectra may not indicate the presence of the two peaks and one is burried in the other or appears only in the shoulder as a small hump.

It is worth noticing that the exponent g^2 has a maximum value at $(\tilde{\varepsilon}_f - \varepsilon_F)/\Gamma_f \simeq -0.5$. This means that the effect of mixing becomes largest when the 3d hole pulls the 4f level into the Fermi sea as much as $\tilde{\varepsilon}_f$ comes close to $\varepsilon_F - 0.5\Gamma_f$. In the limit $V \to 0$, g^2 vanishes and the transition probability gives only the Rayleigh component as given by eq. (22). On the other hand, when V is large enough for Γ_f to be as much as, say, the band width D, it is no longer possible to identify the 4f level as a discrete localized level and a perturbational treatment used in the present work is no longer valid.

The finite life time of the core hole is found to smear out a step-like singularity given by $\theta(-\Omega')$ and also to blur out the diverging singularity below the threshold in the case of $\Omega > 0$. We also studied the effect of γ on the spectra and observed that the increase of γ induces lowering of the peak, broadening of the width and a shift of the peak position towards the lower energy side. This tendency seems to be enhanced as $\Omega(>0)$ approaches to the threshold.

Finally we make some comparison with the NA's work. The most significant difference is that the present study concerns the 3d-4f resonance fluorescence induced by the 4f-conduction band mixing. Therefore, only the resonance diagram shown by Fig. 1 was adopted for the calculation of the transition probability. In the NA's work scattering takes place by first exciting the core electron into the conduction band($\Omega > 0$) and then the conduction electron filling the core hole emitting a photon($\Omega' < 0$). In the process of their consideration the conduction electrons are scattered by the core hole in the intermediate state and electron-hole pairs are left in the final state. Therefore, the mixing strength V appears in second order in the 4f electron propagator(in Γ_f) and in fourth order in the linked cluster part in the present work, whereas the the lowest order contribution of the scattering potential appears in first order as a phase shift and in second order in the linked cluster part in NA's work. The mathematical similarity is due to the fact that asymptotic expression for infinite time was adopted in both cases and logarithmic singularities were sought for the contribution of the core hole.

CONCLUSION

The transition probability was derived for the 3d-4f resonance scattering of lanthanum metal where the 4f level is pulled into the Fermi sea by the 3d core hole and mixes with the conduction band. The life time effect of the 3d and 4f hole is explicitly included in the formalism and the probability was numerically calculated. It was found that the mixing-induced fluorescence component appears slightly below the threshold energy. Further studies are under way for the case of different values of parameters as well as for possible effects expected in other rare-earth metals.

REFERENCES

1. Y. Baer and G. Busch, Phys. Rev. Lett. **31**, 35 (1973).
2. J. K. Lang, Y. Baer, and P. A. Cox, Phys. Rev. Lett. **42**, 74 (1979).
3. G. Crecelius, G. K. Wertheim, and D. N. E. Buchanan, Phys. Rev. **B18**, 6519 (1978).
4. A. Platau, A. Callenas, and S. -E. Karlsson, Sol. St. Comm. 37, 829 (1981).
5. W. F. Egelhoff, Jr., G. G. Tibbetts, M. H. Hecht, and I. Lindau, Phys. Rev. Lett. **46**, 1071 (1981).
6. G. K. Wertheim and G. Crecelius, Phys. Rev. Lett. **40**, 813 (1978).
7. J. W. Allen, L. I. Johansson, R. S. Bauer, I. Lindau, and S. B, M. Hagström, Phys. Rev. Lett. **41**, 1499 (1978).
8. J. W. Allen, L. I. Johansson, I. Lindau, and S. B. Hagström, Phys. Rev. **B21**, 1335 (1980).
9. G. K. Wertheim, "Valence Fluctuations in Solid" edited by L. M. Falicov et al. (North-Holland, Amsterdam, 1981), p. 67.
10. V. F. Demekhin, A. I. Platkov, and M. V. Lyubivaya, Sov. Phys. JETP **35**, 28 (1972).
11. J. M. Mariot and R. C. Karnatak, J. Phys. F **4**, L223 (1974).
12. V. F. Demekhin, Sov. Phys. Solid State **16**, 659 (1974).
13. T. M. Zimkina, A. S. Shulakov, and A. P. Braiko, Sov. Phys. Solid State **23**, 1171 (1981).
14. K. Tsutsumi, O. Aita, and T. Watanabe, Phys. Rev. **B25**, 5415 (1982).
15. L. V. Keldysh, Sov. Phys. JETP **20**, 1018 (1965).
16. P. Nozieres and E. Abrahams, Phys. Rev. **B10**, 3099 (1974).
17. R. J. Liefeld, A. F. Burr, and M. B. Chamberlain, Phys. Rev. **A9**, 316 (1974).
18. J. Kanski, P. O. Nilsson, and I. Curelaru, J. Phys. F **6**, 1073 (1976).
19. G. Wendin and K. Nuroh, Phys. Rev. Lett. **39**, 48 (1977).
20. J. Kanski and P. O. Nilsson, Phys. Rev. Lett. **43**, 1185 (1979).
21. C. Bonnelle, R. C. Karnatak, and J. Suger, Phys. Rev. **A9**, 1920 (1974).
22. A. Kotani and Y. Toyozawa, J. Phys. Soc. Jpn. **35**, 1073 (1973).
23. This expression differs from that of NA(ref. 16) in the exponent g^2. Also, a factor, $|P|^4/_D 2g^2$, is omitted for convenience.

CHANNELING RADIATION: ATOMIC AND MOLECULAR
PHYSICS IN ONE AND TWO DIMENSIONS

Sheldon Datz
Oak Ridge National Laboratory*
Oak Ridge, Tennessee 37830 USA

ABSTRACT

Energetic electrons entering a crystal in low index planar
directions can be trapped into bound states of the potential made up
by the atoms in the planar sheet. This potential has associated
eigenstates analogous to an averaged one dimensional atom. Similarly
electrons injected in an axial direction are bound into eigenstates
of the two dimensional "string" potential characterized by two quantum numbers n, ℓ. In some crystal directions (e.g., diamond <110>)
two strings lie in close proximity and their potentials overlap so
that binding to both strings occurs and two dimensional molecular
levels obtain. When positrons are injected in planar directions they
are bound between the atomic sheets into an almost harmonic potential. In compound crystals the potential can become more complex.
Radiation arising from bound state transitions in all the above con-
figurations has been observed ($Ee^- = 4$-54 MeV, $Ee^+ = 28$-54 MeV, $E_{h\nu} = 4$-150 keV) in a number of materials (C, Si, Ge, LiF). Information
concerning atomic potentials and valence electron densities has been
obtained and is discussed.

INTRODUCTION

Channeling of heavy positively charged particles has been a subject of study since ~ 1963.[1,2] The phenomenon can be most easily
understood if one views a crystal model along low index axial or planar directions (Fig. 1). Viewed along an axis one sees strings of
atomic cores and if a positively charged particle is injected at a
small angle with respect to the row direction it will undergo a set
of small angle correlated collisions which steer the particle away
from the string, i.e., each atom in the row acts to partially shadow
the atom behind it. In 1965 Lindhard[3] showed that for positively
charged projectiles, Z_1, injected with energy E into crystals made up
of atoms Z_2 with spacing d along the rows that for angles

$$\psi_R < (2Z_1Z_2 \, e^2/E \cdot d)^{1/2} \tag{1}$$

this steering effect takes place and that moreover the potential
experienced by such particles is described by a continuum made up
from the atomic potentials in the string. The two dimensional poten-
tial $\overline{V}_R(\rho)$ is cylindrically symmetric and is a function only of ρ the

*Research sponsored by the Office of Basic Energy Sciences, U. S.
Department of Energy, under Contract No. W-7405-eng-26 with the Union
Carbide Corporation.

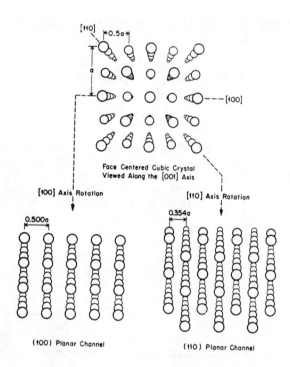

Fig. 1. Face centered cubic crystal, viewed along a <100> axial direction (top) and tilted into (100) planar and (110) planar directions.

distance from the string. For any atomic screened Coulomb potential with screening constant $\underline{a}$

$$\overline{V}_R(\rho) = (2Z_1Z_2\ e^2/d)\ K_R(\rho/a) \qquad (2)$$

If one rotates the crystal about an axis into some arbitrary angle the row symmetry disappears but planar symmetry is retained. Particles injected within an angle

$$\psi_p = (2\pi\ n_p\ Z_1Z_2\ e^2\ a/E)^{1/2}, \qquad (3)$$

where n_p is the areal atomic density in the plane, experience a continuum one dimensional planar potential

$$\overline{V}_p = 2\pi\ n_p\ Z_1Z_2\ e^2\ K_p\ (\rho/a) \qquad (4)$$

where ρ is the distance from the plane.

For massive particles, e.g., $m \gtrsim m_p$, the proton mass, the number of bound states to such potentials is very large and their motion may be treated classically but for lighter particles this is not true and radiative transitions between eigenstates occur and have been observed.

The first reported channeling radiation spectrum, radiation from planar channeled positrons, appeared in 1979.[4] It was soon followed by a paper on electron planar channeling radiation[5] and in 1980 by a paper on axial electron channeling radiation appeared.[6] In the last year these phenomena have come to be clearly understood for pure single crystals and studies of binary systems and higher order effects have been started.

It is not possible within the scope of this paper to describe in detail all of the aspects of these phenomena (line widths, decay lengths, absolute intensities, etc.). An excellent discussion of the theory for electron channeling may be found in a paper by Andersen, Eriksen and Laegsgaard.[7] Here we will confine the discussion to line energies and describe some of the results obtained in our collaborative efforts with groups at Stanford University and Lawrence Livermore National Laboratory (17-54 MeV) and with Aarhus University (4 MeV). The author is thus acting as a raporteur for the many co-workers who are listed in the references.

ELECTRON CHANNELING RADIATION

When electrons are injected into a crystal with small angle with respect to a row or a plane they may be captured into localized bound states which for e.g. row directions are eigenstates of the Hamiltonian

$$[\vec{p}_\perp^2/2m + \overline{V}_R(\vec{\rho})]\psi_i(\vec{\rho}) = E_{\perp,i}\psi_i(\vec{\rho}) \tag{5}$$

where $p_\perp$ and ρ are the projections of the momentum and position on a plane perpendicular to the row and $m \approx \gamma m_0$ with $\gamma = (1-\beta^2)^{-1/2}$ and $\beta = v/c$. Because of the relativistic increase in m the number of bound states increases with γ. (Note that for non-relativistic electrons, $\gamma \approx 1$ the number of bound states is ≤ 1.) Transitions can occur between bound states giving rise to radiation which, for $\beta \approx 1$, have an energy

$$\hbar \omega_L \approx 2\gamma^2(E_{\perp,j} - E_{\perp,i}) \tag{6}$$

when viewed in the forward direction in the laboratory frame. The factor of $2\gamma^2$ arises from two factors; a factor of γ comes from conversion to the rest frame and factor of $\sim 2\gamma$ from the Doppler transformation in the forward direction $(1+\beta)\gamma$.

Electron Planar Channeling Radiation

In this case we have a one dimensional potential and only one quantum number, n, and dipole selection rules dictate Δn odd. An example of the photon spectrum observed in the forward direction for the injection of 54 MeV electrons along the (110) planar direction in Si is shown in Fig. 2. The bound state Δn = 1 transitions are evident up to n = 5→4 and at higher energies the Δn = 3 transitions are also evident. One can now compare the observed spectrum with calculations based on e.g. Hartree-Fock descriptions of the Si atom. This can be done directly through the solution for the one-dimensional Schrödinger equation or one may work in momentum space and use the many-beam formulation of the Schrödinger equation for the transverse motion.[9] The results of the many-beam calculations which use Doyle-Turner scattering factors derived from Hartree-Fock wave functions are compared with experimental results in Table I. It is evident that the agreement is excellent.

Table I. Experimental and calculated line energies for 54 MeV (γ=107) electrons in bound states transitions in {110} of silicon.

Transition i → j	Photon Energy (keV) Calc.	Est.	Transition i → j	Photon Energy (keV) Calc.	Est.
1 → 0	125.3	122.5	4 → 3	49.0	49.1
2 → 1	89.0	88.8	5 → 4	38.4	38.3
3 → 2	64.5	64.3			

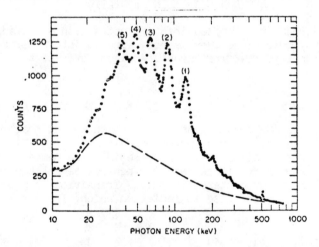

Fig. 2. Photon spectrum from channeling of 54 MeV electrons in the (110) plane of Si. The peaks labeled 1 through 5 correspond to transitions n=1→0, n=2→1, ... respectively. The lower spectrum is that obtained from a nonchanneled electron beam for the same fluence.

Electron Axial Channeling Radiation

In this case we have a two dimensional potential with quantum numbers n and ℓ. We use as an example the channeling radiation spectrum from 16.88 MeV electrons injected along a <100> direction in a diamond crystal which is shown in Fig. 2. Again, as can be seen from Table II, the agreement is satisfactory.

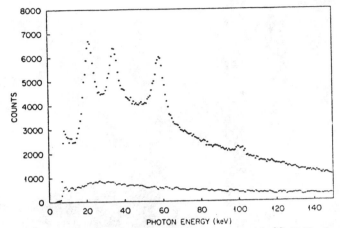

Fig. 3. Photon spectrum from channeling of 16.88 MeV electrons in <100> axial direction. The lower spectrum is that from a non-channeled beam for the same fluence.

Table II. Experimental and calculated line energies for 16.88 MeV electrons for bound state transitions in <100> diamond.

Transition	Experiment	Theory
2p-1s	58.3	58.0
3d-2p	34.4	34.0
2s-2p*	21.5	22.8
3p-1s	100.6	100.8

*Identification of this line is not unambiguous because transitions from higher states also lie in this energy region.

In both the planar and axial case the "nucleus" is extended in space and can be aligned with the incoming electron beam direction. The initial population of states depends upon the transverse momentum and hence impact parameter and incidence angle with respect to the string. Thus by tilting the crystal with respect to the electron beam direction one can selectively populate specific states and essentially map the square of the wave function for that state. This has been elegantly demonstrated by Andersen et al.[9] and an example is shown in Fig. 4 for 4 Mev electrons injected at various angles with respect to the <111> axis in Si. The peaks in order of increasing energy are, as in the diamond case discussed above, due to 2s-2p, 3d-2p, 2p-1s and 3p-1s transitions. Note that the only line which has a maximum in intensity at zero angle with respect to the axis is the 2s-2p because this is the only one which depends on an initial population of an s-state.

728

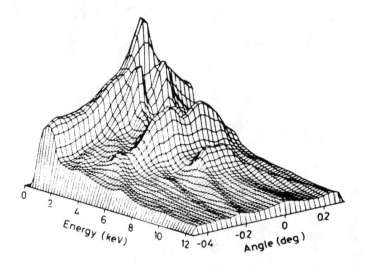

Fig. 4. Experimental photon spectra in the forward direction vs electron angle of incidence to a <111> axis in Si. The electron energy was 4 MeV and the Si crystal thickness was 0.5 μm.

Two-dimensional "Molecular" Bound States for Channeled Electrons[11]

As discussed above axially channeled electrons are captured into bound states of the row or "atomic string" potential. When two rows lie in close proximity as in the case of the <110> direction in diamond the potential overlap, forming a saddle point between the rows. The potential for 4 MeV electrons in <110> diamond is shown in Fig. 5. Solving for the eigenstates of this potential one finds that the 1s states are well localized to a single row. However, the 2p lies above the saddle point and splits into four molecular type levels. The resultant spectrum shown in Fig. 6 gives evidence for this effect.

The qualitative features are obtained in a simple treatment analogous to the linear-combination-of-atomic-orbitals (LCAO) method in chemistry. When the transverse Hamiltonian is diagonalized in the subspace spanned by the four single-string 2p states for a pair of strings, four eigenstates are obtained which may be classified according to their symmetry under reflection in the mid-point between the strings, gerade and ungerade, and under reflection in the line connecting the two strings, σ and π. The line energies for transitions between molecular energy levels obtained from the LCAO treatment are given in Table IV together with the single-string values obtained from a solution to the two-dimensional Schrödinger equation. We have also evaluated the mixing with the near-lying 2s levels, which turns out to be quite strong, in particular for the $\sigma_g 2p$ level, which is lowered considerably when mixing with the $\sigma_g 2s$ state is introduced. The splitting between the lines is seen to increase by ~ 150 eV so that it is now ~ 100 eV larger than the observed splitting. Also, a dipole transition 2s-1s becomes allowed, and, in fact, a corresponding weak line is visible in the spectrum (see Fig. 6). However, while the LCAO-type model is very instructive for qualitative purposes, its accuracy is limited and difficult to assess.

The values obtained for channeling radiation energies from the many-beam formulation are listed in Table III. For the 2p-1s transitions, there is reasonable agreement with the result of the LCAO-type calculations. The upper three lines and the more accurate 1s-2s line are higher than the measured values by ~ 120 eV. However, the most significant deviation is for the separation of the $\sigma_g-\sigma_u$ line which is larger by ~ 130 eV than deduced from the experiment. This we attribute to the accumulation of charge in the tetrahedral bonds in diamond.

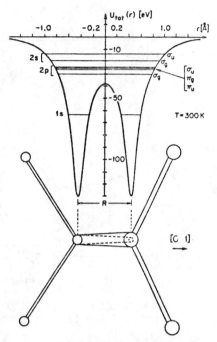

Fig. 5. (Upper) Potential and energy levels for 4 MeV electrons channeled along ⟨110⟩ axis in diamond. (Lower) Atomic core and bond configuration in diamond viewed along the ⟨110⟩ direction.

Table III. Photon energies (in electron volts) for 2s-1s and 2p-1s transitions for 4-MeV electrons in ⟨110⟩ diamond.

Transition	Single string	1st order LCAO	With 2s-2p mixing	Many beam	Experiment
$\sigma_g 2s-\sigma_u 1s$			7253	7103	6933
2p-1s	5801				
$\sigma_u-\sigma_g$		6025	6015	6019	5897
$\pi_g-\sigma_u$		5888		5887	5751
$\pi_u-\sigma_g$		5771		5742	5624
$\sigma_g-\sigma_u$		5279	5115	5090	5084

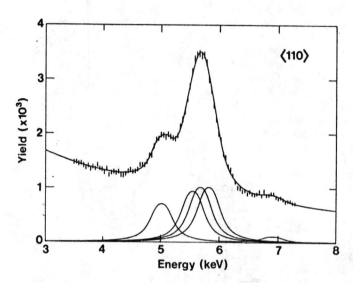

Fig. 6. Decomposition of the <110> 2p-1s and 2s-1s lines (see Table III). See ref. 11 for fitting procedure.

The effect on the <110> potential of a charge accumulation in the bond regions may be understood qualitatively from Fig. 5. Each atom in a <110> row is bonded to two atoms in a neighboring row and to one atom in each of the next-nearest neighboring rows. Hence accumulation of charge in the bonds increases the electron density between close-lying pairs of <110> rows and increases the potential energy of channeled electrons in this region. Since the $\sigma_g 2p$ state has a high density here, this level will increase in energy relative to that of other 2p levels, i.e., the splitting will be reduced, as required by experiment. With use of the many-beam method and the Gaussian parameters obtained from the x-ray analysis, the energy levels have been recalculated as a function of the electron density enhancement and the closest match with the observed splitting is obtained to an electron density of $1.7/Å^3$ in the center of bonds. Although this determination is probably less accurate than that obtained from the x-ray data, it does demonstrate that information on charge distributions and potentials in crystals can be obtained from channeling radiation.

Positron Planar Channeling Radiation

Unlike electrons, which are attracted toward the atomic cores making up the planar potential, positrons are repeled and are caught in the potential made up by two adjacent planes. Here we define the two plane potential $\overline{V}_2$ in terms of the displacement x from the midpoint between the two planes which lies at a distance ℓ from both planes

$$\bar{V}_2(x) = \bar{V}_p (\ell+x) + V_p^-(\ell-x) \qquad (5)$$

Positive particles undergo an oscillatory motion in this potential which, from studies of planar channeled positive ion trajectories,[2,12-14] have been shown to be accurately described in terms of atomic Hartree-Fock potentials.

In single element crystals the interplanar potentials are approximately harmonic and the positron channeling radiation spectrum is nearly monochromatic because of the nearly equal spacing of energy levels, i.e., if the potential were purely harmonic the transition frequency would be

$$\omega_0 = (K_0/m_0)^{1/2} \qquad (6)$$

where the force constant $K_0 = v_2''(o)$ and m_0 is the rest mass of the particle. Since $m = \gamma m_0$, the frequency in the laboratory frame becomes $\omega_L = \omega_0 \gamma^{-1/2}$ and the energy of the photon emitted in the forward direction becomes

$$\omega_{max} \simeq 2\gamma^2 \omega_L = 2 \gamma^{3/2} \omega_0 \qquad (7)$$

This type of radiation was predicted by Kumakhov[15] and was first reported experimentally for positron channeling in silicon by Alguard et al.[4]

In Fig. 7 we show positron channeling potentials and eigenstates which were calculated for the three major planes of LiF.[16] The eigenfunctions and energy levels were computed with the many beam approximation, described by Andersen, Eriksen, and Laegsgaard[7] which takes into account the periodicity of the crystal potential. The Bloch-wave nature of the eigenfunctions becomes apparent only near the tops of the potential wells where the energy levels broaden into bands. The Fourier coefficients U_n of the potentials were derived from the electron scattering factors $f_{el}(s)$ for isolated Li^+ and F^- ions given in Ref. 17. The (110) and (100) planes are made up of equal numbers of Li^+ and F^- ions; the planar average potentials smooths over the individual atom differences and a single two plane potential is obtained (Fig. 7a and 7b). The (111) plane is of particular interest because in this direction the crystal consists of alternating layers of Li^+ and F^- ions. In Fig. 7c the zero position represents a Li^+ sheet of ions which lies half way between two F^- ion planes yielding a double (nearly harmonic) well within a larger (nearly harmonic) well.

The channeling radiation spectra obtained with 54.5 MeV positrons[16] for the (110), (100) and (111) planes of LiF are shown in Fig. 8 together with the theoretical predictions based upon the potentials shown in Fig. 7. The good agreement of the results of this relatively crude calculation with the data, especially for the more complex (111) planar case [Fig. 8(c)], demonstrates the essential correctness of this approach.

732

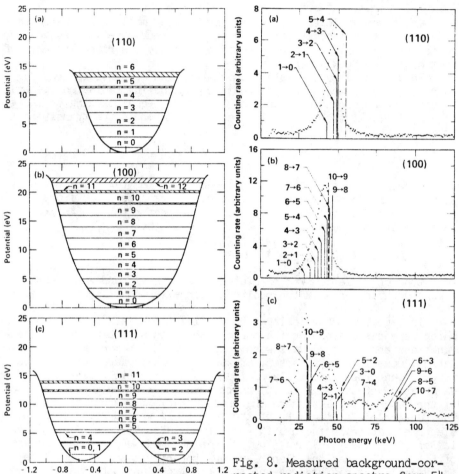

Fig. 7. Calculated interplanar potentials and eigenstates for positron channeling in LiF (a) for the (110) plane, (b) for the (100) plane, (c) for the (111) plane.

Fig. 8. Measured background-corrected radiation spectra from 54-MeV positrons channeled in LiF: (a) along the (110) plane, (b) along the (100) plane, (c) along the (111) plane. The relative intensities of the spectral lines calculated with the potentials of Fig. 7, assuming equal initial level populations, are shown as vertical lines. The dashed lines are somewhat uncertain because of the proximity of the initial level to the top of the well and its consequent band broadening.

For testing descriptions of the potentials of atoms in crystals it is clear that positron and electron channeling spectra give complimentary information. While the same potential must be operative in determining the eigenstates for both particles, positron channeling best tests the region between planes and electron channeling best test the region close to the atomic planes, i.e., for electron channeling, band structure contributions come into play near the midpoint between planes whereas for positrons the opposite is true. Using 54 MeV positron and electron beams from the LLNL linac such complimentary studies have been carried out for silicon, diamond and lithium fluoride and the results and cross comparisons will be published shortly.

1. For a general review of channeling see: D. S. Gemmell, Rev. Mod. Phys. 46, 129 (1974).
2. S. Datz and C. D. Moak, "Heavy Ion Channeling" in "Heavy Ion Physics", D. A. Bromley, ed., Plenum Press, in press.
3. J. Lindhard, K. Dansk. Vidensk. Selsk. Mat. Fys. Medd. 34, No. 14 (1965).
4. M. J. Alguard, R. S. Swent, R. H. Pantell, B. L. Berman, S. D. Bloom and S. Datz, Phys. Rev. Lett. 42, 1148 (1979).
5. R. S. Swent, R. H. Pantell, M. J. Alguard, B. L. Berman, S. D. Bloom and S. Datz, Phys. Rev. Lett. 43, 1723 (1979).
6. J. U. Andersen and E. Laegsgaard, Phys. Rev. Lett. 44, 1079 (1980).
7. J. U. Andersen, K. R. Eriksen and E. Laegsgaard, Physica Scripta 24, 588 (1981).
8. R. L. Swent, R. H. Pantell, S. Datz, M. J. Alguard, B. L. Berman, R. Alverez and D. C. Hamilton, in Physics of Electronic and Atomic Collisions, S. Datz, ed. (Plenum Press, 1981) p. 861.
9. J. U. Andersen, E. Bonderup, E. Laegsgaard, B. B. Marsh and A. H. Sørensen, Nucl. Instrum. Methods 194, 209 (1982).
10. S. Datz, R. Fearick, B. L. Berman, J. Kephart, R. H. Pantell and H. Park, unpublished.
11. J. U. Andersen, S. Datz, E. Laegsgaard, J. F. P. Sellschop and A. H. Sørensen, Phys. Rev. Lett. 49, 215 (1982).
12. S. Datz, C. D. Moak, T. S. Noggle, B. R. Appleton and H. O. Lutz, Phys. Rev. 179, 315 (1969).
13. M. T. Robinson, Phys. Rev. 134, 1461 (1971).
14. C. D. Moak, J. Gomez del Campo, J. A. Biggerstaff, S. Datz, P. F. Dittner, H. F. Krause and P. D. Miller, Phys. Rev. B 25, 4406 (1982).
15. M. A. Kumakhov, Phys. Lett. 57, 17 (1976).
16. B. L. Berman, S. Datz, R. W. Fearick, J. O. Kephart, R. H. Pantell, H. Park and R. L. Swent, Phys. Rev. Lett. 49, 474 (1982).
17. International Tables for X-Ray Crystallography, edited by N.F.M. Henry and K. Lonsdale (Kynoch Press, Birmingham, England, 1959), Vol. II, p. 164.
18. H. Witte and E. Wolfel, Rev. Mod. Phys. 30, 51 (1958).

EQUILIBRATION LENGTHS FOR K-SHELL CHARGE-EXCHANGE PROCESSES FOR FAST IONS TRAVERSING FOILS - EVIDENCE FROM MULTIPLE-SCATTERING DISTRIBUTIONS AND AUGER-YIELD MEASUREMENTS

Elliot P. Kanter, Dieter Schneider and Donald S. Gemmell
Physics Division, Argonne National Laboratory, Argonne, IL 60439

ABSTRACT

Measurements on the angular distributions of fast ions traversing foils show a pronounced dependence of the multiple-scattering widths upon the charge state of the emerging ions. These results can be explained in terms of the large scattering angles achieved by those ions that bear K-vacancies. A quantitative model is described which demonstrates how the "memory" of K-vacancy producing collisions gives rise to large multiple-scattering widths in spite of apparent charge-state equilibration. These data, together with related measurements on projectile Auger-electron yields, permit determination of the equilibration lengths for K-vacancy bearing fractions of ion beams traversing foils.

In studies concerned with energetic heavy-ion beams in solids, it is generally assumed that for target thicknesses larger than those required to achieve "charge-state equilibrium", the multiple-scattering widths of emergent-ion angular distributions are independent of the final charge state. The equilibrium length is usually defined as the minimum target thickness for which the final charge-state distributions remain constant when foil thickness and/or incident beam charge state are varied.[1] In practice, the mean charge state of the emerging ions or the distribution of charge states near the mean charge is used as a quantitative test of this constancy.[2]

The multiple scattering experienced by swift heavy ions traversing matter has received much attention by theorists and experimentalists alike (see, for example, Refs. 3-8, and references contained therein). It has become common practice to compare multiple-scattering distributions measured for projectiles (regardless of their charge state) emerging from thin foils or gaseous targets with predictions based on the theory of Meyer[3] (or the more recent treatment of Sigmund and Winterbon[4]) and agreement is generally excellent (e.g. Refs. 7, 8). These theories assume charge-exchange equilibrium in the foil and use screened ion-atom potentials characterized by a constant mean charge. This therefore results in multiple-scattering distributions that are independent of the final charge state.

Measurements of the angular distributions of ultrahigh-charge-state (charge-states far above the mean) ions produced by stripping of heavy-ion beams in dilute gas targets have demonstrated that these rare charge states are produced primarily by single

violent collisions.[9] We have measured the angular distributions of
ultrahigh-charge-state ions emerging from solid targets and find that
for target thicknesses well in excess of the "normal equilibrium"
length, such ions retain the "memory" of a single small-impact-
parameter collision. We show that this effect leads to a marked
charge-state dependence of measured angular distributions which must
be considered in experimental tests of multiple-scattering
theories. Furthermore, we demonstrate that such angular-distribution
measurements can yield important information about the individual
scattering events experienced by fast ions penetrating condensed
matter.

The apparatus is essentially that used in recent
measurements[10] on the dissociation of molecular-ion beams produced
by Argonne's 4.5-MeV Dynamitron accelerator. Magnetically-analyzed
ion beams were collimated to have a maximum angular divergence of
± .09 mrad at the target position. Sets (both horizontal and
vertical) of "pre-" and "post-deflector" plates permitted
electrostatic deflection of the beam incident upon and emerging from
the target foil. A 25° electrostatic analyzer (ESA) having a
relative energy resolution of 6×10^{-4} (FWHM), and offset by 2 mrad
horizontally from the direction of the undeflected incident beam, was
located several meters downstream from the target. A silicon
surface-barrier detector counted particles emerging from the exit
slit of the analyzer. Distributions in energy and angle were
determined for particles emerging from the target by varying the
voltages on the horizontal pre-deflectors and/or the post-deflectors
in conjunction with that on the ESA. The overall angular resolution
was 0.3 mrad (FWHM). Target thicknesses were measured by energy-loss
techniques with an accuracy of ±10%.

The measurements described here consist of scans of the
horizontal post-deflector voltages with the ESA set to analyze the
most probable final energy for each emergent charge state.
Additional measurements were performed with a fixed voltage on the
post-deflector plates and with a movable silicon surface-barrier
detector collimated to 0.19 mrad. With this latter technique energy-
integrated angular distributions were obtained. These distributions
were consistent with those taken with the ESA.

Two charge-state distributions obtained (at 0°) from the
movable detector measurements are shown in Fig. 1. These
distributions are in accord with measurements for 6- and 16-μg/cm^2
targets.[11,12] Comparisons of the means (3.80 and 3.81), standard
deviations (0.79 and 0.81), and skewnesses (-0.067 and -0.135) of
these distributions measured with 1.8- and 4.9-μg/cm^2 carbon targets
respectively, would lead to the assumption that charge-state
equilibration had already been achieved in the thinner target.

In contrast with the charge-state distributions, the
angular distributions for each charge state display characteristic
non-equilibrium behavior. Figure 2 shows angular distributions for

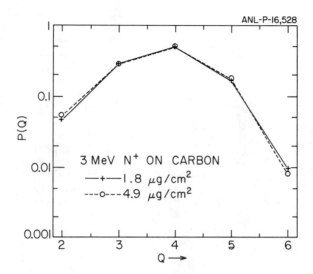

ANL-P-16,528

3 MeV N$^+$ ON CARBON
—+—1.8 μg/cm^2
---o---4.9 μg/cm^2

Fig. 1. Energy- and angle-integrated distributions of final charge
states of nitrogen ions exiting near 0° from 1.8- and 4.9-μg/cm^2
carbon foils bombarded by a 3-MeV N$^+$ beam.

the eight charge states of nitrogen (N^{+q}, q = 0-7) that were observed
after passage of a 3.7-MeV N$^+$ beam through a 1.6-μg/cm^2 carbon
target. The distributions for q = 0-5 are qualitatively similar
(though quantitative comparisons reveal distinct trends with charge
state). The widths are in agreement with theoretical estimates.[3]
The distributions for N^{+6} and N^{+7} are, however, distinctly different
from those for the lower charge states. A significant increase in
the widths of these distributions is readily apparent. Also evident
is a slight "dip" at 0° in the N^{+7} distribution. The N^{+6} and N^{+7}
ions contain K-vacancies upon exiting the target and have, with high
probability, experienced one or more violent small-impact-parameter
collisions in the foil. Such collisions are not very frequent and so
do not significantly affect the angular distributions for the lower
charge states.

These data are representative of more extensive
measurements utilizing 1--4-MeV beams of C$^+$, O$^+$, N$^+$, and Ne$^+$ in
targets of carbon and aluminum ranging in thicknesses from 1-20
μg/cm^2. The results may be summarized as follows:
1. For each combination of beam, energy, and target, there is a
    marked increase in the widths of the angular distributions
    of the two (and sometimes three) highest exiting charge
    states observed, even for the thickest targets used.
2. The increase in width is most dramatic for the lowest
    energies and thinnest targets employed.
3. The second moments of these angular distributions are a
    monotonically increasing function of charge state.

738

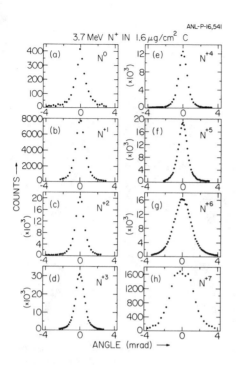

ANL-P-16,541

3.7 MeV N$^+$ IN 1.6 μg/cm^2 C

Fig. 2. Angular distributions, measured with the ESA as described in the text, of nitrogen particles in all final charge states emerging from a 1.6-μg/cm^2 carbon target after impact of 3.7-MeV N$^+$ ions.

We have been able to reproduce these features of the data quantitatively with a simple two-component model for the passage of the ions through the solid. Such models have been successful in explaining the development of projectile K-vacancies in experiments with energetic heavy-ion beams traversing solids.[13-15] In our model, we divide the target into 5-Å thick layers and consider the histories of two beam components during the target dwell time--those ions with either one or no electrons in the K-shell and those with filled K-shells. Small-impact-parameter scattering events leading to the creation of K-shell vacancies are assumed to redistribute the resulting projectile ions into an angular distribution with a hollow center, similar to those observed in single-collision experiments with gaseous targets[16,17] and with solid targets of non-equilibrium thicknesses.[18]

The net effect of all other collisions in the target is assumed (for simplicity) to smear the distributions of both beam components with a Gaussian profile in each 5-Å layer of the target. The width of this Gaussian is chosen to give the measured "normal" multiple-scattering width for the component with a filled K-shell after traversal of the entire target thickness. The final yields of the two components are determined, as are the exchange probabilities in each layer, by the cross sections for vacancy production and destruction.

The angular distributions for the two components are computed by numerical integration for the required target thickness. The two distributions are then combined, with appropriate weights, to produce angular distributions for each final charge state. The weighting factors are chosen to yield the correct final distribution of charge states. The distributions of these weighting factors with respect to charge-state are assumed to be the same for the two components except that the mean charge of the vacancy-bearing

component is shifted upwards by two charge states to account for both
the presence of the K-vacancies and further Auger de-excitation
outside the target. Because of these constraints, the only
adjustable parameters in the calculation are the two cross sections
and the width parameter for the hollow-center single-scattering
distribution. The cross sections determined in this way were
typically in the range 10^{-19}--10^{-17} cm^2. As an example, for 3.7-MeV
N$^+$ in carbon, we find K-vacancy production and destruction cross
sections of 0.2×10^{-18} and 6.3×10^{-18} cm^2, respectively. There are
no reliable previous measurements of such cross sections in solids.
However, there is an abundance of related data available from studies
of single-collision charge-changing processes in gaseous targets.[1]
For comparison Nikolaev et al.[19] report that for 4.6-MeV N$^+$ in dilute
N$_2$ gas, $\sigma(N^{+5} \rightarrow N^{+6}) \approx 3 \times 10^{-18}$ cm^2.

The results of a calculation with this model are compared
in Fig. 3 with the experimental results derived from angular
distributions such as those in Fig. 2. More detailed comparisons
than these do not seem appropriate at present because of the
underlying limitations of this computationally simple model. The

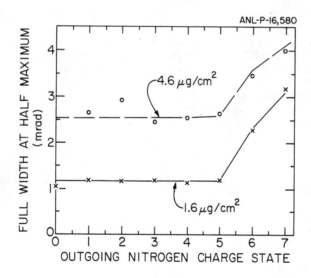

Fig. 3. Full widths at half maximum (FWHM) of angular distributions
of nitrogen ions emerging in various charge states from 1.6-μg/cm^2
and 4.6 μg/cm^2 carbon targets following impact of 3.7-MeV N$^+$ ions.
The points are the FWHMs of best-fit Gaussian profiles to the data.
The lines result from similar analyses of the angular distributions
derived for each charge state from the model described in the text
using the same parameters for both target thicknesses.

long, non-Gaussian tails evident in the data (Fig. 2) strongly influence the second and higher moments of the distributions and cannot be reproduced within the framework of this model. The calculation does, however, produce angular distributions whose second moments increase with charge state, as the presence of the vacancy-bearing component becomes important. The calculation also reproduces the 0° "dips" seen in the angular distributions of the highest charge states.

These results demonstrate that despite the apparent equilibration of the central charge-state distributions, it is possible to observe remnant behavior characteristic of single-scattering events when energetic heavy-ion beams traverse thin solid targets. It is important to recognize the presence of such effects in any experimental tests of heavy-ion multiple-scattering theories. Furthermore, this phenomenon should be considered in the design of heavy-ion stripping systems where the desire for large yields of the high charge states must be balanced against the increased multiple-scattering widths reported here. It is hoped that further refinements to the model described here will yield accurate electron capture and loss cross sections as well as details of the single-scattering angular distributions in the solid medium.

We have also confirmed the non-equilibration of K-shell charge-exchange processes in this thickness regime in measurements of the yield of KLL Auger electrons from 3-MeV N^+ beams exiting solid carbon targets. Previous experimental results from a study of inner-shell vacancy production in heavy ions traversing thin carbon foils (5 and 10 μg/cm^2) have been presented by Garcia et al.[20] That investigation was based on the measurement of absolute Auger-electron yields[21] following the decay of inner-shell vacancies in excited projectile ions. Because of the small fluorescence yields of low-Z ions, measurements of Auger-electron yields can provide reliable information on the formation of excited states[22] as well as on the inner-shell vacancy distribution for low-Z projectiles moving through solids. In particular, the inner-shell excitation and de-excitation processes which are responsible for the formation of equilibrium vacancy fractions and charge state distributions[1,23,24] can be studied.

We report measurements of relative K-Auger-electron yields from .4- and 2.7-MeV N^+ ions excited in thin carbon foils. These yields per ion represent the relative fractions of inner-shell vacancies which are present in the emergent beams. By varying the target thickness we determine the equilibrium length of the inner-shell charge-changing processes.

A well-collimated N^+ beam was passed through thin carbon foils with thicknesses between about 1.5 and 100 μg/cm^2. The target thicknesses were determined both prior to and after the electron measurements by measuring the energy loss of transmitted 2-MeV N^+ ions in the foils. The surface conditions of the target foils were kept constant by first "cleaning" them with an intense heavy-ion beam before the electron spectra were measured and then continuously heating the targets to 60°C to prevent hydrocarbon buildup. The targets were surrounded by a LN_2 cold shield and the vacuum in the scattering chamber was maintained at ~10^{-7} Torr. Electron emission from the target region was observed at an angle of 50° in the forward direction with a 45° parallel-plate electron spectrometer. The electron spectra were normalized to the incoming beam current by counting particles scattered from a rotating chopper in the beam path before the target. The energies of transmitted N^+ ions were measured with a 25° cylindrical electrostatic analyzer. For each target, the incident beam energy was adjusted to give exit ion energies of 0.4 and 2.7 MeV.

Figure 4 shows a typical electron spectrum with the kinematically-shifted projectile Auger structure superimposed on a continuous background. It has been shown previously that K-Auger-electron emission is isotropic in the projectile frame.[25] It was therefore possible to obtain relative Auger yields per projectile ion for the different projectile energies and foil thicknesses after integrating the background-subtracted Auger spectrum over energy and angle. The measured relative Auger electron yields were normalized to the absolute values at 400 keV from earlier measurements.[20,26]

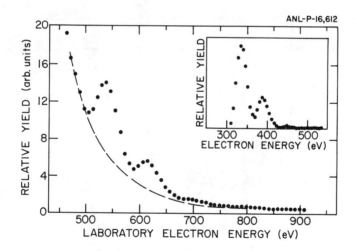

Fig. 4. Electron emission spectrum following 2.7-MeV N^+ ion impact on a 10 $\mu g/cm^2$ C foil. The inset shows a background-subtracted spectrum in the projectile frame. The structure is due to different initial electron configurations formed in the N^{+4} ions.

Figure 5 shows the vacancy fractions on an absolute scale for the two projectile energies 2.7 and 0.4 MeV. We estimate a relative uncertainty of about 5% for these results. The absolute uncertainty is about 25%. It can be seen that the vacancy fractions approach saturation values as the foil thickness increases. The slopes of the vacancy-fraction curves for foil thicknesses below ~ 20 μg/cm^2 are different as are the steady-state saturation values for the two different projectile energies.

A model that describes the inner-shell vacancy fraction production in ion beams penetrating solids was discussed in Ref. 20. As the ions traverse the target foil, excitation and decay of inner-shell vacancies occurs in a rapid sequence. The production of K-shell vacancies in nitrogen ions in this velocity range can be understood within the framework of the molecular orbital model.[27] However, K-shell vacancy production cross sections in solids are available for only a limited number of projectiles.

The decay of inner-shell vacancies is thought to occur via two competing processes; either by spontaneous decay or via collisional de-excitation. The contribution from spontaneous decay depends on the lifetime of the K-shell vacancy as well as the transit time in the foil and it becomes negligible at higher beam velocities. That fact, together with the energy dependence of charge-transfer cross sections, can explain the higher fractions measured for 2.7 MeV in comparison to .4-MeV ion energies.

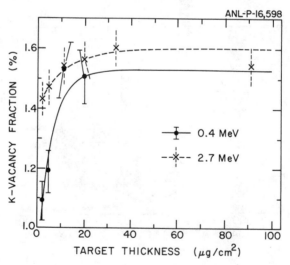

Fig. 5. K-vacancy fractions for nitrogen projectiles leaving carbon foils as a function of foil thickness. Data is shown for 0.4 and 2.7 MeV emergent ions. The solid and dashed curves result from fits to the experimental data with a functional form

$$Y_K(x) = \frac{\sigma_p}{\sigma} (1 - e^{-x/\sigma}).$$

These data show convincingly that the K-vacancy fractions of 0.4--2.7 MeV nitrogen beams do not reach equilibrium values for carbon foils as thick as 10 μg/cm^2. Vacancy production cross sections derived from experiments with such thin targets should be reconsidered in light of this finding.

This work was supported by the U.S. Department of Energy, Office of Basic Energy Sciences, under Contract W-31-109-Eng-38.

REFERENCES

1. H. D. Betz, Rev. Mod. Phys., 44, 465 (1972).
2. W. N. Lennard, T. E. Jackman, and D. Phillips, Phys. Lett. 79A, 309 (1980).
3. L. Meyer, Phys. Stat. Sol. (b) 44, 253 (1971).
4. P. Sigmund and K. B. Winterbon, Nucl. Instrum. Meth. 119, 541 (1974).
5. D. A. Eastham, Nucl. Instrum. Meth. 125, 277 (1975).
6. W. Möller, G. Pospiech, and G. Schrieder, Nucl. Instrum. Meth. 130, 265 (1975).
7. H. H. Andersen, J. Bøttiger, H. Knudsen, P. Møller Petersen, and T. Wohlenberg, Phys. Rev. A 10, 1568 (1974).
8. G. Spahn and K.-O. Groeneveld, Nucl. Instrum. Meth. 123, 425 (1975).
9. G. Ryding, A. Wittkower, and P. H. Rose, Phys. Rev. A 3, 1658 (1971).
10. D. S. Gemmell, Chem. Rev., 80, 301 (1980), and references therein.
11. R. Girardeau, E. J. Knystautas, G. Beauchemin, B. Neveu, and R. Drouin, J. Phys. B 4, 1743 (1971).
12. E. J. Knystautas and M. Jomphe, Phys. Rev. A 23, 679 (1981).
13. T. J. Gray, C. L. Cocke, and R. K. Gardner, Phys. Rev. A 16, 1907 (1977).
14. C. L. Cocke, S. L. Varghese, and B. Carnutte, Phys. Rev. A 15, 874 (1977).
15. F. Hopkins, J. Sokolov, and A. Little, Phys. Rev. A 15, 588 (1977).
16. B. Efken, D. Hahn, D. Hilscher, and G. Wüstefeld, Nucl. Instrum. Meth. 129, 227 (1975).
17. C. D. Moak, IEEE Trans. on Nucl. Sci., NS-23, 1126 (1976).
18. G. D. Alton, J. A. Biggerstaff, L. Bridwell, C. M. Jones, O. Kessel, P. D. Miller, C. D. Moak, and B. Wehring, IEEE Trans. on Nucl. Sci., NS-22, 1685 (1975).
19. V. S. Nikolaev, I. S. Dmitriev, Yu. A. Tashaev, Ya. A. Teplova, and Yu. A. Fainberg, J. Phys. B 8, L58 (1975).
20. J. D. Garcia, R. J. Fortner, H. C. Werner, D. Schneider, N. Stolterfoht, and D. Ridder, Phys. Rev. A 22, 1884 (1980).
21. R. A. Baragiola, P. Ziem, and N. Stolterfoht, J. Phys. B 9, L447 (1976).
22. R. J. Fortner and J. D. Garcia, in Atomic Collisions in Solids, edited by S. Datz, B. R. Appleton, and C. D. Moak (Plenum, New York 1975).

744

23. H. D. Betz, F. Bell, H. Panke, G. Kalkaffen, M. Wetz and D. Evers, Phys. Rev. Lett. 33, 807 (1974).
24. C. L. Cocke, S. L. Vargese and B. Curnutte, Phys. Rev. A 15, 874 (1977).
25. D. Schneider and N. Stolterfoht, Phys. Rev. A 19, 55 (1979).
26. D. Schneider, H. C. Werner, N. Stolterfoht, R. J. Fortner, and D. Ridder, Journal de Physique 40, C1-239 (1979).
27. M. Barat and W. Lichten, Phys. Rev. A6, 211 (1973).

ABSORPTION FINE STRUCTURE FROM SURFACE EXCITATIONS

D. Norman

Science and Engineering Research Council, Daresbury Laboratory,
Daresbury, Warrington WA4 4AD, UK

ABSTRACT

The technique of SEXAFS (Surface Extended X-ray Absorption Fine
Structure) has been developed to the stage where adsorbate-substrate
distances can be determined to within 0.02 Å, and adsorption sites
readily deduced for single crystal surfaces. We discuss several examples of adsorption on semiconductors (Te, I on Si(111), Ge(111) and
Pd, Ag on Si(111)) and on metals (Te on Cu(111) and O on Ni(001))
where the short-range structural parameters offer unusual insights
into details of bonding. Comparison of experimental data with calculations of the X-ray Absorption Near-Edge Structure (XANES) is shown
to hold promise for surface structural determination, and results for
O on Ni(001) are described.

INTRODUCTION

When studying the adsorption of atoms and molecules at solid
surfaces one is interested in obtaining information on the electronic
structure and on the local geometric structure around the adsorbate.
The electronic structure can be deduced by a number of techniques, of
which photoemission has been by far the most successful, but determination of the geometric structure is rather more difficult. Comparison of the results of multiple-scattering calculations with data from
low-energy electron diffraction (LEED) experiments has enabled the solution of over a hundred surface structures[1] over the last few years,
but the application of this technique is limited to single-crystal
substrates with atoms or small molecules adsorbed in a regular, simple pattern possessing long-range order within the adsorbate plane.
Surface EXAFS (SEXAFS), in principle, does not suffer from those limitations and is becoming the technique of choice for the determination
of the local structure of adsorbates at surfaces.

In this paper we describe first, in brief, the basic physics of
the SEXAFS process and its extension to surface studies via electron
and ion yield. We then give a series of examples of adsorption systems studied by SEXAFS, concentrating on some recent applications
where SEXAFS has revealed structural information which could not be
obtained by any other technique. The final section contains the
first results for calculations of XANES of adsorbate systems.

BASIC PHYSICS OF ELECTRON AND ION YIELD SEXAFS

The basic process of photon absorption at a core level, with
backscattering of the photoelectron wave modifying the matrix element
and producing fine structure as a function of energy, has been described in detail elsewhere[2]. A simple expression, valid for absorption at an s-like initial state (K or L_1 edge) gives the EXAFS inten-

sity as a function of the photoelectron wave-vector k ($\chi(k)$):

$$\chi(k) = - \sum_i \left(\frac{N_i^*}{kR_i^2}\right) F_i(k) \, e^{-2\sigma_i^2 k^2} \, e^{-2R_i/\lambda(k)} \quad \sin(2kR_i + \phi_i(k)). \quad (1)$$

The exponential terms contain the Debye-Waller vibrational softening and the damping due to inelastic scattering of the photoelectrons. The summation index i refers to all neighbour shells separated from the central (absorbing) atom by a distance R_i, and ϕ_i is the phase-shift experienced by the photoelectron escaping the potential of the absorbing atom and being backscattered by a neighbour. The backscattering amplitude $F_i(k)$ of a neighbour depends strongly on its atomic number and is, in general, significant for low-Z atoms only at small values of $k (\lesssim 10 \, Å^{-1})$. The effective co-ordination number N_i^* is given by

$$N_i^* = 3 \sum_{j=1,N_i} \cos^2\theta_j \quad (2)$$

where θ_j is the angle between the polarisation vector $\underline{E}$ and the vector $\underline{r}_{ij}$ from the central atom to the jth atom in the ith shell. Thus the EXAFS $\chi(k)$ due to the ith shell of neighbours vanishes if $\underline{E}$ is perpendicular to $\underline{r}_{ij}$, and has maximum amplitude when $\underline{E}$ is parallel to $\underline{r}_{ij}$. This polarisation dependence is of great value in surface studies where $\underline{r}_{ij}$ can often be well-defined.

Implicit in the above description is that EXAFS is a single-scattering process. Although multiple scattering can occasionally cause problems with highly crystalline samples, in general the single-scattering approximation holds good, provided that the near-edge (XANES) structure is discarded from the EXAFS analysis. In the XANES régime (within ca. 50 eV of the edge), electron mean-free-paths are long and scattering is strong, making multiple scattering important and data analysis difficult, but the spectrum in consequence is rich in information on three-body and other correlations. An application of XANES to adsorbates on surfaces is discussed later in this paper.

In conventional transmission EXAFS, the probability of creation of core holes is measured directly, but any process which is linked to the filling of the core hole can also be used to monitor EXAFS. The hole can be annihilated by a radiative transition, with emission of a fluorescent photon, or in a non-radiative process with emission of an Auger electron[3]. For heavy atoms fluorescence is the dominant process, but the probability of Auger emission is greater than 0.5 for the K edges of atoms with Z < 31, and for the L edges of all atoms[4]. The Auger emission is atom-specific, with a fixed kinetic energy characteristic of the binding energies of the levels involved. Adsorbates on surfaces typically contain light atoms (C, N, O, S, etc.) for which Auger emission probabilities are high, and the Auger electrons produced have short mean-free paths, so that most of them emerge only from close to a surface. Put another way, almost all Auger (and photo-) electrons lose energy on their way towards the surface, through electron-electron, electron-plasmon, electron-phonon and other inelastic scattering processes. The total number of electrons (γ) per incident photon created in a depth L of a solid with absorption coefficient μ is $\gamma \propto 1 - \exp(-\mu L)$. But for photon energies $h\nu > 50$ eV, the photon absorption coefficient $\mu > 1000$ Å^{-1}, and L < 50-100 Å so $\mu L \ll 1$ and thus $\gamma \propto \mu L$. Since γ is proportional to μ we can use the total electron yield (γ) as a measure of photon absorption (μ), i.e. EXAFS. But we note that L ~ 50-100 Å for low energy electrons so potentially <u>bulk</u> information is obtained, and <u>surface</u> EXAFS is available only if the surface atoms have an absorption edge within the range of photon energies used. With some systems there may be problems in using total yield caused by direct excitation of photoemitted electrons as $h\nu$ is scanned through substrate core levels, and the technique of partial yield, in which only electrons with energy greater than a set level are detected, has been developed to obviate this difficulty.

Another process which can be linked to absorption at a core level is the desorption of a surface atom as a positive ion[5]. In this case a core hole on an adsorbate or a substrate atom may be filled by an intra- or inter-atomic Auger process which results in holes in the valence shell breaking the bond of the adsorbate complex. This mechanism, first suggested for electron- or photon-stimulated desorption (ESD and PSD) from maximal valence oxides, explains some of the observed features of ESD - such as threshold behaviour - which are inadequately described by earlier theories on Franck-Condon repulsion in valence levels. Although the Auger-initiated mechanism may not be the only process leading to desorption of ions, in some cases it dominates and the resulting ion yield can be used to monitor EXAFS[6].

Details of the experimental arrangements used for SEXAFS measurements can be found elsewhere[7].

Methods of data analysis will not be discussed here, since the basic techniques are the same as for bulk EXAFS[2] (Fourier transformation or curve fitting), but a few comments specific to the case of surfaces are appropriate. Many adsorbate-substrate distances are rather short, giving EXAFS wiggles of a long period, and absorption edges - particularly at low energies - occur fairly close together, leading to a limited photon energy range in the data, and a low information content. The rapid decrease in absorption cross section and the low surface Debye temperature often encountered also conspire to limit the useful range of EXAFS. All of these basic physical problems, coupled with the inherently small numbers of adsorbing atoms, serve to limit the signal-to-noise ratio and the accuracy achievable from many surface EXAFS experiments. However, experimental expertise has now advanced to the state where, with care, adsorbate-substrate distances may be derived to an accuracy of 0.02 Å, and deduction of an adsorption site is usually possible, either from observation of second-nearest-neighbour distances or by calculation of the number of nearest neighbours. The polarization dependence is often very helpful in this, and the symmetry imposed by a surface usually means that there are few structural possibilities, further assisting interpreta-

748

tion of one's data.

ADSORPTION ON SEMICONDUCTORS

It is of vital technological importance that the details of adsorption at semiconductor surfaces are well understood, and it is also of wider theoretical interest, since the bonding mechanisms tend to differ from metal substrates. On metals, bonding is largely nondirectional, but the covalent character and consequent directional bonding of semiconductors suggests that a local orbital approach based on saturation of dangling bonds might be successful in predicting their surface adsorption behaviour. Several experiments have been carried out on reconstructed semiconductors, using photoemission, electron energy loss spectroscopy (EELS) and LEED but no previous sursurface structural experiment has been able reliably to test this idea, in part because no other technique provides the necessary accuracy, but also because all other measurements require long-range order within the overlayer, which may be impossible to achieve. Recently, measurements have been reported[8] for the adsorption of Te and I on clean Si(111)(7×7) and Ge(111)(2×8) surfaces which show that SEXAFS, not being hampered by either of these limitations, enables successful testing of these bond saturation arguments.

The data have been analyzed to give nearest-neighbour and second-nearest-neighbour distances and also effective surface atom coordination numbers. The polarization dependence of this latter parameter is particularly useful in distinguishing between possible adsorption sites. Iodine is found to adsorb, on both Si(111) and Ge(111) in the atop site, (1 in Fig.1) which is what would be expected for a monovalent atom saturating the semiconductor's dangling bond. The second-nearest-neighbour distance on silicon is greater by 0.1 Å than for an unrelaxed (1×1) substrate, suggesting outward relaxation of the surface Si atoms, but this effect is not seen on Ge.

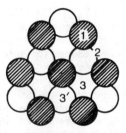

Fig.1. First (shaded) and second layer atoms in unreconstructed Si, Ge(111), showing the four high symmetry adsorption sites.

The case of tellurium adsorption, however, is distinctly different. We should not be too surprised to find different adsorption behaviour for Si and Ge in view of the changed anion coordinations from SiI_4 to GeI_2 and $SiTe_2$ to $GeTe$. However, this would suggest a different site for iodine adsorption on the two surfaces, which does not occur. It is immediately obvious from the experimental data (given in Fig.2 after Fourier transformation and filtering) that the Te site

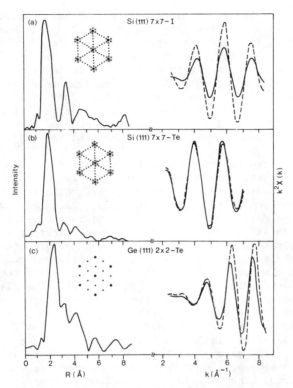

Fig.2. Fourier transformed EXAFS data (left) taken at θ = 90°, with corresponding observed LEED patterns. Polarization dependent filtered data of nearest-neighbours (right) taken at θ = 90° (solid curves) and θ = 35°, 50°, 35° (dashed curves) for (a), (b) and (c) respectively. θ is the angle between the polarization vector E and the surface normal.

cannot be the one-fold position observed for iodine, since the polarization dependence of amplitude and phase is completely different. Inspection of both relative and absolute SEXAFS amplitudes easily rules out both the high symmetry sites 1 and 3 (Fig.1) but the unambiguous distinction between sites 2 and 3' requires the use of amplitude <u>and</u> distance information: this is readily possible with the high quality data obtained and leads to the conclusion that site 2 is occupied for Te on Si(111) but that, on Ge(111), the Te atoms occupy site 3' in alternate rows. Neither of these sites has previously been proposed for a (111) surface so we should consider the implications for bonding on this surface. It is understandable <u>a posteriori</u> that divalent Te should favour a twofold bridging site on Si(111), with one electron per dangling bond, and this again fits in with the directional bond saturation arguments advanced earlier. Te on Ge is more complex, and indeed the model suggested does not explain all of the observations relating to longer-range structure - the higher distance structure in the Fourier transformed data or the (2×2) symmetry of the annealed surface - yet it is clear that the threefold hollow site above the substrate atom (3') is the only one of the four possible high-symmetry sites which fits the short-range SEXAFS data. Reconstructions involving several layers of Ge atoms may have to be invoked to explain the long range structure: such an analysis could well be beyond the capabilities of SEXAFS, which is essentially a short-range, single scattering phenomenon. Finally we note that, out of the four possible high-symmetry sites possible for the (111) surface, three have been identified with Te or I on Si and Ge, and none of them corresponds to the fcc threefold site most commonly observed for (111) metal surfaces.

It is interesting to compare the situation for adsorption of these electronegative atoms with that found for overlayers of metals on silicon. Investigation of these systems could shed light on the mechanisms for Schottky-barrier formation. Different types of interface structure may be envisaged, according to the relative reactivities of the constituent materials: reactive interface formation would involve penetration of the metal atoms into the semiconductor surface and the possible formation of a compound, while non-reactive interface formation would be expected to yield just metal nucleation on top of the semiconductor surface. The adsorption on Si(111)(7×7) of the metals Pd and Ag provides extreme examples of these different effects, as confirmed by a recent SEXAFS study[9].

The spectrum obtained for a 1.5 monolayer thick film of Pd on silicon is compared in Fig.3 with that from a palladium silicide (Pd_2Si) and from palladium metal. Without detailed analysis it is apparent that the initial reaction of Pd with Si at room temperature results in the formation of a silicide-like surface complex. Fourier transformation shows the nearest-neighbour Pd-Si and second-nearest-neighbour Pd-Pd distances to be identical (within 0.02 Å) for the initial overlayer and the silicide, but the Pd-Pd distance is only weakly observed at low coverage: this is attributable to a lower degree of long-range ordering in the Pd_2Si islands initially formed on the surface, and especially to a quicker growth laterally (parallel to the surface) than perpendicular to the surface. This effect would be enhanced by the polarization chosen for this experiment.

The behaviour of silver overlayers, by contrast, is shown in Fig.4, where the Fourier transformations of various spectra are presented. Comparison of Figs.4(a) and (b) reveals that the deposition of a 2.5 monolayer thick layer of Ag results in formation of Ag metal on the surface, without strong metal-semiconductor reaction as was seen for Pd. Peak B corresponds to the nearest-neighbour Ag-Ag distance (2.89 Å) in silver metal and its satellite B' is due to a Ramsauer-Townsend resonance in the Ag backscattering amplitude[10]. The Fourier transforms in Fig.4(c) and (d) for the $\sqrt{3} \times \sqrt{3}$ Ag on Si(111)(7×7) surface (for two different polarization vector directions) are distinctly different from that of Ag metal. Here, in addition to the Ag-Ag distance (peak B), is observed a pronounced peak (A) corresponding to the Ag-Si distance on the surface, i.e. to the Ag atoms forming the $\sqrt{3} \times \sqrt{3}$ overlayer. After correction by the appropriate phaseshifts, a value of 2.50 ± 0.05 Å is found for the Ag-Si distance on the surface. The appearance of peak B for this

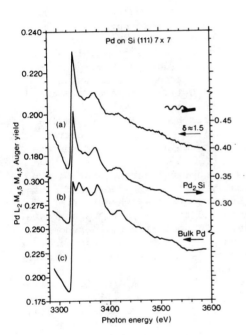

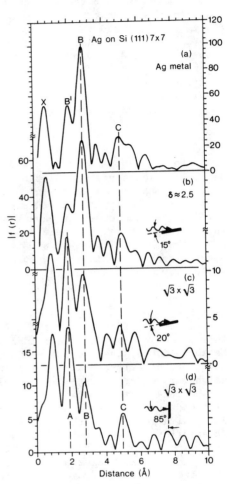

Fig.3. SEXAFS spectra above the Pd L_2 edge for Pd on Si(111)(7×7) (a) 1.5 monolayers of Pd; (b) a thick layer of Pd_2Si; (c) a thick layer of Pd.

Fig.4. Absolute value of Fourier transform of the SEXAFS signal for (a) Ag metal; (b) 2.5 monolayers of Ag on Si(111)(7×7); (c) 1.3 monolayers of Ag on Si(111)(7×7) after heating to produce a √3 × √3 LEED pattern, taken at θ = 20°; (d) as (c) at θ = 85°. Peak X is due to EXAFS from the L_3 edge underlying this spectrum, and should be ignored.

overlayer, as well as peak A, is explicable by part of the Ag atoms (2/3 monolayer) occupying periodic surface sites with an overall √3 × √3 symmetry and the surplus of Ag atoms accumulating in Ag metal-like clusters. Peak B, unlike A, shows no polarization-dependence, exactly as expected for a metal of cubic symmetry. These clusters form in such a way that only a small proportion of the surface is covered, and the LEED pattern is still dominated by the √3 × √3 over-layer. This model agrees with that proposed from ion scattering stud-ies[11], but contradicts other models favoured by LEED/Auger[12] and pho-toemission[13] measurements, and shows the power of SEXAFS in distin-guishing between several proposed surface structural possibilities.

ADSORPTION ON METALS

Atomic adsorption onto clean single crystal surfaces of metals generally yields very predictable results: the adsorbate will sit where the next metal atom would have done in a layer-by-layer exten-sion of the substrate, at a nearest-neighbour distance within about 0.1 Å of the sum of the relevant hard-sphere radii. Exceptions to this rule are rare, and here I describe two such cases in which SEXAFS has recently played a part: Te on Cu(111), where SEXAFS data[14] suggest something close to a substitutional surface site, and O on Ni(001), where SEXAFS data[15] refute earlier proposals that the in-crease in oxygen coverage from the p(2×2) to c(2×2) surface struc-ture involves a change in the spacing between adsorbate and surface layers.

Spectra[14] from Te-covered Cu(001) and (111) surfaces are given in Fig.5. The raw data were identically analyzed by subtraction of a background, to give the curves in the upper panels, then Fourier transformed, filtered and back-transformed to show just the nearest-neighbour contribution (lower panels). The polarization dependences of the SEXAFS amplitudes in these systems are opposite to one another, giving direct evidence that the Te-Cu bonds are oriented very differ-ently with respect to the surface. Analysis of the Cu(001) data straightforwardly gives Te adsorption in the expected fourfold hollow site with a Te-Cu distance of 2.62 ± 0.04 Å. For the (111) surface this distance is analyzed to be 2.69 ± 0.04 Å, and we note from Fig.5 that the amplitude of oscillations is greater for $\theta = 90°$ (normal incidence of photons) than $\theta = 20°$, necessarily implying that the Te-Cu(111) bond is oriented predominantly within the surface plane. The unreconstructed (111) surface cannot accommodate this condition without atomic re-arrangement, and the simplest one appears to be sub-stitutional displacement of one-third of a monolayer of surface Cu atoms by Te. Other, more complicated re-arrangements give unsatisfac-tory agreement with the data, while this simple substitution gives very good agreement with both absolute and relative calculated ampli-tudes (Table I). Note that equation (2), which is valid for absorp-tion at a K or L_1 edge, has to be modified for this case of absorption at an $L_3(2p_{3/2})$ core level. The short-range structural model proposed from these data does not account for the observed (2√3 × √3)R30° sym-metry of the LEED pattern, instead predicting a (√3 × √3)R30° pattern, but this doubled unit cell size could be explained by Te dimerization,

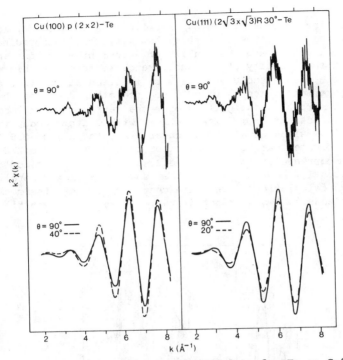

Fig.5. Background-subtracted raw SEXAFS data for Te on Cu(001) and Cu(111), together with polarization-dependent filtered data for the nearest-neighbour Te-Cu scattering only.

Table I. Calculated v. experimental values of N_s, the effective co-ordination number, for Te/Cu(001) and Te/Cu(111).

	θ	atop	bridge	hollow	subst.	experimental
Cu(001)	40°	1.2	2.3	4.3	3.6	3.9 ± 0.7
	90°	0.7	1.6	3.7	4.5	3.2 ± 0.6
	40°/90°	1.7	1.4	1.2	0.8	1.2 ± 0.1*
Cu(111)	20°	1.5	2.7	3.8	4.9	5.4 ± 1.0
	90°	0.7	1.6	2.5	6.6	7.0 ± 1.3
	20°/90°	2.1	1.7	1.5	0.7	0.8 ± 0.1*

*The relative values are derived without reference to a model compound, and thus have much greater accuracy.

surface rumpling or other effects. Despite this lack of a complete picture for the long-range structure, however, the isotropic nearest-neighbour bond length and the quality of agreement shown in Table I both support the validity of the short-range features of this model. This novel proposed structure seems sure to stimulate further study.

The system of oxygen on Ni(001), on the other hand, has been the subject of numerous investigations over the years. It had long been accepted that adsorption is dissociative, with O atoms occupying the fourfold hollow sites at a perpendicular distance ($d_\perp$) of 0.9 Å above the unreconstructed surface Ni layer for both p(2×2) and c(2×2) coverages, which corresponds to an O-Ni nearest-neighbour distance of 1.98 Å. Recently, however, it has been suggested from generalized valence bond calculations[16] that the chemisorption of atomic oxygen on Ni(001) leads to two distinct states. These were identified, by

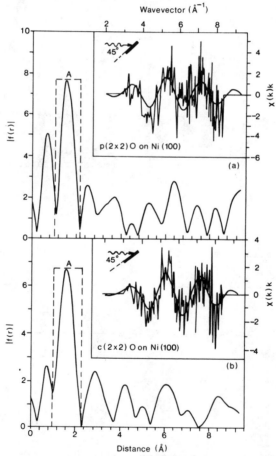

Fig.6. SEXAFS signal and absolute value of Fourier transform for the p(2×2) and c(2×2) coverages of O on Ni(001). The smooth curve through the data corresponds to the back-transformed signal due only to the O-Ni nearest-neighbour peak A.

examination of the wavefunctions and associated charge distributions, as a low-coverage radical state with $d_\perp = 0.88$ Å and a higher-coverage precursor oxide state with $d_\perp = 0.26$ Å. Comparison of calculated vibrational frequencies[17] for the two states with EELS values leads to identification of the $d_\perp = 0.88$ Å state with the p(2×2) structure and the $d_\perp = 0.26$ Å state with the c(2×2) structure. This system, then, seemed ripe for investigation by SEXAFS, and such a study has recently been reported[15].

The SEXAFS spectra, after the usual background subtraction, are given in Fig.6 for the p(2×2) and c(2×2) O/Ni(001) systems, together with their Fourier transforms and the back-transformed curves after filtering for the O-Ni nearest-neighbour signal only. There is no discernible difference between the data for the two structures, and analysis of the spectra, using a phaseshift obtained from bulk NiO standards, gives adsorption in the fourfold hollow site at a distance of 1.96 ± 0.03 Å for both coverages, corresponding to $d_\perp = 0.86$ Å. For $d_\perp = 0.86$ Å only the four nearest-neighbour Ni atoms in the surface plane contribute to peak A (Fig.6). For $d_\perp = 0.26$ Å the distance to the single Ni atom in the second substrate layer directly below the O atom becomes comparable (2.02 Å) to that to the four surface Ni atoms (1.78 Å). Since EXAFS measures just a weighted average distance from a number of similarly spaced scatterers, there remains the possibility of some ambiguity in the structural determination. This is readily removed, however, by the polarization dependence: none is observed, but the apparent O-Ni distance for $d_\perp = 0.26$ Å would be polarization dependent because of the different contributions from the four surface Ni atoms and the one Ni atom directly below the O atom. Further independent proof for the local equivalence of the p(2×2) and c(2×2) structures comes from comparison of the SEXAFS amplitudes, recorded at 45° angle of incidence. The data give an amplitude ratio of 0.96 ± 0.06 indicating that the O-Ni neighbour coordination is identical: the value $d_\perp = 0.26$ Å for the c(2×2) phase would imply an amplitude ratio of 1.28, well outside the experimental error limits.

Thus the local structural equivalence of the p(2×2) and c(2×2) structures of oxygen on Ni(001) is unambiguously established by these SEXAFS measurements. Further evidence comes from comparisons of XANES data with calculations, as described in the following section.

XANES OF ADSORBATES ON SURFACES

The multiple scattering in the XANES region means that analysis cannot proceed by Fourier transformation, as for EXAFS, and must be achieved by comparison with the results of a full multiple scattering calulation. A computational scheme has been developed[18], based on a cluster method, which is thus flexible enough to be applied to crystalline solids[19], molecules[20], amorphous solids[21] and organometallic complexes as well as solid surfaces.

The calculation proceeds by dividing the cluster into shells of atoms (each of whose scattering properties are described by a set of phase-shifts), and the multiple scattering equations are solved within each shell in turn, with the final step being calculation of multiple scattering between shells and the assembly of the whole cluster.

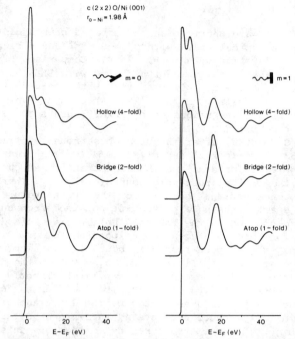

Fig.7. Comparison of the polarization-dependent XANES calculations for c(2×2) O/Ni(001) assuming a nearest-neighbour O-Ni distance of 1.98 Å and adsorption in the atop (1-fold), bridge (2-fold) and hollow (4-fold) high symmetry sites.

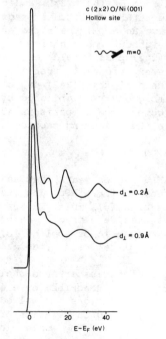

Fig.8. XANES calculations at $\theta = 0°$ (grazing incidence photon beam) for c(2×2) O/Ni(001) at perpendicular O-Ni layer spacings ($d_\perp$) of 0.9 Å and 0.2 Å.

The size of the scattering matrices is reduced by use of any symmetries (mirror or rotational) of the cluster, as we pick out, in the course of the multiple scattering calculation, only those components of the angular momentum expansions which can be coupled by a Hamiltonian possessing these symmetry elements. The calculation can, however, readily deal with systems possessing no symmetry. The reflection matrix thus calculated, when combined with an atomic matrix element linking the core and excited electron states, gives the XANES cross section in one-electron theory. Many-body decay processes which limit the lifetime of the core hole and excited electron are included in our calculation as a complex (absorptive) potential whose effect is to broaden spectral features on the appropriate scale.

One possible use of XANES spectra would be to distinguish between different adsorption sites after the adsorbate-substrate distance had been derived from a SEXAFS analysis. Figure 7 shows the results of such a calculation[22] for absorption at the K edge of oxygen on Ni(001), and depicts the XANES spectra for a c(2×2) symmetry with adsorption in the atop (onefold), bridge (twofold) and hollow (fourfold) high symmetry sites, assuming an O-Ni nearest-neighbour distance of 1.98 Å. Only minor changes are visible for the $m = 1$ case, where the polarization vector is parallel to the surface plane, but major changes are seen with grazing incidence photons. There is satisfactory agreement with experimental data for the hollow site, and convincing evidence that this method may be of use in selecting from a range of coordination sites. As described in the previous section, there was some controversy over the perpendicular spacing of the c(2×2) overlayer of O on Ni(001), and the XANES calculations for the two suggested distances ($d_\perp = 0.9$ Å and 0.2 Å) are shown in Fig.8. The $m = 1$ polarization (not shown) predictably shows little sensitivity to the perpendicular spacing, but the $m = 0$ curves are distinctly different. The experimental spectrum displays peaks at ca. 15 eV and 26 eV above the edge, as calculated for the $d_\perp = 0.9$ Å case, whereas the curve for $d_\perp = 0.2$ Å has valleys at these energies. This is therefore further confirmation of the $d_\perp = 0.9$ Å assignment.

SUMMARY

SEXAFS is now producing very precise results for the short-range structure of adsorbates on surfaces which are often unexpected and seem sure to stimulate further studies using other techniques. XANES, still in its infancy, shows promise for assisting in surface structural determinations.

ACKNOWLEDGEMENTS

I am very grateful to Paul Citrin and Joachim Stöhr for communicating results prior to publication and for many stimulating discussions on SEXAFS.

REFERENCES

1. S.Y. Tong, in "Electron Diffraction", eds. P.J. Dobson, J.B.
    Pendry and C.H.Humphreys, (London: Institute of Physics, 1978)
    270.
2. P.A. Lee, P.H. Citrin, P. Eisenberger and P.M. Kincaid, Rev. Mod.
    Phys. 53, 769 (1981).
3. P.A. Lee, Phys. Rev. B13, 5261 (1976).
4. V.O. Kostroun, M.H. Chen and B. Crasemann, Phys. Rev. A3, 533
    (1971); M.H. Chen, B. Crasemann and V.O. Kostroun, Phys. Rev.
    A4, 1 (1971).
5. M.L. Knotek, these Proceedings.
6. R. Jaeger, J. Feldhaus, J. Haase, J. Stöhr, Z. Hussain, D. Menzel
    and D. Norman, Phys. Rev. Lett. 45, 1870 (1980).
7. J. Stöhr, R. Jaeger, J. Feldhaus, S. Brennan, D. Norman and G.
    Apai, Appl. Optics, 19, 3911 (1980).
8. P.H. Citrin, P. Eisenberger and J.E. Rowe, Phys. Rev. Lett. 48,
    802 (1982).
9. J. Stöhr and R. Jaeger, J. Vac. Sci. Technol., in press.
10. B-K. Teo and P.A. Lee, J. Amer. Chem. Soc. 101, 2815 (1979).
11. M. Saitoh, F. Shoji, K. Oura and T. Hanawa, Jap. J. Appl. Phys.
    19, L421 (1980).
12. F. Wehking, H. Beckerman and R. Niedermayer, Surf. Sci. 71, 364
    (1978).
13. G.V. Hansson, R.Z. Bachrach, R.S. Bauer and P. Chiaradia, Phys.
    Rev. Lett. 46, 1033 (1981).
14. F. Comin, P.H. Citrin, P. Eisenberger and J.E. Rowe, to be
    published.
15. J. Stöhr, R. Jaeger and T. Kendelewicz, Phys. Rev. Lett. 49, 142
    (1982).
16. T.H. Upton and W.A. Goddard III, Phys. Rev. Lett. 46, 1635
    (1981).
17. T.S. Rahman, J.E. Black and D.L. Mills, Phys. Rev. Lett. 46,
    1469 (1981).
18. P.J.Durham, J.B. Pendry and C.H. Hodges, Comput. Phys. Commun.
    25, 193 (1982).
19. G.N. Greaves, P.J. Durham, G. Diakun and P. Quinn, Nature, 294,
    139 (1981).
20. A. Bianconi, M. Del'Ariccia, P.J. Durham and J.B. Pendry, Phys.
    Rev. B, to be published.
21. P.H. Gaskell, D.M. Glover, A.K. Livesey, P.J. Durham and G.N.
    Greaves, J. Phys. C: Solid State Phys. 15, L597 (1982).
22. D. Norman, P.J. Durham, J.B. Pendry, J. Stöhr and R. Jaeger, to
    be published.

PHOTOEMISSION SPECTROSCOPY OF SURFACES AND ADSORBATES

T.-C. Chiang[*]
Department of Physics and Materials Research Laboratory
University of Illinois at Urbana-Champaign
Urbana, IL 61801 USA

G. Kaindl[§]
Institut für Atom-und Festkörperphysik
Freie Universität Berlin, D-1000 Berlin 33, Germany

F. J. Himpsel and D. E. Eastman[+]
IBM, T. J. Watson Research Center
Yorktown Heights, NY 10598 USA

ABSTRACT

Core level photoelectron spectroscopy is providing new information concerning the electronic properties of adsorbates and surfaces. Several examples will be discussed, including studies of adsorbed rare gas submonolayers and multilayers as well as clean metal surfaces. For rare gas multilayers adsorbed on metal surfaces, the photoelectrons and Auger electrons exhibit well-resolved increases in kinetic energy with decreasing distance between the excited atom and the substrate, allowing a direct labeling of the layers. These energy shifts are mainly due to the substrate screening effects, and can be described well by an image-charge model. For a Kr/Xe bilayer system prepared by first coating a Pd substrate with a monolayer of Kr and then overcoating with a layer of Xe, a thermally activated layer inversion process is observed when the temperature is raised, with Xe coming in direct contact with the substrate. For rare gas submonolayers adsorbed on the Al(111) surface, coverage-dependent core level shift and work function measurements provide information about the adatom spatial distributions, polarizabilities, and dipole moments for the ground and excited states. We have also studied the 2p core level shifts for a clean Al(001) surface relative to the bulk. The shifts have a large contribution from the initial-state effects.

INTRODUCTION

When an atom is brought from free space onto a substrate, the atomic core-level binding energies are generally shifted due to the changes in the local atomic environment. From photoemission measurements, such shifts can be determined precisely, which are related to the positions of the adatoms relative to the substrate as well as the

[*]Supported by the U.S. Department of Energy, Division of Materials Sciences, under Contract No. DE-AC02-76ER01198.
[§]Supported by Sonderforschungsbereich-6 of the Deutche Sorschungsgemeinschast.
[+]Supported in part by the U.S. Air Force Office of Scientific Research under Contract No. F49620-81-C-0089.

adatom – adatom and adatom–substrate interactions. These bonding geometries and interactions are subjects of great importance to surface science studies, and much research has been done recently in order to understand the connection between observed adatom core level shifts and the detailed atomic and electronic structures of the adsorbate systems.[1]

Adatom core-level binding-energy shifts as observed in photoemission experiments can often be divided approximately into two types: initial-state (chemical and configurational) shifts and final-state (screening and relaxation) shifts. Roughly speaking, the initial-state shifts are due to charge transfer, change in electronic configuration, formation of chemical bonds, etc., in the ground state during the adsorption process. If a core electron is removed from the adatom by photoemission, other electrons in the system can relax or be polarized to reduce the total energy of the photoexcited system. This effect leads to the final-state shifts, as a result of the many-body response. The division into the initial- and final-state shifts is, generally speaking, somewhat artificial, since the experimentally measured shift is always the total shift. This effect sometimes causes difficulties and ambiguities in interpretation of the experimental results.

In this paper,[2] we will present experimental results on very simple adsorbate systems, i.e., rare gases adsorbed on metal surfaces.[3-6] For these physisorption systems, the bonding between two rare gas adatoms and between adatom and substrate is weak, therefore initial-state effects are much smaller and can often be neglected as a first approximation. The observed shifts for various atomic configurations are mainly due to the final-state effects. We will show that for a rare gas multilayer physisorbed on metal substrates, the core-level binding energies are different for rare gas atoms at different distances from the substrate. The shifts, up to 1 eV or more, can be described quite well using a point-charge image-potential model. The final-state screening effects in this case have the long-range Coulomb form, which is quite different from the expected short-range behavior for the initial-state effects. Furthermore, much larger shifts for doubly-charged Auger final state can also be explained within the same model. Therefore, adatoms in different layers can be distinguished or labeled according to the core binding energies. As an application of this effect, we investigate the thermally induced inversion of Kr/Xe bilayers adsorbed on Pd, prepared by first coating the Pd substrate with a monolayer of Kr and then overcoating with a layer of Xe. Thermal activation energies for inversion are obtained.

Intralayer screening effects from neighboring adatoms can also lead to observable binding-energy shifts for submonolayers of rare gases on metal substrates, although the magnitude of the shifts is much smaller (of the order of 0.1 – 0.2 eV or less). From the measured core-level shifts as functions of coverage for Xe submonolayers adsorbed on Al, the effective charge on a photoionized Xe adatom is determined to be nearly a whole electronic charge; therefore, the screening charge remains in the substrate and hence image-charge-like. From these studies, information on adatom spatial distributions can also be deduced.

Finally, we will discuss the core level binding energy shifts for surface atoms of an Al(001) surface relative to the bulk.[7] The surface Al atoms are tightly bonded to the crystal, therefore, the initial-state shift is significant. The experimental results will be compared with theoretical results.

EXPERIMENT

The photoemission data were taken with a double-cylindrical-mirror-analyzer system and torroidal grating monochromator at the Synchrotron Radiation Center of the University of Wisconsin. The metal substrates, Pd and Al, were prepared by Ar or Ne ion bombard-ment and subsequent annealing. The temperature of the crystal could be varied continuously by use of a sample heater and a closed-cycle helium refrigerator. A thermocouple mechanically attached to the sample was used to monitor the temperature. Rare gases were adsorbed at a substrate temperature of about 40 K. Because of the large dif-ferences in core level binding energies between first- and second-layer adsorbed atoms, full monolayer coverages could be calibrated to within about 5% accuracy in terms of exposure and photoemission in-tensity. Closed-packed monolayers were formed by depositing slightly thicker layers and then annealing at a temperature below the monolay-er desorption temperature but above the second-layer desorption tem-perature. Bilayers were formed by the same method; even thicker layers were formed by controlled exposure. Submonolayer coverages were determined from the angle-integrated photoemission intensity relative to monolayer coverages at high photon energies. The work function was determined ($\pm$ 15 meV for relative measurements; $\pm$ 100 meV for absolute measurements) by subtracting the energy range of photoemitted electrons from the photon energy hν.

RESULTS AND DISCUSSION

For a rare-gas multilayer deposited on a metal substrate, atoms in different layers show different core level binding energies. Figure 1 shows angle-integrated Xe-4d core level photoemission spec-tra and Xe-NOO (4d-5s,p-5s,p) Auger spectra for various configura-tions on a Pd(001) substrate. The 4d core levels show a spin-orbit splitting of 1.98 eV, as seen in Fig. 1(a) for an adsorbed Xe mono-layer. The binding energies referred to the vacuum level are 2.14 eV smaller than the values for the free atoms, mainly due to metallic substrate screening effects to be described below. For an adsorbed Xe bilayer, two sets of Xe 4d spin-orbit-split core levels are clear-ly observed. The less intense set with binding energies roughly the same as for the monolayer originates from the underlayer (first layer), while the more intense set with binding energies higher by 0.72 eV originates from the overlayer (second layer). For a multi-layer (four layers), the 4d spectrum is a superposition of contribu-tions from all four layers, with relative intensities determined by the electron escape depth.

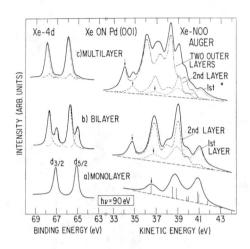

Fig. 1. Xe-4d photo-emission and Xe-NOO Auger spectra for (a) a monolayer, (b) a bilayer, and (c) a multilayer (4 layers) of Xe on Pd(001). The bilayer and multilayer spectra are decomposed into contributions from the first, the second, and the two unresolved outer layers.

The Auger spectra show similar behaviors. The Xe monolayer NOO Auger spectrum is broadened compared with the gas phase spectrum (verticle bars in Fig. 1(a) represent the relative positions and intensities of individual Auger components in the gas phase). This broadening is partly due to inhomogeneous broadening (e.g., binding energy depends on adsorption site) and lifetime effects (lifetime is shortened by coupling to the metal substrate). The arrows in Fig. 1(a) indicate the $^1S_0$ Auger singlet transition which is well separated from all other transitions. For a bilayer coverage, two sets of Auger spectra corresponding to the two layers are observed. The contribution from the overlayer is much sharper compared with the underlayer, because coupling to the metal substrate is reduced. The Auger kinetic energies for the overlayer are shifted to lower values by 2.01 eV with respect to the underlayer contribution. The underlayer kinetic energies are also slightly shifted to higher values by 0.13 eV with respect to the monolayer spectrum, caused by dielectric screening by the overlayer. For four layers of Xe on Pd(001), the contributions from the two outer layers can no longer be resolved. Therefore, the different atomic layers can be easily labeled or distinguished by Auger measurements for a three-layer film of Xe on Pd. From the measured layer to layer shifts in energy, it is seen that the Auger kinetic energy shifts are three times the corresponding 4d binding energy shifts (within 5%).

To understand the origin of the shifts, we first note that a photoionized Xe adatom is fully charged (to be discussed below). Figure 2(a) shows schematically that a free Xe atom is photoionized, and a 4d core electron is removed. The difference between the photon energy hν and the photoelectron kinetic energy is the binding energy of the 4d level. Figure 2(b) shows the same experiment except that the Xe atom is at a finite distance x from the image plane of a metal substrate. After photoionization, the electrons in the metal

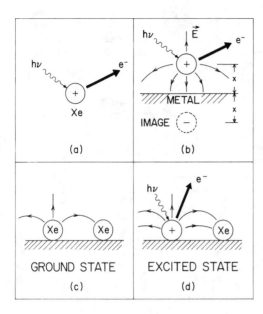

(a)

(b)

GROUND STATE
(c)

EXCITED STATE
(d)

Fig. 2. Schematic drawings for (a) photoionization of a free Xe atom, (b) photo-ionization of a Xe atom at a distance x from the image plane of a metal, (c) a Xe adatom in the ground state, and (d) a photoionized Xe adatom.

substrate tend to screen the electric field produced by the positive charge left on the Xe atom; the total field $\vec{E}$ outside the metal is described quite well by an image-charge model if x is large. The total energy of the Xe ion and metal system is reduced by $e^2/4x$, which is the energy required to move the positive Xe ion to infinity. By energy conservation, the photoelectron kinetic energy must be higher by $e^2/4x$ with respect to the case depicted in Fig. 2(a). Therefore, the 4d binding energy is smaller by $e^2/4x$ when the Xe atom is moved from free space to a distance x from the image plane of the metal substrate.

The experimentally observed binding-energy shifts for the outer-most layers can be explained quite well by an extension of the above model. Consider the photoionized Xe adatom to be a point charge. The underlying layers are approximated by a dielectric slab of dielectric constant $\varepsilon = 2.25$, which is the value for bulk solid Xe. Assume the distance between the point charge and the dielectric boundary be x'. The distance between the point charge and the metal substrate surface atomic plane is determined using a hard-sphere model with the known atomic radii: for a monolayer coverage, the average distance for an incommensurate close-packed layer is used; for thicker layers, the distance is calculated assuming the Xe layers are close-packed and ordered.[8] The location of the image plane for the metal is taken from a local-density-functional calculation.[9] Assuming the lower boundary of the dielectric slab is at the image plane, we obtain the final-state shift from the Coulomb energy of the point charge:

$$\Delta E_B = (-e^2/4) \left\{ \mu/x' + (1 - \mu^2) \sum_{n=0}^{\infty} (-\mu)^n / [(n + 1) x - nx'] \right\}, \quad (1)$$

where $\mu = (\varepsilon - 1)/(\varepsilon + 1)$ and x is the distance between the point charge and the image plane. For $x \to \infty$,

$$\Delta E_B = - \mu e^2/4x'. \quad (2)$$

From measured ΔE_B for a thick layer ($x \rightarrow \infty$), x' is determined using Eq. (2). Initial-state shift ($\lesssim 0.3$ eV)[10] and intralayer screening shift (~ -0.2 eV, see below) are much smaller and not included in the present analysis. The experimental and calculated values of ΔE_B are listed in Table I; the agreement is good within the limitations and accuracies of the simple model.

Table I. Experimental and theoretical Xe 4d binding energy shifts and NOO Auger kinetic energy shifts (all referred to the vacuum level) relative to free Xe atoms.

Xe configuration	Layer	ΔE_B(4d,exp)	ΔE_K(NOO,exp)/3	Theory
monolayer	1st	−2.14	2.19	2.35
bilayer	1st	−2.21	2.23	
	2nd	−1.49	1.56	1.56
multilayer	1st	−2.24	2.26	
	2nd	−1.54	1.58	
	outer	−1.28	1.30	1.28

The same model works well for the Auger-kinetic-energy shifts. The initial state for Auger transition is the final state for 4d photoemission, so the binding energy is shifted by ΔE_B from Eq. (1). The Auger final state is doubly charged; therefore, the binding energy is shifted by 4 ΔE_B from Eq. (1). The difference between the two shifts, −3 ΔE_B, is the Auger-kinetic-energy shift by energy conservation. Table I also lists 1/3 of the measured Auger-kinetic-energy shifts to be compared with − ΔE_B. Again, the agreement is very good.

The ability to label adatoms at different distances from the substrate is useful in studying dynamic processes involving atomic movements. Figure 3 shows one example i.e., the thermally activated inversion of a Kr/Xe bilayer on Pd(111). This bilayer is prepared by coating the substrate with a full monolayer of Kr and then overcoating with a submonolayer of Xe. Since a Xe atom has more electrons than Kr, the van der Waals attraction between Xe and Pd is larger than between Kr and Pd. As the system temperature is raised, second layer Xe atoms tend to move into the first layer and push Kr atoms out. During adsorption, some Xe atoms also had enough thermal kinetic energy to penetrate into the first layer. Since the Xe-4d binding energies are strongly shifted as the adatom moves from the second layer to first layer, this process can be followed quantitatively. In Fig. 3, core-level binding energies for first-layer and second-layer Xe atoms are indicated by arrows. The spectra were obtained after successive annealing at the indicated temperatures for 60 seconds (the base temperature was 49.5 K). The photoemission intensity corrected for attenuation is a direct measure of the degree of inversion. Under various initial conditions, the inversion is

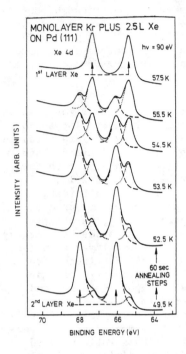

MONOLAYER Kr PLUS 2.5L Xe ON Pd (111)

Xe 4d $h\nu = 90$ eV

1st LAYER Xe

57.5 K

55.5 K

54.5 K

53.5 K

52.5 K

60 sec ANNEALING STEPS

2nd LAYER Xe

49.5 K

INTENSITY (ARB. UNITS)

70   68   66   64
BINDING ENERGY (eV)

Fig. 3. Xe-4d photoemission spectra for a Kr/Xe bilayer on Pd(111) after successive annealing steps as indicated. The Xe-4d doublet components originating from Xe atoms in the 1st and 2nd layers are indicated by dotted and dashed lines, respectively.

described well by a statistical model involving thermal activation. For Kr/Xe on Pd(111), the activation energy is 0.12 ± 0.03 eV.

We now discuss the effects of intralayer screening for rare gas layers, which are small and ignored in the above discussion. Consider a sub-monolayer of Xe adsorbed on an Al(111) substrate. There is a small dipole moment p_1 associated with an isolated Xe adatom due to a very small amount of charge transfer (~ 0.04 e); this is obviously related to the small initial-state shift discussed above. The dipolar field for this adatom can polarize nearby adatoms (schematically shown in Fig. 2(c)); therefore, the net dipole moment at coverages $0 < \theta \leq 1$ has to be calculated self-consistently. This is essentially the Topping model.[11] The dipole layer so produced is related to the work function change $\Delta\phi$ of the system. The result is

$$\Delta\phi = - (8/3^{\frac{1}{2}}) \pi ep_1 d^{-2}\theta(1 + 2\alpha\xi\theta^n d^{-3})^{-1} \qquad (3)$$

where d is the nearest-neighbor distance for a hexagonal close-packed adatom layer, α is the adatom polarizability, $\xi = 11.0$ is a geometrical factor. The index n in Eq. (3) is unity if the submonolayer adatom distribution is random, using a lattice gas mode. At the other extreme, n = 1.5 if the adatom distribution is uniform, namely, the adatom-adatom distnace is a constant equal to $\theta^{-\frac{1}{2}}d$, following the original Topping model (this would be the case if adatom-adatom interaction is highly repulsive). If the adatoms form many large islands, $\Delta\phi$ (averaged over the crystal face) is roughly linearly proportional to θ. The initial stop of $\Delta\phi$ versus θ, from Eq. (3),

$$d\Delta\phi/d\theta\big|_0 = - (8/3^{1/2})\pi ep_1 d^{-2} \qquad (4)$$

and $\Delta\phi$ for full monolayer coverage, given by

$$\Delta\phi\big|_1 = (d\Delta\phi/d\theta\big|_0) (1 + 2 \alpha\xi d^{-3})^{-1}, \qquad (5)$$

are independent of n. The experimental results are shown in Fig. 4

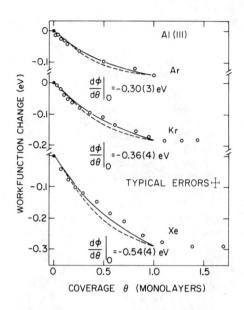

Fig. 4. Measured work function changes as functions of coverage θ for Ar, Kr, and Xe on Al(111) at 40 K. The solid and dashed curves are theoretical curves described in the text.

for Ar, Kr, and Xe adsorbed on Al(111) at 40 K. The solid and dashed cruves are obtained from Eq. (3) with n = 1 and 1.5, respectively; both describe the data well. Clearly, large island formation does not occur in our case. Experimentally, we observe that some Xe atoms are already in the second layer even for θ < 1 at 40 K. These "frozen-on" second layer atoms have weaker dipole moments, and therefore the experimental curve for Xe in Fig. 4 is somewhat rounded at θ ≃ 1. This effect is not observed for Ar and Kr at 40 K. $\Delta\phi$ for full monolayer Xe coverage is determined from a separate experiment with a well annealed monolayer configuration. From measured $\Delta\phi|_1$ and $d\Delta\phi/d\theta|_0$ and Eqs. (4) and (5), α and p_1 are obtained and listed in Table II.

Table II. Polarizability α, dipole moment in the ground state p_1, and dipole moment in the photoionized state p_2 for Ar, Kr, and Xe adsorbed on Al(111).

	Ar	Kr	Xe
$\alpha(\text{Å}^3)$	2.66	2.76	3.20
$p_1(\text{e-Å})$	0.020	0.028	0.049
$p_2(\text{e-Å})$	–	–	1.43

When a Xe adatom is photoionized, a dipole moment p_2 much larger than p_1 is produced (schematically shown in Fig. 2(d)). The dipolar field polarizes nearby adatoms for finite coverages $0 < \theta \leq 1$. The final-state energy is lowered by the depolarization energy which is a function of θ and adatom distribution. By energy conservation, the photoelectron kinetic energy is increased, and the binding energy for the core level is decreased by the depolarization energy. The binding energy shift due to intralayer screening is, to first order, given by

$$\Delta E_B = - 2 \; \alpha p_2^2 \; \eta \theta^m d^{-6} \tag{6}$$

where $\eta = 6.38$ is a geometrical factor. The index m is given by m=3 for a uniform distribution and m = 1 for a random distribution. If large islands are formed, ΔE_B is roughly independent of θ. The shift described here, Eq. (6), is in addition to the shift described by Eq. (1) relative to the free Xe atoms.

The experimental results on the Xe-4d binding-energy shifts versus θ are shown in Fig. 5. The second layer signals shifted by ∼0.6 eV can be observed easily for $\theta > 1$. The data points for $\theta < 1$ were obtained with increasing Xe coverage at 40 K, and therefore were slightly affected by some frozen-on second-layer atoms. The data points at $\theta \simeq 1.1$ were taken after thermal annealing to move some second-layer atoms into the first layer vacancies. This explains the small offset for the first-layer binding energies at $\theta \simeq 1$ in Fig. 5.

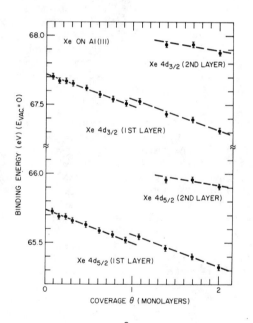

Fig. 5. Xe-4d core level binding energies referred to the vacuum level as functions of coverage on Al(111) at 40 K. The dashed lines are reference straight lines.

Clearly, ΔE_B depends linearly on θ; therefore, a random distribution of adatoms (m = 1 in Eq. (6)) describes the data. A uniform distribution (m = 3) or island formation does not describe the data. From the data in Fig. 5 and using the polarizability α from Table II, we obtain $p_2 = 1.43$ e - Å for Xe on Al(111). The distance between the Xe adatom and the image plane of Al(111) is x = 1.42 Å, where the image plane location is taken from a local-density-functional theory.[9] The effective charge associated with a photoionized Xe adatom is then $p_2/x = 1.0$ e. This result indicates that after photoionization, the Xe adatom is fully ionized and the screening charge remains in the substrate and hence image-like. This is our earlier assumption, and is consistent with other experimental[12] and theoretical[10] results.

In the above experiments, core levels are probed. The experimental results are therefore relatively simple. In contrast, photoemission results from valence levels are more complicated due to the large spatial extent of the valence core hole. The point charge model is obviously not very good, and band dispersion effects[13] have to be considered in general. For example, Fig. 6 shows photoemission

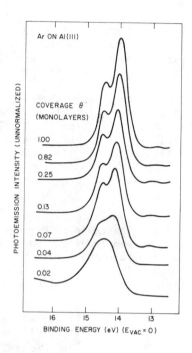

Fig. 6. Photoemission spectra for valence 3p levels of Ar adsorbed on Al(111) at 40 K, taken with a photon energy of 28 eV.

spectra for valence 3p levels of Ar adsorbed on Al(111) at 40 K for $0 < \theta \leq 1$. As θ increases, the lineshape changes due to two-dimensional lateral interactions. The lineshape for a given θ depends also on the adatom spatial distributions and the adatom-substrate interactions (crystal-field effect, etc.). As θ increases, the center of gravity of the 3p levels moves to lower binding energy, similar to the behavior of the core levels. A detailed quantitative understanding of the spectra in Fig. 6, however, requires more detailed experimental and theoretical studies.

Jacobi and Rotermund[14] recently did similar experiments on systems with adsorbed multilayers. Their experimental results are not qualitatively inconsistent with our finding here, yet their interpretation is at variance with ours. These authors ignored final-state screening effects completely. This is an oversimplification and is not justified in general. Recent theoretical results by Lang, et al.[10,15] and our experimental results clearly indicate that for Xe on Al(111), the photoionized adatom lies outside the dipole layer and is fully charged; therefore, the final-state effects are certainly not negligible in determining binding energy shifts.

In the above systems, the initial-state effects are small compared to the final-state effects and therefore neglected, because the bonding between rare gas atoms and the substrates are very weak. This is demonstrated by the very low desorption temperatures for these systems ($\lesssim$ 100 K). We now consider a related simple system, namely, the Al(100) surface. The core level binding energy shifts of surface atoms relative to the bulk atoms are of great interest, because the measured shifts can provide information on surface potentials. Clearly the bonding energy of the surface atoms for Al(100) is much larger than adsorbed rare gas atoms, which is related to the high evaporation temperature of Al. The initial-state shift is not negligible compared to the final-state shift.

The surface shifts for the Al-2p core levels are determined using the partial yield technique to optimize system resolution, and using a sample temperature of 40 K to reduce thermal broadening effects. Results are shown in Fig. 7. The long-dashed and solid curves are secondary electron yields at kinetic energies of 3 ± 0.3 and

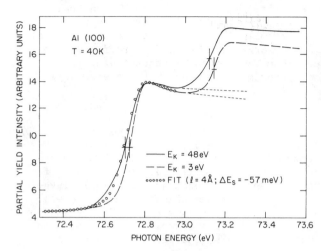

Fig. 7. Photoemission partial yield spectra for a clean Al(100) surface at 40 K at a kinetic energy of 48 eV (solid curve) and 3 eV (long-dashed curve).

48 ± 0.5 eV, respectively, as functions of incident photon energy. The electron escape depth is greater than 50 Å for an electron kinetic energy of 3 eV, and is 4 ± 0.5 Å for an electron kinetic energy of 48 eV; therefore the long-dashed and solid curves in Fig. 7 are bulk- and surface-sensitive, respectively. The partial yield spectra are proportional to optical absorption spectra to a very good approximation. The two edges near 73 eV correspond to transitions from the $2p_{3/2}$ and $2p_{1/2}$ core levels to the Fermi level. Clearly, the edges in the surface-sensitive spectrum are shifted (the half-amplitude points are indicated by crosses; the short dashed lines are estimated backgrounds) relative to the bulk-sensitive spectrum. To estimate the shift, a simple least-square procedure to fit the surface-sensitive spectrum is used. The results (circles in Fig. 7) are obtained by using a linear combination of the bulk spectrum (long-dashed curve) and its replica, with the latter being shifted by the core level shift (fitting parameter) and having the intensity ratio determined by the electron escape depth (taken to be 4 Å at a kinetic energy of 48 eV). From the fit, a surface core level shift of -57 ± 7 meV is obtained, where the uncertainty is estimated from the uncertainty in the electron escape depth (± 0.5 Å).[16] In this simple analysis, subsurface shifts and crystal-field splitting effects are neglected. These quantities cannot be deduced reliably from the data, because the shifts and splittings are very small. Nevertheless, subsurface shifts and crystal-field splitting tend to broaden the absorption edges. The surface-sensitive spectrum is indeed broader than the simple fit in Fig. 7, indicating possible subsurface shifts and/or crystal-field splitting. The experimental value of -57 meV should be regarded as an average value.

At least three theoretical results for the surface shifts of Al(100) are available. From a local-density-functional theory and a jellium model,[17] an initial-state shift of -160 meV is obtained. Wimmer, et al.[18] used a full-potential self-consistent linearized-augmented-plane-wave method to obtain a surface shift of -120 meV and a crystal-field splitting of 38 meV as well as a subsurface layer shift of -50 meV. In this calculation, only initial-state effects

770

are considered. Using the model of Johansson and Martensson, a shift of -120 meV is derived from empirical cohesive energies.[20] These theoretical values are of the same sign as the experimental value, yet they are all too large by about a factor of two. The discrepancy may be due to the final-state effects and/or the various assumptions and approximations made in the theoretical models.

CONCLUSIONS

Core level photoelectron spectroscopy has been used to study several simple adsorbates and surfaces. For physisorbed rare gases, the initial-state effects are small and can be ignored for many applications. The final-state effects are relatively large, and can often be well described by a semiclassical model in which the core level binding energy shifts are due to the depolarization energies of the core hole. The different atomic layers for a rare gas multilayer can be distinguished by the core level binding energies, and the atomic movements can be studied. For rare gas submonolayers, the adatom dipole moment can be determined in the ground and excited states. Results indicate that a photoionized Xe adatom on Al(111) is fully charged, therefore the screening charge is indeed image-like. Information on adatom spatial distribution can also be obtained from the coverage dependence of core level binding energy shifts. To summarize, a simple picture is sufficient to understand many physical effects for these weakly bonded systems. These studies will be the basis for our future studies of more complicated systems involving strong chemisorption bonds. For a related simple system, namely, the clean Al(100) surface, the surface core level shifts have important contributions from the initial-state effects. A full understanding of the shifts, however, requires more detailed theoretical and experimental work.

REFERENCES

1. See, for example, D. Menzel, in Photoemission and the Electronic Properties of Surfaces, edited by Feuerbacher, Fitton, and Willis (Wiley, New York, 1978), pp. 381 - 408.
2. Parts of the results have been published previously, see Refs. 3-7.
3. G. Kaindl, T.-C. Chiang, D. E. Eastman and F. J. Himpsel, Phys. Rev. Lett. 45, 1808 (1980).
4. G. Kaindl, T.-C. Chiang, and D. E. Eastman, Phys. Rev. (in press).
5. T.-C. Chiang, G. Kaindl, and D. E. Eastman, Solid State Commun. 41, 661 (1982).
6. G. Kaindl, T.-C. Chiang, D. E. Eastman, and F. J. Himpsel, in Ordering in Two Dimensions, edited by S. K. Sinha (North Holland, 1980), pp. 97-105.
7. T.-C. Chiang and D. E. Eastman, Phys. Rev. B23, 6836 (1981).

8. This procedure reproduces within 0.05 Å the measured distance for the equivalent case of Xe on Ag(111): N. Stoner, M. A. van Hove, S. Y. Tong, and M. B. Webb, Phys. Rev. Lett. <u>40</u>, 243 (1978).

9. V. L. Morruzi, J. F. Janak, and A. R. Williams, <u>Calculated Electronic Properties of Metals</u>, (Pergamon 1978).

10. N. D. Lang, Phys. Rev. Lett. <u>46</u>, 842 (1981).

11. J. Topping, Proc. Roy. Soc. (London) A<u>114</u>, 67 (1927).

12. J. E. Cunningham, D. K. Greelaw, and C. P. Flynn, Phys. Rev. B<u>22</u>, 717 (1980); C. P. Flynn and Y. C. Chen, Phys. Rev. Lett. <u>46</u>, 447 (1981).

13. K. Horn, M. Scheffler, and A. M. Bradshaw, Phys. Rev. Lett. <u>41</u>, 822 (1978).

14. K. Jacobi and H. H. Rotermund, Surf. Sci. <u>116</u>, 435 (1982).

15. N. D. Lang and A. R. Williams (unpublished).

16. Earlier measurements indicated a broadening with no shift; see, W. Eberhardt, G. Kalkoffen, and C. Kunz, Solid State Commun. <u>32</u>, 901 (1979).

17. N. D. Lang and W. Kohn, Phys. Rev. B<u>1</u>, 4555 (1970).

18. E. Wimmer, M. Weinert, A. J. Freeman, and H. Krakauer, Phys. Rev. B<u>24</u>, 2292 (1981).

19. B. Johansson and N. Martensson, Phys. Rev. B<u>21</u>, 4427 (1980).

20. R. Kammerer, J. Barth, F. Gerken, C. Kunz, S. A. Flodstrom, and L. I. Johansson (unpublished).

Electron- and Photon-Stimulated Desorption†

M. L. Knotek

Sandia National Laboratories, Albuquerque, NM 87185

INTRODUCTION

Electron and Photon Stimulated Desorption (ESD and PSD) have recently received growing attention because of their ability to provide highly specific information on the surface bond. In one sense, stimulated desorption can serve as an electronic and structural probe specific to the desorbed species and its surface bonding environment. This arises from our knowledge of the excitation processes which initiate desorption, and the intrinsic electronic properties of the surface which govern desorption processes. Alternatively the desorption experiment can provide insight into the dynamics of charge motion in the surface bond. The desorption process is a complex interplay between electronic and nuclear motions which occur in response to the highly energetic and localized electronic excitations which are known to induce desorption. A derivation of the details of electronic and nuclear dynamics on the molecular level is at once a very demanding and potentially insightful goal in modeling the surface bond and when coupled with electronic and chemical studies which have long been the mainstay of surface science, will lead to a deeper understanding.

Conceptually the stimulated desorption experiment is quite simple. Ionizing radiation in the form of electrons or photons, of energy E in the range $10eV < E < 10^4eV$, is impinged on a surface. Ionic, neutral or excited atomic and molecular species are desorbed from the surface in response to the deposition of energy into the surface in the form of electronic excitations. The desorbed species mass, energy and angular distributions are rather straightforwardly analyzed. The distribution and relative yields of the various mass fragments from the surface can help determine its chemical makeup.[1,2] Typical desorbed ion kinetic energies range up to 15 eV and the kinetic energy distributions are relatively sharply peaked around a value which can be used as a fingerprint of a given chemical state.[3] The ion energy distributions can also be of great use in deriving an understanding of the precursor or desorptive states, since the kinetic energy distribution is one of the few physical variables we can measure which directly reflect the state of the desorbed species as it leaves the surface. The ion angular distributions can be an important direct probe of bonding geometry.[4,5] Ions are desorbed in well defined cones having both azimuthal and polar structure. The azimuthal patterns directly reflect the symmetry of the bonding site from which the desorption occurs while the polar angle can yield information on the bond angle relative to the surface normal. The derivation of the exact polar angle is complicated by the effect of the ions image charge and the exact nature of the processes leading to desorption.[6] While the theoretical concepts used in understanding desorption have been largely confined

†This work performed at Sandia National Laboratories supported by the U.S. Dept. of Energy under contract # DE-AC04-76DP00789.

electronic structure. The highest occupied level of the Ti^{4+} ion is the Ti(3p) at ~ 34 eV below the conduction band minimum.[27] The valence band is almost exclusively O(2p) in character. If ionizing radiation removes the electron from the Ti atom's shallowest core the predominant core-hole decay will be an interatomic Auger process, due to the absence of higher lying electrons on the Ti atom. A valence electron from the O^{2-} falls into the core-hole and one (or two) electrons will be emitted from the anion to release the energy of the decay as shown in Fig. 1. The loss of the Auger electrons transforms the oxygen to an O^+. The potential in which the O^{2-} sits before the excitation/Auger sequence is the sum of an attractive Madelung term and repulsive core overlap contribution. However, if the charge of the oxygen changes sign, the Madelung term becomes repulsive. Thus the anion is now in a totally repulsive potential and is driven to desorb.

We have singled out the excitation of the cation core-hole because of the unique interatomic charge transfer process which occurs, but excitation of the anion core-level is fully as efficient in causing desorption. Thus in TiO_2 we can excite the Ti M_{23}, M_1, L_{23}, L_1, etc. or the O L_1 or K and will observe desorption of O^+ of varying strengths, depending on the energy available from the core-hole decay. While the original formulation of the Auger decomposition model was to explain the desorption of ions from ionically bonded surfaces, it was soon demonstrated that ions were desorbed from covalently bonded surface complexes by essentially the same mechanism.[26] In addition, recent observation by Niehus and Losch of high O^+ yield from O_2/Ni and Fe while no O^+ yield was observed from their stoichimetric oxides suggest additional factors involved in desorption of covalently bonded oxygen. In the simplest analysis we note that the product of an Auger decay of a core hole in a covalent system is a two- or three-valence-hole final state. Hence, we can envision a "Coulomb explosion,"[26] as observed in gas phase molecules, where the highly repulsive final state results in the production of ionic fragments of the parent species. The presence of multiple valence holes in a covalent system and the possible repulsive reaction between the unscreened nuclei due to them represent sufficient energy to produce ion fragments. In the simplest analysis the multiple holes represent energy stored in the bond. When two holes exist on an atom or in a bond the most obvious way to relieve the large repulsive energy is for one hole to hop into the bulk before nuclear motion can occur. One-hole hopping is often slowed, however, by hole-hole correlations which can block resonant one-hole hopping processes. Typical uncorrelated one-hole hopping times are of the order of 10^{-16} seconds, while desorption times are more of the order of 10^{-13} seconds. The importance of hole-hole correlations was recently pointed out in explaining the existence of atomic-like Auger spectra in narrow d-band metals such as Ni and Co[37,38] in insulators such as oxides and alkali halides[39], and gas phase molecules.[40] This many-body effect makes the Auger-induced desorption process effective not only for covalent materials but for ionic materials as well.[1,40,41]

Consider a simple Hubbard-like Hamiltonian, as in eq. 1, describing an electron moving in a system with bandwidth W. When two holes are created on an atom or in a bond the correlation energy between the two holes is given by U_{eff}, which can have important contributions due to screening.[40]

$$H = \Sigma \; \varepsilon_b c_{1\sigma}^{+} c_{1\sigma} + \Sigma \; W_{1j} c_{1\sigma}^{+} \; c_{j\sigma} + \Sigma \; U_{eff} \; c_{1\downarrow}^{+} c_{1\downarrow} c_{1\uparrow}^{+} c_{1\uparrow} \tag{1}$$

Figure 2 displays the terms in this Hamiltonian schematically. Typical one-hole hopping times in such a system are of the order $T \approx 1/w$ (Fig. 2a). If $U_{eff} > W$, resonant one-hole hopping channels are essentially blocked, $T \gg 1/w$ (Fig. 2b)[37,38] The highly repulsive Auger final state is given an intrinsically long lifetime due to this correlation, of the order of 10^2 to 10^4 times the normal one-hole lifetime, which can then result in nuclear motion to relieve the repulsion. This localization is a necessary but not sufficient condition for desorption to occur. The important point phenomenologically is that we can qualitatively understand the properties of a bond which govern desorption by understanding their effect on U_{eff} or W in the above inequality. Factors which increase U_{eff} and/or decrease W will enhance desorption and vice versa.

U_{eff} contains both an intrasite Coulomb repulsion term and a screening term. The intrasite (or intrabond) repulsion term as above is roughly approximated for atoms by the difference between the second and first ionization potentials and is inversely proportional to the ionic radius. The screening term is enhanced by higher coordination and the presence of unsaturated bonding, which allows very efficient charge-transfer screening to occur as illustrated in Fig. 2c.[40] Thus we can deduce that low coordination and saturated bonding are factors which increase the likelihood of desorption. Increased coordination also increases the effective bandwidth W which will further decrease hole-hole correlation. In general W decreases with increasing ionicity[42] leading to greater desorption in ionic systems. Feibelman[30] points out in addition that due to the substantial reduction in intrasite screening when two electrons are removed from a single atom, the lowest unfilled orbitals have a considerably reduced spatial extent. Thus, the overlap with wavefunctions on neighboring atoms is exponentially reduced, resulting in a like reduction in W. An additional factor in chemisorption systems is the energy difference between one-electron levels in surface and bulk atoms which can further enhance lifetimes.[41]

MODELING DESORPTION IN COVALENT SYSTEMS

Two examples of recent work demonstrate the kinds of processes leading to desorption from covalent systems. The study of condensed species offers the experimenter a rich variation in the choice of the chemical environment of desorbed fragments. A recent ESD study of a series of condensed branched alkanes provides one example.[2] Neopentane ($C(CH_3)_4$) is the first molecule of the series. In neopentane all of the carbons are in a tetrahedral environment with the central carbon having 4 carbon neighbors and the methyl carbons having 3 hydrogen and 1 carbon. The central carbon in neopentane

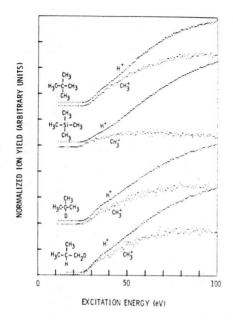

NORMALIZED ION YIELD (ARBITRARY UNITS)

EXCITATION ENERGY (eV)

Fig. 3 H$^+$ and CH$_3^+$ ESD yields from condensed C(CH$_3$)$_4$, Si(CH$_3$)$_4$, DC(CH$_3$)$_3$ and HC(CH$_3$)$_2$(CH$_2$D) as a function of excitation energy. On each molecule H$^+$ and CH$_3^+$ thresholds are equal. Excitation is of methyl group, not central atom. [from ref. 24]

has a delocalized Auger final state while the methyl carbons being effectively lower coordinated than the central carbons and, hence, less efficiently screened have localized final states. This suggests that excitations on the central carbon will be less efficient at causing desorption than will excitations on the methyl group. Shown in Fig. 3 are a series of ESD spectra for 4 condensed molecules C(CH$_3$)$_4$, Si(CH$_3$)$_4$, HC(CH$_3$)$_2$(CH$_2$D), and DC (CH$_3$)$_3$. From these molecules, H$^+$, H$_2^+$ and CH$_3^+$ (or their isotopic analogs) are observed in in desorption. In each of the molecules in Fig. 3 the thresholds for H$^+$ and CH$_3^+$ desorption are equal, indicating that the same excitation breaks C-H bonds in the methyl group or C-CH$_3$ bonds. Changing the central atom from C to Si results in only a small shift in threshold and minor changes in spectral shape, suggesting that the central atom is little involved in the desorption. A test of this is to remove one methyl from the central atom and replace it with a D forming the DC(CH$_3$)$_3$. Here it is found that the D is not efficiently desorbed, supporting the hypothesis that there is no loaclized state in the central carbon. Another isotopic variation is the formation of HC(CH$_3$)$_2$(CH$_2$D) by substitution of a D in one methyl group and an H tp tje central carbon. From this molecule, D$^+$ desorption is observed with a threshold typical of a H$^+$ from methyl group but the yield is lower than an equivalent H due presumably to the istotop effect.

A comparison of the desorption thresholds of Fig. 3 to equivalent gas phase dissociation of methane[43] suggests that the excitation which initiates desorption is of the 3_{a1} (Carbon 2s) orbital on the methyl group, which then decays by autoionization to give a

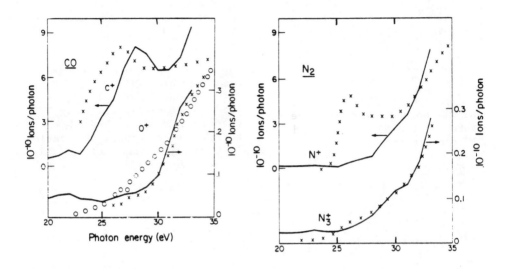

<u>Fig. 4</u> <u>Right</u> N^+ and N_3^+ PSD yield from solid $N_2(1)$ compared to gaseous $N_2(xxx)$ for N^+ and $(N_2)_2$ dimers for $N_3^+(xxx)$. <u>Left</u> C^+ and O^+ PSD yield from solid CO compared to gaseous $CO(xxx)$ and to CO on Ru (001) from ref. 12 (ooo).

localized multi-hole final state on the methyl group. These results can be generalized to suggest that in organic or other solids the highest desorption yield will be observed from terminal groups of low coordination to atoms of the solid. Furthermore, and possibly more important, the desorption spectra reflect the electronic excitation of this terminal group, not the substrate. Hence, the analysis of the spectra can yield information specific to the terminal group in question.

Another example of condensed layer studies are the recent results of Rosenberg et. al.[44] on PSD of condensed N_2 and CO in the photon energy range from 20 to 35 eV shown in Fig. 4. The prominent threshold and peak in the gaseous N^+ yield between 25 and 27 eV is suppressed from solid N_2 whereas the threshold and shape of the N_3^+ curves agree and follow the N^+ yield for solid N_2. For CO the O^+ solid and gaseous curves agree quite well and the C^+ yields have nominally the same shape with the solid shifted up in energy by ~ 1.5 eV. The most dramatic difference between gas and solid the N^+ yield from N_2. The molecular Van der Waals interaction $(N_2)_2$ and $(CO)_2$ are quite small (.007 and .008 eV respectively),[45] and could not be expected to produce such a result. However, the ion molecule interactions are quite large; 0.97 ± 0.04 for $CO^+ \cdot CO$ and 0.90 ± 0.05 eV for $N_2^+ \cdot N_2$.[46] N_2 molecules bond via $3\sigma_g$ orbital, through the end of the molecule. If we consider the top layer of condensed

N_2 in terms of N_2 bonded to a surface, then we find that when the $3\sigma_g$ level is ionized, the hole is stabilized by localization on the N atom nearest the surface.[47] Thus the $3\sigma_g^{-1}$ state of adsorbed N_2^+ is very similar to the $5\sigma^{-1}$ of a CO^+ which has the carbon nearest the surface. Hence, we would expect the desorption of N^+ (the outer N in the adsorbed N_2) to display a behavior similar to O^+ from adsorbed CO, as is seen. It also appears that a similar excitation yields N_3^+ from $(N_2)_2(g)$ and N_2 solid. By contrast, the end of the N_2^+ molecule-ion which is carbon like is always on the attached end of the molecule and won't be seen in desorption, so no C^+-like features in the N^+ are observed from the solid, even though the C^+-like N^+ does appear in the gas. The assignment of the threshold and structure energies to specific excitation of the chemisorbed molecules is not unambiguously determined,[44] but it is noted that in the vapor the dissociative ionization processes in this energy range are thought to be multielectron predissociation transitions.[48] Multielectron excitations are also thought to be involved in the O^+ PSD from adsorbed CO due to both low energy and K-shell excitations.[12,49,50]

While it is known that core-level excitation on adsorbed CO gives desorption,[26] recent results suggest a complicated behavior. Gas phase CO displays a U_{eff} of ~ 15 eV[51] and, in the gas phase, Auger decay of a $C(K)$ or $O(K)$ core-hole is known to produce C^+ and O^+ efficiently.[52] Koel et al's Auger spectrum of CO on $Ni(100)$, however, suggest a U_{eff} of ~ 0 eV, indicating that the charge transfer screening of the surface has totally relieved the repulsive energy of the Auger final state.[53] This may explain the observation by Jaeger et al.[49] that their threshold for desorbing O^+ at the $O(K)$ edge for both CO and NO on $Ni(100)$ is delayed relative to the oxygen KLL Auger yield absorption edges. This they attribute to a complicated $(1\sigma+3\sigma)$ multielectron excitation which the results of Koel et al. suggest may be necessary for efficient desorption. It is important to emphasize that, in detail, the desorption event in such systems can be quite complex and as yet is not fully understood. Similar studies have been carried out on condensed H_2O[15] and CH_3OH.[54]

DESORPTION SPECTROSCOPY

There is a great deal of activity in the area of defining the mechanisms for desorption, both because of the insight to be gained concerning surface bonding and the interaction of radiation with matter and the fact that we have made some important first insights in a largely unexplored area. Quite aside from discussions of detailed desorption mechanisms, there exists a great potential in the use of stimulated desorption to explore surface atomic and electronic structure.[9,13]

The simplest spectra from which to extract information on the excited atom are those for which the desorption yield is directly proportional to the excitation cross-section, e.g. in a system where creation of a core-hole leads to desorption by an Auger decay. In that case, the desorption probability $P(E)$ at an excitation energy E is given by

$$P(E) \ \alpha \ A(1-f)\theta(E) \tag{2}$$

where A is the probability that Auger decay results in a desorptive final state, f is the reneutralization probability, and $\theta(E)$ is the core-hole ionization cross-section at the energy E. Since to a first approximation only θ is a function of E, the desorption spectrum will directly reflect the excited atoms x-ray absorption spectrum. At present we have only a few examples of different desorption spectral phenomena which are more "proof of concept" than finished analytical studies.

When an electron is excited from a core-level, there are a variety of bound or semi-bound and continuum final states to be assessed. When the atomic final state is highly localized, strong resonances can be observed in the near edge region. While these are atomic in nature, they can be affected by the environment as typified in Fig. 5 which shows H^+ from oxidized cerium near the excitation threshold for the Ce(4d) ionization compared to optical absorption in Ce metal and CeO_2.[23] We observe a series of sharp atomic-like resonances which arise from the strong coupling between the Ce 4d and 4f shells.[55,56] Each feature in these spectra is due to a dipole transition of the form

$$4d^{10}4f^n \rightarrow 4d^94f^{n+1} . \tag{3}$$

The exchange interaction between the f electrons themselves and between these and the remaining d electrons splits the $4d^94f^{n+1}$ configuration giving rise to a series of multiplet states. These intermediate bound excited states then autoionize to a final state from which ion desorption occurs. The multiplet structure is critically dependent on the valency of the metal atom. Curve (a) in Fig. 5 shows the H^+ PSD ion-yield spectrum after 750-L O_2 exposure at 300K while curve b is the H^+ desorption after heating to 475K. Curve (c) and (d) are the optical absorption spectra for $CeO_2(Ce^{4+})$ and Ce metal (Ce^{3+}), respectively. The primary difference between (a) and (b) is cross-hatched in (a) and shows that annealing removed Ce^{4+} valency from the surface. Analysis of this experiment and accompaning measurements showed that the Ce oxidizes to a CeO_2 stoichiometry which reduces to Ce_2O_3 on the surface due to the high oxygen conductivity of CeO_2.

In the energy region from the continuum threshold to an energy up to ~ 50 eV there are a variety of very complex structures which are due to the interaction of the electron with the neighboring atoms. These structures are due to density of states structure coupled with scattering from the lattice.

In ionic materials such as Al_2O_3, BeO and SiO_2, the interaction with the anion shell around the excited metal atom leads to peaks inner-well resonances or Tossel structures[62] which can yield information goemetric details of the environment of the excited atom. These structures are considerably more complex to analyze than are EXAFS structures due to multiple scattering processes.[64] In general, to analyze these spectra model site geometries are treated and the

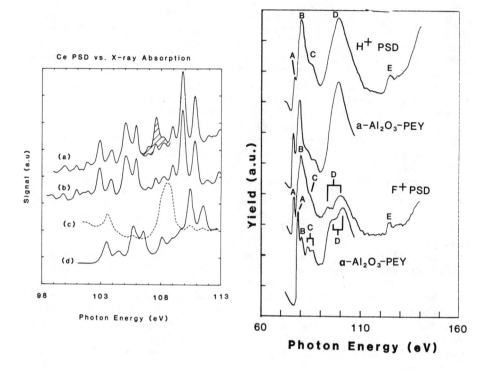

Fig. 5 A comparison of H⁺ PSD from a) freshly oxidized Ce and b) oxidized Ce annealed at 475K compared to x-ray absorption of c) CeO₂ and d) Ce metal. The shaded peak in a) is due to desorption from Ce⁴⁺ sites which are absent after annealing. [from ref. 23]

Fig. 6 A comparison of H⁺ and F⁺ PSD from α-Al₂O₃(0001) and PEY from a-Al₂O₃ (tetrahedral) and α-Al₂O₃ (octahedral). The H bonding site is purely tetrahedral while F shows a mixed signature. [from ref. 57]

derived structures are compared to the experimental data. In principle these structures are more informative than EXAFS in that multiple atom position correlations can be drawn together with detailed information on the bond angles and site symmetries. One promising empirical way to analyze these PSD structures is to make direct comparisons to the x-ray absorption spectra for model compounds when the dominant features in the structure are determined by the first neighbor shell. In Al₂O₃, SiO₂, BeO and other ionic materials where the interaction of the emitted electrons is predominantly with the negative anion shell around the excited atom, the inner-well resonance structures are detemined mainly by the immediate environment of the atom.[62,63] In covalent materials, or in metals, the scattering from several nearest neighbor atomic shells are important in determining the structure, so the use of model compounds to mimic the site environment cannot be so readily employed.[64] Fig. 6 shows H^+ and F^+ PSD from an α-Al_2O_3(0001) surface, and PEY from amorphous Al_2O_3 (a-Al_2O_3) and α-Al_2O_3, indicate H^+ and F^+ bond at quite different sites. The Al_2O_3 PEY has the distinctive tetrahedral signature, similar to SiO_2. The H^+ PSD from a-Al_2O_3 also shows a tetrahedral signature. In α-Al_2O_3 the bulk site is octahedral and the PEY shows a split peak at D. The F^+ PSD shows a similar split peak at D indicating that the molecular field at the Aluminum atom to which the F is bonded is similar to the octahedral site. Note, however, that the peaks A, B and C of the F^+ PSD are not unlike those for the H^+ PSD indicating that the F site has some similarities to the tetrahedral site as well. One important fact pointed out by this data is that further modeling of such systems is needed and would be most fruitful.

Near edge PSD spectra have now been taken for Al_2O_3, BeO and SiO_2,[9,57] O^+ from O/Mo(100),[13] CO and NO on Ni(100),[22] O^+ from O/W, [18,19] O^+ from O/Ti,[10,58] TiO_2,[59] H^+ from Diamond,[60] H_2O/Ti,[61] H^+/ CeO_2.[23]

Several important studies aimed at a theoretical or semiempirical interpretation of x-ray absorption near edge spectra (XANES) have recently appeared.[62-69] The general concepts laid out in these papers will be important in future discussions of PSD data. Noting that the strength of the first prominent peak in x-ray absorption edges is governed to a first approximation by the density of empty states on the excited atom, Lytle et al.[66] use Batsanov's[70] concept of effective "coordination charge"

$$\eta = MI \tag{4}$$

where M is the valence of the absorber atom and I is Paulings ionicity[71]

$$I = 1 - \exp[-1/4(X_A - X_B)^2] \tag{5}$$

where X_A and X_B are the electronegativities of atoms A and B. If the compound were purely ionic then $I = 1$ and $\eta = M$. Lytle et al found a good correlation between the differential area of the first peak in (XANES) and η for a series of Ir, Pt, and Au whose coordination charge was systematically varied. Horsley's correlations confirm this picture.[68] In a study of vanadium in coal Maylotte et al[67] found two distinct vanadium environments. In excitations from the 1s level of vanadium, the octahedral environment permits only a weak quadrupole transition to the empty d-levels while in the tetrahedral site the lack of a center of inversion permits a dipole transition between the 1s and T_2 orbitals which can contain strong 3d contributions, as calculations by Kutzler et al.[69] clearly demonstrate.

In the energy region further above threshold, so-called extended x-ray absorption fine structure (or EXAFS) region, the excited electron's interaction with the environment is simpler to analyze.[21] When the photoemitted electron leaves the excited atom it is scattered by the cores of the atoms in its environment such that the wavefunction at the excited core contains a factor

$$1 + \sum_R A_{k,R} \sin[2kR + \phi(k)] \tag{6}$$

where k is the wavevector of the excited electron, R is the radial distance of a specific coordination shell, and $\phi(k)$ is the scattering phase shift. For a given coordination shell

$$\sum_{shell} A_{k,R} = \frac{N}{kR^2} \; f(k,\pi) \; e^{-2\sigma^2 k^2 - 2R/\lambda(k)} \tag{7}$$

where the first term arises from the spherical wave nature of the outgoing electron, f is a scattering amplitude, the σ^2 term in the exponential is the Debye-Waller factor and λ is the Fourier transform of the radial distribution function of the excited atom. The first example of the use of PSD to obtain EXAFS data analyzed the O^+ PSD yield and the e-yield for an oxygen exposed Mo(100) surface.[13] This data showed a) the Mo-Mo separation for the excited atom is essentially identical to the bulk Mo-Mo separation; b) the coordination of the excited Mo atom is 1/2 that of the bulk; c) from the position of the threshold of this and the other core edges[13] it was determined that the excited Mo was in an "oxidic" state.

CONCLUSION

This paper covers but a few of the salient points which are coming together to suggest that ESD and PSD can contribute greatly to our understanding of surfaces. Specifically, the highly specific local information content of stimulated desorption arises from the intrinsically highly localized processes leading to desorption. Recent examples of near edge and extended range PSD data, together with advances in the interpretation of near edge spectra promise a useful future for these measurements.

REFERENCES

1. M. L. Knotek, Surf. Sci. 91, L17 (1980).
2. M. L. Knotek, Surf. Sci. 101, 334 (1980).
3. M. Nishijima and F. M. Propst, Phys. Rev. B2, 2368 (1970).
4. "The determination of molecular structure at surfaces using angle resolved electron and photon stimulated desorption" T. E. Madey, F. P. Netzer, J. E. Houston, D. M. Hanson, and R. Stockbauer. To be published in the Proceedings of the First International Conference on Desorption Induced by Electronic Transitions, Springer-Verlag.

5. "Angular resolved ESD for surface structure determination," H. Niehus, to be published Appl. Surf. Sci.

6. T. E. Madey, Surf. Sci. 94, 483 (1980).

7. [64][331]

8. P. A. Redhead, Can. J. Phys. 42, 886 (1964).

9. M. L. Knotek, V. O Jones, and V. Rehn, Surf. Sci. 102, 566 (1981).

10. D. M. Hanson, R. Stockbauer, and T. E. Madey, Phys. Rev. B10, 5513 (1981).

11. P. Feulner, R. Treichler and D. Menzel, Phys. Rev. B 24, 7427 (1981).

12. T. E. Madey, R. Stockbauer, S. A. Flodström, J. F. van der Veen, F. J. Himpsel, and D. E. Eastman, Phys. Rev. B 23, 6847 (1981).

13. R. Jaeger, J. Stöhr, J. Feldhaus, S. Brennan and D. Menzel, Phys. Rev. B, 23, 2102 (1981).

14. H. Niehus and W. Losch, Surf. Sci. 111, 344 (1981).

15. R. A. Rosenberg, V. Rehn, V. O. Jones, A. K. Green, C. C. Parks, G. Loubriel, and R. H. Stulen, Chem. Phys. Lett. 80, 488 (1981).

16. S. L. Weng, Phys. Rev. B 23, 3788 (1981).

17. S. L. Weng, Phys. Rev. B 23 1699 (1981).

18. D. P. Woodruff, M. M. Traum, H. H. Farrell, and N. V. Smith, Phys. Rev. B 21, 5642 (1980).

19. T. E. Madey, R. L. Stockbauer, J. F. van der Veen, and D. E. Eastman, Phys. Rev. Lett. 45, 187 (1980).

20. E. Bauer and H. Poppa, Surf. Sci. 99, 341 (1980).

21. R. Jaeger, J. Feldhaus, J. Haase, J. Stöhr, Z. Hussain, D. Menzel, and D. Norman, Phys. Rev. Lett. 45, 1870 (1980).

22. R. Jaeger, R. Treichler and J. Stöhr, Surf. Sci. 117, 533 (1982).

23. B. E. Koel, G. M. Loubriel, M. L. Knotek, R. H. Stulen, R. A. Rosenberg, C. C. Parks, Phys. Rev. B, 25, 5551 (1982).

24. J. A. Kelber and M. L. Knotek, "Electron Stimulated Desorption of Condensed, Branched Alkanes." [Surface Science, In Press]

25. R. H. Stulen, T. E. Felter, R. A. Rosenberg, M. L. Knotek, G. Loubriel and C. C. Parks, Phys. Rev. B, 25, 6530 (1982).

26. R. Franchy and D. Menzel, Phys. Rev. Lett. 43, 865 (1979).

27. M. L. Knotek and P. J. Feibelman, Phys. Rev. Lett. 40, 964 (1979).

28. P. J. Feibelman and M. L. Knotek, Phys. Rev. B 18, 6531 (1978).

29. M. L. Knotek and P. J. Feibelman, Surf. Sci. 90, 78 (1979).

30. P. J. Feibelman, Surf. Sci. 102, L51 (1981).

31. M. L. Knotek, Nature Vol. 291, 452 (1981).

32. D. E. Ramaker, C. T. White and J. S. Murday, J. Vac. Sci. Technol., 18(3) (1981) 748.

33. D. R. Jennison, J. Vac. Sci. Technol. 20, 548 (1982).

34. C. F. Melius, R. H. Stulen, J. O. Poell, Phys. Rev. Lett. 48, 1429 (1982).

35. D. Menzel and R. Gomer, J. Chem. Phys. 41, 3311 (1964).

36. J. Rubio, J. M. Lopez-Sancho and M. P. Lopez-Sancho, J. Vac. Sci. Technol. 20, 217 (1982).

37. G. A. Sawatsky, Phys. Rev. Letters 39, 504 (1977).

38. M. Cini, Solid State Commun. 20, 605 (1976).

39. D. E. Ramaker, Phys. Rev. B21, 4608 (1980).

40. D. R. Jennison, J. A. Kelber and R. R. Rye, Phys. Rev. B 25, 1384 (1982).

41. "On Auger Induced Decomposition/Desorption of Covalent and Ionic Systems" D. E. Ramaker, C. T. White, and J. S. Murday, to be published in Phys. Letters.

42. "Radiation-Induced Decomposition of Inorganic Molecular Ions," E. R. Johnson, Gordon and Breach, N.Y., 1970.

43. R. Locht, J. L. Olivier and J. Momiguy. Chem. Phys. 43, 425 (1979).

44. R. A. Rosenberg, Victor Rehn, A. K. Green, P. R. LaRoe, Michelson Laboratory, Physics Division, NRL, China Lake, CA; C. C. Parks, Lawrence Berkeley Laboratory, Berkeley, CA; "Photon-Stimulated Ion Desorption from Condensed Molecules: N_2, CO C_2H_2, CH_3OH, N_2O, D_2O and NH_3." To be published in the Proceedings of the First International Conference on Desorption Induced by Electronic Transitions, Springer-Verlag.

45. J. O. Hirschfelder, C. F. Curtiss, and R. B. Bird, "Molecular Theory of Gases and Liquids," (Wiley, New York, 1964) p. 1111.

46. S. H. Linn, Y. Ono, and C. Y. Ng, J. Chem. Phys. 74, 3342 (1981).

47. K. Hermann, P. S. Bagus, C. R. Brundle, and D. Menzel, Phys. Rev. B 24, 7025 (1981).

48. G. R. Weight, M. J. Vander Viel and C. E. Brion, J. Phys. B 9, 675 (1976).

49. R. Jaeger, J. Stöhr, R. Treichler, and K. Baberschke, Phys. Rev. Lett. 47, 1300 (1981).

50. J. E. Houston and T. E. Madey, "Core-Level processes in the Electron Stimulated Desorption of CO from the W(110) Surface" To be published in Physical Review.

51. J. A. Kelber, D. R. Jennison and R. R. Rye, J. Chem. Phys. 75, 682 (1981).

52. R. B. Kay, Ph.E. Van der Leeuw and M. J. Van der Weil, J. Phys. B10, 2521 (1977).

53. B. E. Koel, J. M. White, G. M. Loubriel submitted to Surface Science.

54. "The Origin of H^+ in Electron Stimulated Desorption of Condensed CH_3OH." R. Stockbauer, E. Bertel and T. E. Madey, J. Chem. Phys. (in press).

55. J. L. Dehmer, A. F. Starace, U. Fano, J. Sugar and J. W. Cooper, Phys. Rev. Lett. 21, 1521 (1971).

56. J. Sugar, Phys. Rev. B5, 1785 (1972).

57. "Photon Stimulated Desorption of H^+ and F^+ from BeO, Al_2O_3 and SiO_2: Comparison of Near Edge Structure to Photoelectron Yield," M. L. Knotek, R. H. Stulen, G. M. Loubriel, V. Rehn, R. A. Rosenberg, C. C. Parks, submitted to Surface Science.

58. "Photon-Stimulated Desorption from an Oxidized Ti(004) Surface," C. C. Parks, D. A. Shirley, M. L. Knotek, G. M. Loubriel, B. E. Koel, R. A. Rosenberg and R. H. Stulen, submitted to Surface Science.

786

59. M. L. Knotek, V. O. Jones, and V. Rehn, Phys. Rev. Letters $\underline{43}$, 300 (1979).

60. B. B. Pate, M. H. Hecht, C. Binns, I. Lindau and W. E. Spicer, J. Vac. Sci. Technol. Aug. 1982.

61. R. Stockbauer, D. M. Hanson, S. A. Flodstrom, and T. E. Madey, J. Vac. Sci. Technol. $\underline{20}$, 562 (1982).

62. A. Bianconi, Surf. Sci. $\underline{89}$, 41 (1979).

63. A. Bianconi, Appl. of Surf. Sci. $\underline{6}$, 392 (1980).
    to Nature.

64. G. N. Greaves, P. J. Durham, G. Diakun, P. Quinn, submitted to Nature.

65. P. J. Durham, J. B. Pendry, C. H. Hodges, Sol. State Comm. $\underline{38}$, 159 (1981).

66. F. W. Lytle, P. S. P. Wei R. B. Greegor, G. H. Via and J. H. Sinfelt, J. Chem. Phys. $\underline{70}$(11), 4849 (1979).

67. D. H. Maylotte, J. Wong, R. L. St. Peters, F. W. Lytle and R. B. Greegor, Science, $\underline{214}$, 554 (1981).

68. J. A. Horsley, J. Chem. Phys. $\underline{76}$(3), 1451 (1982).

69. F. W. Kutzler, C. R. Natoli, D. K. Misemer, S. Doniach, and K. O. Hodgson, J. Chem. Phys. $\underline{73}$(7), 3274 (1980).

70. S. S. Batsanov, "Electronegativity of Elements and Chemical Bonds," Novosibirsk, 1962, cited in I. A. Ovsyannikova, S. S. Batsanov, L. I. Nasonova, L. R. Batsanova, and E. A. Nekrasova, Bull. Acad. Sci. USSR Phys. Ser. 31, 936 (1976) (English translation)

71. L. Pauling, The Nature of the Chemical Bond, 3rd ed. (Cornell Univ. Press, Ithaca, NY, 1960).

LIST OF PARTICIPANTS

Abe, Schuichi
National Research Institute, Japan

Åberg, Teijo
Helsinki University of Technology, Finland

Ågren, Hans
University of Lund, Sweden

Aiginger, J.
Atominstitut d. Oesterreichischen Universitäten, Austria

Aita, Osamu
University of Osaka Prefecture, Japan

Arber, Judith
Queen Mary College, U. K.

Armen, Brad
University of Oregon, USA

Asaad, W.
The American University in Cairo, Egypt

Bechler, Adam
University of Warsaw, Poland

Becker, Richard L.
Oak Ridge National Lab., USA

Belin, Esther
Laboratoire de Chimie Physique, France

Berényi, D.
Hungarian Academy of Sciences, Hungary

Bernstein, E. M.
Western Michigan University, USA

Berry, Scott
Oak Ridge National Lab., USA

Beyer, H. F.
GSI, West Germany

Blair, John S.
University of Washington, USA

Bleier, Wolfram
Universität Tübingen, West Germany

Bonnet, Jean Jacques
C.N.A.M., France

Borchert, G. L.
KFA Jülich, West Germany

Brandt, Detlev
University of North Carolina, USA

Briand, Jean Pierre
Laboratoire Curie, France

Brion, C. E.
University of British Columbia, Canada

Brown, George
Stanford Synchrotron Radiation Lab., USA

Bruch, Reinhard
Univ. of Freiburg, West Germany

Burr, Alex F.
New Mexico State University, USA

Cardoso, Carlos
University of Strathclyde, Scotland

Cauchois, Y.
Faculté des Sciences de Paris, France

Chang, Chu-Nan
National Taiwan Normal University, Taiwan

Chang, Tu-nan
University of Southern California, USA

Chen, Mau Hsiung
University of Oregon, USA

Chetioui, Annie
Laboratoire Curie, France

Chiang, Tai-Chang
University of Illinois, USA

Cipolla, Sam J.
Creighton University, USA

Cleff, Bernd
Institut für Kernphysik, G.D.R.

Cocke, C. L.
Kansas State University, USA

Cowan, Paul L.
National Bureau of Standards, USA

Cox, Anthony P.
University of Oxford, U.K.

Crasemann, Bernd
University of Oregon, USA

Crisp, R. S.
University of Western Australia, Australia

Curelaru, Irina M.
University of Utah, USA

Dattolo, Evelyne
CEA, France

Dattolo, F.
CEA, France

Datz, Sheldon
Oak Ridge National Lab., USA

Decoster, Alain
Commissariat a l'Energie Atomique, France

Delgrande, Nancy
Lawrence Livermore Laboratory, USA

Deslattes, Richard
National Bureau of Standards, USA

Dicenzo, Stephanie B.
Bell Labs., USA

Dietrich, Daniel
Lawrence Livermore Lab., USA

Drahokoupil, J.
Czechoslavakian Academy of Sciences, Czechoslavakia

Dubois, Robert
Battelle Pacific Northwest Laboratory, USA

Dyall, K. G.
Oxford University, U.K.

Eckart, Mark Joseph
Lawrence Livermore National Lab., USA

Elton, Raymond
Naval Research Laboratory, USA

Fabian, D. J
University of Strathclyde, Scotland

Farley, John
University of Oregon, USA

Feagin, James
University of Nebraska--Lincoln, USA

Feldman, U.
Naval Research Lab., USA

Ferreira, J. Gomes
Faculty of Science-University of Lisbon, Portugal

Florescu, Viorica
Univ. of Bucharest, Rumania

Flynn, C. P.
University of Illinois, USA

Folkmann, Finn
University of Aarhus, Denmark

Ford, A. L.
Texas A & M University, USA

Fortner, Richard
Lawrence Livermore Lab., USA

Freeman, Richard
Bell Labs., USA

Freund, Isaac
Barl-Ilan University, Israel

Fuggle, J. C.
Institut für Festkörperforschung, West Germany

Fujima, K.
Inst. Phys. Chem. Research, Japan

Gavrila, M.
FOM Institute, The Netherlands

Gemmell, Donald
Argonne National Lab., USA

Genz, Harald
T.H. Darmstadt, West Germany

Gerken, Friedrich
Universität Hamburg, West Germany

Ghatikar, M. N.
Indian Institute of Technology, Bombay, India

Goldberg, Isser
University of Pittsburgh, USA

Graeffe, Gunnar
Tampere University, Finland

Greegor, Robert B.
The Boeing Company, USA

Greenberg, Jack
Yale University, USA

Gregory, Donald
Oak Ridge National Lab., USA

Gunnarsson, Olle
Max Planck Institut, West Germany

Hagmann, S.
Kansas State University, USA

Harte, William E.
College Park, Maryland, USA

Haynes, Sherwood
Michigan State University, USA

Henins, Albert
National Bureau of Standards, USA

Hink, Wolfgang
Univ. Würzburg, West Germany

Holmberg, Lennart
University of Stockholm, Sweden

Huang, Keh-Ning
Argonne National Lab., USA

Ice, Gene
Oak Ridge National Lab., USA

Ilakovac, K.
Zagreb University, Yugoslavia

Intemann, Robert
Temple University, USA

Ishii, Keizo
Tohoku University, Japan

Iwinski, Zbigniew
New York University, USA

Jitschin, W.
University of Bielefeld, West Germany

Jones, Keith
Brookhaven National Laboratory, USA

Jossem, E. Leonard
Ohio State University, USA

Källne, E.
Harvard Smithsonian Astrophyical Observatory, USA

Källne, Jan C.
Harvard Smithsonian Astrophysical Observatory, USA

Kanter, Elliot
Argonne National Lab., USA

Karim, Kh Rezaul
University of Oregon, USA

Karnatak, Ramesh
LURE, France

Kerkhoff, Hans
Lawrence Berkeley Laboratory, USA

Kessel, Quentin
University of Connecticut, USA

Kessler, Ernest
National Bureau of Standards, USA

Khalaff, Mansour
Emory University, USA

Kissel, Lynn
Sandia National Labs., USA

Knotek, M.
Sandia National Labs., USA

Koch, Peter
Yale University, USA

Kodre, Alojz
University of E. Kardelj, Yugoslavia

Komma, M.
Phys. Inst. Univ. Tübingen, West Germany

Kostroun, V.
Cornell University, USA

Krause, M. O.
Oak Ridge National Lab., USA

Kuhn, Kelin J.
Stanford University, USA

Kumbartzki, W. Gerfried
University Bonn, West Germany

Lähdeniemi, Matti
University of Turku, Finland

Land, David
Naval Surface Weapons Center, USA

Lapicki, G.
East Carolina University, USA

Laramore, G. E.
University of Washington, USA

Larkins, Frank
Monash University, Australia

Lavilla, Robert
National Bureau of Standards, USA

Law, J.
University of Guelph, Ontario, Canada

Lee, Yim Tin
Lawrence Livermore National Lab., USA

Levin, Jon
University of Oregon, USA

Lin, Chii-Dong
Kansas State University, USA

Lindau, Ingolf
Stanford Synchrotron Radiation Lab., USA

Lindle, Dennis W.
Lawrence Berkeley Lab., USA

Logan, B. A.
University of Ottawa, Canada

Lucatorto, Thomas B.
National Bureau of Standards, USA

Mallett, John
Lawrence Livermore Lab., USA

Mandé, Chintamani
Nagpur University, India

Manson, Steven T.
Georgia State University, USA

Mark, Hans
NASA, USA

Mårtensson, Nils
Linköping University, Sweden

Martinez, J. V.
U. S. Department of Energy, USA

Martins, Maria Da Conceiçao
CFFII, Portugal

Mäusli, Pierre-Alain
University of Lausanne, Switzerland

McGuire, Eugene
Sandia National Labs., USA

Mehlhorn, Werner
University of Freiburg, West Germany

Meisel, A.
Karl-Marx-Universität, G.D.R.

Meron, Mati
Brookhaven National Laboratory, USA

Merzbacher, Eugen
University of North Carolina, USA

Meyerhof, W. E.
Stanford University, USA

Moncton, D. E.
Brookhaven National Laboratory, USA

Morenzoni, Elvezio
Stanford University, USA

Moseley, John T.
University of Oregon, USA

Mowat, Richard
N. Carolina State University, USA

Mukoyama, Takeshi
Kyoto University, Japan

Müller, Alfred
Univ. Giessen, West Germany

Müller, Berndt
University Frankfurt, West Germany

Nagel, David J.
Naval Research Lab., USA

Nakel, W.
University of Tübingen, West Germany

Neifert, Richard
Lawrence Livermore Lab., USA

Neumann, Walter
Techn. Universität München, West Germany

Noguera, Claudine
Université Paris-Sud, France

Nolte, Günther
Hahn-Meitner Institut, West Germany

Nordgren, Joseph
Institute of Physics, Sweden

Nordling, C.
University of Uppsala, Sweden

Norman, David
Daresbury, Lab., U.K.

Ohmura, Yoshihiro
Tokyo Institute of Technology, Japan

Ojala, Eero
University of Turku, Finland

Olabanji, S. O.
Centre for Energy Research, Nigeria

Oshima, Masaharu
Stanford University, USA

Parente, Fernando
University of Lisbon, Portugal

Paul, H.
University of Linz, Austria

Pedersen, Erik
Kansas State University, USA

Pedrazzini, Greg
Texas A & M University, USA

Pellegrini, C.
Brookhaven National Lab., USA

Pengra, James
Whitman College, USA

Petrasso, Richard
American Sc. & Eng. USA

Pétroff, Y.
LURE, France

Pianetta, Piero
Stanford University, USA

Pindzola, Michael
Oak Ridge National Lab., USA

Pratt, Richard H.
University of Pittsburgh, USA

Putila-Mantyla, Pirjo
Tampere University of Technology, Finland

Quarles, Carrol
Texas Christian University, USA

Rao, Venugopala
Emory University, USA

Reading, John
Texas A & M University, USA

Richard, Patrick
Kansas State University, USA

Richards, J. A.
Monash University, Australia

Ron, Akiva
Hebrew University of Jerusalem, Israel

Saethre, Leif
University of Tromsö, Norway

Sauder, William C.
Virginia Military Institute, USA

Schappert, G. T.
Los Alamos Lab., USA

Scheer, J.
University of Bremen, West Germany

Schmidt, V.
University of Freiburg, West Germany

Schneider, Dieter
Argonne National Lab., USA

Schneider, Stanley
McDonnell Douglas Astro. Co., USA

Schuch, R.
University Heidelberg, West Germany

Scofield, James
Lawrence Livermore National Lab., USA

Sénémaud, C.
Lab. de Chimie Physique, France

Sevier, Kenneth
Université Louis Pasteur, France

Shafroth, S. M.
University of North Carolina, USA

Shirley, David A.
Lawrence Berkeley Lab., USA

Simons, Donald
Naval Surface Weapons Center, USA

Sinclair, Rolf
National Science Foundation, USA

Smith, W. W.
University of Connecticut, USA

Sorek, Yacov
Nuclear Research Centre-Negev, Israel

Southworth, Stephen
National Bureau of Standards, USA

Stein, Josef
Hebrew University, Israel

Stewart, Richard
Lawrence Livermore Lab., USA

Stiebing, Kurt-Ernst
Univ. of Frankfurt, West Germany

Stöckli, Martin P.
Kansas State University, USA

Stoller, Christian
ETH, Switzerland

Stratton, Thomas
Los Alamos National Lab., USA

Taniguchi, Kazuo
Osaka Electro-Communication University, Japan

Tanis, John
Western Michigan University, USA

Theodosiou, C. E.
University of Toledo, USA

Thomas, Darrah
Oregon State University, USA

Toburen, L. H.
Battelle Pacific Northwest Lab., USA

Tsutsumi, Kenjiro
University of Osaka Prefecture, Japan

Uda, M.
Inst. Phys. Chem. Research, Wako, Japan

Ungier, Leon
Oregon State University, USA

Urch, David
Queen Mary College, U.K.

Van Nordstrand, Robert A.
Chevron Research Co., USA

Väyrynen, Juhani
University of Oulu, Finland

Vernazza, Jorge E.
Lawrence Livermore Lab., USA

Wang, Duan-Wei
Texas A & M University USA

Watanabe, Takeshi
Tohoku University, Japan

Wendin, Göran
LURE, France

Wiech, Gerhard
Sektion Physik der Universität München, West Germany

Wilkins, John
Cornell University, USA

Wuilleumier, François
LURE, France

Yamaka, Naohito
Osaka Electro-Communication University, Japan

Zhang, An-Dong
University of Maryland, USA

Ziggel, Heiko
University of Bremen, West Germany

AIP Conference Proceedings

		L.C. Number	ISBN
No.1	Feedback and Dynamic Control of Plasmas	70-141596	0-88318-100-2
No.2	Particles and Fields - 1971 (Rochester)	71-184662	0-88318-101-0
No.3	Thermal Expansion - 1971 (Corning)	72-76970	0-88318-102-9
No.4	Superconductivity in d-and f-Band Metals (Rochester, 1971)	74-18879	0-88318-103-7
No.5	Magnetism and Magnetic Materials - 1971 (2 parts) (Chicago)	59-2468	0-88318-104-5
No.6	Particle Physics (Irvine, 1971)	72-81239	0-88318-105-3
No.7	Exploring the History of Nuclear Physics	72-81883	0-88318-106-1
No.8	Experimental Meson Spectroscopy - 1972	72-88226	0-88318-107-X
No.9	Cyclotrons - 1972 (Vancouver)	72-92798	0-88318-108-8
No.10	Magnetism and Magnetic Materials - 1972	72-623469	0-88318-109-6
No.11	Transport Phenomena - 1973 (Brown University Conference)	73-80682	0-88318-110-X
No.12	Experiments on High Energy Particle Collisions - 1973 (Vanderbilt Conference)	73-81705	0-88318-111-8
No.13	π-π Scattering - 1973 (Tallahassee Conference)	73-81704	0-88318-112-6
No.14	Particles and Fields - 1973 (APS/DPF Berkeley)	73-91923	0-88318-113-4
No.15	High Energy Collisions - 1973 (Stony Brook)	73-92324	0-88318-114-2
No.16	Causality and Physical Theories (Wayne State University, 1973)	73-93420	0-88318-115-0
No.17	Thermal Expansion - 1973 (lake of the Ozarks)	73-94415	0-88318-116-9
No.18	Magnetism and Magnetic Materials - 1973 (2 parts) (Boston)	59-2468	0-88318-117-7
No.19	Physics and the Energy Problem - 1974 (APS Chicago)	73-94416	0-88318-118-5
No.20	Tetrahedrally Bonded Amorphous Semiconductors (Yorktown Heights, 1974)	74-80145	0-88318-119-3
No.21	Experimental Meson Spectroscopy - 1974 (Boston)	74-82628	0-88318-120-7
No.22	Neutrinos - 1974 (Philadelphia)	74-82413	0-88318-121-5
No.23	Particles and Fields - 1974 (APS/DPF Williamsburg)	74-27575	0-88318-122-3
No.24	Magnetism and Magnetic Materials - 1974 (20th Annual Conference, San Francisco)	75-2647	0-88318-123-1
No.25	Efficient Use of Energy (The APS Studies on the Technical Aspects of the More Efficient Use of Energy)	75-18227	0-88318-124-X